Essential Genetics
A GENOMICS PERSPECTIVE
Fourth Edition

Daniel L. Hartl, Harvard University
Elizabeth W. Jones, Carnegie Mellon University

JONES AND BARTLETT PUBLISHERS
Sudbury, Massachusetts
BOSTON   TORONTO   LONDON   SINGAPORE

For Christopher, Ted and Dana
DLH

For Nick, Evan and Freddy
EWJ

About the Authors

Daniel L. Hartl is Higgins Professor of Biology at Harvard University. He has published more than 300 research papers and 25 books in genetics, genomics, molecular evolution, and population genetics. He has served as president of both the Genetics Society of America and the Society for Molecular Biology and Evolution. He is an elected member of the National Academy of Sciences, as well as the American Academy of Arts and Sciences.

Elizabeth W. Jones is Schwertz University Professor of Life Sciences, Howard Hughes Medical Institute Professor, and Head of Biological Sciences at Carnegie Mellon University. She has published more than 80 papers and edited more than 20 books and monographs in genetics and cell biology. She has served as President of the Genetics Society of America and is currently Editor-in-Chief of GENETICS. She is a member of the American Academy of Microbiology.

About the Cover

Despite appearances, the fur of the polar bear *Ursus maritimus* is not white. Each hair is actually transparent, but has a hollow shaft that scatters visible light, much as ice and snow does. Some years ago, three polar bears at the San Diego Zoo turned green because algae had invaded the hollow hair shafts. (Treatment with an algaecide restored their normal color.) Male polar bears, weighing up to 1,500 pounds, are the world's largest land predators. The females are only about half as large, and newborn cubs weigh only about a pound. The world population of polar bears currently numbers about 25,000. (Photo © Thomas Brakefield/Digital Vision/Getty Images)

World Headquarters

Jones and Bartlett Publishers	Jones and Bartlett Publishers Canada	Jones and Bartlett Publishers International
40 Tall Pine Drive	6339 Ormindale Way	Barb House, Barb Mews
Sudbury, MA 01776	Mississauga, ON L5V 1J2	London W6 7PA
978-443-5000	CANADA	UK
info@jbpub.com		
www.jbpub.com		

Jones and Bartlett's books and products are available through most bookstores and online booksellers. To contact Jones and Bartlett Publishers directly, call 800-832-0034, fax 978-443-8000, or visit our website www.jbpub.com.

Substantial discounts on bulk quantities of Jones and Bartlett's publications are available to corporations, professional associations, and other qualified organizations. For details and specific discount information, contact the special sales department at Jones and Bartlett via the above contact information or send an email to specialsales@jbpub.com.

Production Credits
Chief Executive Officer: Clayton Jones
Chief Operating Officer: Don W. Jones, Jr.
President, Higher Education and Professional Publishing: Robert W. Holland, Jr.
V.P., Design and Production: Anne Spencer
V.P., Sales and Marketing: William Kane
V.P., Manufacturing and Inventory Control: Therese Connell
Executive Editor, Science: Stephen Weaver
Managing Editor, Science: Dean W. DeChambeau
Associate Editor: Rebecca Seastrong
Senior Photo Researcher: Kimberly Potvin
Photo Researcher: Christine McKeen
Marketing Manager: Andrea DeFronzo
Associate Marketing Manager: Laura Kavigian
Production, Text, and Cover Design: Anne Spencer
Composition: Graphic World
Recurring chapter opening image: © Photos.com
Printing and Binding: Courier Kendallville
Cover Printing: Courier Companies

Library of Congress Cataloging-in-Publication Data
Hartl, Daniel L.
  Essential genetics : a genomics perspective / Daniel L. Hartl, Elizabeth
W. Jones. -- 4th ed.
    p. ; cm.
  Includes bibliographical references and index.
  ISBN 0-7637-3527-2
  1. Genetics. 2. Genomics.   I. Jones, Elizabeth W. II. Title.
  [DNLM: 1. Genetics, Medical. 2. Genomics.   QZ 50 H331e 2006]
  QH430.H3732 2006
  572.8'6--dc22
                                    2005013813
Printed in the United States of America
09 08 07 06 05   10 9 8 7 6 5 4 3 2 1

Brief Contents

1 The Genetic Code of Genes and Genomes 1

2 Transmission Genetics: Heritage from Mendel 32

3 The Chromosomal Basis of Heredity 72

4 Gene Linkage and Genetic Mapping 120

5 Human Chromosomes and Chromosome Behavior 162

6 DNA Structure, Replication, and Manipulation 202

7 The Genetics of Bacteria and Their Viruses 240

8 The Molecular Genetics of Gene Expression 280

9 Molecular Mechanisms of Gene Regulation 316

10 Genomics, Proteomics, and Genetic Engineering 356

11 The Genetic Control of Development 390

12 Molecular Mechanisms of Mutation and DNA Repair 420

13 Molecular Genetics of the Cell Cycle and Cancer 458

14 Molecular Evolution and Population Genetics 492

15 The Genetic Basis of Complex Inheritance 530

Answers to Even-Numbered Problems   561

Word Roots: Prefixes, Suffixes, and Combining Forms   571

Glossary   575

Index   590

Contents

1 The Genetic Code of Genes and Genomes 1

1.1 DNA is the molecule of heredity 3

Genetic traits can be altered by treatment with pure DNA 3

Transmission of DNA is the link between generations 4

1.2 The structure of DNA is a double helix composed of two intertwined strands 7

1.3 In replication, each parental DNA strand directs the synthesis of a new partner strand 8

1.4 Genes code for proteins 9

Enzyme defects result in inborn errors of metabolism 9

the human connection **Black Urine 12**
Archibald E. Garrod

A defective enzyme results from a mutant gene 14

One of the DNA strands directs the synthesis of a molecule of RNA 14

A molecule of RNA directs the synthesis of a polypeptide chain 16

The genetic code is a triplet code 18

1.5 Genes change by mutation 20

1.6 Traits are affected by environment as well as by genes 22

1.7 Evolution means continuity of life with change 22

Groups of related organisms descend from a common ancestor 23

The molecular unity of life is seen in comparisons of genomes and proteomes 25

2 Transmission Genetics: Heritage from Mendel 32

2.1 Mendel took a distinctly modern view of transmission genetics 34

Mendel was careful in his choice of traits 35

Reciprocal crosses yield the same types of offspring 36

The wrinkled mutation causes an inborn error in starch synthesis 38

Analysis of DNA puts Mendel's experiments in a modern context 38

2.2 Genes come in pairs, separate in gametes, and join randomly in fertilization 39

Genes are physical entities that come in pairs 40

The paired genes separate (segregate) in the formation of reproductive cells 41

Gametes unite at random in fertilization 41

Genotype means genetic endowment; phenotype means observed trait 42

The progeny of the F_2 generation support Mendel's hypothesis 43

The progeny of testcrosses also support Mendel's hypothesis 44

2.3 The alleles of different genes segregate independently 45

The F_2 genotypes in a dihybrid cross conform to Mendel's prediction 45

The progeny of testcrosses show the result of independent assortment 47

2.4 Chance plays a central role in Mendelian genetics 49

The addition rule applies to mutually exclusive possibilities 49

The multiplication rule applies to independent possibilities 50

2.5 The results of segregation can be observed in human pedigrees 51

2.6 Dominance is a property of a pair of alleles in relation to a particular attribute of phenotype 53

Flower color in snapdragons illustrates incomplete dominance 54

The human ABO blood groups illustrate both dominance and codominance 55

The Human Connection Blood Feud 57
Karl Landsteiner

A mutant gene is not always expressed in exactly the same way 58

2.7 Epistasis can affect the observed ratios of phenotypes 59

2.8 Complementation between mutations of different genes is a fundamental principle of genetics 62

Complementation means that two mutations are in different genes 63

Lack of complementation means that two mutations are alleles of the same gene 63

The complementation test enables one to group mutations into allelic classes 63

3 The Chromosomal Basis of Heredity 72

3.1 Each species has a characteristic set of chromosomes 74

3.2 The daughter cells of mitosis have identical chromosomes 75

In mitosis, the replicated chromosomes align on the spindle, and the sister chromatids pull apart 76

3.3 Meiosis results in gametes that differ genetically 78

The first meiotic division reduces the chromosome number by half 84

The second meiotic division is equational 87

3.4 Eukaryotic chromosomes are highly coiled complexes of DNA and protein 88

Chromosome-sized DNA molecules can be separated by electrophoresis 88

The nucleosome is the basic structural unit of chromatin 88

Chromatin fibers form discrete chromosome territories in the nucleus 92

The metaphase chromosome is a hierarchy of coiled coils 92

Heterochromatin is rich in satellite DNA and low in gene content 94

3.5 The centromere and telomeres are essential parts of chromosomes 95

The centromere is essential for chromosome segregation 95

The telomere is essential for the stability of the chromosome tips 96

The Human Connection Sick of Telomeres 99
William C. Hahn, Christopher M. Counter, Ante S. Lundberg, Roderick L. Beijersbergen, Mary W. Brooks, and Robert A. Weinberg

3.6 Genes are located in chromosomes 99

Special chromosomes determine sex in many organisms 99

X-linked genes are inherited according to sex 100

Hemophilia is a classic example of human X-linked inheritance 102

In birds, moths, and butterflies, the sex chromosomes are reversed 103

Experimental proof of the chromosome theory came from nondisjunction 103

3.7 Genetic data analysis makes use of probability and statistics 105

Progeny of crosses are predicted by the binomial probability formula 105

Chi-square tests goodness of fit of observed to expected numbers 107

Are Mendel's data a little too good to be true? 110

4 Gene Linkage and Genetic Mapping 120

4.1 Linked alleles tend to stay together in meiosis 122

The degree of linkage is measured by the frequency of recombination 122

The frequency of recombination is the same for coupling and repulsion heterozygotes 124

The frequency of recombination differs from one gene pair to the next 125

Recombination does not occur in *Drosophila* males 125

4.2 Recombination results from crossing-over between linked alleles 126

Physical distance is often—but not always—correlated with map distance 132

One crossover can undo the effects of another 132

4.3 Double crossovers are revealed in three-point crosses 134

The Human Connection Count Your Blessings 135
Joe Hin Tijo and Albert Levan

Interference decreases the chance of multiple crossing-over 138

4.4 Polymorphic DNA sequences are used in human genetic mapping 140

Single-nucleotide polymorphisms (SNPs) are abundant in the human genome 143

4.5 Tetrads contain all four products of meiosis 143

Unordered tetrads have no relation to the geometry of meiosis 144

Tetratype tetrads demonstrate that crossing-over takes place at the four-strand stage of meiosis and is reciprocal 145

Tetrad analysis affords a convenient test for linkage 145

The geometry of meiosis is revealed in ordered tetrads 148

Gene conversion suggests a molecular mechanism of recombination 151

4.6 Recombination is initiated by a double-stranded break in DNA 152

5 Human Chromosomes and Chromosome Behavior 162

5.1 Human beings have 46 chromosomes in 23 pairs 164

The standard human karyotype consists of 22 pairs of autosomes and two sex chromosomes 165

Chromosomes with no centromere, or with two centromeres, are genetically unstable 166

Dosage compensation adjusts the activity of X-linked genes in females 168

The calico cat shows visible evidence of X-chromosome inactivation 170

Some genes in the X chromosome are also present in the Y chromosome 170

The pseudoautosomal region of the X and Y chromosomes has gotten progressively shorter in evolutionary time 170

The history of human populations can be traced through studies of the Y chromosome 172

5.2 Chromosome abnormalities are frequent in spontaneous abortions 174

Down syndrome results from three copies of chromosome 21 175

The Human Connection **Catch-21 176**
Jerome Lejeune, Marthe Gautier, and Raymond Turpin

Trisomic chromosomes undergo abnormal segregation 177

An extra X or Y chromosome usually has a relatively mild effect 179

The rate of nondisjunction can be increased by chemicals in the environment 179

5.3 Chromosome rearrangements can have important genetic effects 179

A chromosome with a deletion has genes missing 180

Rearrangements are apparent in giant polytene chromosomes 180

A chromosome with a duplication has extra genes 182

Human color-blindness mutations result from unequal crossing-over 182

A chromosome with an inversion has some genes in reverse order 184

Reciprocal translocations interchange parts between nonhomologous chromosomes 186

5.4 Polyploid species have multiple sets of chromosomes 189

Polyploids can arise from genome duplications occurring before or after fertilization 190

Polyploids can include genomes from different species 192

Plant cells with a single set of chromosomes can be cultured 194

5.5 The grass family illustrates the importance of polyploidy and chromosome rearrangements in genome evolution 195

6 DNA Structure, Replication, and Manipulation 202

6.1 Genome size can differ tremendously, even among closely related organisms 204

6.2 DNA is a linear polymer of four deoxyribonucleotides 205

6.3 Duplex DNA is a double helix in which the bases form hydrogen bonds 208

6.4 Replication uses each DNA strand as a template for a new one 210

Nucleotides are added one at a time to the growing end of a DNA strand 210

DNA replication is semiconservative: The parental strands remain intact 211

DNA strands must unwind to be replicated 213

Eukaryotic DNA molecules contain multiple origins of replication 216

6.5 Many proteins participate in DNA replication 217

Each new DNA strand or fragment is initiated by a short RNA primer 218

DNA polymerase has a proofreading function that corrects errors in replication 220

The Human Connection **Sickle-Cell Anemia: The First "Molecular Disease" 221**
Vernon M. Ingram

One strand of replicating DNA is synthesized in pieces 222

Precursor fragments are joined together when they meet 223

6.6 Knowledge of DNA structure makes possible the manipulation of DNA molecules 224

Single strands of DNA or RNA with complementary sequences can hybridize 224

Restriction enzymes cleave duplex DNA at particular nucleotide sequences 225

Specific DNA fragments are identified
by hybridization with a probe   228

**6.7   The polymerase chain reaction makes
possible the amplification of a particular DNA
fragment   228**

**6.8   Chemical terminators of DNA synthesis are
used to determine the base sequence   230**

The incorporation of a dideoxynucleotide
terminates strand elongation   231
Dideoxynucleoside analogs are also used in the
treatment of diseases   232

7   The Genetics of Bacteria and Their Viruses   240

**7.1   Many DNA sequences in bacteria are mobile
and can be transferred between individuals
and among species   242**

A plasmid is an accessory DNA molecule, often a
circle   242
The F plasmid is a conjugative plasmid   243
Insertion sequences and transposons play a key
role in bacterial populations   245
Nonconjugative plasmids can be mobilized by
cointegration into conjugative plasmids   247
Integrons have special site-specific recombinases
for acquiring antibiotic-resistance cassettes   247
Bacteria with resistance to multiple antibiotics are
an increasing problem in public health   250

**7.2   Mutations that affect a cell's ability
to form colonies are often used in bacterial
genetics   250**

**7.3   Transformation results from the uptake of
DNA and recombination   251**

**7.4   In bacterial mating, DNA transfer is
unidirectional   253**

The F plasmid can integrate into the bacterial
chromosome   253

Chromosome transfer begins at F and proceeds in
one direction   255
The unit of distance in the *E. coli* genetic map is
the length of chromosomal DNA transferred in
one minute   255
Some F plasmids carry bacterial genes   259

**7.5   Some phages can transfer small pieces of
bacterial DNA   259**

The Human Connection **One Gene, One
Enzyme   261**
George W. Beadle and Edward L. Tatum

**7.6   Bacteriophage DNA molecules in the same
cell can recombine   263**

Bacteriophages form plaques on a lawn of
bacteria   264
Infection with two mutant bacteriophages yields
recombinant progeny   265
Recombination occurs within genes   265

**7.7   Lysogenic bacteriophages do not necessarily
kill the host   267**

Specialized transducing phages carry
a restricted set of bacterial genes   273

8   The Molecular Genetics of Gene Expression   280

**8.1   Polypeptide chains are linear polymers of
amino acids   282**

Human proteins, and those of other vertebrates,
have a more complex domain structure than do
the proteins of invertebrates   284

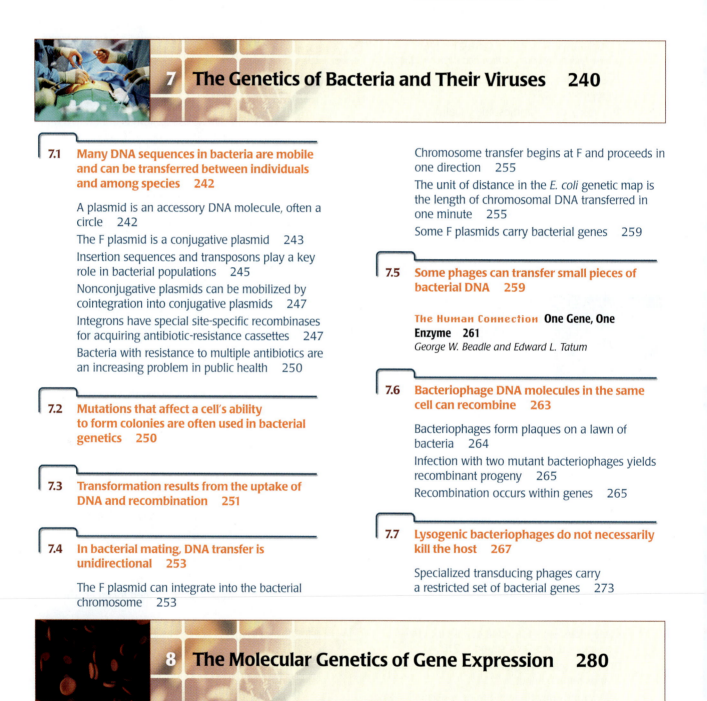

8.2 The linear order of amino acids is encoded in a DNA base sequence 284

8.3 The base sequence in DNA specifies the base sequence in an RNA transcript 284

The chemical synthesis of RNA is similar to that of DNA 284

Eukaryotes have several types of RNA polymerase 286

Particular nucleotide sequences define the beginning and end of a gene 286

Messenger RNA directs the synthesis of a polypeptide chain 289

8.4 RNA processing converts the original RNA transcript into messenger RNA 289

Splicing removes introns from the RNA transcript 291

Human genes tend to be very long even though they encode proteins of modest size 292

Many exons code for distinct protein-folding domains 293

8.5 Translation into a polypeptide chain takes place on a ribosome 293

In eukaryotes, initiation takes place by scanning the mRNA for an initiation codon 294

Elongation takes place codon by codon 294

A termination codon signals release of the finished polypeptide chain 298

Most polypeptide chains fold correctly as they exit the ribosome 299

Prokaryotes often encode multiple polypeptide chains in a single mRNA 301

8.6 The genetic code for amino acids is a triplet code 302

Genetic evidence for a triplet code came from three-base insertions and deletions 302

Most of the codons were determined from *in vitro* polypeptide synthesis 304

The Human Connection Poly-U 305
Marshall W. Nirenberg and J. Heinrich Matthaei

Redundancy and near-universality are principal features of the genetic code 305

An aminoacyl-tRNA synthetase attaches an amino acid to its tRNA 306

Much of the code's redundancy comes from wobble in codon–anticodon pairing 308

8.7 Several ribosomes can move in tandem along a messenger RNA 308

9 Molecular Mechanisms of Gene Regulation 316

9.1 Regulation of transcription is a common mechanism in prokaryotes 318

In negative regulation, the default state of transcription is "on" 318

In positive regulation, the default state of transcription is "off" 319

9.2 In prokaryotes, groups of adjacent genes are often transcribed as a single unit 319

The first regulatory mutations that were discovered affected lactose metabolism 319

Lactose-utilizing enzymes can be inducible (regulated) or constitutive 320

Repressor shuts off messenger RNA synthesis 321

The lactose operator is an essential site for repression 321

The lactose promoter is an essential site for transcription 322

The lactose operon contains linked structural genes and regulatory sequences 322

The Human Connection X-ing Out Gene Activity 323
Mary F. Lyon

The lactose operon is also subject to positive regulation 324

Tryptophan biosynthesis is regulated by the tryptophan operon 326

9.3 Gene activity can be regulated through transcriptional termination 328

Attenuation allows for fine-tuning of transcriptional regulation 328

Riboswitches combine with small molecules to control transcriptional termination 330

9.4 Eukaryotes regulate transcription through transcriptional activator proteins, enhancers, and silencers 330

Galactose metabolism in yeast illustrates transcriptional regulation 332

Transcription is stimulated by transcriptional activator proteins 334

Enhancers increase transcription; silencers decrease transcription 335

The eukaryotic transcription complex includes numerous protein factors 335

Chromatin-remodeling complexes prepare chromatin for transcription 338

Some eukaryotic genes have alternative promoters 339

9.5 Gene expression can be affected by heritable chemical modifications in the DNA 340

Transcriptional inactivation is associated with heavy DNA methylation 340

In mammals, some genes are imprinted by methylation in the germ line 340

9.6 Regulation also takes place at the levels of RNA processing and decay 342

The primary transcripts of many genes are alternatively spliced to yield different products 342

The coding capacity of the human genome is enlarged by extensive alternative splicing 343

Different messenger RNAs can differ in their persistence in the cell 343

RNA interference (RNAi) results in the destruction of RNA transcripts 343

9.7 Regulation can also take place at the level of translation 344

Small regulatory RNAs can control translation by base-pairing with the messenger RNA 344

9.8 Some developmental processes are controlled by programmed DNA rearrangements 345

Programmed deletions take place in the development of the mammalian immune system 345

Programmed transpositions take place in the regulation of yeast mating type 348

10 Genomics, Proteomics, and Genetic Engineering 356

10.1 Cloning a DNA molecule takes place in several steps 358

Restriction enzymes cleave DNA into fragments with defined ends 358

Restriction fragments are joined end to end to produce recombinant DNA 360

A vector is a carrier for recombinant DNA 361

Specialized vectors can carry very large DNA fragments 362

Vector and target DNA fragments are joined with DNA ligase 363

A recombinant cDNA contains the coding sequence of a eukaryotic gene 364

Loss of β-galactosidase activity is often used to detect recombinant vectors 365

Recombinant clones are often identified by hybridization with labeled probe 368

10.2 A genomic sequence is like a book without an index, and identifying genes and their functions is a major challenge 368

10.3 Genomics and proteomics reveal genome-wide patterns of gene expression and networks of protein interactions 371

The Human Connection The Human Genome Sequence 371
Eric S. Lander and 248 other investigators

DNA microarrays are used to estimate the relative level of gene expression of each gene in the genome 372

Microarrays reveal groups of genes that are coordinately expressed during development 374

Yeast two-hybrid analysis reveals networks of protein interactions 375

10.4 Reverse genetics creates an organism with a designed mutation 377

Recombinant DNA can be introduced into the germ line of animals 377

Recombinant DNA can also be introduced into plant genomes 379

Transformation rescue is used to determine experimentally the physical limits of a gene 380

10.5 Genetic engineering is applied in medicine, industry, agriculture, and research 381

Animal growth rate can be genetically engineered 381

Crop plants with improved nutritional qualities can be created 382

The production of useful proteins is a primary impetus for recombinant DNA 382

Animal viruses may prove useful vectors for gene therapy 384

11 The Genetic Control of Development 390

11.1 Mating type in yeast illustrates transcriptional control of development 392

11.2 The determination of cell fate in *C. elegans* development is largely autonomous 393

Development in *C. elegans* exhibits a fixed pattern of cell divisions and cell lineages 394

Cell fate is determined by autonomous development and/or intercellular signaling 394

Developmental mutations often affect cell lineages 395

Transmembrane receptors often mediate signaling between cells 395

Cells can determine the fate of other cells through ligands that bind with their transmembrane receptors 397

11.3 Development in *Drosophila* illustrates progressive regionalization and specification of cell fate 399

The Human Connection Maternal Impressions 400
David H. Skuse, Rowena S. James, Dorothy V. M. Bishop, Brian Coppin, Paola Dalton, Gina Aamodt-Leeper, Monique Barcarese-Hamilton, Catharine Creswell, Rhona McGurk, and Patricia A. Jacobs

Mutations in a maternal-effect gene result in defective oocytes 401

Embryonic pattern formation is under genetic control 401

Coordinate genes establish the main body axes 403

Gap genes regulate other genes in broad anterior–posterior regions 404

Pair-rule genes are expressed in alternating segments or parasegments 404

Segment-polarity genes govern differentiation within segments 405

Interactions among genes in the regulatory hierarchy ensure an orderly progression of developmental events 407

Homeotic genes function in the specification of segment identity 407

HOX genes function at many levels in the regulatory hierarchy 409

11.4 Floral development in *Arabidopsis* illustrates combinatorial control of gene expression 411

Flower development in *Arabidopsis* is controlled by MADS box transcription factors 412

Flower development in *Arabidopsis* is controlled by the combination of genes expressed in each concentric whorl 413

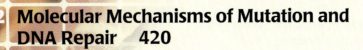

12 Molecular Mechanisms of Mutation and DNA Repair 420

12.1 Mutations are classified in a variety of ways 422

Mutagens increase the chance that a gene undergoes mutation 422

Germ-line mutations are inherited; somatic mutations are not   422

Conditional mutations are expressed only under certain conditions   423

Mutations can affect the amount or activity of the gene product, or the time or tissue specificity of expression   423

12.2   Mutations result from changes in DNA sequence   424

A base substitution replaces one nucleotide pair with another   424

Mutations in protein-coding regions can change an amino acid, truncate the protein, or shift the reading frame   425

Sickle-cell anemia results from a missense mutation that confers resistance to malaria   426

In the human genome, some trinucleotide repeats have high rates of mutation   427

12.3   Transposable elements are agents of mutation   430

Some transposable elements transpose via a DNA intermediate, others via an RNA intermediate   430

Transposable elements can cause mutations by insertion or by recombination   433

Almost 50 percent of the human genome consists of transposable elements, most of them no longer able to transpose   434

12.4   Mutations are statistically random events   435

Mutations arise without reference to the adaptive needs of the organism   435

The discovery of mutagens made use of special strains of *Drosophila*   437

Mutations are nonrandom with respect to position in a gene or genome   438

12.5   Spontaneous and induced mutations have similar chemistries   439

Purine bases are susceptible to spontaneous loss   439

Some weak acids are mutagenic   440

A base analog masquerades as the real thing   440

Highly reactive chemicals damage DNA   441

Some agents cause base-pair additions or deletions   442

Ultraviolet radiation absorbed by DNA is mutagenic   442

Ionizing radiation is a potent mutagen   442

The Chernobyl nuclear accident had unexpectedly large genetic effects   445

The Human Connection   Damage Beyond Repair   446
Frederick S. Leach and 34 other investigators

12.6   Many types of DNA damage can be repaired   447

Mismatch repair fixes incorrectly matched base pairs   447

The AP endonuclease system repairs nucleotide sites at which the base has been lost   448

Special enzymes repair damage caused to DNA by ultraviolet light   450

Excision repair works on a wide variety of damaged DNA   450

Postreplication repair skips over damaged bases   450

12.7   Genetic tests are useful for detecting agents that cause mutations and cancer   451

13   Molecular Genetics of the Cell Cycle and Cancer   458

13.1   The cell cycle is under genetic control   460

Many genes are transcribed during the cell cycle just before their product is needed   462

Mutations affecting the cell cycle have helped to identify the key regulatory pathways   463

Cyclins and cyclin-dependent protein kinases propel the cell through the cell cycle   464

The retinoblastoma protein controls the initiation of DNA synthesis   466

Protein degradation also helps regulate the cell cycle   467

13.2   Checkpoints in the cell cycle allow damaged cells to repair themselves or to self-destruct   467

The p53 transcription factor is a key player in the DNA damage checkpoint 468

The centrosome duplication checkpoint and the spindle checkpoint function to maintain the normal complement of chromosomes 472

13.3 Cancer cells have a small number of mutations that prevent normal checkpoint function 474

Proto-oncogenes normally function to promote cell proliferation or to prevent apoptosis 476

Tumor-suppressor genes normally act to inhibit cell proliferation or to promote apoptosis 477

13.4 Mutations that predispose to cancer can be inherited through the germ line 478

Cancer initiation and progression occurs through mutations that allow affected cells to evade normal cell-cycle checkpoints 478

Retinoblastoma is an inherited cancer syndrome associated with loss of heterozygosity in the tumor cells 480

Some inherited cancer syndromes result from defects in processes of DNA repair 482

13.5 Acute leukemias are proliferative diseases of white blood cells and their precursors 483

Some acute leukemias result from a chromosomal translocation that fuses a transcription factor with a leukocyte regulatory sequence 483

The Human Connection Two Hits, Two Errors 484
Alfred G. Knudson

Other acute leukemias result from a chromosomal translocation that fuses two genes to create a novel chimeric gene 485

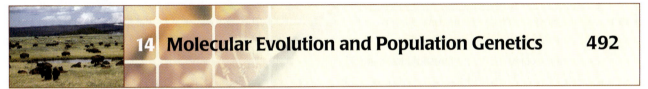

14 Molecular Evolution and Population Genetics 492

14.1 DNA and protein sequences contain information about the evolutionary relationships among species 494

A gene tree is a diagram of the inferred ancestral history of a group of sequences 494

Bootstrapping is a method of assigning a level of confidence to each node in a gene tree 496

A gene tree does not necessarily coincide with a species tree 496

Rates of evolution can differ dramatically from one protein to another. 497

Rates of evolution of nucleotide sites differ according to their function 498

New genes usually evolve through duplication and divergence 499

14.2 Genotypes may differ in frequency from one population to another 500

Allele frequencies are estimated from genotype frequencies 500

The allele frequencies among gametes equal those among reproducing adults 502

14.3 Random mating means that mates pair without regard to genotype 502

The Hardy–Weinberg principle has important implications for population genetics 504

If an allele is rare, it is found mostly in heterozygous genotypes 505

Hardy–Weinberg frequencies can be extended to multiple alleles 506

X-linked genes are a special case because males have only one X-chromosome 506

14.4 Highly polymorphic sequences are used in DNA typing 507

DNA exclusions are definitive 510

14.5 Inbreeding means mating between relatives 510

Inbreeding results in an excess of homozygotes compared with random mating 512

14.6 Evolution is accompanied by genetic changes in species 513

14.7 Mutation and migration bring new alleles into populations 513

14.8 Natural selection favors genotypes that are better able to survive and reproduce 514

Fitness is the relative ability of genotypes to survive and reproduce 514

Allele frequencies change slowly when alleles are either very rare or very common 515

Selection can be balanced by new mutations 516

Occasionally the heterozygote is the superior genotype 516

The Human Connection **Resistance in the Blood 517**
Anthony C. Allison

14.9 Some changes in allele frequency are random 518

14.10 Mitochondrial DNA is maternally inherited 521

Human mtDNA evolves changes in sequence at an approximately constant rate 521

Modern human populations originated in subsaharan Africa approximately 100,000 years ago 522

15 The Genetic Basis of Complex Inheritance 530

15.1 Multifactorial traits are determined by multiple genes and the environment 532

Continuous, categorical, and threshold traits are usually multifactorial 533

The distribution of a trait in a population implies nothing about its inheritance 534

15.2 Variation in a trait can be separated into genetic and environmental components 536

The genotypic variance results from differences in genotype 538

The environmental variance results from differences in environment 539

Genotype and environment can interact, or they can be associated 541

There is no genotypic variance in a genetically homogeneous population 542

The number of genes affecting a quantitative trait need not be large 542

The broad-sense heritability includes all genetic effects combined 543

Twin studies are often used to assess genetic effects on variation in a trait 544

15.3 Artificial selection is a form of "managed evolution" 544

The narrow-sense heritability is usually the most important in artificial selection 545

There are limits to the improvement that can be achieved by artificial selection 547

Inbreeding is generally harmful, and hybrids may be the best 547

15.4 Genetic variation is revealed by correlations between relatives 548

Covariance is the tendency for traits to vary together 548

The additive genetic variance is transmissible; the dominance variance is not 549

The most common disorders in human families are multifactorial 549

15.5 Pedigree studies of genetic polymorphisms are used to map loci for quantitative traits 551

The Human Connection **Win, Place, or Show? 552**
Hugh E. Montgomery and 18 other investigators

Some QTLs have been identified by examining candidate genes 553

Answers to Even-Numbered Problems 561
Word Roots: Prefixes, Suffixes, and Combining Forms 571
Glossary 575
Index 590

Papers excerpted in Connections in chronological order

Karl Landsteiner 1901
Anatomical Institute,
Vienna, Austria
*On Agglutination Phenomena
in Normal Blood*

Archibald E. Garrod 1908
St. Bartholomew's Hospital,
London, England
Inborn Errors of Metabolism

**George W. Beadle and
Edward L. Tatum** 1941
Stanford University, Stanford, California
Genetic Control of Biochemical Reactions in Neurospora

Anthony C. Allison 1954
Radcliffe Infirmary,
Oxford, England
*Protection Afforded by Sickle-Cell Trait Against Subtertian
Malarial Infection*

Joe Hin Tijo[1] and Albert Levan[2] 1956
[1]Estacion Experimental de Aula Dei, Zaragoza, Spain
[2]Institute of Genetics, Lund, Sweden
The Chromosome Number in Man

Vernon M. Ingram 1957
Cavendish Laboratory, University of Cambridge, England
*Gene Mutations in Human Hemoglobin: The Chemical
Difference Between Normal and Sickle-Cell Hemoglobin*

**Jerome Lejeune, Marthe Gautier,
and Raymond Turpin** 1959
National Center for Scientific Research, Paris, France
*Study of the Somatic Chromosomes
of Nine Down Syndrome Children*
(original in French)

Mary F. Lyon 1961
Medical Research Council,
Harwell, England
Gene Action in the X Chromosome of the Mouse
(Mus musculus L.)

Marshall W. Nirenberg and Heinrich Matthaei 1961
National Institutes of Health, Bethesda, Maryland
The Dependence of Cell-Free Protein Synnthesis in E. Coli
upon Naturally Occurring or Synthetic Polyribonucleotides

Alfred G. Knudson 1971
M. D. Anderson Hospital
The University of Texas
Houston, Texas
*Mutation and Cancer: Statistical Study
of Retinoblastoma*

**Frederick S. Leach and
34 other investigators** 1993
Johns Hopkins University,
Baltimore, MD, and ten other
research institutions
*Mutations of a mutS Homolog
in Hereditary Nonpolyposis
Colorectal Cancer*

**David H. Skuse[1],
Rowena S. James[2],
Dorothy V. M. Bishop[3],
Brian Coppin[4],
Paola Dalton[2],
Gina Aamodt-Leeper[1],
Monique Barcarese-Hamilton[1],
Catharine Creswell[1],
Rhona McGurk[1], and
Patricia A. Jacobs[2]** 1997
[1]Institute of Child Health,
 London, England
[2]Salisbury District Hospital,
 Salisbury, Wiltshire, UK
[3]Medical Research Council
 Applied Psychology Unit,
 Cambridge, UK
[4]Princess Anne Hospital,
 Southampton, UK
*Evidence from Turner's Syndrome of an Imprinted
X-linked Locus Affecting Cognitive Function*

**Hugh E. Montgomery and
18 other investigators** 1998
University College, London,
and 6 other research institutions
Human Gene for Physical Performance

**William C. Hahn,[1,2]
Christopher M. Counter,[3]
Ante S. Lundberg,[1,2]
Roderick L. Beijersbergen,[1]
Mary W. Brooks,[1] and
Robert A. Weinberg[1]** 1999
[1]Massachusetts Institute of Technology,
 Cambridge, Massachusetts
[2]Harvard Medical School, Boston,
 Massachusetts
[3]Duke University Medical Center,
 Durham, North Carolina
*Creation of Human Tumor Cells with Defined
Genetic Elements*

**Eric S. Lander and 248 other
investigators** 2001
The Whitehead Institute for Biomedical Research,
Massachusetts Institute of Technology,
Cambridge, Massachusetts, and
23 other research institutions
*Initial Sequencing and Analysis of the
Human Genome*

Preface

Biology undergraduates often find genetics difficult for two reasons. First, certain aspects of genetics are abstract and include unfamiliar concepts from probability and statistics. Second, genetics covers a broad territory, ranging from molecular biology to population genetics. Yet undergraduates are motivated to study genetics because of media reports about the human genome and genetic risk factors for disease, as well as many social and ethical controversies related to genetics such as genetic privacy, cloning, and stem-cell research. The challenges for the instructor are to sustain this motivation, to kindle a desire to understand the principles of genetics and genomics, and to help students integrate their knowledge into a wider social and ethical context. We have written *Essential Genetics* to help instructors meet these challenges. It is designed for the shorter, less comprehensive introductory course. The brevity of the text fits the pace of what can be covered in a typical one-semester or one-quarter course. The topics have been carefully chosen to help students achieve the following learning objectives:

- Understand the basic processes of gene transmission, mutation, expression, and regulation
- Learn to formulate genetic hypotheses, work out their consequences, and test the results against observed data
- Develop basic skills in problem solving, including single-concept exercises, those requiring the application of several concepts in logical order, and numerical problems requiring some arithmetic for solution
- Gain some sense of the social and historical context in which genetics has developed as well as an appreciation of current trends
- Become aware of some of the genetic resources and information that are available through the World Wide Web

What's New in the 4th Edition?

The complete integration of "molecular genetics" and "classical genetics" makes *Essential Genetics* unique. They are recognized as merely different aspects of the same things—the transmission, mutation, and function of the genetic material. These processes are manifested at the molecular level and can be studied by such techniques as DNA electrophoresis, as well as at the organismic level through their association with visible traits. These levels of genetic analysis are integrated even in our discussion of Mendel's experiments.

The fourth edition includes major changes. A great deal of new material has been added to incorporate new discoveries, and a considerable amount has been rewritten to reflect current thinking in the field. *Over 100 of the major illustrations are completely new or significantly revised to increase clarity.* We have also gone over the text, word by word and line by line, to condense, clarify, and update. To keep the book at a reasonable size, whenever new material has been added, we have tried to delete or abbreviate an equivalent amount that is less relevant, of secondary importance, or no longer needed.

The fourth edition includes dramatic new results on the evolutionary relationships in the **tree of life** (Chapter 1). In addition to the recent major reorganization of the phylogenetic tree of eukaryotes, we examine the similarities and differences in the genetic systems of bacteria, achaeans, and eukaryotes and in the bacterial origins of mitochondria and chloroplasts. Our emphasis is on the molecular unity of life as seen in comparisons of genomes and proteomes.

Chapter 3 includes the correct structure of the **30-nanometer chromatin fiber,** which has recently been shown to consist of a string of nucleosomes formed into a series of stacked right-handed coils in which the path of the linker DNA repeatedly traces out a seven-pointed star as it travels down the length of the fiber. The fourth edition also includes a discussion of how the chromatin in nondividing cells is organized into **chromosome territories** and of how these might contribute to genetic regulation. The DNA sequence organization of **human centromeres** has also been included based on new data from the Human Genome Project.

An updated model for **homologous recombination** is included for the first time (Chapter 4). This model is based on compelling evidence that recombination is initiated by a double-stranded break and that the process includes two Holliday junctions. Furthermore, we distinguish between the noncrossover pathway for break repair and the crossover pathway, since it now appears that these pathways are distinct.

Chapter 5 includes new data showing that the **inactive X chromosome** in mammals is not completely silenced. About 15% of the genes in the "inactive X" escape inactivation to some degree. This chapter also includes a major new section on tracing human history through **Y-chromosome haplotypes.** It gives several examples in detail, including the legacy of Genghis Kahn, the so-called black Jews of Africa, and the origin of European Gypsies. New data are also included on the effects of chemicals in the environment on **nondisjunction.** Strange as it may seem, new data have also required a major revision in the way we think about the origin of polyploidy. In particular, it has become clear that chromosomally **unreduced gametes** are a significant contributor to the formation of new polyploid species.

Even DNA synthesis has had a makeover (Chapter 6) in light of new findings that the mechanism of **primer removal** and replacement is rather different in eukaryotes than it is in prokaryotes.

Chapter 8 includes new discussions of transcription and RNA processing as **coupled processes,** in which each step in the process recruits the proteins needed to accomplish the next step along the way. A new section on **protein folding** is also included, emphasizing the role of trigger factor, chaperones, and chaperonins in helping large or slowly folding polypeptides to find their correct three-dimensional structure. There are also new illustrations of polypeptide chain elongation and termination, showing more clearly how the small and large ribosomal subunits move along the mRNA in a ratchet-like mechanism.

Hardly any field of genetics and genomics moves as rapidly as that of **gene regulation,** and much of this new information in incorporated into Chapter 9. Regulation by means of RNA has become a particularly active field, and we include major new discussions of **riboswitches,** small regulatory RNAs that control translation (**"kissing complexes"**), and the mechanism of RNA interference (**RNAi**).

The chapter on genomics, proteomics, and genetic engineering (Chapter 10) has been completely reorganized and expanded. There are new sections on the use of **DNA microarrays** to study the coordinate regulation of genes in development, the yeast **two-hybrid system** to identify protein that interact, and the **network of protein-protein interactions** among nuclear proteins in yeast.

The genetic control of development (Chapter 11) is also a rapidly changing field. The fourth edition includes a discussion of interactions among genes in the **regulatory hierarchy** that ensure an orderly progression of events during *Drosophila* development, as well as new illustrations of **floral ABC model** as it functions in *Arabidopsis.*

Chapter 12 deals with mutation. It has been relocated to be nearer the chapter on cancer, which is preeminently a disease of somatic mutation. There is a major new discussion of **trinucleotide expansion diseases** with emphasis on the fragile-X syndrome, including the molecular mechanism by which the fragile-X mental retardation protein (FMRP) exerts its effects.

The fourth edition has a major new section on **genetic control of the cell cycle** through cyclins, cyclin-dependent protein kinases, and other mechanisms such as proteolysis (Chapter 13). The emphasis is on **cell-cycle checkpoints** that prevent runaway growth or that promote programmed

cell death, and this sets the stage for a discussion of cancer as a series of mutations that subvert these control mechanisms.

Chapter 14 has a major new section on **molecular evolution** that focuses on how inferences about gene evolution, protein evolution, and organismic evolution can be gleaned from DNA or protein sequences. We discuss methods of molecular phylogenetic trees and bootstrap support for the nodes, gene trees and species trees, and rates and patterns of protein and DNA evolution including the molecular clock and summarize current thinking on how new genes come into being.

Much progress has also been made in using methods of genetic mapping and candidate genes to identify the genes involved in complex multifactorial traits. Chapter 15 has a thoroughly updated section on the polymorphism in the **serotonin reuptake transporter** and its relation to risk of **severe depression** in response to stressful life events.

Understanding genetics and genomics includes a significant component of simply learning the language of genetics and how to use it. To aid in this process, we have added a new type of problem at the end of each chapter that helps students to learn the **proper use of key terms** in their usual context. These exercises are in addition to the multiplicity of different types of chapter-end problems updated from the previous edition.

Chapter Organization

To help the student keep track of the main concepts without being distracted by details, each chapter begins with a list of **Key Concepts** written in simple declarative sentences, highlighting the most important concepts presented in the chapter. An **Outline** shows the principal subjects to be discussed. The body of each chapter provides more detailed information and experimental evidence. An opening paragraph gives an overview of the chapter, illustrates the subject with some specific examples, and shows how the material is connected to genetics as a whole. The section and subsection **Headings** are in the form of complete sentences that encapsulate the main message. The text makes liberal use of **Numbered** and **Bulleted Lists** and **Bullets** to aid students in organizing their learning, as well as **Summary Statements** set apart from the main text in order to emphasize important principles. A feature introduced in the third edition, designed to reinforce understanding, called **A Moment to Think,** appears again in this edition. This is a problem integrated into the text in which the student is asked to interrupt studying to think about the concept just described and to use it in solving an actual problem. Each chapter also includes **The Human Connection.** This special feature highlights a research paper in human genetics that reports a key experiment or raises important social, ethical, or legal issues. Each Human Connection has a brief introduction of its own, explaining the importance of the experiment and the context in which it was carried out. At the end of each chapter is a complete **Chapter Summary, Key Terms, GeNETics on the Web** exercises that guide students in the use of Internet resources in genetics, and several different types and levels of **Problems.** At the back of the book are **Answers** to even-numbered problems and a complete **Glossary** as well as a list of frequently used **Word Roots** that will help students to understand key genetic terms and make them part of their vocabulary.

Contents

The organization of the chapters is that favored by the majority of instructors who teach genetics. It is the organization we use in our own courses. An important feature is the presence of an introductory chapter providing a broad overview of the gene: what it is, what it does, how it changes, how it evolves. Today, most students learn about DNA in grade school or high school. In our teaching, we have found it rather artificial to pretend that DNA does not exist until the middle of the term. The introductory chapter therefore serves to connect the more advanced concepts that students are about to learn with what they already know. It also serves to provide each student with a solid framework for integrating the material that comes later.

Throughout each chapter, there is a balance between challenge and motivation, between observation and theory, and between principle and concrete example. Molecular and classical genetics are integrated throughout, and the principles of human genetics are interwoven into the entire fabric of the book. On the other hand, the book is also liberally supplied with examples from animals and plants, especially model organisms.

A number of points related to organization and coverage should be noted:

Chapter 1 is an overview of genetics designed to bring students with disparate backgrounds to a common level of understanding. This chapter enables classical, molecular, and evolutionary genetics to be integrated throughout the rest of the book. Included in Chapter 1 are the basic concepts of genetics: genes as DNA that function through transcription and translation, that change by mutation, and that affect organisms through inborn errors of metabolism. Chapter 1 also includes a discussion of the classical experiments demonstrating that DNA is the genetic material.

Chapters 2 through 5 are the core of Mendelian genetics, including segregation and independent assortment, the chromosome theory of heredity, mitosis and meiosis, linkage and chromosome mapping, tetrad analysis in fungi, and chromosome mechanics. An important principle of genetics, too often ignored or given inadequate treatment, is that of the complementation test and how complementation differs from segregation or other genetic principles. Chapter 4 expands on the use of molecular markers in genetics, because these are the principal types of genetic markers in use today.

Chapter 6 deals with DNA, including the details of DNA structure and replication. It also discusses how basic research that revealed the molecular mechanisms of DNA replication ultimately led to such important practical applications as DNA hybridization analysis, DNA sequencing, and the polymerase chain reaction. These examples illustrate the value of basic research in leading, often quite unpredictably, to practical applications.

Chapter 7 deals with the principles of genetics in prokaryotes, beginning with the genetics of mobile DNA, plasmids, and integrons, and their relationships to the evolution of multiple antibiotic resistance. There is a thorough discussion of mechanisms of genetic recombination in microbes, including transformation, conjugation, and transduction, as well as discussion of temperate and virulent bacteriophages.

Chapters 8 through 12 focus on molecular genetics in the strict sense. Chapter 8 examines the details of gene expression, including transcription, RNA processing, and translation. Chapter 9 is an integrative chapter that deals with genetic mechanisms of regulation, with examples of mechanisms of gene regulation in prokaryotes as well as eukaryotes. We include broader aspects of gene regulation that are topics of much current research, such as chromatin remodeling complexes, imprinting, and RNAi. Chapter 10 deals with recombinant DNA and genome analysis. Included are the use of restriction enzymes and vectors in recombinant DNA, cloning strategies, transgenic animals and plants, and applications of genetic engineering. Chapter 11 examines the genetic control of development with emphasis on models in *C. elegans*, *D. melanogaster*, and *A. thaliana*. Chapter 12 focuses on mechanisms of mutation and DNA repair, including chemical mutagens and information on the genetic effects of the Chernobyl nuclear accident.

Chapter 13 stresses cancer from the standpoint of the genetic control of cell division, with emphasis on the checkpoints that, in normal cells, result either in inhibition of cell division or in programmed cell death (apoptosis). Cancer results from a series of successive mutations, usually in somatic cells, that overcome the normal checkpoints that control cellular proliferation.

Chapters 14 and 15 deal with molecular evolution and population genetics. The discussion includes gene trees and species trees and the population genetics of the CCR5 receptor mutation that confers resistance to infection by HIV virus. It also includes DNA typing in criminal investigations, paternity testing, the effects of inbreeding, and the evolutionary mechanisms that drive changes in allele frequency. The approach to quantitative genetics includes a discussion of how particular genes influencing quantitative traits (QTLs, or quantitative-trait loci) may be identified and mapped by linkage analysis. There is also a section on the genetic determinants of human behavior with examples of the approach using "candidate" genes that led to the identification of the "natural Prozac" polymorphism in the human serotonin transporter gene.

Special Features

A Moment to Think

A unique feature of this book is found in boxes called **A Moment to Think.** These are problems that ask a student to pause and think about a concept and apply it to an actual situation. Often these problems use the results of classical experiments to help the student transform a concept from abstract to concrete, and carry it from thought to action. The answer is provided on a different page.

The Human Connection

The Human Connection in each chapter is our way of connecting to the world of human genetics outside the classroom. All of the Connections include short excerpts from the original literature of genetics, usually papers, each introduced with a short explanatory passage. Many of the Connections are excerpts from classic materials, such as Garrod's book on inborn errors of metabolism, but by no means are all of the "classic" papers are old papers, as you will see by examining the publication dates.

The pieces are called The Human Connection because each connects the material to something that broadens or enriches its implications for human beings. Some of the Connections raise issues of ethics in the application of genetic knowledge, social issues that need to be addressed, or issues related to laboratory animals. They illustrate other things as well. Because each Connection names the place where the research was carried out, the student will learn that great science is done in many universities and research institutions throughout the world. In papers that use outmoded or unfamiliar terminology, or archaic gene symbols, we have substituted the modern equivalent to make the material more accessible to the student.

GeNETics on the Web

The World Wide Web is a rich source of information on all aspects of genetics. To make genetic information on the Internet available to the beginning student, we have developed GeNETics on the Web, links that make use of Internet resources related to human genetics. The relevant genetics sites are accessed through the use of key words that are highlighted in each exercise. The key words are maintained as hot links at the publisher's Web site (http://www.jbpub.com/genetics) and are kept up to date should the address of each site change.

Solutions Step by Step

Each chapter contains a section entitled **Solutions Step by Step** that demonstrates problems worked in full, explaining step by step a path of logical reasoning that can be followed to analyze the problem. The Solutions Step by Step serve as another level of review of the important concepts used in working problems. The solutions also emphasize some of the most common mistakes made by beginning students and give pointers on how the student can avoid falling into these conceptual traps.

Levels and Types of Problems

Each chapter provides numerous problems for solution, graded in difficulty, so students can test their understanding. The problems are of two different types:

- **Issues and Ideas** ask for genetic principles to be restated in the student's own words; some are matters of definition or call for the application of elementary principles.

- **Concepts in Action** are problems that require the student to reason using genetic concepts. The problems make use of a variety of formats, including true or false, multiple choice, matching, and traditional types of word problems. Many of the Concepts in Action require some numerical calculation. The level of mathematics is that of arithmetic and elementary probability as it pertains to genetics. None of the problems uses mathematics beyond elementary algebra.

Answers to Problems

The answers to the even-numbered Concepts in Action are included in the **Answer** section at the end of the book. The answers are complete; they explain the logical foundation of the solution and lay out the methods. The answers to the rest of the Concepts in Action problems are available for the instructor on the Instructor's ToolKit CD-ROM.

Word Roots and Glossary

We have included a compilation of **Word Roots** that students find helpful in interpreting and remembering the meaning of technical terms. This precedes the **Glossary** of Key Words.

Further Reading

Each chapter also includes recommendations for **Further Reading** for the student who either wants more information or who needs an alternative explanation for the material presented in the book. Some additional "classic" papers and historical perspectives are included.

Illustrations

Every chapter is richly illustrated with beautiful graphics in which color is used functionally to enhance the value of each illustration as a learning aid. The illustrations are also heavily annotated with "process labels" explaining step-by-step what is happening at each level of the illustration. These labels make the art user-friendly, inviting, and maximally informative.

Adaptability and Flexibility

There is no necessary reason to start at the beginning and proceed straight to the end. Each chapter is a self-contained unit that stands on its own. This feature gives the book the flexibility to be used in a variety of course formats. Throughout the book, we have integrated classical and molecular principles, so you can begin a course with almost any of the chapters. Most teachers will prefer starting with the overview in Chapter 1, possibly as suggested reading, because it brings every student to the same basic level of understanding. Teachers preferring the Mendel-early format should continue with Chapter 2; those preferring to teach the details of DNA early should continue with Chapter 6. Some teachers are partial to a chromosomes-early format, which would suggest continuing with Chapter 3, followed by Chapters 2 and 4. A novel approach would put genetic engineering first, which could be implemented by continuing with Chapter 10. The writing and illustration program was designed to accommodate a variety of formats, and we encourage teachers to take advantage of this flexibility in order to meet their own special needs.

Instructor and Student Supplements

An unprecedented offering of traditional and interactive multimedia supplements is available to assist instructors and aid students in mastering genetics. Additional information and review copies of any of the following items are available through your Jones and Bartlett Sales Representative.

For the Instructor

Instructor's ToolKit CD-ROM

This CD-ROM provides the instructor with a powerful set of programs that can easily be integrated into your daily routine to help save time, while making classroom presentations more educational for students. The programs include:

The Kaleidoscope Media Viewer is an easy-to-use multimedia tool containing over 300 figures from the text specially enhanced for classroom presentation. You select the images you need by chapter, topic, or figure number to create your own lecture aid.

PowerPoint Image Bank provides all of the illustrations, photos, and tables (to which Jones and Bartlett owns the copyright or has permission to reprint digitally), inserted into PowerPoint slides. With the Microsoft PowerPoint program you can quickly and easily copy individual image slides into your existing slides.

PowerPoint Presentations, revised by Elena R. Lozovsky of Harvard University, provides outline summaries of each chapter. The slide set can be customized to meet your classroom needs.

The Computerized Test Bank, prepared by Elena R. Lozovsky, contains over 700 test items. There is a mix of factual, descriptive, analytical, and quantitative question types. A typical chapter file contains 20 multiple-choice objective questions, 15 fill-ins, and 15 quantitative. The Computerized Test Bank allows the instructor to easily generate quizzes and tests from the complete set of over 700 questions.

A Solutions Manual, authored by Elena R. Lozovsky, contains worked solutions for all the Concepts in Action problems found at the end of each chapter in the main text. Only solutions to even-numbered problems are provided in the back of the main text. This allows the instructor to control access to solutions for odd-numbered problems. The solutions are supplied as a Microsoft Word document. Solutions to the *Student Study Guide* problems can be found here.

An Electronic Companion to Genetics™ Version 2.0 © 2001, Cogito Learning Media, Inc.

This Mac/IBM CD-ROM, by Philip Anderson and Barry Ganetzky of the University of Wisconsin, Madison, reviews important genetics concepts using state of the art interactive multimedia. It consists of hundreds of animations, diagrams, and videos that dynamically explain difficult concepts to students. New to this version are review screens and self-tests on genomics and improved navigation.

For the Student

Study Guide and Solutions Manual

Authored by Elena R. Lozovsky, this study aid contains chapter outlines, key terms, and study questions to enhance student learning. This essential study tool also includes worked solutions for all the Concepts in Action problems found at the end of each chapter in the main text. Only solutions to even-numbered problems are provided in the back of the main text.

The Gist of Genetics: Guide to Learning and Review

Written by Rowland H. Davis and Stephen G. Weller of the University of California, Irvine, this study aid uses illustrations, tables and text outlines to review all of the fundamental elements of genetics. It includes extensive practice problems and review questions with solutions for self-check. The Gist helps students formulate appropriate questions and generate hypothesis that can be tested with classical principles and modern genetic techniques.

GeNETics on the Web

GeNETics on the Web provides animated flashcards, practice questions, and an online glossary. Corresponding to the end-of-chapter GeNETics on the Web exercises, the site offers genetics-related links, articles and monthly updates to other genetics sites on the Web. Material for this site is carefully selected and updated by the authors, and Jones and Bartlett Publishers ensures that links for the site are regularly maintained. Visit the GeNETics on the Web site at http://www.jbpub.com/genetics.

An Electronic Companion to Genetics™ Version 2.0 © 2001, Cogito Learning Media, Inc.

This Mac/IBM CD-ROM, by Philip Anderson and Barry Ganetzky of the University of Wisconsin, Madison, reviews important genetics concepts covered in class using state of the art interactive multimedia. It consists of hundreds of animations, diagrams, and videos that dynamically explain difficult concepts to students. In addition, it contains over 400 interactive multiple choice, "drop and drag," true/false, and fill-in problems. New to this version are review screens and self-tests on genomics and improved navigation. These resources will prove invaluable to students in a self-study environment and to instructors as a lecture enhancement tool. This CD-ROM is available for packaging exclusively with Jones and Bartlett Publishers texts.

Acknowledgments

We are indebted to our colleagues whose advice and thoughts were immensely helpful throughout the preparation of this book. These colleagues range from specialists in various aspects of genetics who checked for accuracy or suggested improvement to instructors who evaluated the material for suitability in teaching or sent us comments on the text as they used it in their courses.

Laura Adamkewitz, George Mason University
Jeremy C. Ahouse, Brandeis University
Mary Alleman, Duquesne University
Peter D. Ayling, University of Hull
John C. Bauer, Stratagene, Inc., La Jolla CA
Anna C. Berkowitz, Purdue University
Mary K. B. Berlyn, Yale University
Thomas A. Bobik, University of Florida
Colin G. Brooks, The Medical School, Newcastle
Pierre Carol, Université Joseph Fourier
Domenico Carputo, University of Naples
Sean Carroll, University of Wisconsin
Chris Caton, University of Birmingham
John Celenza, Boston University
Alan C. Christiensenm University of Nebraska-Lincoln
Christoph Cremer, University of Heidelberg
Marion Cremer, Ludwig Maximilians University
Thomas Cremer, Ludwig Maximilians University
Leslie Dendy, University of New Mexico
Stephen J. D'Surney, University of Mississippi
John W. Drake, National Institute of Environmental Health Sciences, Research Triangle Park, NC
Kathleen Dunn, Boston College
Chris Easton, State University of New York
Wolfgang Epstein, University of Chicago
Brian E. Fee, Manhattan College
Gyula Ficsor, Western Michigan University
Robert G. Fowler, San Jose State University
David W. Francis, University of Deleware

Gail Gasparich, Towson University
Ellion S. Goldsteinm Arizona State University
Patrick Guilfoile, Bemidji State University
Jeffrey C. Hall, Brandeis University
Mark L. Hammond, Campbell University
Steven Henikoff, Fred Hutchinson Cancer Research Center, Seattle WA
Charles Hoffman, Boston College
Ivan Huber, Fairleigh Dickenson University
Kerry Hull, Bishop's University
Lynn A, Hunter, University of Pittsburgh
Richard Imberski, University of Maryland
Joyce Katich, Monsanto, Inc., St. Louis MO
Jeane M. Kennedy, Monsanto, Inc., St. Louis MO
Jeffrey King, University of Berne
Tobias A. Knoch, German Cancer Research Center, Heidelberg, Germany
Yan B. Linhart, University of Colorado
K. Brooks Low, Yale University
Sally A. MacKenzie, Purdue University
Gustavo Maroni, University of North Carolina
Jeffrey Mitton, University of Colorado
Robert K. Mortimer, University of California
Gisela Mosig, Vanderbilt University
Steve O'Brien, National Cancer Institute
Kevin O'Hare, Imperial College, London
Ronald L. Phillips, University of Minnesota
Robert Pruitt, Purdue University
Peggy Redshaw, Austin College

Pamela Reinagel, California Institute of Technology
Susanne Renner, University of Missouri
Lynn S. Ripley, University of Medicine and Dentistry of New Jersey
Andrew J. Roger, Dalhousie University, Halifax
Kenneth E. Rudd, National Library of Medicine, Bethesda MD
Thomas F. Savage, Oregon State University
Joseph Schlammadinger, University of Debrecen
David Shepard, University of Delaware
Alastair G.B. Simpson, Dalhousie University
Leslie Smith, National Institute of Environmental Health Sciences, Research Triangle Park NC
Charles Staben, University of Kentucky
Johan H. Stuy, Florida State University
David T. Sullivan, Syracuse University
Jeanne Sullivan, West Virginia Wesleyan College
Millard Susman, University of Wisconsin
Irwin Tessman, Purdue University
James H. Thomas, University of Washington
Michael Tully, University of Bath
David Ussery, The Technical University of Denmark
George von Dassow, Friday Harbor Laboratories, Friday Harbor WA
Denise Wallack, Muhlenberg College
Kenneth E. Weber, University of Southern Maine
Tamara Western, Okanagan University College

Finally, a very special thanks goes to geneticist Elena R. Lozovsky of Harvard University, who contributed substantially to the instructional materials at the end of each chapter, as well as to the instructor and student supplements. Her help and support is very gratefully acknowledged.

We also wish to acknowledge the superb art, production, and editorial staff at Jones and Bartlett who helped make this book possible: Stephen Weaver, Dean DeChambeau, Rebecca Seastrong, Anne Spencer, Kimberly Potvin, Shellie Newell, Judy May, Trudi Elliott, and Karen Crowley. We are also grateful to the many people, acknowledged in the legends of the illustrations, who contributed photographs, drawings, and micrographs from their own research and publications, especially those who provided color photographs for this edition. Every effort has been made to obtain permission to use copyrighted material and to make full disclosure of its source. We are grateful to the authors, journal editors, and publishers for their cooperation. Any errors or omissions are wholly inadvertant and will be corrected at the first opportunity.

This grid shows the relative levels of gene expression of all 5538 genes in the genome of two strains of budding yeast. Each spot corresponds to one gene. Red indicates a much higher level of expression in one strain, green much higher level of expression in the other strain. [Courtesy of Dr. Jason Kang/National Cancer Institute.]

key concepts

- Inherited traits are affected by genes.
- Genes are composed of the chemical deoxyribonucleic acid (DNA).
- DNA replicates to form (usually identical) copies of itself.
- DNA contains a code specifying what types of enzymes and other proteins are made in cells.
- DNA occasionally mutates, and the mutant forms specify altered proteins.
- A mutant enzyme is an "inborn error of metabolism" that blocks one step in a biochemical pathway for the metabolism of small molecules.
- Traits are affected by environment as well as by genes.
- Organisms change genetically through generations in the process of biological evolution.
- Because of their common descent, organisms share many features of their genetics and biochemistry.

1

The Genetic Code of Genes and Genomes

chapter organization

1.1 DNA is the molecule of heredity.

1.2 The structure of DNA is a double helix composed of two intertwined strands.

1.3 In replication, each parental DNA strand directs the synthesis of a new partner strand.

1.4 Genes code for proteins.

1.5 Genes change by mutation.

1.6 Traits are affected by environment as well as by genes.

1.7 Evolution means continuity of life with change.

the human connection Black Urine

Chapter Summary
Issues & Ideas
Key Terms & Concepts
Solutions: Step by Step
Concepts in Action: Problems for Solution
Further Readings

geNETics on the web

Every science occasionally undergoes a major advance that completely changes perspectives on the field. This happened in physics with the discovery of subatomic particles and in chemistry with the understanding of the nature of the chemical bond. In genetics it has happened most recently with the development of *genomics.*

Genetics is the study of biologically inherited traits. Each species of living organism is united by a common set of inherited traits, observable characteristics that set it apart from all other species of organisms. For example, a human being habitually stands upright and has long legs, relatively little body hair, a large brain, and a flat face with a prominent nose, jutting chin, distinct lips, and small teeth. Some traits present in human beings we share with other animals to whom we are more distantly related. In common with other mammals, human beings are warmblooded, and human mothers feed their young with milk secreted by mammary glands. In common with other vertebrates, human beings have a backbone and a spinal cord. Every normal human being exhibits these biological characteristics, and all of these traits are inherited.

© Photodisc.

The fundamental concept of genetics is that

> **key concept**
>
> Inherited traits are determined by the elements of heredity that are transmitted from parents to offspring in reproduction; these elements of heredity are called **genes.**

What makes genomics so important to the study of genetics is that it represents significant advances in discovering genes and analyzing their functions. One noteworthy application of genomics led to the development of methods to determine the complete sequence of the constituents that make up the DNA in an organism. A second noteworthy application provided the means to study the patterns of expression of all of an organism's genes simultaneously, to learn which groups of genes act together to result in normal function or disease.

Genomics is merely the latest in a continuing series of major advances in *molecular genetics,* the study of the chemical nature of genes and of how genes function to affect the traits of living organisms. This chapter serves as an introduction to molecular genetics. First we discuss key experiments showing that DNA is the genetic material. Then we provide an overview of how DNA is duplicated in going from one cell generation to the next, how it functions to determine the chemical

makeup of enzymes and other proteins in the cell, and how it undergoes mutation to produce defective proteins that are often associated with inherited diseases. Later in this chapter we return to genomics and give an overview of the extent to which living organisms share similar proteins and cellular processes.

The existence of genes and the rules governing their transmission from generation to generation were discovered by Gregor Mendel in the 1860s. His work with garden peas represents the inauguration of what would become the science of genetics. Mendel's formulation of inheritance was in terms of the abstract rules by which hereditary elements are transmitted from parents to offspring. The approach to the study of genetics through the analysis of offspring from matings is sometimes referred to as *classical* genetics. (Mendel's experiments are the main subject of Chapter 2.)

Molecular genetics got its start only three years after Mendel reported his experiments. In 1869 Friedrich Miescher discovered a new type of weak acid, abundant in the nuclei of white blood cells, that turned out to be the chemical substance of which genes are made. Miescher's weak acid is now called **deoxyribonucleic acid** or **DNA.** Nevertheless, even though the two main pieces of the puzzle of heredity—genes and DNA—had been discovered, the pieces were not put together until about the middle of the twentieth century when the chemical identity between genes and DNA was conclusively demonstrated. The next section shows how this connection was made.

© Photodisc.

EACH SPECIES of organism has its own unique hereditary endowment different from that of other species.

1.1

DNA is the molecule of heredity.

The importance of the cell nucleus in inheritance became clear in the 1870s when the nuclei of the male and female reproductive cells were observed to fuse in the process of fertilization. The next major advance was the discovery of **chromosomes,** thread-like objects inside the nucleus that become visible in the light microscope when

©Andrew S. Bajer.

stained with certain dyes. Chromosomes exhibit a characteristic "splitting" behavior, in which each daughter cell formed by cell division receives an identical complement of chromosomes. How this happens is taken up in Chapter 3. More evidence for the importance of chromosomes was provided by the observation that, whereas the number of chromosomes in each cell differs from one biological species to the next, the number of chromosomes is nearly always constant within the cells of any particular species. These features of chromosomes were well understood by about 1900, and they made it seem likely that chromosomes were the carriers of the genes.

By the 1920s, several lines of indirect evidence suggested a close relationship between chromosomes and DNA. Microscopic studies with special stains showed that DNA is present in chromosomes. Various types of proteins are present in chromosomes too. But whereas most of the DNA in cells of higher organisms is present in chromosomes, and the amount of DNA per cell is constant, the amount and kinds of proteins and other large molecules differ greatly from one cell type to another. The indirect evidence for DNA as the genetic material was unconvincing, because crude chemical analyses had suggested (erroneously, as it turned out) that DNA lacked the chemical diversity needed in a genetic substance. The favored candidate for the genetic material was protein, because proteins were known to be an exceedingly diverse collection of molecules. Proteins therefore became widely accepted as the genetic material, and DNA was thought to provide only the structural framework of chromosomes. Any researcher who hoped to demonstrate that DNA was the genetic material had a double handicap. Such experiments had to demonstrate not only that DNA *is* the genetic material but also that proteins are *not* the genetic material. Some of the experiments regarded as decisive in implicating DNA are described in this section.

■ Genetic traits can be altered by treatment with pure DNA.

One type of bacterial pneumonia in mammals is caused by strains of *Streptococcus pneumoniae* able to synthesize a gelatinous capsule composed of polysaccharide (complex carbohydrate). This capsule surrounds the bacterium and protects it from the defense mechanisms of the infected animal; thus it enables the bacterium to cause disease. When a bacterial cell is grown on solid medium, it undergoes repeated cell divisions to form a visible clump of cells called a **colony.** The enveloping capsule makes the size of each colony large and gives it a glistening or smooth (S) appearance (Figure 1.1). Certain strains of *S. pneumoniae*, however, are unable to synthesize the capsular polysaccharide, and they form small colonies that have a rough (R) surface. The R strains do not cause pneumonia; lacking the capsule, these bacteria are inactivated by the immune system of the host. Both types of bacteria "breed true" in the sense that the progeny formed by cell division have the capsular type of the parent, whether S or R.

When mice are injected either with living R cells or with heat-killed S cells, they remain healthy. However, in 1928 Frederick Griffith

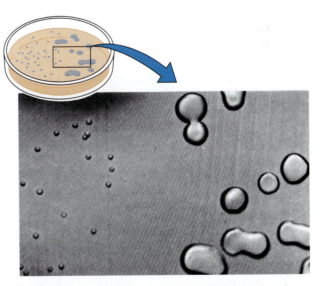

R colonies, breed true   S colonies, breed true

Figure 1.1 Colonies of *Streptococcus pneumoniae.* The small colonies on the left are from a rough (R) strain, and the large colonies on the right are from a smooth (S) strain. The S colonies are larger because of the capsule on the S cells. [Photograph from O. T. Avery, C. M. MacLeod, and M. McCarty. 1944. *J. Exp. Med.* 79: 137.]

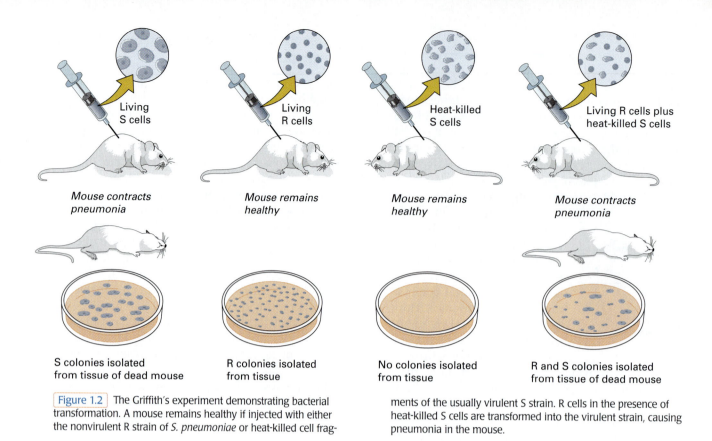

Living S cells	Living R cells	Heat-killed S cells	Living R cells plus heat-killed S cells
Mouse contracts pneumonia	Mouse remains healthy	Mouse remains healthy	Mouse contracts pneumonia
S colonies isolated from tissue of dead mouse	R colonies isolated from tissue	No colonies isolated from tissue	R and S colonies isolated from tissue of dead mouse

Figure 1.2 The Griffith's experiment demonstrating bacterial transformation. A mouse remains healthy if injected with either the nonvirulent R strain of *S. pneumoniae* or heat-killed cell fragments of the usually virulent S strain. R cells in the presence of heat-killed S cells are transformed into the virulent strain, causing pneumonia in the mouse.

showed that when mice are injected with a *mixture* of living R cells and heat-killed S cells, they often die of pneumonia (Figure 1.2). Bacteria isolated from blood samples of the dead mice produce S cultures with a capsule typical of the injected S cells, even though the injected S cells had been killed by heat. Evidently, the injected material from the dead S cells includes a substance that can enter living R bacterial cells and give them the ability to synthesize the S-type capsule. In other words, the R bacteria can be changed—or undergo **transformation**—into S bacteria, and the new characteristics are inherited by descendants of the transformed bacteria.

Griffith's transformation of *Streptococcus* was not in itself definitive; in 1944 the chemical substance responsible for changing the R cells into S cells was identified as DNA. In a milestone experiment, Oswald Avery, Colin MacLeod, and Maclyn McCarty showed that the substance causing the transformation of R cells into S cells was DNA. In preparation for the experiment, they had to develop chemical procedures for obtaining DNA in almost pure form from bacterial cells, which had not been done before. When they added DNA isolated from S cells to growing cultures of R cells, they observed that a few type-S cells were pro-

duced. Although the DNA preparations contained traces of protein and RNA (ribonucleic acid, an abundant cellular macromolecule chemically related to DNA), the transforming activity was not altered by treatments that destroy either protein or RNA. However, treatments that destroy DNA eliminated the transforming activity (Figure 1.3). These experiments implied that the substance responsible for genetic transformation was the DNA of the cell—and hence that DNA is the genetic material.

■ **Transmission of DNA is the link between generations.**

A second pivotal finding was reported by Alfred Hershey and Martha Chase in 1952. They studied cells of the intestinal bacterium *Escherichia coli* after infection by the virus T2. A virus that attacks bacterial cells is called a **bacteriophage,** often shortened to **phage.** (*Bacteriophage* means "bacteria eater.") The T2 particle is exceedingly small, yet it has a complex structure composed of a head containing the phage DNA, collar, tail, and tail fibers. (The head of a human sperm is about 30–50 times larger in both length and width than the head of T2.) Hershey and Chase were already

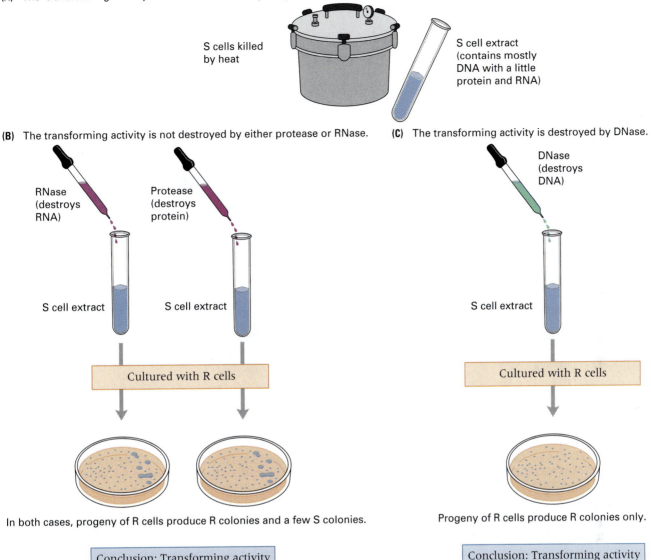

(A) The transforming activity in S cells is not destroyed by heat.

S cells killed by heat

S cell extract (contains mostly DNA with a little protein and RNA)

(B) The transforming activity is not destroyed by either protease or RNase.

(C) The transforming activity is destroyed by DNase.

RNase (destroys RNA)

Protease (destroys protein)

DNase (destroys DNA)

S cell extract

S cell extract

S cell extract

Cultured with R cells

Cultured with R cells

In both cases, progeny of R cells produce R colonies and a few S colonies.

Progeny of R cells produce R colonies only.

Conclusion: Transforming activity is not protein or RNA.

Conclusion: Transforming activity is most likely DNA.

Figure 1.3 A diagram of the experiment demonstrating that DNA is the active material in bacterial transformation. (A) Purified DNA extracted from heat-killed S cells can convert some living R cells into S cells, but the extract may still contain undetectable traces of protein and/or RNA. (B) The transforming activity is not destroyed by either protease or RNase. (C) The transforming activity is destroyed by DNase and so probably consists of DNA.

aware that T2 infection proceeds via the attachment of a phage particle by the tip of its tail to the bacterial cell wall, entry of phage material into the cell, multiplication of this material to form a hundred or more progeny phage, and release of the progeny phage by bursting (lysis) of the bacterial host cell. They also knew that T2 particles are composed of DNA and protein in approximately equal amounts.

Because DNA contains phosphorus but no sulfur, whereas most proteins contain sulfur but no phosphorus, it is possible to label DNA and proteins differentially by the use of radioactive isotopes of the two elements. Hershey and Chase produced particles containing radioactive DNA by infecting $E.$ $coli$ cells that had been grown for several generations in a medium that included ^{32}P (a radioactive isotope of phosphorus) and then collecting the phage progeny. They obtained other particles containing labeled proteins in the same way, using medium that included ^{35}S (a radioactive isotope of sulfur).

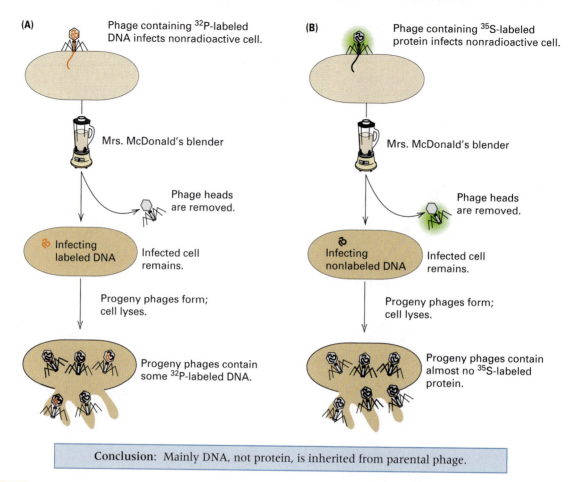

(A) Phage containing ^{32}P-labeled DNA infects nonradioactive cell.

Mrs. McDonald's blender

Phage heads are removed.

Infecting labeled DNA — Infected cell remains.

Progeny phages form; cell lyses.

Progeny phages contain some ^{32}P-labeled DNA.

(B) Phage containing ^{35}S-labeled protein infects nonradioactive cell.

Mrs. McDonald's blender

Phage heads are removed.

Infecting nonlabeled DNA — Infected cell remains.

Progeny phages form; cell lyses.

Progeny phages contain almost no ^{35}S-labeled protein.

Conclusion: Mainly DNA, not protein, is inherited from parental phage.

Figure 1.4 The Hershey–Chase ("blender") experiment, which demonstrated that DNA, not protein, is responsible for directing the reproduction of phage T2 in infected *E. coli* cells. (A) Radioactive DNA is transmitted to progeny phage in substantial amounts. (B) Radioactive protein is transmitted to progeny phage in negligible amounts.

In the experiments summarized in Figure 1.4, nonradioactive *E. coli* cells were infected with phage labeled with *either* ^{32}P (part A) *or* ^{35}S (part B) in order to follow the DNA and proteins separately. Infected cells were separated from unattached phage particles by centrifugation, resuspended in fresh medium, and then swirled violently in a kitchen blender to shear attached phage material from the cell surfaces. This treatment was found to have no effect on the subsequent course of the infection, which implies that the genetic material must enter the infected cells very soon after phage attachment (Figure 1.5). The kitchen blender turned out to be the critical piece of equipment. Other methods had been tried to tear the phage heads from the bacterial cell surface, but nothing had worked reliably. Hershey later explained, "We tried various grinding arrangements, with results that weren't very encouraging. When Margaret McDonald loaned us her kitchen blender, the experiment promptly succeeded."

After the phage heads were removed by blending, the infected bacteria were examined. Most of the radioactivity from ^{32}P-labeled phage was found

to be associated with the bacteria, whereas only a small fraction of the ^{35}S radioactivity was present in the infected cells. The retention of most of the labeled DNA, contrasted with the loss of most of

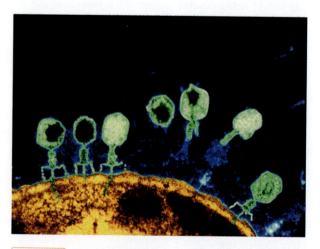

Figure 1.5 T2 phages infecting a cell of *E. coli*. Each phage attaches to the bacterial cell wall and injects its DNA into the host. The image has been color-enhanced to show the injected phage DNA in green. [© Oliver Meckes/E.O.S./MPI Tubingen/Photo Researchers, Inc.]

the labeled protein, implied that a T2 phage transfers most of its DNA, but very little of its protein, to the cell it infects. The critical finding (Figure 1.4) was that about 50 percent of the transferred ^{32}P-labeled DNA, but less than 1 percent of the transferred ^{35}S-labeled protein, was inherited by the *progeny* phage particles. Hershey and Chase interpreted this result to mean that the genetic material in T2 phage is DNA.

The transformation experiment and the Hershey–Chase experiment are regarded as classics in the demonstration that genes consist of DNA. At the present time, the equivalent of the transformation experiment is carried out daily in many research laboratories throughout the world, usually with bacteria, yeast, or animal or plant cells grown in culture. These experiments indicate that DNA is the genetic material in these organisms as well as in phage T2.

key concept

There are no known exceptions to the generalization that DNA is the genetic material in all cellular organisms.

It is worth noting, however, that in a few types of viruses, the genetic material consists of another kind of nucleic acid called RNA.

1.2

The structure of DNA is a double helix composed of two intertwined strands.

Even after it was shown that genes consist of DNA, many questions remained. How is the DNA in a gene duplicated when a cell divides? How does the DNA in a gene control a hereditary trait? What happens to the DNA when a mutation (a change in the DNA) takes place in a gene? In the early 1950s, a number of researchers began to try to understand the detailed molecular structure of DNA in hopes that the structure alone would suggest answers to these questions. The first essentially correct three-dimensional structure of the DNA molecule was proposed in 1953 by James Watson and Francis Crick at Cambridge University. The structure was dazzling in its elegance and revolutionary in suggesting how DNA duplicates itself, controls hereditary traits, and undergoes mutation. Even while the tin sheet and wire model of the DNA molecule was still incomplete, Crick visited his favorite pub and exclaimed, "We have discovered the secret of life."

In the Watson–Crick structure, DNA consists of two long chains of subunits twisted around one another to form a double-stranded helix. The double helix is right-handed, which means that as one looks along the barrel, each chain follows a clockwise path as it progresses. You can see the right-

handed coiling in part A of Figure 1.6 if you imagine yourself looking up into the structure from the bottom: The smaller spheres outline the "backbone" of each individual strand, and they coil in a clockwise direction. The subunits of each strand are **nucleotides,** each of which contains any one of four chemical constituents called **bases.** The four bases in DNA are

$$\text{Adenine (A)} \qquad \text{Guanine (G)}$$
$$\text{Thymine (T)} \qquad \text{Cytosine (C)}$$

The chemical structures of the nucleotides and bases need not concern us at this point. They are examined in Chapter 6. A key point for our present purposes is that the bases in the double helix are paired as shown in Figure 1.6, part B. That is,

key concept

At any position on the paired strands of a DNA molecule, if one strand has an A, then the partner strand has a T; and if one strand has a G, then the partner strand has a C.

The base pairing between A and T and between G and C is said to be **complementary base pairing;** the complement of A is T, and the complement of G

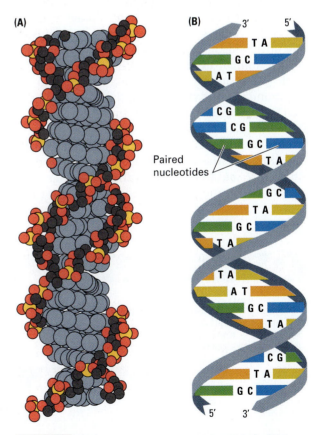

Figure 1.6 Molecular structure of a DNA double helix. (A) A "space-filling" model, in which each atom is depicted as a sphere. (B) A diagram highlighting the helical backbones on the outside of the molecule and the stacked A–T and G–C base pairs inside.

is C. The complementary pairing in the duplex molecule means that each base along one strand of the DNA is matched with a base in the opposite position on the other strand. Furthermore,

This principle explains how only four bases in DNA can code for the huge amount of information needed to make an organism. It is the linear order or *sequence* of bases along the DNA that encodes the genetic information, and the sequence is completely unrestricted.

The complementary pairing is also called *Watson–Crick base pairing.* In the three-dimensional structure (Figure 1.6, part A), the base pairs are represented by the larger spheres filling the interior of the double helix. The base pairs lie almost flat, stacked on top of one another perpendicular to the long axis of the double helix, like pennies in a roll. When discussing a DNA molecule, biologists frequently refer to the individual strands as **single-stranded DNA** and to the double helix as **double-stranded DNA** or **duplex DNA.**

Each DNA strand has a **polarity,** or directionality, like a chain of circus elephants linked trunk to tail. In this analogy, each elephant corresponds to one nucleotide along the DNA strand. The polarity is determined by the direction in which the nucleotides are pointing. The "trunk" end of the strand is called the *5' end* of the strand, and the "tail" end is called the *3' end.* In double-stranded DNA, the paired strands are oriented in opposite directions: The 5' end of one strand is aligned with the 3' end of the other. The oppositely oriented strands are said to be **antiparallel.** The molecular basis of the strand polarity, and the reason for the antiparallel orientation of the strands in duplex DNA, are explained in Chapter 6. In illustrating DNA molecules, we use an arrow-like ribbon to represent the backbone, and we use tabs jutting off the ribbon to represent the nucleotides. The polarity of a DNA strand is indicated by the direction of the arrow-like ribbon. The tail of the arrow represents the 5' end of the DNA strand, the head the 3' end.

Beyond the most optimistic hopes, knowledge of the structure of DNA immediately gave clues to its function:

1. The sequence of bases in DNA could be copied by using each of the separate "partner" strands as a pattern for the creation of a new partner strand with a complementary sequence of bases.

2. The DNA could contain genetic information in coded form in the sequence of bases, analogous to letters printed on a strip of paper.

3. Changes in genetic information (mutations) could result from errors in copying in which the base sequence of the DNA became altered.

In the remainder of this chapter, we discuss some of the implications of these clues.

1.3

In replication, each parental DNA strand directs the synthesis of a new partner strand.

"It has not escaped our notice," wrote Watson and Crick, "that the specific base pairing we have postulated immediately suggests a copying mechanism for the genetic material." The copying process in which a single DNA molecule becomes two identical molecules is called **replication.** The replication

A Moment to Think

Problem: When the DNA sequence of the bacteriophage λ (lambda), which infects *E. coli,* was first determined, geneticists were surprised to find a 12-base single-stranded overhang at the 5' end of each strand. The structure of the ends is diagrammed below. The dots represent the remaining 48,501 base pairs of phage sequence that are not shown.

```
GGGCGGCGACCTCGCG . . . TACG
        GCGC . . . ATGCCCCGCCGCTGGA
```

Using your knowledge that, in double-stranded DNA, A pairs with T and G pairs with C, and that the paired strands in duplex DNA are antiparallel, identify what is special about these single-stranded ends and suggest a manner in which they might interact. *Hint:* The single-stranded regions are called cohesive ends. (The answer can be found on page 10.)

mechanism that Watson and Crick had in mind is illustrated in Figure 1.7 . The strands of the original (parent) duplex separate, and each individual strand serves as a pattern, or **template,** for the synthesis of a new strand (replica). The replica strands are synthesized by the addition of successive nucleotides in such a way that each base in the replica is complementary (in the Watson–Crick pairing sense) to the base across the way in the template strand. Although the mechanism in Figure 1.7 is simple in principle, it is a complex process that is fraught with geometric problems and requires a variety of enzymes and other proteins. The details are examined in Chapter 6. The end result of replication is that a single double-stranded molecule becomes replicated into two copies with identical sequences:

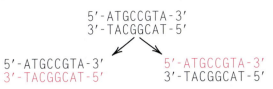

Here the bases in the newly synthesized strands are shown in red. In the duplex on the left, the top strand is the template from the parental molecule and the bottom strand is newly synthesized; in the duplex on the right, the bottom strand is the template from the parental molecule and the top strand is newly synthesized.

1.4

Genes code for proteins.

One of the important principles of molecular genetics is that genes exert their effects on organisms indirectly. For most genes, the genetic information contained in the nucleotide sequence specifies a particular type of *protein*. Proteins control the chemical and physical processes of cells known as **metabolism.** Many proteins are **enzymes,** a term introduced in 1878 to refer to the biological catalysts that accelerate biochemical reactions. Enzymes are essential for the breakdown of organic molecules, generating the chemical energy needed for cellular activities; they are also essential for the assembly of small molecules into large molecules and complex cellular structures.

Although the fundamental connection between genes and proteins was not widely appreciated until the 1940s, the first evidence for a relationship came much earlier. The pioneering observations were made by Archibald Garrod, a British physician, who studied genetic diseases caused by inherited defects in metabolism. He concluded that an inherited defect in metabolism results from an inherited defect in an enzyme. The key observations on which Garrod based this conclusion are summarized in the following sections.

■ Enzyme defects result in inborn errors of metabolism.

In 1908 Garrod gave a series of lectures in which he proposed this fundamental hypothesis about the relationship between enzymes and disease:

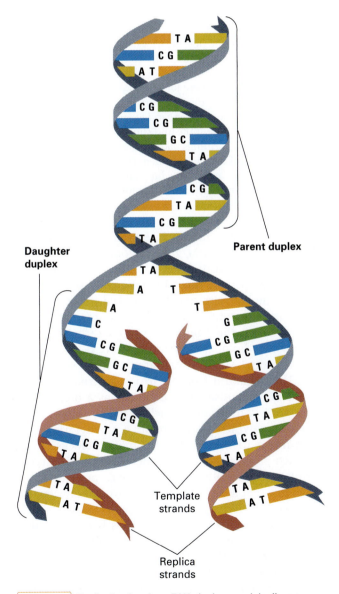

Figure 1.7 Replication in a long DNA duplex as originally proposed by Watson and Crick. The parental strands separate, and each parental strand serves as a template for the formation of a new daughter strand by means of A—T and G—C base pairing.

key concept

Any hereditary disease in which cellular metabolism is abnormal results from an inherited defect in an enzyme.

1.4 Genes code for proteins   9

A Moment to Think

Answer to Problem: The single-stranded regions are complementary in sequence and polarity, so each end can loop around and form base pairs with the other end. The result is that the phage DNA can form a circle, as illustrated in the diagram. (The term *cohesive ends* comes from the fact that the ends can stick together, or cohere.)

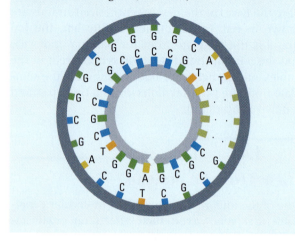

Figure 1.8 Urine from a person with alkaptonuria turns black because of the oxidation of the homogentisic acid that it contains. [Courtesy of Daniel De Aguiar.]

Such diseases became known as **inborn errors of metabolism,** a term still in use today.

Garrod studied a number of inborn errors of metabolism in which the patients excreted abnormal substances in the urine. One of these was **alkaptonuria.** In this case, the abnormal substance excreted is **homogentisic acid:**

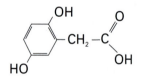

This is a conventional chemical representation in which each corner of the hexagon represents a carbon atom, and hydrogen atoms attached to the ring are not shown. The six-carbon ring is called a *phenyl* ring. An early name for homogentisic acid was *alkapton*—hence the name *alkaptonuria.* Even though alkaptonuria is rare, with an incidence of about one in 200,000 people, it was well known even before Garrod studied it. The disease itself is relatively mild, but it has one striking symptom: The urine of the patient turns black because of the oxidation of homogentisic acid (Figure 1.8). This is why alkaptonuria is also called *black urine disease.* The passing of black urine can hardly escape being noticed. One case was described in the year 1649:

The patient was a boy who passed black urine and who, at the age of fourteen years, was submitted to a

drastic course of treatment that had for its aim the subduing of the fiery heat of his viscera, which was supposed to bring about the condition in question by charring and blackening his bile. Among the measures prescribed were bleedings, purgation, baths, a cold and watery diet, and drugs galore. None of these had any obvious effect, and eventually the patient, who tired of the futile and superfluous therapy, resolved to let things take their natural course. None of the predicted evils ensued. He married, begat a large family, and lived a long and healthy life, always passing urine black as ink. (Quotation from Garrod, 1908.)

Garrod was primarily interested in the biochemistry of alkaptonuria, but he took note of family studies that indicated that the disease was inherited as though it were due to a defect in a single gene. As to the biochemistry, he deduced that the problem in alkaptonuria was the patients' inability to break down the phenyl ring of six carbons that is present in homogentisic acid. Where does this ring come from? Most animals are unable to synthesize it. They obtain it from their diet. Garrod proposed that homogentisic acid originates as a breakdown product of two amino acids, phenylalanine and tyrosine, which also contain a phenyl ring. An **amino acid** is one of the "building blocks" from which proteins are made. Phenylalanine and tyrosine are constituents of normal proteins. The scheme that illustrates the relationship between the molecules is shown in Figure 1.9 . Any such sequence of biochemical reactions is called a **biochemical pathway** or a **metabolic pathway.** Each arrow in the pathway represents a single step depicting the transition from the "input" or **substrate molecule,** shown at the tail of the arrow, to the "output" or **product molecule,** shown at the tip. Biochemical pathways are usually

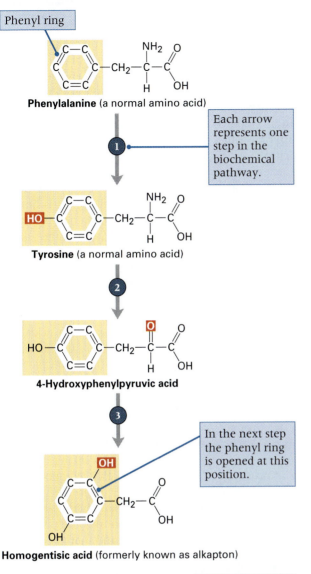

Phenyl ring

Phenylalanine (a normal amino acid)

Each arrow represents one step in the biochemical pathway.

Tyrosine (a normal amino acid)

4-Hydroxyphenylpyruvic acid

In the next step the phenyl ring is opened at this position.

Homogentisic acid (formerly known as alkapton)

This is the step that is blocked in alkaptonuria; homogentisic acid accumulates.

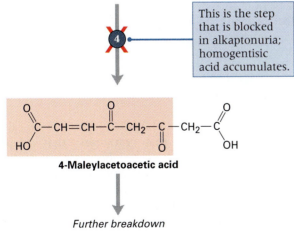

4-Maleylacetoacetic acid

Further breakdown

Figure 1.9 Metabolic pathway for the breakdown of phenylalanine and tyrosine. Each step in the pathway, represented by an arrow, requires a particular enzyme to catalyze the reaction. The key step in the breakdown of homogentisic acid is the breaking open of the phenyl ring.

oriented either vertically with the arrows pointing down, as in Figure 1.9, or horizontally, with the arrows pointing from left to right. Garrod did not know all of the details of the pathway in Figure 1.9, but he did understand that the key step in the breakdown of homogentisic acid is the breaking open of the phenyl ring and that the phenyl ring in homogentisic acid comes from dietary phenylalanine and tyrosine.

What allows each step in a biochemical pathway to occur? Garrod's insight was to see that each step requires a specific enzyme to catalyze the reaction and allow the chemical transformation to take place. Persons with an inborn error of metabolism, such as alkaptonuria, have a defect in one step of a metabolic pathway because they lack a functional enzyme for that step. When an enzyme in a pathway is defective, the pathway is said to have a **block** at that step. One frequent result of a blocked pathway is that the substrate of the defective enzyme accumulates. Observing the accumulation of homogentisic acid in patients with alkaptonuria, Garrod proposed that there must be an enzyme whose function is to open the phenyl ring of homogentisic acid and that this enzyme is missing in these patients. Discovery of all the enzymes in the pathway in Figure 1.9 took a long time. The

This highly trained Golden Retriever is a canine companion helping to keep its blind owner safe in a dangerous urban environment.

the human connection

Black Urine

Archibald E. Garrod 1908
St. Bartholomew's Hospital,
London, England
Inborn Errors of Metabolism

Although he was a distinguished physician, Garrod's lectures on the relationship between heredity and congenital defects in metabolism had no impact when they were delivered. The important concept that one gene corresponds to one enzyme (the "one gene–one enzyme hypothesis") was developed independently in the 1940s by George W. Beadle and Edward L. Tatum, who used the bread mold *Neurospora crassa* as their experimental organism. When Beadle finally became aware of *Inborn Errors of Metabolism,* he was generous in praising it. This excerpt shows Garrod at his best, interweaving history, clinical medicine, heredity, and biochemistry in his account of alkaptonuria. The excerpt also illustrates how the severity of a genetic disease depends on its social context. Garrod writes as though alkaptonuria were a harmless curiosity. This is indeed largely true when the life expectancy is short. With today's longer life span, however, alkaptonuria patients accumulate the dark pigment

in their cartilage and joints and can eventually develop severe arthritis.

. . . .

To students of heredity the inborn errors of metabolism offer a promising field of investigation.... It was pointed out [by others] that the mode of incidence of alkaptonuria finds a ready explanation if the anomaly be regarded as a rare recessive character in the Mendelian sense.... Of the cases of alkaptonuria a very large proportion have been in the children of first cousin marriages.... It is also noteworthy that, if one takes families with five or more children [with both parents normal and at least one child affected with alkaptonuria], the totals work out in strict conformity to Mendel's law, i.e. 57 [normal children] : 19 [affected children] in the proportions 3 : 1.... Of inborn errors of metabolism, alkaptonuria is that of which we know most. In itself it is a trifling matter, inconvenient rather than harmful....

> **We may further conceive that the splitting of the phenyl ring in normal metabolism is the work of a special enzyme and that in congenital alkaptonuria this enzyme is wanting.**

Indications of the anomaly may be detected in early medical writings, such as that in 1584 of a schoolboy who, although he enjoyed good health, continuously excreted black urine; and that in 1609 of a monk who exhibited a similar peculiarity and stated that he had done so all his life.... There are no sufficient grounds [for doubting that the blackening substance in the urine originally called alkapton] is homogentisic acid, the excretion of which is the essential feature of the alkaptonuric.... Homogentisic acid is a product of normal metabolism.... The most likely sources of the phenyl ring in homogentisic acid are phenylananine and tyrosine, [because when these amino acids are administered to an alkaptonuric] they cause a very conspicuous increase in the output of homogentisic acid.... Where the alkaptonuric differs from the normal individual is in having no power of destroying homogentisic acid when formed—in other words of breaking up the phenyl ring of that compound.... We may further conceive that the splitting of the phenyl ring in normal metabolism is the work of a special enzyme and that in congenital alkaptonuria this enzyme is wanting.

Source: Originally published in London, England, by the Oxford University Press. Excerpts from the reprinted edition in Harry Harris. 1963. *Garrod's Inborn Errors of Metabolism.* London, England: Oxford University Press.

enzyme that opens the phenyl ring of homogentisic acid was not actually isolated until 50 years after Garrod's lectures. In normal people it is found in cells of the liver. Just as Garrod had predicted, the enzyme is defective in patients with alkaptonuria.

The pathway for the breakdown of phenylalanine and tyrosine, as it is understood today, is shown in Figure 1.10 . In this figure the emphasis is on the enzymes rather than on the structures of the **metabolites,** or small molecules, on which the enzymes act. As Garrod would have predicted, each step in the pathway requires the presence of a

particular enzyme that catalyzes that step. Although Garrod knew only about alkaptonuria, in which the defective enzyme is homogentisic acid 1,2-dioxygenase, we now know the clinical consequences of defects in the other enzymes. Unlike alkaptonuria, which is a relatively benign inherited disease, the others are very serious. The condition known as **phenylketonuria (PKU)** results from the absence of (or a defect in) the enzyme **phenylalanine hydroxylase (PAH).** When this step in the pathway is blocked, phenylalanine accumulates. The excess phenylalanine is

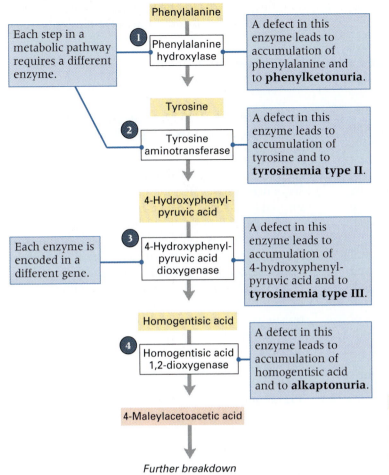

Each step in a metabolic pathway requires a different enzyme.

Phenylalanine

1 Phenylalanine hydroxylase

A defect in this enzyme leads to accumulation of phenylalanine and to **phenylketonuria**.

Tyrosine

2 Tyrosine aminotransferase

A defect in this enzyme leads to accumulation of tyrosine and to **tyrosinemia type II**.

4-Hydroxyphenyl-pyruvic acid

Each enzyme is encoded in a different gene.

3 4-Hydroxyphenyl-pyruvic acid dioxygenase

A defect in this enzyme leads to accumulation of 4-hydroxyphenyl-pyruvic acid and to **tyrosinemia type III**.

Homogentisic acid

4 Homogentisic acid 1,2-dioxygenase

A defect in this enzyme leads to accumulation of homogentisic acid and to **alkaptonuria**.

4-Maleylacetoacetic acid

Further breakdown

Figure 1.10 Inborn errors of metabolism in the breakdown of phenylalanine and tyrosine. A different inherited disease results when each of the enzymes is missing or defective. Alkaptonuria results from a defective homogentisic acid 1,2-dioxygenase, phenylketonuria from a defective phenylalanine hydroxylase.

broken down into harmful metabolites that cause defects in myelin formation that damage a child's developing nervous system and lead to severe mental retardation.

If PKU is diagnosed in children soon enough after birth, they can be placed on a specially formulated diet low in phenylalanine (Figure 1.11). The child is allowed only as much phenylalanine as can be used in the synthesis of proteins, so excess phenylalanine does not accumulate. The special diet is very strict. It excludes meat, poultry, fish, eggs, milk and milk products, legumes, nuts, and bakery goods manufactured with regular flour. These foods are replaced by a synthetic formula that is very expensive. With the special diet, however, the detrimental effects of excess phenylalanine on mental development can largely be avoided. In many countries, including the United States, all newborn babies have their blood tested for chemical signs of PKU. Routine screening is cost-effective because PKU is relatively common. In the United States, the incidence is about one in 8000 among Caucasian births. The disease is less common in other ethnic groups.

In the metabolic pathway in Figure 1.10, defects in the breakdown of tyrosine or of 4-hydroxy-phenylpyruvic acid lead to types of tyrosinemia. These are also severe diseases. Type II is associated with skin lesions and mental retardation, type III with severe liver dysfunction.

Figure 1.11 The women in this photograph are sisters. Both are homozygous for the same mutant phenylalanine hydroxylase (*PAH*) gene. The bride is the younger of the two. She was diagnosed just three days after birth and put on the PKU diet soon thereafter. Her maid of honor, the older sister, was diagnosed too late for the diet to be effective. [Courtesy of Charles R. Scriver.]

■ A defective enzyme results from a mutant gene.

It follows from Garrod's work that a defective enzyme results from a mutant gene. How does a mutant gene result in a defective enzyme? Garrod did not speculate. For all he knew, genes *were* enzymes. This would have been a logical hypothesis at the time. We now know that the relationship between genes and enzymes is somewhat indirect. With a few exceptions, each enzyme is *encoded* in a particular sequence of nucleotides present in a region of DNA. The DNA region that codes for the enzyme, as well as adjacent regions that regulate when and in which cells the enzyme is produced, make up the "gene" that encodes the enzyme.

The genes for the enzymes in the biochemical pathway in Figure 1.10 have all been identified and the nucleotide sequence of the DNA determined. In the following list, and throughout this book, we use the typographical convention that the names of *genes* are printed in *italic* type, whereas gene products are printed in regular type:

- The gene *PAH* on the long arm of chromosome 12 encodes phenylalanine hydroxylase (PAH).
- The gene *TAT* on the long arm of chromosome 16 encodes tyrosine aminotransferase (TAT).
- The gene *HPD* on the long arm of chromosome 12 encodes 4-hydroxyphenylpyruvic acid dioxygenase (HPD).

- The gene *HGD* on the long arm of chromosome 3 encodes homogentisic acid 1,2-dioxygenase (HGD).

Next we turn to the issue of *how* genes code for enzymes and other proteins.

■ One of the DNA strands directs the synthesis of a molecule of RNA.

Watson and Crick were correct in proposing that the genetic information in DNA is contained in the sequence of bases in a manner analogous to letters printed on a strip of paper. In a region of DNA that directs the synthesis of a protein, the genetic code for the protein is contained in only one strand, and it is decoded in a linear order. The result of protein synthesis is a **polypeptide chain,** which consists of a linear sequence of amino acids connected end to end. Each polypeptide chain folds into a characteristic three-dimensional configuration that is determined by its particular sequence of amino acids. A typical protein is made up of one or more polypeptide chains. For example, the enzyme PAH consists of four identical polypeptide chains, each 452 amino acids in length. Figure 1.12 shows the complex folding of the polypeptide chains in the active molecule. (Only two of the four subunits are shown.) The polypeptide chains are held together by weak chemical interactions between the regions shown in green.

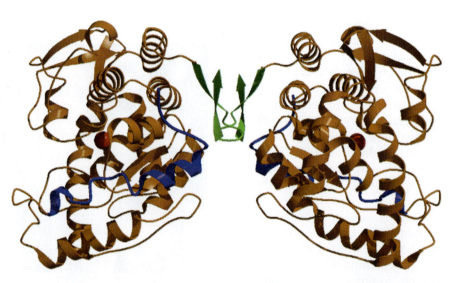

Figure 1.12 Three-dimensional structure of two of the four protein subunits present in phenylalanine hydroxylase. The chains of amino acids are represented as a sequence of curls, loops, and flat arrows, which represent different types of local structure. The regions that hold the subunits together are shown in green.

[Courtesy of R. C. Stevens, T. Flatmark, and H. Erlandsen. Reprinted by permission from *Nature Structural Biology* 4:995, H. Erlandsen, F. Fusetti, A Martinez, E. Hough, T. Flatmark, and R. C. Stevens. Copyright 1997 Macmillan Magazines Ltd.]

In the decoding of DNA, each successive "code word" in the DNA specifies the next amino acid to be added to the polypeptide chain as it is being made. The amount of DNA required to code for the polypeptide chain of PAH is therefore $452 \times 3 = 1356$ nucleotide pairs. The entire gene is very much longer—90,000 nucleotide pairs. Only 1.5 percent of the gene is devoted to coding for the amino acids. The noncoding part includes sequences that control the activity of the gene, but it is not known how much of the gene is involved in regulation.

There are 20 different amino acids. How can four bases code for 20 amino acids? Because each "word" in the genetic code consists of three adjacent bases. For example, the base sequence ATG specifies the amino acid methionine (Met), TCC specifies serine (Ser), ACT specifies threonine (Thr), and GCG specifies alanine (Ala). There are 64 possible three-base combinations but only 20 amino acids, because some combinations code for the same amino acid. For example, TCT, TCC, TCA, TCG, AGT, and AGC all code for serine (Ser), and CTT, CTC, CTA, CTG, TTA, and TTG all code for leucine (Leu). An example of the relationship between the base sequence in a DNA duplex and the amino acid sequence of the corresponding protein is shown in Figure 1.13. This particular DNA duplex is the human sequence that codes for the first seven amino acids in the polypeptide chain of PAH.

The scheme outlined in Figure 1.13 indicates that DNA codes for protein not directly but indirectly through the processes of *transcription* and *translation*. The indirect route of information transfer,

$$DNA \rightarrow RNA \rightarrow Protein$$

is known as the **central dogma** of molecular genetics. The term *dogma* means "set of beliefs"; it dates from the time the idea was put forward first as a theory. Since then the "dogma" has been confirmed experimentally, but the term persists. The main concept in the central dogma is that DNA does not code for protein directly but rather acts through an intermediary molecule of **ribonucleic acid (RNA).** The structure of RNA is similar to, but not identical with, that of DNA. There is a difference in the sugar (RNA contains the sugar **ribose** instead of deoxyribose), RNA is usually single-stranded (not a duplex), and RNA contains the base **uracil (U)** instead of thymine (T), which is present in DNA. Three types of RNA take part in the synthesis of proteins:

- A molecule of **messenger RNA (mRNA),** which carries the genetic information from DNA and is used as a template for polypeptide synthesis. In most mRNA molecules, a relatively high proportion of the nucleotides actually code for amino acids. For example, the mRNA for PAH is 2400 nucleotides in length and codes for a polypeptide of 452 amino acids; in this case, more than 50 percent of the length of the mRNA codes for amino acids.

- Four types of **ribosomal RNA (rRNA),** which are major constituents of the cellular particles called **ribosomes** on which polypeptide synthesis takes place.

- A set of about 45 **transfer RNA (tRNA)** molecules, each of which carries a particular amino acid as well as a three-base recognition region that base-pairs with a group of three adjacent bases in the mRNA. As each tRNA participates in translation, its amino acid becomes the terminal subunit of the growing polypeptide chain. A tRNA that carries methionine is denoted $tRNA^{Met}$, one that carries serine is denoted $tRNA^{Ser}$, and so forth. (Because there are more than 20 different tRNAs, but only 20 amino acids, some amino acids can be attached to any of several tRNAs.)

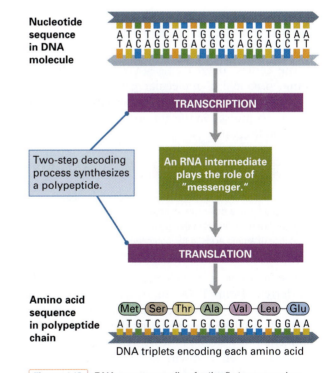

Figure 1.13 DNA sequence coding for the first seven amino acids in a polypeptide chain. The DNA sequence specifies the amino acid sequence through a molecule of RNA that serves as an intermediary "messenger." Although the decoding process is indirect, the net result is that each amino acid in the polypeptide chain is specified by a group of three adjacent bases in the DNA. In this example, the polypeptide chain is that of phenylalanine hydroxylase (PAH).

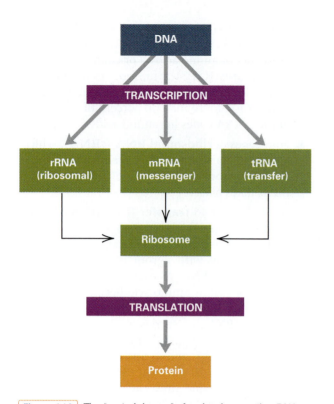

Figure 1.14 The "central dogma" of molecular genetics: DNA codes for RNA, and RNA codes for proteins. The DNA → RNA step is transcription, and the RNA → protein step is translation.

place of T. The rules of base pairing between DNA and RNA are summarized below. Like DNA, an RNA strand also exhibits polarity, its 5′ and 3′ ends determined by the orientation of the nucleotides. The 5′ end of the RNA transcript is synthesized first, and in the RNA–DNA duplex formed in transcription, the polarity of the RNA strand is opposite to that of the DNA strand. Each gene includes particular nucleotide sequences that initiate and terminate transcription. The RNA transcript made from any gene begins at an initiation site in the template strand, which is located "upstream" from the amino-acid coding region, and ends at a termination site, which is located "downstream" from the amino-acid coding region. For any gene, the length of the RNA transcript is very much smaller than the length of the DNA in the entire chromosome. For example, the transcript of the *PAH* gene for phenylalanine hydroxylase is 90,000 nucleotides in length, but the DNA in chromosome 12 is about 130,000,000 nucleotide pairs. In this case, the length of the *PAH* transcript is less than 0.1 percent of the length of the DNA in the chromosome. A different gene in chromosome 12 would be transcribed from a different region of the DNA molecule in chromosome 12, but the transcribed region would again be small in comparison with the total length of the DNA in the chromosome.

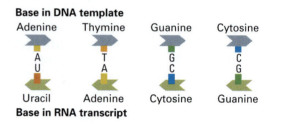

The central dogma (Figure 1.14) is the fundamental principle of molecular genetics because it summarizes how the genetic information in DNA becomes expressed in the amino acid sequence in a polypeptide chain.

key concept

The sequence of nucleotides in a gene specifies the sequence of nucleotides in a molecule of messenger RNA; in turn, the sequence of nucleotides in the messenger RNA specifies the sequence of amino acids in the polypeptide chain.

The manner in which genetic information is transferred from DNA to RNA is shown in Figure 1.15. The DNA opens up, and one of the strands is used as a template for the synthesis of a complementary strand of RNA. (How the template strand is chosen is discussed in Chapter 8.) The process of making an RNA strand from a DNA template is **transcription,** and the RNA molecule that is made is the **transcript.** The base sequence in the RNA is complementary (in the Watson–Crick pairing sense) to that in the DNA template, except that U (which pairs with A) is present in the RNA in

▪ A molecule of RNA directs the synthesis of a polypeptide chain.

The synthesis of a polypeptide under the direction of an mRNA molecule is known as **translation.** Although the sequence of bases in the mRNA codes for the sequence of amino acids in a polypeptide, the molecules that actually do the "translating" are the tRNA molecules. The mRNA molecule is translated in nonoverlapping groups of three bases called **codons.** For each codon in the mRNA that specifies an amino acid, there is one tRNA molecule containing a complementary group of three adjacent bases that can pair with the bases in the codon. The correct amino acid is attached to the other end of the tRNA, and when this tRNA comes into line, the amino acid attached to it becomes the new terminal end of the growing polypeptide chain.

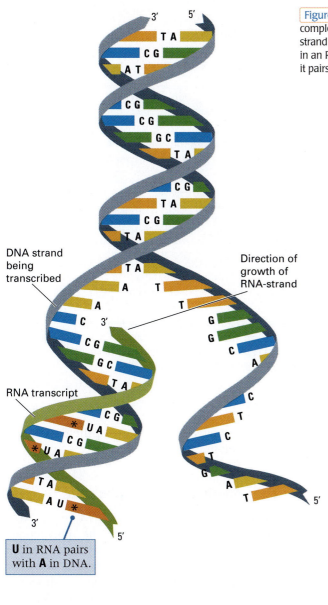

Figure 1.15 Transcription is the production of an RNA strand that is complementary in base sequence to a DNA strand. In this example, a DNA strand is being transcribed into an RNA strand at the bottom left. Note that in an RNA molecule, the base U (uracil) plays the role of T (thymine) in that it pairs with A (adenine). Each A—U pair is marked with an asterisk.

DNA strand being transcribed

Direction of growth of RNA-strand

RNA transcript

U in RNA pairs with A in DNA.

Q A Moment to Think

Problem: A 1975 Nobel Prize was awarded to David Baltimore and Howard Temin for their discovery of an enzyme called *reverse transcriptase,* which can produce a complementary DNA strand from a template of RNA. Immediately recognized as one of the most important enzymes ever discovered, reverse transcriptase could account for the ability of certain animal viruses whose genetic material is RNA to create a complementary DNA molecule that can become inserted into the genetic material of an infected cell. Reverse transcription is similar to ordinary transcription in that the RNA template and DNA transcript are antiparallel and that the DNA transcript grows by the addition of successive nucleotides to the 3′ end. Shown here is part of the RNA sequence of the virus HIV-1 (human immunodeficiency virus-1), the causative agent of AIDS (acquired immune deficiency syndrome), isolated from an infected child in Italy. The sequence encodes part of the HIV-1 reverse transcriptase and is itself reverse-transcribed.

5′-UCCUAUUGAAACUGUACCAGUAAAAUU-3′

What single-stranded DNA sequence would be reverse-transcribed from this stretch of RNA? Would the DNA transcript strand grow from left to right or from right to left? (The answer can be found on page 19.)

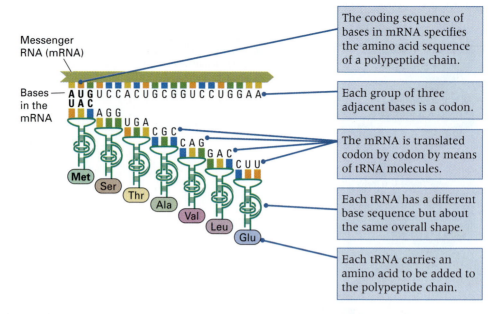

Messenger RNA (mRNA)

Bases in the mRNA

AUG U C C A C U G C G G U C C U G G A A
UAC
A G G
U G A
C G C
C A G
G A C
C U U

Met Ser Thr Ala Val Leu Glu

The coding sequence of bases in mRNA specifies the amino acid sequence of a polypeptide chain.

Each group of three adjacent bases is a codon.

The mRNA is translated codon by codon by means of tRNA molecules.

Each tRNA has a different base sequence but about the same overall shape.

Each tRNA carries an amino acid to be added to the polypeptide chain.

Figure 1.16 The role of messenger RNA in translation is to carry the information contained in a sequence of DNA bases to a ribosome, where it is translated into a polypeptide chain. Translation is mediated by transfer RNA (tRNA) molecules, each of which can base-pair with a group of three adjacent bases in the mRNA. Each tRNA also carries an amino acid, and when it is brought to the ribosome by base pairing, its amino acid becomes the growing end of the polypeptide chain.

The role of tRNA in translation is illustrated in Figure 1.16 and can be described as follows:

key concept

The mRNA is read codon by codon. Each codon that specifies an amino acid matches with a complementary group of three adjacent bases in a single tRNA molecule. One end of the tRNA is attached to the correct amino acid, so the correct amino acid is brought into line.

The tRNA molecules used in translation do not line up along the mRNA simultaneously as shown in Figure 1.16. The process of translation takes place on a ribosome, which combines with a single mRNA and moves along it in steps, three nucleotides at a time (codon by codon). As each new codon comes into place, the next tRNA binds with the ribosome, and the growing end of the polypeptide chain becomes attached to the amino acid on the tRNA. In this way, each tRNA in turn serves temporarily to hold the polypeptide chain as it is being synthesized. As the polypeptide chain is transferred from each tRNA to the next in line, the tRNA that previously held the polypeptide is released from the ribosome. The polypeptide chain elongates one amino acid at a step until any one of three particular codons specifying "stop" is encountered. At this point, synthesis of the chain of amino acids is finished, and the polypeptide chain is released from the ribosome. (This brief description

of translation glosses over many of the details that are presented in Chapter 8.)

■ The genetic code is a triplet code.

Figure 1.16 indicates that the mRNA codon AUG specifies methionine (Met) in the polypeptide chain, UCC specifies Ser (serine), ACU specifies Thr (threonine), and so on. The complete decoding table is called the **genetic code,** and it is shown in **Table 1.1.** For any codon, the column on the left corresponds to the first nucleotide in the codon (reading from the 5′ end), the row across the top corresponds to the second nucleotide, and the column on the right corresponds to the third nucleotide. The complete codon is given in the body of the table, along with the amino acid (or "stop") that the codon specifies. Each amino acid is designated by its full name as well as by a three-letter abbreviation and a single-letter abbreviation. Both types of abbreviations are used in molecular genetics. The code in Table 1.1 is the "standard" genetic code used in translation in the cells of nearly all organisms.

In addition to the 61 codons that code only for amino acids, there are 4 codons that have specialized functions:

- The codon AUG, which specifies Met (methionine), is also the "start" codon for polypeptide synthesis. The positioning of a tRNA^{Met} bound

Table 1.1

The standard genetic code

		Second nucleotide in codon			
	U	**C**	**A**	**G**	

	U	C	A	G	
U	UUU Phe F *Phenylalanine*	UCU Ser S *Serine*	UAU Tyr Y *Tyrosine*	UGU Cys C *Cysteine*	U
	UUC Phe F *Phenylalanine*	UCC Ser S *Serine*	UAC Tyr Y *Tyrosine*	UGC Cys C *Cysteine*	C
	UUA Leu L *Leucine*	UCA Ser S *Serine*	UAA Termination	UGA Termination	A
	UUG Leu L *Leucine*	UCG Ser S *Serine*	UAG Termination	UGG Trp W *Tryptophan*	G
C	CUU Leu L *Leucine*	CCU Pro P *Proline*	CAU His H *Histidine*	CGU Arg R *Arginine*	U
	CUC Leu L *Leucine*	CCC Pro P *Proline*	CAC His H *Histidine*	CGC Arg R *Arginine*	C
	CUA Leu L *Leucine*	CCA Pro P *Proline*	CAA Gln Q *Glutamine*	CGA Arg R *Arginine*	A
	CUG Leu L *Leucine*	CCG Pro P *Proline*	CAG Gln Q *Glutamine*	CGG Arg R *Arginine*	G
A	AUU Ile I *Isoleucine*	ACU Thr T *Threonine*	AAU Asn N *Asparagine*	AGU Ser S *Serine*	U
	AUC Ile I *Isoleucine*	ACC Thr T *Threonine*	AAC Asn N *Asparagine*	AGC Ser S *Serine*	C
	AUA Ile I *Isoleucine*	ACA Thr T *Threonine*	AAA Lys K *Lysine*	AGA Arg R *Arginine*	A
	AUG Met M *Methionine*	ACG Thr T *Threonine*	AAG Lys K *Lysine*	AGG Arg R *Arginine*	G
G	GUU Val V *Valine*	GCU Ala A *Alanine*	GAU Asp D *Aspartic acid*	GGU Gly G *Glycine*	U
	GUC Val V *Valine*	GCC Ala A *Alanine*	GAC Asp D *Aspartic acid*	GGC Gly G *Glycine*	C
	GUA Val V *Valine*	GCA Ala A *Alanine*	GAA Glu E *Glutamic acid*	GGA Gly G *Glycine*	A
	GUG Val V *Valine*	GCG Ala A *Alanine*	GAG Glu E *Glutamic acid*	GGG Gly G *Glycine*	G

First nucleotide in codon (5′ end) — *Third nucleotide in codon (3′ end)*

Codon — Three-letter and single-letter abbreviations

to AUG is one of the first steps in the initiation of polypeptide synthesis, so all polypeptide chains begin with Met. In most organisms, the tRNA^{Met} used for initiation of translation is the same tRNA^{Met} used to specify methionine at internal positions in a polypeptide chain.

- The codons UAA, UAG, and UGA, each of which is a "stop," specify the termination of translation and result in release of the completed polypeptide chain from the ribosome. These codons do not have tRNA molecules that recognize them but are instead recognized by protein factors that terminate translation.

How the genetic code table is used to infer the amino acid sequence of a polypeptide chain may be illustrated using PAH again, in particular the DNA sequence coding for amino acid numbers 1 through 7. The DNA sequence is

5′-ATGTCCACTGCGGTCCTGGAA-3′
3′-TACAGGTGACGCCAGGACCTT-5′

This region is transcribed into RNA in a left-to-right direction, and because RNA grows by the addition of

successive nucleotides to the 3′ end (Figure 1.15), it is the bottom strand that is transcribed. The nucleotide sequence of the RNA is that of the top strand of the DNA, except that U replaces T, so the mRNA for amino acids 1 through 7 is

5′-AUGUCCACUGCGGUCCUGGAA-3′

The codons are read from left to right according to the genetic code shown in Table 1.1. Codon AUG

A Moment to Think

Answer to Problem: The RNA and the reverse-transcribed DNA match as follows:

5′-UCCUAUUGAAACUGUACCAGUAAAAUU-3′
3′-AGGATAACTTTGACATGGTCATTTTAA-5′

Because the DNA strand grows by the addition of nucleotides to the 3′ end, this strand would grow from right to left. (By a mechanism described in Chapter 11, reverse transcriptase also uses the first DNA strand as a template to produce a second DNA strand; the result is the production of a double-stranded DNA molecule from a single strand of RNA.)

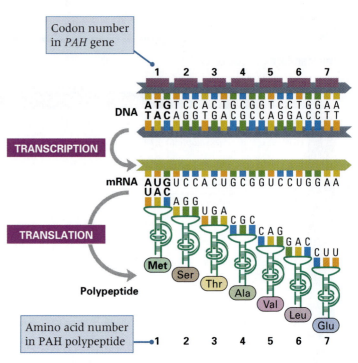

Codon number in *PAH* gene

DNA
mRNA

TRANSCRIPTION

TRANSLATION

Polypeptide

Amino acid number in PAH polypeptide

Figure 1.17 The central dogma in action. The DNA that encodes PAH serves as a template for the production of a messenger RNA, and the mRNA, in turn, serves to specify the sequence of amino acids in the PAH polypeptide chain through interactions with the tRNA molecules.

codes for Met (methionine), UCC codes for Ser (serine), and so on. Altogether, the amino acid sequence of this region of the polypeptide is

5'-AUGUCCACUGCGGUCCUGGAA-3'
MetSerThrAlaValLeuGlu

or, in terms of the single-letter abbreviations,

5'-AUGUCCACUGCGGUCCUGGAA-3'
M S T A V L E

The full decoding operation for this region of the *PAH* gene is shown in Figure 1.17. In this figure, the initiation codon AUG is highlighted because some patients with PKU have a mutation in this particular codon. As might be expected from the fact that methionine is the initiation codon for polypeptide synthesis, cells in patients with this particular mutation fail to produce any of the PAH polypeptide. Mutation and its consequences are considered next.

1.5

Genes change by mutation.

The term **mutation** refers to any heritable change in a gene (or, more generally, in the genetic material); the term also refers to the process by which such a change takes place. One type of mutation results in a change in the sequence of bases in DNA. The change may be simple, such as the substitution of one pair of bases in a duplex molecule for a different pair of bases. For example, a C—G pair in a duplex molecule may mutate to T—A, A—T, or G—C. The change in base sequence may also be more complex, such as the deletion or addition of base pairs. Geneticists also use the term **mutant,** which refers to the result of a mutation. A mutation yields a mutant gene, which in turn produces a mutant mRNA, a mutant protein, and finally a mutant organism that exhibits the effects of the mutation—for example, an inborn error of metabolism.

DNA from patients from all over the world who have phenylketonuria has been studied to determine what types of mutations are responsible for the inborn error. There are a large variety of mutant types. More than 400 different mutations have been described. In some cases part of the gene is missing, so the genetic information to make a complete PAH enzyme is absent. In other cases the genetic defect is more subtle, but the result is still either the failure to produce a PAH protein or the production of a PAH protein that is inactive. In the mutation shown in Figure 1.18, substitution of a G—C base pair for the normal A—T base pair at the very first position in the coding sequence changes the normal codon AUG (Met) used for the initiation of translation into the codon GUG, which normally specifies valine (Val) and cannot be used as a "start" codon. The result is that translation of the PAH mRNA cannot occur, so no PAH polypeptide is made. This mutant is designated M1V because the codon for M (methionine) at amino acid position 1

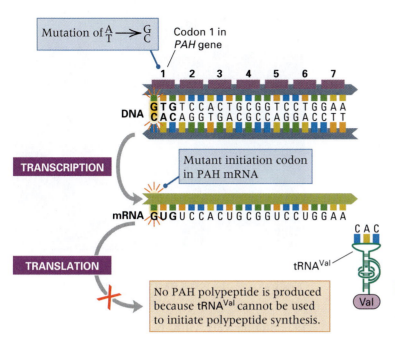

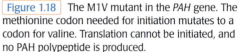

in the PAH polypeptide has been changed to a codon for V (valine). Although the M1V mutant is quite rare worldwide, it is common in some localities, such as in Québec Province in Canada.

One PAH mutant that is quite common is designated R408W, which means that codon 408 in the PAH polypeptide chain has been changed from one coding for arginine (R) to one coding for tryptophan (W). This mutant is one of the four most common in cases of PKU among European Caucasians. The molecular basis of the mutation is

shown in Figure 1.19. In this case, the first base pair in codon 408 is changed from a C—G base pair into a T—A base pair. The result is that the PAH mRNA has a mutant codon at position 408; specifically, it has UGG instead of CGG. Translation does occur in this mutant because everything else about the mRNA is normal, but the result is that the mutant PAH carries a tryptophan (Trp) instead of an arginine (Arg) at position 408 in the polypeptide chain. The consequence of the seemingly minor change of one amino acid is very drastic, because

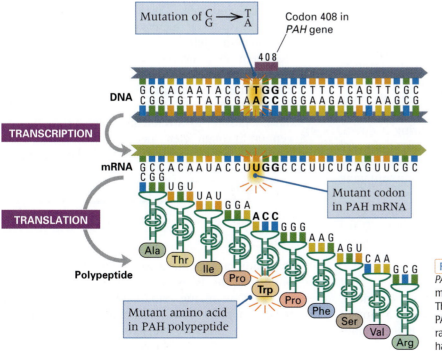

Figure 1.19 The R408W mutant in the *PAH* gene. Codon 408 for arginine (R) is mutated into a codon for tryptophan (W). The result is that position 408 in the mutant PAH polypeptide is occupied by tryptophan rather than by arginine. The mutant protein has no PAH enzyme activity.

the mutant PAH has no enzyme activity and so is unable to catalyze its metabolic reaction. In other words, the mutant PAH protein is complete but inactive. With PAH, as with other proteins, some amino acid replacements result in a polypeptide chain that is unable to fold properly. The incorrectly folded polypeptides are digested by proteases in the cell, which recycles the amino acids for use in the synthesis of other proteins.

1.6

Traits are affected by environment as well as by genes.

Inborn errors of metabolism illustrate the general principle that genes code for proteins and that mutant genes code for mutant proteins. In cases such as PKU, mutant proteins cause such a drastic change in metabolism that a severe genetic defect results. But biology is not necessarily destiny. Organisms are also affected by the environment. PKU serves as an example of this principle, because patients who adhere to a diet restricted in the amount of phenylalanine develop mental capacities within the normal range. What is true in this example is true in general. Most traits are determined by the interaction of genes and environment.

It is also true that most traits are affected by multiple genes. No one knows how many genes are involved in the development and maturation of the brain and nervous system, but the number must be in the thousands. This number is in addition to the genes that are required in all cells to carry out metabolism and other basic life functions. It is easy to lose sight of the multiplicity of genes when considering extreme examples, such as PKU, in which a single mutation can have such a drastic effect on mental development. The situation is the same as that with any complex machine. An airplane can function only if thousands of parts are working together in harmony, but it takes only one defective part, if it affects a vital system, to bring it down. Likewise, the development and functioning of every trait require a large number of genes working in harmony, but in some cases a single mutant gene can have catastrophic consequences.

In other words, the relationship between a gene and a trait is not necessarily a simple one. Three principles govern the relationships between genes and traits:

1. One gene can affect more than one trait through secondary or indirect effects. The various, sometimes seemingly unrelated effects of a mutant gene are called **pleiotropic effects,** and the phenomenon itself is known as **pleiotropy.** For exam-

Figure 1.20 Cats with white fur and blue eyes have a high risk of being born deaf, a pleiotropic effect. [© Medioimages/Alamy Images]

ple, among cats with white fur and blue eyes, about 40 percent are born deaf (Figure 1.20). The reason for the defective hearing is not known, nor why it is so often associated with hair and eye color. This form of deafness can be regarded as a pleiotropic effect of white fur and blue eyes.

2. Any trait can be affected by more than one gene. We discussed this principle earlier in connection with the large number of genes that are required for the normal development and functioning of the brain and nervous system. Multiple genes affect even simpler traits such as hair color and eye color.

3. Many traits are affected by environmental factors as well as by genes. Consider again the low-phenylalanine diet. Children with PKU are not "doomed" to severe mental deficiency. Their capabilities can be brought into the normal range by dietary treatment. PKU serves as an example of what motivates geneticists to search to discover the molecular basis of inherited disease. The hope is that knowing the metabolic basis of the disease will eventually make it possible to develop methods for clinical intervention through diet, medication, or other treatments that will reduce the severity of the disease.

1.7

Evolution means continuity of life with change.

The pathway for the breakdown and excretion of phenylalanine is by no means unique to human beings. One of the remarkable generalizations to have emerged from molecular genetics is that organisms that are very distinct—for example, plants and animals—share many features in their genetics and biochemistry. These similarities indicate a fundamental "unity of life":

■ **Groups of related organisms descend from a common ancestor.**

Why is there unity of life? Because all creatures share a common origin. The process of **evolution** takes place whenever a population of organisms

gradually changes in genetic composition through time. From an evolutionary perspective, the unity of fundamental molecular processes is derived by inheritance from a distant common ancestor in which the mechanisms were already in place.

Not only the unity of life but also many other features of living organisms become comprehensible from an evolutionary perspective. For example, the interposition of an RNA intermediate in the basic flow of genetic information from DNA to RNA to protein makes sense if the earliest forms of life used RNA for both genetic information and enzyme catalysis. The importance of the evolutionary perspective in understanding aspects of biology that seem pointless or needlessly complex is summed up in the now famous aphorism of the evolutionary biologist Theodosius Dobzhansky: "Nothing in biology makes sense except in the light of evolution."

One indication of the common ancestry among Earth's creatures is illustrated in Figure 1.21 . The

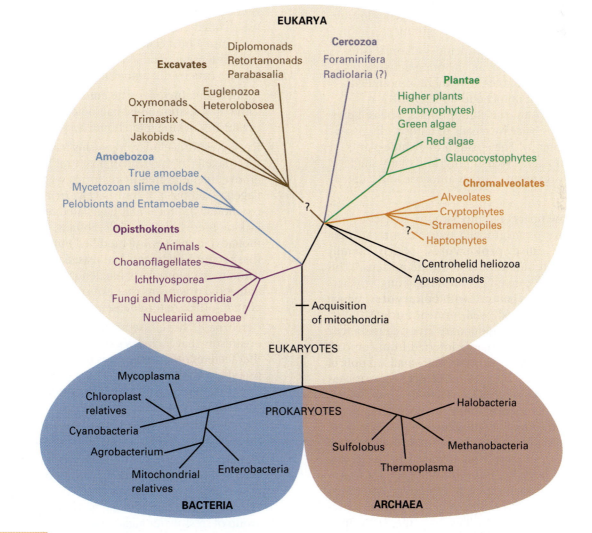

Figure 1.21 Evolutionary relationships among the major life forms as inferred from similarities in DNA sequence. The three major groups—Bacteria, Archaea, and Eukarya—are evident. Although the great diversity among eukaryotic forms is illustrated, there is also much more diversity among bacterial and archaeal forms than indicated here. [Courtesy of Andrew J. Roger and Alastair G. B. Simpson (eukaryotes) and Mitchell L. Sogin (prokaryotes).]

tree of relationships was inferred from similarities in nucleotide sequence in an RNA molecule found in the small subunit of the ribosome. Three major groups of organisms are distinguished.

1. **Bacteria**   This group includes most bacteria and cyanobacteria (formerly called blue-green algae). Cells of these organisms lack a membrane-bounded nucleus and mitochondria, are surrounded by a cell wall, and divide by binary fission.

2. **Archaea**   This group was initially discovered among microorganisms that produce methane gas or that live in extreme environments, such as hot springs or high salt concentrations. They are widely distributed in more normal environments as well. Like those of bacteria, the cells of archaea lack internal membranes. DNA sequence analysis indicates that the machinery for DNA replication and transcription in archaea resembles that in eukarya, whereas their metabolism strongly resembles that of bacteria. About half of the genes found in archaea are unique to this group.

3. **Eukarya**   This group includes all organisms whose cells contain an elaborate network of internal membranes, a membrane-bounded nucleus, and mitochondria. Their DNA is organized into true chromosomes, and cell division takes place by means of mitosis (discussed in Chapter 3). The eukaryotes include plants and animals as well as fungi and many single-celled organisms, such as amoebae and ciliated protozoa.

The members of the groups Bacteria and Archaea are often grouped together into a larger assemblage called **prokaryotes,** which literally means "before [the evolution of] the nucleus." This terminology is convenient for designating prokaryotes as a group in contrast with **eukaryotes,** which literally means "good [well-formed] nucleus."

How are such evolutionary trees constructed? The tree in Figure 1.21 was inferred from comparisons of the RNA sequence of the small subunit of the ribosome as well as sequence comparisons among a sample of protein-coding genes of diverse function. Among eukaryotes, the deepest branches are based on certain unique genetic features that are shared among organisms within each branch, such as the fusion of two genes involved in nucleotide synthesis, a large insertion into the gene for a protein elongation factor used in translation, and transfer of a gene for glyceraldehyde-3-phosphate dehydrogenase from the nucleus into the DNA in a plastid. Methods for inferring evolutionary relationships among organisms based on comparisons of

their DNA or protein sequences are discussed further in Chapter 14.

In Figure 1.21, the trees for bacteria and achaea are greatly simplified in order to emphasize the bacterial origins of mitochondria and chloroplasts, as well as the unusual environments in which many archaeans live. The tree for eukaryotes is more detailed but still excludes some important groups whose positions in the tree are unclear. Many of the groups of eukaryotic organisms in Figure 1.21 have odd and unfamiliar names—even to most professional biologists. Some of the defining features of these organisms are listed below. These descriptions are not intended to be memorized, but to demonstrate the astonishing diversity among eukaryotic organisms. The known eukaryotic organisms, for the most part, fall into six major groups:

- Opisthokonts: This group includes human beings and other animals; fungi; singe-celled organisms with a flagellum surrounded by a collar-like structure (choanoflagellates); obligate intracellular parasites (microsporidia) that use a coiled filament to inject their spore contents into a host cell; and amoeboid organisms with fine pseudopodia unsupported by microtubules (nucleariid amoebae).

- Amoebozoa: Includes true amoebae, certain slime molds, and some anaerobic amoeboid organisms lacking mitochondria.

- Excavates: Unicellular organisms featuring multiple flagella at the head of a ventral feeding groove; includes free-living forms with two flagella (jakobids) or four flagella (trimastix, oxymonads); unicellular flagellated organisms lacking mitochondria (diplomonads, retortamonads), or possessing modified mitochondria that produce hydrogen (parabasalia); forms containing chloroplasts (euglenozoa); and some amoeboid organisms that can convert into flagellated forms (heterolobosea).

- Cercozoa: Diverse assemblage of unicellular organisms that includes free-living flagellates that feed on other organisms using fine pseudopodia; includes amoeboid protozoa with fine pseudopodia and an external mineral shell (foraminiferans) or internal mineral skeleton (radiolarians); photosynthetic amoebae; and some plant pathogens resembling fungi.

- Plantae: Includes higher plants; green algae with chloroplasts containing chlorophylls a and b; red algae with chloroplasts containing fluorescent phycobiliprotein pigments; and a small group of freshwater algae (glaucocystophytes) with an unusual plastid that retains a cell wall chemically like those in bacteria.

• Chromalveolates: Organisms derived from a common ancestor whose cells incorporated a symbiotic red alga; includes the alveolates comprising ciliated protozoans, dinoflagellates, and apicomplexans such as the malaria parasite; the cryptophytes, predominantly unicellular water-dwelling organisms with chloroplasts bounded by four membranes; stramenopiles such as brown algae and diatoms; and the photosynthetic flagellated haptophytes whose resting stages are covered with a shell of calcareous plates.

The question marks in Figure 1.21 serve to emphasize that understanding the true evolutionary tree of eukaryotes is an ongoing process that periodically requires updating as new data become available. For example, the grouping *excavates* is somewhat controversial in that it includes many flagellated forms without mitochondria that were previously thought to branch off near the base of the tree.

■ **The molecular unity of life is seen in comparisons of genomes and proteomes.**

The totality of DNA in a single cell is called the **genome** of the organism. In sexual organisms, the genome is usually regarded as the DNA present in a reproductive cell. Modern methods for sequencing DNA are so rapid and efficient that the complete DNA sequence is known for more than 1000 genomes and large DNA molecules. These include the genomes of several species each of mammals (including the human genome), fish, fruit flies, nematodes, yeast and higher plants, as well as hundreds of bacteria, archaea, cellular organelles, and viruses.

Table 1.2 gives some examples of organisms whose genomes have been completely sequenced. Genome size is given in megabases (Mb), or millions of base pairs. The genome of the bacteria *Hemophilus influenzae*, like that of most bacteria, is very compact in that most of it codes for proteins. A high density of genes, relative to the amount of DNA, is also found in the budding yeast (5538 genes), the nematode worm (18,250 genes), the fruit fly (13,350 genes), and the diminutive flowering plant *Arabidopsis thaliana* (about 25,000 genes). The human genome, by contrast, contains large amounts of noncoding DNA. Comparison with the nematode is illuminating. Whereas the human genome is about 30 times larger than that of the worm, the number of genes is less than 2 times larger. This discrepancy reflects that fact that only about 1.5 percent of the human genome sequence codes for protein. (About 27 percent of the human genome is present in genes, but much of the DNA sequence present in genes is not protein coding.)

Table 1.2

Comparisons of genomes and proteomes

Organism	Genome size, Mb^a (approximate)	Number of genes (approximate)	Number of distinct proteins in proteome^b (approximate)	Shared protein families
Hemophilus influenzae (causes bacterial meningitis)	4.6	1700	1400	
Saccharomyces cerevisiae (budding yeast)	13	5538	4400	3000
Caenorhabditis elegans (soil nematode)	100	18,250	9500	5000
Drosophila melanogaster (fruit fly)	120^c	13,350	8000	7000^d
Mus musculus (laboratory mouse)	2500	30,000	10,000	9900^f
Homo sapiens (human being)	2900^e	30,000	10,000	

^{a} Millions of base pairs.
^{b} Excludes "families" of proteins with similar sequences (and hence related functions).
^{c} Excludes 60 Mb of specialized DNA ("heterochromatin") that has a very low content of genes.
^{d} Based on similarity with sequences in messenger RNA (mRNA).
^{e} For convenience, this estimate is rounded to 3000 Mb elsewhere in this book.
^{f} Based on the observation that only about 1% of mouse genes lack a similar gene in the human genome, and *vice versa*.

The complete set of proteins encoded in the genome is known as the **proteome.** In less complex genomes, such as the yeast, worm, and fruit fly, the number of proteins in an organism's proteome is approximately the same as the number of genes. However, as we shall see in Chapter 9, some genes encode two or more proteins through a process called *alternative splicing* in which segments of the original RNA transcript are joined together in a variety of combinations to produce different messenger RNAs. Alternative splicing is especially prevalent in the human genome. At least one-third of human genes, and possibly as many as two-thirds, undergo alternative splicing, and among the genes that undergo alternative splicing, the number of different messenger RNAs per gene ranges from 2 to 7. Hence, with its seemingly limited repertoire of 30,000 genes, the human genome can create approximately 60,000–90,000 different mRNAs. The widespread use of alternative splicing to multiply the coding capacity of genes is one source of human genetic complexity.

Most eukaryotic organisms contain *families* of related proteins that can be grouped according to similarities in their amino acid sequence. These families exist because the evolution of a new gene function is typically preceded by the duplication of an already existing gene, followed by changes in nucleotide sequence in one of the copies that gives rise to the new function. The new function is usually similar to the previous one (for example, a

change in the substrate specificity of a transporter protein), so that the new protein retains enough similarity in amino acid sequence to the original that their common ancestry can be recognized.

The molecular unity of life can be seen in the similarity of proteins in the proteome among diverse types of organisms. Such comparisons are shown in the right-hand column in Table 1.2. In this tabulation, each family of related proteins is counted only once, in order to estimate the number of proteins in the proteome that are "distinct" in the sense that their sequences are dissimilar. In yeast, worms, and flies, the number of distinct proteins is approximately 4400, 9500, and 8000, respectively. The brackets in Table 1.2 indicate the number of distinct proteins that share sequence similarity between species. From these comparisons, it appears that most multicellular animals share 5,000–10,000 proteins that are similar in sequence and function. Approximately 3000 of these are shared with eukaryotes as distantly related as yeast, and approximately 1000 with prokaryotes as distantly related as bacteria. What these comparisons among proteomes imply is that biological systems are based on protein components numbering in the thousands. This is a challenging level of complexity to understand, but the challenge is much less intimidating than it appeared to be at an earlier time when human cells were thought to produce as many as a million different proteins.

chapter summary

1.1 DNA is the molecule of heredity.

- Genetic traits can be altered by treatment with pure DNA.
- Transmission of DNA is the link between generations.

Organisms of the same species have many traits (characteristics) in common but may differ in other traits. Many of the differences between organisms result from genetic differences, the effects of the environment, or both. Genetics is the study of inherited traits, including those influenced in part by the environment. The elements of heredity consist of genes, which are transmitted from parents to offspring in reproduction. Although the sorting of genes in successive generations was first expressed numerically by Mendel, the chemical basis of genes was discovered by Miescher in the form of a weak acid—deoxyribonucleic acid (DNA). However, experimental proof that DNA is the genetic material did not come until about the middle of the twentieth century.

The first convincing evidence of the role of DNA in heredity came from the experiments of Avery, MacLeod, and McCarty, who showed that genetic characteristics in bacteria could be altered from one type to another by treatment with purified DNA. In studies of *Streptococcus pneumoniae,* they transformed mutant cells unable to cause pneumonia into cells that could do so by treating them with pure DNA from disease-causing forms. A second important line of evidence was the Hershey–Chase experiment. Hershey and Chase showed that the T2 bacterial virus injects primarily DNA into the host bacterium (*Escherichia coli*) and that a much higher proportion of parental DNA, compared with parental protein, is found among the progeny phage.

1.2 The structure of DNA is a double helix composed of two intertwined strands.

1.3 In replication, each parental DNA strand directs the synthesis of a new partner strand.

The three-dimensional structure of DNA, proposed in 1953 by Watson and Crick, gave many clues to the manner in which DNA functions as the genetic material. A molecule of DNA consists of two long chains of nucleotide subunits twisted around each other to form a right-handed helix. Each nucleotide subunit contains any one of four bases: A (adenine), T (thymine), G (guanine), or C (cytosine). The bases are paired in the two strands of a DNA molecule: Wherever one strand has an A, the partner strand has a T; and wherever one strand has a G, the partner strand has a C. The base pairing means that the two paired strands in a DNA duplex molecule have complementary base sequences along their lengths. The structure of the DNA molecule suggested that genetic information could be encoded in DNA in the sequence of bases. Mutations—changes in the genetic material—could result from changes in the sequence of bases, such as the substitution of one nucleotide for another or the insertion or deletion of one or more nucleotides. The structure of DNA also suggested a mode of replication: The two strands of the parental DNA molecule separate, and each individual strand serves as a template for the synthesis of a new, complementary strand.

1.4 Genes code for proteins.

- Enzyme defects result in inborn errors of metabolism.
- A defective enzyme results from a mutant gene.
- One of the DNA strands directs the synthesis of a molecule of RNA.
- A molecule of RNA directs the synthesis of a polypeptide chain.
- The genetic code is a triplet code.

Most genes code for proteins. More precisely stated, most genes specify the sequence of amino acids in a polypeptide chain. The transfer of genetic information from DNA into protein is a multistep process that includes several types of RNA (ribonucleic acid). Structurally, an RNA strand is similar to a DNA strand except that the "backbone" contains a different sugar (ribose instead of deoxyribose) and RNA contains the base uracil (U) instead of thymine (T). Also, RNA is usually present in cells in the form of single, unpaired strands. The initial step in gene expression is transcription, in which a molecule of RNA is synthesized that is complementary in base sequence to whichever DNA strand is being transcribed. In polypeptide synthesis, which takes place on a ribosome, the base sequence in the RNA transcript is translated in groups of three adjacent bases (codons). The codons are recognized by different types of transfer RNA (tRNA) through base pairing. Each type of tRNA is attached to a particular amino acid, and when a tRNA base-pairs with the proper codon on the mRNA, the growing end of the polypeptide chain is transferred to the amino acid on the tRNA. The table of all codons and the amino acids they specify is called the genetic code. Special codons specify the "start" (AUG, Met) and "stop" (UAA, UAG, and UGA) of polypeptide synthesis. The reason why various types of RNA are an intimate part of transcription and translation is proba-

bly that the earliest forms of life used RNA for both genetic information and enzyme catalysis.

1.5 Genes change by mutation.

A mutation that alters one or more codons in a gene can change the amino acid sequence of the resulting polypeptide chain synthesized in the cell. Often the altered protein is functionally defective, so an inborn error of metabolism results. One of the first inborn errors of metabolism studied was alkaptonuria; it results from the absence of an enzyme for cleaving homogentisic acid, which accumulates and is excreted in the urine, turning black upon oxidation. Phenylketonuria (PKU) is an inborn error of metabolism that affects the same metabolic pathway. The enzyme defect in PKU results in an inability to convert phenylalanine to tyrosine. Phenylalanine accumulation has catastrophic effects on the development of the brain. Children with the disease have severe mental deficits unless they are immediately placed on a special diet low in phenylalanine.

1.6 Traits are affected by environment as well as by genes.

Most visible traits of organisms result from many genes acting together in combination with environmental factors. The relationship between genes and traits is often complex because (1) every gene potentially affects many traits (pleiotropy), (2) every trait is potentially affected by many genes, and (3) many traits are significantly affected by environmental factors as well as by genes.

1.7 Evolution means continuity of life with change.

- Groups of related organisms descend from a common ancestor.
- The molecular unity of life is seen in comparisons among genomes and proteomes.

All living creatures are united by sharing many features of the genetic apparatus (for example, transcription and translation) and many aspects of metabolism. The unity of life results from the common ancestry of all living things. The great diversity among species results from the process of evolution. The three major groups of organisms are the Bacteria (which lack a membrane-bounded nucleus), the Archaea (which share features with both Bacteria and Eukarya but form a distinct group), and Eukarya (all organisms whose cells have a membrane-bounded nucleus that contains DNA organized into discrete chromosomes). Members of the groups Bacteria and Archaea collectively are often called prokaryotes.

The genomes of eukaryotes contain about 3000 genes or gene families that are shared by organisms as dissimilar as human beings and budding yeast. Eukaryotes share about 1000 genes or gene families with bacteria. Gene families are initially created by gene duplications that take place within a species to create extra copies. Through the process of evolution, the extra copies can change in nucleotide sequence and acquire modified functions.

issues & ideas

- What does it mean to say that DNA is the genetic material?
- What special feature of the structure of DNA allows each strand to be replicated separately?
- How does a strand of DNA specify the structure of a molecule of RNA?
- What types of RNA participate in protein synthesis, and what is the role of each type of RNA?
- What is meant by the phrase *the genetic code*, and how is the genetic code relevant to the translation of a polypeptide chain from a molecule of messenger RNA?

- How does the "central dogma" explain Garrod's discovery that nonfunctional enzymes result from mutant genes?
- In what way does phenylketonuria demonstrate the importance of the environment even for traits that are "determined" by genes?
- What is meant by the phrase *gene family*? How do gene families originate and proliferate?

key terms & concepts

adenine (A)
alkaptonuria
amino acid
antiparallel
Archaea
Bacteria
bacteriophage
base (in DNA or RNA)
biochemical pathway
block (in a biochemical pathway)
central dogma
chromosome
codon
colony
complementary base pairing

cytosine (C)
deoxyribonucleic acid (DNA)
double-stranded DNA
duplex DNA
enzyme
Eukarya
eukaryote
evolution
gene
genetic code
genetics
genome
guanine (G)
homogentisic acid
inborn error of metabolism
messenger RNA (mRNA)

metabolic pathway
metabolism
metabolite
mutant
mutation
nucleotide
phenylalanine hydroxylase (PAH)
phenylketonuria (PKU)
pleiotropic effect
pleiotropy
polarity (of DNA or RNA)
polypeptide chain
product molecule
prokaryote
proteome

replication
ribonucleic acid (RNA)
ribose
ribosomal RNA (rRNA)
ribosome
single-stranded DNA
substrate molecule
template
thymine (T)
transcript
transcription
transfer RNA (tRNA)
transformation
translation
uracil (U)

1. _____ The chemical interaction that holds the two strands of a DNA double helix together.

2. _____ This term means that the two strands in a DNA double helix have opposite polarity.

3. _____ DNA → RNA → protein.

4. _____ Production of a complementary strand of RNA from a template strand of DNA.

5. _____ The cellular particle, composed of RNA and protein, in which a messenger RNA is translated into a polypeptide chain.

6. _____ An "adaptor" type of RNA molecule that carries both an amino acid and a set of three nucleotides capable of binding with the corresponding codon in the mRNA.

7. _____ The complete set of proteins encoded in an organism's genome.

8. _____ In a metabolic pathway, the small molecule "input" that an enzyme acts upon.

9. _____ The earliest recognized inborn error of metabolism, this condition is characterized by black urine owing to the oxidation of homogentisic acid.

10. _____ Severe form of mental disability caused by mutant forms of the enzyme phenylalanine hydroxylase (PAH).

11. _____ Term referring to a secondary, often seemingly unrelated, effect of a mutant gene, such as deafness associated with white fur in cats.

12. _____ Eukarya is to eukaryote as Bacteria and Archaea are to this term.

solutions: step by step

Problem 1

In the human gene for the β chain of hemoglobin (the oxygen-carrying protein in the red blood cells), the first 30 nucleotides in the amino-acid–coding region have the sequence

3'-TACCACGTGGACTGAGGACTCCTCTTCAGA-5'

What is the sequence of the partner strand?

■ Solution The base pairing between the strands is A with T and G with C, but it is equally important that the strands in a

DNA duplex have opposite polarity. The partner strand is therefore oriented with its 5′ end at the left, and the base sequence is

```
5'-ATGGTGCACCTGACTCCTGAGGAGAAGTCT-3'
```

Problem 2

If the DNA duplex for the β chain of hemoglobin in Step-by-Step Problem 1 were transcribed from left to right, deduce the base sequence of the RNA in this coding region.

■ **Solution** To deduce the RNA sequence, we must apply three concepts. First, in the transcription of RNA, the base pairing is such that an A, T, G, or C in the DNA template strand is transcribed as U, A, C, or G, respectively, in the RNA strand. Second, the RNA transcript and the DNA template strand have opposite polarity. Third (and critically for this problem), the RNA transcript is always transcribed in the 5′-to-3′ direction, so the 5′ end of the RNA is the end synthesized first. This being the case, and considering the opposite polarity, the 3′ end of the template strand must be transcribed first. Because we are told that transcription takes place from left to right, we can deduce that the transcribed strand is that shown in Step-by-Step Problem 1. The RNA transcript therefore has the base sequence

```
5'-AUGGUGCACCUGACUCCUGAGGAGAAGUCU-3'
```

Problem 3

Given the RNA sequence coding for part of human β hemoglobin deduced in Step-by-Step Problem 2, what is the amino acid sequence in this part of the β polypeptide chain?

■ **Solution** The polypeptide chain is translated in successive groups of three nucleotides (each group constituting a codon),

starting at the 5′ end of the coding sequence and moving in the 5′-to-3′ direction. The amino acid corresponding to each codon can be found in the genetic code table. The first ten amino acids in the polypeptide chain are therefore

```
5'-AUGGUGCACCUGACUCCUGAGGAGAAGUCU-3'
   MetValHisLeuThrProGluGluLysSer
```

Problem 4

A very important mutation in human hemoglobin occurs in the DNA sequence shown in Step-by-Step Problem 1. In this mutation, the red T at nucleotide position 20 is replaced with an A. The mutant hemoglobin is called *sickle-cell hemoglobin*, and it is associated with a severe anemia known as *sickle-cell anemia*. Severe as the genetic disease is, the mutant gene is present at relatively high frequency in some human populations because carriers of the gene, who have only a mild anemia, are more resistant to falciparum malaria than are noncarriers. What is the nucleotide sequence of this region of the DNA duplex in sickle-cell hemoglobin (both strands) and that of the messenger RNA, and what is the amino acid replacement that results in sickle-cell hemoglobin?

■ **Solution** The mutation is already given as a T → A substitution at position 20. The sequence of the DNA duplex is obtained as in Step-by-Step Problem 1, that of the RNA as in Problem 2, and that of the mutant polypeptide chain as in Problem 3, except that at each step there is one nucleotide (or one amino acid) that differs from the nonmutant. The DNA, RNA, and polypeptide in this region of sickle-cell hemoglobin are shown below, where the differences from the nonmutant gene are in red. The amino acid replacement is glutamic acid → valine.

Problem 4 solution—Nucleotide sequences

DNA (transcribed strand)	3'-TACCACGTGGACTGAGGACACCTCTTCAGA-5'
DNA (nontranscribed strand)	5'-ATGGTGCACCTGACTCCTGTGGAGAAGTCT-3'
	↓
RNA coding region	5'-AUGGUGCACCUGACUCCUGUGGAGAAGUCU-3'
Polypeptide chain	MetValHisLeuThrProValGluLysSer

concepts in action: problems for solution

1.1 Classify each of the following statements as true or false.
(a) Each gene is responsible for only one visible trait.
(b) Every trait is potentially affected by many genes.
(c) The sequence of nucleotides in a gene specifies the sequence of amino acids in a protein encoded by the gene.
(d) There is one-to-one correspondence between the set of codons in the genetic code and the set of amino acids encoded.

1.2 Prior to the Avery, MacLeod, and McCarty experiment, what features of cells and chromosomes were already known that could have been interpreted as evidence that DNA is an important constituent of the genetic material?

1.3 In the early years of the twentieth century, why did most biologists and biochemists believe that proteins were probably the genetic material?

1.4 Like DNA, molecules of RNA contain large amounts of phosphorus. When Hershey and Chase grew their T2 phage in bacterial cells in the presence of radioactive phosphorus, the RNA must also have incorporated the labeled phosphorus, and yet the experimental result was not compromised. Why not?

1.5 From their examination of the structure of DNA, what were Watson and Crick able to infer about the probable mechanisms of DNA replication, coding capability, and mutation?

concepts in action: problems for solution *continued*

1.6 What are three principal structural differences between RNA and DNA?

1.7 When the base composition of a DNA sample from *Micrococcus luteus* was determined, 37.5 percent of the bases were found to be cytosine. The DNA of this organism is known to be double-stranded. What is the percentage of adenine in this DNA?

1.8 DNA extracted from a certain virus has the following base composition: 20 percent adenine, 40 percent thymine, 25 percent guanine, and 15 percent cytosine. How would you interpret this result in terms of the structure of the viral DNA?

1.9 A duplex DNA molecule contains 642 occurrences of the dinucleotide 5′-GT-3′ in one or the other of the paired strands. What other dinucleotide is also present exactly 642 times?

1.10 A repeating polymer with the sequence

$$5′\text{-}AAUAAUAAUAAU \cdots \text{-}3′$$

was found to produce only two types of polypeptides in a translation system that uses cellular components but not living cells (called an *in vitro* translation system). One polypeptide consisted of repeating Asn and the other of repeating Ile. How can you explain this result?

1.11 Analysis of the DNA of the *PAH* gene in a patient with phenylketonuria revealed a mutation in the protein-coding region whose predicted effect would be to replace the amino acid aspartic acid with histidine. Nevertheless, no mutant protein could be found in the patient's cells. Explain how this could happen.

1.12 In the course of evolution, genes can become longer or shorter via the insertion or deletion of nucleotides in the DNA sequence. Such changes that take place in coding regions are always exactly three nucleotides in length or an exact multiple of three. Explain why this is the case.

1.13 Part of the protein-coding region in a gene has the base sequence 3′-ACAGCATAAACGTTC-5′. What is the sequence of the partner DNA strand?

1.14 If the DNA sequence in Problem 1.13 is the template strand that is transcribed in the synthesis of messenger RNA, would it be transcribed from left to right or from right to left? What base sequence would this region of the RNA contain?

1.15 What amino acid sequence would be synthesized from the messenger RNA region in Problem 1.14?

1.16 If a mutation occurs in the DNA sequence in Problem 1.15 in which the red C is replaced with T, what amino acid sequence would result?

1.17 A polymer is made that has a random sequence consisting of 75 percent Gs and 25 percent Us. Among the amino acids in the polypeptide chains resulting from *in vitro* translation, what is the expected frequency of Trp? of Val? of Phe?

1.18 The coding sequence in the messenger RNA for amino acids 1 through 10 of human phenylalanine hydroxylase is

$$5′\text{-}AUGUCCACUGCGGUCCUGGAAAACCCAGGC\text{-}3′$$

(a) What are the first ten amino acids?
(b) What sequence would result from a mutant RNA in which the red A was changed to G?
(c) What sequence would result from a mutant RNA in which the red C was changed to G?
(d) What sequence would result from a mutant RNA in which the red U was changed to C?
(e) What sequence would result from a mutant RNA in which the red G was changed to U?

1.19 Shown here is the terminal part of a metabolic pathway in a bacterium in which a substrate metabolite (small molecule) W is converted into a final product metabolite Z through a sequence of three steps catalyzed by the enzymes A, B, and C. Each of the enzymes is the product of a different gene.

$$W \xrightarrow{A} X \xrightarrow{B} Y \xrightarrow{C} Z$$

Which metabolites would be expected to be missing, and which present in excess, in cells that are mutant for:
(a) Enzyme A?
(b) Enzyme B?
(c) Enzyme C?

1.20 A mutation isolated in the bacterium discussed in Problem 19 affects one of the enzymes in the pathway shown, but it is not known which step (A, B, or C) is blocked. The final product Z of the pathway is essential for growth. When mutant cells are placed in cultures lacking Z, they cannot grow. If Z is added to the medium, they can grow. Experiments are carried out to determine whether any of the intermediates can substitute for Z in supporting growth. It is found that mutant cells can grow in the presence of Y but not in the presence of W or X. Deduce from these data what step in the pathway is blocked in the mutant.

geNETics on the web

GeNETics on the Web will introduce you to some of the most important sites for finding genetic information on the Internet. To explore these sites, visit the Jones and Bartlett home page at

http://www.jbpub.com/genetics

For the book *Essential Genetics: A Genomics Perspective,* choose the link that says Enter **GeNETics on the Web**. You will be presented with a chapter-by-chapter list of highlighted keywords. Select any highlighted keyword and you will be linked to a Web site containing genetic information related to the keyword.

- James D. Watson once said that he and Francis Crick had no doubt that their proposed DNA structure was essentially correct, because the structure was so beautiful it had to be true! At an Internet site accessed by the keyword **DNA**, you can view a large collection of different types of models of DNA structure. Some models highlight the sugar–phosphate backbones, others the A–T and G–C base pairs, still others the helical structure of double-stranded DNA.

- With proper dietary control of blood phenylalanine, patients with PKU can develop normally and lead normal lives. When dietary control is relaxed, however, blood phenylalanine returns to high levels. This situation is extremely dangerous for a developing fetus, resulting in high risk of congenital heart disease, small head size, mental retardation, and slow growth. Affected children are said to have **maternal PKU**. They are affected, not because of their own inability to metabolize phenylalanine, but because of high levels of phenyalanine in their mothers' blood. The risk can be reduced, but not entirely eliminated, if PKU mothers plan their pregnancies and adhere to a strict dietary regimen prior to and during pregnancy. To learn more about this unanticipated consequence of dietary treatment of PKU, log onto the phenylalanine hydroxylase knowledge database at the keyword site.

further readings

Abedon, S. T. 2000. The murky origin of Snow White and her T-even dwarfs. *Genetics* 155: 481.

Bearn, A. G. 1994. Archibald Edward Garrod, the reluctant geneticist. *Genetics* 137: 1.

Calladine, C. R. 1997. *Understanding DNA: The Molecule and How It Works.* New York: Academic Press.

Ciechanover, A., A. Orian, and A. L. Schwartz. 2000. Ubiquitin-mediated proteolysis: Biological regulation via destruction. *Bioessays* 22: 442.

Doolittle, W. F. 2000. Uprooting the tree of life. *Scientific American,* February.

Gehrig, A., S. R. Schmidt, C. R. Muller, S. Srsen, K. Srsnova, and W. Kress. 1997. Molecular defects in alkaptonuria. *Cytogenetics & Cell Genetics* 76: 14.

Gould, S. J. 1994. The evolution of life on the earth. *Scientific American,* October.

Horowitz, N. H. 1996. The sixtieth anniversary of biochemical genetics. *Genetics* 143: 1.

Judson, H. F. 1996. *The Eighth Day of Creation: The Makers of the Revolution in Biology.* Cold Spring Harbor, NY: Cold Spring Harbor Laboratory Press.

Kellis, M., N. Patterson, M. Endrizzi, B. Birren, and E. S. Lander. 2003. Sequencing and comparison of yeast sequences to identify genes and regulatory elements. *Nature* 423: 241.

Mirsky, A. 1968. The discovery of DNA. *Scientific American,* June.

Olson, G. J., and C. R. Woese. 1997. Archaeal genomics: An overview. *Cell* 89: 991.

Radman, M., and R. Wagner. 1988. The high fidelity of DNA duplication. *Scientific American,* August.

Rennie, J. 1993. DNA's new twists. *Scientific American,* March.

Scazzocchio, C. 1997. Alkaptonuria: From humans to moulds and back. *Trends in Genetics* 13: 125.

Scriver, C. R., and P. J. Waters. 1999. Monogenic traits are not simple: Lessons from phenylketonuria. *Trends in Genetics* 15: 267.

Simpson, A. G. B., and A. J. Roger. 2002. Eukaryote evolution: Getting to the root of the problem. *Current Biology* 12: R691.

Stadler, D. 1997. Ultraviolet-induced mutation and the chemical nature of the gene. *Genetics* 145: 863.

Susman, M. 1995. The Cold Spring Harbor phage course (1945–1970): A 50th anniversary remembrance. *Genetics* 139: 1101.

Watson, J. D. 1968. *The Double Helix.* New York: Atheneum.

Zhang, J. Z. 2000. Protein-length distributions for the three domains of life. *Trends in Genetics* 16: 107.

For centuries, flower-loving horticulturalists have nurtured rare and interesting "sports," or mutant forms, such as this yellow snapdragon. [© Photos.com]

key concepts

- Inherited traits are determined by the genes present in the reproductive cells united in fertilization.

- Genes are usually inherited in pairs, one from the mother and one from the father.

- The genes in a pair may differ in DNA sequence and in their effect on the expression of a particular inherited trait.

- The maternally and paternally inherited genes are not changed by being together in the same organism.

- In the formation of reproductive cells, the paired genes separate again into different cells.

- Random combinations of reproductive cells containing different genes result in Mendel's ratios of traits appearing among the progeny.

- The ratios actually observed for any traits are determined by the types of dominance and gene interaction.

2

Transmission Genetics: Heritage from Mendel

chapter organization

2.1 Mendel took a distinctly modern view of transmission genetics.

2.2 Genes come in pairs, separate in gametes, and join randomly in fertilization.

2.3 The alleles of different genes segregate independently.

2.4 Chance plays a central role in Mendelian genetics.

2.5 The results of segregation can be observed in human pedigrees.

2.6 Dominance is a property of a pair of alleles in relation to a particular attribute of phenotype.

2.7 Epistasis can affect the observed ratios of phenotypes.

2.8 Complementation between mutations of different genes is a fundamental principle of genetics.

the human connection **Blood Feud**

Chapter Summary
Issues & Ideas
Key Terms & Concepts
Solutions: Step by Step
Concepts in Action: Problems for Solution
Further Readings

geNETics on the web

Selective breeding has long been a practice of farmers and herders, dating back thousands of years to the early domestication of plants and animals. It was generally understood that desirable traits could be passed from one generation to the next, with some degree of predictability if not certainty. The study of the patterns of inheritance from generation to generation is known as **transmission genetics.** In eukaryotic organisms, transmission genetics is often called **Mendelian genetics** after Gregor Mendel, himself the son of a farmer. His story is one of the inspiring legends in the history of science.

While serving as a monk at the distinguished monastery of St. Thomas in the town of Brno (Brünn), in what is now the Czech Republic, Mendel taught physics and natural history at a local secondary school. His teaching was said to be "clear, logical, and well suited to the needs of his students." He also carried out biological experiments, the most important being his study of crosses of the common garden pea (*Pisum sativum*). These were carried out from 1856 to 1863 in a small garden plot nestled in a corner of the monastery grounds. He reported his experiments to a local natural history society, published the results and his interpretation in its scientific journal in 1866, and began exchanging letters with Carl Nägeli in Munich, one of the leading botanists of the time. No one understood the significance of the experiments.

In 1868 Mendel was elected abbot of the monastery, and his scientific work effectively came to an end. Shortly before his death in 1884, Mendel is said to have remarked to one of the younger monks, "My scientific work has brought me a great deal of satisfaction, and I am convinced that it will be appreciated before long by the whole world."

The prophecy was fulfilled 16 years later when Hugo de Vries, Carl Correns, and Erich von Tschermak, each working independently and in a different European country, published the results of experiments similar to Mendel's, drew attention to Mendel's paper, and attributed priority of discovery to him.

Although some modern historians of science disagree over Mendel's intentions in carrying out his work, everyone concedes that Mendel was a first-rate experimenter who performed careful and exceptionally well-documented experiments. His paper contains the first clear exposition of the statistical rules governing the transmission of hereditary elements from generation to generation. The elegance of his experiments explains why they were embraced as the foundation of modern genetics. Mendel's breakthrough experiments and concepts are the subject of this chapter.

2.1

Mendel took a distinctly modern view of transmission genetics.

Mendel's name will forever be associated with peas: round or wrinkled, yellow or green, tall or short. But it was not only his choice of experimental organism and his choice of traits that made Mendel's success possible. The basic premise underlying Mendel's experiments represented an important shift in approach. Although he didn't know about DNA or chromosomes, he came to realize that each parent contributed to its progeny a number of separate and distinct elements of heredity ("factors" as he called them—in modern terms, genes). More important still, he realized that each

GREGOR MENDEL (above) and the small monastery garden (right) in which Mendel grew and classified more than 33,500 pea plants.

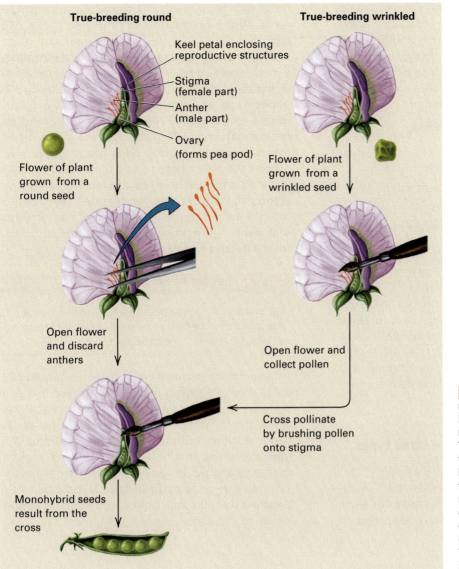

True-breeding round

Keel petal enclosing reproductive structures

Stigma (female part)

Anther (male part)

Ovary (forms pea pod)

Flower of plant grown from a round seed

Open flower and discard anthers

Monohybrid seeds result from the cross

True-breeding wrinkled

Flower of plant grown from a wrinkled seed

Open flower and collect pollen

Cross pollinate by brushing pollen onto stigma

Figure 2.1 Crossing pea plants requires some minor surgery in which the anthers of a flower are removed before they produce pollen. The stigma, the female part of the flower, is not removed. It is fertilized by brushing with mature pollen grains taken from another plant. Each pollinated flower has a single ovary that develops into the seed pod. The ovary contains as many as ten ovules, which develop into seeds upon fertilization. These seeds represent the second-generation, in this instance, a hybrid.

of these parental factors remained unchanged as it was passed from one generation to the next.

Given the unchanging nature of these factors, Mendel set out to track their movement through generations of pea plants by observing the appearance of the traits associated with them, such as round or wrinkled seeds. He thought in quantitative, numerical terms. Mendel did not ask merely "What types of peas are present?" in the progeny of a cross, but also "What are their numerical ratios?" He proceeded by carrying out simple crossing experiments and then looked for statistical regularities that might identify general rules. In his own words, he wanted to "determine the number of different forms in which hybrid progeny appear" and, among these, to "ascertain their numerical interrelationships."

Mendel selected peas for his experiments for two reasons. First, he had access to varieties that differed in observable alternative characteristics, such as round versus wrinkled seeds, or yellow versus green

seeds. Second, his preliminary studies had indicated that peas normally reproduce by self-fertilization, in which pollen produced in a flower is used to fertilize the eggs in the same flower. Left alone, pea flowers always self-fertilize. Carrying out a cross between two different varieties is actually very tedious. One must open the keel petal (which encloses the reproductive structures), remove the immature anthers (the pollen-producing structures) before they shed pollen, and dust the stigma (part of the female structure) with mature pollen taken from a flower on a different plant (Figure 2.1).

■ **Mendel was careful in his choice of traits.**

Mendel recognized the need to study traits that were uniform within any given variety of peas but different between varieties. For this reason, at the beginning of his experiments, he established **true-breeding** varieties, in which the plants produced only progeny like themselves when allowed to self-fertilize. For example, one true-breeding variety always yielded round seeds, whereas another true-breeding variety always yielded wrinkled seeds. For his experiments, Mendel chose seven pairs of varieties, each of which was true-breeding for a different trait. The contrasting traits affected

- seed shape (round versus wrinkled)
- seed color (yellow versus green)
- flower color (purple versus white)
- pod shape (smooth versus constricted)
- pod color (green versus yellow)
- flower and pod position (axial versus terminal)
- stem length (standard versus dwarf)

When two varieties that differ in one or more traits are crossed, the progeny constitute a **hybrid** between the parental varieties. Crosses in which the parental varieties differ in one, two, or three traits of interest are called *monohybrid, dihybrid,* or *trihybrid* crosses, respectively. Unless a trait is relevant to the experiment under consideration, it is normally ignored even if the parental varieties happen to differ in regard to this trait.

■ Reciprocal crosses yield the same types of offspring.

It is worthwhile to examine a few of Mendel's original experiments to learn what his methods were and how he interpreted his results. One pair of traits that he studied was round versus wrinkled seeds. When pollen from a variety of plants with wrinkled seeds was used to cross-pollinate plants from a variety with round seeds, all of the resulting hybrid seeds were round (Figure 2.2, cross A). Geneticists call the true-breeding parents the P_1 **generation** and the hybrid *filial* seeds or plants the F_1 **generation.** Mendel also performed the **reciprocal cross** (Figure 2.2, cross B), in which plants from the variety with round seeds were used as the pollen parents and those from the variety with wrinkled seeds as the female parents. As before, all of the F_1 seeds were round. The reciprocal crosses in Figure 2.2 illustrate the principle that, with a few important exceptions to be discussed in later chapters,

key concept

The outcome of a genetic cross does not depend on which trait is present in the male and which is present in the female; reciprocal crosses yield the same result.

Equivalent results were obtained when Mendel made crosses between plants that differed in any of the pairs of alternative characteristics. In each case, all of the F_1 progeny exhibited only one of the parental traits, and the other trait was absent. The trait expressed in the F_1 generation in each of the monohybrid crosses is shown in Figure 2.3. The trait expressed in the hybrids Mendel called the *dominant* trait; the trait not expressed in the hybrids he called *recessive*.

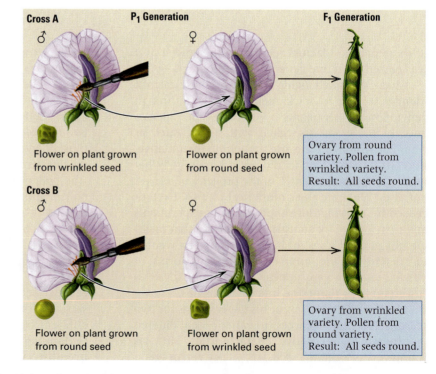

Figure 2.2 Reciprocal crosses of true-breeding pea plants. In this example, the hybrid seeds are round, irrespective of the direction of the cross.

Cross A

Cross B

P_1 Generation

F_1 Generation

Flower on plant grown from wrinkled seed

Flower on plant grown from round seed

Ovary from round variety. Pollen from wrinkled variety. Result: All seeds round.

Flower on plant grown from round seed

Flower on plant grown from wrinkled seed

Ovary from wrinkled variety. Pollen from round variety. Result: All seeds round.

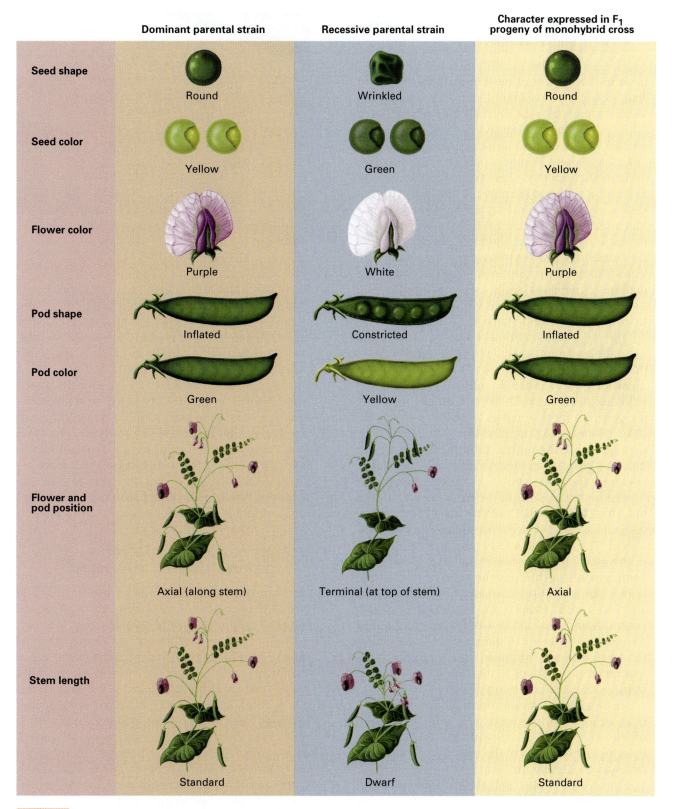

	Dominant parental strain	Recessive parental strain	Character expressed in F_1 progeny of monohybrid cross
Seed shape	Round	Wrinkled	Round
Seed color	Yellow	Green	Yellow
Flower color	Purple	White	Purple
Pod shape	Inflated	Constricted	Inflated
Pod color	Green	Yellow	Green
Flower and pod position	Axial (along stem)	Terminal (at top of stem)	Axial
Stem length	Standard	Dwarf	Standard

Figure 2.3 The seven character differences in peas studied by Mendel. The character considered dominant is the trait that appears in the hybrid produced by crossing. Which of each of the pairs of contrasting characters is dominant is revealed only after the F_1 progeny are formed.

■ The wrinkled mutation causes an inborn error in starch synthesis.

Let us now consider Mendel's round and wrinkled seeds in the context of modern methods of genetic analysis and what we know today. Although most of Mendel's original experimental material has been lost, a strain of peas bearing what is thought to be the original wrinkled mutation was perpetuated by seed dealers in Eastern Europe. Analysis of this mutation using modern methods has revealed the function of the normal gene and the molecular basis of the wrinkled mutation. The normal gene encodes an enzyme, starch-branching enzyme I (SBEI), required to synthesize a branched-chain form of starch known as *amylopectin.* As pea seeds dry, they lose water and shrink. Round seeds contain amylopectin and shrink uniformly; wrinkled seeds lack amylopectin and shrink irregularly. In other words, wrinkled peas have an inborn error in starch metabolism.

The most common form of a trait occurring in a natural population is considered the **wildtype,** in this case the round pea. Any form that differs from the wildtype is considered a mutant, in this case the wrinkled pea. The wildtype and mutant forms of the gene are represented as *W* and *w,* respectively. (It is customary to print gene symbols in italic type. Geneticists use the mutant form to name the trait. A capital letter often identifies the dominant form, lowercase the recessive.)

The molecular basis of the wrinkled mutation is that the *SBEI* gene has become interrupted by the insertion of a DNA sequence called a *transposable element.* These are DNA sequences that are capable of moving (*transposition*) from one location to another within a chromosome or between chromosomes. The molecular mechanisms of transposition are discussed in Chapter 12, but for present purposes it is necessary to know only that transposable elements are present in most genomes, especially the large genomes of eukaryotes, and that many spontaneous mutations result from the insertion of transposable elements into a gene.

Figure 2.4, part A, is a simplified diagram of the DNA structure of the wildtype (nonmutant) form of the *SBEI* gene, along with the mutant form showing the insertion of the transposable element. One way to identify the *W* and *w* forms of the gene is a procedure called **gel electrophoresis.** It is used for separating DNA molecules of different sizes. In this procedure, samples containing relatively small fragments of duplex DNA are placed into slots near one edge of a slab of a jelly-like material (usually agarose), which is then submerged in a buffer solution and subjected to an electric field (Figure 2.4, part B). DNA fragments in the samples move in response to the electric field in accordance with their lengths. Shorter fragments

FRAGMENTS OF DNA can be separated by size in a thin slab of agarose gel.

move faster and farther than long fragments. In the case of DNA fragments corresponding to the *W* and *w* forms of the *SBEI* gene, the *W* fragment moves farther than the *w* fragment because the *w* fragment is larger (owing to the insertion of the transposable element). The separation of the *W* and *w* fragments is indicated by the dark rectangles, called *bands,* shown in the gel. As noted, a sample containing a mixture of both *W* and *w* fragments yields two bands, one corresponding to *W* and the other to *w.* There's much more to tell about obtaining and isolating the DNA fragments; these procedures are discussed in Chapter 6. For the moment we will focus only on how the direct analysis of DNA helps one to understand Mendel's experiments.

■ Analysis of DNA puts Mendel's experiments in a modern context.

In discussing Mendel's results with round and wrinkled peas from a modern point of view, we must be careful to specify how the trait is examined. To avoid confusion, we use the terms morphological trait and molecular trait. A *morphological trait* is one that is manifest, plainly shown, and readily perceived by the senses. A *molecular trait* is one that can be perceived only by means of special methods, such as gel electrophoresis, that enable differences between molecules to be visualized. Classical geneticists studied primarily morphological traits (although their observations were sometimes aided by instruments such as the microscope). Modern geneticists study morphological traits too, but they usually supplement this with molecular analysis using techniques such as gel electrophoresis and DNA sequencing. With regard to round and wrinkled peas, the morphological trait corresponds to whether the shape of a seed is manifestly round or wrinkled. The molecular trait corresponds to the pattern of bands in an electrophoresis gel: whether the DNA extracted from a

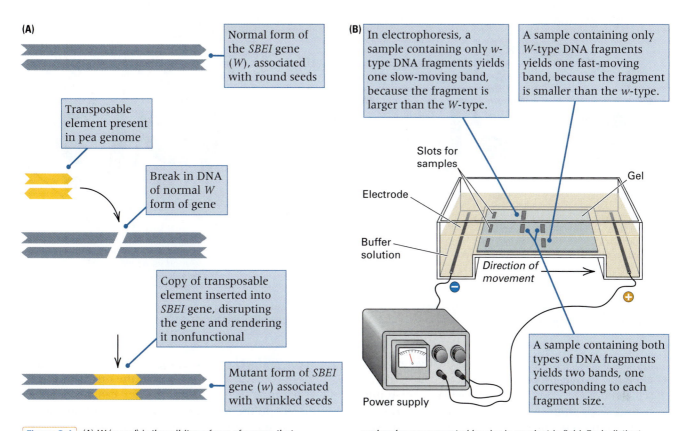

(A)

Normal form of the *SBEI* gene (*W*), associated with round seeds

Transposable element present in pea genome

Break in DNA of normal *W* form of gene

Copy of transposable element inserted into *SBEI* gene, disrupting the gene and rendering it nonfunctional

Mutant form of *SBEI* gene (*w*) associated with wrinkled seeds

(B)

In electrophoresis, a sample containing only *w*-type DNA fragments yields one slow-moving band, because the fragment is larger than the *W*-type.

A sample containing only *W*-type DNA fragments yields one fast-moving band, because the fragment is smaller than the *w*-type.

Slots for samples

Electrode

Gel

Buffer solution

Direction of movement

A sample containing both types of DNA fragments yields two bands, one corresponding to each fragment size.

Power supply

Figure 2.4 (A) *W* (round) is the wildtype form of a gene that specifies the amino acid sequence of starch-branching enzyme I (SBEI). The allele *w* (wrinkled) encodes an inactive form of the enzyme, inactive because its DNA sequence is interrupted by the insertion of a transposable element. (B) The molecular difference between *W* and *w* can be detected using electrophoresis. The DNA molecules are separated by size in an electric field. Each distinct size of DNA molecule produces a band at a characteristic position in the gel. A DNA molecule from the *w* gene, because it includes the transposable element, is larger than a molecule from the *W* gene and will migrate more slowly in the gel. A DNA sample containing both types of molecules will yield two bands in the gel.

seed yields one rapidly migrating band, one slowly migrating band, or two bands.

Morphological traits are frequently dominant or recessive, but this is not necessarily true of molecular traits. In Figure 2.4, part B, for example, consider the molecular trait defined by the distance traveled by each DNA band from its starting position in the gel. The true-breeding strain with round seeds has a single rapidly migrating band, the true-breeding strain with wrinkled seeds has a single slowly migrating band, and the progeny of the cross (which has round seeds) exhibit both bands.

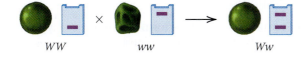

WW × ww → Ww

In other words, the progeny of the cross between the true-breeding strains show the molecular trait associated with both forms of the gene (in this case, a rapidly migrating DNA band along with a slowly migrating DNA band). In situations in which alternative forms of a gene (in this case, *W* and *w*) can both be detected when they are present in a cell or organism, we say that the forms of the gene are **codominant.** Molecular traits are often (but by no means always) codominant.

In the next section we will use the gel symbol to show the molecular traits whose existence Mendel could only infer as he followed the morphological traits of pea plants through many generations. This approach puts Mendel's experiments in the context of modern molecular genetics.

2.2

Genes come in pairs, separate in gametes, and join randomly in fertilization.

The prevailing concept of heredity in Mendel's time was that the traits of the parents became blended in the hybrid, as though the hereditary material consisted of fluids that became permanently mixed when combined. Following this logic to its natural conclusion, one would expect to see generations of offspring that move toward a set of shared traits, with little to distinguish one individual from another. This did not happen with Mendel's monohybrid peas. In the first generation of hybrids, the recessive visible trait "dis-

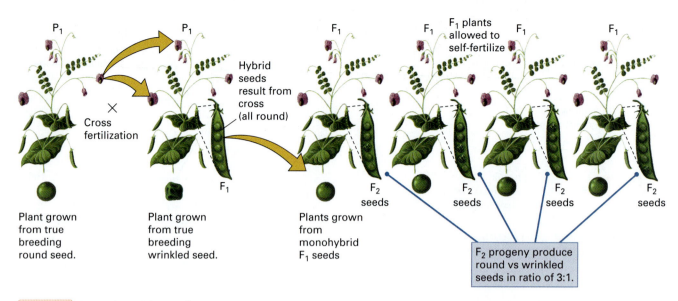

Plant grown from true breeding round seed.

× Cross fertilization

Plant grown from true breeding wrinkled seed.

Hybrid seeds result from cross (all round)

F_1

Plants grown from monohybrid F_1 seeds

F_2 seeds

F_1 plants allowed to self-fertilize

F_2 seeds

F_2 seeds

F_2 seeds

F_2 progeny produce round vs wrinkled seeds in ratio of 3:1.

Figure 2.5 Some of Mendel's traits (such as flower color and plant height) are visible only in the mature plants grown from seeds, but other traits (such as seed shape and seed color) are visible in the seeds themselves.

appeared," only to reappear in the next generation, after the hybrid progeny were allowed to undergo self-fertilization. For example, when the round hybrid seeds from the round × wrinkled cross—the F_1 generation—were grown into plants and allowed to self-fertilize, some of the resulting seeds were round and others wrinkled (Figure 2.5). The progeny seeds produced by self-fertilization of the F_1 generation constitute the **F_2 generation.** Mendel found that the dominant and recessive traits appear in the F_2 progeny in the proportions 3 round : 1 wrinkled.

Similar results were obtained in the F_2 generation of crosses between plants that differed in any of the pairs of alternative characteristics (Table 2.1). Note that the first two traits (round versus wrinkled

seeds and yellow versus green seeds) have many more observations than any of the other traits; this is because seed shape and color can be classified directly in the seeds, whereas the other traits can be classified only in the mature plants. Relative to the inheritance of visible traits, the principal conclusions from the data in Table 2.1 were as follows:

- The F_1 hybrids express only the dominant trait.
- In the F_2 generation, plants with either the dominant or the recessive trait are present.
- In the F_2 generation, there are approximately three times as many plants with the dominant trait as plants with the recessive trait. In other words, the F_2 ratio of dominant : recessive is approximately 3 : 1.

In the remainder of this section, we will see how Mendel deduced from these basic observations his hypothesis of discrete genetic units and the principles governing their inheritance. We shall also see how he used statistical analysis to support it.

■ Genes are physical entities that come in pairs.

Important to Mendel's formulation of his hypothesis was the fact that in his monohybrid crosses, the recessive trait that seemingly disappeared in the F_1 generation reappeared again in the F_2 generation. Not only did the recessive trait reappear, it was in no way different from the trait present in the recessive P_1 plants. In a letter describing this finding, Mendel noted that in the F_2 generation, "the two parental traits appear, separated and unchanged, and there is nothing to indicate that one of them has either inherited or taken over anything from

Table	2.1

Results of Mendel's monohybrid experiments

Parental traits	F_1 trait	Number of F_2 progeny	F_2 ratio
round × wrinkled (seeds)	round	5474 round 1850 wrinkled	2.96 : 1
yellow × green (seeds)	yellow	6022 yellow 2001 green	3.01 : 1
purple × white (flowers)	purple	705 purple 224 white	3.15 : 1
inflated × constricted (pods)	inflated	882 inflated 299 constricted	2.95 : 1
green × yellow (unripe pods)	green	428 green 152 yellow	2.82 : 1
axial × terminal (flower position)	axial	651 axial 207 terminal	3.14 : 1
long × short (stems)	long	787 long 277 short	2.84 : 1

the other." From this finding, Mendel concluded that the hereditary determinants for the traits in the parental lines were transmitted as two different elements that retain their purity in the hybrids. In other words, the hereditary determinants do not "mix" or "contaminate each other." The implication of this conclusion is that a plant with the dominant trait might carry, in unchanged form, a hereditary determinant for the recessive trait.

The hypothesis of genetic transmission that Mendel developed to explain the reappearance of the recessive trait is outlined in Figure 2.6 . The first element of the hypothesis is that each reproductive cell, or **gamete,** contains one representative of each kind of hereditary determinant in the plant. Mendel proposed that in the true-breeding variety with round seeds, all of the reproductive cells would contain the "round factor" (W) and that in the true-breeding variety with wrinkled seeds, all of the reproductive cells would contain the "wrinkled factor" (w). When the varieties are crossed, the F_1 hybrid should receive one each of W and w and so have the genetic constitution Ww (Figure 2.6). Because, with respect to seed shape, round (W) is dominant to wrinkled (w), the presence of w in the F_1 seeds is concealed, and so the seeds are round. Although the mutant w form of the gene is concealed with regard to the visible trait, it is not concealed with regard to the molecular trait. This is signified by the gel icon for the F_1 progeny in Figure 2.6, in which the DNA band corresponding to the mutant w form of the gene is clearly present. Whereas Mendel had to infer the presence of w from the progeny of crosses involving the F_1 progeny, the analysis of DNA allows the presence of w in the F_1 progeny to be detected directly.

■ The paired genes separate (segregate) in the formation of reproductive cells.

The second key feature of Mendel's hypothesis in Figure 2.6 is that when an F_1 plant is self-fertilized (denoted by the encircled cross sign), the W and w determinants separate from one another and are included in the gametes in equal numbers. This separation of the hereditary elements is the heart of Mendelian genetics. The principle is called **segregation.**

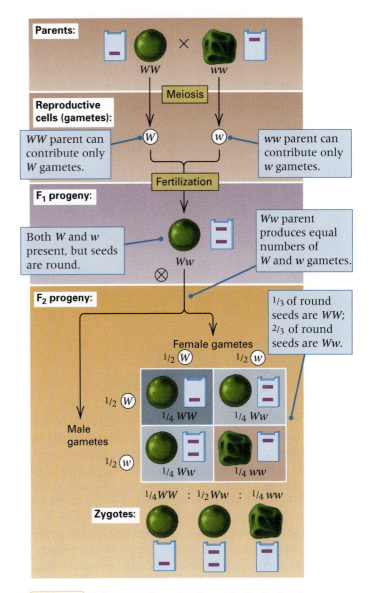

Figure 2.6 A diagrammatic explanation of the 3 : 1 ratio of dominant : recessive visible traits observed in the F_2 generation of a monohybrid cross. While a 3 : 1 ratio of visible traits is observed, the ratio of molecular traits ($WW : Ww : ww$) in the F_2 generation is 1 : 2 : 1, as depicted by the bands in the DNA gels.

> **key concept**
>
> **The Principle of Segregation:** In the formation of gametes, the paired hereditary determinants (genes) separate (segregate) in such a way that each gamete is equally likely to contain either member of the pair.

The principle of segregation implies not only that the hereditary determinants separate in the formation of gametes but also that, when separated, the hereditary determinants are completely unaltered by their having been paired in the previous generation. In Mendel's words, neither of them has "inherited or taken over anything from the other."

■ Gametes unite at random in fertilization.

The third key feature of Mendel's hypothesis is that the gametes produced by segregation come together in pairs *at random* to yield the progeny of the next generation. The assumption of random fertilization means that the result of self-fertilization of the F_1 plants in Figure 2.6 can be deduced by cross-multiplication in a square grid, as shown for the F_2

progeny. Each square within the grid represents the result of fertilization of one type of pollen with one type of egg. When a W-bearing pollen fertilizes a W-bearing egg, the result is a WW fertilized egg, or **zygote.** Similarly, when a W-bearing pollen fertilizes a w-bearing egg, the result is a Ww zygote. The consequence of fertilization by w-bearing pollen is either a Ww zygote (if the egg carries W) or a ww zygote (if the egg carries w). These possibilities are shown in the grid in Figure 2.6.

The critical point is that the outcome of fertilization is a chance—or random—event. The probability of a particular gene combination occurring in the zygote is directly related to the frequency with which a particular gamete occurs. In this instance there is a 1-in-2 chance that a gamete will bear either the W or w gene (written as a probability of 1/2 along the top and left side of the grid).

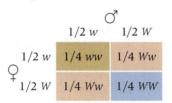

The probability of two such chance events occurring together—one particular female gamete combining with a particular male gamete—is calculated as the product of their individual probabilities. In this case, each zygote combination within the grid has a probability of $1/2 \times 1/2 = 1/4$ (a 1-in-4 chance of occurring). Because, in a Ww seed, it does not matter whether the W came through the pollen or the egg, random combinations of the gametes result in an F_2 generation with the genetic composition $1/4\ WW$, $1/2\ Ww$, and $1/4\ ww$. This is the ratio of genetic types that would be observed via electrophoretic analysis of the DNA in the seeds, owing to the codominance of W and w at the molecular level. However, because of dominance at the level of the visible trait, the underlying $1/4\ WW$: $1/2\ Ww$: $1/4\ ww$ ratio is concealed, and instead one observes a ratio of 3/4 round : 1/4 wrinkled. (The 3/4 comes from the fact that round seeds include $1/4\ WW + 1/2\ Ww = 3/4$ altogether.)

In summary, Mendel's key observation and the inference he made from it are as follows:

> **key concept**
>
> In the F_2 progeny of a monohybrid cross, the observed ratio of visible traits is 3/4 dominant : 1/4 recessive (or 3 : 1), but the dominance expressed at the level of the visible trait conceals the fact that the underlying ratio of genetic types is
>
> $$1/4 : 1/2 : 1/4 \ (\text{or } 1 : 2 : 1)$$
>
> For example, $1/4\ WW : 1/2\ Ww : 1/4\ ww$.

■ **Genotype means genetic endowment; phenotype means observed trait.**

The genetic hypothesis outlined in Figure 2.6 also illustrates another of Mendel's important deductions: Two plants with the same outward appearance—for example, with round seeds—might nevertheless differ in their hereditary makeup. One of the handicaps under which Mendel wrote was the absence of an established vocabulary of terms suitable for describing his concepts. Hence he made a number of seemingly elementary mistakes, such as occasionally confusing the outward appearance of an organism with its hereditary constitution. The necessary vocabulary was developed only after Mendel's work was rediscovered, and it includes the following essential terms.

1. A hereditary determinant of a trait is called a **gene.**

2. The different forms of a particular gene are called **alleles.** In Figure 2.6, the alleles of the gene for seed shape are W for round seeds and w for wrinkled seeds. W and w are alleles because they are alternative forms of the gene for seed shape. Alternative alleles are typically represented by the same letter or combination of letters, distinguished either by upper case versus lower case or by means of superscripts or subscripts or some other typographic identifier.

3. The **genotype** is the genetic constitution of an organism or cell—its molecular makeup. With respect to seed shape in peas, WW, Ww, and ww are examples of the possible genotypes for the W and w alleles. Because gametes contain only one allele of each gene, W and w are examples of genotypes of gametes.

4. A genotype in which the members of a pair of alleles are different, as in the Ww hybrids in Figure 2.6, is said to be **heterozygous;** a genotype in which the two alleles are alike is said to be **homozygous.** A homozygous organism may be homozygous dominant (WW) or homozygous recessive (ww). The terms *homozygous* and *heterozygous* cannot apply to gametes, because gametes contain only one allele of each gene.

5. The observable properties of an organism—including its visible traits—constitute its **phenotype.** Round seeds and wrinkled seeds are phenotypes. So are yellow seeds and green seeds. The phenotype of an organism does not necessarily imply anything about its genotype. For example, a seed with the phenotype "round" could have either the genotype WW or the genotype Ww.

6. A **dominant trait** is that expressed in the phenotype when the genotype is either heterozygous or homozygous. A **recessive trait** is that expressed in the phenotype when a genotype is homozygous for the alternative allele. The presence of a dominant trait masks a recessive trait.

■ The progeny of the F₂ generation support Mendel's hypothesis.

Mendel realized that the key to proving the genetic hypothesis outlined in Figure 2.6 lay with the round seeds in the F_2 generation. If his hypothesis was correct, then **1**/3 of the round seeds should have the genetic composition WW and **2**/3 of the round seeds should have the genetic composition Ww. The reason for the **1 : 2** ratio is shown in Figure 2.7 . The ratio of $WW : Ww : ww$ in the F_2 generation is 1 : 2 : 1, but if we disregard the ww seeds, then the ratio of $WW : Ww$ is 1 : 2. In other words, 1/3 of the round seeds are WW and 2/3 are Ww. These ratios are apparent from the molecular analysis of the round seeds, but Mendel had to identify the genotypes of the seeds on the basis of the breeding behavior of the plants that grew out of them. He realized that, upon self-fertilization, the WW genotypes should be true-breeding for round seeds, whereas the Ww genotypes should yield round and wrinkled seeds in the ratio 3 : 1. Furthermore, among the wrinkled seeds in the F_2 generation, all should have the genetic composition ww, and so, upon self-fertilization, they should be true-breeding for wrinkled seeds.

For several of his traits, Mendel carried out self-fertilization of the F_2 plants in order to test these predictions. His results for round versus wrinkled seeds are summarized in Figure 2.8 . As predicted from Mendel's genetic hypothesis, the plants grown

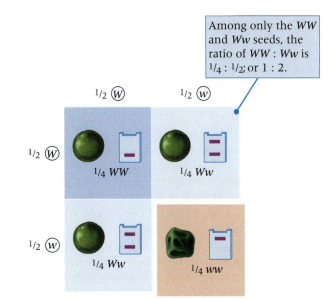

Among only the WW and Ww seeds, the ratio of $WW : Ww$ is $1/4 : 1/2$, or 1 : 2.

Figure 2.7   In the F_2 generation, the ratio of $WW : Ww : ww$ is 1 : 2 : 1. However, *among those seeds that are round,* the ratio of $WW : Ww$ is 1 : 2, hence 1/3 of the round seeds are WW and 2/3 are Ww.

from F_2 wrinkled seeds were true-breeding for wrinkled seeds. They produced only wrinkled seeds in the F_3 generation. Moreover, among 565 plants grown from F_2 round seeds, 193 were true-breeding, producing only round seeds in the F_3 generation, whereas the other 372 plants produced both round and wrinkled seeds in a proportion very close to 3 : 1. The ratio 193 : 372 equals 1 : 1.93, which is very close to the ratio 1 : 2 of $WW : Ww$ genotypes predicted theoretically from the genetic hypothesis in Figure 2.7. Overall, taking all of the F_2 plants into account, the ratio of genotypes observed was very close to the predicted 1 : 2 : 1 of $WW : Ww : ww$ expected from Figure 2.7.

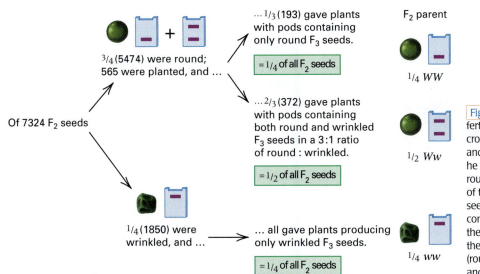

Figure 2.8   Mendel's results of self-fertilization of the F_2 progeny of a cross between plants with round seeds and plants with wrinkled seeds. When he self-fertilized F_2 plants grown from round seeds (the dominant trait), 1/3 of the progeny consisted of all round seeds and 2/3 of the progeny consisted of round : wrinkled seeds in the ratio 3 : 1. The result shows that the F_2 seeds with the dominant trait (round) include two genetic types, WW and Ww, in a ratio of 1 : 2.

■ The progeny of testcrosses also support Mendel's hypothesis.

Mendel devised a second way to test the genetic makeup of the F_1 seeds, lending further support to his hypothesis (Figure 2.6). By crossing the plants grown from F_1 seeds with plants that were homozygous recessive, the genotype of the F_1 seeds would be revealed. Such a cross, between an organism of dominant phenotype (genotype unknown) and an organism of recessive phenotype (genotype known to be homozygous recessive), is called a **testcross.** If the parent with the dominant phenotype is homozygous, then the cross will produce progeny with the dominant phenotype. If the parent with the dominant phenotype is heterozygous (for example, *Ww*), then the result of the testcross will be progeny with both dominant and recessive phenotypes, as shown in Figure 2.9. Because of segregation, the heterozygous parent is expected to produce *W* and *w* gametes in equal numbers. When these gametes combine at random with the *w*-bearing gametes produced by the homozygous recessive parent, the expected progeny are 1/2 with the genotype *Ww* and 1/2 with the genotype *ww*. The former have the dominant visible phenotype (round, because *W* is dominant to *w*), whereas the latter have the recessive visible phenotype (wrinkled), and so the expected ratio of dominant to recessive phenotypes is 1 : 1. If the organism being tested in a testcross is heterozygous, the ratio of visible phenotypes will be the same as the ratio of molecular phenotypes. Both ratios are 1 : 1, as indicated by the gel icons in Figure 2.9. This is why a testcross is often extremely useful in genetic analysis.

key concept

In a testcross, the relative proportion of the different gametes produced by the heterozygous parent can be observed directly in the proportion of phenotypes of the progeny, because the recessive parent contributes only recessive alleles.

Mendel carried out a series of testcrosses with the genes for round versus wrinkled seeds, yellow versus green seeds, purple versus white flowers, and long versus short stems. The results are shown in Table 2.2. In all cases, the ratio of phenotypes among the progeny is very close to the 1 : 1 ratio expected from segregation of the alleles in the heterozygous parent.

Q A Moment to Think

Problem: Mendel was the first to consider the implications of segregation for an entire population of organisms. He deduced the expected proportions of each genotype in a population undergoing successive generations of self-fertilization in a population so large that it is effectively infinite in size. In the initial generation (generation 0), the genotype of each individual in the population is heterozygous *Aa* (these are the allele symbols that Mendel used). Self-fertilization of these individuals implies that in generation 1, the genotypic composition of the population must be 1/4 *AA*, 1/2 *Aa*, and 1/4 *aa*. Now the situation becomes slightly more complex, because the homozygous genotypes breed true, whereas the heterozygous genotypes continue to segregate. Using the principle of segregation, calculate the expected genotypic composition of the population after 2, 3, 4, and 5 generations of self-fertilization. What does this suggest about the genotype of hybrid organisms left to self-fertilize over a number of generations? (The answer can be found on page 48.)

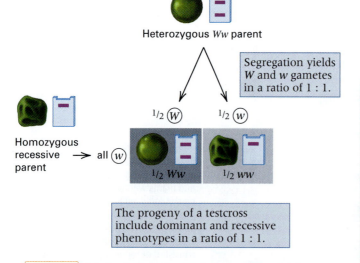

Figure 2.9 A testcross shows the result of segregation directly in the phenotypes of the progeny. This example illustrates a testcross of a *Ww* heterozygous parent with a *ww* homozygous recessive. The expected progeny are *Ww* and *ww* in a ratio of 1 : 1.

Table 2.2

Results of Mendel's testcross experiments

Testcross (F_1 heterozygote × homozygous recessive)	Progeny from testcross	Ratio
round × wrinkled seeds	193 round 192 wrinkled	1.01:1
yellow × green seeds	196 yellow 189 green	1.04:1
purple × white flowers	85 purple 81 white	1.05:1
long × short stems	85 long 79 short	1.01:1

Another valuable type of cross is a **backcross,** in which hybrid organisms are crossed with one of the parental genotypes. Backcrosses are commonly used by geneticists and by plant and animal breeders, as we will see in later chapters. Note that the testcrosses in Table 2.2 are also, in effect, backcrosses. The F_1 heterozygous parent that came from a cross between the homozygous dominant and the homozygous recessive is backcrossed with a homozygous recessive.

2.3

The alleles of different genes segregate independently.

In the experiments described so far, Mendel was concerned with successive generations of progeny from parents that differed in a single contrasting trait, such as round seeds versus wrinkled seeds. In each case, he observed the result of segregation of the pair of alleles determining the trait. Mendel also carried out experiments that tracked two or more traits simultaneously. He wanted to determine whether the traits would be inherited together or whether they would be inherited independently of each other. For example, plants from a true-breeding variety with round and yellow seeds were crossed with plants from a true-breeding variety with wrinkled and green seeds. The F_1 progeny were hybrid for both characteristics (dihybrid), and the phenotype of the seeds was round (which is dominant to wrinkled) and yellow (which is dominant to green, see Figure 2.3). Then Mendel self-fertilized the F_1 progeny to obtain seeds in the F_2 generation. He observed four types of seed phenotypes in the progeny, and in counting the seeds, he obtained the following numbers:

round, yellow	315
round, green	108
wrinkled, yellow	101
wrinkled, green	32
Total	556

In these data, Mendel noted the presence of the expected monohybrid 3 : 1 ratio for each trait separately. With respect to each trait, the progeny were

round : wrinkled
$$= (315 + 108) : (101 + 32)$$
$$= 423 : 133$$
$$= 3.18 : 1$$

yellow : green
$$= (315 + 101) : (108 + 32)$$
$$= 416 : 140$$
$$= 2.97 : 1$$

Furthermore, in the F_2 progeny of the dihybrid cross, the separate 3 : 1 ratios for the two traits

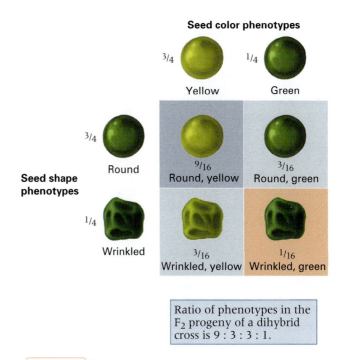

Ratio of phenotypes in the F_2 progeny of a dihybrid cross is 9 : 3 : 3 : 1.

Figure 2.10 The 3 : 1 ratio of round : wrinkled, when combined at random with the 3 : 1 ratio of yellow : green, yields the 9 : 3 : 3 : 1 ratio that Mendel observed in the F_2 progeny of the dihybrid cross.

were combined at random, as shown in Figure 2.10. That is, among the 3/4 of the progeny that are round, 3/4 are yellow and 1/4 green; similarly, among the 1/4 of the progeny that are wrinkled, 3/4 are yellow and 1/4 green. The overall proportions of round–yellow to round–green to wrinkled–yellow to wrinkled–green are therefore expected to be

$$3/4 \times 3/4 \text{ to } 3/4 \times 1/4 \text{ to } 1/4 \times 3/4 \text{ to } 1/4 \times 1/4$$

or

$$9/16 : 3/16 : 3/16 : 1/16$$

The observed ratio of 315 : 108 : 101 : 32 equals 9.84 : 3.38 : 3.16 : 1, which is reasonably close to the 9 : 3 : 3 : 1 ratio expected from the cross multiplication of the separate 3 : 1 ratios in Figure 2.10.

■ The F_2 genotypes in a dihybrid cross conform to Mendel's prediction.

Mendel carried out similar experiments with other combinations of traits. For each pair of traits, he consistently observed the 9 : 3 : 3 : 1 ratio. He also deduced the biological reason for the observation. To illustrate his explanation using the dihybrid round × wrinkled cross, we can represent the dominant and recessive alleles of the pair affecting seed shape as W and w, respectively, and the allelic pair affecting seed

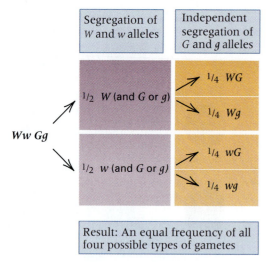

Segregation of W and w alleles	Independent segregation of G and g alleles
$1/2$ W (and G or g)	$1/4$ WG $1/4$ Wg
$1/2$ w (and G or g)	$1/4$ wG $1/4$ wg

$Ww\ Gg$

Result: An equal frequency of all four possible types of gametes

Figure 2.11 Independent segregation of the Ww and Gg allele pairs means that among each of the W and w classes, the ratio of $G : g$ is 1 : 1. Likewise, among each of the G and g classes, the ratio of $W : w$ is 1 : 1.

color as G (yellow) and g (green). Mendel proposed that the underlying reason for the 9 : 3 : 3 : 1 ratio in the F_2 generation is that the segregation of the alleles W and w for round or wrinkled seeds has no effect on the segregation of the alleles G and g for yellow or green seeds. Each pair of alleles undergoes segregation into the gametes independently of the segregation of the other pair of alleles. The parental genotypes in the P_1 generation are $WW\ GG$ (round, yellow seeds) and $ww\ gg$ (wrinkled, green seeds). When these are crossed, the genotype of the F_1 hybrid is the double heterozygote $Ww\ Gg$.

The result of independent segregation in the F_1 plants is that the W allele is just as likely to be included in a gamete with G as with g, and the w allele is just as likely to be included in a gamete with G as with g. The independent segregation is illustrated in Figure 2.11. The independent segregation of the W, w and the G, g allele pairs implies that the gametes produced by the double heterozygote $Ww\ Gg$ are

$$1/4\ W\,G \quad 1/4\ W\,g \quad 1/4\ w\,G \quad 1/4\ w\,g$$

When the four types of gametes combine at random to form the zygotes of the next generation, the result of independent segregation is as shown in Figure 2.12. Again, we use cross multiplication to show

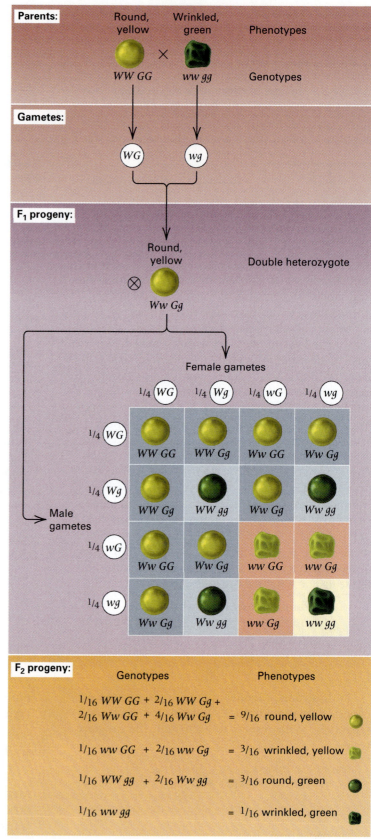

Figure 2.12 Diagram showing the basis for the 9 : 3 : 3 : 1 ratio of F_2 phenotypes resulting from a cross in which the parents differ in two traits determined by genes that undergo independent segregation.

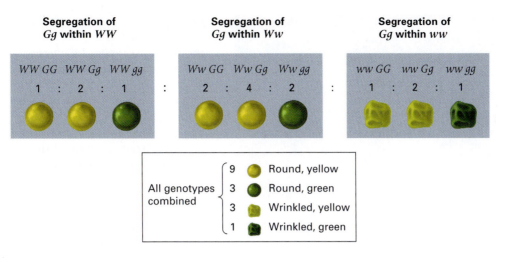

Segregation of Gg within WW			Segregation of Gg within Ww			Segregation of Gg within ww		
$WW\,GG$	$WW\,Gg$	$WW\,gg$	$Ww\,GG$	$Ww\,Gg$	$Ww\,gg$	$ww\,GG$	$ww\,Gg$	$ww\,gg$
1 :	2 :	1	2 :	4 :	2	1 :	2 :	1

All genotypes combined		
9		Round, yellow
3		Round, green
3		Wrinkled, yellow
1		Wrinkled, green

Figure 2.13 | In the F_2 progeny of the dihybrid cross for seed shape and seed color, in any of the genotypes for one of the allele pairs, the ratio of homozygous dominant, heterozygous, and homozygous recessive genotypes for the other allele pair is 1 : 2 : 1.

how the F_1 female and male gametes combine at random to produce the F_2 genotypes. This format is called a **Punnett square.** In the Punnett square, the combinations of seed shape and color phenotypes of the F_2 progeny are indicated. Note that the ratio of phenotypes is 9 : 3 : 3 : 1 for round yellow : wrinkled yellow : round green : wrinkled green.

The Punnett square in Figure 2.12 also shows that the ratio of *genotypes* in the F_2 generation is not 9 : 3 : 3 : 1. With independent segregation, the ratio of genotypes in the F_2 generation is

$$1 : 2 : 1 : 2 : 4 : 2 : 1 : 2 : 1$$

The reason for this ratio is shown in Figure 2.13. Among seeds with the WW genotype, the ratio of $GG : Gg : gg$ is 1 : 2 : 1. Among seeds with the Ww genotype, the ratio is 2 : 4 : 2 (the 1 : 2 : 1 is multiplied by 2 because there are twice as many Ww genotypes as either WW or ww). And among seeds with the ww genotype, the ratio of $GG : Gg : gg$ is 1 : 2 : 1. The phenotypes of the seeds are shown beneath the genotypes. The combined ratio of phenotypes is 9 : 3 : 3 : 1.

Mendel tested the hypothesis of independent segregation by ascertaining whether the predicted genotypes were actually present in the expected proportions. He did the tests by growing plants from the F_2 seeds and obtaining F_3 progeny by self-pollination. To illustrate the tests, consider one series of crosses in which he grew plants from F_2 seeds that were round, green. Note in Figures 2.12 and 2.13 that round, green F_2 seeds are expected to have either the genotype $Ww\,gg$ or the genotype

$WW\,gg$ in the ratio 2 : 1. Mendel grew 102 plants from such seeds and found that 67 of them produced pods containing both round, green and wrinkled, green seeds (indicating that the parental plants must have been $Ww\,gg$) and 35 of them produced pods containing only round, green seeds (indicating that the parental genotype was $WW\,gg$). The ratio 67 : 35 is in good agreement with the expected 2 : 1 ratio of genotypes.

Mendel's observation of independent segregation of two pairs of alleles has come to be known as the principle of **independent assortment:**

key concept

The Principle of Independent Assortment: Segregation of the members of any pair of alleles is independent of the segregation of other pairs in the formation of reproductive cells.

Although the principle of independent assortment is fundamental to Mendelian genetics, in later chapters we will see that there are important exceptions.

■ The progeny of testcrosses show the result of independent assortment.

A second way in which Mendel tested the hypothesis of independent assortment was by carrying out a testcross with the F_1 genotypes that were heterozygous for both genes ($Ww\,Gg$). In this testcross, one parental genotype has to be multiply

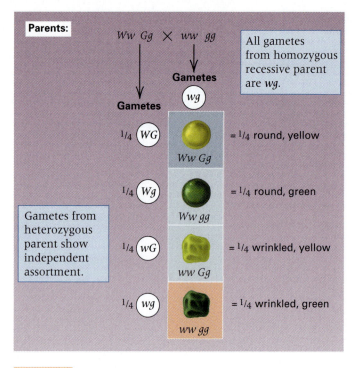

Figure 2.14 Genotypes and phenotypes resulting from a testcross of a *Ww Gg* double heterozygote.

homozygous recessive—in this case, *ww gg*. As shown in Figure 2.14, the double heterozygotes produce four types of gametes—*W G*, *W g*, *w G*, and *w g*—in equal proportions, whereas the *ww gg* plants produce only *w g* gametes. Thus the progeny phenotypes are expected to consist of round yellow, round green, wrinkled yellow, and wrinkled green in a ratio of 1 : 1 : 1 : 1. As in a testcross of a monohybrid, the ratio of phenotypes in the progeny is a direct demonstration of the ratio of gametes produced by the heterozygous parent, because no dominant alleles are contributed by the homozygous recessive parent to obscure the results. In the actual cross, Mendel obtained 55 round yellow, 51 round green, 49 wrinkled yellow, and 53 wrinkled green, which is in good agreement with the predicted 1 : 1 : 1 : 1 ratio. The results were the same in the reciprocal cross with *Ww Gg* as the female parent and *ww gg* as the male parent. This observation confirmed Mendel's assumption that the gametes of both sexes included all possible genotypes in approximately equal proportions.

A | A Moment to Think

Answer to Problem: The self-fertilization of monohybrid *Aa* plants will yield in generation 1 seeds with the genotypic composition 1/4 *AA*, 1/2 *Aa*, and 1/4 *aa*. The 1/4 of the population that consist of *AA* genotypes produce all *AA* progeny; the 1/2 that consist of *Aa* genotypes produce 1/4 *AA*, 1/2 *Aa*, and 1/4 *aa* progeny; and the 1/4 that consist of *aa* genotypes produce all *aa* progeny. As shown in the graphic below, the genotypic composition in generation 2 is therefore, for *AA*, 1/4 × 1 + 1/2 × 1/4 = 3/8. In this expression, the first term (1/4 × 1) counts the *AA* progeny from *AA* parents, and the second term (1/2 × 1/4) counts the *AA* progeny from *Aa* parents. (There is no term corresponding to *AA* progeny from *aa* parents, because *aa* parents cannot have *AA* progeny.) The proportion of *Aa* genotypes in generation 2 is calculated as 1/2 × 1/2 = 1/4. And the proportion of *aa* genotypes in generation 2 is 1/4 × 1 + 1/2 × 1/4 = 3/8. For generation 3 the genotypic composition is obtained similarly: For *AA* the proportion is 3/8 × 1 + 1/4 × 1/4 = 7/16; for *Aa* it is 1/4 × 1/2 = 1/8; and for *aa* it is 3/8 × 1 + 1/4 × 1/4 = 7/16. For generation 4 we proceed as above: *Aa*, 7/16 × 1 + 1/8 × 1/4 = 15/32; *Aa*, 1/8 × 1/2 = 1/16; *aa*, 7/16 × 1 + 1/8 × 1/4 = 15/32. And for generation 5 we have *Aa*, 15/32 × 1 + 1/16 × 1/4 = 31/64; *Aa*, 1/16 × 1/2 = 1/32; *aa*, 15/32 × 1 + 1/16 × 1/4 = 31/64. The pattern is evident from the table shown above: The proportion of heterozygous genotypes is reduced by one-half in each generation. What this means is that the population eventually comes to consist of two homozygous, true-breeding types.

	Genotype		
Generation	***AA***	***Aa***	***aa***
0		1	
1	1/4	1/2	1/4
2	3/8	1/4	3/8
3	7/16	1/8	7/16
4	15/32	1/16	15/32
5	31/64	1/32	31/64

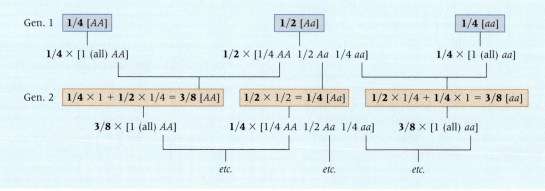

An interesting historical note: Mendel's paper does not explicitly state either the principle of segregation (sometimes called Mendel's first law) or the principle of independent assortment (sometimes called Mendel's second law). On this basis, one could argue that Mendel did not discover Mendel's laws! On the other hand, Mendel did seem to have a pretty clear idea of what was going on. Six times in his relatively short paper, he repeated what he evidently thought was the main message: "Pea hybrids form germinal and pollen cells that in their composition correspond in equal numbers to all the constant [true-breeding] forms resulting from the combination of traits united through fertilization." One could not draw this conclusion without invoking both segregation and independent assortment.

2.4

Chance plays a central role in Mendelian genetics.

As we have seen, chance plays a central role in Mendelian genetics. In the creation and then fertilization of gametes, the particular combination of alleles that occurs is random and subject to chance variation. In a genetic cross, the proportions of the different types of offspring obtained are the cumulative result of numerous individual events of fertilization. It is for this reason that a working knowledge of the rules of probability is basic to understanding the transmission of hereditary characteristics.

In the analysis of genetic crosses, the probability of a particular outcome may be considered equivalent to the number of times that an outcome is expected to occur over a large number of repeated trials. This number, expressed as a ratio, is also considered equivalent to the *probability* that this particular outcome will occur in a single trial. For example, in the F_2 generation of the hybrid between pea varieties with round seeds and those with wrinkled seeds, Mendel observed 5474 round seeds and 1850 wrinkled seeds (Table 2.1). In this case, the proportion of wrinkled seeds was $1850/(1850 + 5474) = 1/3.96$, or very nearly $1/4$. We may therefore regard $1/4$ as the approximate proportion of wrinkled seeds to be expected among a large number of progeny from this cross. Completely equivalently, we can regard $1/4$ as the probability that any particular seed chosen at random will be wrinkled.

Evaluating the probability of any possible outcome of a genetic cross usually requires an understanding of the mechanism of inheritance and knowledge of the particular cross. For example, in evaluating the probability of obtaining a round seed

from a particular cross, one needs to know that there are two alleles, W and w, with W dominant to w. One also needs to know the particular cross, because the probability of round seeds is determined by whether the cross is

$WW \times ww$, in which case all the progeny seeds are expected to be round,

$Ww \times Ww$, in which case $3/4$ of the progeny seeds are expected to be round, or

$Ww \times ww$, in which case $1/2$ of the progeny seeds are expected to be round.

■ The addition rule applies to mutually exclusive possibilities.

Sometimes an outcome of interest can be expressed in terms of two or more possibilities. For example, a seed with the phenotype "round" may have either of two genotypes, WW or Ww. A seed that is round cannot have both genotypes at the same time. Only one possibility, such as the presence of the WW or the Ww genotype, can be realized in any one organism, and the realization of one such possibility precludes the realization of others. In this example, the realization of the genotype WW in a seed precludes the realization of the genotype Ww in the same seed, and the other way around. Outcomes that exclude each other in this manner are said to be *mutually exclusive*. When the possible outcomes are mutually exclusive, their probabilities are combined according to the addition rule.

> **key concept**
>
> **Addition Rule:** The probability of the realization of one or the other of two mutually exclusive possibilities, A or B, is the sum of their separate probabilities.

In symbols, where Prob is used to mean *probability*, the addition rule is written

$$\text{Prob \{A or B\}} = \text{Prob \{A\}} + \text{Prob \{B\}}$$

The addition rule can be applied to determine the proportion of round seeds expected from the cross $Ww \times Ww$, which is illustrated in Figure 2.7. The round-seed phenotype results from the expression of either of two genotypes, WW and Ww, and these possibilities are mutually exclusive. In any particular progeny seed, the probability of genotype WW is $1/4$ and that of Ww is $1/2$. Hence the overall probability of either WW or Ww is

$$\text{Prob \{WW or Ww\}} = \text{Prob \{WW\}} + \text{Prob \{Ww\}}$$
$$= 1/4 + 1/2 = 3/4$$

Because $3/4$ is the probability of an individual seed being round, it is also the expected proportion of round seeds among a large number of progeny.

■ The multiplication rule applies to independent possibilities.

Possible outcomes that are not mutually exclusive may be *independent,* which means that the realization of one outcome has no influence on the possible realization of any others. For example, in Mendel's crosses for seed shape and color, the two traits are independent, and the proportions of phenotypes in the F_2 generation are expected to be 9/16 round yellow, 3/16 round green, 3/16 wrinkled yellow, and 1/16 wrinkled green. These proportions can be obtained by considering the traits separately, because they are independent. Considering only seed shape, we can expect the F_2 generation to consist of 3/4 round and 1/4 wrinkled seeds. Considering only seed color, we can expect the F_2 generation to consist of 3/4 yellow and 1/4 green. Because the traits are inherited independently, among the 3/4 of the seeds that are round, there should be 3/4 that are yellow, and so the overall proportion of round yellow seeds is expected to be $3/4 \times 3/4 = 9/16$ (Figure 2.10). Likewise, among the 3/4 of the seeds that are round, there should be 1/4 green, yielding $3/4 \times 1/4 = 3/16$ as the expected proportion of round, green seeds. The proportions of the other phenotypic classes can be deduced in a similar way using the cross multiplication method illustrated in Figure 2.10. The principle is that when outcomes are independent, the probability that they are realized together is obtained by multiplication.

Successive offspring from a cross are also independent outcomes, which means that the genotypes of early progeny have no influence on the relative proportions of genotypes in later progeny. The independence of successive offspring contradicts the widespread belief that in each human family, the ratio of girls to boys must "even out" at approximately 1 : 1 such that if parents already have, say, four girls, then they are somehow more likely to have a boy the next time around. But this belief is not supported by theory, and it is also contradicted by actual data on the sex ratios in human sibships. (The term **sibship** refers to a group of offspring from the same parents.) The data indicate that parents are no more likely to have a girl on the next birth if they already have five boys than if they already have five girls. The statistical reason is that although the sex ratios tend to balance out when they are averaged across a large number of sibships, they do not need to balance within individual sibships. Thus, among families in which there are five children, the sibships consisting of five boys balance those consisting of five girls, for an overall sex ratio of 1 : 1. However, both of these sibships are unusual in their sex distribution.

When the possible outcomes are independent (such as independent traits or successive offspring from a cross), the probabilities are combined by means of the multiplication rule.

key concept

Multiplication Rule: The probability of two independent possibilities, A and B, being realized simultaneously is given by the product of their separate probabilities.

In symbols, the multiplication rule is

$$\text{Prob }\{A \text{ and } B\} = \text{Prob }\{A\} \cdot \text{Prob }\{B\}$$

The multiplication rule can be used to answer questions like the following: Of two offspring from the mating $Aa \times Aa$, what is the probability that both have the dominant phenotype? Because the mating is $Aa \times Aa$, the probability that any particular offspring has the dominant phenotype equals 3/4. Using the multiplication rule, we find that the probability that both of two offspring have the dominant phenotype is $3/4 \times 3/4 = 9/16$.

Here is a typical genetic question that can be answered by using the addition and multiplication rules together: Of two offspring from the mating $Aa \times Aa$, what is the probability of one dominant phenotype (probability of 3/4) and one recessive (probability of 1/4)? Sibships of one dominant phenotype and one recessive can come about in two different ways, with the dominant born first or with the dominant born second, and these outcomes are mutually exclusive. The probability of the first case is $3/4 \times 1/4$ and that of the second is $1/4 \times 3/4$; because the outcomes are mutually exclusive, the probabilities are added. The answer is therefore

$$(3/4 \times 1/4) + (1/4 \times 3/4) = 2(3/16) = 3/8$$

The addition and multiplication rules are very powerful tools for calculating the probabilities of genetic events. Figure 2.15 shows how the rules are applied to determine the expected proportions of the nine different genotypes possible among the F_2 progeny produced by self-pollination of a $Ww\ Gg$ dihybrid.

In genetics, independence applies not only to the successive offspring formed by a mating but also to genes that segregate according to the principle of independent assortment (Figure 2.16). The independence means that the multiplication rule can be used to determine the probability of the various types of progeny from a cross in which there is independent assortment among numerous pairs of alleles. This principle is the theoretical basis for the expected progeny types from the dihybrid cross shown in Figure 2.15. One can also use the multiplication rule to calculate the probability of a spe-

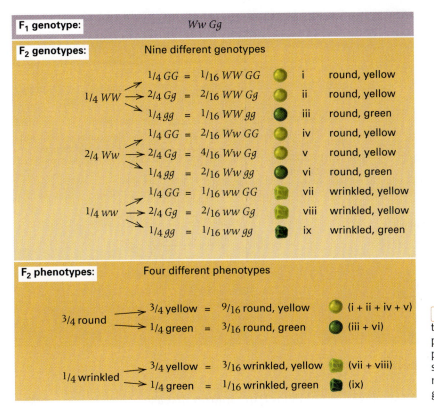

F₁ genotype:	$Ww\ Gg$				
F₂ genotypes:	Nine different genotypes				

$1/4\ WW$ → $1/4\ GG$ = $1/16\ WW\ GG$ — i — round, yellow
$2/4\ Gg$ = $2/16\ WW\ Gg$ — ii — round, yellow
$1/4\ gg$ = $1/16\ WW\ gg$ — iii — round, green

$2/4\ Ww$ → $1/4\ GG$ = $2/16\ Ww\ GG$ — iv — round, yellow
$2/4\ Gg$ = $4/16\ Ww\ Gg$ — v — round, yellow
$1/4\ gg$ = $2/16\ Ww\ gg$ — vi — round, green

$1/4\ ww$ → $1/4\ GG$ = $1/16\ ww\ GG$ — vii — wrinkled, yellow
$2/4\ Gg$ = $2/16\ ww\ Gg$ — viii — wrinkled, yellow
$1/4\ gg$ = $1/16\ ww\ gg$ — ix — wrinkled, green

F₂ phenotypes: Four different phenotypes

$3/4$ round → $3/4$ yellow = $9/16$ round, yellow (i + ii + iv + v)
→ $1/4$ green = $3/16$ round, green (iii + vi)

$1/4$ wrinkled → $3/4$ yellow = $3/16$ wrinkled, yellow (vii + viii)
→ $1/4$ green = $1/16$ wrinkled, green (ix)

Figure 2.15 Example of the use of the addition and multiplication rules to determine the probabilities of the nine genotypes and four phenotypes in the F₂ progeny obtained from self-pollination of a dihybrid F₁. The roman numerals are arbitrary labels identifying the F₂ genotypes.

(A)

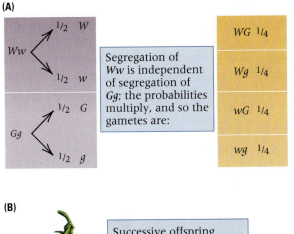

Ww → $1/2\ W$, $1/2\ w$
Gg → $1/2\ G$, $1/2\ g$

Segregation of Ww is independent of segregation of Gg; the probabilities multiply, and so the gametes are:

WG $1/4$
Wg $1/4$
wG $1/4$
wg $1/4$

(B)

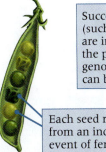

Successive offspring (such as peas in a pod) are independent, and so the probabilities of genotypes or phenotypes can be multiplied.

Each seed results from an independent event of fertilization.

Figure 2.16 In genetics, two important types of independence are independent segregation of alleles that show independent assortment (A) and independent fertilizations resulting in successive offspring (B). In these cases, the probabilities of each of the individual outcomes of segregation or fertilization are multiplied to obtain the overall probability.

cific genotype among the progeny of a cross. For example, if a quadruple heterozygote of genotype $Aa\ Bb\ Cc\ Dd$ is self-fertilized, the probability of a quadruple heterozygote $Aa\ Bb\ Cc\ Dd$ offspring is

$$(1/2)(1/2)(1/2)(1/2) = (1/2)^4 = 1/16$$

assuming independent assortment of all four pairs of alleles.

2.5

The results of segregation can be observed in human pedigrees.

Determining the genetic basis of a trait from the kinds of crosses that we have considered requires controlled matings and large numbers of offspring. The analysis of segregation by this method is not possible in human families, and it is usually not feasible for traits in large domestic animals. However, the mode of inheritance of a trait can sometimes be determined by examining the appearance of the phenotypes that reflect the segregation of alleles in several generations of related individuals. This is typically done with a family tree that shows the phenotype of each individual; such a diagram is called a **pedigree.** An important application of probability in genetics is its use in pedigree analysis.

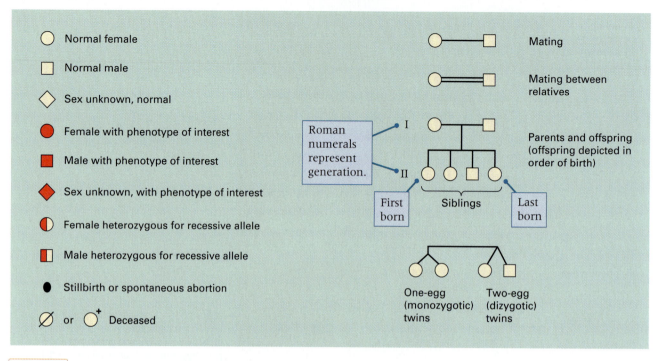

Figure 2.17 depicts most of the standard symbols used in drawing a human pedigree. Females are represented by circles and males by squares. (A diamond is used if the sex of an individual is unknown.) Persons with the phenotype of interest are indicated by colored or shaded symbols. For recessive alleles, heterozygous carriers are sometimes depicted with half-filled symbols. A mating between a female and a male is indicated by joining their symbols with a horizontal line, which is connected vertically to a second horizontal line running beneath that connects the symbols for their offspring. The offspring within a sibship, called **siblings** or **sibs** regardless of sex, are represented from left to right in order of their birth.

A typical pedigree for a trait due to a dominant allele is shown in Figure 2.18 . In this example the trait is **Huntington disease,** which is a progressive nerve degeneration, usually beginning about middle age, that results in severe physical and mental disability and ultimately in death. The numbers in the pedigree are added for convenience in referring to particular persons. The successive generations are designated by Roman numerals. Within any generation, all of the persons are numbered consecutively from left to right. The pedigree starts with the woman I-1 and the affected man I-2. The pedigree shows the characteristic features of inheritance due to a simple Mendelian dominant allele:

- The trait affects both sexes.
- Every affected person has an affected parent.
- Approximately 1/2 of the offspring of affected persons are affected.

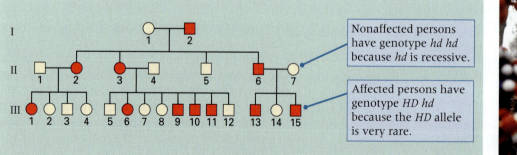

Figure 2.18 Pedigree of a human family showing the inheritance of the dominant gene for Huntington disease, a dominant genetic disorder. The condition does not exhibit itself until individuals are 30 years of age, or older—after they may have had children. Females and males are represented by circles and squares, respectively. Red symbols indicate persons affected with the disease. In the photograph is Nancy Wexler, a pioneering researcher in Huntington disease, whose mother was afflicted. Courtesy of Nancy Wexler/Hereditary Disease Foundation (www.hdfoundation.org)

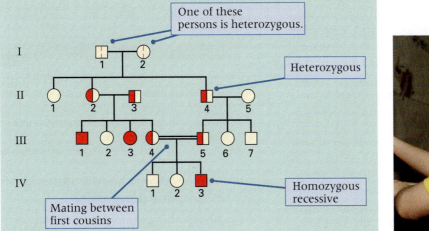

Because the dominant allele, *HD*, that causes Huntington disease is very rare, all affected persons in the pedigree have the heterozygous genotype *HD hd*. Nonaffected persons have the homozygous normal genotype *hd hd*.

A typical pedigree pattern for a trait due to a homozygous recessive allele is shown in Figure 2.19. The trait is **albinism**, absence of pigment in the skin, hair, and iris of the eyes. The pedigree characteristics of recessive inheritance are as follows:

- The trait affects both sexes.
- Most affected persons have parents who are not themselves affected; the parents are heterozygous for the recessive allele and are called **carriers.**
- Approximately 1/4 of the children of heterozygous parents are affected.
- The parents of affected individuals are often relatives.

The reason for the 1/4 ratio is that in a mating between carriers ($Aa \times Aa$), each offspring has a 1/4 chance of being homozygous *aa* and hence being affected. The reason why mating between relatives is important, particularly with traits due to rare recessive alleles, is that when a recessive allele is rare, it is more likely to become homozygous through inheritance from a common ancestor than from parents who are completely unrelated. When a common ancestor of an individual's parents is a carrier, the recessive allele may, by chance, be transmitted down both sides of the pedigree to the parents of the individual. That allele then has a 1/4 chance of becoming homozygous when the relatives mate. Mating between relatives constitutes *inbreeding;* the consequences of inbreeding are discussed further in Chapter 14.

2.6

Dominance is a property of a pair of alleles in relation to a particular attribute of phenotype.

In Mendel's experiments, all visible traits had clear dominant–recessive patterns. This was fortunate, because otherwise he might not have made his discoveries. However, departures from strict dominance are frequently observed. In fact, even for such a classic trait as round versus wrinkled seeds in peas, it is an oversimplification to say that round is dominant. At the level of whether a seed is round or wrinkled, round is dominant in the sense that the genotypes *WW* and *Ww* cannot be distinguished by the outward appearance of the seeds. However, as we noted in Chapter 1, every gene potentially affects many traits. It often happens that the same pair of alleles show complete dominance for one trait but not complete dominance for another trait. For example, in the case of round versus wrinkled seeds, the biochemical defect in wrinkled seeds is the absence of the active form of the enzyme starch-branching enzyme I (SBEI), which is needed for the synthesis of amylopectin, a branched-chain form of starch. Seeds that are heterozygous *Ww* have only half as much SBEI as homozygous *WW* seeds, and seeds that are homozygous *ww* have virtually none (Figure 2.20, part A). Homozygous *WW* peas contain large, well-rounded starch grains. As a result, the seeds retain water and shrink uniformly as they ripen, and so they do not become wrinkled. In homozygous *ww* seeds, the starch grains lack amylopectin; they are irregular in shape. When these seeds ripen, they lose water too rapidly and shrink unevenly, resulting in the wrinkled phenotype (Figure 2.20, parts B and C).

The *w* allele also affects the shape of the starch grains in *Ww* heterozygotes. In heterozygous seeds, the starch grains are intermediate in shape (Figure 2.20, part B). Nevertheless, their amylopectin content is high enough to result in uniform shrinking of the seeds and no wrinkling (Figure 2.20, part C). Thus there is an apparent paradox of dominance. If we consider only the overall shape of the seeds, round is dominant over wrinkled. If we examine the shape of the starch grains with a microscope, however, all three genotypes can be distinguished from each other: large rounded starch grains in *WW*, large irregular grains in *Ww*, and small irregular grains in *ww*.

The example in Figure 2.20 makes it clear that "dominance" is not simply a property of a particular pair of alleles independent of which aspect of the phenotype is observed. When a gene affects multiple traits (as most genes do), a particular pair of alleles might show complete dominance for some

traits but not others. The general principle illustrated in Figure 2.20 is that

key concept

The total phenotype of an organism consists of many different physical and biochemical attributes, and dominance may be observed for some of these attributes and not for others; thus dominance is a property of a pair of alleles in relation to a particular attribute of phenotype.

■ Flower color in snapdragons illustrates incomplete dominance.

When the phenotype of the heterozygous genotype is intermediate between the phenotypes of the homozygous genotypes, there is said to be **incomplete dominance.** As shown in Figure 2.20, the *W* and *w* alleles for SBEI show incomplete dominance for the traits "amount of active SBEI enzyme" and "microscopic shape of starch grains." Incomplete dominance sometimes occurs for visible traits, too. An example concerns flower color in the snapdragon *Antirrhinum* (Figure 2.21). In wildtype flowers, a red type of anthocyanin pigment is formed by a sequence of enzymatic reactions. A wildtype enzyme, encoded by the *I* allele, affects the rate of the overall reaction—the more enzyme available, the more red pigment produced.

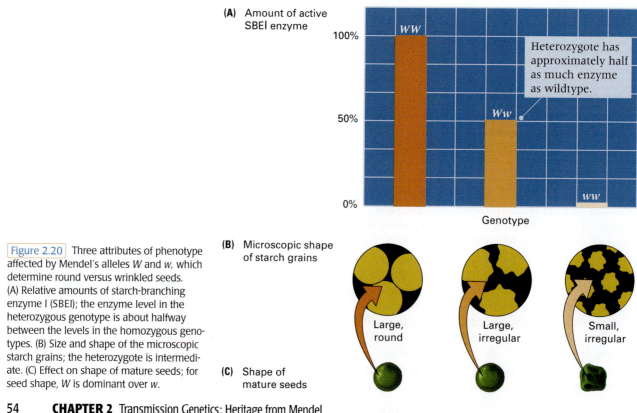

Figure 2.20 Three attributes of phenotype affected by Mendel's alleles *W* and *w*, which determine round versus wrinkled seeds. (A) Relative amounts of starch-branching enzyme I (SBEI); the enzyme level in the heterozygous genotype is about halfway between the levels in the homozygous genotypes. (B) Size and shape of the microscopic starch grains; the heterozygote is intermediate. (C) Effect on shape of mature seeds; for seed shape, *W* is dominant over *w*.

(A) Amount of active SBEI enzyme

Heterozygote has approximately half as much enzyme as wildtype.

Genotype

(B) Microscopic shape of starch grains

Large, round

Large, irregular

Small, irregular

(C) Shape of mature seeds

The alternative *i* allele codes for an inactive enzyme, and *ii* flowers, which have no red pigment, are ivory in color. Because the amount of the critical enzyme is reduced in *Ii* heterozygotes, the amount of red pigment in the flowers is reduced also, and the effect of the dilution is to make the flowers pink.

The result of Mendelian segregation is observed directly when snapdragons that differ in flower color are crossed. For example, a cross between plants from a true-breeding red-flowered variety and a true-breeding ivory-flowered variety results in F_1 plants with pink flowers. In the F_2 progeny obtained via self-fertilization of the F_1 hybrids, one experiment resulted in 22 plants with red flowers, 52 with pink flowers, and 23 with ivory flowers. These numbers agree fairly well with the Mendelian ratio of 1 dominant homozygote : 2 heterozygotes : 1 recessive homozygote. In agreement with the predictions from simple Mendelian inheritance, when self-fertilized, the red-flowered F_2 plants produced only red-flowered progeny; the ivory-flowered plants produced only ivory-flowered progeny; and the pink-flowered plants produced red, pink, and ivory progeny in the proportions 1/4 red : 1/2 pink : 1/4 ivory.

Incomplete dominance is often observed when the phenotype is quantitative rather than discrete. A trait that is *quantitative* can be measured on a continuous scale and is sometimes referred to as a *continuous* trait. Examples include height, weight, number of eggs laid by a hen, time of flowering of a plant, and amount of enzyme in a cell or organism. A trait that is *discrete* is all-or-nothing, for example yellow versus green seeds. We consider round versus wrinkled seeds to be discrete, even though the shape and size of their starch grains fall under the category continuous. With a phenotype that is quantitative, the measured value of a heterozygote is usually intermediate between the homozygotes, and thus there is incomplete dominance.

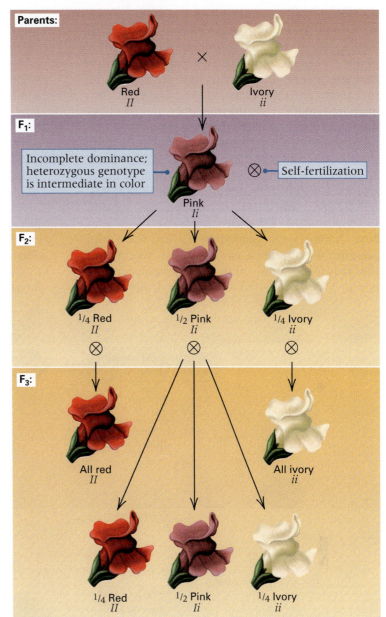

Figure 2.21 Incomplete dominance in the inheritance of flower color in snapdragons.

■ The human ABO blood groups illustrate both dominance and codominance.

Beginning students are often confused by the difference between incomplete dominance and codominance. Incomplete dominance means that the phenotype of the heterozygous genotype is *intermediate* between those of the homozygous genotypes. Incomplete dominance is more frequent for morphological traits than for molecular traits. For example, in the snapdragon in Figure 2.21, the color pink is intermediate between red and white. Codominance means that the heterozygous genotype exhibits the traits associated with *both* homozygous genotypes. Codominance is more frequent for molecular traits than for morphological traits. In Figure 2.7, for example, the gel pattern of the heterozygous *Ww* genotype shows both the small DNA fragment associated with *WW* and the large DNA fragment associated with *ww,* and therefore *W* and *w* are regarded as codominant.

An example illustrating both dominance and codominance is found in the familiar A, B, AB, and O human blood groups determined by polysaccharides (polymers of sugars) present on the surface of red blood cells. Both the A and B polysaccharides are formed from a precursor substance that is modified by the enzyme product of either the I^A or the I^B allele. The gene products are transferase enzymes that attach either of two types of sugar units to the precursor (Figure 2.22). People of genotype $I^A I^A$ produce red blood cells having only the A polysaccharide and are said to have blood type A. Those of genotype $I^B I^B$ have red blood cells with only the B polysaccharide and have blood type B. Heterozygous $I^A I^B$ people have red cells with both the A and the B polysaccharide and have blood type AB. The $I^A I^B$ genotype illustrates codominance, because the heterozygous genotype has the characteristic of both homozygous genotypes—in this case, the presence of both the A and the B carbohydrate on the red blood cells. Although the polypeptides encoded by I^A and I^B differ in only 4 out of 355

amino acids, these differences are at strategic positions in the molecules and change their substrate specificity.

Both I^A and I^B are dominant to the recessive allele I^O. The I^O allele has a single-base deletion in codon 86 that shifts the translational reading frame of the mRNA, resulting in an incomplete, inactive enzyme. The precursor substrate remains unchanged, and neither the A nor the B type of polysaccharide is produced. Homozygous $I^O I^O$ persons therefore lack both the A and the B polysaccharide; they are said to have blood type O. In $I^A I^O$ heterozygotes, presence of the I^A allele results in production of the A polysaccharide; and in $I^B I^O$ heterozygotes, presence of the I^B allele results in production of the B polysaccharide. The result is that $I^A I^O$ persons have blood type A and $I^B I^O$ persons have blood type B, and so I^O is recessive to both I^A

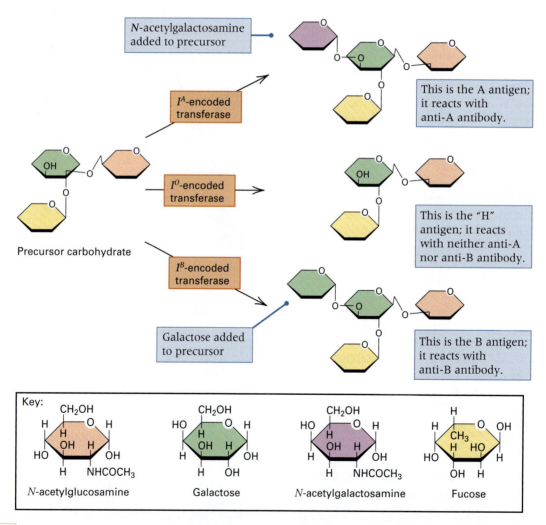

Figure 2.22 The ABO antigens on the surface of human red blood cells are carbohydrates. They are formed from a precursor carbohydrate by the action of transferase enzymes encoded by alleles of the I gene. Allele I^O codes for an inactive enzyme and leaves the precursor unmodified. The unmodified form is called the H substance. The I^A allele encodes an enzyme that adds

N-acetylgalactosamine (purple) to the precursor. The I^B allele encodes an enzyme that adds galactose (green) to the precursor. The other colored sugar units are N-acetylglucosamine (orange) and fucose (yellow). The sugar rings also have side groups attached to one or more of their carbon atoms; these are shown in the detailed structures inside the box.

the human connection

Blood Feud

Karl Landsteiner 1901
Anatomical Institute,
Vienna, Austria
*On Agglutination Phenomena
in Normal Blood*

Early blood transfusions were extremely hazardous. The patient receiving the blood often went into shock and died. This outcome was caused by massive clumping (agglutination) of red blood cells in the recipient, leading to blockage of the oxygen supply to many vital organs. In this paper, Landsteiner demonstrates that the clumping reaction can be observed in the test tube and that blood cells from each person can be classified as type A, type B, or type O, according to whether the cells are agglutinated by blood sera from other persons. The blood samples were taken from volunteers at the Institute at which Landsteiner worked ("Dr. St.", "Dr. Plecn.", and so forth; the abbreviation "Landst." refers to the author himself). In this excerpt, we have preserved the terms *agglutinin* (antibody) and *corpuscle* (red blood cell) as in the

original but have replaced the blood type that Landsteiner called type C with its modern equivalent, type O. (Blood type AB was not found in these experiments, because the number of persons tested was too small.) Landsteiner's discovery led quickly to the matching of donor and recipient for the ABO blood groups in blood transfusions, and the disastrous incompatibility reactions were almost completely eliminated. As an interesting exercise, you may wish to deduce the blood type of "Landst."

....

Some time ago I observed and reported that blood serum of normal human beings is often capable of agglutinating red blood corpuscles of other healthy individuals.... I will mention in the following the results obtained in some recent experiments.... The table is self-explanatory. About equal amounts of serum and approximately 5 percent blood suspension were mixed in 0.6 percent saline solution and observed in test tubes. The plus sign denotes agglutination.

> **The reaction may possibly be suitable for forensic purposes of identification in some cases.**

The experiment demonstrates that my data require no correction. All examined sera [22 altogether] from healthy persons gave the reaction. The result obviously would have been different had I not used a number of different corpuscles for the test.... In several cases (group A) the serum reacted on the corpuscles of another group (B), but not on those of group A, whereas the A corpuscles are again influenced in the same manner by serum B. In the third group (O) the serum aggregates the corpuscles of A and B, while the O corpuscles are not affected by sera of A and B. In ordinary speech, it can be said that in these cases at least two different kinds of agglutinins are present: some in A, others in B, and both together in O. The corpuscles are naturally to be considered as insensitive for the agglutinins which are present in the same serum.... I also did the agglutination successfully with blood which had been dried on linen and preserved for 14 days. Thus the reaction may possibly be suitable for forensic purposes of identification in some cases.... Finally, it must be mentioned that the reported observations allow us to explain the variable results in therapeutic transfusions of human blood.

Source: Wiener Klinische Wochenschrift
14: 1132–1134. Original in German. Excerpt from translation in S. H. Boyer, IV. 1963. *Papers on Human Genetics.* Englewood Cliffs, NJ: Prentice-Hall, pp. 27–31.

Sera	Blood corpuscles of					
	Dr. St.	Dr. Plecn.	Dr. Sturl.	Dr. Erdh.	Zar.	Landst.
Dr. St.	−	+	+	+	+	−
Dr. Plecn.	−	−	+	+	−	−
Dr. Sturl.	−	+	−	−	+	−
Dr. Erdh.	−	+	−	−	+	−
Zar.	−	−	+	+	−	−
Landst.	−	+	+	+	+	−

2.6 Dominance is a property of a pair of alleles in relation to a particular attribute of phenotype   **57**

Table 2.3

Genetic control of the human ABO blood groups

Genotype	Antigens present on red blood cells	ABO blood group phenotype	Antibodies present in blood fluid	Blood types that can be tolerated in transfusion	Blood types that can accept blood for transfusion
$I^A I^A$	A	Type A	Anti-B	A & O	A & AB
$I^A I^O$	A	Type A	Anti-B	A & O	A & AB
$I^B I^B$	B	Type B	Anti-A	B & O	B & AB
$I^B I^O$	B	Type B	Anti-A	B & O	B & AB
$I^A I^B$	A & B	Type AB	Neither anti-A nor anti-B	A, B, AB & O	AB only
$I^O I^O$	Neither A nor B	Type O	Anti-A & anti-B	O only	A, B, AB & O

and I^B. The genotypes and phenotypes of the ABO blood-group system are summarized in the first three columns of **Table 2.3**.

The ABO blood groups are critical in medicine because of the frequent need for blood transfusions. An important feature of the ABO system is that most human blood contains antibodies to either the A or the B polysaccharide. An **antibody** is a protein that is made by the immune system in response to a stimulating molecule called an **antigen** and is capable of binding to the antigen. An antibody is usually specific in that it recognizes only one antigen. Some antibodies combine with antigen and form large molecular aggregates that may precipitate.

Antibodies act in the body's defense against invading viruses and bacteria, as well as other cells, and help remove such invaders from the body. Although antibodies do not normally form without prior stimulation by the antigen, people capable of producing anti-A and anti-B antibodies do produce them. Production of these antibodies may be stimulated by antigens similar to polysaccharides A and B present on the surfaces of many common bacteria. However, a mechanism called *tolerance* prevents an organism from producing antibodies against its own antigens. This mechanism ensures that A-antigen or B-antigen elicits antibody production only in people whose own red blood cells do not contain A or B, respectively. The end result:

key concept

People of blood type O make both anti-A and anti-B antibodies; those of blood type A make anti-B antibodies; those of blood type B make anti-A antibodies; and those of blood type AB make neither type of antibody.

The antibodies found in the blood fluid of people with each of the ABO blood types are shown in the fourth column in Table 2.3. The clinical significance

of the ABO blood groups is that transfusion of blood containing A or B red-cell antigens into persons who make antibodies against these antigens results in an agglutination reaction in which the donor red blood cells are clumped. In this reaction, the anti-A antibody will agglutinate red blood cells of either blood type A or blood type AB, because both carry the A antigen (Figure 2.23). Similarly, anti-B antibody will agglutinate red blood cells of either blood type B or blood type AB. When the blood cells agglutinate, many blood vessels are blocked, and the recipient of the transfusion goes into shock and may die. Incompatibility in the other direction, in which the donor blood contains antibodies against the recipient's red blood cells, is usually acceptable because the donor's antibodies are diluted so rapidly that clumping is avoided. The types of compatible blood transfusions are shown in the last two columns of Table 2.3. Note that a person of blood type AB can receive blood from a person of any other ABO type; type AB is called a *universal recipient*. Conversely, a person of blood type O can donate blood to a person of any ABO type; type O is called a *universal donor*.

■ **A mutant gene is not always expressed in exactly the same way.**

Simple Mendelian ratios are not always observed even when a trait is determined by the alleles of a single gene. The reason is that the same genotype may be expressed in different individuals in different ways. Variation in the phenotypic expression of a particular genotype may happen because other genes modify the phenotype or because the biological processes that produce the phenotype are sensitive to environmental conditions.

The types of variable gene expression are usually grouped into two categories:

• **Variable expressivity** refers to genes that are expressed to different degrees in different

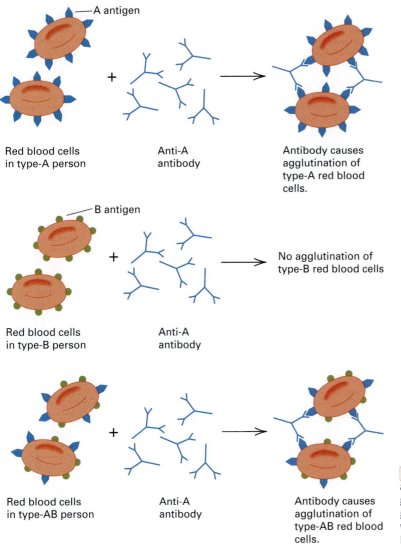

Red blood cells in type-A person

Anti-A antibody

Antibody causes agglutination of type-A red blood cells.

B antigen

Red blood cells in type-B person

Anti-A antibody

No agglutination of type-B red blood cells

Red blood cells in type-AB person

Anti-A antibody

Antibody causes agglutination of type-AB red blood cells.

A antigen

Figure 2.23 Antibody against type-A antigen will agglutinate red blood cells carrying the type-A antigen, whether or not they also carry the type-B antigen. Hence, blood fluid containing anti-A antibody will agglutinate red blood cells of type A and type AB, but not red blood cells of type B or type O.

organisms. For example, inherited genetic diseases in human beings are often variable in expression from one person to the next. One patient may be very sick, whereas another with the same disease is less severely affected. Variable expressivity means that the same mutant gene can cause a severe form of the disease in one person but a mild form in another. The different degrees of expression often form a continuous series from full expression to almost no expression of the expected phenotypic characteristics.

- **Penetrance** refers to the proportion of organisms whose phenotype matches their genotype for a given trait. A genotype that is always expressed has a penetrance of 100 percent. A penetrance of less than 100 percent (*incomplete penetrance*) is the extreme of variable expressivity in which the genotype is not expressed to any detectable degree in some individuals. For example, people with a genetic predisposition

to lung cancer may not get the disease if they don't smoke tobacco. A lack of gene expression may result from environmental conditions, such as in the example of not smoking, or from the effects of other genes.

2.7

Epistasis can affect the observed ratios of phenotypes.

In Chapter 1 we saw that the products of several genes may be necessary to carry out all the steps in a biochemical pathway. In genetic crosses in which two mutations that affect different steps in a single pathway are both segregating, the typical F_2 dihybrid ratio of $9 : 3 : 3 : 1$ is not observed. One example is found in the interaction of two recessive mutations, each in a different gene, that affect flower coloration in peas. Plants of genotypes CC and Cc have purple flower color, which is the normal or wildtype expression of the trait, whereas homozygous cc plants have white flowers. For the other gene, plants of genotypes PP and Pp have wildtype purple flowers, whereas homozygous pp plants have white flowers. Geneticists often use a dash to indicate an allele whose identity is not specified; for example, the symbol $C-$ means that in this genotype, one allele is known to be C and the other (unspecified) allele, indicated by the dash, may be either C or c. The symbol $C-$ is therefore a shorthand designation meaning "either CC or Cc." Using this type of symbolism, we could say that genotypes $C-$ and $P-$ have wildtype purple flowers, whereas genotypes cc and pp have white flowers. Homozygous recessive cc or pp plants have white flowers regardless of the genotype of the other gene.

Figure 2.24 shows a cross between plants of genotype $CC\,pp$ and plants of genotype $cc\,PP$. Both plants have white flowers because they are homozygous for either pp or cc. Even though both parental plants have white flowers, the flower-color phenotype of the plants in the F_1 generation is purple because the genotype of the F_1 progeny is $Cc\,Pp$ and heterozygous for both recessive alleles. Self-fertilization of the F_1 plants results in the F_2 progeny genotypes shown in the Punnett square. Because only the progeny with at least one C allele $(C-)$ and at least one P allele $(P-)$ have purple flowers, and all the rest have white flowers, the ratio of purple flowers to white flowers in the F_2 generation is $9:7$.

The $9:7$ ratio of purple : white flowers is a modified form of the $9:3:3:1$ ratio in which the "9" class has purple flowers and the "$3:3:1$" classes all have white flowers. This is an example of **epistasis,** a term that refers to any type of gene interaction that results in the F_2 dihybrid ratio of $9:3:3:1$ being modified into some other ratio. In a more general sense, what this means is that one gene is masking the expression of the other.

For a trait determined by the interaction of two genes, each with a dominant allele, there are only a limited number of ways in which the $9:3:3:1$ dihybrid ratio can be modified. The possibilities are illustrated in Figure 2.25 . The pie chart at the top shows the $9:3:3:1$ ratio expected of independent assortment when there is no epistasis. Each of the other pie charts shows a different modification of the $9:3:3:1$ ratio, depending on which genotypes have the same phenotype (indicated by sectors of the same color).

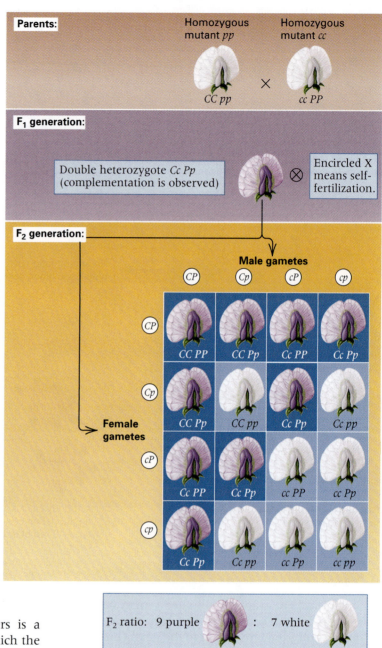

Figure 2.24 A cross showing epistasis in the determination of flower color in peas. Formation of the purple pigment requires the dominant allele of both the C and P genes. With this type of epistasis, the dihybrid F_2 ratio is modified to 9 purple : 7 white.

Taking all the possible modified ratios in Figure 2.25 together, there are nine possible dihybrid ratios when both genes show complete dominance. Examples of each of the modified ratios are known. Some of the most frequently encountered modified ratios are illustrated in the following examples, which are taken from a variety of organisms. Other examples can be found in the problems at the end of the chapter.

9 : 7 This is the ratio observed when a homozygous recessive mutation in either or both of two

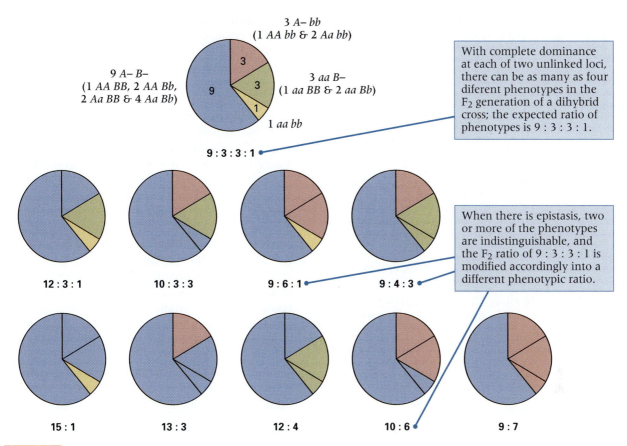

$3\ A-\ bb$
$(1\ AA\ bb\ \&\ 2\ Aa\ bb)$

$9\ A-\ B-$
$(1\ AA\ BB,\ 2\ AA\ Bb,$
$2\ Aa\ BB\ \&\ 4\ Aa\ Bb)$

$3\ aa\ B-$
$(1\ aa\ BB\ \&\ 2\ aa\ Bb)$

$1\ aa\ bb$

9 : 3 : 3 : 1

With complete dominance at each of two unlinked loci, there can be as many as four diferent phenotypes in the F_2 generation of a dihybrid cross; the expected ratio of phenotypes is 9 : 3 : 3 : 1.

12 : 3 : 1   10 : 3 : 3   9 : 6 : 1   9 : 4 : 3

When there is epistasis, two or more of the phenotypes are indistinguishable, and the F_2 ratio of 9 : 3 : 3 : 1 is modified accordingly into a different phenotypic ratio.

15 : 1   13 : 3   12 : 4   10 : 6   9 : 7

Figure 2.25 Modified F_2 dihybrid ratios. In each pie chart, different colors indicate different phenotypes.

different genes results in the same mutant phenotype. It is exemplified by the segregation of purple and white flowers in Figure 2.24. Genotypes that are $C-$ for the C gene and $P-$ for the P gene have purple flowers; all other genotypes have white flowers. Recall that the dash in $C-$ means that the unspecified allele could be either C or c, and so $C-$ means "either CC or Cc." Similarly, the dash in $P-$ means that the unspecified allele could be either P or p.

12 : 3 : 1 A modified dihybrid ratio of the 12 : 3 : 1 variety results when the presence of a dominant allele of one gene masks the genotype of a different gene. For example, if the $A-$ genotype renders the $B-$ and bb genotypes indistinguishable, then the dihybrid ratio is 12 : 3 : 1 because the $A-\ B-$ and $A-\ bb$ genotypes are expressed as the same phenotype.

In a genetic study of the color of the hull in oat seeds, a variety having white hulls was crossed with a variety having black hulls. The F_1 hybrid seeds had black hulls. Among 560 progeny in the F_2 generation produced by self-fertilization of the F_1, the following seed phenotypes were observed in the indicated numbers:

418 black hulls   106 gray hulls   36 white hulls

The observed ratio of phenotypes is 11.6 : 2.9 : 1, or very nearly 12 : 3 : 1. These results can be explained by a genetic hypothesis in which the

black-hull phenotype results from the presence of a dominant allele (say, A) and the gray-hull phenotype results from another dominant allele (say, B) whose effect is apparent only in the aa homozygotes. On the basis of this hypothesis, the original true-breeding varieties must have had genotypes $aa\ bb$ (white) and $AA\ BB$ (black). The F_1 has genotype $Aa\ Bb$ (black). If the A, a allele pair and the B, b allele pair undergo independent assortment, then the F_2 generation is expected to have the following composition of genotypes:

9/16	$A-\ B-$	(black hull)
3/16	$A-\ bb$	(black hull)
3/16	$aa\ B-$	(gray hull)
1/16	$aa\ bb$	(white hull)

This type of epistasis accounts for the 12 : 3 : 1 ratio.

13 : 3 This type of epistasis is illustrated by the difference between White Leghorn chickens (genotype $CC\ II$) and White Wyandotte chickens (genotype $cc\ ii$). The C allele is responsible for colored feathers but in White Leghorns the I allele is a dominant inhibitor of feather coloration. The F_1 generation of a dihybrid cross between these breeds has the genotype $Cc\ Ii$, which results in the presence of white feathers because of the inhibitory effects of the I allele. In the F_2 generation, only the $C-ii$ genotype has colored feathers; hence there is a 13 : 3 ratio of white : colored.

9 : 4 : 3 This dihybrid ratio (often stated as 9 : 3 : 4) is observed when homozygosity for a recessive allele with respect to one gene masks the expression of the genotype of a different gene. For example, if the aa genotype has the same phenotype regardless of whether the genotype is $B-$ or bb, then the 9 : 4 : 3 ratio results.

In the mouse, the grayish coat color called agouti is produced by the presence of a horizontal band of yellow pigment just beneath the tip of each hair. The agouti pattern results from the presence of a dominant allele A, and in aa animals the coat color is black. A second dominant allele, C, is necessary for the formation of hair pigments of any kind, and cc animals are albino (white fur). In a cross of $AA\ CC$ (agouti) $\times aa\ cc$ (albino), the F_1 progeny are $Aa\ Cc$ and agouti. Crosses between F_1 males and females produce F_2 progeny in the following proportions:

$$9/16 \quad A-C- \quad \text{(agouti)}$$
$$3/16 \quad A-cc \quad \text{(albino)}$$
$$3/16 \quad aa\ C- \quad \text{(black)}$$
$$1/16 \quad aa\ cc \quad \text{(albino)}$$

The dihybrid ratio is therefore 9 agouti : 4 albino : 3 black.

9 : 6 : 1 This dihybrid ratio is observed when homozygosity for a recessive allele of either of two genes results in the same phenotype but the phenotype of the double homozygote is distinct. For example, red coat color in Duroc–Jersey pigs requires the presence of two dominant alleles R and S. Pigs of genotype $R-ss$ and $rr\ S-$ have sandy-colored coats, and $rr\ ss$ pigs are white. The F_2 dihybrid ratio is therefore

$$9/16 \quad R-S- \quad \text{(red)}$$
$$3/16 \quad R-ss \quad \text{(sandy)}$$
$$3/16 \quad rr\ S- \quad \text{(sandy)}$$
$$1/16 \quad rr\ ss \quad \text{(white)}$$

The 9 : 6 : 1 ratio results from the fact that both single recessives have the same phenotype.

2.8

Complementation between mutations of different genes is a fundamental principle of genetics.

As we have just seen, one gene can mask the effect of another, a situation illustrated for purple (wildtype) and white (mutant) flower color in pea flow-

ers (Figure 2.24). Epistatic interaction of mutant recessive alleles p and c masks the effect of the dominant alleles P and C, creating white flowers in plants that have at least one homozygous recessive pair in their genotype: $C-pp$, $cc\ P-$, and $cc\ pp$. An important feature of this interaction is that a cross between a white-flowered plant of genotype $CC\ pp$ and a white-flowered plant of genotype $cc\ PP$ would result in offspring with genotype $Cc\ Pp$, which would have the normal purple flower color. In this cross, the individual that inherits the c allele from one parent and the p allele from the other parent has the wildtype phenotype. This is because the gamete carrying the c allele also carries P and that carrying the p allele also carries C, and both c and p are recessive. The mutations c and p are said to exhibit **complementation** because the F_1 progeny of the cross between homozygous mutant genotypes has the wildtype phenotype. Complementation occurs because c and p are mutations in different genes. They are not mutant alleles of the same gene.

Because complementation between c and p shows that they are mutations in different genes, these mutations identify two genes that are necessary for purple flower coloration. Suppose that a pea geneticist isolates more recessive mutations that cause white flower color. Mutant genes might be recovered in an experiment in which peas are exposed to radiation or to a chemical agent known to cause mutations. Such experiments are done routinely in genetics to isolate and identify any and all genes that affect key steps in a biological process—in this example, the production of color pigment in flowers. The first thing to do is to compare each new mutation with every other one, and to ask, for each pair of mutations being compared: "Are these mutations in the same gene, or are they mutations in different genes?"

If two mutations are in the same gene, it does not imply that they are identical in DNA sequence. We saw in Chapter 1 that genes typically consist of thousands of base pairs, many of which are so critical to function that a single-base change results in a visible mutant phenotype. Mutations can also be more drastic changes, such as a deletion of part of the gene. Because there are so many different ways in which a gene can mutate to impair or eliminate its function, it is very unlikely that two independent mutations are identical in DNA sequence, even if they are alleles of the same gene. The various possible forms of a gene are called **multiple alleles.** (It is important to note that the concept of multiple alleles, sometime referred to as *allelism*, includes not only mutant alleles. In natural populations of organisms there are usually also multiple *wildtype* alleles, even though we often refer to wildtype as "the" wildtype allele. The multiple wildtype alleles differ in nucleotide sequence, but they all encode a functional gene product.)

■ Complementation means that two mutations are in different genes.

First, how can the geneticist find out whether two new mutations are (or are not) alleles of the same gene? The approach is extremely simple: The mutant homozygous genotypes are crossed and the phenotype of the progeny examined. The key to interpreting the result is illustrated in Figure 2.24, in which the cross of $CC\ pp \times cc\ PP$ yields F_1 progeny whose flowers are wildtype. The reason for the wildtype phenotype is that the genotype of the F_1 progeny is $Cc\ Pp$. The F_1 progeny are heterozygous for a mutation in each of two different genes. Each mutation is recessive, and its presence is concealed by the dominant allele in the F_1 progeny. This finding of complementation between two recessive mutations means that the mutations are in different genes, as shown for the genes C and P in Figure 2.24.

■ Lack of complementation means that two mutations are alleles of the same gene.

Lack of complementation between mutant recessive alleles in the same gene is illustrated in part A of Figure 2.26 for mutant strains designated 1 and 2.

The finding of a mutant phenotype in the F_1 generation (lack of complementation) indicates that the mutant gene in strain 1 is an allele of the mutant gene in strain 2. The expected result of a cross when the mutations are in different genes is shown in Figure 2.26, part B. In this case the finding of a wildtype phenotype in the F_1 generation (complementation) means that the mutant allele in strain 1 must be in a different gene from the mutant allele in strain 3.

■ The complementation test enables one to group mutations into allelic classes.

The kind of cross illustrated in Figure 2.26 is a **complementation test.** As we have seen, it is used to determine whether recessive mutations in each of two different strains are alleles of the same gene. Because the result indicates presence or absence of allelism, the complementation test is one of the key experimental operations in genetics.

To illustrate application of the complementation test in practice, suppose a mutant screen were carried out to isolate new mutants with white flowers. A **mutant screen** is a large-scale, systematic experiment designed to isolate multiple new mutations affecting a particular trait. For a mutant screen affecting flower color, we start with a strain with purple flowers that is completely homozygous for the wildtype allele of every gene affecting the

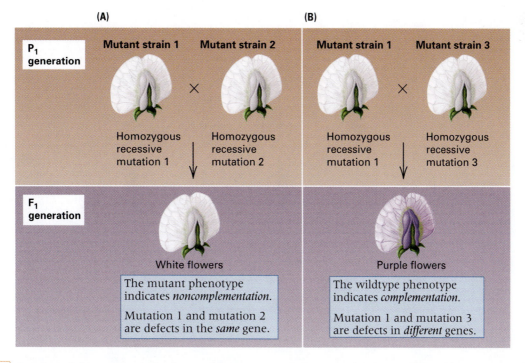

Figure 2.26 | Complementation reveals whether two recessive mutations are alleles of different genes. To test for complementation, homozygous recessive genotypes are crossed. If the phenotype of the F_1 progeny is mutant (A), it means that the mutations in the parental strains are alleles of the same gene. If the phenotype of the F_1 progeny is wildtype (B), it means that the mutations in the parental strains are mutations in different genes.

trait. The complete homozygosity is important, otherwise pre-existing recessive mutations hidden by heterozygosity could affect the result. Completely homozygous strains are usually obtained in plants by repeated generations of self-fertilization, or in animals by repeated generations of brother-sister mating.

Starting with a completely homozygous strain with purple flowers, we treat pollen with x rays, and use the irradiated pollen to fertilize ovules to obtain seeds. The F_1 seeds are grown and the resulting plants allowed to self-fertilize, after which the F_2 plants are grown. A few of the F_1 seeds may contain a new mutation that causes white flowers, but because the white phenotype is recessive, the flower will be purple. However, the resulting F_1 plant will be heterozygous for the new white mutation, so self-fertilization will result in the formation of F_2 plants with a 3 : 1 ratio of purple : white flowers. Because mutations that result in a particular phenotype are quite rare, even when induced by radiation, among many thousands of self-fertilized plants only a few will be found to be a new mutant with white flowers. Furthermore, because mutations are rare even in the presence of agents that cause mutations, any mutant strain is extremely unlikely to carry a mutation in more than one gene

affecting flower color. Therefore, we may safely assume that each new mutant strain is homozygous recessive for one, and only one, mutation affecting flower color.

Let us suppose that we are lucky enough to obtain four new mutants, in addition to the p and c mutants already identified. How are we going to name these four new mutations? We can make no assumptions about the number of genes represented. All four could be new mutations in either P or C, and then they would be alleles of the already known mutations p and c. On the other hand, each of the four could be the result of a new mutation in a different gene needed for flower color. For the moment, let us call the new mutant genes $x1$, $x2$, $x3$, and $x4$, where the x does not imply a gene but rather indicates that the mutant gene was obtained with x irradiation. Each mutant gene is recessive and was identified through the white flowers of the homozygous recessive F_2 seeds (for example, $x1\ x1$).

Now the complementation test is used to classify the "x" mutations, as well as the p and c mutations, into groups. Figure 2.27 shows that the results of a complementation test can be reported in a triangular array of + and − signs. The crosses that yield F_1 progeny with the wildtype phenotype (in this case, purple flowers) are denoted with a + in

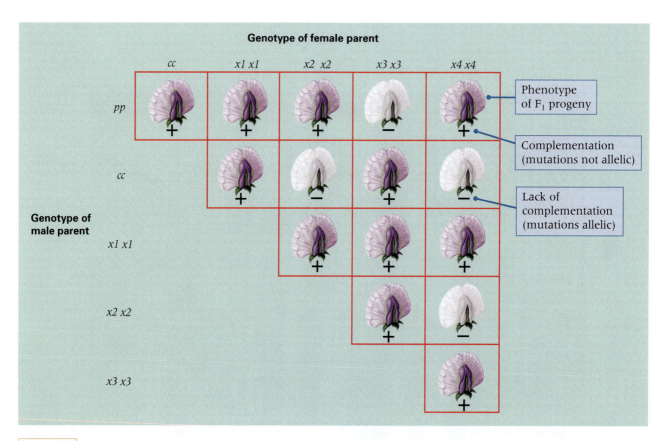

Figure 2.27 Results of complementation tests among six mutant strains of peas, each homozygous for a recessive allele resulting in white flowers. Each box gives the phenotype of the F_1 progeny of a cross between the male parent whose genotype is indicated in the far-left column and the female parent whose genotype is indicated in the top row.

the box where imaginary lines from the male parent and the female parent intersect. The crosses that yield F_1 progeny with the mutant phenotype (white flowers) are denoted with a $-$ sign. The $+$ signs indicate complementation between the mutant alleles in the parents; the $-$ signs indicate lack of complementation. The bottom half of the triangle is unnecessary, because the reciprocal of each cross produces F_1 progeny with the same genotype and phenotype as the cross that is shown. The diagonal showing a cross between any two organisms that carry the same mutant gene—for example, *x1 x1* $\times$ *x1 x1*—is also unnecessary. Such a cross must yield homozygous recessive *x1 x1* progeny, which will have the mutant phenotype. As we saw in Figure 2.26, complementation in a cross means that the parental strains are mutant for different genes. Lack of complementation means that the parental strains are mutant for the same gene. The principle underlying the complementation test is as follows:

key concept

The Principle of Complementation: If two recessive mutations are alleles of the same gene, then the phenotype of an organism that contains one copy of each mutation is mutant; if they are alleles of different genes, then the phenotype of an organism that contains one copy of each mutation is wildtype (nonmutant).

In the interpretation of complementation data such as those in Figure 2.27, the principle is actually applied the other way around. Examination of the phenotype of the F_1 progeny of each possible cross reveals which of the mutations are alleles of the same gene.

key concept

In a complementation test, if the combination of two recessive mutations results in a mutant phenotype, then the mutations are regarded as alleles of the *same* gene; if the combination results in a wildtype phenotype, then the mutations are regarded as alleles of *different* genes.

A convenient way to analyze the data in Figure 2.27 is to arrange the alleles in a circle as shown in Figure 2.28, part A. Then, for each possible pair of mutations, connect the pair by a straight line if the mutant genes *fail* to complement (Figure 2.28, part B). According to the principle of complementation, the lines must connect mutant genes that are alleles of each other, because in a comple-

mentation test, lack of complementation means that the mutant genes are alleles. In this example, mutant *x3* is an allele of *p*, so *x3* and *p* are different mutant alleles of the gene *P*. Similarly, the mutants *x2*, *x4*, and *c* are different mutant alleles of the gene *C*. The mutant *x1* complements all of the other alleles. It represents a third gene, different from *P* and *C*, that affects flower coloration.

In an analysis like that in Figure 2.28, the mutant alleles belonging to the same gene make up a distinct **complementation group.** Thus each complementation group defines a gene.

key concept

A *gene* is defined experimentally as a set of mutant alleles that make up one complementation group. Any pair of mutant alleles in such a group fail to complement one another and result in an organism with a mutant phenotype.

The mutations in Figure 2.28 therefore represent three genes, a mutation in any one of which results in white flowers. The gene *P* is represented by the alleles *p* and *x3*; the gene *C* is represented by the alleles *c*, *x2*, and *x4*; and the allele *x1* represents a third gene different from either *P* or *C*. Each gene coincides with one of the complementation groups.

At this point in a genetic analysis, it is good practice to rename the mutant genes to indicate which ones are true alleles. Because the *p* allele

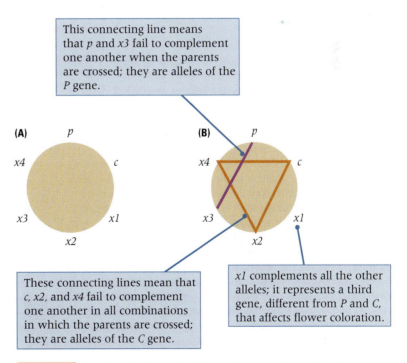

This connecting line means that *p* and *x3* fail to complement one another when the parents are crossed; they are alleles of the *P* gene.

These connecting lines mean that *c*, *x2*, and *x4* fail to complement one another in all combinations in which the parents are crossed; they are alleles of the *C* gene.

x1 complements all the other alleles; it represents a third gene, different from *P* and *C*, that affects flower coloration.

Figure 2.28 A method for interpreting the results of complementation tests. (A) Arrange the mutations in a circle. (B) Connect by a straight line any pair of mutations that fail to complement (that yield a mutant phenotype); any pair of mutations so connected are alleles of the same gene. In this example, there are three complementation groups, each of which represents a different gene needed for purple flower coloration.

already had its name before the mutant screen was carried out to obtain more flower-color mutants, the new allele of p that was previously called $x3$ should be renamed to reflect its allelism with p. For example, we might rename $x3$ as p_3 to emphasize its allelism with p. For similar reasons, we might rename $x2$ and $x4$ as c_2 and c_4 to emphasize their allelism with the original c and to convey their independent origins. The mutation $x1$ represents an allele of a new gene to which we can assign a name arbitrarily. For example, we might call the mutant allele *albus* (Latin for "white") and assign the $x1$ allele the new name *alb*. The wild-type dominant allele of *alb*, which is necessary for purple coloration, would then be symbolized as *Alb* or as alb^+. The procedure of sorting new mutations into complementation groups and renaming them according to their allelism is an example of how geneticists identify genes and name alleles. Such renaming of alleles is the typical manner in which genetic terminology evolves as knowledge advances.

chapter summary

2.1 Mendel took a distinctly modern view of transmission genetics.

- Mendel was careful in his choice of traits.
- Reciprocal crosses yield the same types of offspring.
- The wrinkled mutation causes an inborn error in starch synthesis.
- Analysis of DNA puts Mendel's experiments in a modern context.

2.2 Genes come in pairs, separate in gametes, and join randomly in fertilization.

- Genes are physical entities that come in pairs.
- The paired genes separate (segregate) in the formation of reproductive cells.
- Gametes unite at random in fertilization.
- Genotype means genetic endowment; phenotype means observed trait.
- The progeny of the F_2 generation support Mendel's hypothesis.
- The progeny of testcrosses also support Mendel's hypothesis.

Inherited traits are determined by particulate elements called genes. In a higher plant or animal, the genes are present in pairs. One member of each gene pair is inherited from the maternal parent, the other member from the paternal parent.

A gene can have different forms that result from differences in DNA sequence. The different forms of a gene are called alleles. The particular combination of alleles present in an organism constitutes its genotype. The observable characteristics of an organism constitute its phenotype. In an organism, if the two alleles of a gene pair are the same (for example, AA or aa), the genotype is homozygous for the A or a allele; if the alleles are different (Aa), the genotype is heterozygous. When the phenotype of a heterozygote is the same as that of one of the homozygous genotypes, the allele that is expressed is called dominant and the hidden allele is called recessive.

2.3 The alleles of different genes segregate independently.

- The F_2 genotypes in a dihybrid cross conform to Mendel's prediction.

- The progeny of testcrosses show the result of independent assortment.

In genetic studies, the organisms produced by a mating constitute the F_1 generation. Matings between members of the F_1 generation produce the F_2 generation. In a cross such as $AA \times aa$, in which only one gene is considered (a monohybrid cross), the ratio of genotypes in the F_2 generation is 1 dominant homozygote (AA) : 2 heterozygotes (Aa) : 1 recessive homozygote (aa). The phenotypes in the F_2 generation appear in the ratio 3 dominant : 1 recessive. The Mendelian ratios of genotypes and phenotypes result from segregation in gamete formation (when the members of each allelic pair segregate into different gametes) and random union of gametes in fertilization.

2.4 Chance plays a central role in Mendelian genetics.

- The addition rule applies to mutually exclusive possibilities.
- The multiplication rule applies to independent possibilities.

2.5 The results of segregation can be observed in human pedigrees.

The processes of segregation, independent assortment, and random union of gametes follow the rules of probability, which provide the basis for predicting outcomes of genetic crosses. Two basic rules for combining probabilities are the addition rule and the multiplication rule. The addition rule applies to mutually exclusive possibilities: it states that the probability of observing one or the other of two possible outcomes equals the sum of the respective probabilities. The multiplication rule applies to independent possibilities: it states that the probability of observing both of two possible outcomes is equal to the product of the respective probabilities. In some organisms—for example, human beings—it is not possible to perform controlled crosses, and genetic analysis is accomplished through the study of several generations of a family tree, called a pedigree. Pedigree analysis is determination of the possible genotypes of the family members in a pedigree and of the probability that an individual member has a particular genotype.

2.6 Dominance is a property of a pair of alleles in relation to a particular attribute of phenotype.

- Flower color in snapdragons illustrates incomplete dominance.
- The human ABO blood groups illustrate both dominance and codominance.
- A mutant gene is not always expressed in exactly the same way.

In heterozygous genotypes, complete dominance of one allele over the other is not always observed. In most cases, a heterozygote for a wildtype allele and a mutant allele that encodes a defective gene product produces less gene product than does the wildtype homozygote. If the phenotype is determined by the amount of wildtype gene product rather than by its mere presence, the heterozygote will have an intermediate phenotype. This situation is called incomplete dominance. Codominance means that both alleles in a heterozygote are expressed, and so the heterozygous genotype exhibits the phenotypic characteristics of both homozygous genotypes. Codominance is exemplified by the I^A and I^B alleles in persons with blood-group AB. Codominance is often observed for DNA fragments that differ in size among alleles or for proteins when each alternative allele codes for a different amino acid replacement, because the alternative forms of the protein may be able to be distinguished by chemical or physical means. Genotypes are not always expressed in the same way in different individuals; this phenomenon is called variable expressivity. A genotype that is not expressed at all in some individuals is said to have incomplete penetrance.

2.7 Epistasis can affect the observed ratios of phenotypes.

Dihybrid crosses differ in two genes—for example, $AA\ BB \times aa\ bb$. The phenotypic ratios in the dihybrid F_2 are 9 : 3 : 3 : 1, provided that both the A and the B alleles are dominant and that the genes undergo independent assortment. The 9 : 3 : 3 : 1 ratio can be modified in various ways by interaction between the genes (epistasis). Different types of epistasis may result in dihybrid ratios such as 9 : 7 or 12 : 3 : 1.

2.8 Complementation between mutations of different genes is a fundamental principle of genetics.

- Complementation means that two mutations are in different genes.
- Lack of complementation means that two mutations are alleles of the same gene.
- The complementation test enables one to group mutations into allelic classes.

The complementation test is the functional definition of a gene. Two recessive mutations are considered alleles of different genes if a cross between the homozygous recessives results in nonmutant progeny. Such alleles are said to complement each other. On the other hand, two recessive mutations are considered alleles of the same gene if a cross between the homozygous recessives results in mutant progeny. Such alleles are said to fail to complement. For any group of recessive mutant alleles, a complete complementation test entails crossing the homozygous recessives in all pairwise combinations.

issues & ideas

- What constitutes the genotype of an organism? What constitutes the phenotype? Why is it important in genetics that genotype and phenotype be distinguished?
- What is the difference between a gene and an allele? How can a gene have more than two alleles? Give an example of multiple alleles of a gene.
- What is the principle of segregation, and how is this principle demonstrated in the results of Mendel's monohybrid crosses?
- What is the principle of independent assortment, and how is this principle demonstrated in the results of Mendel's dihybrid crosses?
- Explain why random union of male and female gametes is necessary for Mendelian segregation and independent assortment to occur.

- What is the difference between mutually exclusive possibilities and independent possibilities? How are the probabilities of these two types of possible outcomes combined? Give two examples of genetic possibilities that are mutually exclusive and two examples of genetic possibilities that are independent.
- When two pairs of alleles show independent assortment, under what conditions will a 9 : 3 : 3 : 1 ratio of phenotypes in the F_2 generation *not* be observed?
- Explain this statement: In genetics, a gene is identified experimentally by a set of mutant alleles that fail to show complementation.

key terms & concepts

addition rule	backcross	complementation test	gamete
albinism	carrier	dominant trait	gel electrophoresis
allele	codominant	epistasis	gene
antibody	complementation	F_1 generation	genotype
antigen	complementation group	F_2 generation	heterozygous

homozygous
Huntington disease
hybrid
incomplete dominance
independent assortment
Mendelian genetics
multiple alleles

multiplication rule
mutant screen
P_1 generation
pedigree
penetrance
phenotype
Punnett square

recessive trait
reciprocal cross
segregation
sib
sibling
sibship

testcross
transmission genetics
true-breeding
variable expressivity
wildtype
zygote

1. _____ One of possibly several alternative forms (DNA sequences) of a gene.

2. _____ Refers to the "normal" or non-mutant form of a gene.

3. _____ The genetic endowment of an individual.

4. _____ Mendel's greatest discovery: the random separation of alleles into different gametes.

5. _____ A cross to identify the genotype of an individual, in which one parent is homozygous recessive for all of the genes of interest.

6. _____ Another term for the independent segregation of the alleles of two genes, which with a dominant allele of each gene results in a 9 : 3 : 3 : 1 ratio of phenotypes in the F_2 generation.

7. _____ Refers to a situation in which the phenotype of the heterozygous genotype is intermediate between those of the homozygous genotypes.

8. _____ Diagram of a family history.

9. _____ General term meaning either a brother or a sister.

10. _____ Interaction between the alleles of different genes that can result in modification of the 9 : 3 : 3 : 1 dihybrid ratio in the F_2 generation.

11. _____ A type of cross used to determine whether two mutant recessive alleles are, or are not, alleles of the same gene.

12. _____ Incomplete penetrance is an extreme form of this phenomenon, in which the same genotype is expressed to differing degrees in different individuals.

solutions: step by step

Problem 1

In garden peas, an allele T for axial flowers (positioned along the stem) is dominant to an allele t for terminal flowers (positioned at the tips of the branches).
(a) In the F_2 generation of a monohybrid cross, what is the expected ratio of axial : terminal?
(b) Among the F_2 progeny, what proportion are heterozygous?
(c) Among the F_2 progeny with axial flowers, what proportion are heterozygous?
(d) In a testcross of the F_1 progeny of a monohybrid cross, what is the expected ratio of axial : terminal?

■ Solution (a) The monohybrid cross is between the genotypes $TT \times tt$, which yields F_1 plants with genotype Tt. Crossing these among themselves ($Tt \times Tt$) yields F_2 plants given by the accompanying Punnett square. Therefore, the expected ratio of axial (TT or Tt) to terminal (tt) is $1/4 + 1/2 = 3/4$ to $1/4$, or $3 : 1$.

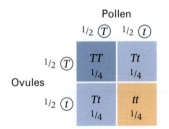

Pollen

(b) The proportion of heterozygous Tt progeny in the F_2 generation is evident from the Punnett square as $1/4 + 1/4 = 1/2$.

(c) The critical phrase in this question is *with axial flowers*, because it means that in considering the Punnett square, we are supposed to disregard the tt plants with terminal flowers. When these are disregarded, there remain $1/3$ TT plants and $2/3$ Tt plants, so the proportion of F_2 plants *with axial flowers* that are heterozygous is $2/3$.

(d) By definition, the testcross is with a homozygous recessive parent, so the parents in the testcross are $Tt \times tt$. The Punnett square is indicated, and it shows that the expected ratio of axial (genotype Tt) to terminal (genotype tt) plants is $1/2 : 1/2$, or $1 : 1$.

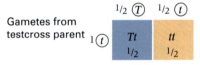

Problem 2

In some people, the carbohydrate A and B antigens of the ABO blood groups not only are found on the surface of the red blood cells but also are secreted into the saliva and other body fluids. Secretion of the antigens is due to a dominant allele of another gene called *secretor*. People of genotype Se Se or Se se do secrete the A and B antigens, but those of genotype se se do not. There is epistasis between the genetic systems, because people of blood group O are nonsecretors regardless of their secretor genotype. The ABO and secretor genes are in different chromosomes and therefore show independent assortment. In crosses of the following type, what is the expected ratio of secretors to nonsecretors?

$$I^A\,I^O\ Se\ se \times I^B\,I^O\ Se\ se$$

■**Solution** One way to work this problem is to make a 4×4 Punnett square with the gametes $I^A\ Se$, $I^A\ se$, $I^O\ Se$, and $I^O\ se$ (1/4 for each) along one axis and $I^B\ Se$, $I^B\ se$, $I^O\ Se$, and $I^O\ se$ (again 1/4 for each) along the other; then fill in all the boxes and tabulate which are secretors and which are not. But because the genes segregate independently, the approach can be simplified by considering each gene separately. The ABO segregation produces the genotypes $1/4\ I^A\,I^B$, $1/4\ I^A\,I^O$, $1/4\ I^B\,I^O$, and $1/4\ I^O\,I^O$. The first three are potential secretors, depending on the *secretor* genotype, whereas the $I^O\,I^O$ genotype is a nonsecretor no matter what. Segregation of the *secretor* gene produces $3/4\ Se-$ (where the dash indicates *either Se or se*) and $1/4\ se\ se$. The former are potential secretors, depending on the ABO genotype, whereas the *se se* genotype is a nonsecretor no matter what. Because these segregations are independent, we can make the sort of Punnett square illustrated, which shows that the ratio of secretor : nonsecretor is $9/16 : 7/16$, or $9 : 7$.

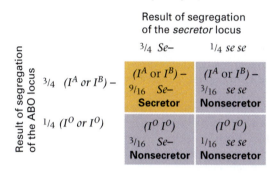

Problem 3

Mutant genes *a* through *f*, all recessive, are identified in a mutant screen for genes affecting pharynx development in the nematode worm *Caenorhabditis elegans*. Complementation tests are carried out between the homozygous mutant strains, with the results shown in the accompanying matrix. A + sign indicates complementation, and a − sign indicates lack of complementation. Configure the mutant genes in

the form of a circle, and use straight lines to connect the alleles that are in the same complementation group. How many different genes affecting pharynx development do these data indicate?

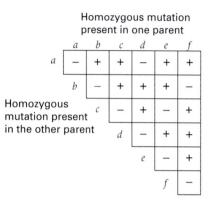

■**Solution** This problem uses the key concept of a *complementation test*, which is a cross between two genotypes, each homozygous for an independently identified mutant gene. The mutations may or may not be alleles of the same gene. If they *are* alleles of the same gene, then all of the progeny will have the mutant phenotype. If they are *not* alleles of the same gene, then all of the progeny will have the nonmutant phenotype. A set of recessive mutations in which all pairwise crosses between the homozygous mutants yield progeny with the mutant phenotype constitutes a *complementation group*, which is the geneticist's experimental definition of a *gene*. The circular pattern called for in the problem is indicated. The lines connect mutants that, when crossed, yield only mutant progeny (−sign in the complementation matrix). The circular analysis of the complementation data indicates that mutations *a* and *d* are alleles of one gene affecting pharynx development, *b* and *f* are alleles of a different gene affecting pharynx development, and *c* and *e* are alleles of a third such gene.

---

concepts in action: problems for solution

2.1 How many different alleles of a particular gene may exist in the population? How many alleles of a particular gene can be present in a single individual?

2.2 Complete the chart below by placing a 0, 1/4, 1/2, or 1 in each box, corresponding to the proportion of each genotype of offspring expected from each type of mating.

Parents	Progeny		
	AA	**Aa**	**aa**
$AA \times AA$			
$AA \times Aa$			
$AA \times aa$			
$Aa \times Aa$			
$Aa \times aa$			
$aa \times aa$			

2.3 A round pea seed is germinated and the mature plant self-fertilized. It produces some wrinkled seeds. What was the genotype of the original seed? What is the expected proportion of wrinkled seeds produced by the mature plant?

2.4 The recurrence risk of a genetic disorder is the probability that the next child born into a sibship will be affected, given that one or more previous children is affected. What is the recurrence risk for:
(a) A dominant trait in which one parent is affected?
(b) A recessive trait in which neither parent is affected?
(c) A recessive trait in which one parent is affected?

2.5 With independent assortment, how many different types of gametes are possible from the genotype *Aa Bb Cc Dd*, and in what proportions are they expected?

2.6 The pedigree illustrated here shows individual II-2 affected with a recessive trait. Let A and a represent the dominant and recessive alleles.

(a) What is the genotype of II-2 ?

(b) What are the genotypes of I-1 and I-2?

(c) What are the possible genotypes of II-1 and II-3?

(d) What is the probability that II-3 is a heterozygous "carrier" of the a allele?

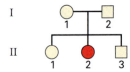

2.7 In Problem 2.6, what is the probability that both II-1 and II-3 are carriers? that neither is a carrier? that at least one is a carrier?

2.8 The accompanying diagram shows an electrophoresis gel in which DNA samples are placed ("loaded") in the depressions ("wells") at the top of the gel and electrophoresis is in the downward direction. The dashed lines on the right denote the positions to which DNA fragments of various sizes would migrate. The fragment sizes are given in kilobase pairs (kb); 1 kb refers to a duplex DNA molecule 1000 base pairs in length. Also shown is the position of a DNA fragment corresponding to part of the coding region of a gene in DNA extracted from a homozygous wildtype (AA) organism. Assuming that a_1 is a mutant allele that has a 2-kb insertion of DNA into the wildtype fragment, and that a_2 is a mutant allele that has a 1-kb deletion within the wildtype fragment, show the positions at which DNA bands would be expected in each of the other genotypes shown.

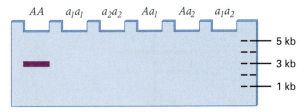

2.9 A woman who is homozygous recessive for a mutation that results in deafness marries an unrelated man who is also deaf because of homozygosity for a recessive mutation. They have a child whose hearing is normal. Explain how this can happen. What genetic principle does this situation exemplify?

2.10 Complementation tests of the recessive mutant genes a through f produced the data in the accompanying matrix. The circles represent missing data. Assuming that all of the missing mutant combinations would yield data consistent with the entries that are known, complete the table by filling each circle with a + or − as needed.

	a	b	c	d	e	f
a	○	+	−	○	+	○
b	○	○	○	○	○	−
c	○	−	○	○		
d	○	○	○			
e	○	+				
f	○					

2.11 In the shepherd's purse, *Capsella bursapastoris*, the capsule containing the seeds can be either triangular or ovoid. A cross between certain true-breeding strains with triangular capsules yielded an F_1 with triangular capsules. The observed F_2 ratio was 15 triangular : 1 ovoid. What genetic hypothesis can explain these results? What crosses would you carry out to test this hypothesis?

2.12 The dominant allele *Cy* (*Curly*) in *Drosophila* results in curly wings. The cross $Cy/+ \times Cy/+$ (where + represents the wildtype allele of *Cy*) results in a ratio of 2 curly : 1 wildtype F_1 progeny. The cross between curly F_1 progeny also gives a ratio of 2 curly : 1 wildtype F_2 progeny. How can this result be explained?

2.13 A trihybrid cross $A/A; B/B; r/r \times a/a; b/b; R/R$ is made in a plant species in which A and B are dominant to their respective alleles but there is incomplete dominance between R and r. Assume independent assortment, and consider the F_2 progeny from this cross.

(a) How many phenotypic classes are expected?

(b) What is the probability of the parental $a/a; b/b; R/R$ genotype?

(c) What proportion of the progeny would be expected to be homozygous for all three genes?

2.14 In the summer squash *Curcurbita pepo*, the shape of the fruit in wildtype genotypes is a Frisbee-like flattened shape known as disc. Either of two recessive mutations, when homozygous, result in sphere-shaped fruit. The doubly homozygous genotype has a football-shaped fruit known as elongate. Assuming independent assortment, what ratio of disc : sphere : elongate would be expected in the F_2 generation of a cross between homozygous disc and homozygous elongate?

2.15 A man and a woman each have a 50 percent chance of being a carrier (heterozygous) for a recessive allele associated with a genetic disease. If they have one child, what is the chance that the child will be homozygous recessive?

2.16 Huntington disease is a rare degenerative human disease determined by a dominant allele, *HD*. The disorder is usually manifested after the age of 45. A young man has learned that his father has developed the disease.

(a) What is the probability that the young man will later develop the disorder?

(b) What is the probability that a child of the young man carries the *HD* allele?

2.17 Consider human families with four children, and assume that each birth is equally likely to result in a boy (B) or a girl (G).

(a) What proportion will include at least one boy?

(b) What fraction will have the gender order GBGB?

2.18 In plants, certain mutant genes are known that affect the ability of gametes to participate in fertilization. Suppose that an allele A is such a mutation and that pollen cells bearing the A allele are only half as likely to survive and participate in fertilization as pollen cells bearing the a allele. What is the expected ratio of $AA : Aa : aa$ plants in the F_2 generation in a monohybrid cross? (*Hint:* Use a Punnett square.)

2.19 Assume that the trait in the accompanying pedigree is due to simple Mendelian inheritance.

(a) Is it likely to be due to a dominant allele or a recessive allele? Explain.

geNETics on the web

(b) What is the meaning of the double horizontal line connecting III-1 with III-2?

(c) What is the biological relationship between III-1 and III-2?

(d) If the allele responsible for the condition is rare, what are the most likely genotypes of all of the persons in the pedigree in generations I, II, and III? (Use A and a for the dominant and recessive alleles, respectively.)

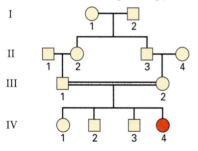

2.20 *Meiotic drive* is a phenomenon observed occasionally in which a heterozygous genotype does not produce a 1: 1 proportion of functional gametes, usually because one of the gametic types is not formed or fails to function. Suppose that an allele D shows meiotic drive such that heterozygous Dd genotypes form 3/4 D-bearing and 1/4 d-bearing functional gametes. What is the expected ratio of genotypes in the F_2 generation of a monohybrid cross under the assumptions stipulated below? (*Hint:* Use Punnett squares.)

(a) The meiotic drive occurs equally in both sexes.

(b) The meiotic drive occurs only in females.

further readings

Bowler, P. J. 1989. *The Mendelian Revolution.* Baltimore, MD: Johns Hopkins University Press.

Carlson, E. A. 1987. *The Gene: A Critical History.* 2d ed. Philadelphia: Saunders.

Dunn, L. C. 1965. *A Short History of Genetics.* New York: McGraw-Hill.

Hartl, D. L., and V. Orel. 1992. What did Gregor Mendel think he discovered? *Genetics* 131: 245.

Hawley, R. S., and C. A. Mori. 1999. *The Human Genome: A User's Guide.* San Diego: Academic Press.

Henig, R. M. 2000. *The Monk in the Garden.* Boston, MA: Houghton Mifflin.

Judson, H. F. 1996. *The Eighth Day of Creation: The Makers of the Revolution in Biology.* Cold Spring Harbor, NY: Cold Spring Harbor Laboratory Press.

Lewontin, R. C. 2000. *It Ain't Necessarily So: The Dream of the Human Genome and Other Illusions.* New York: New York Review of Books.

Maroni, G. 2000. *Molecular and Genetic Analysis of Human Traits.* Malden, MA: Blackwell Science.

Mendel, G. 1866. Experiments in plant hybridization. (Translation.) In *The Origins of Genetics: A Mendel Source Book,* ed. C. Stern and E. Sherwood. 1966. New York: Freeman.

Olby, R. C. 1966. *Origins of Mendelism.* London: Constable.

Orel, V. 1996. *Gregor Mendel: The First Geneticist.* Oxford, England: Oxford University Press.

Orel, V., and D. L. Hartl. 1994. Controversies in the interpretation of Mendel's discovery. *History and Philosophy of the Life Sciences* 16: 423.

Sandler, I. 2000. Development: Mendel's legacy to genetics. *Genetics* 154: 7.

Stern, C., and E. Sherwood. 1966. *The Origins of Genetics: A Mendel Source Book.* New York: Freeman.

Sturtevant, A. H. 1965. *A Short History of Genetics.* New York: Harper & Row.

Unlike Snow White's Seven Dwarfs, these seven little tykes (from left to right, Curious, Cautious, Touchy, Weepy, Crawly, Bawler, and Watcher) will grow bigger. Their differences, already evident, will magnify as each baby's unique genetic constitution unfolds its developmental plan in the context of each baby's unique environment. [© Comstock Images/Alamy Images].

key concepts

- Chromosomes in eukaryotic cells are usually present in pairs.
- The chromosomes of each pair separate in meiosis, one going to each gamete.
- In meiosis, the chromosomes of different pairs undergo independent assortment.
- Chromosomes consist largely of DNA combined with histone proteins.
- In many animals, sex is determined by a special pair of chromosomes, the X and Y.

- Irregularities in the inheritance of an X-linked gene in *Drosophila* gave experimental proof of the chromosomal theory of heredity.
- The progeny of genetic crosses follow the binomial probability formula.
- The chi-square statistical test is used to determine how well observed genetic data agree with expectations from a hypothesis.

3

The Chromosomal Basis of Heredity

chapter organization

3.1 Each species has a characteristic set of chromosomes.

3.2 The daughter cells of mitosis have identical chromosomes.

3.3 Meiosis results in gametes that differ genetically.

3.4 Eukaryotic chromosomes are highly coiled complexes of DNA and protein.

3.5 The centromere and telomeres are essential parts of chromosomes.

3.6 Genes are located in chromosomes.

3.7 Genetic data analysis makes use of probability and statistics.

the human connection
Sick of Telomeres

Chapter Summary
Issues & Ideas
Key Terms & Concepts
Solutions: Step by Step
Concepts in Action: Problems for Solution
Further Readings

geNETics  the web

Gregor Mendel's experiments made it clear that, in heterozygous genotypes, neither allele is altered by the presence of the other. The hereditary units remain stable and unchanged in passing from one generation to the next. Mendel emphasized this finding in a long letter to Carl Nägeli sent on April 18, 1867: "I have never observed gradual transitions between the parental traits or a progressive approach toward one of them. . . . In each generation, the two parental traits appear, separated and unchanged, and there is nothing to indicate that one of them has either inherited or taken over anything from the other." Nevertheless, at the time, the biological basis of the transmission of the hereditary factors from one generation to the next was quite mysterious. Neither the role of the nucleus in reproduction nor the details of cell division had been discovered. Once these phenomena were understood, and when microscopy had improved enough that the chromosomes could be observed and were finally realized to be the bearers of the genes, new understanding came at a rapid pace. This chapter examines the mechanism of chromosome segregation in cell division and the relationship between DNA and chromosomes.

3.1

Each species has a characteristic set of chromosomes.

The importance of the cell nucleus and its contents was suggested as early as the 1840s when Carl Nägeli observed that in dividing cells, the nucleus divided first. This was the same man with whom Mendel would later correspond. Nägeli failed to understand Mendel's findings, and he also failed to see the importance of his own discovery of nuclear division. He regarded the cells in which he saw nuclear division as aberrant. Nevertheless, by the 1870s it had become clear that nuclear division is a universal attribute of cell division. The importance of the nucleus in inheritance was reinforced by the nearly simultaneous discovery that the nuclei of two gametes fuse in the process of fertilization. The next major advance came in the 1880s with the discovery of **chromosomes,** which had been made visible by light microscopy when stained with basic dyes. A few years later, chromosomes were found to segregate by an orderly process into the daughter cells formed by cell division, as well as into the gametes formed by the division of reproductive cells. Finally, three important regularities were observed about the **chromosome complement** (the complete set of chromosomes) of plants and animals:

1. The nucleus of each **somatic cell** (a cell of the body, in contrast with a **germ cell,** or gamete) contains a fixed number of chromosomes typical of the particular species. However, the numbers vary tremendously among species and have little relationship to the complexity of the organism (Table 3.1).

2. The chromosomes in the nuclei of somatic cells are usually present in pairs. For example, the 46 chromosomes of human beings consist of 23 pairs (Figure 3.1). Similarly, the 14 chromosomes of peas consist of 7 pairs. Cells with nuclei of this sort, containing two similar sets of chromo-

Table 3.1

Somatic chromosome numbers of some plant and animal species

Organism	Chromosome number	Organism	Chromosome number
Field horsetail	216	Yeast (*Saccharomyces cerevisiae*)	32
Bracken fern	116	Fruit fly (*Drosophila melanogaster*)	8
Giant sequoia	22	Nematode (*Caenorhabditis elegans*)	11 ♂, 12 ♀
Macaroni wheat	28	House fly	12
Bread wheat	42	Scorpion	4
Fava bean	12	Geometrid moth	224
Garden pea	14	Common toad	22
Mustard cress (*Arabidopsis thaliana*)	10	Chicken	78
Corn (*Zea mays*)	20	Mouse	40
Lily	24	Gibbon	44
Snapdragon	16	Human being	46

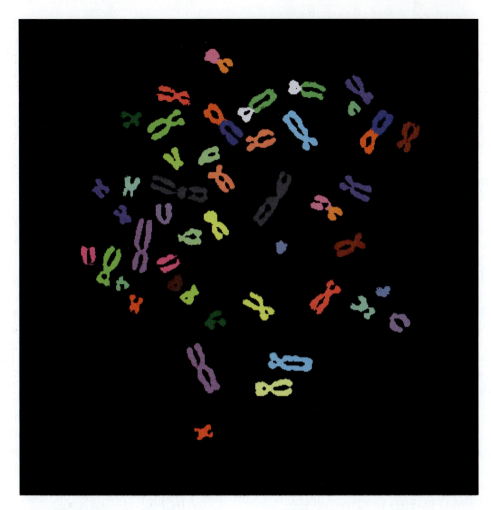

Figure 3.1 Chromosome complement of a human male. There are 46 chromosomes, present in 23 pairs. At the stage of the division cycle in which these chromosomes were observed, each chromosome consists of two identical halves lying side by side longitudinally. Except for the members of one chromosome pair (the XY pair that determines sex), the members of all of the chromosome pairs are the same color because they contain DNA molecules that were labeled with the same mixture of fluorescent dyes. The colors differ from one pair to the next because the dye mixtures differ in color. In some cases, the long and the short arm have been labeled with different colors. [Courtesy of David C. Ward and Michael R. Speicher.]

somes, are called **diploid.** The chromosomes are present in pairs because one chromosome of each pair derives from the maternal parent of the organism and the other from its paternal parent.

3. The germ cells, or gametes, contain only one set of chromosomes, consisting of one member of each of the pairs. The gamete nuclei are said to be **haploid.** The haploid gametes unite in fertilization to produce the diploid state of somatic cells.

In a multicellular organism, which develops from a single fertilized egg, the presence of the diploid chromosome number in somatic cells and the haploid chromosome number in germ cells indicates that there are *two* different processes of nuclear division. One of these, mitosis, maintains chromosome number, whereas the other, meiosis, halves the number. These two processes are examined in the following sections.

3.2

The daughter cells of mitosis have identical chromosomes.

Mitosis is a precise process of nuclear division that ensures that each of two daughter cells receives a diploid complement of chromosomes identical with the diploid complement of the parent cell. Mitosis is usually accompanied by **cytokinesis,** the process in which the cell itself divides to yield two daughter cells. The essential details of mitosis are the same in all organisms, and the basic process is remarkably uniform:

1. Each chromosome is already present as a duplicated structure at the beginning of nuclear division. (The duplication of each chromosome coincides with the replication of the DNA molecule contained within it.)

2. Each chromosome divides longitudinally into identical halves that become separated from each other.

3. The separated chromosome halves move in opposite directions, and each becomes included in one of the two daughter nuclei that are formed.

In a cell that is not undergoing mitosis, the chromosomes are not visible with a light microscope. This stage of the cell cycle is called **interphase.** In preparation for mitosis, the genetic material (DNA) in the chromosomes is replicated during a period of interphase called **S** (Figure 3.2). (The S stands for *synthesis* of DNA.) DNA replication is accompanied by chromosome duplication. Before and after S, there are periods, called G_1 and G_2, respectively, in which DNA replication does not take place. The **cell cycle,** or the life cycle of a cell, is commonly described in terms of these three interphase periods followed by mitosis, **M.** The order of events is therefore $G_1 \rightarrow S \rightarrow G_2 \rightarrow M$, as shown in Figure 3.2. In this representation, the M period includes cytokinesis, which is the division of the cytoplasm into two approximately equal parts, each containing one daughter nucleus. The length of time required for a complete life cycle varies with cell type. In higher eukaryotes, the majority of cells require from 18 to 24 hours. The relative duration of the different periods in the cycle also varies considerably with cell type. Mitosis is usually the shortest period, requiring from 1/2 hour to 2 hours.

The cell cycle itself is controlled by mechanisms that are essentially identical in all eukaryotes. These are discussed in detail in Chapter 13. The transitions from G_1 into S and from G_2 into M

are called **checkpoints** because the transitions are delayed unless key processes have been completed (Figure 3.2). For example, for DNA replication to be initiated at the G_1/S checkpoint, some cell types require that sufficient time must have elapsed since the preceding mitosis, whereas other cell types require that the cell must have attained a particular size. Similarly, for the M phase to begin at the G_2/M checkpoint, DNA replication and repair of any DNA damage must be completed.

■ In mitosis, the replicated chromosomes align on the spindle, and the sister chromatids pull apart.

The diagram in Figure 3.3 shows the essential features of chromosome behavior in mitosis. The process is conventionally divided into four stages: **prophase, metaphase, anaphase,** and **telophase.** The stages have the following characteristics.

1. Prophase In interphase, the chromosomes have the form of extended filaments and cannot be seen as discrete bodies with a light microscope. Except for the presence of one or more conspicuous dark bodies, each called a **nucleolus,** the nucleus has a diffuse, granular appearance. The beginning of prophase is marked by the condensation of chromosomes to form visibly distinct, thin threads within the nucleus. Each chromosome is already longitudinally double, consisting of two closely associated subunits called **chromatids.** The longitudinally bipartite nature of each chromosome is readily seen later in prophase. Each pair of chromatids is the product of the duplication of one chromosome in the S period of interphase. The chromatids in a pair are held together at a specific region of the chromosome called the **centromere.** As prophase progresses, the chromosomes become shorter and thicker, as a result of further coiling. At the end of prophase, the nucleoli disappear and the nuclear envelope, a membrane surrounding the nucleus, abruptly disintegrates.

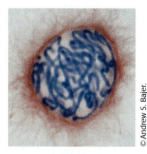

PROPHASE of *Haemanthus.*

© Andrew S. Bajer.

2. Metaphase At the beginning of metaphase, the mitotic **spindle** forms. The spindle is a bipolar structure arching between the centrosomes that consists of fiber-like bundles of microtubules. The spindle fibers attach to each chromosome in the region of the centromere at a structure technically

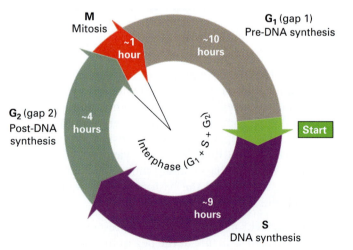

Figure 3.2 The cell cycle of a typical mammalian cell growing in tissue culture with a generation time of 24 hours.

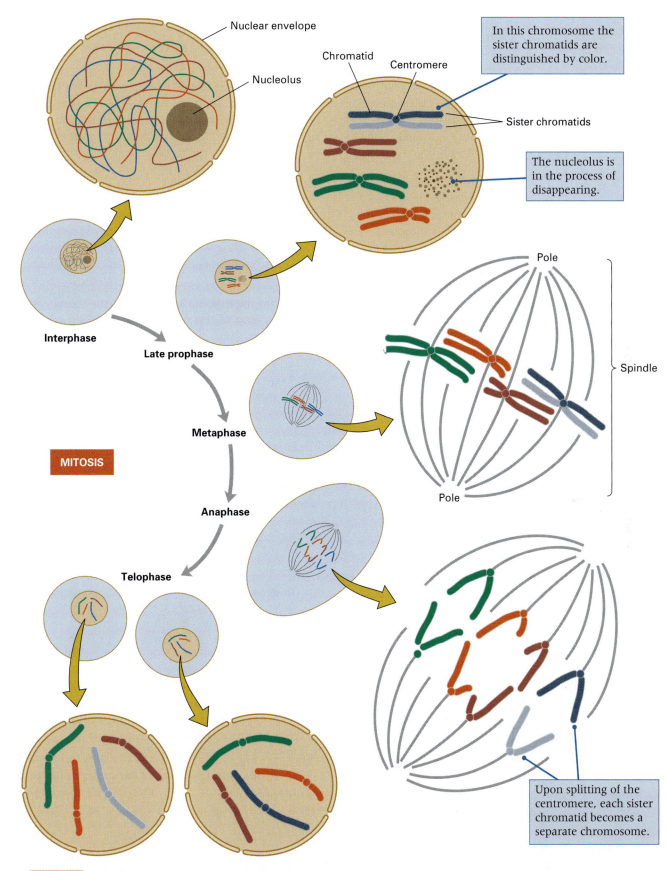

In this chromosome the sister chromatids are distinguished by color.

Nuclear envelope

Nucleolus

Chromatid   Centromere

Sister chromatids

The nucleolus is in the process of disappearing.

Pole

Spindle

Pole

Interphase

Late prophase

Metaphase

MITOSIS

Anaphase

Telophase

Upon splitting of the centromere, each sister chromatid becomes a separate chromosome.

Figure 3.3 Chromosome behavior during mitosis in an organism with two pairs of chromosomes (red/rose versus green/blue). At each stage, the smaller, inner diagram represents the entire cell, and the larger diagram is an exploded view showing the chromosomes at that stage.

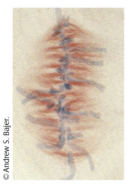

METAPHASE of *Haemanthus.*

known as the **kinetochore.** After the chromosomes are attached to spindle fibers, they move toward the center of the cell until all the kinetochores lie on an imaginary plane equidistant from the spindle poles. This imaginary plane is called the **metaphase plate.** Aligned on the metaphase plate, the chromosomes reach their maximum contraction and are easiest to count and examine for differences in morphology. Proper chromosome alignment is an important checkpoint for controlling the cell cycle at metaphase in both mitosis and meiosis.

3. Anaphase In anaphase, the centromeres divide longitudinally, and the two **sister chromatids** of each chromosome move toward opposite poles of the spindle. Once the centromeres divide, each sister chromatid is regarded as a separate chromosome in its own right. Chromosome movement results in part

ANAPHASE of *Haemanthus.*

from progressive shortening of the spindle fibers attached to the centromeres, which pulls the chromosomes in opposite directions toward the poles. At the completion of anaphase, the chromosomes lie in two groups near opposite poles of the spindle. Each group contains the same number of chromosomes that was present in the original interphase nucleus.

4. Telophase In telophase, a nuclear envelope forms around each compact group of chromosomes, nucleoli are formed, and the spindle disappears. The chromosomes undergo a reversal of condensation until they are no longer visible as discrete entities. The two daughter nuclei slowly assume a typical interphase appearance as the cytoplasm of the cell divides in two by means of a gradually deepening furrow around the periphery. (In plants, a new cell wall is synthesized between the daughter cells and separates them.)

TELOPHASE of *Haemanthus.*

3.3

Meiosis results in gametes that differ genetically.

Meiosis is a mode of cell division in which cells are created that contain only one member of each pair of chromosomes present in the premeiotic cell. When a diploid cell with two sets of chromosomes undergoes meiosis, the result is four daughter cells, each genetically different and each containing one haploid set of chromosomes.

Meiosis consists of two successive nuclear divisions. The essentials of chromosome behavior during meiosis are outlined in Figure 3.4. This outline affords an overview of meiosis as well as an introduction to the process as it takes place in a cellular context.

1. Prior to the first nuclear division, the members of each pair of chromosomes become closely associated along their length (part A). The chromosomes that pair with each other are said to be **homologous** chromosomes. Because each member of a pair of homologous chromosomes is already replicated, it consists of a duplex of two sister chromatids joined at the centromere. The pairing of the homologous chromosomes therefore produces a four-stranded structure.

2. In the first nuclear division, the homologous chromosomes are separated from each other, one member of each pair going to opposite poles of the spindle (part B). Two nuclei are formed, each containing a haploid set of duplex chromosomes (part C).

3. The second nuclear division loosely resembles a mitotic division, *but there is no DNA replication.* At metaphase, the chromosomes align on the metaphase plate, and at anaphase, the chromatids of each chromosome are separated into opposite daughter nuclei (part D). The net effect of the two divisions in meiosis is the creation of four haploid daughter nuclei, each containing the equivalent of a single sister chromatid from each pair of homologous chromosomes (part E).

Figure 3.4 does not show that at the time of chromosome pairing, the homologous chromosomes can exchange genes. The exchanges result in the formation of chromosomes that consist of segments from one homologous chromosome intermixed with segments from the other. In Figure 3.4, the exchanged chromosomes would be depicted as segments of alternating color. The exchange process is one of the critical feature of meiosis, and it will be examined in the next section.

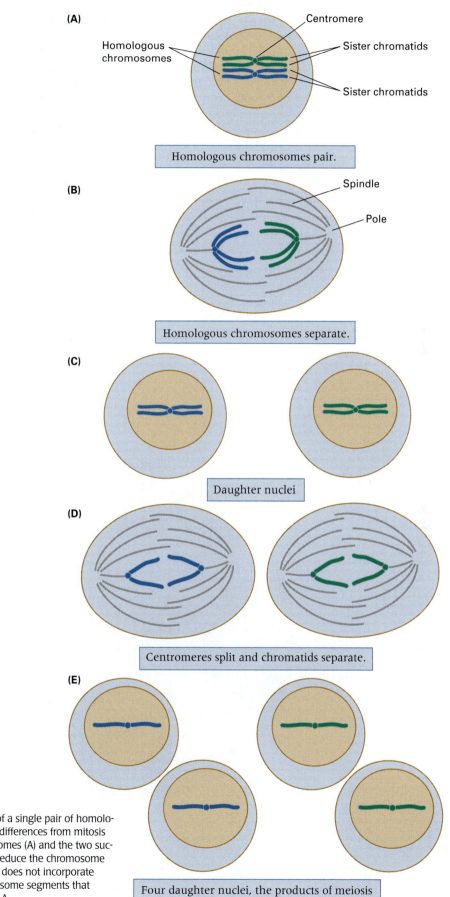

(A)

Centromere

Homologous chromosomes

Sister chromatids

Sister chromatids

Homologous chromosomes pair.

(B)

Spindle

Pole

Homologous chromosomes separate.

(C)

Daughter nuclei

(D)

Centromeres split and chromatids separate.

(E)

Four daughter nuclei, the products of meiosis

Figure 3.4 Overview of the behavior of a single pair of homologous chromosomes in meiosis. The key differences from mitosis are the pairing of homologous chromosomes (A) and the two successive nuclear divisions (B and D) that reduce the chromosome number by half. For clarity, this diagram does not incorporate crossing-over, an interchange of chromosome segments that takes place at the stage depicted in part A.

In animals, meiosis takes place in specific cells called *meiocytes,* a general term for the primary oocytes and spermatocytes in the gamete-forming tissues (Figure 3.5). The oocytes form egg cells and the spermatocytes form sperm cells. Although the process of meiosis is similar in all sexually reproducing organisms, in the female of both animals and plants, only one of the four products develops into a functional cell (the other three disintegrate). In animals, the products of meiosis form gametes (sperm or eggs).

In plants, the situation is slightly more complicated:

1. The products of meiosis typically form *spores,* which undergo one or more mitotic divisions to produce a haploid *gametophyte* organism. The gametophyte produces gametes by mitotic division of a haploid nucleus (Figure 3.6).

The world has about 1500 named species of scorpions, a few with a genome consisting of only two pairs of chromosomes. Yet they all have about the same number of genes.

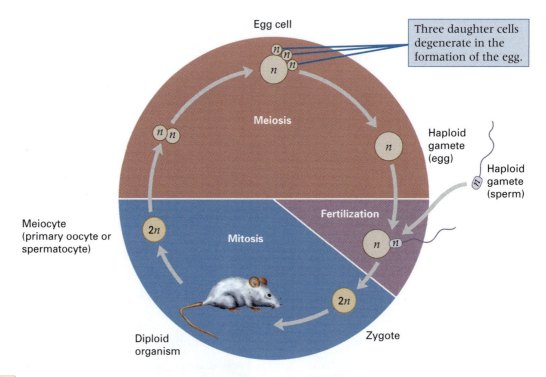

Figure 3.5 The life cycle of a typical animal. The number n is the number of chromosomes in the haploid chromosome complement. In males, the four products of meiosis develop into functional sperm; in females, only one of the four products develops into an egg.

2. Fusion of haploid gametes creates a diploid zygote that develops into the *sporophyte* plant, which undergoes meiosis to produce spores and so restarts the cycle.

Meiosis is a more complex and considerably longer process than mitosis and usually requires days or even weeks. The entire process of meiosis is illustrated in its cellular context in Figure 3.7 (pages 82–83). The essence is that *meiosis consists of two divisions of the nucleus but only one replication of the chromosomes.* The nuclear divisions—called the *first meiotic division* and the *second meiotic division*—can be separated into a sequence of stages similar to those used to describe mitosis. The distinctive events of this important process occur during the first division of the nucleus; these events are described in the following section.

© Ablestock

Amphibian eggs are large, and embryonic development takes place outside the mother's body. These features rank amphibians among the best organisms for studies of the genetic control of development.

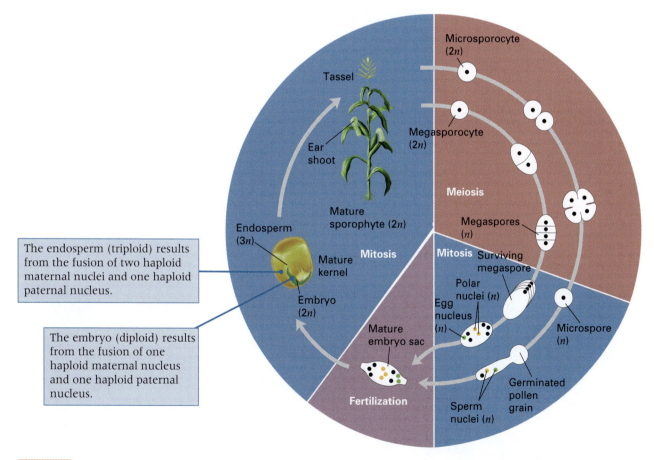

The endosperm (triploid) results from the fusion of two haploid maternal nuclei and one haploid paternal nucleus.

The embryo (diploid) results from the fusion of one haploid maternal nucleus and one haploid paternal nucleus.

Figure 3.6 The life cycle of corn, *Zea mays.* As is typical in higher plants, the diploid spore-producing (sporophyte) generation is conspicuous, whereas the gamete-producing (gametophyte) generation is microscopic. The egg-producing spore is the *mega-spore* and the sperm-producing spore is the *microspore.* Nuclei participating in meiosis and fertilization are shown in yellow and green.

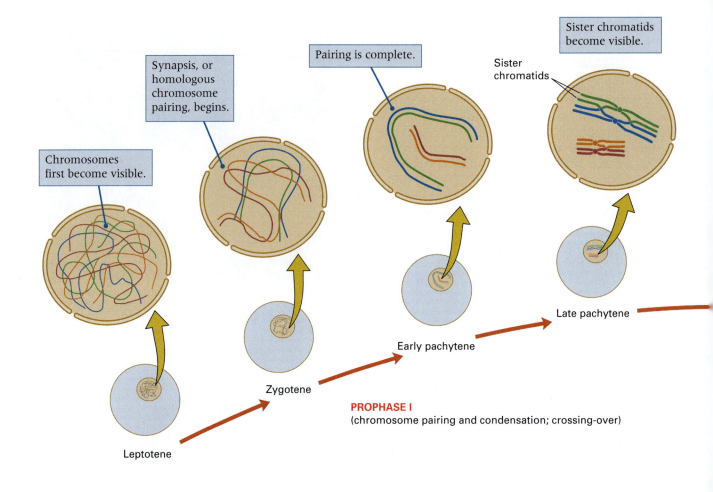

Chromosomes first become visible.

Synapsis, or homologous chromosome pairing, begins.

Pairing is complete.

Sister chromatids become visible.

Sister chromatids

Leptotene

Zygotene

Early pachytene

Late pachytene

PROPHASE I
(chromosome pairing and condensation; crossing-over)

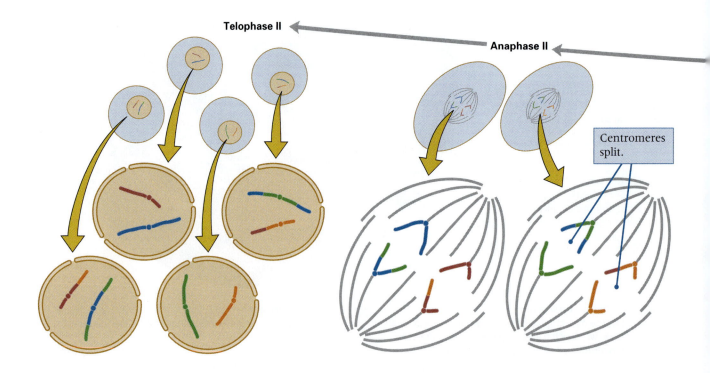

Telophase II

Anaphase II

Centromeres split.

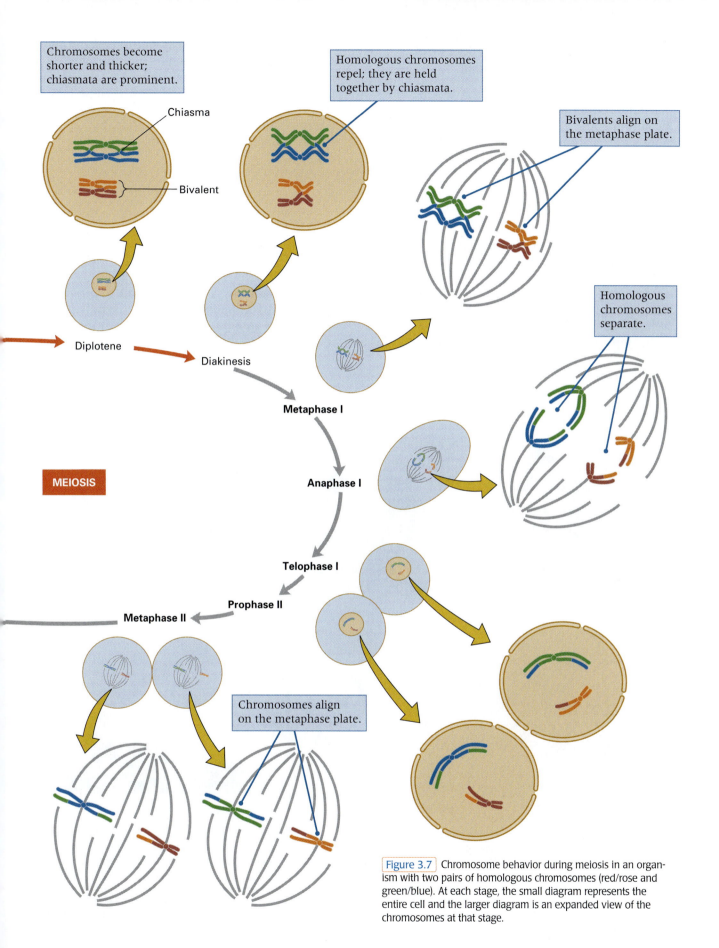

Chromosomes become shorter and thicker; chiasmata are prominent.

Chiasma

Bivalent

Homologous chromosomes repel; they are held together by chiasmata.

Bivalents align on the metaphase plate.

Homologous chromosomes separate.

Diplotene

Diakinesis

Metaphase I

Anaphase I

MEIOSIS

Telophase I

Prophase II

Metaphase II

Chromosomes align on the metaphase plate.

Figure 3.7 Chromosome behavior during meiosis in an organism with two pairs of homologous chromosomes (red/rose and green/blue). At each stage, the small diagram represents the entire cell and the larger diagram is an expanded view of the chromosomes at that stage.

■ The first meiotic division reduces the chromosome number by half.

The first meiotic division (meiosis I) is sometimes called the **reductional division** because it divides the chromosome number in half. By analogy with mitosis, the first meiotic division can be split into the four stages of **prophase I, metaphase I, anaphase I,** and **telophase I.** These stages are generally more complex than their counterparts in mitosis. The stages and substages can be visualized with reference to Figure 3.7 and Figure 3.8 .

1. Prophase I   This long stage lasts several days in most higher organisms and is commonly divided into five substages: *leptotene, zygotene, pachytene, diplotene,* and *diakinesis.* These are descriptive terms that indicate the appearance of the chromosomes at each substage.

In **leptotene,** which literally means "thin thread," the chromosomes first become visible as long, thread-like structures. The pairs of sister chro-

matids can be distinguished by electron microscopy. In this initial phase of condensation of the chromosomes, numerous dense granules appear at irregular intervals along their length. These localized contractions, called *chromomeres,* have a characteristic number, size, and position in a given chromosome (Figure 3.8, part A).

Leptotene

The **zygotene** ("paired thread") period is marked by the lateral pairing, or **synapsis,** of homologous chromosomes, beginning at the chromosome tips. As the pairing process proceeds along the length of the chromosomes, it results in a precise chromomere-by-chromomere association

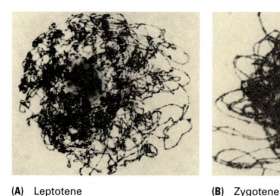

(A) Leptotene   **(B)** Zygotene   **(C)** Early pachytene

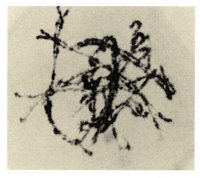

(D) Late pachytene   **(E)** Diplotene   **(F)** Detail of synapsis

Figure 3.8   Substages of prophase of the first meiotic division in microsporocytes of a lily (*Lilium longiflorum*). (A) Leptotene, in which condensation of the chromosomes is initiated and bead-like chromomeres are visible along the length of the chromosomes. (B) Zygotene, in which pairing (synapsis) of homologous chromosomes occurs (paired and unpaired regions can be seen particularly at the lower left in this photograph). (C) Early pachytene, in which synapsis is completed and crossing-over between homologous chromosomes occurs. (D) Late pachytene, showing the continued shortening and thickening of the chromosomes. (E) Diplotene, characterized by mutual repulsion of the paired homologous chromosomes, which remain held together at one or more cross points (chiasmata) along their length; diplotene is followed by diakinesis (not shown), in which the chromosomes reach their maximum contraction. (F) Zygotene (at higher magnification in another cell) showing paired homologs and matching of chromomeres during synapsis. [Courtesy of Marta Walters (A, B, C, E, F) and Herbert Stern (D).]

(Figure 3.8, part B and part F). Each pair of synapsed homologous chromosomes is referred to as a **bivalent.**

Zygotene

During **pachytene** (Figure 3.8, part C and part D), condensation of the chromosomes continues. The term literally means "thick thread," and throughout this period, the chromosomes continue to shorten and thicken (Figure 3.7). By late pachytene, it can sometimes be seen that each bivalent (that is, each set of paired chromosomes) actually consists of a **tetrad** of four chromatids, but the two sister chromatids of each chromosome are usually juxtaposed very tightly. The important event of genetic exchange, **crossing-over,** takes place during pachytene, but crossing-over does not become apparent until the transition to diplotene. In Figure 3.7, the sites of exchange are indicated by the points where chromatids of different colors cross over each other.

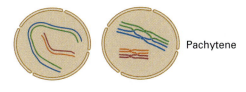

Pachytene

At the onset of **diplotene** ("double thread"), the synapsed chromosomes begin to separate, and the diplotene chromosomes are clearly double (Figure 3.8, part E). However, the homologous chromosomes remain held together at intervals along their length by cross-connections resulting from crossing-over. Each cross-connection, called a **chiasma** (plural, *chiasmata*), is formed by a breakage and rejoining between nonsister chromatids. As shown in the chromosome and diagram in Figure 3.9, *a chiasma results from physical exchange between chromatids of homologous chromosomes*. In normal meiosis, each bivalent usually has at least one chiasma, and bivalents of long chromosomes often have three or more.

Diplotene

The final period of prophase I is **diakinesis,** in which the homologous chromosomes seem to repel each other and the segments not connected by chiasmata move apart. (Diakinesis means "moving apart.") It is at this substage that the chromosomes attain their maximum condensation. The homologous chromosomes in a bivalent remain connected by at least one chiasma, which persists until the first meiotic anaphase. Near the end of diakinesis, the formation of a spindle is initiated and the nuclear envelope breaks down.

Diakinesis

(A)

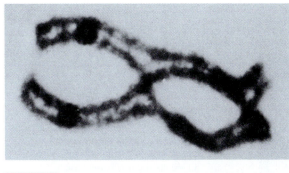

(B)

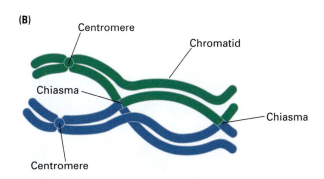

Figure 3.9 Light micrograph (A) and interpretative drawing (B) of a bivalent consisting of a pair of homologous chromosomes. This bivalent was photographed at late diplotene in a spermatocyte of the salamander *Oedipina poelzi*. It shows two chiasmata where the chromatids of the homologous chromosomes appear to exchange pairing partners. [From F. W. Stahl. 1964. *The Mechanics of Inheritance.* Englewood Cliffs, NJ: Prentice-Hall; courtesy of James Kezer.]

2. Metaphase I The bivalents become positioned with the centromeres of the two homologous chromosomes on opposite sides of the metaphase plate (part A, Figure 3.10). As each bivalent moves onto the metaphase plate, its centromeres are oriented at random with respect to the poles of the spindle. As shown in Figure 3.11, the bivalents formed from nonhomologous pairs of chromosomes can be oriented on the metaphase plate in either of two ways. The orientation of the centromeres determines which member of each bivalent will subsequently move to each pole. If each of the nonhomologous chromosomes is heterozygous for a pair of alleles, one type of alignment results in $A\ B$ and $a\ b$ gametes, and the other type results in $A\ b$ and $a\ B$ gametes (Figure 3.11). Because the metaphase alignment takes place at random, the two types of alignment—and therefore the four types of gametes—are equally frequent. The ratio of the four types of gametes is $1:1:1:1$, which means that the A, a and B, b pairs of alleles undergo independent assortment. In other words,

Metaphase I

key concept

Genes on different chromosomes undergo independent assortment because nonhomologous chromosomes align at random on the metaphase plate in meiosis I.

The experimental demonstration of this principle in 1913 gave strong support to the idea, already accepted by many geneticists, that the chromosomes were the cellular objects that contained the genetic material. These studies were carried out by Eleanor Carothers working with the grasshopper, *Brachystola magna.*

3. Anaphase I In this stage, homologous chromosomes, each composed of two chromatids joined at an undivided centromere, separate from one another and move to opposite poles of the spindle (Figure 3.10, part B). Chromosome separation at anaphase is the cellular basis of the segregation of alleles.

Anaphase I

key concept

The physical separation of homologous chromosomes in anaphase is the physical basis of Mendel's principle of segregation.

Note, however, that the centromeres of the sister chromatids are stuck together tightly and behave as a single unit, owing to the presence of a protein "glue" that holds them together until the onset of anaphase II.

4. Telophase I At the completion of anaphase I, a haploid set of chromosomes consisting of one

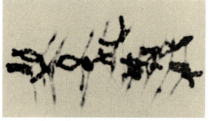

(A) Metaphase I

(B) Anaphase I

(C) Metaphase II (telophase I and prophase II not shown)

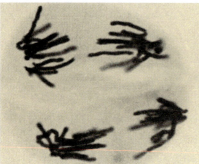

(D) Anaphase II

(E) Telophase II

Figure 3.10 Later meiotic stages in microsporocytes of the lily *Lilium longiflorum.* (A) Metaphase I. (B) Anaphase I. (C) Metaphase II. (D) Anaphase II. (E) Telophase II. Cell walls have begun to form in telophase, which will lead to the formation of four pollen grains. [Courtesy of Herbert Stern.]

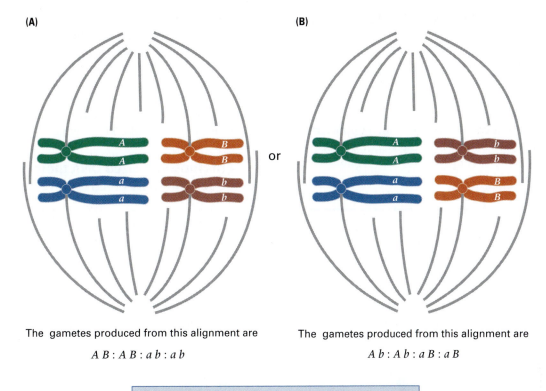

(A) The gametes produced from this alignment are

$A\,B : A\,B : a\,b : a\,b$

or

(B) The gametes produced from this alignment are

$A\,b : A\,b : a\,B : a\,B$

Because the alignments are equally likely, the overall ratio of gametes is

$A\,B : A\,b : a\,B : a\,b = 1 : 1 : 1 : 1$

This ratio is characteristic of independent assortment.

Figure 3.11 Random alignment of nonhomologous chromosomes at metaphase I results in the independent assortment of genes on nonhomologous chromosomes.

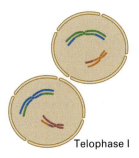

Telophase I

homolog from each bivalent is located near each pole of the spindle (Figure 3.9). In telophase, the spindle breaks down, and, depending on the species, either a nuclear envelope briefly forms around each group of chromosomes or the chromosomes enter the second meiotic division after only a limited uncoiling.

■ The second meiotic division is equational.

The second meiotic division (meiosis II) is sometimes called the **equational division** because the chromosome number remains the same in each cell before and after the second division. In some species, the chromosomes pass directly from telophase I to **prophase II** without loss of condensation; in others, there is a brief pause between the two meiotic divisions and the chromosomes may "decondense" (uncoil) somewhat. *Chromosome replication never takes place between the two divisions;* the chromosomes present at the beginning of the second division are identical with those present at the end of the first division.

After a short prophase (prophase II) and the formation of second-division spindles, the centromeres of the chromosomes in each nucleus become aligned on the central plane of the spindle at **metaphase II** (Figure 3.10, part C). In **anaphase II,** the centromeres divide longitudinally and the chromatids of each chromosome move to opposite poles of the spindle (Figure 3.10, part D). Once the centromere has split at anaphase II, each chromatid is considered a separate chromosome.

Telophase II (Figure 3.10, part E) is marked by a transition to the interphase condition of the chromosomes in the four haploid nuclei, accompanied by division of the cytoplasm. Thus the second meiotic division superficially resembles a mitotic division. However, there is an important difference:

key concept

The chromatids of a chromosome are usually not genetically identical along their entire length because of crossing-over associated with the formation of chiasmata during prophase of the first division.

This principle has consequences of such importance for genetic analysis that it is the major subject of Chapter 4.

3.4

Eukaryotic chromosomes are highly coiled complexes of DNA and protein.

A eukaryotic chromosome contains a single DNA molecule of enormous length. For example, the largest chromosome in the *D. melanogaster* genome has a DNA content of about 65,000 kb (6.5×10^7

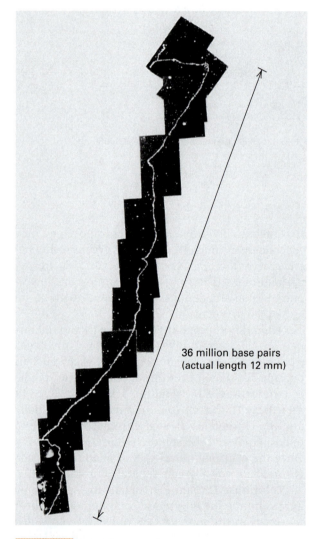

36 million base pairs
(actual length 12 mm)

Figure 3.12 Autoradiogram of a DNA molecule from *D. melanogaster.* The molecule is 12 mm long (approximately 36,000 kb). [From R. Kavenoff, L. C. Klotz, and B. H. Zimm. 1974. *Cold Spring Harbor Symp. Quant. Biol.* 38: 4.]

nucleotide pairs), which is equivalent to a continuous linear duplex about 22 mm long. (Recall that the abbreviation *kb* stands for *kilobase pairs;* 1 kb equals 1000 base pairs.) These long molecules usually fracture during isolation, but some fragments that are recovered are still very long. Figure 3.12 is an autoradiograph of radioactively labeled *Drosophila* DNA more than 36,000 kb in length.

■ Chromosome-sized DNA molecules can be separated by electrophoresis.

As we saw in Chapter 2, DNA molecules move in response to an electric field and can be separated according to size by means of electrophoresis, smaller fragments moving faster. Conventional electrophoresis, in which the orientation and strength of the electric field are held constant, can separate molecules smaller than about 20 kb. Under these conditions, all molecules larger than about 20 kb have the same electrophoretic mobility and so form a single band in the gel.

Simple modifications of the electrophoresis procedure result in separation among much larger DNA molecules, including those present in small chromosomes. The most common modifications alter the geometry of the electric field at periodic intervals during the course of the electrophoretic separation. In one type of large-fragment electrophoresis, the apparatus is similar to that for conventional electrophoresis, but the orientation of the electric field is reversed periodically. The improved separation of large DNA molecules apparently results from the additional time that it takes for large molecules to reorient themselves when the orientation of the electric field is changed. Figure 3.13 illustrates the separation of the 16 chromosomes from budding yeast, *Saccharomyces cerevisiae*, by means of this procedure. The chromosomes range in size from approximately 200 kb to 2.2 Mb. (The abbreviation *Mb* stands for *megabase pairs;* 1 Mb equals 1,000,000 base pairs.) Electrophoretic separation yields a visual demonstration that each of the chromosomes contains a single DNA molecule that runs continuously throughout its length.

■ The nucleosome is the basic structural unit of chromatin.

The DNA of all eukaryotic chromosomes is associated with numerous protein molecules in a stable ordered aggregate called **chromatin.** Some of the proteins present in chromatin determine chromosome structure and the changes in structure that occur during the division cycle of the cell. Other chromatin proteins appear to have important roles in regulating chromosome functions.

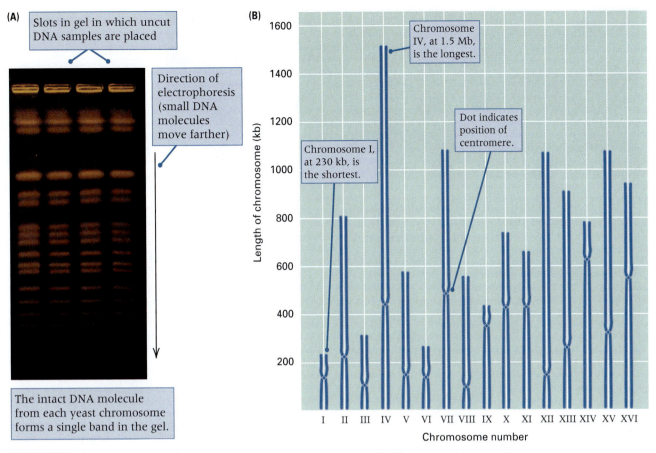

(A)

Slots in gel in which uncut DNA samples are placed

Direction of electrophoresis (small DNA molecules move farther)

The intact DNA molecule from each yeast chromosome forms a single band in the gel.

(B)

Length of chromosome (kb)

Chromosome IV, at 1.5 Mb, is the longest.

Chromosome I, at 230 kb, is the shortest.

Dot indicates position of centromere.

I II III IV V VI VII VIII IX X XI XII XIII XIV XV XVI

Chromosome number

Figure 3.13 (A) Separation of chromosomes of the yeast *Saccharomyces cerevisiae* by pulsed-field gel electrophoresis, in which there is regular change in the orientation of the electric field. (B) Histogram of sizes of the 16 yeast chromosomes. [B ©1988 BioRad Laboratories, Inc.; permission to use this image has been granted by Bio-Rad Laboratories Inc.]

The simplest form of chromatin is present in nondividing eukaryotic cells, when chromosomes are not sufficiently condensed to be visible by light microscopy. Chromatin isolated from such cells is a complex aggregate of DNA and proteins. The major class of proteins comprises the **histone** proteins. Histones are largely responsible for the structure of chromatin. Five major types—H1, H2A, H2B, H3, and H4—are present in the chromatin of nearly all eukaryotes in amounts about equal in mass to that of the DNA. Histones are small proteins (100 to 200 amino acids) that differ from most other proteins in that from 20 to 30 percent of the amino acids are lysine and arginine, both of which have a positive charge. (Only a few percent of the amino acids of a typical protein are lysine and arginine.) The positive charges enable histone molecules to bind to DNA, primarily by electrostatic attraction to the negatively charged phosphate groups in the sugar–phosphate backbone of DNA. Placing chromatin in a solution with a high salt concentration (for example, 2 molar NaCl) to eliminate the elec-

trostatic attraction causes the histones to dissociate from the DNA. Histones also bind tightly to each other; both DNA–histone and histone–histone binding are important for chromatin structure.

The histone molecules from different organisms are remarkably similar to one another, with the exception of H1. In fact, the amino acid sequences of H3 molecules from widely different species are almost identical. For example, the sequences of H3 of cow chromatin and pea chromatin differ by only 4 of 135 amino acids. The H4 proteins of all organisms also are quite similar; cow and pea H4 differ by only 2 of 102 amino acids. There are few other proteins whose amino acid sequences vary so little from one species to the next. When the variation is very small between organisms, one says that the sequence is highly *conserved*. The extraordinary conservation in histone composition through hundreds of millions of years of evolutionary divergence is consistent with the important role of these proteins in the structural organization of eukaryotic chromosomes.

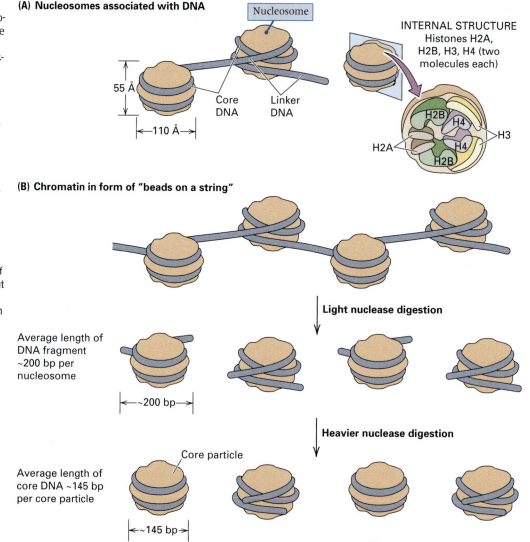

Figure 3.14 Dark-field electron micrograph of chromatin showing the beaded structure at low salt concentration. The beads have diameters of about 100 Å. [Courtesy of Ada Olins and Donald E. Olins.]

In the electron microscope, chromatin resembles a regularly beaded thread (Figure 3.14). The bead-like units in chromatin are called **nucleosomes.** The organization of the nucleosomes in chromatin is illustrated in part A of Figure 3.15. Each unit has a definite composition, consisting of two molecules each of H2A, H2B, H3, and H4; a segment of DNA containing about 200 nucleotide pairs; and one molecule of histone H1. The complex of two subunits each of H2A, H2B, H3, and H4, as well as part

of the DNA, forms each "bead," and the remaining DNA bridges between the beads. Histone H1 also appears to play a role in bridging between the beads, but it is not shown in Figure 3.15, part A.

Brief treatment of chromatin with certain DNases yields a collection of small particles of quite uniform size consisting only of histones and DNA. The DNA fragments in these particles are of lengths equal to about 200 nucleotide pairs or small multiples of that unit size (the precise size varies with

Figure 3.15
(A) Organization of nucleosomes. The DNA molecule is wound one and three-fourths turns around a histone octamer called the core particle. If H1 were present, it would bind to the octamer surface and to the linkers, causing the linkers to cross. (B) Effect of treatment with micrococcal nuclease. Brief treatment cleaves the DNA between the nucleosomes and results in core particles associated with histone H1 and approximately 200 base pairs of DNA. More extensive treatment results in loss of H1 and digestion of all but 145 base pairs of DNA in intimate contact with each core particle

(A) Nucleosomes associated with DNA

Nucleosome

INTERNAL STRUCTURE
Histones H2A, H2B, H3, H4 (two molecules each)

H2B, H4, H3, H2A, H4, H2B

55 Å
110 Å
Core DNA
Linker DNA

(B) Chromatin in form of "beads on a string"

Light nuclease digestion

Average length of DNA fragment ~200 bp per nucleosome

~200 bp

Heavier nuclease digestion

Core particle

Average length of core DNA ~145 bp per core particle

~145 bp

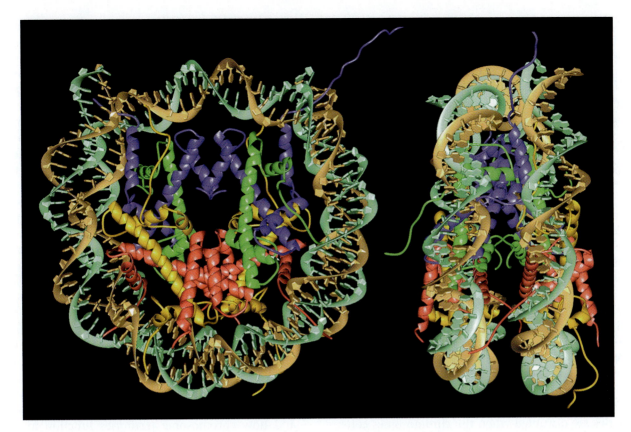

Figure 3.16 Subunit structure of the nucleosome core particle with the DNA duplex wrapped around it. The color coding for the histone monomers is H2A, yellow; H2B, red; H3, blue; and H4, green. [Courtesy of Timothy J. Richmond. Reprinted by permission from *Nature* 389: 251, K. Luger, A. W. Mäder, R. K. Richmond, D. F. Sargent, and T. J. Richmond. © 1997 Macmillan Magazines Ltd.]

species and tissue). These particles result from cleavage of the linker DNA segments between the beads (Figure 3.15, part B). More extensive treatment with DNase results in loss of the H1 histone and digestion of all the DNA except that protected by the histones in the bead. The resulting structure is called a **core particle,** the detailed structure of which is shown in Figure 3.16. It consists of an octamer of pairs of H2A, H2B, H3, and H4, around which the remaining DNA, approximately 145 base pairs, is wound in about one and three-fourths turns. Each nucleosome is composed of a core particle, additional DNA called *linker DNA* that links adjacent core particles (the linker DNA is removed by extensive nuclease digestion), and one molecule of H1 that binds to the histone octamer and to the linker DNA.

At the salt concentration present in living cells, the nucleosome fiber compacts into a shorter, thicker fiber with an average diameter ranging from 300 to 350 Å; this is called the **30-nm chromatin fiber** (Figure 3.17 , part A). In forming the 30-nm chromatin fiber, the string of nucleosomes forms a series of stacked right-handed coils (part B) in which each nucleosome is attached to its neighbor

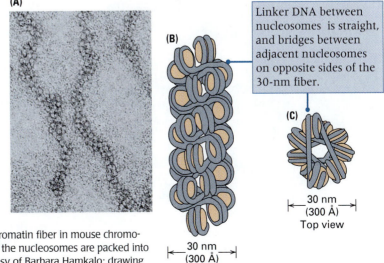

Linker DNA between nucleosomes is straight, and bridges between adjacent nucleosomes on opposite sides of the 30-nm fiber.

Figure 3.17 (A) Electron micrograph of the 30-nm chromatin fiber in mouse chromosomes. (B) and (C) Model showing the manner in which the nucleosomes are packed into the 30-nm chromatin fiber. [Electron micrograph courtesy of Barbara Hamkalo; drawing after B. Dorigo et al. 2004. *Science* 306: 1571.]

3.4 Eukaryotic chromosomes are highly coiled complexes of DNA and protein   91

by linker DNA that stretches nearly linearly across to the opposite side of the coil. Looking down at the 30-nm fiber from the top (part C), one can trace the path of the linker DNA as it travels down the length of the fiber. In each revolution around the fiber axis, the path of the linker DNA closely approximates the shape of a seven-pointed star.

■ Chromatin fibers form discrete chromosome territories in the nucleus.

In the nucleus of a nondividing cell, the 30-nm chromatin fiber is organized into higher order structures that can be visualized using modern methods of optical sectioning and image reconstruction. Figure 3.18 shows a computer-generated image of 30-nm chromatin fibers within the nucleus of a nondividing cell. The chromatin fibers are folded into small *chromatin loops* with a DNA content of approximately 100 kb each, and these are further organized into *chromatin domains* with a DNA content of approximately 1 Mb each. Each chromosome arm occupies a discrete **chromosome territory,** denoted by the different colors. In cells cycling through mitosis, the chromosome territories are disrupted when the chromosomes condense and the cell divides, but they are reconstituted again in the next interphase. However, the chromosome territories may differ in position in different cell types as well as in the same cell type at different times in development.

Chromosome territories are correlated with gene densities. The territories of chromatin domains con-

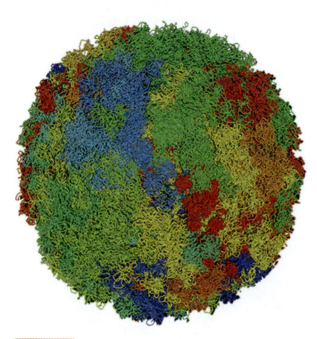

Figure 3.18 Computer-generated image of chromosome territories formed by 30-nm chromatin fibers within the nucleus of a nondividing cell. [Courtesy of Dr. Tobias A. Knoch.]

taining relatively few genes tend to be located near the periphery of the nucleus or near the nucleolus, whereas the territories of chromosome domains that are relatively gene rich tend to be located toward the interior of the nucleus. For example, human chromosome 18 (85 Mb in size) is relatively gene poor whereas chromosome 19 (67 Mb in size) is relatively gene rich. In the nucleus, chromosome 18 territories tend to be at the nuclear periphery whereas those of chromosome 19 tend to be in the interior.

The spaces between the chromatin domains form a network of channels, like the holes permeating through a sponge. The channels are large enough to allow passage of the molecular machinery for replication, transcription, and RNA processing. Evidence suggests that these molecules gain access to chromatin by means of passive diffusion. Replication, transcription, and RNA processing are all ordered processes. DNA replication takes place in small discrete regions that exhibit a reproducible temporal and spatial pattern, and transcription takes place in a few hundred discrete locations. However, many important details are still unknown about the organization of chromatin in the nucleus and how chromosome territories function in the coordination of the central molecular processes of replication, transcription, and RNA processing.

■ The metaphase chromosome is a hierarchy of coiled coils.

The hierarchical nature of chromosome structure is illustrated in Figure 3.19 . Assembly of DNA and histones is the first level, resulting in a sevenfold reduction in length of the DNA and the the formation of a beaded flexible fiber 110 Å (11 nm) wide (part B), roughly five times the width of free DNA (part A). The structure of chromatin varies with the concentration of salts, and the 110-Å fiber is present only when the salt concentration is quite low. In the living cell, this is usually compacted into the 30-nm chromatin fiber (part C), which in the interphase nucleus is folded into 100-kb chromatin loops that are organized into 1-Mb chromatin domains that form the chromosome territories.

In cells cycling through mitosis, the interphase chromatin organization is replaced by a more compact organization in which the 30-nm chromatin fiber condenses into a chromatid of the metaphase chromosome (Figure 3.19, parts D through F). Little is known about this process other than that it seems to proceed in stages, and there is no strong evidence supporting any of the particular coiled structures greater than the 30-nm chromatin fiber. In electron micrographs of isolated metaphase chromosomes from which histones have been removed, the partly unfolded

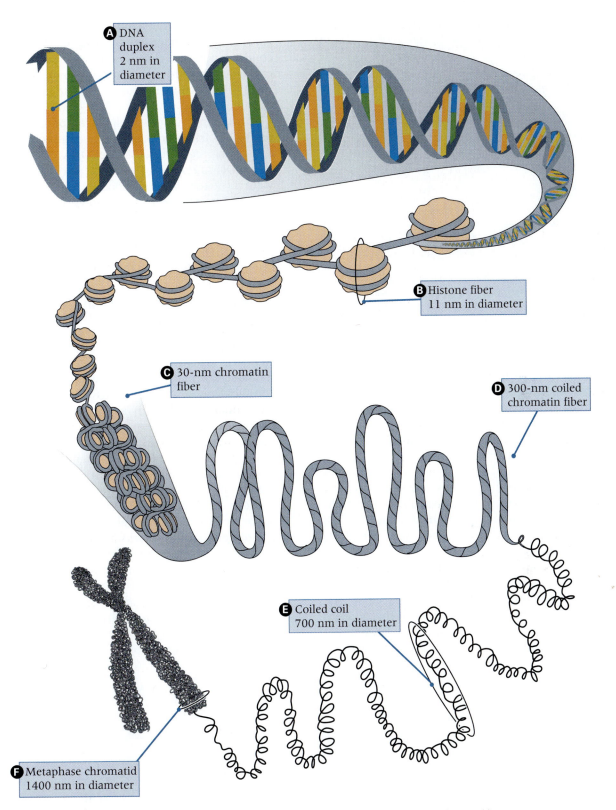

A DNA duplex 2 nm in diameter

B Histone fiber 11 nm in diameter

C 30-nm chromatin fiber

D 300-nm coiled chromatin fiber

E Coiled coil 700 nm in diameter

F Metaphase chromatid 1400 nm in diameter

Figure 3.19 Condensation of DNA (A) and chromatin (B through E) to form a metaphase chromosome (F). The details of the structures in D–F are hypothetical.

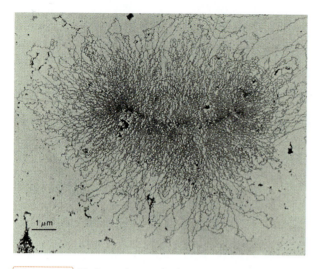

Figure 3.20 Electron micrograph of a partially disrupted anaphase chromosome of the milkweed bug *Oncopeltus fasciatus,* showing multiple loops of 30-nm chromatin at the periphery. [From V. Foe, H. Forrest, L. Wilkinson, and C. Laird. 1982. *Insect Ultrastructure* 1: 222.]

DNA has the form of an enormous number of loops that seem to extend from a central core, or **scaffold,** composed of nonhistone chromosomal proteins (Figure 3.20). Electron microscopic studies of chromosome condensation in mitosis and meiosis suggest that the scaffold extends along the chromatid and that the 30-nm fiber becomes arranged into a helix of loops radiating from the scaffold. Details are not known about the additional folding that is required of the fiber in each loop to produce the fully condensed metaphase chromosome.

The genetic significance of the compaction of DNA and protein into chromatin and ultimately into the chromosome is that it greatly facilitates the movement of the genetic material during nuclear division. Relative to a fully extended DNA molecule, the length of a metaphase chromosome is

reduced by a factor of approximately 10^4 because of chromosome condensation. Without chromosome condensation, the chromosomes would become so entangled that there would be many more abnormalities in the distribution of genetic material into daughter cells.

An analogy may be helpful in appreciating the prodigious feat of packaging that chromosome condensation represents. If the DNA molecule in human chromosome 1 (the longest chromosome) were a cooked spaghetti noodle 1 mm in diameter, it would stretch for 25 miles; in chromosome condensation, this noodle is gathered together, coil upon coil, until at metaphase it is a canoe-sized tangle of spaghetti 16 feet long and 2 feet wide. After cell division, the noodle is unwound again.

■ **Heterochromatin is rich in satellite DNA and low in gene content.**

Certain regions of the chromosome have a dense, compact structure in interphase and are darkly stainable by many standard dyes used to make chromosomes visible. Regions of chromatin that are compact and heavily stained in interphase are known as **heterochromatin.** The rest of the chromatin, which becomes visible only after chromosome condensation in mitosis or meiosis, is called **euchromatin.** Sometimes the heterochromatin remains highly condensed throughout the cell cycle and can be distinguished even at metaphase. The major heterochromatic regions are adjacent to the centromere; smaller blocks are present at the ends of the chromosome arms (the telomeres) and interspersed with the euchromatin (Figure 3.21). At the DNA level, a substantial part of the heterochromatin consists of long tracts of relatively short base sequences, typically from 5 to 500 base pairs in length, each repeated in tandem. The highly repeated sequences are often called **satellite DNA**

(A)

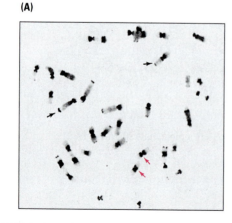

(B)

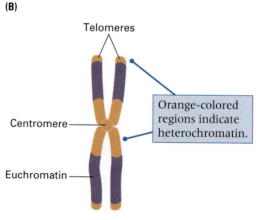

Telomeres

Centromere

Euchromatin

Orange-colored regions indicate heterochromatin.

Figure 3.21 (A) Metaphase chromosomes of the ground squirrel *Ammospermophilus harrissi,* stained to show the heterochromatic regions near the centromere of most chromosomes (red arrows)

and the telomeres of some chromosomes (black arrows). (B) An interpretive drawing. [Micrograph courtesy of T. C. Hsu and Sen Pathak.]

for reasons related to the original method of their isolation. Each satellite sequence has its own distinctive distribution in the heterochromatin. In many species, an entire chromosome—such as the Y chromosome in *Drosophila*—is almost completely heterochromatic.

The genetic content of heterochromatin is summarized in the following generalization:

key concept

The number of genes located in heterochromatin is small relative to the number in euchromatin.

The relatively small number of genes means that many large blocks of heterochromatin are genetically almost inert, or devoid of function. Indeed, heterochromatic blocks can often be rearranged in the genome, duplicated, or even deleted without major phenotypic consequences.

3.5

The centromere and telomeres are essential parts of chromosomes.

Eukaryotic chromosomes contain regions specialized for maneuvering the chromosomes in cell division and for capping the ends. These regions are discussed next.

■ The centromere is essential for chromosome segregation.

The *centromere* is a specific region of the eukaryotic chromosome that becomes visible as a distinct morphological entity along the chromosome during condensation. It serves as a central component of the *kinetochore*, the complex of DNA and proteins to which the spindle fibers attach and move the chromosomes in both mitosis and meiosis. The kinetochore is also the site at which the spindle fibers shorten, causing the chromosomes to move toward the poles.

Figure 3.22 A yeast centromere. (A) Diagram of centromeric DNA showing the major regions (CDE1 through CDE4) common to all yeast centromeres. The letter R stands for any purine (A or G), and the letter N indicates any nucleotide. Inverted-repeat segments in region 3 are indicated by arrows. The sequence of region CDE4 varies from one centromere to the next. (B) Positions of the centromere core and the nucleosomes on the DNA. The DNA is wrapped around histones in the nucleosomes, but the detailed organization and composition of the centromere core are unknown. [After K. S. Bloom, M. Fitzgerald-Hayes, and J. Carbon. 1982. *Cold Spring Harbor Symp. Quant. Biol.* 47: 1175.]

Electron microscopic analysis has shown that in some organisms—for example, the budding yeast *Saccharomyces cerevisiae*—a single spindle-fiber protein is attached to centromeric chromatin. Most other organisms have multiple spindle fibers attached to each centromeric region. The centromeres of *S. cerevisiae* are also unusual in that they are relatively small in terms of DNA content. In most other eukaryotes, including higher eukaryotes, the centromere may contain hundreds of kilobases of DNA.

The chromatin segment of the centromeres of *Saccharomyces cerevisiae* has a unique structure exceedingly resistant to the action of various DNases; it has been isolated as a protein–DNA complex containing from 220 to 250 base pairs. The nucleosomal constitution and DNA base sequences of all of the yeast centromeres have been determined. Several common features of the base sequences are shown in part A of Figure 3.22 . There are four regions, labeled CDE1, CDE2, CDE3, and CDE4. All yeast centromeres have sequences highly similar to those indicated for regions 1, 2, and 3, but the sequence of region CDE4 varies from one centromere to another. Region 2 is noteworthy in

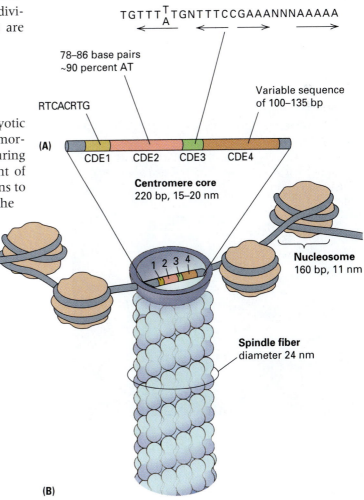

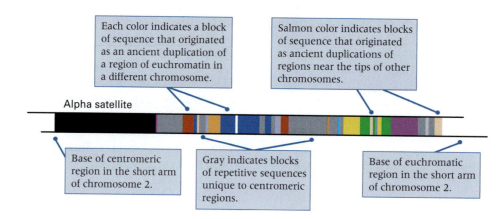

Each color indicates a block of sequence that originated as an ancient duplication of a region of euchromatin in a different chromosome.

Salmon color indicates blocks of sequence that originated as ancient duplications of regions near the tips of other chromosomes.

Alpha satellite

Base of centromeric region in the short arm of chromosome 2.

Gray indicates blocks of repetitive sequences unique to centromeric regions.

Base of euchromatic region in the short arm of chromosome 2.

Figure 3.23 Mosaic ancestry of duplicated sequences present in the centromeric region of the short arm of chromosome 2. The total length of the region shown is approximately 750 kb. [After Xinwei She et al. 2004. *Nature* 430: 857.]

that approximately 90 percent of the base pairs are A–T pairs. The centromeric DNA is contained in a structure (the centromeric core particle) that contains more DNA than a typical yeast nucleosome core particle (which contains 160 base pairs) and is larger. This structure is responsible for the resistance of centromeric DNA to DNase. The spindle fiber is believed to be attached directly to this particle (Figure 3.22, part B).

The centromeres of budding yeast are unusually small and simple. In higher eukaryotes, each centromeric region encompasses a million base pairs or more, to which numerous spindle fibers become attached. Figure 3.23 illustrates the organization of

DNA sequences in the centromeric region of the short arm of human chromosome 2. The organization is typical in that it includes a patchwork of DNA sequences derived from duplicated regions of euchromatin from different chromosomes. How these patchworks are put together is not clear, but the duplications occur at an estimated rate of six to seven events per million years. Chromosome 2 is also typical in containing a large fraction of repetitive satellite DNA sequences. The region at the left (nearest the centromere) consists of tandem repeats of a 170-bp DNA sequence called **alpha satellite.** Most human chromosomes contain 100 to 1000 copies of this sequence (Figure 3.24). DNA sequences needed for spindle fiber attachment are interspersed among the alpha-satellite repeats, and the alpha-satellite repeats themselves appear to contribute to centromere activity.

■ The telomere is essential for the stability of the chromosome tips.

Each end of a linear chromosome is composed of a special DNA–protein structure called a **telomere** that is essential for chromosome stability. Genetic and microscopic observations first indicated that telomeres are special structures. In *Drosophila,* chromosomes without ends formed by x-ray breakage cannot be recovered; in maize, broken chromosome ends frequently fuse with one another and form new chromosomes with abnormal structures (often having two centromeres). As we shall see in Chapter 6, DNA replication cannot begin precisely at the 3' end of a template strand, so the 3' end of a replicated duplex must terminate in a short stretch in which the DNA is single-stranded. This single-stranded overhang is subject to degradation by nucleases. Without some mechanism to restore the digested end, the DNA molecule in a chromosome would become slightly shorter with each replication. There is such a mechanism, and in mutant

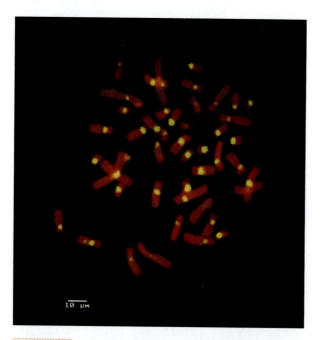

Figure 3.24 Hybridization of human metaphase chromosomes (red) with alpha-satellite DNA. The yellow areas result from hybridization with the labeled DNA. The sites of hybridization of the alpha satellite coincide with the centromeric regions of all 46 chromosomes. [Courtesy of Paula Coelho and Claudio E. Sunkel.]

cells in which the mechanism is defective, each chromosome end does become shorter in each replication until, eventually, there is so much degradation that the cell dies.

The mechanism of restoring the ends of a DNA molecule in a chromosome relies on an enzyme called **telomerase.** This enzyme works by adding tandem repeats of a simple sequence to the 3′ end of a DNA strand. In the ciliated protozoan *Tetrahymena,* in which the enzyme was first discovered, the simple repeating sequence is —TTGGGG-3′, and in humans and other vertebrate organisms, it is —TTAGGG-3′. The tandem repeats of these sequences constitute the telomere. As the repeating telomere sequence is being elongated, DNA replication occurs to synthesize a partner strand, and so, for example, the telomere sequence of the right-

hand end of any *Tetrahymena* chromosome would be a DNA duplex of the form

```
-TTGGGGTTGGGGTTGGGGTTGGGGTTGGGGTTGGGGTTGGGG-3′
-AACCCCAACCCCAACCCCAACCCC-5′
```

with a single-stranded overhang at the 3′ end that can be elongated further by the telomerase.

The role of telomerase in the replication of chromosomal DNA is illustrated in Figure 3.25. Part A represents the duplex DNA in a chromosome, with the telomere sequences shown in red. Because DNA replication cannot start precisely at the 3′ end, the 5′ end of each daughter strand in part B is a little shorter than the template strand from which it was replicated. The unreplicated part of the telomere sequence is subject to degradation by nucleases. The 3′ end of each daughter molecule also has

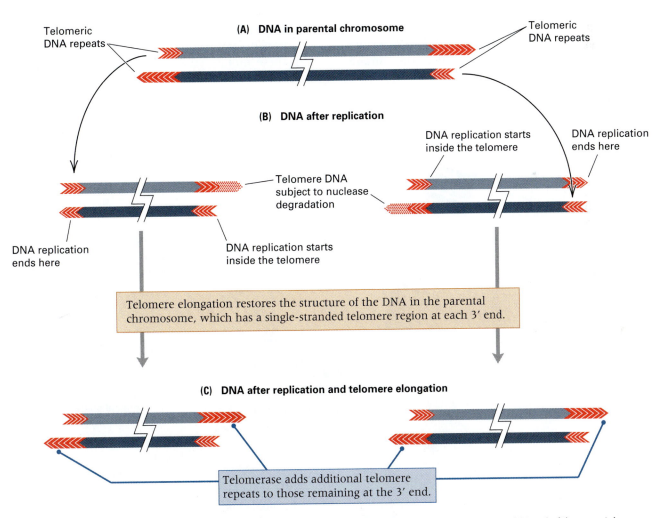

Figure 3.25 The function of telomerase. (A) Chromosomal DNA is double-stranded. Each end of each strand terminates in a set of telomere repeats, but the 3′ end of each strand is longer as a result of telomerase action after the previous replication. (B) In the replication of each parental DNA strand, the new daughter DNA strand is initiated within the telomere repeat at the 3′ end; the telomere at the 5′ end in the new strand is shorter than that in the parental strand. The unreplicated 3′ end of the parental strand is vulnerable to digestion by nucleases. (C) In the daughter DNA duplexes formed by replication, the 3′ strand at each end is elongated via the addition of telomere repeats by the telomerase. The length and 3′ overhang of the telomeres are restored to the state that was present in the original parental molecule.

a short telomere, because this end is replicated from the underhanging 5' end of the telomere in the parental strand. The shortened telomere remaining at each 3' end is the substrate of the telomerase, which elongates each 3' end by the addition of more repeating telomere units (in the case of *Tetrahymena*, −TTGGGG-3'). Telomere elongation restores the structure of the original parental chromosome in which each end has a larger number of telomere repeats at the 3' end and a smaller number of repeats at the 5' end.

Relatively few copies of the telomere repeat are necessary to prime the telomerase to add additional copies and form a telomere. Remarkably, the telomerase enzyme incorporates an essential RNA molecule, called a *guide RNA*, that contains sequences complementary to the telomere repeat and that serves as a template for telomere synthesis and elongation. For example, the *Tetrahymena* guide RNA contains the sequence 3'-AACCCCAAC-5'. The guide RNA undergoes base-pairing with the telomere repeat and serves as a template for telomere elongation by the addition of more repeating units (Figure 3.26). The complementary DNA strand of the telomere is synthesized by cellular DNA repli-

cation enzymes. In the telomeric regions of most eukaryotic chromosomes, there are also longer, moderately repetitive DNA sequences just preceding the terminal repeats. These sequences differ among organisms and even among different chromosomes in the same organism.

What limits the length of a telomere? In most organisms the answer is unknown. In yeast, however, a protein called Rap1p has been identified that appears to be important in regulating telomere length. The Rap1p protein binds to the yeast telomere sequence. Molecules of Rap1p bind to the telomere sequence as it is being elongated until about 17 Rap1p molecules have been bound. At this point, telomere elongation stops, probably because the accumulation of Rap1p inhibits telomerase activity. Because each Rap1p molecule binds to about 18 base pairs of the telomere, the predicted length of a yeast telomere is $17 \times 18 = 306$ base pairs, which is very close to the value observed. Additional evidence for the role of Rap1p comes from mutations in the *RAP1* gene producing a protein that cannot bind to telomere sequences; in these mutant strains, massive telomere elongation is observed.

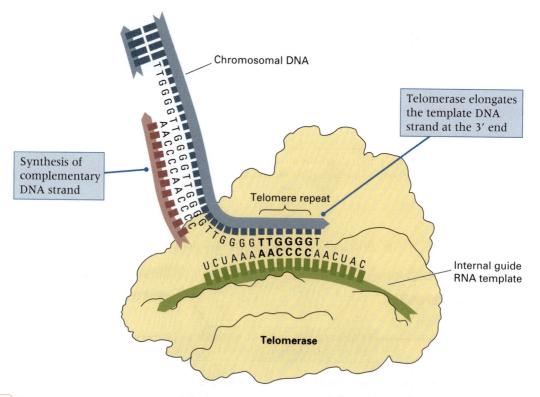

Figure 3.26 Telomere formation in *Tetrahymena*. The telomerase enzyme contains an internal guide RNA with a sequence complementary to the telomere repeat. The RNA undergoes base pairing with the telomere repeat and serves as a template for telomere elongation. The newly forming DNA strand is produced by DNA polymerase.

Sick of Telomeres

William C. Hahn,[1,2] Christopher M. Counter,[3] Ante S. Lundberg,[1,2] Roderick L. Beijersbergen,[1] Mary W. Brooks,[1] and Robert A. Weinberg[1] 1999

[1]Massachusetts Institute of Technology, Cambridge, Massachusetts

[2]Harvard Medical School, Boston, Massachusetts

[3]Duke University Medical Center, Durham, North Carolina

Creation of Human Tumor Cells with Defined Genetic Elements

Some years ago at a well-known midwestern medical school, the head of the Department of Medicine confronted the head of the Department of Genetics and said, "Your Professor Blank up the hall studies telomeres. Who cares about telomeres? Nobody ever gets sick because of their telomeres!" We recount this story to emphasize that the directions of basic research that prove most important in the long run cannot usually be predicted, even by experts. As this important paper shows, lots of people get sick—very sick—because of their telomeres.

During malignant transformation, cancer cells acquire genetic mutations that override the normal mechanisms controlling cellular proliferation. Primary rodent cells are efficiently converted into tumorigenic cells by the coexpression of cooperating oncogenes [genes that cause cancer when mutated]. However, similar experiments with human cells have consistently failed to yield tumorigenic transformants, indicating a fundamental difference in the biology of human and rodent cells. . . . One ostensibly important difference between rodent and human cells derives from their telomere biology. Murine somatic cells express telomerase activity and have much longer telomeres than their human counterparts, which lack telomerase activity. Because normal human cells progressively lose telomeric DNA with passage in culture, telomeric erosion is thought to limit cellular life span. Ectopic expression of the *hTERT* gene, which encodes the catalytic subunit of the telomerase, enables some . . . pre-senescent primary human cells to multiply indefinitely. . . . To determine whether

hTERT human cells were tumorigenic, . . . we also introduced large-T antigen [which inhibits both the retinoblastoma and p53 tumor-suppressor proteins] and an oncogenic *ras* [signal transduction] mutation. . . . We observed efficient colony formation in soft agar [an assay for the tumorigenic state] only with cells expressing the combination of large-T, *ras,* and *hTERT.* . . .

It is now highly likely that telomere maintenance contributes directly to oncogenesis [cancer].

When these cells were introduced into [immunologically deficient] mice, rapidly growing tumors were repeatedly observed with high efficiency. . . . We conclude that ectopic expression of a defined set of genes . . . suffices to convert normal human cells into tumorigenic cells. . . . It is now highly likely that telomere maintenance contributes directly to oncogenesis by allowing pre-cancerous cells to proliferate beyond the number of replicative doublings allotted to their normal precursors.

Source: Nature 400: 464–468.

3.6
Genes are located in chromosomes.

Not long after the rediscovery of Mendel's paper, it became widely assumed that genes were physically located in the chromosomes. The strongest evidence was that Mendel's principles of segregation and independent assortment paralleled the behavior of chromosomes in meiosis. But the first indisputable proof that genes are parts of chromosomes was obtained in experiments concerned with the pattern of transmission of the **sex chromosomes,** the chromosomes responsible for determination of

the separate sexes in some plants and in almost all higher animals. We will examine these results in this section.

■ Special chromosomes determine sex in many organisms.

The sex chromosomes are an exception to the rule that all chromosomes of diploid organisms are present in pairs of morphologically similar homologs. As early as 1891, microscopic analysis had shown that one of the chromosomes in males of some insects, such as grasshoppers, does not have a homolog. This unpaired chromosome was called

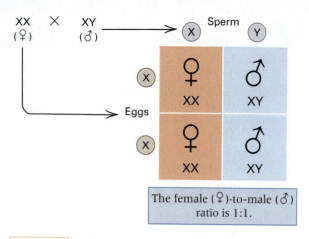

The female ($♀$)-to-male ($♂$) ratio is 1:1.

Figure 3.27 In chromosomal sex determination as found in humans and many other animals, each son gets his X chromosome from his mother and his Y chromosome from his father.

the **X chromosome,** and it was present in all somatic cells of the males but in only half the sperm cells. The biological significance of these observations became clear when females of the same species were shown to have two X chromosomes.

In other species in which the females have two X chromosomes, the male has one X chromosome along with a morphologically different chromosome. This different chromosome is referred to as the **Y chromosome,** and it pairs with the X chromosome during meiosis in males because the X and Y share a small region of homology. The difference in the chromosomal constitution of males and females is a chromosomal mechanism for determining sex at the time of fertilization (Figure 3.27). Whereas every egg cell contains an X chromosome, half the sperm cells contain an X chromosome and the rest contain a Y chromosome. Fertilization of an X-bearing egg by an X-bearing sperm results in an XX zygote, which normally develops into a female; and fertilization by a Y-bearing sperm results in an XY zygote, which normally develops into a male. The result is a criss-cross pattern of inheritance of the X chromosome in which a male receives his X chromosome from his mother and transmits it only

to his daughters. The XX-XY type of chromosomal sex determination is found in mammals, including human beings, in many insects, and in other animals, as well as in some flowering plants.

The X and Y chromosomes together constitute the sex chromosomes; this term distinguishes them from other pairs of chromosomes, which are called **autosomes.** Although the sex chromosomes control the developmental switch that determines the earliest stages of female or male development, the developmental process itself requires many genes scattered throughout the chromosome complement, including genes on the autosomes. The X chromosome also contains many genes with functions unrelated to sexual differentiation. In most organisms, the Y chromosome carries few genes other than those related to male determination. In human beings, for example, the Y chromosome is about 51 Mb in length and contains an estimated 128 genes, many of which are thought to be nonfunctional remnants of genes whose functional counterparts are in the X chromosome.

■ X-linked genes are inherited according to sex.

The compelling evidence that genes are in chromosomes came from the study of a *Drosophila* gene for white eyes, which proved to be present in the X chromosome. Recall that in Mendel's crosses, reciprocal crosses gave the same result; it did not matter which trait was present in the male parent and which in the female parent. One of the earliest exceptions to this rule was found by Thomas Hunt Morgan in 1910, in an early study of a mutation in the fruit fly *Drosophila melanogaster* that had white eyes. The wildtype eye color is a brick-red combination of red and brown pigments (Figure 3.28). Although white eyes can result from certain combinations of autosomal genes that eliminate the pigments individually, the white-eye mutation that Morgan studied results in a metabolic block that knocks out both pigments simultaneously.

(A)

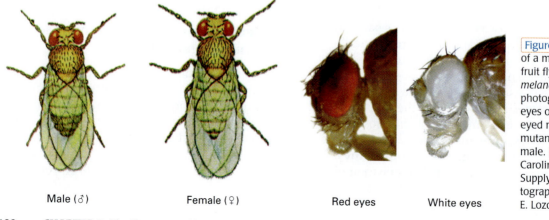

Male ($♂$)   Female ($♀$)

(B)

Red eyes   White eyes

Figure 3.28 Drawings of a male and a female fruit fly, *Drosophila melanogaster.* The photographs show the eyes of a wildtype red-eyed male and a mutant white-eyed male. [Drawings © Carolina Biological Supply Company; photographs courtesy of E. Lozovsky.]

Morgan's study started with a single male with white eyes that appeared in a wildtype laboratory population that had been maintained for many generations. In a mating of this male with wildtype females (cross A, Figure 3.29), all of the F_1 progeny of both sexes had red eyes, showing that the allele for white eyes is recessive. In the F_2 progeny from the mating of F_1 males and females, Morgan observed 2459 red-eyed females, 1011 red-eyed males, and 782 white-eyed males. The white-eyed phenotype was somehow connected with sex, because all of the white-eyed flies were males.

On the other hand, white eyes were not restricted to males. For example, when red-eyed F_1 females from the cross of wildtype ♀♀ × white ♂♂ were backcrossed with their white-eyed fathers, the progeny consisted of both red-eyed and white-eyed females and red-eyed and white-eyed males in approximately equal numbers.

A key observation came from the mating of white-eyed females with wildtype males (cross B, Figure 3.29). All of the female progeny had wildtype eyes, but all of the male progeny had white eyes. This is the reciprocal of the cross A of wildtype ♀♀ × white ♂♂, which had yielded only wildtype females and wildtype males, and so the reciprocal crosses gave different results.

Morgan realized that reciprocal crosses would yield different results if the allele for white eyes were present in the X chromosome. This is because the X chromosome is transmitted in a different pattern by males and females, and the Y chromosome does not contain a counterpart of the *white* gene. Figure 3.29 shows that a male transmits his X chromosome only to his daughters, whereas a female transmits one of her X chromosomes to the offspring of both sexes. A gene located in the X chromosome is said to be **X-linked.**

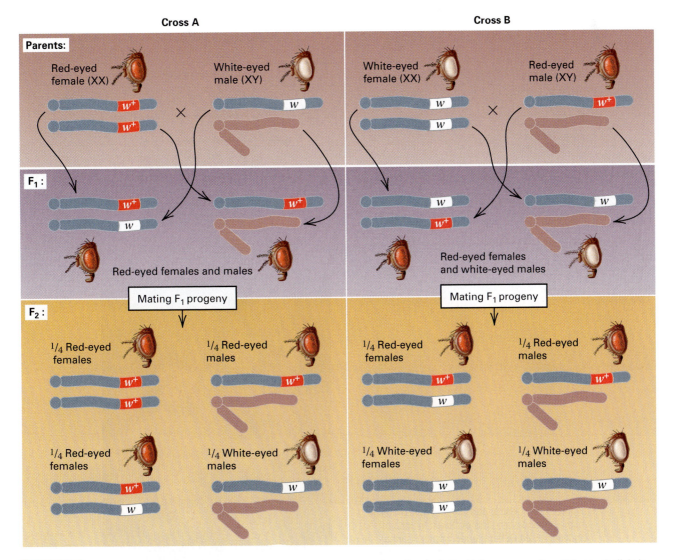

Figure 3.29 | A chromosomal interpretation of the results obtained in F_1 and F_2 progenies in crosses of *Drosophila*. Cross A is a mating of a wildtype (red-eyed) female with a white-eyed male. Cross B is the reciprocal mating of a white-eyed female with a red-eyed male. In the X chromosome, the wildtype w^+ allele is shown in red and the mutant w allele in white. The Y chromosome does not carry either allele of the w gene.

■ Hemophilia is a classic example of human X-linked inheritance.

A classic example of a human trait with an X-linked pattern of inheritance is **hemophilia A,** a severe disorder of blood clotting determined by a recessive allele. Affected persons lack a blood-clotting protein called factor VIII that is needed for normal clotting, and they suffer excessive, often life-threatening bleeding after injury. A famous pedigree of hemophilia starts with Queen Victoria of England (Figure 3.30). One of her sons, Leopold, was hemophilic, and three of her daughters were heterozygous carriers of the gene. Two of Victoria's granddaughters were also carriers, and by marriage they introduced the gene into the royal families of Russia and Spain. The heir to the Russian throne of

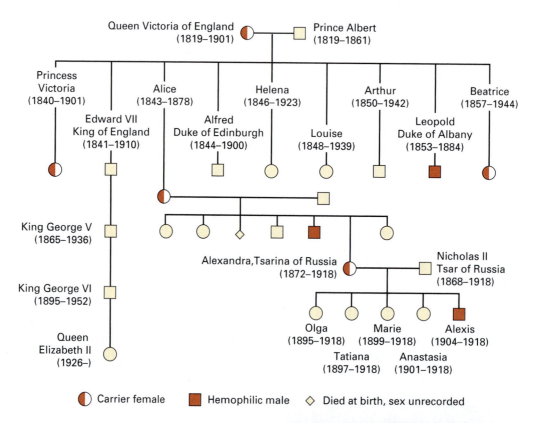

Carrier female ◼ Hemophilic male ◇ Died at birth, sex unrecorded

Figure 3.30 Genetic transmission of hemophilia A among the descendants of Queen Victoria of England, including her granddaughter, Tsarina Alexandra of Russia, and Alexandra's five children. The photograph is that of Tsar Nicholas II, Tsarina Alexandra, and the Tsarevich Alexis, who was afflicted with hemophilia. [Courtesy of Boston Public Library, Print Department.]

the Romanoffs, Tsarevich Alexis, was afflicted with the condition. He inherited the gene from his mother, Tsarina Alix, one of Victoria's granddaughters. The Tsar, the Tsarina, Alexis, and his four sisters were all executed by the Bolsheviks in the 1918 Russian revolution. The present royal family of England is descended from a normal son of Victoria and is free of the disease.

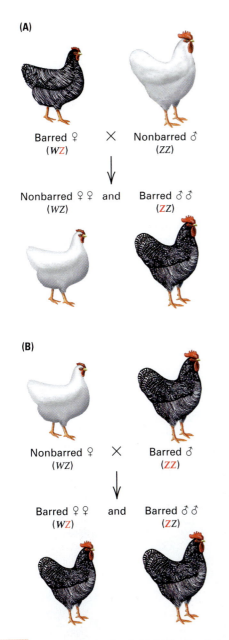

Figure 3.31 In sex determination in birds, the female has the unmatched sex chromosomes (called W and Z), whereas the male has the matched sex chromosomes (ZZ). The recessive mutant gene for nonbarred feathers is in the Z chromosome. (A) A cross of barred females with nonbarred males yields nonbarred female and barred male progeny. (B) A cross of nonbarred females with barred males yields barred female and barred male progeny. These results are the opposite of those observed with the white-eye *Drosophila* mutant in Figure 3.29.

X-linked inheritance in human pedigrees shows several characteristics that distinguish it from other modes of genetic transmission:

1. For any rare trait due to an X-linked recessive allele, the affected individuals are exclusively, or almost exclusively, male. There is an excess of males because females carrying the rare X-linked recessive allele are almost exclusively heterozygous and so do not express the mutant phenotype.

2. Affected males who reproduce have normal sons. This follows from the fact that a male transmits his X chromosome only to his daughters.

3. A woman whose father was affected has normal sons and affected sons in the ratio 1 : 1. This is true because any daughter of an affected male must be heterozygous for the recessive allele.

■ In birds, moths, and butterflies, the sex chromosomes are reversed.

In some organisms, sex is determined by sex chromosomes, but the mammalian situation is reversed: The males are XX and the females are XY. This type of sex determination is found in birds, in some reptiles and fish, and in moths and butterflies. The reversal of XX and XY in the sexes results in an opposite pattern of nonreciprocal inheritance of X-linked genes. For example, some breeds of chickens have feathers with alternating transverse bands of light and dark color, resulting in a phenotype referred to as barred. The feathers are uniformly colored in the nonbarred phenotypes of other breeds. Reciprocal crosses between true-breeding barred and nonbarred types give the results shown in Figure 3.31. The results indicate that the gene determining barring is in the chicken X chromosome and is dominant. To distinguish sex determination in birds, butterflies, and moths from the usual XX-XY mechanism, in these organisms the sex chromosome constitutions are usually designated WZ for the female and ZZ for the male.

■ Experimental proof of the chromosome theory came from nondisjunction.

The parallel between the inheritance of the *Drosophila white* mutation and the genetic transmission of the X chromosome supported the chromosome theory of heredity that genes are parts of chromosomes. Other experiments with *Drosophila* provided the definitive proof.

One of Morgan's students, Calvin Bridges, discovered rare exceptions to the expected pattern of inheritance in crosses with several X-linked genes.

To understand these experiments, it is necessary to know that *Drosophila* is unusual among organisms with an XX-XY type of sex determination in that the Y chromosome, although it is associated with maleness, is not male-determining. In *Drosophila*, XXY embryos develop into morphologically normal, fertile females, whereas XO embryos develop into morphologically normal, but sterile, males. (The O is written in the formula XO to emphasize that a sex chromosome is missing.) The sterility of XO males shows that the Y chromosome, though not necessary for male development, is essential for male fertility.

When Bridges crossed white-eyed *Drosophila* females with red-eyed males, most of the progeny consisted of the expected red-eyed females and white-eyed males. However, about one in every 2000 F_1 flies was an exception: either a white-eyed female or a red-eyed male. Bridges showed that these rare exceptional offspring resulted from occasional failure of the two X chromosomes in the mother to separate from each other during meiosis—a phenomenon called **nondisjunction.** The consequence of nondisjunction of the X chromosomes is the formation of some eggs with two X chromosomes and others with none. Four classes of zygotes are expected from the fertiliza-

tion of these abnormal eggs (Figure 3.32). Animals with no X chromosome are not detected because embryos that lack an X chromosome die early in development; likewise, most progeny with three X chromosomes are not viable. Microscopic examination of the chromosomes of the exceptional progeny from the cross white ♀♀ × wild-type ♂♂ showed that the exceptional white-eyed females had two X chromosomes *plus* a Y chromosome and that the exceptional red-eyed males had a single X but were *lacking* a Y. The latter were sterile XO males.

These and related experiments demonstrated conclusively the validity of the chromosome theory of heredity.

key concept

Chromosome Theory of Heredity: Genes are contained in the chromosomes.

Bridges's evidence for the chromosome theory was that exceptional behavior of chromosomes is precisely paralleled by exceptional inheritance of their genes. This proof of the chromosome theory ranks among the most important and elegant experiments in genetics.

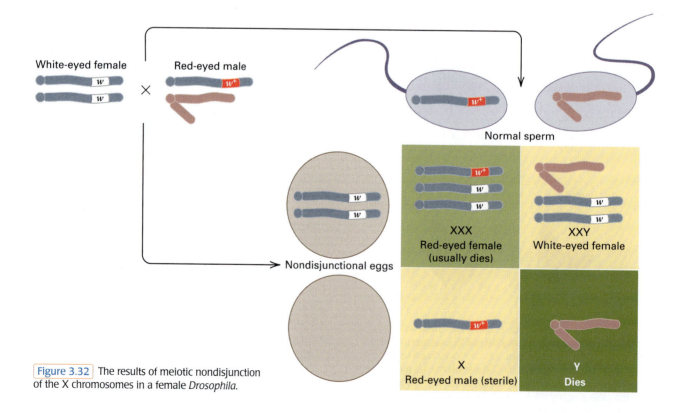

Figure 3.32 The results of meiotic nondisjunction of the X chromosomes in a female *Drosophila*.

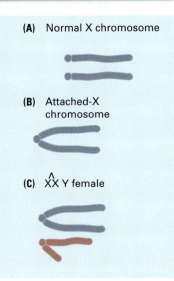

A Moment to Think

Problem: In 1922 Lilian V. Morgan, wife of Thomas Hunt Morgan and a first-rate geneticist in her own right, discovered a remarkable female of *Drosophila* in which the pattern of X-linked inheritance seemed to be reversed. The fly was homozygous for an X-linked mutation y for yellow body. When this fly was mated with a wildtype y^+ male with a gray body, the females resulting from the cross were all yellow and the males all gray. She realized immediately that this female was extremely important. (There is an old story, of uncertain validity, that the original female escaped temporarily, before she could be crossed, causing consternation in the laboratory and occasioning a mad search until finally she was found resting on a window pane.) Further genetic analysis indicated that the female had two X chromosomes that were physically joined together, united by a single centromere. This configuration is known as an *attached-X chromosome.* The illustration here shows the normal X chromosomes in a female (A) and an attached-X chromosome (B). As shown in (C), *Drosophila* females with an attached-X chromosome usually also carry a Y chromosome, which they acquire from their fathers. In female meiosis, the attached-X chromosome segregates from the Y chromosome. Using this information, draw a Punnett square showing the result of a cross of a yellow-body attached-X/Y female with a wildtype XY male. Indicate which progeny fail to survive; and show the body-color phenotype and sex of the offspring that do survive. In what sense do X chromosomes that are attached *always* undergo nondisjunction? (The answer can be found on page 110.)

(A) Normal X chromosome

(B) Attached-X chromosome

(C) X̂X̂ Y female

3.7

Genetic data analysis makes use of probability and statistics.

Genetic transmission includes a large component of chance. A particular gamete from an *Aa* organism might or might not include the *A* allele, depending on chance. A particular gamete from an *Aa Bb* organism might or might not include both the *A* and *B* alleles, depending on the chance orientation of the chromosomes on the metaphase I plate. Genetic ratios result not only from the chance assortment of genes into gametes but also from the chance combination of gametes into zygotes. Although exact predictions are not possible for any particular event, it is possible to determine the probability that a particular event might be realized, as we have seen in Chapter 2. In this section, we consider some of the probability methods used in interpreting genetic data.

■ **Progeny of crosses are predicted by the binomial probability formula.**

The addition rule of probability deals with possible outcomes of a genetic cross that are *mutually exclusive.* Outcomes are mutually exclusive if they are incompatible in the sense that they cannot occur at the same time. For example, there are four mutually exclusive outcomes of the sex distribution of sibships with three children—namely, the inclusion of 0, 1, 2, or 3 girls. These have the probabilities 1/8, 3/8, 3/8, and 1/8, respectively. The addition rule states that the overall probability of any combination of mutually exclusive outcomes is equal to the sum of the probabilities of the outcomes taken separately. For example, the probability that a sibship of size 3 contains *at least one girl* includes the outcomes 1, 2, and 3 girls, and so the overall probability of at least one girl equals $3/8 + 3/8 + 1/8 = 7/8$.

The multiplication rule of probability deals with outcomes of a genetic cross that are independent. Any two outcomes are independent if the knowledge that one outcome is actually realized provides no information about whether the other is realized also. For example, in a sequence of births, the sex of any one child is not affected by the sex distribution of any children born earlier and has no influence whatsoever on the sex distribution of any siblings born later. Each successive birth is independent of all the others. When possible outcomes are independent, the multiplication rule states that the probability of any combination of outcomes being realized equals the product of the probabilities of each of the outcomes taken separately. For example, the probability that a sibship of three children will consist of three girls equals $1/2 \times 1/2 \times 1/2$, because the probability of each birth resulting in a girl is 1/2, and the successive births are independent.

Probability calculations in genetics frequently use the addition and the multiplication rules together. For example, to find the probability that each of three children in a family will be of the same sex, we use both the addition and the multiplication rules. The probability that all three will be girls is $(1/2)(1/2)(1/2) = 1/8$, and the probability that all three will be boys is also $1/8$. Because these outcomes are mutually exclusive (a sibship of size three cannot include three boys *and* three girls), the probability of either three girls or three boys is the sum of the two probabilities, or $1/8 + 1/8 = 1/4$. The other possible outcomes for sibships of size three are that two of the children will be girls and the other a boy, or that two will be boys and the other a girl. For each of these outcomes, three different orders of birth are possible—for example, GGB, GBG, and BGG—each having a probability of $1/2 \times 1/2 \times 1/2 = 1/8$. The probability of two girls and a boy, disregarding birth order, is the sum of the probabilities for the three possible orders, or $3/8$; likewise, the probability of two boys and a girl is also $3/8$. Therefore, the distribution of probabilities for the sex ratio in families with three children is

GGG	BGG	GBB	BBB
	GBG	BGB	
	GGB	BBG	

$$(1/2)^3 + 3(1/2)^2(1/2)^1 + 3(1/2)^1(1/2)^2 + (1/2)^3$$

$$1/8 \quad + \quad 3/8 \quad + \quad 3/8 \quad + 1/8 \quad = 1$$

The sex-ratio probabilities can be obtained by expanding the binomial expression $(p + q)^n$, in which p is the probability of the birth of a girl $(1/2)$, q the probability of the birth of a boy $(1/2)$, and n the number of children. In the present example,

$$(p + q)^3 = 1p^3 + 3p^2q + 3pq^2 + 1q^3$$

in which the red numerals are the possible number of birth orders for each sex distribution. Similarly, the binomial distribution of probabilities for the sex ratios in families of five children is

$$(p + q)^5 = 1p^5 + 5p^4q + 10p^3q^2 + 10p^2q^3 + 5pq^4 + 1q^5$$

Each term tells us the probability of a particular combination. For example, the third term is the probability of three girls (p^3) and two boys (q^2) in a family having five children:

$$10(1/2)^3(1/2)^2 = 10/32 = 5/16$$

There are $n + 1$ terms in a binomial expansion. The exponents of p decrease by one from n in the first term to 0 in the last term, and the exponents of q increase by one from 0 in the first term to n in the last term. The coefficients generated by successive values of n can be arranged in a regular triangle known as **Pascal's triangle** (Figure 3.33). Note that the horizontal rows of the triangle are symmetrical and that each number is the sum of the two numbers on either side of it in the row above.

In general, if the probability of a possible outcome A is p and that of B is q, and the two events are independent and mutually exclusive (see Chapter 2), then the probability that A will be realized four times and B two times—in a specific order—is p^4q^2, by the multiplication rule. However, suppose that we are interested in the combination of events "four of A and two of B," irrespective of order. In that case, we multiply the probability that the combination 4A : 2B will be realized in any one specific order by the number of possible orders. The number of different combinations of six things, four of one kind and two of another, is

$$\frac{6!}{4!2!} = \frac{1 \times 2 \times 3 \times 4 \times 5 \times 6}{(1 \times 2 \times 3 \times 4) \times (1 \times 2)} = 15$$

The symbol ! stands for **factorial,** or the product of all positive integers from 1 through a given number. Except for $n = 0$, the formula for factorial is $n! = 1 \times 2 \times 3 \times 4 \times \cdots \times n - 1 \times n$. The case $n = 0$ is an exception because 0! is defined as equal to 1. The values of the first few factorials are given in **Table 3.2**. The value of $n!$ increases very rapidly as n increases; 15! is more than a trillion.

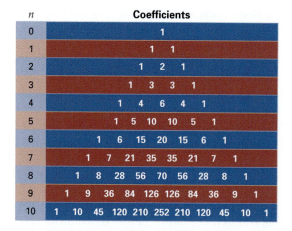

Figure 3.33 Pascal's triangle. The numbers are the coefficients of the terms obtained by multiplying out the expression $(p + q)^n$ for successive values of n from 0 through 10.

Table 3.2

Factorials

n	$n!$	n	$n!$
0	1	8	40,320
1	1	9	362,880
2	2	10	3,628,800
3	6	11	39,916,800
4	24	12	479,001,600
5	120	13	6,227,020,800
6	720	14	87,178,291,200
7	5040	15	1,307,674,368,000

The factorial formula $6!/(4! \times 2!) = 720/(24 \times 2) = 15$ is the coefficient of the term p^4q^2 in the expansion of the binomial $(p + q)^6$. Therefore, the probability that outcome A will be realized four times and outcome B two times is $15p^4q^2$.

The general rule for repeated trials of events with constant probabilities is as follows:

<table>
<tr><td>

key concept

If the probability of possibility A is p and the probability of the alternative possibility B is q, then the probability that, in n trials, A is realized s times and B is realized t times is

$$\frac{n!}{s!t!}p^sq^t \qquad (3.1)$$

</td></tr>
</table>

in which $s + t = n$ and $p + q = 1$. Equation (3.1) applies even when either s or t equals 0, because 0! is defined to equal 1. (Remember also that any number raised to the zero power equals 1; for example, $2^0 = 1$.) Any individual term in the expansion of the binomial $(p + q)^n$ is given by Equation (3.1) for the appropriate values of s and t.

In Equation (3.1), $n!/(s!\ t!)$ enumerates all possible ways in which s elements of one kind and t elements of another kind can be arranged in order, provided that the s elements and the t elements are not distinguished among themselves. A specific example might include s yellow peas and t green peas. Although the yellow peas and the green peas can be distinguished from each other because they have different colors, the yellow peas are not distinguishable from one another (because they are all yellow), and the green peas are not distinguishable from one another (because they are all green). Altogether there are $n!/(s!\ t!)$ different orders in which the yellow and green peas can be arranged in a row.

Let us use Equation (3.1) to calculate the probability that a mating between two heterozygous parents yields exactly the expected $3:1$ ratio of the dominant and recessive traits among sibships of a particular size. The probability p of a child showing the dominant trait is $3/4$, and the probability q of a child showing the recessive trait is $1/4$. Suppose that we wanted to know how often families with eight children would contain exactly six children with the dominant phenotype and two with the recessive phenotype. This is the "expected" Mendelian ratio. In this case, $n = 8$, $s = 6$, $t = 2$, and the probability of this combination of events is

$$\frac{8!}{6!2!}p^6q^2 = \frac{6! \times 7 \times 8}{6! \times 2!}(3/4)^6(1/4)^2 = 0.31$$

That is, in only 31 percent of the families with eight children would the offspring exhibit the expected $3:1$ phenotypic ratio; the other sibships would deviate in one direction or the other because of chance variation. The importance of this example is in demonstrating that although a $3:1$ ratio is the "expected" outcome (and also the single most probable outcome), the majority of the families (69 percent) actually have a distribution of offspring different from $3:1$.

■ **Chi-square tests goodness of fit of observed to expected numbers.**

Geneticists often need to decide whether an observed ratio is in satisfactory agreement with a theoretical prediction. Mere inspection of the data is unsatisfactory because different investigators may disagree. Suppose, for example, that we crossed a plant having purple flowers with a plant having white flowers and, among the progeny, observed 14 plants with purple flowers and 6 with white flowers. Is this result close enough to be accepted as a $1:1$ ratio? What if we observed 15 plants with purple flowers and 5 with white flowers? Is this result consistent with a $1:1$ ratio? There is bound to be statistical variation in the observed results from one experiment to the next. Who is to say what results are consistent with a particular genetic hypothesis? In this section, we describe a test of whether observed results deviate too far from a theoretical expectation. The test is called a test for **goodness of fit,** where the word *fit* means how closely the observed results "fit," or agree with, the expected results.

A conventional measure of goodness of fit is a value called **chi-square** (symbol, χ^2), which is calculated from the number of progeny observed in each of various classes, compared with the number expected in each of the classes on the basis of some genetic hypothesis. For example, in a cross between plants with purple flowers and those with white flowers, we may be interested in testing the hypothesis that the parent with purple flowers is heterozygous for a pair of alleles determining flower color and that the parent with white flowers is homozygous recessive. Suppose further that we examine 20 progeny plants from the mating and find that 14 are purple and 6 are white. The procedure to be followed in testing this genetic hypothesis (or any other genetic hypothesis) by means of the chi-square method is as follows:

1. *State the genetic hypothesis in detail, specifying the genotypes and phenotypes of the parents and the possible progeny.* In the example using flower color, the genetic hypothesis implies that the genotypes in the cross purple $\times$ white could be represented as $Pp \times pp$. The possible progeny genotypes are either Pp or pp.

2. *Use the rules of probability to make explicit predictions of the types and proportions of progeny that*

should be observed if the genetic hypothesis is true. Convert the proportions to numbers of progeny (percentages are not allowed in a χ^2 test). If the hypothesis about the flower-color cross is true, then we expect the progeny genotypes Pp and pp in a ratio of 1 : 1. Because the hypothesis is that Pp flowers are purple and pp flowers are white, we expect the phenotypes of the progeny to be purple or white in the ratio 1 : 1. Among 20 progeny, the expected numbers are 10 purple and 10 white.

3. *For each class of progeny in turn, subtract the expected number from the observed number. Square this difference and divide the result by the expected number.* In our example, the calculation for the purple progeny is $(14 - 10)^2/10 = 1.6$, and that for the white progeny is $(6 - 10)^2/10 = 1.6$.

4. *Sum the result of the numbers calculated in step 3 for all classes of progeny. The summation is the value of χ^2 for these data.* The sum for the purple and white classes of progeny is $1.6 + 1.6 = 3.2$, and this is the value of χ^2 for the experiment, calculated on the assumption that our genetic hypothesis is correct.

In symbols, the calculation of χ^2 can be represented by the expression

$$\chi^2 = \sum \frac{(observed - expected)^2}{expected}$$

in which Σ means the summation over all the classes of progeny. Note that χ^2 is calculated using the observed and expected *numbers,* not the proportions, ratios, or percentages. Using something other than the actual numbers is the beginner's most common mistake in applying the χ^2 method. The χ^2 value is reasonable as a measure of goodness of fit, because the closer the observed numbers are to the expected numbers, the smaller the value of χ^2. A value of $\chi^2 = 0$ means that the observed numbers fit the expected numbers perfectly.

As another example of the calculation of χ^2, suppose that the progeny of an $F_1 \times F_1$ cross include two contrasting phenotypes observed in the numbers 99 and 45. The genetic hypothesis might be that the trait is determined by a pair of alleles of a single gene, in which case the expected ratio of dominant : recessive phenotypes among the F_2 progeny is 3 : 1. Considering the data, the question is whether the observed ratio of 99 : 45 is in satisfactory agreement with the expected 3 : 1. Calculation of the value of χ^2 is illustrated in **Table 3.3**. The total number of progeny is $99 + 45 = 144$. The *expected* numbers in the two classes, on the basis of the genetic hypothesis that the true ratio is 3 : 1, are calculated as $(3/4) \times 144 = 108$ and $(1/4) \times 144 = 36$. Because there are two classes of data, there are two terms in the χ^2:

$$\chi^2 = \frac{(99 - 108)^2}{108} + \frac{(45 - 36)^2}{36}$$

$$= 0.75 + 2.25 = 3.00$$

Once the χ^2 value has been calculated, the next step is to interpret whether this value represents a good fit or a bad fit to the expected numbers. This assessment is done with the aid of the graphs in Figure 3.34. The x-axis gives the χ^2 values measuring goodness of fit, and the y-axis gives the probability P that a worse fit (or one equally bad) would be obtained by chance, assuming that the genetic hypothesis is true. If the genetic hypothesis is true, then the observed numbers should be reasonably close to the expected numbers. Suppose that the observed χ^2 is so large that the probability of a fit as bad or worse is very small. Then the observed results do *not* fit the theoretical expectations. This means that the genetic hypothesis used to calculate the expected numbers of progeny must be rejected, because the observed numbers of progeny deviate too much from the expected numbers.

In practice, the critical values of P are conventionally chosen as 0.05 (the 5 percent level) and 0.01 (the 1 percent level). For P values ranging from 0.01 to 0.05, the probability that chance alone would lead to a fit as bad or worse is between 1 in 20 experiments and 1 in 100. This is the purple region in Figure 3.34; if the P value falls in this range, the correctness of the genetic hypothesis is considered very doubtful. The result

Table 3.3

Calculation of χ^2 for a monohybrid ratio

Phenotype (class)	Observed number	Expected number	Deviation from expected (obs − exp)	$\frac{(obs - exp)^2}{exp}$
Wildtype	99	108	−9	0.75
Mutant	45	36	+9	2.25
Total	144	144		$\chi^2 = 3.00$

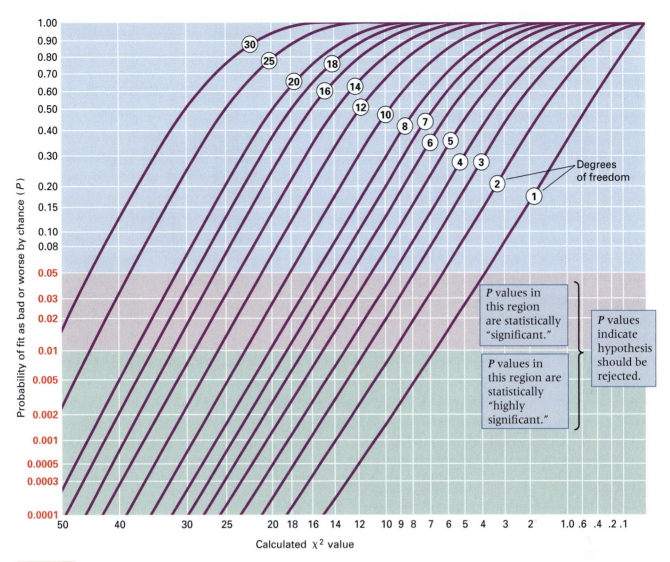

Calculated χ^2 value

Probability of fit as bad or worse by chance (P)

Degrees of freedom

P values in this region are statistically "significant."

P values in this region are statistically "highly significant."

P values indicate hypothesis should be rejected.

Figure 3.34 Graphs for interpreting goodness of fit to genetic predictions using the chi-square test. For any calculated value of χ^2 along the x-axis, the y-axis corresponding to the curve with the appropriate number of degrees of freedom gives the probability P that chance alone would produce a fit as bad as or worse than that actually observed, when the genetic predictions are correct. Tests with P in the purple region (less than 5 percent) or in the green region (less than 1 percent) are regarded as statistically significant and normally require rejection of the genetic hypothesis that led to the prediction.

is said to be **statistically significant** at the 5 percent level. For P values smaller than 0.01, the probability that chance alone would lead to a fit as bad or worse is less than 1 in 100 experiments. This is the green region in Figure 3.34; in this case, the result is said to be **statistically highly significant** at the 1 percent level, and the genetic hypothesis is rejected outright. If the terminology of statistical significance seems backwards, it is because the term *significant* refers to the magnitude of the difference between the observed and the expected numbers; in a result that is statistically significant, there is a large ("significant") difference between what is observed and what is expected.

To use Figure 3.34 to determine the P value corresponding to a calculated χ^2, we need the number

of **degrees of freedom** of the particular χ^2 test. For the type of χ^2 test illustrated in Table 3.3, the number of degrees of freedom equals the number of classes of data minus 1. Table 3.3 contains two classes of data (wildtype and mutant), so the number of degrees of freedom is $2 - 1 = 1$. The reason for subtracting 1 is that, in calculating the expected numbers of progeny, we make sure that the total number of progeny is the same as that actually observed. For this reason, one of the classes of data is not really "free" to contain any number we might specify. Because the expected number in one class must be adjusted to make the total come out correctly, one "degree of freedom" is lost. Analogous χ^2 tests with three classes of data have 2 degrees of freedom, and those with four classes of data have 3 degrees of freedom.

Answer to Problem: The Punnett square shows the outcome of a cross between a female with an attached-X chromosome and a normal male. The eggs contain either the attached-X or the Y chromosome; they combine at random with X-bearing or Y-bearing sperm. Genotypes with either three X chromosomes or no X chromosomes do not survive. Note that a male fly receives its X chromosome from its father and its Y chromosome from its mother, which is the opposite of the usual situation in *Drosophila*. The X chromosomes that are attached always undergo nondisjunction because the arms stay together in anaphase I, rather than being separated.

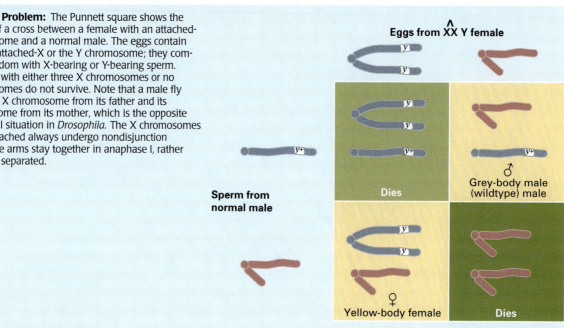

Once we have determined the appropriate number of degrees of freedom, we can interpret the χ^2 value in Table 3.3. Refer to Figure 3.34, and observe that each curve is labeled with its degrees of freedom. To determine the P value for the data in Table 3.3, in which the χ^2 value is 3.00, first find the location of $\chi^2 = 3.00$ along the x-axis in Figure 3.34. Trace vertically from 3.00 until you intersect the curve with 1 degree of freedom. Then trace horizontally to the left until you intersect the y-axis and read the P value—in this case, $P = 0.08$. This means that chance alone would produce a χ^2 value as great as or greater than 3 in about 8 percent of experiments of the type in Table 3.3; and because the P value is in the blue region, the goodness of fit to the hypothesis of a 3 : 1 ratio of wildtype : mutant is judged to be satisfactory.

As a second illustration of the χ^2 test, we will determine the goodness of fit of Mendel's round-versus-wrinkled data to the expected 3 : 1 ratio. Among the 7324 seeds that he observed, 5474 were round and 1850 were wrinkled. The expected numbers are $(3/4) \times 7324 = 5493$ round and $(1/4) \times 7324 = 1831$ wrinkled. The χ^2 value is calculated as

$$\chi^2 = \frac{(5474 - 5493)^2}{5493} + \frac{(1850 - 1831)^2}{1831}$$

$$= 0.26$$

The fact that the χ^2 is less than 1 already implies that the fit is very good. To find out how good, note that the number of degrees of freedom equals $2 - 1 = 1$ because there are two classes of data (round and wrinkled). From Figure 3.34, the P value for $\chi^2 = 0.26$ with 1 degree of freedom is approximately 0.65. This means that in about 65 percent of all experiments of this type, a fit as bad or worse would be expected simply because of chance; only about 35 percent of all experiments would yield a better fit.

■ Are Mendel's data a little too good to be true?

Many of Mendel's experimental results are very close to the expected values. For the ratios listed in Table 2.1 in Chapter 2, the χ^2 values are 0.26 (round versus wrinkled seeds), 0.01 (yellow versus green seeds), 0.39 (purple versus white flowers), 0.06 (inflated versus constricted pods), 0.45 (green versus yellow pods), 0.35 (axial versus terminal flowers), and 0.61 (long versus short stems). (As an exercise in χ^2, you should confirm these calculations for yourself.) All of the χ^2 tests have P values of 0.45 or greater (Figure 3.34), which means that the reported results are in excellent agreement with the theoretical expectations.

The statistician Ronald Fisher pointed out in 1936 that Mendel's results are *suspiciously* close to the theoretical expectations. In a large number of

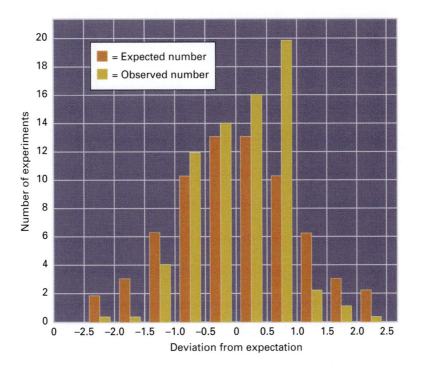

Legend:
= Expected number
= Observed number

Y-axis: Number of experiments
X-axis: Deviation from expectation

Figure 3.35 Distribution of deviations observed in 69 of Mendel's experiments (yellow bars) compared with expected values (orange bars). There is no suggestion that the data in the middle have been adjusted to improve the fit. However, several experiments with large deviations may have been discarded or repeated, because there are not as many experiments with large deviations as might be expected.

experiments, some experiments can be expected to yield fits that appear doubtful simply because of chance variation from one experiment to the next. In Mendel's data, the doubtful values that are to be expected by chance appear to be missing. Figure 3.35 shows the observed deviations in Mendel's experiments compared with the deviations expected to occur by chance. (The measure of deviation is the square root of the χ^2, which is assigned either a plus or a minus sign according to whether the dominant or the recessive phenotypic class was in excess of the expected number.) For each magnitude of deviation, the height of the yellow bar gives the number of experiments that Mendel observed with that magnitude of deviation, and the orange bar gives the number of experiments expected to deviate by this amount as a result of chance alone. There are clearly too few experiments with deviations smaller than -1 or larger than $+1$. This type of discrepancy could be explained if Mendel discarded or repeated a few experiments with large deviations that made him suspect that the results were not to be trusted.

Did Mendel cheat? Did he deliberately falsify his data to make them appear better than they were? Mendel's paper reports extremely deviant ratios from individual plants, as well as experiments repeated a second time when the first results were doubtful. These are not the kinds of things that a dishonest person would admit. Only a small bias is necessary to explain the excessive goodness of fit in

Figure 3.34. In a count of seeds or individual plants, only about 2 phenotypes per 1000 would need to be assigned to the wrong category to account for the bias in the 91 percent of the data related to the testing of monohybrid ratios. The excessive fit could also be explained if three or four entire experiments were discarded or repeated because deviant results were attributed to pollen contamination or other accident. After careful reexamination of Mendel's data in 1966, the evolutionary geneticist Sewall Wright concluded,

> Mendel was the first to count segregants at all. It is rather too much to expect that he would be aware of the precautions now known to be necessary for completely objective data. . . . Checking of counts that one does not like, but not of others, can lead to systematic bias toward agreement. I doubt whether there are many geneticists even now whose data, if extensive, would stand up wholly satisfactorily under the χ^2 test. . . . Taking everything into account, I am confident that there was no deliberate effort at falsification.

Mendel's data are some of the most extensive and complete "raw data" ever published in genetics. Additional examinations of the data will surely be carried out as new statistical approaches are developed. However, the point to be emphasized is that up to the present time, no reputable statistician has alleged that Mendel knowingly and deliberately adjusted his data in favor of the theoretical expectation.

chapter summary

3.1 Each species has a characteristic set of chromosomes.

The chromosomes in somatic cells of higher plants and animals are present in pairs. The members of each pair are homologous chromosomes. Pairs of homologs are usually identical in appearance, whereas nonhomologous chromosomes often show differences in size and structural detail that make them visibly distinct from each other. A cell whose nucleus contains two sets of homologous chromosomes is diploid. One set of chromosomes comes from the maternal parent and the other from the paternal parent. Gametes are haploid. A gamete contains only one set of chromosomes, consisting of one member of each pair of homologs.

3.2 The daughter cells of mitosis have identical chromosomes.

- In mitosis, the replicated chromosomes align on the spindle, and the sister chromatids pull apart.

Mitosis is the process of nuclear division that maintains the chromosome number when a somatic cell divides. Before mitosis, each chromosome replicates, forming a two-part structure that consists of two sister chromatids joined at the centromere (kinetochore). At the onset of mitosis, the chromosomes become visible and, at metaphase, become aligned on the metaphase plate perpendicular to the spindle. At anaphase, the centromere of each chromosome divides, and the sister chromatids are pulled by spindle fibers to opposite poles of the cell. The separated sets of chromosomes present in telophase nuclei are genetically identical.

3.3 Meiosis results in gametes that differ genetically.

- The first meiotic division reduces the chromosome number by half.
- The second meiotic division is equational.

Meiosis is the type of nuclear division that takes place in germ cells, and it reduces the diploid number of chromosomes to the haploid number. The genetic material is replicated before the onset of meiosis, so each chromosome consists of two sister chromatids. The first meiotic division is the reduction division, which reduces the chromosome number by half. The homologous chromosomes first pair (synapsis) and then, at anaphase I, separate. The resulting products contain chromosomes that still consist of two chromatids attached to a common centromere. However, as a result of crossing-over, which takes place in prophase I, the chromatids may not be genetically identical along their entire length. In the second meiotic division, the centromeres divide and the homologous chromatids separate. The end result of meiosis is the formation of four genetically different haploid nuclei.

A distinctive feature of meiosis is the synapsis, or side-by-side pairing, of homologous chromosomes in the zygotene substage of prophase I. During the diplotene sub-stage, the paired chromosomes are seen to be connected by chiasmata (the physical manifestations of crossing-over), and they do not separate until anaphase I. This separation is called disjunction (unjoining), and failure of chromosomes to separate is called nondisjunction. Nondisjunction results in a gamete that contains either two copies or no copies of a particular chromosome. Meiosis is the physical basis of the segregation and independent assortment of genes. In *Drosophila*, an unexpected pattern of inheritance of the *white* gene was shown to be accompanied by nondisjunction of the X chromosome; these observations gave experimental proof to the chromosome theory of heredity.

3.4 Eukaryotic chromosomes are highly coiled complexes of DNA and protein.

- Chromosome-sized DNA molecules can be separated by electrophoresis.
- The nucleosome is the basic structural unit of chromatin.
- Chromatin fibers form discrete territories in the nucleus.
- The metaphase chromosome is a hierarchy of coiled coils.
- Heterochromatin is rich in satellite DNA and low in gene content.

In eukaryotes, the DNA is compacted into chromosomes that are thick enough to be visible by light microscopy during the mitotic phase of the cell cycle. The DNA–protein complex of eukaryotic chromosomes is called chromatin. The protein component of chromatin consists primarily of five distinct proteins: histones H1, H2A, H2B, H3, and H4. The last four histones aggregate to form an octameric protein containing two molecules of each. DNA is wrapped around the histone octamer, forming a particle called a nucleosome. This wrapping is the first level of compaction of the DNA in chromosomes. Each nucleosome unit contains about 200 nucleotide pairs of which about 145 are in contact with the protein. The remaining 55 nucleotide pairs link adjacent nucleosomes. Histone H1 binds to the linker segment and draws the nucleosomes nearer to one another. The DNA in its nucleosome form is further compacted into a helical fiber, the 30-nm fiber. In the nucleus of a nondividing cell, the 30-nm fiber corresponding to each chromosome arm occupies a distinct chromosome territory. In dividing cells, the 30-nm fiber forms a visible chromosome by undergoing several additional levels of coiling, producing a highly compact visible chromosome. The result is that a eukaryotic DNA molecule, whose length and width are about 50,000 and 0.002 μm, respectively, is folded to form a chromosome with a length of about 5 μm and a width of about 0.5 μm.

3.5 The centromere and telomeres are essential parts of chromosomes.

- The centromere is essential for chromosome segregation.
- The telomere is essential for the stability of the chromosome tips.

Centromeres and telomeres are regions of eukaryotic chromosomes specialized for spindle-fiber attachment and stabilization of the tips, respectively. The centromeres of most higher eukaryotes are associated with localized, highly repeated, satellite DNA sequences. Telomeres are formed by a telomerase enzyme that contains a guide RNA that serves as a template for the addition of nucleotides to the $3'$ end of a telomerase addition site. The complementary strand is synthesized by the normal DNA replication enzymes. In mammals and other vertebrates, the $3'$ strand of the telomere terminates in tandem repeats of the simple sequence $5'$-TTAGGG-$3'$. Relatively few copies of this sequence are needed to prime the telomerase.

3.6 Genes are located in chromosomes.

- Special chromosomes determine sex in many organisms.
- X-linked genes are inherited according to sex.
- Hemophilia is a classic example of human X-linked inheritance.
- In birds, moths, and butterflies, the sex chromosomes are reversed.
- Experimental proof of the chromosome theory came from nondisjunction.

The X and Y sex chromosomes differ from other chromosome pairs in that they are visibly different and contain different genes. In mammals and in many insects and other animals, as well as in some flowering plants, the female contains two X chromosomes (XX) and the male contains one X chromosome and one Y chromosome (XY). In birds, moths, butterflies, and some reptiles, the situation is the reverse. The Y chromosome in many species contains only a few genes, but in human beings and other mammals it includes a male-determining factor. In most organisms, the X chromosome contains many genes unrelated to sexual differentiation. These X-linked genes show a characteristic pattern of inheritance that is due to their location in the X chromosome.

3.7 Genetic data analysis makes use of probability and statistics.

- Progeny of crosses are predicted by the binomial probability formula.
- Chi-square tests goodness of fit of observed to expected numbers.
- Are Mendel's data a little too good to be true?

The progeny of genetic crosses often conform to the theoretical predictions of the binomial probability formula. The degree to which the observed numbers of different genetic classes of progeny fit theoretically expected numbers is usually ascertained with a chi-square (χ^2) test. On the basis of the criterion of the χ^2 test, Mendel's data fit the expectations somewhat more closely than chance would dictate. However, the bias in the data is relatively small and is unlikely to be due to anything more than his recounting or repeating certain experiments whose results were regarded as unsatisfactory.

issues & ideas

- What is the genetic significance of the fact that gametes contain half the chromosome complement of somatic cells?
- The term *mitosis* derives from the Greek *mitos,* which means "thread." The term *meiosis* derives from the Greek *meioun,* which means "to make smaller." What feature, or features, of these types of nuclear division might have led to the choice of these terms?
- Explain the meaning of the terms *reductional division* and *equational division.* What is "reduced" or "kept equal"? To which nuclear divisions do the terms refer?
- How is independent assortment of genes on different chromosomes related physically to the process of chromosome alignment on the metaphase plate in meiosis I?

- What are some of the important differences between the first meiotic division and the second meiotic division?
- Why is X-linked inheritance often called "criss-cross inheritance"? How can this term be misleading in regard to the genetic transmission of the X chromosome?
- In what ways is the inheritance of Y-linked genes different from that of X-linked genes?
- How did nondisjunction "prove" the chromosome theory of heredity?
- Why is a statistical test necessary to determine whether an observed set of data yields an acceptable fit to the result expected from a particular genetic hypothesis? What statistical test is conventionally used for this purpose?

key terms & concepts

alpha satellite	centromere	chromosome complement	degrees of freedom
anaphase	checkpoint	chromosome territory	diakinesis
anaphase I	chiasma	chromosome theory of	diploid
anaphase II	chi-square (χ^2)	heredity	diplotene
autosomes	chromatid	crossing-over	equational division
bivalent	chromatin	core particle	euchromatin
cell cycle	chromosome	cytokinesis	factorial

G_1 period	meiosis	prophase I	synapsis
G_2 period	metaphase	prophase II	telomerase
germ cell	metaphase I	reductional division	telomere
goodness of fit	metaphase II	satellite DNA	telophase
haploid	metaphase plate	S period	telophase I
hemophilia A	mitosis	scaffold	telophase II
heterochromatin	nondisjunction	sex chromosome	tetrad
histone	nucleosome	sister chromatid	30-nm fiber
homologous chromosomes	nucleolus	somatic cell	X chromosome
interphase	pachytene	spindle	X-linked gene
kinetochore	Pascal's triangle	statistically highly significant	Y chromosome
leptotene	prophase	statistically significant	zygotene
M period			

1. _____ Term for the pairing of homologous chromosomes that takes place during zygotene of meiosis.

2. _____ The region of a chromosome to which the spindle fibers become attached; geneticists usually refer to it as the centromere.

3. _____ This is the "double thread" substage of meiosis I at which bivalents are clearly visible.

4. _____ Any of the cross-shaped structures that bridge between pairs of nonsister chromatids in a bivalent.

5. _____ Enzyme containing a guide RNA that is responsible for maintaining the length of the chromosome tips.

6. _____ This anaphase separation marks the equational division.

7. _____ The protein component of nucleosomes.

8. _____ In a cell not cycling through mitosis, the region of the nucleus in which the 30-nm nucleosome fiber corresponding to a given chromosome or chromosome arm is located.

9. _____ Regions of chromosomes, often adjacent to the centromere, that differ from euchromatin in taking up chromosome dyes during interphase; these regions contain a high proportion of repetitive DNA and a low density of genes.

10. _____ Failure of anaphase separation of a pair of chromosomes or chromatids during meiosis, resulting in a gamete that lacks a chromosome or that contains an extra copy of a chromosome.

11. _____ Test for goodness of fit in which the test statistic is the sum over all classes of data of *(observed number − expected number)^2/expected number*.

12. _____ Result of a statistical test in which the *P*-value is smaller than 0.05 but greater than 0.01.

solutions: step by step

Problem 1

The accompanying diagrams show the appearance of a pair of homologous chromosomes in prophase I of meiosis. Arrange the diagrams in chronological order, and identify each stage as leptotene, zygotene, pachytene, diplotene, or diakinesis.

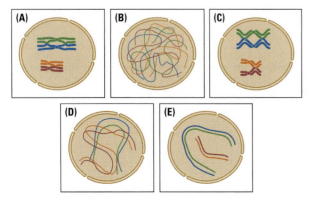

Solution First recall what distinguishes one stage of prophase I from the next. *Leptotene* literally means "thin thread," when each chromosome is in an extended, thread-like condition prior to synapsis; this stage corresponds to diagram B. *Zygotene* means "paired threads," and the pairing begins at the chromosome tips; this is configuration D. *Pachytene* means "thick thread"; it commences when pairing is completed and the homologous chromosomes still appear to be single, which corresponds to diagram E. *Diplotene* means "double thread"; at this time each homologous chromosome clearly consists of two sister chromatids, and chiasmata are apparent, which is shown in diagram A. *Diakinesis* means "moving apart," and in this stage the synapsed homologous chromosomes begin to repel one another, being held together by the chiasmata, producing configuration C. Therefore, the order of the stages is BDEAC; that is, leptotene (B), zygotene (D), pachytene (E), diplotene (A), and diakinesis (C).

Problem 2

Most color blindness in people is due to relatively common X-linked recessive alleles. A woman with normal color vision whose father was color blind marries a normal man. What types of color vision are expected in the offspring, and in what frequencies?

■ Solution In these kinds of problems it is helpful to draw a pedigree, showing the information given, and to identify the genotypes of the persons in the pedigree insofar as possible. In this case the pedigree is as indicated; the woman whose father was color blind is number II-1. (Her father's genotype must be as shown, because he was color blind.) We are told nothing about the mother's genotype, but because II-1 has normal color vision, the mother I-1 must have at least one nonmutant allele (designated cb^+). The normal male II-2 must have a nonmutant allele in his X chromosome, as shown.

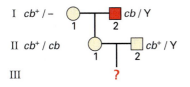

The progeny in question are those in generation III, and their expected composition is shown in the Punnett square that follows. The expected offspring are 1/2 normal females, 1/4 normal males, and 1/4 color-blind males. Note that half of the female offspring are carriers of the recessive allele (heterozygous).

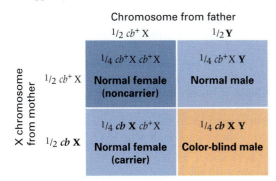

Problem 3

Suppose that a cell undergoes meiosis in a normal human male and nondisjunction of the sex chromosomes takes place. Determine what chromosome constitution would result in a zygote formed from a normal egg and either of the abnormal gametes resulting from nondisjunction, under the following conditions:

(a) The nondisjunction takes place in meiosis I.
(b) Nondisjunction happens to the X chromosome in meiosis II.
(c) Nondisjunction happens to the Y chromosome in meiosis II.

Problem 3 Solution—Diagrams of Meiotic Divisions

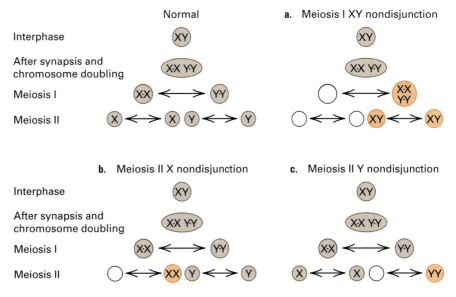

■ Solution In approaching problems like this, it is essential to draw diagrams of the meiotic divisions, showing the nondisjunctions that are postulated to happen. The consequences then become clear. The diagrams shown above illustrate the normal situation, along with the three types of nondisjunction stipulated in the problem. The raised dot in the symbols $X \cdot X$ and $Y \cdot Y$ serves to indicate that at this stage, each chromosome consists of two chromatids attached to a single centromere. The consequences of the nondisjunction events are clear from the diagrams.

(a) The abnormal gametes resulting from XY nondisjunction in meiosis I carry either no sex chromosome ("nullo-X") or both an X and a Y. A zygote from a nullo-X gamete will have 45 chromosomes, with a missing X (this chromosome constitution is designated 45,X); a zygote from an XY gamete will have 47 chromosomes with an XXY sex-chromosome constitution (designated 47,XXY). **(b)** The abnormal gametes resulting from X nondisjunction in meiosis II are either nullo-X or XX. The nullo-X gamete yields a 45,X zygote, and the XX gamete yields a 47,XXX zygote (47 chromosomes total, with three X chromosomes). **(c)** The abnormal gametes resulting from Y nondisjunction in meiosis II are either nullo-Y or YY. The nullo-Y gamete yields a 45,X zygote, and the YY gamete yields a 47,XYY zygote (47 chromosomes total, with an XYY sex-chromosome constitution).

Problem 4

Certain people are able to taste the chemical phenylthiocarbamide (PTC) when it is present in wet filter paper. These people are called "tasters." Others are unable to detect PTC and have the "nontaster" phenotype. PTC tasting is an example of variation in sensory perception, and the genetic basis is thought to be rather simple. The ability to taste PTC is attributed to a dominant allele, denoted T, located in chromosome 7. The recessive allele is designated t. The genotypes TT and Tt are tasters of PTC, and tt genotypes are nontasters. From 204 matings of heterozygous tasters with nontasters, 259 taster and 278 nontaster progeny were observed. Use a chi-square test to determine whether these numbers give a satisfactory fit to the Mendelian expectation.

■Solution First you need to determine the Mendelian expectation. The problem specifies that the matings are heterozygous tasters with nontasters, or $Tt \times tt$. The Mendelian expectation is therefore $1/2$ Tt (taster) : $1/2$ tt (nontaster) progeny. The total number of progeny observed was $259 + 278 = 537$, so the expected number of tasters : nontasters is $268.5 : 268.5$. The chi-square value is given by Σ ($observed - expected$)^2/$expected$, where the sum is over all classes of data. In this case,

$$\chi^2 = \frac{(259 - 268.5)^2}{268.5} + \frac{(278 - 268.5)^2}{268.5}$$

$$= 0.336 + 0.336$$

$$= 0.672$$

This chi-square has 1 degree of freedom, because there are two classes of data. The corresponding P value from Figure 3.34 is about 0.47, which means that there is about a 47 percent chance of obtaining a fit as bad or worse than $259 : 278$. This means that there is no reason, on the basis of these data, to reject the hypothesis of simple Mendelian inheritance of tasting ability.

concepts in action: problems for solution

3.1 Explain the statement "The physical separation of homologous chromosomes in anaphase is the physical basis of Mendel's principle of segregation."

3.2 The chemical colchicine is a "spindle poison" that interferes with the organization of the spindle. Somatic cells undergoing division in the presence of colchicine arrest at metaphase. Eventually the splitting of the centromeres that is characteristic of anaphase occurs, but cell division does not take place. If a human cell with a normal diploid complement of 46 chromosomes undergoes one round of the cell cycle in the presence of colchicine, what is the expected number of chromosomes in the resulting cell?

3.3 Explain the statement "Genes on different chromosomes undergo independent assortment because nonhomologous chromosomes align at random on the metaphase plate in meiosis I."

3.4 A somatic cell has 24 chromosomes aligned at metaphase. How many chromosomes are present at anaphase, after the centromeres have split?

3.5 Emmer wheat (*Triticum dicoccum*) has a somatic chromosome number of 28, and rye (*Secale cereale*) has a somatic chromosome number of 14. Hybrids produced by crossing these cereal grasses are highly sterile and have many characteristics intermediate between the parental species. How many chromosomes do the hybrids possess?

3.6 What would you expect to happen to the chromosomes in successive cell cycles in a cell lineage that had a nonfunctional telomerase? Explain your answer.

3.7 It is often advantageous to be able to determine the sex of newborn chickens from their plumage. How could this be done by using the Z-linked dominant allele S for silver plumage and the recessive allele s for gold plumage?

(Remember that, in chickens, the homogametic and heterogametic sexes are the reverse of those in mammals: Females are WZ and males are ZZ.)

3.8 The diagrams shown here depict anaphase in cell division in a cell of a hypothetical organism with two pairs of chromosomes. Identify the panels as being anaphase of mitosis, anaphase I of meiosis, or anaphase II of meiosis, stating on what basis you reached your conclusions.

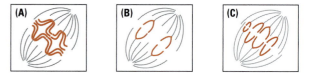

3.9 A cytogeneticist examining cells in *Tradescantia* stamen hairs is trying to determine the length of the various stages in mitosis and the cell cycle. She examines 2000 cells and finds 320 cells in prophase, 150 cells in metaphase, 80 cells in anaphase, and 120 cells in telophase. Assuming that the cells are sampled in proportion to the duration of each stage in the cell cycle, what conclusion can be drawn about the relative length of each stage of the cell cycle, including the time spent in interphase? Express each answer as a percentage of the total cell-cycle time.

3.10 In *Drosophila*, sex is determined by the ratio of X chromosomes to autosomes, called the X/A ratio, where A indicates one complete haploid set of autosomes. Embryos with X/A of 1 or greater develop as females, those with X/A of 1/2 or less develop as males, and those with intermediate values develop as intersexes. Using this information, deduce the expected sex of each of the following genotypes of flies.
(a) X/AA   **(b)** XX/AA
(c) XXX/AA   **(d)** XX/AAA

3.11 The most common form of color blindness in human beings results from X-linked recessive alleles. One type of allele, call it cb^r, results in defective red perception, whereas another type of allele, call it cb^g, results in defective green perception. A woman who is heterozygous cb^r/cb^g and a normal male produce a son whose chromosome constitution is XXY. What are the possible genotypes of this boy under each of the following circumstances?

(a) The nondisjunction took place in meiosis I in the mother.

(b) The nondisjunction took place in the cb^r-bearing chromosome in meiosis II in the mother.

(c) The nondisjunction took place in the cb^g-bearing chromosome in meiosis II in the mother.

3.12 A recessive mutation of an X-linked gene in human beings results in hemophilia, marked by a prolonged increase in the time needed for blood clotting. Suppose that a phenotypically normal couple produces two normal daughters and a son affected with hemophilia.

(a) What is the probability that both of the daughters are heterozygous carriers?

(b) If one of the daughters and a normal man produce a son, what is the probability that the son will be affected?

3.13 People with the chromosome constitution 47,XXY are phenotypically males. A normal woman whose father had hemophilia mates with a normal man and produces an XXY son who also has hemophilia. What kind of nondisjunction can explain this result?

3.14 Mendel studied the inheritance of phenotypic characters determined by seven pairs of alleles. It is an interesting coincidence that the pea plant also has seven pairs of chromosomes. What is the probability that no two of the traits studied by Mendel were determined by genes located on the same pair of chromosomes?

3.15 Duchenne-type muscular dystrophy is an inherited disease of muscle due to a mutant form of a protein called dystrophin. The pattern of inheritance of the disease has these characteristics: (1) affected males have unaffected children, (2) the unaffected sisters of affected males often have affected sons, and (3) the unaffected brothers of affected males have unaffected children. What type of inheritance do these findings suggest? Explain your reasoning.

3.16 Tall, red-flowered hibiscus is mated with short, white-flowered hibiscus. Both varieties are true-breeding. All the F_1 plants are backcrossed with the short, white-flowered variety. This backcross yields 188 tall red, 203 tall white, 175 short red, and 178 short white plants. Does the observed result fit the genetic hypothesis of $1 : 1 : 1 : 1$ segregation as assessed by a chi-square test?

3.17 What are the values of chi-square that yield P values of 5 percent (statistically significant) when there are 1, 2, 3, 4, and 5 degrees of freedom? For the types of chi-square tests illustrated in this chapter, how many classes of data do these degrees of freedom represent? Because the significant chi-square values increase with the number of degrees of freedom, does this mean that it becomes increasingly "hard" (less likely) to obtain a statistically significant chi-square value when the genetic hypothesis is true?

3.18 In *D. melanogaster*, the alleles dp^+ and dp determine long versus short wings, and e^+ and e determine gray versus ebony body. A dihybrid cross was carried out to produce flies homozygous for both dp and e. The following phenotypes were obtained in the F_2 generation:

long wing, gray body	462
long wing, ebony body	167
short wing, gray body	127
short wing, ebony body	44

Test these data for agreement with the $9 : 3 : 3 : 1$ ratio expected if the two pairs of alleles undergo independent assortment.

3.19 Are the observed progeny numbers of 11, 10, 22, and 23 consistent with a genetic hypothesis that predicts a $1 : 1 : 1 : 1$ ratio?

3.20 The accompanying pedigree and gel diagram show the molecular phenotypes obtained from genomic DNA samples. The bands are characteristic DNA fragments that distinguish two alleles of a single gene. What mode of inheritance does the pedigree suggest? On the basis of this hypothesis, and using A_1 to represent the allele associated with the 4-kb band and A_2 to represent that associated with the 9-kb band, deduce the genotype of each individual in the pedigree.

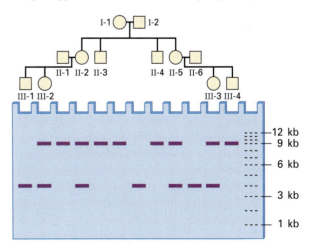

further readings

Allen, G. E. 1978. *Thomas Hunt Morgan: The Man and His Science*. Princeton, NJ: Princeton University Press.

Chandley, A. C. 1988. Meiosis in man. *Trends in Genetics* 4: 79.

Cohen, J. S., and M. E. Hogan. 1994. The new genetic medicine. *Scientific American,* December.

Cremer, T., and C. Cremer. 2001. Chromosome territories, nuclear architecture and gene regulation in mammalian cells. *Nature Reviews Genetics* 2: 292.

Ingber, D. E. 1998. The architecture of life. *Scientific American*, January.

McIntosh, J. R., and K. L. McDonald. 1989. The mitotic spindle. *Scientific American,* October.

McKusick, V. A. 1965. The royal hemophilia. *Scientific American,* August.

Miller, O. J. 1995. The fifties and the renaissance of human and mammalian genetics. *Genetics* 139: 484.

Page, A. W., and T. L. Orr-Weaver. 1997. Stopping and starting the meiotic cell cycle. *Current Opinion in Genetics & Development* 7: 23.

Prescott, J. C., and E. H. Blackburn. 1999. Telomerase: Dr. Jekyll or Mr. Hyde? *Current Opinion in Genetics & Development* 9: 368.

She, X. et al. 2004. The structure and evolution of centromeric transition regions within the human genome. *Nature* 430: 857.

Sokal, R. R., and F. J. Rohlf. 1969. *Biometry.* New York: Freeman.

Sturtevant, A. H. 1965. *A Short History of Genetics.* New York: Harper & Row.

Voeller, B. R., ed. 1968. *The Chromosome Theory of Inheritance: Classical Papers in Development and Heredity.* New York: Appleton-Century-Crofts.

Welsh, M. J., and A. E. Smith. 1995. Cystic fibrosis. *Scientific American*, December.

Zielenski, J., and L. C. Tsui. 1995. Cystic fibrosis: Genotypic and phenotypic variations. *Annual Review of Genetics* 29: 777.

geNETics on the web

GeNETics on the Web will introduce you to some of the most important sites for finding genetic information on the Internet. To explore these sites, visit the Jones and Bartlett home page at

http://www.jbpub.com/genetics

For the book *Essential Genetics: A Genomics Perspective,* choose the link that says Enter **GeNETics on the Web**. You will be presented with a chapter-by-chapter list of highlighted keywords. Select any highlighted keyword and you will be linked to a Web site containing genetic information related to the keyword.

- A thorough command of chromosome behavior during cell division is an absolute prerequisite to understanding transmission genetics. The keyword **meiosis** links to a site that reviews key features of mitosis and meiosis and reinforces this information with animations.

- Duchenne muscular dystrophy (**DMD**) is a wasting disease of muscle that results from an X-linked mutation in one of the largest genes in the human genome, spanning more than 2.3 million nucleotide pairs. It affects about 1 in every 3500 newborn boys and has no cure. Early signs usually occur between the ages of 2 and 6 when affected boys are observed to fall frequently, have difficulty getting up from a reclining position, and have a waddling gait. Log onto this keyword site to learn more about Duchenne muscular dystrophy and other inherited diseases of muscle.

Busy bees. The insect order Hymenoptera is genetically unique in that the females are diploid but the males are haploid. The haploid males arise from the development of unfertilized eggs. This mode of inheritance has important implications for the genetic relationship between a mother and her offspring. [Courtesy of Jack Dykinga/USDA].

key concepts

- Genes that are located in the same chromosome and that do not show independent assortment are said to be linked.
- The alleles of linked genes present together in the same chromosome tend to be inherited as a group.
- Crossing-over between homologous chromosomes results in recombination, which breaks up combinations of linked alleles.
- A genetic map depicts the relative positions of genes along a chromosome.

- The map distance between genes in a genetic map is related to the rate of recombination between the genes.
- Physical distance along a chromosome is often, but not always, correlated with map distance.
- Tetrads are sensitive indicators of linkage because they include all the products of meiosis.
- At the DNA level, recombination is initiated by a double-stranded break in a DNA molecule. Use of the homologous DNA molecule as a template for repair can result in a crossover, in which both strands of the participating DNA molecules are broken and rejoined.

4

Gene Linkage and Genetic Mapping

chapter organization

4.1 Linked alleles tend to stay together in meiosis.

4.2 Recombination results from crossing-over between linked alleles.

4.3 Double crossovers are revealed in three-point crosses.

4.4 Polymorphic DNA sequences are used in human genetic mapping.

4.5 Tetrads contain all four products of meiosis.

4.6 Recombination is initiated by a double-stranded break in DNA.

the human connection
Count Your Blessings

Chapter Summary
Issues & Ideas
Key Terms & Concepts
Solutions: Step by Step
Concepts in Action: Problems for Solution
Further Readings

geNETics on the web

Genetic mapping means determining the relative positions of genes along a chromosome. It is one of the main experimental tools in genetics. This may seem odd in organisms in which the DNA sequence of the genome has been determined. If every gene in an organism is already sequenced, then what is the point of genetic mapping? The answer is that a gene's sequence does not always reveal its function, nor does a genomic DNA sequence reveal which genes interact in a complex biological process. When a new mutant gene is discovered, the first step in genetic analysis is usually genetic mapping to determine its position in the genome. It is at this point that the genomic sequence, if known, becomes useful, because in some cases the position of the mutant gene coincides with a gene whose sequence suggests a role in the biological process being investigated. For example, in the case of flower color, a new mutation may map to a region containing a gene whose sequence suggests that it encodes an enzyme in anthocyanin synthesis. But the function of a gene is not always revealed by its DNA sequence, and so in some cases, further genetic or molecular analysis is necessary to sort out which one of the genes in a sequenced region corresponds to a mutant gene mapped to that region. In human genetics, genetic mapping is important because it enables genes associated with hereditary diseases, such as those that predispose to breast cancer, to be localized and correlated with the genomic sequence in the region.

4.1

Linked alleles tend to stay together in meiosis.

As we saw in Chapter 3, homologous chromosomes form pairs in prophase I of meiosis by undergoing synapsis, and the individual members of each pair separate from one another at anaphase I. Genes that are close enough together in the same chromosome might therefore be expected to be transmitted together. Thomas Hunt Morgan examined this issue using two genes present in the X chromosome of *Drosophila*. One was a mutation for white eyes, the other a mutation for miniature wings. Morgan found that the *white* and *miniature* alleles present in each X chromosome of a female do tend to remain together in inheritance, a phenomenon known as **linkage.** Nevertheless, the linkage is incomplete. Some gametes are produced that have different combinations of the *white* and *miniature* alleles than those in the parental chromosomes. The new combinations are produced because homologous chromosomes can exchange segments when they are paired. This process

(crossing-over) results in **recombination** of alleles between the homologous chromosomes. The probability of recombination between any two genes serves as a measure of genetic distance between the genes and allows the construction of a **genetic map,** which is a diagram of a chromosome showing the relative positions of the genes. The linear order of genes along a genetic map is consistent with the conclusion that each gene occupies a well-defined position, or **locus,** in the chromosome, with the alleles of a gene in a heterozygote occupying corresponding locations in the pair of homologous chromosomes.

In discussing linked genes, it is necessary to distinguish which alleles are present together in the parental chromosomes. This is done by means of a slash ("/", also called a virgule). The alleles in one chromosome are depicted to the left of the slash, and those in the homologous chromosome are depicted to the right of the slash. For example, in the cross $AA\ BB \times aa\ bb$, the genotype of the doubly heterozygous progeny is denoted $A\ B/a\ b$ because the A and B alleles were inherited in one parental chromosome and the alleles a and b were inherited in the other parental chromosome. In this genotype the A and B alleles are said to be in the **coupling** or *cis* **configuration;** likewise, the a and b alleles are in coupling. Among the four possible types of gametes, the $A\ B$ and $a\ b$ types are called **parental combinations** because the alleles are in the same configuration as in the parental chromosomes, and the $A\ b$ and $a\ B$ types are called **recombinants** (Figure 4.1, part A).

Another possible configuration of the A, a and B, b allele pairs is $A\ b/a\ B$. In this case the A and B alleles are said to be in the **repulsion** or *trans* **configuration.** Now the parental and recombinant gametic types are reversed (Figure 4.1, part B). The $A\ b$ and $a\ B$ types are the parental combinations, and the $A\ B$ and $a\ b$ types are the recombinants.

■ The degree of linkage is measured by the frequency of recombination.

In his early experiments with *Drosophila*, Morgan found mutations in each of several X-linked genes that provided ideal materials for studying linkage. One of these genes, with alleles w^+ and w, determines normal red eye color versus white eyes, as discussed in Chapter 3; another such gene, with the alleles m^+ and m, determines whether the size of the wings is normal or miniature. The initial cross is shown as Cross 1 in Figure 4.2. It was a cross between females with white eyes and normal wings and males with red eyes and miniature wings:

$$\frac{w\ m^+}{w\ m^+}\ ♀♀ \times \frac{w^+\ m}{Y}\ ♂♂$$

(A) Parental alleles in coupling or *cis* configuration

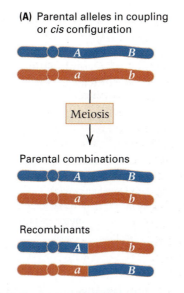

(B) Parental alleles in repulsion or *trans* configuration

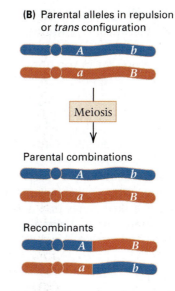

Figure 4.1 For any pair of alleles, the gametes produced through meiosis have the alleles either in a parental configuration or in a recombinant configuration. Which types are parental and which recombinant depends on whether the configuration of the alleles in the parent is (A) coupling or (B) repulsion.

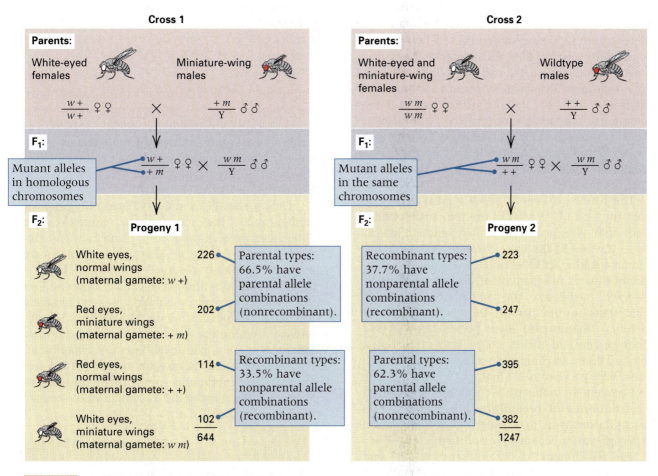

Figure 4.2 An experiment demonstrating that the frequency of recombination between two mutant alleles is independent of whether they are present in the same chromosome or in homologous chromosomes. (A) Cross 1 produces F_1 females with the genotype $w+/+m$, and the $w-m$ recombination frequency is 33.5 percent. (B) Cross 2 produces F_1 females with the genotype $wm/++$, and the $w-m$ recombination frequency is 37.7 percent. These values are within the range of variation expected to occur by chance.

In this way of writing the genotypes, the horizontal line replaces the slash. Alleles written above the line are present in one chromosome, and those written below the line are present in the homologous chromosome. In the females, both X chromosomes carry w and m^+. In males, the X chromosome carries the alleles w^+ and m. (The Y written below the line denotes the Y chromosome in the male.) Figure 4.2 illustrates a simplified symbolism, commonly used in *Drosophila* genetics, in which a wildtype allele is denoted by a + sign in the appropriate position. The + symbolism is unambiguous because the linked genes in a chromosome are always written in the same order. Using the + notation,

$$\frac{w\,+}{w\,+} \quad \text{means} \quad \frac{w\,m^+}{w\,m^+}$$

and

$$\frac{+\,m}{Y} \quad \text{means} \quad \frac{w^+\,m}{Y}$$

The resulting F_1 female progeny from Cross 1 have the genotype $w\,+/+\,m$ (or, equivalently, $w\,m^+/w^+\,m$). In this genotype, the w^+ and m^+ alleles are in repulsion. When these females were mated with $w\,m/Y$ males, the offspring denoted as Progeny 1 in Figure 4.2 were obtained. In each class of progeny, the gamete from the female parent is shown in the column at the left, and the gamete from the male parent carries either $w\,m$ or the Y chromosome. The cross is equivalent to a testcross, and so the phenotype of each class of progeny reveals the alleles present in the gamete from the mother.

The results of Cross 1 show a great departure from the $1:1:1:1$ ratio of the four male phenotypes that is expected with independent assortment. If genes in the same chromosome tended to remain together in inheritance but were not completely linked, this pattern of deviation might be observed. In this case, the combinations of phenotypic traits in the parents of the original cross (parental phenotypes) were present in 428/644 (66.5 percent) of the F_2 males, and nonparental combinations (recombinant phenotypes) of the traits were present in 216/644 (33.5 percent). The 33.5 percent recombinant X chromosomes is called the **frequency of recombination,** and it should be contrasted with the 50 percent recombination expected with independent assortment.

The recombinant X chromosomes $w^+\,m^+$ and $w\,m$ result from crossing-over in meiosis in F_1 females. In this example, the frequency of recombination between the linked w and m genes was 33.5 percent. With other pairs of linked genes,

the frequency of recombination ranges from near 0 to 50 percent. Even genes in the same chromosome can undergo independent assortment (frequency of recombination equal to 50 percent) if they are sufficiently far apart. This implies the following principle:

key concept

Genes with recombination frequencies smaller than 50 percent are present in the same chromosome (linked). Two genes that undergo independent assortment, indicated by a recombination frequency equal to 50 percent, either are in nonhomologous chromosomes or are located far apart in a single chromosome.

■ The frequency of recombination is the same for coupling and repulsion heterozygotes.

Morgan also studied progeny from the coupling configuration of the w^+ and m^+ alleles, which results from the mating designated as Cross 2 in Figure 4.2. In this case, the original parents had the genotypes

$$\frac{w\,m}{w\,m} \,♀♀ \times \frac{+\,+}{Y}\,♂♂$$

The resulting F_1 female progeny from Cross 2 have the genotype $w\,m/+\,+$ (equivalently, $w\,m/w^+\,m^+$). In this case the wildtype alleles are in the same chromosome. When these F_1 female progeny were crossed with $w\,m/Y$ males, they yielded the types of progeny tabulated as Progeny 2 in Figure 4.2.

Because the alleles in Cross 2 are in the coupling configuration, the parental-type gametes carry either $w\,m$ or $+\,+$, and the recombinant gametes carry either $w\,+$ or $+\,m$. The types of gametes are the same as those observed in Cross 1, but the parental and recombinant types are opposite. Yet the frequency of recombination is approximately the same: 37.7 percent versus 33.5 percent. The difference is within the range expected to result from random variation from experiment to experiment. The consistent finding of equal recombination frequencies in experiments in which the mutant alleles are in the *trans* or the *cis* configuration leads to the following conclusion:

key concept

Recombination between linked genes takes place with the same frequency whether the alleles of the genes are in the repulsion (*trans*) configuration or in the coupling (*cis*) configuration; it is the same no matter how the alleles are arranged.

over (part A), the alleles present in each homologous chromosome remain in the same combination. When a crossover does take place (part B), the outermost alleles in two of the chromatids are interchanged (recombined).

The unit of distance in a genetic map is called a **map unit;** one map unit is equal to 1 percent recombination. For example, two genes that recombine with a frequency of 3.1 percent are said to be located 3.1 map units apart. One map unit is also called a **centimorgan,** abbreviated cM, in honor of T. H. Morgan. A distance of 3.1 map units therefore equals 3.1 centimorgans and indicates 3.1 percent recombination between the genes. An example is shown in part A of Figure 4.5, which deals with the *Drosophila* mutants *w* for white eyes and *dm (diminutive)* for small body. The female parent in the testcross is the *trans* heterozygote, but as we have seen, this configuration is equivalent in frequency of recombination to the *cis* heterozygote. Among 1000 progeny there are 31 recombinants. Using this estimate, we can express the genetic distance between *w* and *dm* in four completely equivalent ways:

- As the *frequency of recombination*—in this case 0.031
- As the *percent recombination,* or 3.1 percent
- As the distance in *map units*—in this example, 3.1 map units
- As the distance in *centimorgans,* or 3.1 centimorgans (3.1 cM)

A genetic map based on these data is shown in Figure 4.5, part B. The chromosome is represented as a horizontal line, and each gene is assigned a position on the line according to its genetic distance from other genes. In this example, there are only two genes, *w* and *dm,* and they are separated by a distance of 3.1 centimorgans (3.1 cM), or 3.1 map units. Genetic maps are usually truncated to show only the genes of interest. The full genetic map of the *Drosophila* X chromosome extends considerably farther in both directions than indicated in this figure.

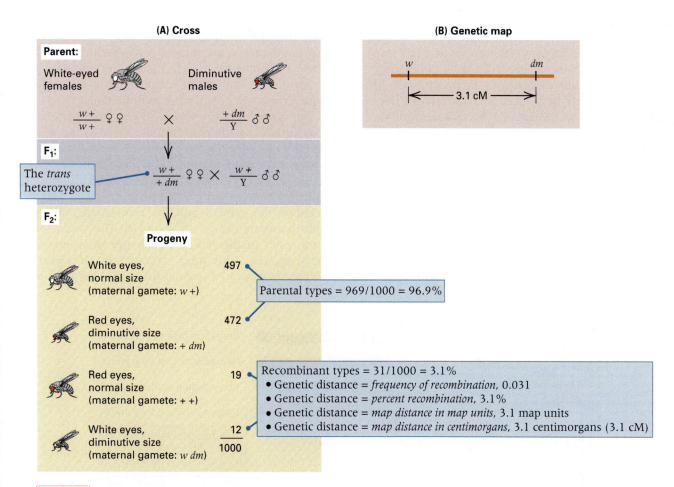

Figure 4.5 An experiment illustrating how the frequency of recombination is used to construct a genetic map. (A) There is 3.1 percent recombination between the genes *w* and *dm*. (B) A genetic map with *w* and *dm* positioned 3.1 map units (3.1 centi-morgans, cM) apart, corresponding to 3.1 percent recombination. The map distance equals frequency of recombination only when the frequency of recombination is sufficiently small.

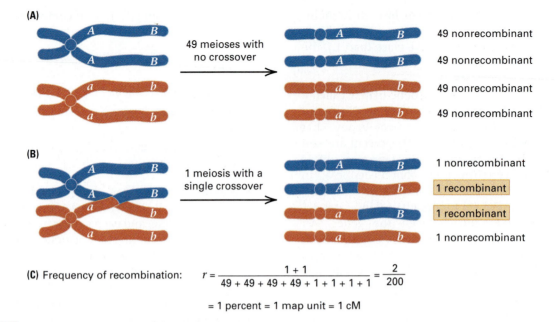

(A)

49 meioses with no crossover

49 nonrecombinant
49 nonrecombinant
49 nonrecombinant
49 nonrecombinant

(B)

1 meiosis with a single crossover

1 nonrecombinant
1 recombinant
1 recombinant
1 nonrecombinant

(C) Frequency of recombination:

$$r = \frac{1 + 1}{49 + 49 + 49 + 49 + 1 + 1 + 1 + 1} = \frac{2}{200}$$

$$= 1 \text{ percent} = 1 \text{ map unit} = 1 \text{ cM}$$

Figure 4.6 Diagram of chromosomal configurations in 50 meiotic cells, in which 1 has a crossover between two genes. (A) The 49 cells without a crossover result in 98 $A\,B$ and 98 $a\,b$ chromosomes; these are all nonrecombinant. (B) The cell with a crossover yields chromosomes that are $A\,B$, $A\,b$, $a\,B$, and $a\,b$, of which the middle two types are recombinant chromosomes. (C) The recombination frequency equals 2/200, or 1 percent, also called 1 map unit or 1 cM. Hence, 1 percent recombination means that 1 meiotic cell in 50 has a crossover in the region between the genes.

Physically, one map unit corresponds to a length of the chromosome in which, on the average, one crossover is formed in every 50 cells undergoing meiosis. This principle is illustrated in Figure 4.6 . If one meiotic cell in 50 has a crossover, the frequency of *crossing-over* equals 1/50, or 2 percent. Yet the frequency of *recombination* between the genes is 1 percent. The correspondence of 1 percent recombination with 2 percent crossing-over is a little confusing until you consider that a crossover results in two recombinant chromatids and two nonrecombinant chromatids (Figure 4.6). A crossover frequency of 2 percent means that of the 200 chromosomes that result from meiosis in 50 cells, exactly 2 chromosomes (those involved in the crossover) are recombinant for genetic markers spanning the particular chromosome segment. To put the matter in another way, 2 percent crossing-over corresponds to 1 percent recombination because only half of the chromatids in each cell with a crossover are actually recombinant.

In situations in which there are **genetic markers** along the chromosome, such as the A, a and B, b pairs of alleles in Figure 4.6, recombination between the marker genes takes place only when a crossover occurs *between* the genes. Figure 4.7 illustrates a case in which a crossover takes place between the gene A and the centromere, rather than between the genes A and B. The crossover does result in the physical exchange of segments between the innermost chromatids. However, because it is located outside the region between A

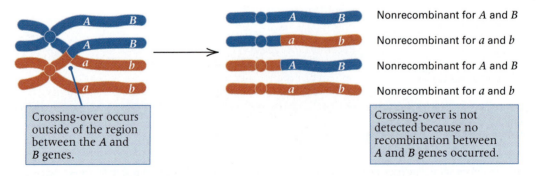

Nonrecombinant for A and B
Nonrecombinant for a and b
Nonrecombinant for A and B
Nonrecombinant for a and b

Crossing-over occurs outside of the region between the A and B genes.

Crossing-over is not detected because no recombination between A and B genes occurred.

Figure 4.7 Crossing-over outside the region between two genes is not detectable through recombination. Although a segment of chromosome is exchanged, the genetic markers outside the region of the crossovers stay in the nonrecombinant configurations, in this case $A\,B$ and $a\,b$.

and B, all of the resulting gametes must carry either the A B or the a b allele combination. These are nonrecombinant chromosomes. The presence of the crossover is undetected because it is not in the region between the genetic markers.

In some cases, the region between genetic markers is large enough that two (or even more) crossovers can be formed in a single meiotic cell. One possible configuration for two crossovers is shown in Figure 4.8. In this example, both crossovers are between the same pair of chromatids. The result is that there is a physical exchange of a segment of chromosome between the marker genes, but the double crossover remains undetected because the markers themselves are not recombined. The absence of recombination results from the fact that the second crossover reverses the effect of the first, insofar as recombination between A and B is concerned. The resulting chromosomes are either A B or a b, both of which are nonrecombinant.

Because double crossovers in a region between two genes can remain undetected (this happens when they do not result in recombinant chromosomes), there is an important distinction between the distance between two genes as measured by the recombination frequency and as measured in map units:

- The *map distance* between two genes equals one-half of the average number of crossovers that take place in the region per meiotic cell; it is a measure of crossing-over.

- The *recombination frequency* between two genes indicates how much recombination is actually observed in a particular experiment; it is a measure of recombination.

The difference between map distance and recombination frequency arises because double crossovers that do not yield recombinant gametes, like the one depicted in Figure 4.8, *do* contribute to the map distance but *do not* contribute to the recombination

Studies of laboratory mice are essential in discovering the functions of human genes and understanding the organization of the human genome.

frequency. The distinction is important only when the region in question is large enough for double crossing-over to occur. If the region between the genes is short enough that no more than one crossover can occur in the region in any one meiosis, then map units and recombination frequencies are the same (because there are no multiple crossovers that can undo each other). This is the basis for defining a map unit as being equal to 1 percent recombination:

key concept

Over an interval so short that multiple crossovers are precluded (typically yielding 10 percent recombination or less), the map distance equals the recombination frequency because all crossovers result in recombinant gametes.

Furthermore, when adjacent chromosome regions separating linked genes are so short that

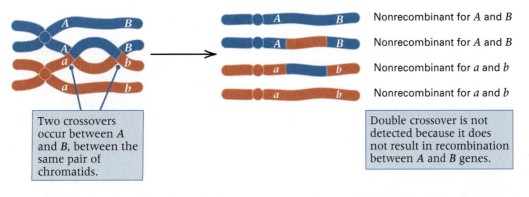

Two crossovers occur between A and B, between the same pair of chromatids.

Nonrecombinant for A and B
Nonrecombinant for A and B
Nonrecombinant for a and b
Nonrecombinant for a and b

Double crossover is not detected because it does not result in recombination between A and B genes.

Figure 4.8 | If two crossovers take place between marker genes A and B, and both involve the same pair of chromatids, then neither crossover is detected because all of the resulting chromosomes are nonrecombinant A B or a b.

multiple crossovers are not formed, the recombination frequencies (and hence the map distances) between the genes are additive. This important feature of recombination, as well as the logic used in genetic mapping, is illustrated by the example in Figure 4.9 . The genes are located in the X chromosome of *Drosophila*—*y* for yellow body, *rb* for ruby eye color, and *cv* for shortened wing crossvein. The experimentally measured recombination frequency between genes *y* and *rb* is 7.5 percent, and that between *rb* and *cv* is 6.2 percent. The genetic map might be any one of three possibilities, depending on which gene is in the middle (*y*, *cv*, or *rb*). Map C, which has *y* in the middle, can be excluded because it implies that the recombination frequency between *rb* and *cv* should be greater than that between *rb* and *y*, and this contradicts the observed data. In other words, map C can be excluded because it implies that the frequency of recombination between *y* and *cv* must be negative.

Maps A and B are both consistent with the observed recombination frequencies. They differ in their predictions regarding the recombination frequency between *y* and *cv*. Using the principle of additivity of map distances, the predicted *y*−*cv* map distance in A is 13.7 map units, whereas the predicted *y*−*cv* map distance in B is 1.3 map units. In fact, the observed recombination frequency between *y* and *cv* is 13.3 percent. Map A is therefore correct. However, there are actually two genetic maps corresponding to

map A. They differ only in whether *y* is placed at the left or at the right. One map is *y*−*rb*−*cv*, which is the one shown in Figure 4.9; the other is *cv*−*rb*−*y*. The two ways of depicting the genetic map are completely equivalent because there is no way of knowing from the recombination data whether *y* or *cv* is closer to the telomere. (Other data indicate that *y* is, in fact, near the telomere.)

A genetic map can be expanded by this type of reasoning to include all of the known genes in a chromosome; these genes constitute a **linkage group.** The number of linkage groups is the same as the haploid number of chromosomes of the species. For example, cultivated corn (*Zea mays*) has ten pairs of chromosomes and ten linkage groups. A partial genetic map of chromosome 10 is shown in Figure 4.10 , along with the dramatic phenotypes caused by some of the mutations. The ears of corn shown in parts C and F demonstrate the result of Mendelian segregation. The ear in part C shows a 3 : 1 segregation of yellow : orange kernels produced by the recessive *orange pericarp-2* (*orp-2*) allele in a cross between two heterozygous genotypes.

Courtesy of M. G. Neuffer.

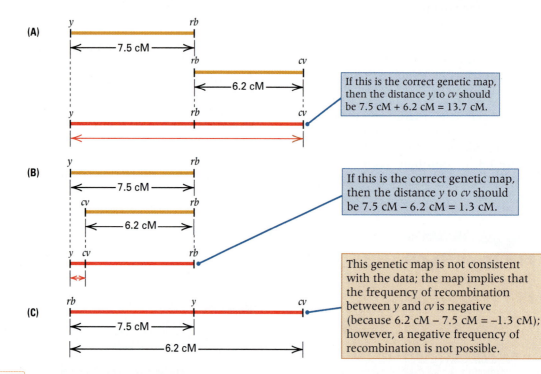

Figure 4.9 In *Drosophila*, the genes *y* (yellow body) and *rb* (ruby eyes) have a recombination frequency of 7.5 percent, and *rb* and *cv* (shortened wing crossvein) have a recombination frequency of 6.2 percent. There are three possible genetic maps, depending on whether *rb* is in the middle (part A), *cv* is in the middle (part B), or

y is in the middle (part C). Map (C) can be excluded because it implies that *rb* and *y* should be closer than *rb* and *cv*, whereas the observed recombination frequency between *rb* and *y* is actually greater than that between *rb* and *cv*. Maps (A) and (B) are compatible with the data given.

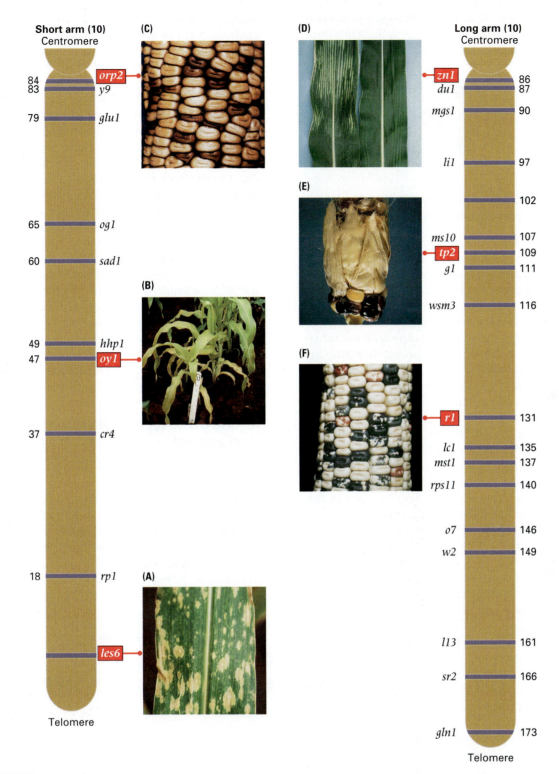

Short arm (10)
Centromere

84 — orp2
83 — y9
79 — glu1
65 — og1
60 — sad1
49 — hhp1
47 — oy1
37 — cr4
18 — rp1
— les6

Telomere

(C)
(B)
(A)

Long arm (10)
Centromere

zn1 — 86
du1 — 87
mgs1 — 90
li1 — 97
— 102
ms10 — 107
tp2 — 109
g1 — 111
wsm3 — 116
r1 — 131
lc1 — 135
mst1 — 137
rps11 — 140
o7 — 146
w2 — 149
l13 — 161
sr2 — 166
gln1 — 173

Telomere

(D)
(E)
(F)

Figure 4.10 Genetic map of chromosome 10 of corn, *Zea mays*. The map distance to each gene is given in map units (centimorgans) relative to a position 0 for the telomere of the short arm (lower left). (A) Mutations in the gene *lesion-6 (les6)* result in many small to medium-sized, irregularly spaced, discolored spots on the leaf blade and sheath. (B) Mutations in the gene *oil yellow-1 (oy1)* result in a yellow-green plant. In the photograph, the plant in front shows the mutant phenotype; behind it is a normal plant. (C) The *orp2* allele is a recessive expressed as orange pericarp, a maternal tissue that surrounds the kernels. The photograph shows segrega-

tion in the F_2, yielding a 3 : 1 ratio of yellow : orange seeds. (D) In *zn1* (*zebra necrotic-1*) mutants, stripes of leaf tissue die. In the photograph, the left leaf is homozygous *zn1*, and the right is wildtype. (E) Mutants for the gene *teopod-2* (*tp2*) have small, partially podded ears and a simple tassle. (F) The mutation *R1-mb* is an allele of the *r1* gene resulting in red or purple color in the aleurone layer of the seed. The photograph shows the marbled color in kernels of an ear segregating for *R1-mb*. [Photographs courtesy of M. G. Neuffer; genetic map courtesy of E. H. Coe.]

The ear in part F shows a 1 : 1 segregation of marbled : white kernels produced by the dominant allele *R1-mb* in a cross between a heterozygous genotype and a homozygous wildtype.

Courtesy of M. G. Neuffer.

■ Physical distance is often—but not always—correlated with map distance.

Generally speaking, the greater the physical separation between genes along a chromosome, the greater the map distance between them. Physical distance and genetic map distance are usually correlated, because a greater distance between genetic markers affords a greater chance for a crossover to take place; crossing-over is a physical exchange between the chromatids of paired homologous chromosomes.

On the other hand, the general correlation between physical distance and genetic map distance is by no means absolute. We have already noted that the frequency of recombination between genes may differ in males and females. An unequal frequency of recombination means that the sexes can have different map distances in their genetic maps, although the physical chromosomes of the two sexes are the same and the genes must have the same linear order. For example, because there is no recombination in male *Drosophila*, the map distance between any pair of genes located in the same chromosome, when measured in the male, is 0. (On the other hand, genes on different chromosomes do undergo independent assortment in males.)

The general correlation between physical distance and genetic map distance can even break down in a single chromosome. For example, crossing-over is much less frequent in heterochromatin, which consists primarily of gene-poor regions near the centromeres, than in euchromatin. Consequently, a given length of heterochromatin will appear much shorter in the genetic map than an equal length of euchromatin. In heterochromatic regions, therefore, the genetic map gives a distorted picture of the physical map. An example of such distortion is illustrated in Figure 4.11, which compares the physical map and the genetic map of chromosome 2 in *Drosophila*. The physical map depicts the appearance of the chromosome in metaphase of mitosis. Two genes near the tips and two near the euchromatin–heterochromatin junction are indicated in the genetic map. The map distances across the euchromatic arms are 54.5 and 49.5 map units, respectively, for a total euchromatic map distance of 104.0 map units. However, the heterochromatin, which constitutes approximately 25 percent of the entire chromosome, has a genetic length in map units of only 3.0 percent. The distorted length of the heterochromatin in the genetic map results from the reduced frequency of crossing-over in the heterochromatin. In spite of the distortion of the genetic map across the heterochromatin, in the regions of euchromatin there is a good correlation between the physical distance between genes and their distance, in map units, in the genetic map.

■ One crossover can undo the effects of another.

When two genes are located far apart along a chromosome, more than one crossover can be formed between them in a single meiosis, and this complicates the interpretation of recombination data. The probability of multiple crossovers increases with the distance between the genes. Multiple crossing-over complicates genetic mapping because map distance is based on the number of physical exchanges that are formed, and some of the multi-

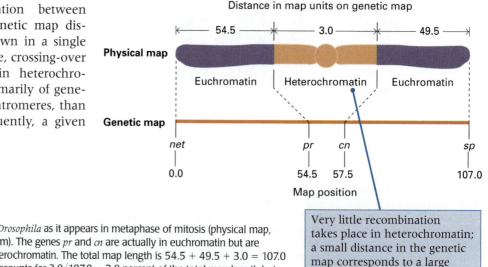

Figure 4.11 | Chromosome 2 in *Drosophila* as it appears in metaphase of mitosis (physical map, top) and in the genetic map (bottom). The genes *pr* and *cn* are actually in euchromatin but are located near the junction with heterochromatin. The total map length is 54.5 + 49.5 + 3.0 = 107.0 map units. The heterochromatin accounts for 3.0/107.0 = 2.8 percent of the total map length but constitutes approximately 25 percent of the physical length of the metaphase chromosome.

ple exchanges between two genes do not result in recombination of the genes and hence are not detected. As we saw in Figure 4.8, the effect of one crossover can be canceled by another crossover further along the way. If two exchanges between the same two chromatids take place between the genes A and B, then their net effect will be that all chromosomes are nonrecombinant, either $A\ B$ or $a\ b$. Two of the products of this meiosis have an interchange of their middle segments, but the chromosomes are not recombinant for the genetic markers and so are genetically indistinguishable from noncrossover chromosomes. The possibility of such canceling events means that the observed recombination value is an *underestimate* of the true exchange frequency and the map distance between the genes. In higher organisms, double crossing-over is effectively precluded in chromosome segments that are sufficiently short, usually about 10 map units or less. Therefore, multiple crossovers that cancel each other's effects can be avoided by using recombination data for closely linked genes to build up genetic linkage maps.

The minimum recombination frequency between two genes is 0. The recombination frequency also has a maximum:

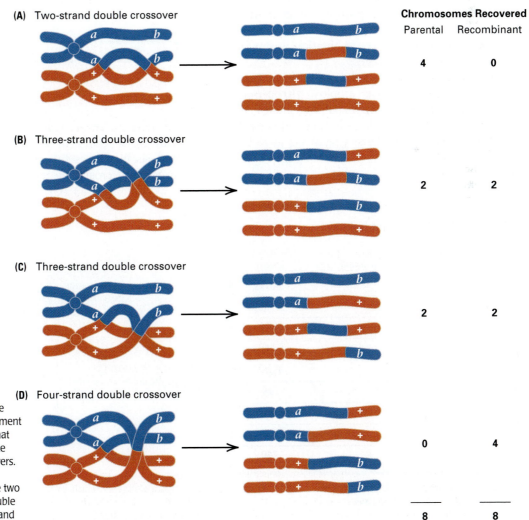

> **key concept**
>
> No matter how far apart two genes may be, the maximum frequency of recombination between any two genes is 50 percent.

Fifty percent recombination is the same value that would be observed if the genes were on nonhomologous chromosomes and assorted independently. The maximum frequency of recombination is observed when the genes are so far apart in the chromosome that at least one crossover is almost always formed between them. In part B of Figure 4.6, it can be seen that a single exchange in every meiosis would result in half of the products having parental combinations and the other half having recombinant combinations of the genes. The occurrence of two exchanges between two genes has the same effect, as shown in Figure 4.12. Part A

Figure 4.12 Diagram showing that the result of two crossovers in the interval between two genes is indistinguishable from independent assortment of the genes, provided that the chromatids participate at random in the crossovers. (A) A two-strand double crossover. (B) and (C) The two types of three-strand double crossovers. (D) A four-strand double crossover.

shows a two-strand double crossover, in which the same chromatids participate in both exchanges; no recombination of the marker genes is detectable. When the two exchanges have one chromatid in common (three-strand double crossover, parts B and C), the result is indistinguishable from that of a single exchange; two products with parental combinations and two with recombinant combinations are produced. Note that there are two types of three-strand doubles, depending on which three chromatids participate. The final possibility is that the second exchange connects the chromatids that did not participate in the first exchange (four-strand double crossover, part D), in which case all four products are recombinant.

In most organisms, when double crossovers are formed, the chromatids that take part in the two exchange events are selected at random. In this case, the expected proportions of the three types of double exchanges are 1/4 four-strand doubles, 1/2 three-strand doubles, and 1/4 two-strand doubles. This means that on the average, $(1/4)(0) + (1/2)(2) + (1/4)(4) = 2$ recombinant chromatids will be found among the 4 chromatids produced from meioses with two exchanges between a pair of genes. This is the same proportion obtained with a single exchange between the genes. Moreover, a maximum of 50 percent recombination is obtained for any number of exchanges.

Double crossing-over is detectable in recombination experiments that employ **three-point crosses,** which include three pairs of alleles. If a third pair of alleles, c^+ and c, is located between the outermost genetic markers (Figure 4.13), double

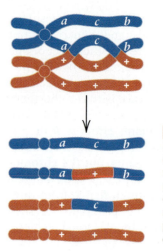

Figure 4.13 Diagram showing that two crossovers that occur between the same chromatids and span the middle pair of alleles in a triple heterozygote will result in a reciprocal exchange of the middle pair of alleles between the two chromatids.

exchanges in the region can be detected when the crossovers flank the c gene. The two crossovers, which in this example take place between the same pair of chromatids, would result in a reciprocal exchange of the c^+ and c alleles between the chromatids. A three-point cross is an efficient way to obtain recombination data; it is also a simple method for determining the order of the three genes, as we will see in the next section.

4.3

Double crossovers are revealed in three-point crosses.

The data in Table 4.1 result from a testcross in corn with three genes in a single chromosome. The analysis illustrates the approach to interpreting a three-point cross. The recessive alleles of the genes

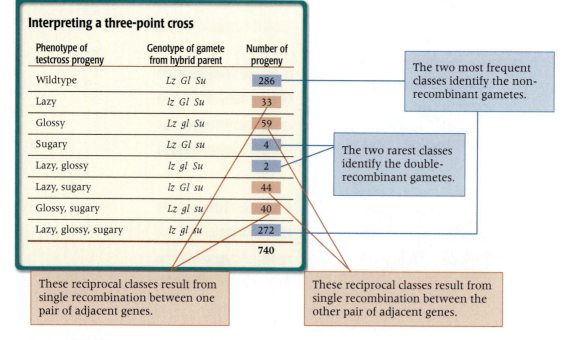

Table 4.1

Interpreting a three-point cross

Phenotype of testcross progeny	Genotype of gamete from hybrid parent	Number of progeny
Wildtype	Lz Gl Su	286
Lazy	lz Gl Su	33
Glossy	Lz gl Su	59
Sugary	Lz Gl su	4
Lazy, glossy	lz gl Su	2
Lazy, sugary	lz Gl su	44
Glossy, sugary	Lz gl su	40
Lazy, glossy, sugary	lz gl su	272
		740

The two most frequent classes identify the non-recombinant gametes.

The two rarest classes identify the double-recombinant gametes.

These reciprocal classes result from single recombination between one pair of adjacent genes.

These reciprocal classes result from single recombination between the other pair of adjacent genes.

the human connection

Count Your Blessings

Joe Hin Tijo[1] and Albert Levan[2] 1956
[1]Estacion Experimental de Aula Dei,
Zaragoza, Spain
[2]Institute of Genetics, Lund, Sweden
The Chromosome Number in Man

This paper marks the beginning of modern human cytogenetics. Tijo and Levan made a technical improvement for spreading chromosomes on a microscope slide, which for the first time allowed accurate chromosome counts to be made. Previous methods ran up against the problem that human chromosomes are small, relatively numerous, and bunched together at metaphase. Until 1956 it was widely believed that the human chromosome number was 48, as it is in the chimpanzee and gorilla. The number 48 became dogma, so much so that other counts were disbelieved, as illustrated in this excerpt by the reference to a previous researcher who repeatedly obtained a chromosome number of 46 in liver cells but abandoned the study because she was unable to find the two "missing" chromosomes. What worked for Tijo and Levan was the simple trick of soaking the cells in a hypotonic solution (a solution with a lower concentration of charged inorganic ions than of the cells themselves). When cells are bathed in a hypotonic solution, water surges through the cell membrane into the cells and causes them to swell. The nucleus becomes considerably enlarged, thereby spread-

ing out the chromosomes. Once this technique was in use, the discovery of many human chromosomal abnormalities followed quickly, such as the finding of three copies of chromosome 21 in Down syndrome. (See "The Human Connection" in Chapter 5.)

••••••••••••••••••••••••••••••••••••

While staying last summer at the Sloan-Kettering Institute, New York, one of us tried out hypotonic treatment on various human tissue cultures.... The results were promising inasmuch as some fairly satisfactory chromosome analyses were obtained.... The treatment [had] a tendency to make the chromosome outlines somewhat blurred and vague. We consequently tried to abbreviate the treatment to a minimum, hoping to induce the scattering of the chromosomes without unfavorable effects on the chromosome surface.... Treatment with hypotonic solution for only one or two minutes gave good results.... Ordinary squash preparations were made. For chromosome counts the squashing was made very mild in order to keep the chromosomes in the metaphase group. For studies of chromosome morphology a more thorough squashing was preferable. In many cases single cells were squashed under the microscope by a slight pressure of a needle. In such cases it was directly observed that no chromosomes escaped.... We were surprised to find that [among 261 cells] the chromosome

number 46 predominated in the tissue cultures from embryonic cells.... Lower numbers were frequent, of course, but always in cells that seemed damaged. These were consequently disregarded.... The chromosomes are easily arranged in pairs, but only certain of these pairs are individually distinguishable.... The almost exclusive occurrence of the chromosome number 46 in embryonic cell cultures is a very unexpected finding.... After the conclusion had been drawn that the tissue studied by us had 46 as a chromosome

> **We were surprised to find that the chromosome number 46 predominated.**

number, Dr. Eva Hansen-Melander kindly informed us that during last Spring she had studied the chromosomes of embryonic liver mitosis. This study, however, was temporarily discontinued because the workers were unable to find all the 48 human chromosomes in their material; as a matter of fact, the number 46 was repeatedly counted in their slides. This finding suggests that 46 may be the correct chromosome number for human liver tissue, too.... We do not wish to generalize our present findings into a statement that the chromosome number of human beings is $2n = 46$, but it is hard to avoid the conclusion that this would be the most natural explanation of our data.

Source: Hereditas 42: 1–6.

in this cross are lz (for lazy or prostrate growth habit), gl (for glossy leaf), and su (for sugary endosperm), and the multiply heterozygous parent in the cross had the genotype

$$\frac{Lz \ Gl \ Su}{lz \ gl \ su}$$

where each symbol with an initial capital letter represents the dominant allele. (The use of this type of symbolism is customary in corn genetics.) The two classes of progeny that inherit noncrossover (parental-type) gametes are therefore the wildtype plants and those with the lazy-glossy-sugary phenotype. The number of progeny in these classes is far larger than the number in any of the crossover classes. Because the frequency of recombination is never greater than 50 percent, the very fact that these progeny are the most numerous indicates that the gametes that gave rise to them have the parental allele configurations, in this case *Lz Gl Su* and *lz gl su*. Using this principle, we could have inferred the genotype of the heterozygous parent even if the genotype had not been stated. This is a point important enough to state more generally:

key concept

In any genetic cross, no matter how complex, the two most frequent types of gametes with respect to any pair of genes are *nonrecombinant*; these provide the linkage phase (*cis* versus *trans*) of the alleles of the genes in the multiply heterozygous parent.

In mapping experiments, the gene sequence is usually not known. In this example, the order in which the three genes are shown is entirely arbitrary. However, there is an easy way to determine the correct order from three-point data. The gene order can be deduced by identifying the genotypes of the double-crossover gametes produced by the heterozygous parent and comparing these with the nonrecombinant gametes. Because the probability

of two simultaneous exchanges is considerably smaller than that of either single exchange, the double-crossover gametes will be the least frequent types. It is clear in Table 4.1 that the classes composed of four plants with the sugary phenotype and two plants with the lazy-glossy phenotype (products of the *Lz Gl su* and *lz gl Su* gametes, respectively) are the least frequent and therefore constitute the double-crossover progeny. Now we apply another principle:

key concept

The effect of double crossing-over is to interchange the members of the *middle* pair of alleles between the chromosomes.

This principle is illustrated in Figure 4.14. With three genes there are three possible orders, depending on which gene is in the middle. If *gl* were in the middle (part A), the double-recombinant gametes would be *Lz gl Su* and *lz Gl su*, which is inconsistent with the data. Likewise, if *lz* were in the middle (part C), the double-recombinant gametes would be *Gl lz Su* and *gl Lz su*, which is also inconsistent with the data.

The correct order of the genes, *lz−su−gl*, is given in part B, because in this case, the double-recombinant gametes are *Lz su Gl* and *lz Su gl*,

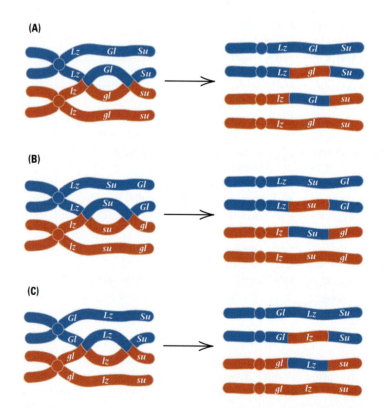

Figure 4.14 The order of genes in a three-point testcross may be deduced from the principle that double recombination interchanges the middle pair of alleles. For the genes *Lu*, *Gl*, and *Su*, there are three possible orders (parts A, B, and C), each of which predicts a different pair of gametes as the result of double recombination. Only the order in part B is consistent with the finding that *Lz Gl su* and *lz gl Su* are the double-recombinant gametes.

which Table 4.1 indicates is actually the case. Although one can always infer which gene is in the middle by going through all three possibilities, there is a shortcut. Each double-recombinant gamete will always match one of the parental gametes in two of the alleles. In Table 4.1, for example, the double-recombinant gamete *Lz Gl su* matches the parental gamete *Lz Gl Su* except for the allele *su*. Similarly, the double-recombinant gamete *lz gl Su* matches the parental gamete *lz gl su* except for the allele *Su*. The middle gene can be identified because the "odd man out" in the comparisons—in this case, the alleles of *Su*—is always the gene in the middle. The reason is that only the middle pair of alleles is interchanged by double crossing-over.

Taking the correct gene order into account, the genotype of the heterozygous parent in the cross yielding the progeny in Table 4.1 should be written as

$$\frac{Lz \; Su \; Gl}{lz \; su \; gl}$$

The consequences of single crossing-over in this genotype are shown in Figure 4.15. A single crossover in the *lz–su* region (part A) yields the reciprocal recombinants *Lz su gl* and *lz Su Gl*, and a single crossover in the *su–gl* region (part B) yields the reciprocal recombinants *Lz Su gl* and *lz su Gl*. The consequences of double crossing-over are illustrated in Figure 4.16. There are four different types of double crossovers: a two-strand double (part A), two types of three-strand doubles (parts B and C), and a four-strand double (part D). These types were illustrated earlier in Figure 4.12, where the main point was that with two genetic markers flanking the crossovers, the occurrence of double crossovers cannot be detected genetically. The difference in the present case is that, here, the genetic marker *su* is located in the middle between the two crossovers, so some of the double crossovers can be detected genetically. On the right in Figure 4.16, the asterisks mark the sites of crossing-over between nonsister chromatids. In terms of recombination, the result is that

- A two-strand double crossover (part A) yields the reciprocal double-recombinant products *Lz su Gl* and *lz Su gl*.
- One three-strand double crossover (part B) yields the double-recombinant product *Lz su Gl* and two single-recombinant products, *Lz Su gl* and *lz Su Gl*.
- The other three-strand double crossover (part C) yields the double-recombinant product *lz Su gl* and two single-recombinant products, *Lz su gl* and *lz su Gl*.

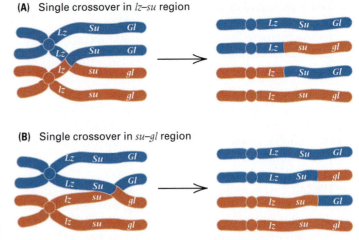

Figure 4.15 Result of single crossovers in a triple heterozygote, using the *Lz−Su−Gl* region as an example. (A) A crossover between *Lz* and *Su* results in two gametes that show recombination between *Lz* and *Su* and two gametes that are nonrecombinant. (B) A crossover between *Su* and *Gl* results in two gametes that show recombination between *Su* and *Gl* and two gametes that are nonrecombinant.

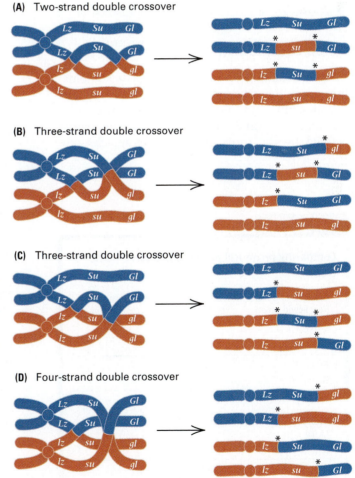

Figure 4.16 Result of double crossovers in a triple heterozygote, using the *Lz−Su−Gl* region as an example. Note that chromosomes showing double recombination derive from the two-strand double crossover (A) or from either type of three-strand double crossover (B and C). The four-strand double crossover (D) results only in single-recombinant chromosomes.

- The four-strand double crossover (part C) yields reciprocal single recombinants in the *lz–su* region, namely *Lz su gl* and *lz Su Gl*, and reciprocal single recombinants in the *su–gl* region, namely *Lz Su gl* and *lz su Gl*.

Note that the products of recombination in the three-strand double crossovers (parts B and C) are the reciprocals of each other. Because these two types of double crossovers are equally frequent, the reciprocal products of recombination are expected to appear in equal numbers.

We can now summarize the data in Table 4.1 in a more informative way by writing the genes in the correct order and grouping reciprocal gametic genotypes together. This grouping is shown in **Table 4.2**. Note that each class of single recombinants consists of two reciprocal products and that these are found in approximately equal frequencies (40 versus 33 and 59 versus 44). This observation illustrates an important principle:

key concept

The two reciprocal products resulting from any crossover, or any combination of crossovers, are expected to appear in approximately equal frequencies among the progeny.

In calculating the frequency of recombination from the data, remember that the double-recombinant chromosomes result from *two* exchanges, one in each of the chromosome regions defined by the three genes. Therefore, chromo-

somes that are recombinant between *lz* and *su* are represented by the following chromosome types:

$$
\begin{array}{ll}
Lz\ su\ gl & 40 \\
lz\ Su\ Gl & 33 \\
Lz\ su\ Gl & 4 \\
lz\ Su\ gl & \underline{2} \\
& 79
\end{array}
$$

The total implies that 79/740, or 10.7 percent, of the chromosomes recovered in the progeny are recombinant between the *lz* and *su* genes, so the map distance between these genes is 10.7 map units, or 10.7 centimorgans. Similarly, the chromosomes that are recombinant between *su* and *gl* are represented by

$$
\begin{array}{ll}
Lz\ Su\ gl & 59 \\
lz\ su\ Gl & 44 \\
Lz\ su\ Gl & 4 \\
lz\ Su\ gl & \underline{2} \\
& 109
\end{array}
$$

In this case the recombination frequency between *su* and *gl* is 109/740, or 14.7 percent, so the map distance between these genes is 14.7 map units, or 14.7 centimorgans. The genetic map of the chromosome segment in which the three genes are located is therefore

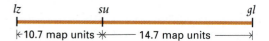

The most common error in learning how to interpret three-point crosses is to forget to include the double recombinants when calculating the recombination frequency between adjacent genes. You can keep from falling into this trap by remembering that the double-recombinant chromosomes have single recombination in *both* regions.

■ **Interference decreases the chance of multiple crossing-over.**

The detection of double crossing-over makes it possible to determine whether exchanges in two different regions of a pair of chromosomes are formed independently of each other. Using the information from the example with corn, we know from the recombination frequencies that the probability of recombination is 0.107 between *lz* and *su* and 0.147 between *su* and *gl*. If recombination is independent in the two regions (which means that the formation of one crossover does not alter the probability of the second crossover), the probability of a single recombination in both regions is the product of these separate probabilities, or $0.107 \times 0.147 = 0.0157$ (1.57 percent). This implies that in a sample of 740 gametes, the expected number of double recombinants would be 740×0.0157, or 11.6, whereas the

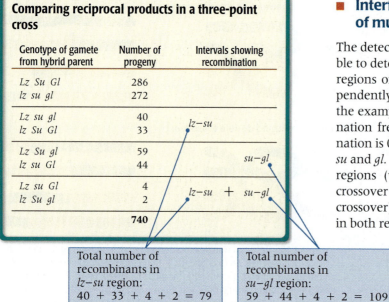

Table 4.2

Comparing reciprocal products in a three-point cross

Genotype of gamete from hybrid parent	Number of progeny	Intervals showing recombination
Lz Su Gl *lz su gl*	286 272	
Lz su gl *lz Su Gl*	40 33	*lz–su*
Lz Su gl *lz su Gl*	59 44	*su–gl*
Lz su Gl *lz Su gl*	4 2	*lz–su* + *su–gl*
	740	

Total number of recombinants in *lz–su* region: 40 + 33 + 4 + 2 = 79

Total number of recombinants in *su–gl* region: 59 + 44 + 4 + 2 = 109 number actually observed was only 6 (Table 4.2). Such deficiencies in the observed number of double recombinants are common; they reflect a phenomenon called chromosome **interference,** in which a crossover in one region of a chromosome reduces the probability of a second crossover in a nearby region. Over short genetic distances, chromosome interference is nearly complete.

The **coefficient of coincidence** is the observed number of double-recombinant chromosomes divided by the expected number. Its value provides a quantitative measure of the degree of interference, which is defined as

i = Interference

$= 1 - $ (Coefficient of coincidence)

From the data in the corn example, the coefficient of coincidence is calculated as follows:

- Observed frequency of double recombinants = 6
- Expected frequency of double recombinants = $0.107 \times 0.147 \times 740 = 11.6$
- Coefficient of coincidence $= 6/11.6 = 0.52$

The 0.52 means that the observed number of double recombinants was only about half of the number expected if crossing-over in the two regions were independent. The value of the interference depends on the distance between the genetic markers and on the species. In some species, the interference increases as the distance between the two outside markers becomes smaller, until a point is reached at which double crossing-over is eliminated; that is, no double recombinants are found, and the coefficient of coincidence equals 0 (or, to say the same thing, the interference equals 1). In *Drosophila* this distance is about 10 map units.

The effect of interference on the relationship between genetic map distance and the frequency of recombination is illustrated in Figure 4.17 . Each

curve is an example of a *mapping function,* which is the mathematical relation between the genetic distance across an interval in map units (centimorgans) and the observed frequency of recombination across the interval. In other words, a mapping function tells one how to convert a *map distance* between genetic markers into a *recombination frequency* between the markers. As we have seen, when the map distance between the markers is small, the recombination frequency equals the map distance. This principle is reflected in the curves in Figure 4.17 in the region in which the map distance is smaller than about 10 cM. At less than this distance, all of the curves are nearly straight lines, which means that map distance and recombination frequency are equal; 1 map unit equals 1 percent recombination, and 10 map units equals 10 percent recombination. For distances greater than 10 map units, the recombination frequency becomes smaller than the map distance according to the pattern of interference along the chromosome. Each pattern of interference yields a different mapping function, as shown by the three examples in Figure 4.17.

A Moment to Think

Problem: In his pioneering 1913 studies in *Drosophila* that resulted in the first genetic linkage map, A. H. Sturtevant included three genetic markers now known to cover almost the entirety of the euchromatin of the X chromosome. The marker *w* (white eyes) is near the tip, *m* (miniature body) near the middle, and *r* (rudimentary wings) near the centromere. In two-point crosses, Sturtevant obtained recombination frequencies of 0.32 for the interval *w−m*, 0.25 for the interval *m−r*, and 0.45 for the interval *w−r*. He noted that 0.45 is smaller than the sum of 0.32 + 0.25 = 0.57, which is the value expected if the frequencies of recombination over such large distances were additive. He commented that the discrepancy "is probably due to the occurrence of two breaks in the same chromosome, or double crossing-over." From Sturtevant's data, calculate the coincidence and the interference across the region. (The answer can be found on page 143.)

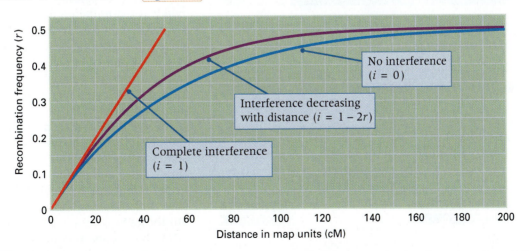

Figure 4.17 A mapping function is the relation between genetic map distance across an interval and the observed frequency of recombination across the interval. Map distance is defined as one-half the average number of crossovers converted into a percentage. The three mapping functions correspond to different assumptions about interference, *i*.

4.4

Polymorphic DNA sequences are used in human genetic mapping.

Until quite recently, mapping genes in human beings was very tedious and slow. Numerous practical obstacles complicated genetic mapping in human pedigrees:

1. Most genes that cause genetic diseases are rare, so they are observed in only a small number of families.

2. Many mutant genes of interest in human genetics are recessive, so they are not detected in heterozygous genotypes.

3. The number of offspring per human family is relatively small, so segregation cannot usually be detected in single sibships.

4. The human geneticist cannot perform testcrosses or backcrosses, because human matings are not dictated by an experimenter.

Human genetics has been revolutionized by the use of techniques for manipulating DNA. These techniques have enabled investigators to carry out genetic mapping in human pedigrees primarily by using genetic markers present in the DNA itself, rather than through the phenotypes produced by mutant genes. There are many minor differences in DNA sequence from one person to the next. On the average, the DNA sequences at corresponding positions in any two chromosomes, taken from any two people, differ at approximately one in every thousand base pairs. A genetic difference that is relatively common in a population is called a **polymorphism.** Most polymorphisms in DNA sequence are not associated with any inherited disease or disability; many occur in DNA sequences that do not code for proteins. Nevertheless, each of the polymorphisms serves as a convenient genetic marker, and those genetically linked to genes that cause hereditary diseases are particularly important. Some polymorphisms in DNA sequence are detected by means of a type of enzyme called a **restriction endonuclease,** which cleaves double-stranded DNA molecules wherever a particular, short sequence of bases is present. For example, the restriction enzyme *Eco*RI cleaves DNA wherever the sequence GAATTC appears in either strand, as illustrated in Figure 4.18 . Restriction enzymes are considered in detail in Chapter 6. For present purposes, their significance is related to the fact that a difference in DNA sequence that eliminates a cleavage site can be detected because the region lacking the cleavage site will be cleaved into one larger fragment instead of two smaller ones (Figure 4.19). More rarely, a mutation in the DNA sequence will create a new site rather than destroy one already present. The main point is that any difference in DNA sequence that alters a cleavage site also changes the length of the DNA fragments produced by cleavage with the corresponding restriction enzyme. The different DNA fragments can be separated by size by an electric field in a supporting gel and detected by various means. Differences in DNA fragment length produced by presence or absence of the cleavage sites in DNA molecules are known as **restriction fragment length polymorphisms (RFLPs).**

RFLPs are typically formed in one of two ways. A mutation that changes a base sequence may result in loss or gain of a cleavage site that is recognized by the restriction endonuclease in use. Figure 4.20 , part A, gives an example. On the left is shown the relevant region in the homologous DNA molecules in a person who is heterozygous for such a sequence polymorphism. The homologous chromosomes in the person are distinguished by the letters *a* and *b*. In the region of interest, chromosome *a* contains two cleavage sites and chromosome *b* contains three. On the right is shown the position of the DNA fragments produced by cleavage after separation in an electric field. Each fragment appears as a discrete band in the gel. The fragment from chromosome *a*

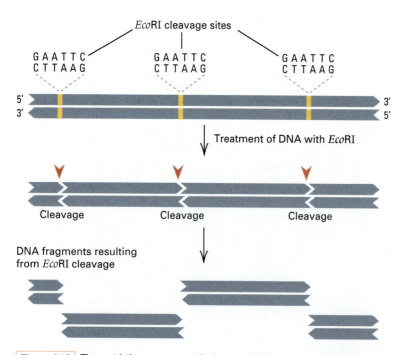

Figure 4.18 The restriction enzyme *Eco*RI cleaves double-stranded DNA wherever the sequence 5'-GAATTC-3' is present. In the example shown here, the DNA molecule contains three *Eco*RI cleavage sites, and it is cleaved at each site, producing a number of fragments.

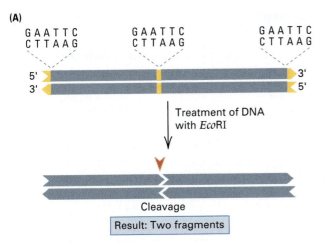

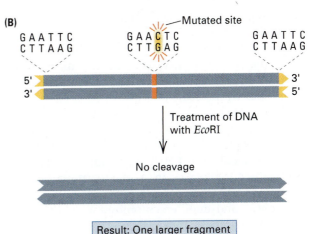

Figure 4.19 A minor difference in the DNA sequence of two molecules can be detected if the difference eliminates a restriction site. (A) This molecule contains three restriction sites for EcoRI, including one at each end. It is cleaved into two fragments by the enzyme. (B) This molecule has a mutant base sequence in the EcoRI site in the middle. It changes 5'-GAATTC-3' into 5'-GAACTC-3', which is no longer cleaved by EcoRI. Treatment of this molecule with EcoRI results in one larger fragment.

migrates more slowly than those from chromosome b because it is longer, and longer fragments move more slowly through the gel. In this example, DNA from a person heterozygous for the a and b types of chromosomes (genotype ab) would yield three

bands in a gel. Similarly, DNA from homozygous aa would yield one band, and that from homozygous bb would yield two bands.

A second type of DNA polymorphism results from differences in the number of copies of a short

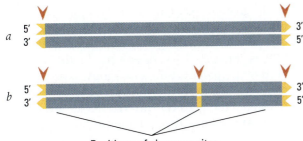

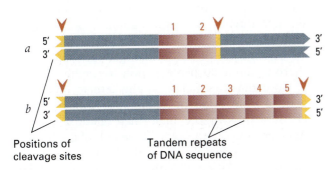

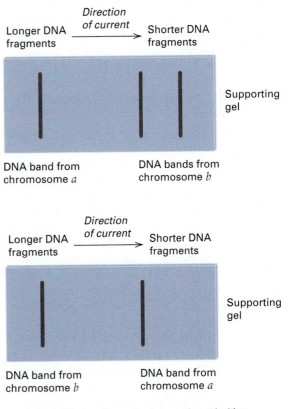

Figure 4.20 Two types of genetic variation that are widespread in most natural populations of animals and plants. (A) RFLP (restriction fragment length polymorphism), in which alleles differ in the presence or absence of a cleavage site in the DNA. The different alleles yield different fragment lengths (shown in the gel pattern at the right) when the molecules are cleaved with a restriction enzyme. (B) SSR (simple sequence repeat), in which alleles differ in the number of repeating units present between two cleavage sites.

DNA sequence that may be repeated many times in tandem at a particular site in a chromosome (Figure 4.20, part B). In a particular chromosome, the tandem repeats may contain any number of copies, typically ranging from ten to a few hundred. When a DNA molecule is cleaved with a restriction endonuclease that cleaves at sites flanking the tandem repeat, the size of the DNA fragment produced is determined by the number of repeats present in the molecule. Figure 4.20, part B, illustrates homologous DNA sequences in a heterozygous person containing one chromosome *a* with two copies of the repeat and another chromosome *b* with five copies of the repeat. When cleaved and separated in a gel, chromosome *a* yields a shorter fragment than that from chromosome *b*, because *a* contains fewer copies of the repeat. A genetic polymorphism resulting from a tandemly repeated short DNA sequence is called a **simple sequence repeat (SSR).** An example of an SSR is the repeating sequence

$$5'-\ldots TGTGTGTGTGTG\ldots-3'$$

and the polymorphism consists of differences in the number of TG repeats. A particular "allele" of the SSR is defined by the number of TG repeats it includes.

One source of the utility of SSRs in human genetic mapping is the high density of SSRs across the genome. There is an average of one SSR per 2 kb of human DNA. Some examples are shown in **Table 4.3.** The prevalence of different SSRs differs. Some dinucleotide repeats, such as 5'-AC-3' and 5'-AT-3', are far more frequent than others, such as 5'-GC-3'. Overall, dinucleotide repeats are much more abundant than trinucleotide repeats. The second source of utility of SSRs in genetic mapping is the large number of alleles that can be present in any human population. The large number of alleles also implies that most people will be heterozygous, and so their DNA will yield two bands upon cleavage with the appropriate restriction endonuclease. Because of their high degree of variation among people, DNA polymorphisms are also widely used in DNA typing in criminal investigations (Chapter 14).

In genetic mapping, the phenotype of a person with respect to a DNA polymorphism is a pattern of bands in a gel. As with any other type of genetic marker, the genotype of a person with respect to the polymorphism is inferred, insofar as it is possible, from the phenotype. Linkage between different polymorphic loci is detected through lack of independent assortment of the alleles in pedigrees, and recombination and genetic mapping are carried out using the same principles as apply in other organisms, except that in human beings, because of the small family size, different pedigrees are pooled for analysis. Primarily through the use of DNA poly-

<table>
<tr><td colspan="2">Table   4.3</td></tr>
<tr><td colspan="2">**Some simple sequence repeats in the human genome**</td></tr>
<tr><td>SSR
repeat unit</td><td>Number of SSRs in
the human genome</td></tr>
<tr><td>5'-AC-3'</td><td>80,330</td></tr>
<tr><td>5'-AT-3'</td><td>56,260</td></tr>
<tr><td>5'-AG-3'</td><td>23,780</td></tr>
<tr><td>5'-GC-3'</td><td>290</td></tr>
<tr><td>5'-AAT-3'</td><td>11,890</td></tr>
<tr><td>5'-AAC-3'</td><td>7,540</td></tr>
<tr><td>5'-AGG-3'</td><td>4,350</td></tr>
<tr><td>5'-AAG-3'</td><td>4,060</td></tr>
<tr><td>5'-ATG-3'</td><td>2,030</td></tr>
<tr><td>5'-CGG-3'</td><td>1,740</td></tr>
<tr><td>5'-ACC-3'</td><td>1,160</td></tr>
<tr><td>5'-AGC-3'</td><td>870</td></tr>
<tr><td>5'-ACT-3'</td><td>580</td></tr>
</table>

Data from Lander et al. 2001. *Nature* 409: 889.

morphisms, genetic mapping in humans has progressed rapidly.

To give an example of the type of data used in human genetic mapping, a three-generation pedigree of a family segregating for several alleles of an SSR is illustrated in Figure 4.21 . In this example, each of the parents is heterozygous, as are all of the children. Yet every person can be assigned his or her genotype because the SSR alleles are codominant. At present, DNA polymorphisms are the principal types of genetic markers used in genetic mapping in human pedigrees. Such polymorphisms are prevalent, are located in virtually all regions of the chromosome set, and have multiple alleles and so yield a high proportion of heterozygous genotypes. Furthermore, only a small amount of biological material is needed to perform the necessary tests.

One feature of the human genetic map is that there is about 60 percent more recombination in females than in males, so the female and male genetic maps differ in length. The female map is about 4400 cM, the male map about 2700 cM. Averaged over both sexes, the length of the human genetic map for all 23 pairs of chromosomes is about 3500 cM. Because the total DNA content per haploid set of human chromosomes is 3200 million base pairs, there is, very roughly, 1 cM per million base pairs in the human genome.

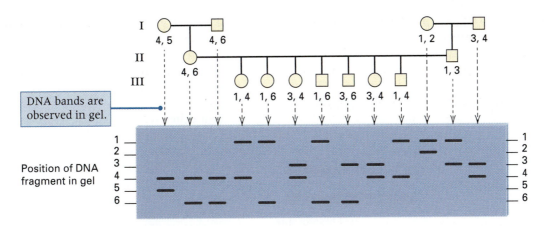

DNA bands are observed in gel.

Position of DNA fragment in gel

Figure 4.21 Human pedigree showing segregation of SSR alleles. Six alleles (1–6) are present in the pedigree, but any one person can have only one allele (if homozygous) or two alleles (if heterozygous).

■ Single-nucleotide polymorphisms (SNPs) are abundant in the human genome.

The most prevalent type of polymorphism in the human genome, as well as in those of most other organisms, is the **single-nucleotide polymorphism (SNP),** in which a single base pair at a particular nucleotide site may differ from one individual to the next. Rare mutants of virtually every nucleotide can probably be found, if one searches hard enough. But rare variants are not generally useful for family studies of heritable variation in susceptibility to disease or for reconstructing historical patterns of human migration. For this reason, in order for a difference in nucleotide sequence to be considered as an SNP, the less-frequent base must have a frequency of greater than about 5 percent in the human population. By this definition, the density of SNPs in the human genome averages about one per 1300 bp. A catalog of SNPs is regarded as the ultimate compendium of DNA

markers, because SNPs are the most common form of genetic differences among people and because they are distributed approximately uniformly along the chromosomes.

Fluorescent dyes are often used to label DNA so that the positions of DNA fragments in a gel can be identified. In this gel the labeled bands are fluorescing a pink color against the blue background.

Courtesy of National Cancer Institute.

A Moment to Think

Answer to Problem: The coincidence is equal to the observed number of double crossovers divided by the expected number, but the observed and expected frequencies (proportions) can be used as well, because in the conversion to frequencies, both numerator and denominator are divided by the same number. Set $r_1 = 0.32$ and $r_2 = 0.25$. We know that r_1 includes the proportion of gametes with single recombination in region 1 (call this s_1) plus the proportion of gametes with double recombination (call this d). Hence $r_1 = s_1 + d$. Similarly, $r_2 = s_2 + d$, where s_2 is the proportion of gametes with single recombination in region 2. In the two-point cross between w and r, the observed frequency of recombination, 0.45, equals $s_1 + s_2$ because none of the double recombinants are detected. Now we can solve for d because $r_1 + r_2 = (s_1 + d) + (s_2 + d) = s_1 + s_2 + 2d = 0.57$, whereas $s_1 + s_2 = 0.45$. Subtracting, we obtain $2d = 0.57 - 0.45 = 0.12$, or $d = 0.06$. This is the "observed" (inferred in this case) frequency of double recombinants. The expected frequency of double recombinants equals $r_1 \times r_2 = 0.32 \times 0.25 = 0.08$. Thus the coincidence across the region is $0.06/0.08 = 0.75$, and the interference equals $1 - 0.75 = 0.25$. In other words, there is about a 25% deficit in double recombinants from the frequency that would be expected with independence.

4.5

Tetrads contain all four products of meiosis.

In some species of fungi, each meiotic tetrad is contained in a sac-like structure, called an **ascus,** and can be recovered as an intact group. Each product of meiosis is included in a reproductive cell called an **ascospore,** and all of the ascospores formed from one meiotic cell remain together in the ascus (Figure 4.22). The advantage of these organisms for the study of recombination is the potential for analyzing all of the products from each meiotic division. For example, one can see immediately from the diagram in Figure 4.22 that a tetrad containing the products of a single meiosis in a heterozygous Aa organism contains 2 A ascospores and 2 a ascospores. The 2 A : 2 a segregation means that the Mendelian ratio of 1 : 1 is realized in the products of each individual meiotic division and is not merely an average over a large number of meioses.

Two other features of ascus-producing organisms are especially useful for genetic analysis: (1) They are haploid, so dominance is not a complicating factor because the genotype is expressed directly in the phenotype. (2) They produce very large numbers of progeny, making it possible to detect rare events and to estimate their frequencies accurately. Furthermore, the life cycles of the organisms tend to be short. The only diploid stage is the zygote, which undergoes meiosis soon after it is formed; the resulting haploid meiotic products (which form the ascospores) germinate to regenerate the vegetative stage (Figure 4.23). In most of the organisms, the meiotic products, or their derivatives, are not arranged in any particular order in the ascus. However, bread molds of the genus *Neurospora* and related organisms have the useful characteristic that the meiotic products are arranged in a definite order directly related to the planes of the meiotic divisions. In *Neurospora*, each of the four products of meiosis also undergoes a mitotic division, with the result that each member of the tetrad yields a *pair* of genetically identical ascospores. We will examine the ordered system after first looking at unordered tetrads.

■ Unordered tetrads have no relation to the geometry of meiosis.

In the tetrads when two pairs of alleles are segregating, three patterns of segregation are possible. For example, in the cross $A\ B \times a\ b$, the three types of tetrads are

(AB) (AB) (ab) (ab) referred to as **parental ditype,** or **PD.** Only two genotypes are represented, and their alleles have the same combinations found in the parents.

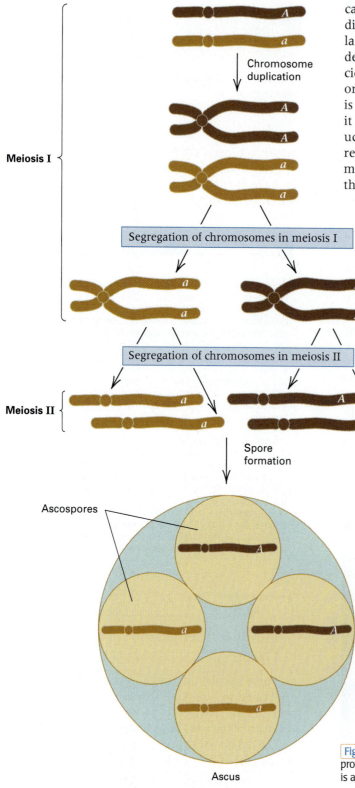

Figure 4.22 Formation of an ascus containing all of the four products of a single meiosis. Each ascospore present in the ascus is a reproductive cell formed from one of the products of meiosis.

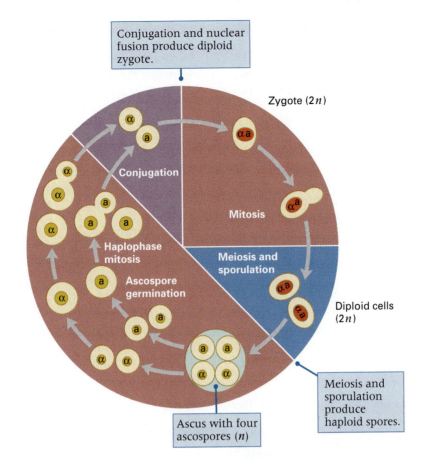

Conjugation and nuclear fusion produce diploid zygote.

Zygote (2*n*)

Conjugation

Mitosis

Haplophase mitosis

Meiosis and sporulation

Ascospore germination

Diploid cells (2*n*)

Meiosis and sporulation produce haploid spores.

Ascus with four ascospores (*n*)

Figure 4.23 Life cycle of the yeast *Saccharomyces cerevisiae.* Mating type is determined by the alleles **a** and **α**. Both haploid and diploid cells normally multiply by mitosis (budding). Depletion of nutrients in the growth medium induces meiosis and sporulation of cells in the diploid state. Diploid nuclei are shown in red, haploid nuclei in yellow.

(Ab) (Ab) (aB) (aB) referred to as **nonparental ditype,** or **NPD.** Only two genotypes are represented, but their alleles have nonparental combinations.

(AB) (Ab) (aB) (ab) referred to as **tetratype,** or **TT.** All four of the possible genotypes are present.

■ **Tetratype tetrads demonstrate that crossing-over takes place at the four-strand stage of meiosis and is reciprocal.**

We noted earlier that tetrads from heterozygous organisms regularly contain 2 *A* and 2 *a* ascospores, which implies that Mendelian segregation takes place in each meiosis. The existence of tetratype tetrads for linked genes demonstrates two features about crossing-over that we have assumed, so far without proof.

1. The exchange of segments between parental chromatids takes place in the first meiotic prophase, *after the chromosomes have duplicated.* Tetratype tetrads demonstrate this assertion

because only two of the four products of meiosis show recombination. This would not be possible unless crossing-over took place at the four-strand stage.

2. The exchange process consists of the breaking and rejoining of the two chromatids, resulting in the *reciprocal* exchange of equal and corresponding segments. Tetratype tetrads demonstrate this point because they contain the reciprocal products (*A b* and *a B* if the parental alleles were in coupling, *A B* and *a b* if they were in repulsion).

■ **Tetrad analysis affords a convenient test for linkage.**

Tetrad analysis is an effective way to determine whether two genes are linked, because

key concept

When genes are *unlinked,* the parental ditype tetrads and the nonparental ditype tetrads are expected in equal frequencies (PD = NPD).

4.5 Tetrads contain all four products of meiosis   **145**

The reason for the equality PD = NPD for unlinked genes is shown in part A of Figure 4.24, where the two pairs of alleles A, a and B, b are located in different chromosomes. In the absence of crossing-over between either gene and its centromere, the two chromosomal configurations are equally likely at metaphase I, and so PD = NPD. When there is a crossover between either gene and its centromere (Figure 4.24, part B), a tetratype tetrad results, but this does not change the fact that PD = NPD.

In contrast, when genes are linked, parental ditypes are far more frequent than nonparental ditypes. To see why, assume that the genes are linked and consider the events required for the production of the three types of tetrads. Figure 4.25 shows that when no crossing-over takes place between the genes, a PD tetrad is formed. Single crossover between the genes results in a TT tetrad. The formation of a two-strand, three-strand, or four-strand double crossover results in a PD, TT, or NPD tetrad, respectively. With linked genes, meiotic cells with no crossovers always outnumber those with four-strand double crossovers. Therefore,

The relative frequencies of the different types of tetrads can be used to determine the map distance between two linked genes. The simplest case is one in which the genes are sufficiently close that double and higher levels of crossing-over can be neglected. In this case, tetratype tetrads arise only from meiotic cells in which a single crossover occurs between the genes (Figure 4.25, part A and part B). As we saw in Figure 4.6, the genetic map distance across an interval is defined as one-half the proportion of cells with a crossover in the interval, so the map distance implied by the tetrads is given by

Map distance =

$$\frac{1}{2} \times \frac{\text{Number of tetratype tetrads}}{\text{Total number of tetrads}} \times 100 \quad (4.1)$$

To take a specific example, suppose 100 tetrads are analyzed from the cross $A\,B \times a\,b$, and the result is that 91 are PD and 9 TT. The finding that NPD $<<$ PD means that the genes are linked, and the fact that NPD = 0 means that the genes are so closely

> **key concept**
>
> Linkage is indicated when nonparental ditype tetrads appear with a much lower frequency than parental ditype tetrads (NPD $<<$ PD).

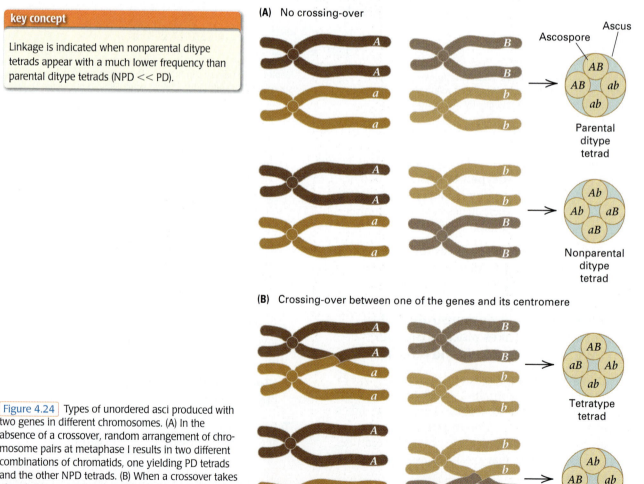

Figure 4.24 Types of unordered asci produced with two genes in different chromosomes. (A) In the absence of a crossover, random arrangement of chromosome pairs at metaphase I results in two different combinations of chromatids, one yielding PD tetrads and the other NPD tetrads. (B) When a crossover takes place between one gene and its centromere, the two chromosome arrangements yield TT tetrads. If both genes are closely linked to their centromeres (so that crossing-over is rare), few TT tetrads are produced.

(A) No crossing-over

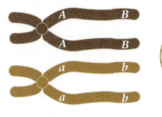

Parental
ditype (PD)

(B) Single crossover

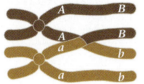

Tetratype (TT)

(C) 2-strand double crossover

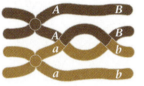

Parental
ditype (PD)

(D) 3-strand double crossover

Tetratype (TT)

(E) 3-strand double crossover

Tetratype (TT)

(F) 4-strand double crossover

Nonparental
ditype (NPD)

Figure 4.25 Types of tetrads produced with two linked genes. (A) In the absence of a crossover, a PD tetrad is produced. (B) With a single crossover between the genes, a TT tetrad is produced. (C–F) Among the four possible types of double crossovers between the genes, only the four-strand double crossover in part F yields an NPD tetrad.

linked that double crossing-over does not occur between them. The map distance between A and B is calculated as follows:

$$\text{Map distance} = \frac{1}{2} \times \frac{9}{100} \times 100 = 4.5 \text{ cM}$$

We must emphasize that Equation (4.1) is valid only when NPD = 0, so that interference across the region prevents the occurrence of double crossing-over. When double crossovers do take place in the interval, then NPD ≠ 0, and the formula for map distance has to be modified to take the double crossovers into account.

The mapping procedure using tetrads differs from that presented earlier in the chapter in that the map distance is not calculated directly from the number of recombinant and nonrecombinant chromatids. Instead, the map distance is calculated directly from the tetrads and the inferred crossovers that give rise to each type of tetrad. However, it is not necessary to carry out a full tetrad analysis for estimating linkage. The alternative is to examine spores chosen at random after allowing the tetrads to break open and disseminate their spores. This procedure is called *random-spore analysis,* and the linkage relationships are determined exactly as described earlier for *Drosophila* and corn. In particular, the frequency of recombination equals the number of spores that are recombinant for the genetic markers divided by the total number of spores.

© Photos.com.

Zebras can differ in hair color, just as other mammals can, owing to genetic differences among individuals and between different species.

■ The geometry of meiosis is revealed in ordered tetrads.

In the bread mold *Neurospora crassa,* a species used extensively in genetic investigations, the products of meiosis are contained in an *ordered* array of ascospores (Figure 4.26). A zygote nucleus, contained in a sac-like ascus, undergoes meiosis almost immediately after it is formed. The four nuclei produced by meiosis are in a linear, ordered sequence in the ascus, and each of them undergoes a mitotic division to form two genetically identical and adjacent ascospores. Each mature ascus contains eight ascospores arranged in four pairs, each pair derived from one of the products of meiosis. The ascospores can be removed one by one from an ascus and each germinated in a culture tube to determine its genotypes.

Ordered asci also can be classified as PD, NPD, or TT with respect to two pairs of alleles, which makes it possible to assess the degree of linkage between the genes. The fact that the arrangement of meiotic products is ordered also makes it possible to determine the recombination frequency between any particular gene and its centromere. The logic of the mapping technique is based on the feature of meiosis shown in Figure 4.27 :

key concept

Homologous centromeres of parental chromosomes separate at the first meiotic division; the centromeres of sister chromatids separate at the second meiotic division.

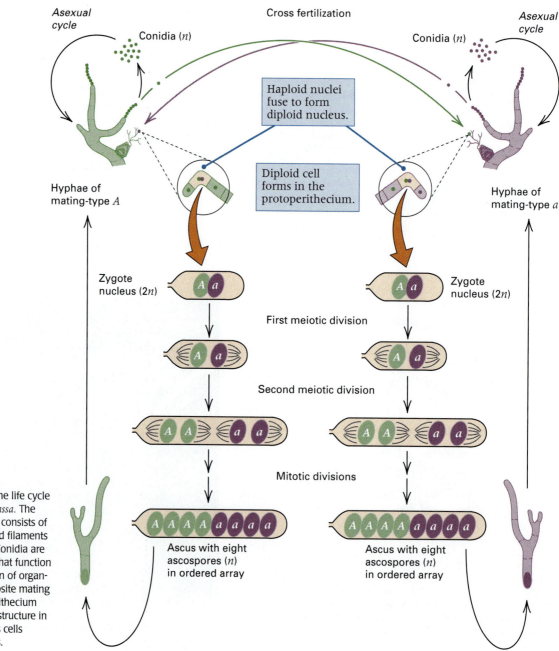

Figure 4.26 The life cycle of *Neurospora crassa.* The vegetative body consists of partly segmented filaments called hyphae. Conidia are asexual spores that function in the fertilization of organisms of the opposite mating type. A protoperithecium develops into a structure in which numerous cells undergo meiosis.

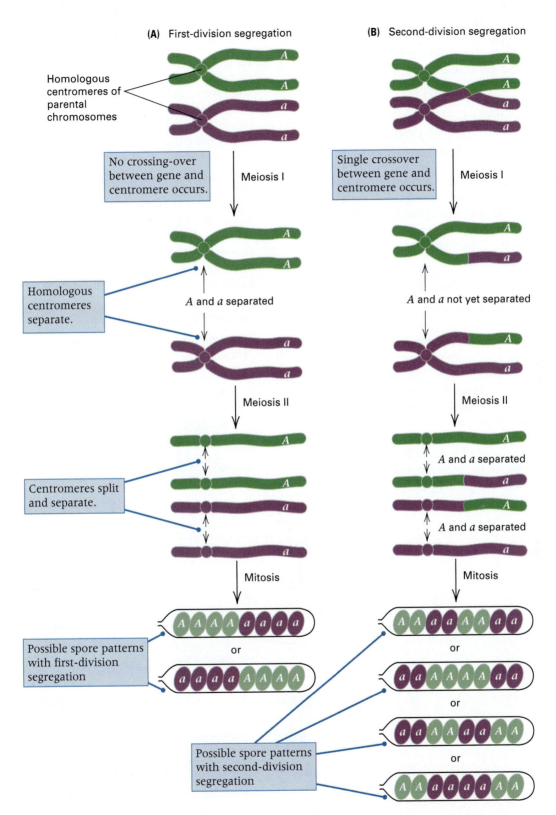

(A) First-division segregation

Homologous centromeres of parental chromosomes

A
A
a
a

No crossing-over between gene and centromere occurs.

Meiosis I

A
A

Homologous centromeres separate.

A and a separated

a
a

Meiosis II

A
A

Centromeres split and separate.

a
a

Mitosis

A A A A a a a a

or

a a a a A A A A

Possible spore patterns with first-division segregation

(B) Second-division segregation

A
A
a
a

Single crossover between gene and centromere occurs.

Meiosis I

A
a

A and a not yet separated

A
a

Meiosis II

A
A and a separated
a
A
A and a separated
a

Mitosis

A A a a A A a a

or

a a A A A A a a

or

a a A A a a A A

or

A A a a a a A A

Possible spore patterns with second-division segregation

Figure 4.27 First- and second-division segregation in *Neurospora*. (A) First-division segregation patterns are found in the ascus when a crossover between the gene and centromere does not take place. The alleles separate (segregate) in meiosis I. Two spore patterns are possible, depending on the orientation of the pair of chromosomes on the first-division spindle. The orientation shown results in the pattern in the upper ascus. (B) Second-division segregation patterns are found in the ascus when a crossover between the gene and the centromere delays separation of A from a until meiosis II. Four patterns of spores are possible, depending on the orientation of the pair of chromosomes on the first-division spindle and that of the chromatids of each chromosome on the second-division spindle. The orientation shown results in the pattern in the top ascus.

Thus, in the absence of crossing-over between a gene and its centromere, the alleles of the gene (for example, A and a) must separate in the first meiotic division; this separation is called **first-division segregation.** If, instead, a crossover is formed between the gene and its centromere, the A and a alleles do not become separated until the second meiotic division; this separation is called **second-division segregation.** The distinction between first-division and second-division segregation is shown in Figure 4.27. As shown in part A, only two possible arrangements of the products of meiosis can yield first-division segregation—$A A a a$ or $a a A A$. However, four patterns of second-division segregation are possible because of the random arrangement of homologous chromosomes at metaphase I and of the chromatids at metaphase II. These four arrangements, which are shown in part B, are

$$A a A a, \quad a A a A, \quad A a a A, \quad \text{and} \quad a A A a$$

The percentage of asci with second-division segregation patterns for a gene can be used to map the gene with respect to its centromere. For example, let us assume that 30 percent of a sample of asci from a cross have a second-division segregation pattern for the A and a alleles. This means that 30 percent of the cells undergoing meiosis had a crossover between the A gene and its centromere. Because the map distance between two genes is, by definition, equal to one-half times the proportion of cells with a crossover between the genes, the map distance between a gene and its centromere is given by the equation

Map distance =
$$\frac{1}{2} \times \frac{\text{Number of asci with second division segregation}}{\text{Total number of asci}} \times 100 \quad (4.2)$$

Equation (4.2) is valid as long as the gene is close enough to the centromere for us to neglect multiple crossovers. Reliable linkage values are best determined for genes that are near the centromere. The location of more distant genes is then accomplished by mapping these genes relative to genes nearer the centromere.

If a gene is far from its centromere, crossing-over between the gene and its centromere will be so frequent that the A and a alleles become randomized with respect to the four chromatids. The result is that the six possible spore arrangements shown in Figure 4.27 are all equally frequent. Therefore, when the chromatids participating in each crossover are chosen at random,

key concept

The maximum frequency of second-division segregation asci is 2/3.

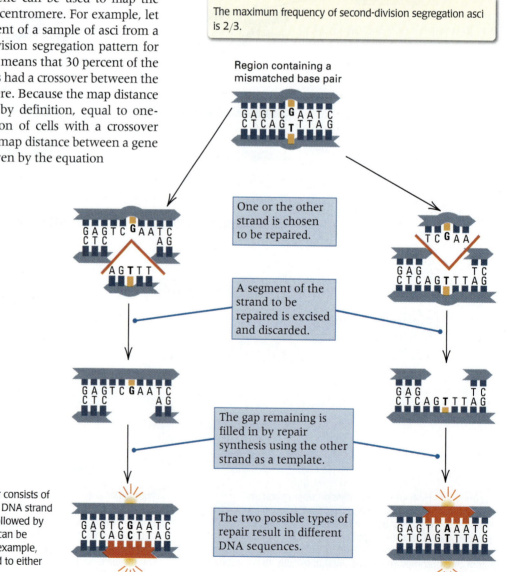

Figure 4.28 Mismatch repair consists of the excision of a segment of a DNA strand containing a base mismatch followed by repair synthesis. Either strand can be excised and corrected. In this example, the G—T mismatch is corrected to either G—C (left) or A—T (right).

Gene conversion suggests a molecular mechanism of recombination.

Genetic recombination may be regarded as a process of breakage and repair between two DNA molecules. In eukaryotes, the process takes place early in meiosis after each molecule has replicated, and with respect to genetic markers, it results in two molecules of the parental type and two recombinants (Chapter 3). For genetic studies of recombination, fungi such as yeast or *Neurospora* are particularly useful, because all four products of any meiosis can be recovered in a four-spore (yeast) or eight-spore (*Neurospora*) ascus. As we have noted, most asci from heterozygous *Aa* diploids contain ratios of

2 *A* : 2 *a* in four-spored asci, or

4 *A* : 4 *a* in eight-spored asci

demonstrating normal Mendelian segregation. Occasionally, however, aberrant ratios are also found, such as

3 *A* : 1 *a* or 1 *A* : 3 *a* in four-spored asci, and

5 *A* : 3 *a* or 3 *A* : 5 *a* in eight-spored asci

Different types of aberrant ratios can also occur. The aberrant asci are said to result from **gene conversion** because it appears as if one allele has "converted" the other allele into a form like itself. Gene conversion is frequently accompanied by recombination between genetic markers on either side of the conversion event, even when the flanking markers are tightly linked. This implies that gene conversion can be one consequence of the recombination process.

Gene conversion results from a normal DNA repair process in the cell known as **mismatch repair.** In this process, an enzyme recognizes any base pair in a DNA duplex in which the paired bases are mismatched—for example, G paired with T, or A paired with C. When such a mismatch is found in a molecule of duplex DNA, a small segment of one strand is excised and replaced with a new segment synthesized using the remaining strand as a template. In this manner the mismatched base pair is replaced. Figure 4.28 shows an example in which a mismatched G—T pair is being repaired. The strand that is excised could be either the strand containing T or the one containing G, and the newly synthesized (repaired) segment, shown in red, would contain either a C or an A, respectively. The two possible products of repair differ in DNA sequence.

The role of mismatch repair in gene conversion is illustrated in Figure 4.29 . The pair of DNA duplexes across the top represents the DNA molecules of two alleles in a cell undergoing meiosis. One duplex contains a G–C base pair highlighted in color; this corresponds to the *A* allele. The other duplex contains an A–T base pair at the same position, which corresponds to the *a* allele. In the process of recombination, the participating DNA duplexes can exchange pairing partners. The result is shown in the second row. The exchange of pairing partners creates a **heteroduplex** region in which any bases that are not identical in the parental duplexes become mismatched. In this example, one heteroduplex contains a G–T base pair and the other an A–C base pair. At this point, the mismatch repair system comes into play and corrects the mismatches. Each mismatch can be

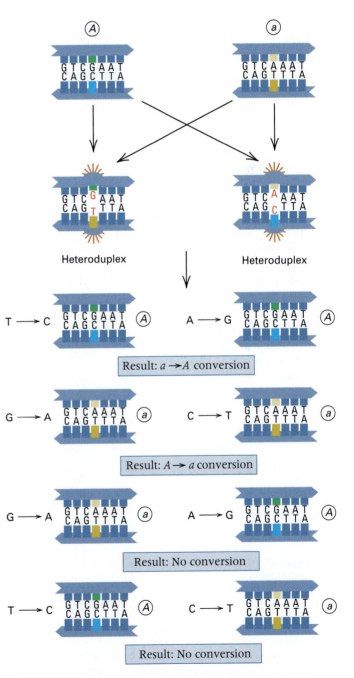

Figure 4.29 Mismatch repair resulting in gene conversion. Only a small part of the heteroduplex region is shown.

corrects the mismatches. Each mismatch can be repaired in either of two ways, so there are four possible ways in which the mismatches can be repaired. One type of repair results in gene conversion of *a* to *A*, another results in gene conversion of *A* to *a*, and the remaining two restore the sequences of the original duplexes and so do not result in gene conversion.

4.6

Recombination is initiated by a double-stranded break in DNA.

As emphasized in Chapter 3, chiasmata are the physical manifestations of crossing-over between DNA molecules. These structures bridge between pairs of sister chromatids in a bivalent and are important in the proper alignment of the bivalent at the metaphase plate in preparation for anaphase I. Bivalents that lack chiasmata to help hold them together are prone to undergo nondisjunction.

The crossovers needed for the chiasmata to form are initiated by programmed double-stranded breaks in DNA (Figure 4.30, part A). In a double-stranded break, the size of the gap is usually increased by nuclease digestion of the broken ends, with greater degradation of the 5' ends leaving overhanging 3' ends as shown in the illustration. These gaps are repaired using the unbroken homologous DNA molecule as a template, but in meiosis the repair process can result in crossovers that yield chiasmata between nonsister chromatids. These crossovers are also the phys-

ical basis of what is observed genetically as recombination. In some organisms, including humans and other mammals, the programmed DNA breaks are much more likely to occur at certain positions in the genome than others. Crossovers resulting in recombination are much more likely to occur at these positions, and so they are referred to as *hot spots* of recombination.

A double-stranded break does not necessarily result in a crossover, however. Repair of the double-stranded break by the noncrossover pathway is illustrated in Figure 4.30, part B. The first step in repair is that a broken 3' end invades the homologous unbroken DNA duplex, forming a short heteroduplex region with one strand and a looped-out region of the other strand called a **D loop**. (Specific proteins are required to mediate strand invasion; in

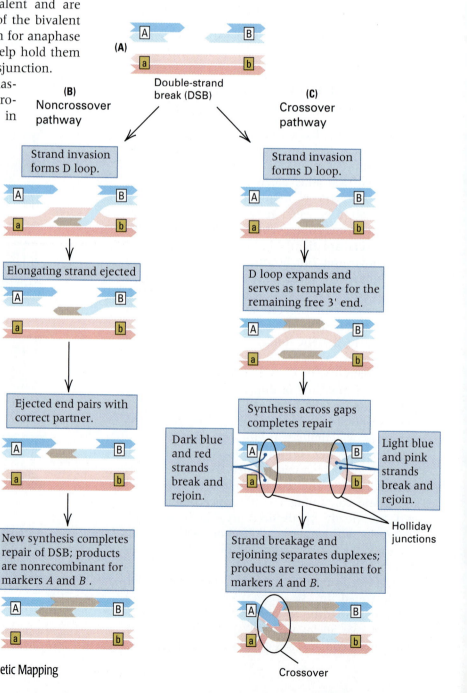

Figure 4.30 (A) Double-strand break in a duplex DNA molecule with overhanging 3' ends facing the gap. Repair of the break makes use of the nonbroken homologous duplex. (B) Repair pathway that does not result in crossing-over, although the heteroduplex regions can undergo gene conversion. (C) Repair pathway that does result in crossing-over, also with possible gene conversion in heteroduplex regions. [After D. K. Bishop and D. Zickler. 2004. *Cell* 117: 9.]

152   **CHAPTER 4** Gene Linkage and Genetic Mapping

RecA.) In the illustration, the heteroduplex region is the region where the light blue strand is paired with the red strand. Because it is a heteroduplex, any base-pair mismatches in this region could be corrected by mismatch repair in such a way as to result in gene conversion. Such heteroduplex regions are typically only a few hundred base pairs in length. They are much shorter than a gene and vastly shorter than a chromosome, and so gene conversions are rare events except for short regions very near the site of a double-stranded break.

At one end of the heteroduplex, the free 3' end of the broken DNA strand is extended (brown), but after a time it is ejected from the template, and the strands of the unbroken duplex are able to come together again. At this point, the extension of the 3' end is long enough that pairing can take place with the complementary strand in the broken duplex. At the same time, this pairing provides a template for the 3' end of the other broken strand. Extension of the 3' ends across the remaining gaps completes the repair of the double-stranded break. Note that although gene conversion can occur in the non-crossover pathway, the resulting duplex DNA molecules are nonrecombinant.

The crossover pathway for repairing a double-stranded break is illustrated in Figure 4.30, part C. Again invasion of the unbroken duplex forms a D loop and a short heteroduplex region in which gene conversion can occur. As in the noncrossover pathway, the free 3' end of the broken DNA strand is extended (brown), but in this case it continues until it displaces the partner strand (pink) of the template strand (red). The displaced strand can then serve as a template for the elongation of the 3' end of the other broken strand. Eventually, the extensions of the broken strands become long enough that they can be attached to the broken 5' ends. This completes the repair of the double-stranded break, but note that the resulting structure includes two places where the strands have exchanged pairing partners. Each of the structures where pairing partners are switched is called a **Holliday junction,** named after Robin Holliday, who first predicted that such structures would be involved in recombination.

The problem with Holliday junctions is that they are places where DNA strands from different duplex molecules are interconnected. How the strands are interconnected is shown for the DNA double helices in Figure 4.31, part A. Resolution of the Holliday junctions is necessary for the DNA molecules to become free of one another. This requires breakage and rejoining of one pair of DNA strands at each Holliday junction. The breakage and rejoining is an enzymatic function carried out by an enzyme called the **Holliday junction-resolving enzyme.**

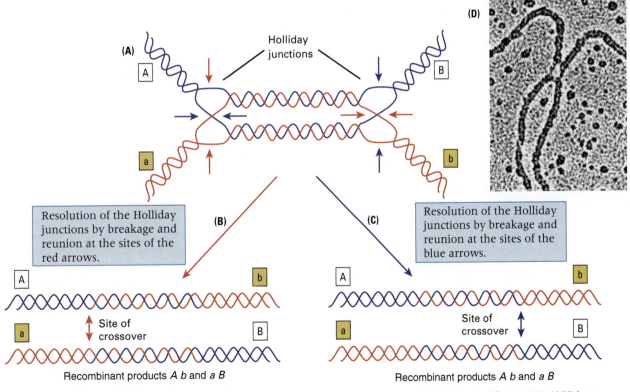

Figure 4.31 (A) Two Holliday junctions in a pair of DNA molecules undergoing recombination; (B and C) two modes of resolution depending which strands are broken and rejoined. Part D is an electron micrograph showing a single Holliday junction between a pair of DNA molecules. [Adapted © 1997 from *Essential Cell Biology, 1st edition* by Dr. Bruce Alberts. Reproduced by permission of Garland Science/Taylor & Francis Books, Inc.

Parts B and C in Figure 4.31 show two ways in which the Holliday structures can be resolved. Breakage and rejoining of the strands indicated by the red arrows results in a crossover at the site of the left-hand Holliday junction, whereas breakage and rejoining of the strands indicated by the blue arrows results in a crossover at the site of the right-hand Holliday junction. In both cases, the resulting DNA molecules have a crossover that yields reciprocal recombinant $A\ b$ and $a\ B$ products. (In principle, resolution could also take place at the red arrows in one Holliday junction and the blue arrows in the other, but these resolutions result in noncrossover products. It is unclear how often these noncrossover types of resolution take place.)

chapter summary

4.1 Linked alleles tend to stay together in meiosis.

- The degree of linkage is measured by the frequency of recombination.
- The frequency of recombination is the same for coupling and repulsion heterozygotes.
- The frequency of recombination differs from one gene pair to the next.
- Recombination does not occur in *Drosophila* males.

Nonallelic genes located in the same chromosome tend to remain together in meiosis rather than undergoing independent assortment. This phenomenon is called linkage. The indication of linkage is deviation from the 1 : 1 : 1 : 1 ratio of phenotypes in the progeny of a mating of the form $Aa\ Bb \times aa\ bb$. When alleles of two linked genes segregate, more than 50 percent of the gametes produced have parental combinations of the segregating alleles, and fewer than 50 percent have nonparental (recombinant) combinations of the alleles. The recombination of linked genes results from crossing-over, a process in which nonsister chromatids of the homologous chromosomes exchange corresponding segments in the first meiotic prophase.

4.2 Recombination results from crossing-over between linked alleles.

- Physical distance is often—but not always— correlated with map distance.
- One crossover can undo the effects of another.

The frequencies of crossing-over between different genes can be used to determine the relative order and locations of the genes in chromosomes. This is called genetic mapping. Distance between adjacent genes in such a map (a genetic or linkage map) is defined in map units. Across regions in which multiple crossing-over does not take place, the map distance between two genes is proportional to the frequency of recombination between them. In this case, one unit of map distance corresponds to 1 percent recombination. Across longer regions in which multiple crossovers are possible, map distance equals one-half the average number of crossovers expressed as a percentage. One map unit therefore corresponds to a physical length of the chromosome in which a crossover event occurs, on the average, once in every 50 meioses. For short distances, map units are additive. (For example, for three genes with order $a\ b\ c$, if the map distances $a-b$ and $b-c$ are 2 and 3 map units, respectively, then the map distance $a-c$ is $2 + 3 = 5$ map units.) The recombination frequency underestimates the map distance between genes if the length of the region is too great. This discrepancy results from multiple crossovers, which yield either no recombinants or the same number produced by a single event. For example, two crossovers in the region between two genes may yield no recombinants, and three crossover events may yield recombinants of the same type as that from a single crossover.

4.3 Double crossovers are revealed in three-point crosses.

- Interference decreases the chance of multiple crossing-over.

When many genes are mapped in a particular species, they form linkage groups equal in number to the haploid chromosome number of the species. The maximum frequency of recombination between any two genes in a mating is 50 percent; this happens when the genes are in nonhomologous chromosomes and assort independently or when the genes are sufficiently far apart in the same chromosome that at least one crossover is formed between them in every meiosis.

4.4 Polymorphic DNA sequences are used in human genetic mapping.

- Single-nucleotide polymorphisms are abundant in the human genome.

In many organisms, some of the most useful genetic markers are polymorphisms in nucleotide sequence that are not associated with any phenotypic abnormalities. Prominent among these are nucleotide substitutions that create or destroy a particular cleavage site recognized by a restriction endonuclease. Such mutations can be detected because different chromosomes yield restriction fragments that differ in size according to the positions of the cleavage sites. Genetic variation of this type is called restriction fragment length polymorphism (RFLP). Most species also have considerable genetic variation in which one allele differs from the next according to the number of copies it contains of a short tandemly repeated DNA sequence (simple sequence repeat, or SSR).

4.5 Tetrads contain all four products of meiosis.

- Unordered tetrads have no relation to the geometry of meiosis.
- Tetratype tetrads demonstrate that crossing-over takes place at the four-stand stage of meiosis and is reciprocal.
- The geometry of meiosis is revealed in ordered tetrads.
- Gene conversion suggests a molecular mechanism of recombination.

The four haploid products of individual meiotic divisions can be used to analyze linkage and recombination in some species of fungi and unicellular algae. The method is called tetrad analysis. In *Neurospora* and related fungi, the meiotic tetrads are contained in a tubular sac, or ascus, in a linear order, making it possible to determine whether a pair of alleles segregated in the first or the second meiotic division. Linkage analysis in unordered tetrads is based on the frequencies of parental ditype (PD), nonparental ditype (NPD), and tetratype (TT) tetrads. The observation that NPD << PD is a sensitive indicator of linkage between two genetic markers. Ordered tetrads are convenient for genetic analysis because the dis-tance between a gene and its centromere is related to the frequency of asci showing first-division segregation.

4.6 Recombination is initiated by a double-stranded break in DNA.

Genetic recombination is intimately connected with DNA repair because the process always includes breakage and rejoining of DNA molecules. In gene conversion, one allele becomes converted into a homologous allele, which is detected by aberrant segregation in fungal asci, yielding ratios of alleles such as 3 : 1 or 1 : 3. Gene conversion is the outcome of mismatch repair in heteroduplexes. At the DNA level, the process of recombination is initiated by a double-stranded break and includes the creation of heteroduplexes in the region of the exchange, so gene conversion is often accompanied by recombination of genetic markers flanking the conversion event. Repair of double-stranded breaks results in two Holliday junctions that join homologous DNA duplexes together. Resolution of the Holliday junctions by strand breakage and rejoining results in crossing-over and genetic recombination.

issues & ideas

- Distinguish between genetic recombination and genetic complementation. Is it possible for two mutant genes to show complementation but not recombination? Is it possible for two mutant genes to show recombination but not complementation?
- In genetic analysis, why is it important to know the position of a gene along a chromosome?
- What is the maximum frequency of recombination between two genes? Is there a maximum map distance between two genes?
- Why is the frequency of recombination over a long interval of a chromosome always smaller than the map distance over the same interval?

- What is meant by the term *chromosome interference?*
- In human genetics, why are molecular variations in DNA sequence, rather than phenotypes such as eye color or blood-group differences, used for genetic analysis?
- In genetic analysis, what is so special about the ability to examine tetrads in certain fungi?
- Explain how tetratype tetrads demonstrate that recombination takes place at the four-strand stage of meiosis and is reciprocal.
- Explain why the observation PD >> NPD with respect to tetrads is a sensitive indicator of linkage.

key terms & concepts

ascospore
ascus
centimorgan
chromosome map
cis configuration
coefficient of coincidence
coupling configuration
D loop
first-division segregation
frequency of recombination
gene conversion

genetic map
genetic marker
heteroduplex
Holliday junction
Holliday junction-resolving
 enzyme
interference
linkage
linkage group
linkage map
locus

map distance
map unit
mismatch repair
nonparental ditype (NPD)
parental combination
parental ditype (PD)
polymorphism
recombinant
recombination
repulsion configuration

restriction endonuclease
restriction fragment length
 polymorphism (RFLP)
second-division segregation
simple sequence repeat (SSR)
single-nucleotide
 polymorphism (SNP)
tetratype (TT)
three-point cross
trans configuration

key terms & concepts
continued

1. _____ Another term for lack of independent assortment.

2. _____ Unit of distance in a genetic map equal to 1 percent recombination. This unit is also called a centimorgan.

3. _____ Defined in technical terms as one-half of the average number of chiasmata between two genes multiplied by 100.

4. _____ Another term for the *trans* configuration of A and B in the genotype $A\ b\ /\ a\ B$.

5. _____ Ratio of the observed number of double crossovers to the expected number.

6. _____ The tendency for a crossover occurring at one location to inhibit the formation of crossovers in nearby regions.

7. _____ Type of enzyme used in the study of restriction fragment length polymorphisms.

8. _____ Unexpected change in allelic state resulting from mismatch repair in a heteroduplex.

9. _____ Structure formed in the process of recombination in which strands from two DNA duplexes switch pairing partners.

10. _____ Type of segregation yielding a *Neurospora* ordered ascus in which the order of A spores and a spores along the ascus is $A\ A\ a\ a\ a\ a\ A\ A$.

11. _____ Among unordered asci produced from the fungal cross $A\ b \times a\ B$, the type of ascus containing the spore genotypes $A\ B$, $A\ B$, $a\ b$, and $a\ b$.

12. _____ Physical relationship between two genes indicated by the expression NPD << PD.

solutions: step by step

Problem 1

The mutations *cinnabar* (*cn*, bright red eyes) and *vestigial* (*vg*, malformed wings) are linked in the second chromosome of *Drosophila*. Among 1000 progeny of the cross *cn vg* / + + females × *cn vg* / *cn vg* males, the following genotypes of progeny were observed. From these data, estimate the frequency of recombination between the *cn* and *vg* genes.

```
cn vg / cn vg    445
cn +  / cn vg     45
+  vg / cn vg     55
+  +  / cn vg    455
```

■ **Solution** In linkage problems of this type, you must first classify the progeny in terms of whether they are recombinant or nonrecombinant. Because the parental genotype is given as *cn vg* / + +, the nonrecombinant progeny are *cn vg* / *cn vg* (445) and + + / *cn vg* (455), and the recombinant progeny are *cn* + / *cn vg* (45) and + *vg* / *cn vg* (55). If the parental genotypes had not been given, it would still have been possible to classify the progeny using the principle that when there is linkage, the recombinant progeny are always less frequent than the nonrecombinant progeny. Because the frequency of recombination is determined by the frequency of progeny carrying recombinant chromosomes, the frequency of recombination between *cn* and *vg* is estimated as $r = (45 + 55)/1000 = 0.10$. The frequency of recombination can also be expressed as a percentage. For recombination frequencies smaller than about 10 percent, the percent recombination equals the number of map units between the genes; in this case, $r = 10$ percent recombination, or 10 map units.

Problem 2

In addition to *cn* and *vg*, the genes *curved* (*c*, curved wings) and *plexus* (*px*, extra wing veins) are linked in the second chromosome of *Drosophila*. In a cross of *cn c px* / + + + females × *cn c px* / *cn c px* males, the following progeny were counted:

```
cn  c  px  / cn c px    296
cn  c  +   / cn c px     63
cn  +  +   / cn c px    119
cn  +  px  / cn c px     10
+   c  px  / cn c px     86
+   c  +   / cn c px     15
+   +  +   / cn c px    329
+   +  px  / cn c px     82
        Total          1000
```

(a) What is the frequency of recombination between *cn* and *c*?

(b) What is the frequency of recombination between *c* and *px*?

(c) What is the frequency of recombination between *cn* and *px*?

(d) Why is the frequency of recombination between *cn* and *px* smaller than the sum of that between *cn* and *c* and that between *c* and *px*?

(e) What is the coefficient of coincidence across this region? What is the value of the interference?

(f) Draw a genetic map of the region, showing the locations of *cn*, *c*, and *px* and the map distances between the genes.

(g) From the data in Step-by-Step Problem 1, where in this map would you put the gene *vg*?

■Solution Do not try to hurry through linkage problems! You will be rewarded by taking time to organize the information in the optimal manner. First, group the progeny types into reciprocal pairs—*cn c px* with + + +, *cn c* + with + + *px*, and so forth—and make a new list organized as shown here. (Ignore the *cn c px* chromosome from the father because it contributes no information about recombination.)

cn c px	296	}	625	
+ + +	329			
cn c +	63	}	145	
+ + *px*	82			
cn + +	119	}	205	
+ *c px*	86			
cn + *px*	10	}	25	
+ *c* +	15			
Total			1000	

In this tabulation, a space has been inserted between the pairs of reciprocal products in order to keep the groups separate. The number next to each brace is the total number of chromosomes in the group. The most numerous group of reciprocal chromosomes (in this case, *cn c px* and + + +) consists of the nonrecombinants, and the least numerous group of reciprocal chromosomes (in this case, *cn* + *px* and + *c* +) consists of the double recombinants. Rearrange the order of the groups, if necessary, so that the nonrecombinants are at the top of the list and the double recombinants are at the bottom. (In the present example, rearrangement is not necessary.) At this point, also make sure that the order of the genes is correct as given, by comparing the genotypes of the double recombinants with those of the nonrecombinants. If the gene order is correct, then it will require two recombination events (one in each interval) to derive the double-recombinant chromosomes from the non-recombinants. If this is not the case, rearrange the order of the genes. (The "odd man out" in comparing the double recombinants with the nonrecombinants is always the gene in the middle.) In this particular example, the gene order is correct as given. Finally, with this preliminary bookkeeping done, we can proceed to tackle the questions. **(a)** The frequency of recombination between *cn* and *c* is given by the totals of all classes of progeny showing recombination in the *cn−c* interval, in this case $(205 + 25)/1000 = 0.23$. **(b)** The frequency of recombination between *c* and *px* equals $(145 + 25)/1000 = 0.17$. **(c)** The frequency of recombination between *cn* and *px* equals $(145 + 205)/1000 = 0.35$. (Note that the double recombinants are not included in this total, because the double recombinants are not recombined for *cn* and *px*; their allele combinations for *cn* and *px* are the same as in the nonrecombinants.) **(d)** The frequency of recombination between *cn* and *px* (0.35) is smaller than the sum of that between *cn* and *c* and that between *c* and *px* $(0.23 + 0.17 = 0.40)$ because of double recombination. **(e)** The coefficient of coincidence equals the observed number of double recombinants divided by the expected number. The observed number is 25 and the expected number is

$0.23 \times 0.17 \times 1000 = 39.1$; the coefficient of coincidence therefore equals $25/39.1 = 0.64$. The interference equals $1 - $ *coefficient of coincidence*, so the interference equals $1 - 0.64 = 0.36$. **(f)** The genetic map is shown in the accompanying diagram. The distances are in map units (centimorgans). However, the map distances of 23 and 17 map units are based on the 23 percent and 17 percent recombination observed between *cn* and *c* and between *c* and *px*, respectively; the actual distances in map units are probably a little greater than these estimates because of a small amount of double recombination within each of the intervals. **(g)** The position of *vg* is located 10 map units from *cn* because there is 10 percent recombination observed between *vg* and *cn*. However, there is no way of knowing from the information given whether *vg* is to the left of *cn* or the right, so both possible positions for *vg* are indicated. (In fact, *vg* is located between *cn* and *c*.)

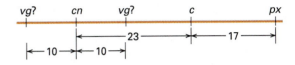

Problem 3

Genes called *spore killers* are relatively frequent in natural populations of *Neurospora crassa*. In asci produced by a heterozygous spore-killer genotype, all spores that do not carry the spore-killer allele are killed, and they are unable to be germinated. For example, in a cross of *Sk* × *sk*, where *Sk* represents the spore-killer allele, all spores carrying *sk* are killed. For one spore-killer allele, a cross of *Sk* × *sk* produced 125 asci. All of the asci had four dead ascospores and four live ascospores. In 95 of the asci, the four dead spores were all adjacent at either the top end of the ascus or the bottom end. What is the map distance between the *Sk* allele and its centromere?

■Solution Centromere mapping in *Neurospora* is based on the relative frequencies of first-division and second-division segregation. One feature that distinguishes these two types of asci is that in first-division segregation, the four spores carrying a particular allele are at one end of the ascus or the other. In the *Sk* × *sk* cross, the first-division asci would have one of the configurations shown in the diagram. Because the *sk*-bearing asci die, in these asci (and only in these) the dead ascospores are all at either the top or the bottom of the ascus. The observation is that 95/125 asci are of this type, which implies that $95/125 = 76$ percent of the asci show first-division segregation. Because the distance between a gene and its centromere is estimated as 1/2 times the frequency of *second*-division segregation, this spore-killer gene is estimated to be at a distance of $(1 - 0.76)/2 = 12$ map units from its centromere.

concepts in action: problems for solution

4.1 What is the principal difference between the genotype $A\,B/a\,b$ and the genotype $A\,b/a\,B$? What is the role of the slash (/, technically called a virgule) in such an expression? In which genotype are the A and B alleles in coupling (*cis*)? In which genotype are they in repulsion (*trans*)?

4.2 What gametes, and in what frequencies, are produced by a female *Drosophila* of genotype $A\,B/a\,b$ when the genes are present in the same chromosome and the frequency of recombination between them is 8 percent? What gametes, and in what frequencies, are produced by a male of the same genotype?

4.3 The genetic map of the X chromosome of *Drosophila melanogaster* has a length of 73.1 map units. The X chromosome of the related species *Drosophila virilis* is much longer (170.5 map units). In view of the fact that the recombination rate between two genes cannot exceed 50 percent, how is it possible for the map distance between genes at opposite ends of a chromosome to exceed 50 map units?

4.4 Which of the following ways to complete this statement are correct? A coefficient of coincidence of 0.25 means that:
(a) The frequency of double crossovers was 25 percent.
(b) The frequency of double crossovers was 25 percent of the number that would be expected if there were no interference.
(c) There were four times as many single crossovers as double crossovers.
(d) There were four times as many single crossovers in one region as there were in an adjacent region.
(e) There were four times as many parental as recombinant progeny.

4.5 In the case of independently assorting genes in fungi, why are parental ditype (PD) and nonparental ditype (NPD) tetrads formed in equal numbers? Why are NPD tetrads rare for linked genes?

4.6 In *Drosophila*, the eye-color mutation scarlet (*st*) and the bristle mutation spineless (*ss*) are located in chromosome 3 at a distance of 14 map units. What phenotypes, and in what proportions, would you expect in the progeny from the mating of $st^+\,ss^+ / st\,ss$ females with $st\,ss / st\,ss$ males?

4.7 Construct a map of a chromosome from the following recombination frequencies between individual pairs of genes: $r-c$, 10; $c-p$, 12; $p-r$, 3; $s-c$, 16; $s-r$, 8. You will discover that the distances are not strictly additive. Why aren't they?

4.8 A *Drosophila* cross is carried out with a female that is heterozygous for both the y (yellow body) and bb (bobbed bristles) mutations. Both genes are located in the X chromosome. Among 200 male progeny, there were 59 wildtype for both traits, 46 with yellow body, 43 with bobbed bristles, and 52 mutant for both genes. Do these genes show evidence for linkage? [*Note:* The appropriate chi-square test is a test for a 1 : 1 ratio of parental : recombinant gametes.]

4.9 Two genes in chromosome 7 of corn are identified by the recessive alleles gl (glossy), determining glossy leaves,

and ra (ramosa), determining branching of ears. When a plant heterozygous for each of these alleles was crossed with a homozygous recessive plant, the progeny consisted of the following genotypes with the numbers of each indicated:

$Gl\,ra/gl\,ra$	88	$gl\,Ra/gl\,ra$	103
$Gl\,Ra/gl\,ra$	6	$gl\,ra/gl\,ra$	3

Calculate the frequency of recombination between these genes.

4.10 In the yellow-fever mosquito, *Aedes aegypti*, a dominant gene *DDT* for DDT resistance (DDT is dichlorodiphenyltrichloroethane, an insecticide) and a dominant gene *Dl* for Dieldrin resistance (Dieldrin is another long-lasting insecticide) are known to be in the same chromosome. A cross was carried out between a DDT-resistant strain and a Dieldrin-resistant strain, and female progeny resistant to both insecticides were testcrossed with wildtype males. The progeny were 99 resistant to both insecticides, 88 resistant to DDT only, 89 resistant to Dieldrin only, and 106 sensitive to both insecticides.
(a) Are *DDT* and *Dl* alleles of the same gene? How can you tell?
(b) Are *DDT* and *Dl* linked?
(c) What can you say about the positions of *DDT* and *Dl* along the chromosome?

4.11 In a testcross of an individual heterozygous for each of three linked genes, the most frequent classes of progeny were $A\,B\,c/a\,b\,c$ and $a\,b\,C/a\,b\,c$, and the least frequent classes were $A\,B\,C/a\,b\,c$ and $a\,b\,c/a\,b\,c$. What was the genotype of the triple heterozygote parent, and what is the order of the genes?

4.12 In corn, the genes v (virescent seedlings), pr (red aleurone), and bm (brown midrib) are all on chromosome 5, but not necessarily in the order given. The cross

$$v^+\,pr\,bm / v\,pr^+\,bm^+ \times v\,pr\,bm / v\,pr\,bm$$

produces 1000 progeny with the following phenotypes:

v^+	pr	bm	209
v	pr^+	bm^+	213
v^+	pr	bm^+	175
v	pr^+	bm	181
v^+	pr^+	bm	69
v	pr	bm^+	76
v^+	pr^+	bm^+	36
v	pr	bm	41

(a) Determine the gene order, the recombination frequencies between adjacent genes, the coefficient of coincidence, and the interference.
(b) Explain why, in this example, the recombination frequencies are not good estimates of map distance.

4.13 The man I-2 in the accompanying pedigree is affected with Huntington disease, a neuromuscular degeneration caused by a rare autosomal dominant mutation *HD* with complete penetrance. The wildtype allele is denoted *hd*. The woman II-1 is also affected. The RFLP alleles A and a yielding the bands in the gel are linked to the Huntington locus with a recombination frequency of 10 percent.

(a) Is the genotype of II-1 HD *A/hd a* or is it *HD a/hd A*?

(b) Given the pattern of bands in the gel, what is the probability that III-1 will be affected?

(c) Given the pattern of bands in the gel, what is the probability that III-2 will be affected?

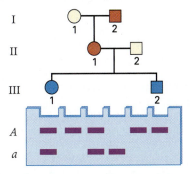

4.14 A human geneticist discovers the molecular variation in DNA sequence illustrated in the accompanying diagrams of electrophoresis gels. In the human population as a whole, she finds any of four phenotypes, shown in panel A. She believes that this may be a simple genetic polymorphism with three alleles, like the ABO blood groups. There are two alleles that yield DNA fragments of different sizes, fast (F) or slow (S) migration, and a "null" allele (O) in which the DNA fragment is deleted. The genotypes in panel A would therefore be, from left to right, FF or FO, SS or SO, FS, and OO. In the population as a whole, the putative OO genotype is extremely common, and the FS genotype is quite rare. To investigate this hypothesis further, the geneticist studies offspring of matings between parents who have the putative FS genotype (panel B). The types of progeny, and their numbers, are shown in panel C.

(a) What result would be expected from the three-allele hypothesis?

(b) Are the observed data consistent with this result? Why or why not?

(c) Suggest a genetic hypothesis that can explain the data in panel C.

(d) Are the data consistent with your hypothesis?

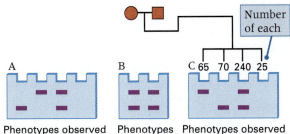

4.15 The following classes and frequencies of ordered tetrads were obtained from the cross $a^+ b^+ \times a\, b$ in *Neurospora*. (Only one member of each pair of spores is shown.) What is the order of the genes in relation to the centromere?

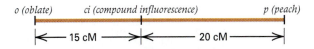

	Spore pair			Number of asci
1–2	3–4	5–6	7–8	
$a^+ b^+$	$a^+ b^+$	$a\ b$	$a\ b$	1766
$a^+ b^+$	$a\ b$	$a^+ b^+$	$a\ b$	220
$a^+ b^+$	$a\ b^+$	$a^+ b$	$a\ b$	14

4.16 A portion of the linkage map of chromosome 2 in the tomato is illustrated here. The oblate phenotype has flattened fruit, peach results in hairy fruit (like a peach), and compound influorescence means clustered flowers.

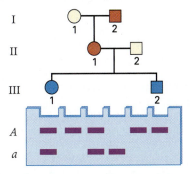

o (oblate)   *ci (compound influorescence)*   *p (peach)*

← 15 cM → ← 20 cM →

Among 1000 gametes produced by a plant of genotype $o\ ci + / + + p$, what types of gametes would be expected, and what number would be expected of each? Assume that the chromosome interference across this region is 80 percent but that interference within each region is complete.

4.17 The yeast *Saccharomyces cerevisiae* has unordered tetrads. In a cross made to study the linkage relationships among three genes, the tetrads in the accompanying table were obtained. The cross was between a strain of genotype $+ b\, c$ and one of genotype $a + +$.

Tetrad type	Genotypes of spores in tetrads				Number of tetrads
1	$a + +$	$a + +$	$+ b\ c$	$+ b\ c$	132
2	$a\ b +$	$a\ b +$	$+ + c$	$+ + c$	124
3	$a + +$	$a + c$	$+ b +$	$+ b\ c$	64
4	$a\ b +$	$a\ b\ c$	$+ + +$	$+ + c$	80
Total					400

(a) From these data determine which, if any, of the genes are linked.

(b) For any linked genes, determine the map distances.

4.18 A small portion of the genetic map of *Neurospora crassa* chromosome VI is illustrated here. The *cys-1* mutation blocks cysteine synthesis, and the *pan-2* mutation blocks pantothenic acid synthesis. Assuming complete chromosome interference, determine the expected frequencies of the following types of asci in a cross of *cys-1 pan-2* × *CYS-1 PAN-2*.

(a) First-division segregation of *cys-1* and first-division segregation of *pan-2*.

(b) First-division segregation of *cys-1* and second-division segregation of *pan-2*.

(c) Second-division segregation of *cys-1* and first-division segregation of *pan-2*.

(d) Second-division segregation of *cys-1* and second-division segregation of *pan-2*.

(e) Parental ditype, tetratype, and nonparental ditype tetrads.

cys-1   *pan-2*

← 7 cM → ← 3 cM →

4.19 The accompanying gel diagram shows the DNA-band phenotypes associated with the *A, a* and *B, b* allele pairs for two linked genes. On the left are the phenotypes of the parents, and on the right are the phenotypes of the progeny and the number of each observed. Is the linkage phase of *A* and *B* in the doubly heterozygous parent coupling or repulsion? What is the frequency of recombination between these genes?

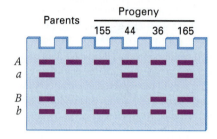

4.20 The gel diagram below shows the DNA-band phenotypes associated with allele pairs for three linked genes. On the left are the phenotypes of the parents, and on the right are the phenotypes of the progeny and the number of each observed.

(a) What is the order of the genes?

(b) What is the genotype of the triply heterozygous parent?

(c) What is the frequency of recombination in the smaller of the two regions?

(d) What is the frequency of recombination in the larger of the two regions?

(e) What is the value of the coincidence across the region?

(f) What is the interference across the region?

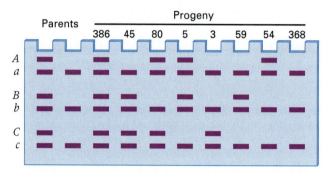

further readings

Botstein, D., R. L. White, M. Skolnick, and R. W. Davis. 1980. Construction of a genetic linkage map in man using restriction fragment length polymorphisms. *American Journal of Human Genetics* 32: 314.

Carlson, E. A. 1987. *The Gene: A Critical History.* 2d ed. Philadelphia: Saunders.

Fincham, J. R. S., P. R. Day, and A. Radford. 1979. *Fungal Genetics.* Oxford, England: Blackwell.

Green, M. M. 1996. The "Genesis of the White-Eyed Mutant" in *Drosophila melanogaster*: A reappraisal. *Genetics* 142: 329.

Kohler, R. E. 1994. *Lords of the Fly.* Chicago: University of Chicago Press.

Levine, L. 1971. *Papers on Genetics.* St. Louis, MO: Mosby.

Lewis, E. B. 1995. Remembering Sturtevant. *Genetics* 141: 1227.

Morton, N. E. 1995. LODs past and present. *Genetics* 140: 7.

Risch, N. J. 2000. Searching for genetic determinants in the new millennium. *Nature* 405: 847.

Stewart, G. D., T. J. Hassold, and D. M. Kurnit. 1988. Trisomy 21: Molecular and cytogenetic studies of nondisjunction. *Advances in Human Genetics* 17: 99.

Sturtevant, A. H. 1965. *A History of Genetics.* New York: Harper & Row.

Sturtevant, A. H., and G. W. Beadle. 1962. *An Introduction to Genetics.* New York: Dover.

Voeller, B. R., ed. 1968. *The Chromosome Theory of Inheritance: Classical Papers in Development and Heredity.* New York: Appleton-Century-Crofts.

Wang, D. G., J.-B. Fan, C.-J. Siao, A. Berno, P. Young, R. Sapolsky et al. 1998. Large-scale identification, mapping and genotyping of single-nucleotide polymorphisms in the human genome. *Science* 280: 1077.

White, R., and J.-M. Lalouel. 1988. Chromosome mapping with DNA markers. *Scientific American*, February.

geNETics on the web

GeNETics on the Web will introduce you to some of the most important sites for finding genetic information on the Internet. To explore these sites, visit the Jones and Bartlett home page at

http://www.jbpub.com/genetics

For the book *Essential Genetics: A Genomics Perspective,* choose the link that says **Enter GeNETics on the Web**. You will be presented with a chapter-by-chapter list of highlighted keywords. Select any highlighted keyword and you will be linked to a Web site containing genetic information related to the keyword.

- The linkage map of the **human chromosomes** currently includes more than 5000 genetic markers. Detailed linkage maps of the human chromosomes are available in graphical format at this keyword site. The site also includes tables of the physical length of each chromosome in nucleotide pairs, the genetic length in centimorgans in males and females, and the number of informative meioses that are necessary to yield statistically significant lod scores for any specified recombination fraction.

- One of the earliest cases of genetic linkages in a human autosome, reported in 1965, was that between the ABO blood-group gene and a gene that, when mutated, causes nail–patella syndrome (**NPS**). The frequency of recombination is about 10 percent but is higher in females than in males. At this keyword site, you can learn more about NPS and find the more recent linkage data.

A human chromosome at metaphase of mitosis. The DNA molecule folded inside this structure is about ten thousand times longer than the structure itself. How this prodigious feat of packaging is accomplished is one of the many remaining mysteries in molecular genetics.

key concepts

- The standard human karyotype consists of 22 pairs of autosomes and two sex chromosomes.

- Chromosome abnormalities are a major factor in human spontaneous abortions and an important cause of genetic disorders, such as trisomy 21 (Down syndrome).

- Dosage compensation in mammals results from genetic inactivation (silencing) of all but one X chromosome at an early stage in embryonic development.

- Aneuploid (unbalanced) chromosome rearrangements usually have greater phenotypic effects than euploid (balanced) chromosome rearrangements.

- Reciprocal translocations result in abnormal gametes because they upset segregation.

- The genetic imbalance caused by a single chromosome that is extra or missing may have a more serious phenotypic effect than an entire extra set of chromosomes.

- Duplication of the entire chromosome complement that is present in a species, or in a hybrid between species, is a major factor in the evolution of higher plants.

5

Human Chromosomes and Chromosome Behavior

chapter organization

5.1 Human beings have 46 chromosomes in 23 pairs.

5.2 Chromosome abnormalities are frequent in spontaneous abortions.

5.3 Chromosome rearrangements can have important genetic effects.

5.4 Polyploid species have multiple sets of chromosomes.

5.5 The grass family illustrates the importance of polyploidy and chromosome rearrangements in genome evolution.

the human connection
Catch-21

Chapter Summary
Issues & Ideas
Key Terms & Concepts
Solutions: Step by Step
Concepts in Action: Problems for Solution
Further Readings

geNETics on the web

In most species, the chromosome complement is virtually identical from one individual to the next, with a few exceptions such as the XX—XY sex difference in many animals. As might be expected, organisms with an extra chromosome or a missing chromosome usually have developmental or other types of abnormalities that result from the increase or decrease in copy number (*dosage*) of the genes in this chromosome. Some organisms, usually rare, are found to have a variation in chromosome structure. The abnormal chromosome may have a particular segment missing, duplicated, reversed in orientation, or attached to a different chromosome. Each of these structural abnormalities has different genetic implications. In this chapter we consider the human chromosome complement and some of the major chromosomal abnormalities encountered in human populations. We also examine chromosome abnormalities in other organisms. Generally speaking, animals are much less tolerant of chromosomal changes than are plants. As we shall see, the acquisition of entire extra sets of chromosomes is not necessarily harmful, especially in plants. In some lineages of plants, the duplication of entire chromosome sets has figured prominently in genome evolution and the origin of species.

5.1

Human beings have 46 chromosomes in 23 pairs.

The normal chromosome complement of a cell in mitotic metaphase from a human male is illustrated in Figure 5.1. The chromosomes have been labeled via a technique called **chromosome painting,** in which different colors are "painted" on each chromosome by hybridization (formation of duplex molecules) with DNA strands labeled with different fluorescent dyes. Individual chromosomes are first isolated by any of a variety of techniques, and then the chromosome-specific DNA samples are labeled with fluorescence. A mixture of differently labeled strands from all the chromosomes is used in hybridization with metaphase chromosomes squashed onto a glass slide, allowing the fluorescent strands to hybridize with complementary strands present in the chromosomes. Unhybridized DNA is washed from the slide, and the preparation is examined through a confocal microscope to read the fluorescent signals for conversion into visible colors. (A confocal microscope produces images of a single region in a single focal plane, because it is able to reject scattered and extraneous light.)

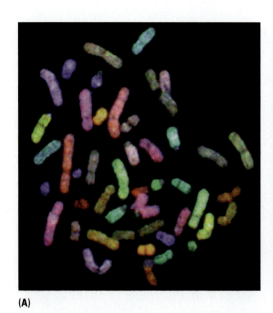

(A)

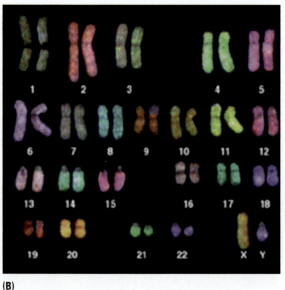

(B)

Figure 5.1 Human chromosome painting, in which each pair of chromosomes is labeled by hybridization with a different fluorescent probe. (A) Metaphase spread showing the chromosomes in a random arrangement as they were squashed onto the slide. (B) A karyotype, in which the chromosomes have been grouped in pairs and arranged in conventional order. Chromosomes 1–20 are arranged in order of decreasing size, but for historical reasons, chromosome 21 precedes chromosome 22, even though chromosome 21 is smaller. [Courtesy of Johannes Wienberg and Thomas Ried.]

The standard human karyotype consists of 22 pairs of autosomes and two sex chromosomes.

Chromosome painting dramatically identifies the pairs of homologous chromosomes. The presentation shown in part A of Figure 5.1 is a *metaphase spread,* in which the chromosomes are arranged just as they appear in the cytological preparation. A more conventional representation, called a **karyotype,** is shown in part B of Figure 5.1. In a karyotype, the autosomes in the metaphase spread are rearranged systematically in pairs, from longest to shortest, and numbered from 1 (the longest) through 22. In this example, the sex chromosomes are set off at the bottom right. The single X and Y chromosomes are evident. The karyotype of a normal human female has a pair of X chromosomes, instead of an X and a Y, in addition to the 22 pairs of autosomes. Chromosome painting is of considerable utility in human cytogenetics because even complex chromosome rearrangements can be detected rapidly and easily.

Another, less colorful metaphase spread and karyotype are shown in Figure 5.2 . In this case the chromosomes have been treated with a staining reagent called Giemsa, which causes the chromo-

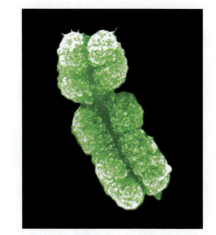

SCANNING ELECTRON MICROGRAPH of a human chromosome at mitotic metaphase.

somes to exhibit transverse bands (*G-bands*). The bands form in large regions in which the base composition of the DNA has a relatively low abundance of G–C base pairs, and the banding pattern is specific for each pair of homologs. These bands permit smaller segments of each chromosome arm to be identified. The chromosomes are grouped into seven sets denoted by the letters A through G. (The X chromosome is included in group C, the Y in

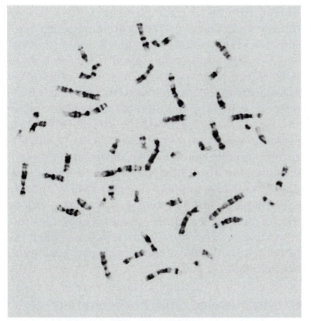

(A) Photograph of metaphase chromosomes

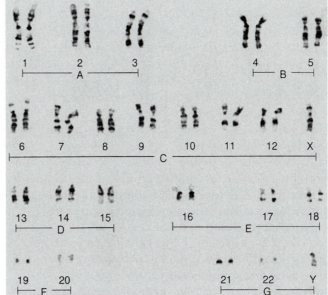

(B) Karyotype

Figure 5.2 A karyotype of a normal human male. Blood cells arrested in metaphase were stained with Giemsa and photographed with a microscope. (A) The chromosomes as seen in the cell by microscopy. (B) The chromosomes have been cut out of the photograph and paired with their homologs. [Courtesy of Patricia Jacobs.]

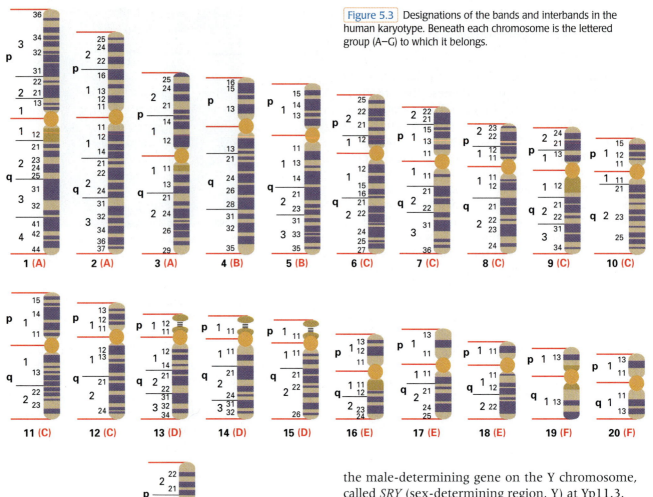

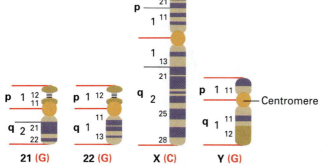

the male-determining gene on the Y chromosome, called *SRY* (sex-determining region, Y) at Yp11.3.

Figure 5.4 shows a chromosome painting of a human chromosome complement at metaphase of mitosis. Only one of each pair of homologous autosomes is shown, along with the X and Y chromosome. Below each chromosome is the amount of DNA in the chromosome in megabase pairs (Mb), an estimate of the number of genes from the human genome sequence, and the approximate gene density. Gene density is not highly correlated with chromosome size, and it can differ greatly from one chromosome to the next. Two of the smallest chromosomes, 19 and 22, have the highest gene densities (27 and 23 genes per Mb, respectively), and two of the largest chromosomes (4 and 5) have among the smallest gene densities (8 and 9 genes per Mb, respectively). The total genome size is 3000 Mb (3 billion base pairs), and the estimated gene number is 25,000–30,000.

group G.) These conventional groupings date from a time prior to G-banding and chromosome painting, when the chromosomes could be sorted only by size and centromere position.

The nomenclature of the banding patterns in human chromosomes is shown in Figure 5.3, where the red letter beneath each chromosome indicates its group. For each chromosome, the short arm is designated with the letter p, which stands for "petite," and the long arm by the letter q, which stands for "not-p." Within each arm, the regions are numbered according to standard conventions. Some familiar genetic landmarks in the human genome are the ABO blood-group locus at 9q34; the red–green color-blindness genes at Xq28, and

■ Chromosomes with no centromere, or with two centromeres, are genetically unstable.

As is true in nearly all eukaryotic organisms, each human chromosome is linear and has a single centromere. Chromosomes are often classified according to the relative position of their cen-

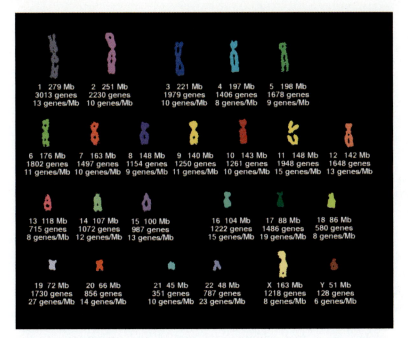

1 279 Mb 3013 genes 13 genes/Mb	2 251 Mb 2230 genes 10 genes/Mb	3 221 Mb 1979 genes 10 genes/Mb	4 197 Mb 1406 genes 8 genes/Mb	5 198 Mb 1678 genes 9 genes/Mb

| 6 176 Mb
1802 genes
11 genes/Mb | 7 163 Mb
1497 genes
10 genes/Mb | 8 148 Mb
1154 genes
9 genes/Mb | 9 140 Mb
1250 genes
11 genes/Mb | 10 143 Mb
1261 genes
10 genes/Mb | 11 148 Mb
1948 genes
15 genes/Mb | 12 142 Mb
1648 genes
13 genes/Mb |

| 13 118 Mb
715 genes
8 genes/Mb | 14 107 Mb
1072 genes
12 genes/Mb | 15 100 Mb
987 genes
13 genes/Mb | 16 104 Mb
1222 genes
15 genes/Mb | 17 88 Mb
1486 genes
19 genes/Mb | 18 86 Mb
580 genes
8 genes/Mb |

| 19 72 Mb
1730 genes
27 genes/Mb | 20 66 Mb
856 genes
14 genes/Mb | 21 45 Mb
351 genes
10 genes/Mb | 22 48 Mb
787 genes
23 genes/Mb | X 163 Mb
1218 genes
8 genes/Mb | Y 51 Mb
128 genes
6 genes/Mb |

Figure 5.4 The human chromosome complement at metaphase of mitosis showing the amount of DNA in each chromosome, the estimated number of genes, and the approximate gene density. For the autosomes, only one of each homologous pair is shown. [Genome sequence data from E. S. Lander et al. 2001. *Nature* 409: 860 and J. C. Venter et al. 2001. *Science* 291: 1304. Chromosome images courtesy of Michael Speicher.]

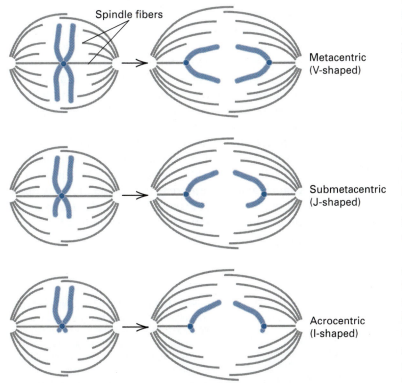

Spindle fibers

Metacentric (V-shaped)

Submetacentric (J-shaped)

Acrocentric (I-shaped)

Figure 5.5 Three possible shapes of monocentric chromosomes in anaphase as determined by the position of the centromere (shown in dark blue).

tromeres, which determines the appearance of the daughter chromosomes as they separate from each other in anaphase (Figure 5.5). A chromosome with its centromere about in the middle is a *metacentric chromosome;* the arms are of approximately equal length, and each daughter chromosome forms a V shape at anaphase. When the centromere is somewhat off center, the chromosome is a *submetacentric chromosome,* and each daughter chromosome forms a J shape at anaphase. A chromosome with the centromere very close to one end appears I-shaped at anaphase because the arms are grossly unequal in length; such a chromosome is *acrocentric.*

Chromosomes with a single centromere are usually the only ones that are reliably transmitted from parental cells to daughter cells or from parental organisms to their progeny. When a cell divides, spindle fibers attach to the centromere of each chromosome and pull the sister chromatids to opposite poles. A chromosome that lacks a centromere is an **acentric chromosome.** Acentric chromosomes are genetically unstable because they cannot be maneuvered properly during cell division and are lost. Occasionally, a chromosome arises that has two centromeres and is said to be **dicentric.** A dicentric chromosome is also genetically unstable because it is not transmitted in a predictable fashion. The dicentric chromosome is frequently lost from a cell when the two centromeres proceed to opposite poles in the course of cell division; in this case, the chromosome is stretched and forms a *bridge* between the daughter cells. This bridge may not be included in either daughter nucleus, or it may break, with the result that each daughter nucleus receives a broken chromosome. We will consider one mechanism by which dicentric and acentric chromosomes are formed when we discuss inversions.

Although most dicentric chromosomes are genetically unstable, if the two centromeres are close enough together, they can frequently behave as a single unit and be transmitted normally. This principle was important in the evolution of human chromosome 2. Among higher primates, chimpanzees and human beings have 23 pairs of chromosomes that are similar in morphology and G-banding pattern, but chimpanzees have no obvious homolog of

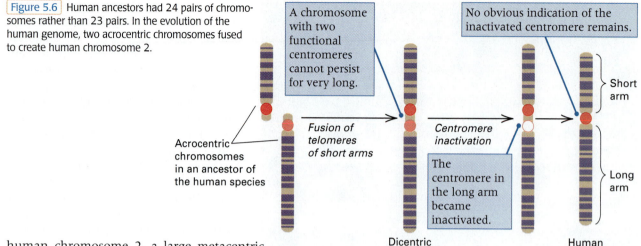

Figure 5.6 Human ancestors had 24 pairs of chromosomes rather than 23 pairs. In the evolution of the human genome, two acrocentric chromosomes fused to create human chromosome 2.

A chromosome with two functional centromeres cannot persist for very long.

No obvious indication of the inactivated centromere remains.

Acrocentric chromosomes in an ancestor of the human species

Fusion of telomeres of short arms

Centromere inactivation

The centromere in the long arm became inactivated.

Dicentric chromosome

Human chromosome 2

Short arm

Long arm

human chromosome 2, a large metacentric chromosome. Instead, chimpanzees have two medium-sized acrocentric chromosomes not found in the human genome. The cause of this situation is shown in Figure 5.6. The G-banding patterns indicate that human chromosome 2 was formed by fusion of the telomeres between the short arms of two acrocentric chromosomes that, in chimpanzees, remain acrocentrics. The fusion created a dicentric chromosome with its two centromeres close together. The new chromosome must have been sufficiently stable to be retained in the human lineage. The present-day human chromosome 2 has only a single functional centromere. The centromere on the slightly longer arm is inactive. Whether this inactivation occurred simultaneously with the chromosome fusion or evolved gradually is unknown. In any case, this chromosome fusion

reduced the chromosome number in the human lineage from 48, which is characteristic of the great apes (chimpanzee, gorilla, and orangutan), to the number 46.

■ **Dosage compensation adjusts the activity of X-linked genes in females.**

For all organisms with XX–XY sex determination, there is a problem of the dosage of genes on the X and Y chromosome, because females have two copies of this chromosome whereas males have only one. (There is less of a problem with Y-linked genes, because the Y chromosome is largely heterochromatic and carries relatively few genes.) In most organisms, a mechanism of **dosage compensation** has evolved in which the unequal dosage in the sexes is corrected either by increasing the activity of genes in the X chromosome in males or by reducing the activity of genes in the X chromosome in females. The mechanism of dosage compensation in mammals is seemingly simple: In the early cleavage divisions of the embryo, all X chromosomes except one, chosen at random, are genetically inactivated. Different tissues undergo **X inactivation** at different times, and an X chromosome that is inactivated in a particular somatic cell remains inactive in all the descendants of that cell (Figure 5.7).

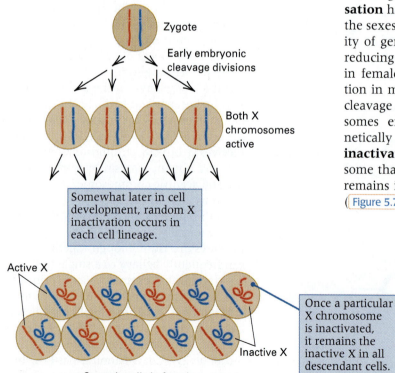

Zygote

Early embryonic cleavage divisions

Both X chromosomes active

Somewhat later in cell development, random X inactivation occurs in each cell lineage.

Active X

Once a particular X chromosome is inactivated, it remains the inactive X in all descendant cells.

Inactive X

Somatic cells in female

Figure 5.7 Schematic diagram of somatic cells of a normal female showing that the female is a mosaic for X-linked genes. The two X chromosomes are shown in red and blue. An active X is depicted as a straight chromosome, an inactive X as a tangle. Each cell has just one active X, but the particular X that remains active is a matter of chance. In human beings, the inactivation includes all but a few genes in the tip of the short arm.

The X-chromosome inactivation takes place in all embryos with two or more X chromosomes, including normal XX females. The inactivation process is one of chromosome condensation initiated at a site called *XIC* (for *X-inactivation center*) near the centromere on the long arm between Xq11.2 and Xq21.1. Cytologically, this site marks a "bend" in the chromosome in mitotic metaphase, which is thought to be a visible manifestation of the X inactivation. The *XIC* includes a transcribed region in band Xq13 designated *Xist* (for *X-inactivation–specific transcript*). Transcription of *Xist* is the earliest event observed in X inactivation, and *Xist* is the only gene known to be transcribed only from the inactive X chromosome. Remarkably, the spliced transcript of *Xist* does not contain an open reading frame encoding a protein. It appears to function as a structural RNA, and as transcription of *Xist* continues, the spliced transcript progressively coats the inactive X chromosome, spreading outward from the *XIC*. Thereafter, other molecular changes take place along the inactive X chromosome that are typically associated with gene silencing. In mouse embryos in which the *Xist* homolog has been disrupted, X inactivation does not take place, which demonstrates that *Xist* is essential for inactivation. On the other hand, a study of the level of transcription of 624 genes along the human X chromosome has shown that about 15 percent of the X-linked genes escape inactivation to some degree. The "inactive X" is therefore not completely silenced.

In certain cell types, the inactive X chromosome in females can be observed microscopically as a densely staining heterochromatic body in the nucleus of interphase cells. This is called a **Barr body** (arrow in Figure 5.8). Although cells of normal females have one Barr body, cells of normal males have none. Persons with two or more X chromosomes have all but one X chromosome per cell inactivated, and the number of Barr bodies equals the number of inactivated X chromosomes. Hybridization *in situ* of labeled *Xist* RNA with chromosomes from a normal female demonstrates that the *Xist* RNA is localized within the nucleus at the same position as the Barr body.

X-chromosome inactivation has two consequences. First, it results in dosage compensation. It equalizes the number of active copies of X-linked genes in females and males. Although a female has two X chromosomes and a male has only one, because of inactivation of one X chromosome in each of the somatic cells of a female, both sexes have the same number of active X chromosomes. The mechanism of dosage compensation by means of X inactivation was originally proposed by Mary Lyon and is called the **single-active-X principle.**

The second consequence of X-chromosome inactivation is that a normal female is a mosaic for the expression of X-linked genes. A genetic *mosaic* is an individual that contains cells of two or more different genotypes. A normal female is a mosaic for gene expression, because the X chromosome that is genetically active can differ from one cell to the next. The mosaicism can be observed directly in females that are heterozygous for X-linked alleles that determine different forms of an enzyme, A and B: When cells from a heterozygous female are individually cultured in the laboratory, half of the clones are found to produce only the A form of the enzyme and the other half to produce only the B form. Mosaicism can also be observed directly in women who are heterozygous for an X-linked recessive mutation that results in the absence of sweat glands; these women exhibit large patches of skin in which sweat glands are present (these patches are derived from embryonic cells in which the normal X chromosome remained active and the mutant X was inactivated) and other large patches of skin in which sweat glands are absent (these patches are derived from embryonic cells in which the normal X chromosome was inactivated and the mutant X remained active.)

Although all mammals use X-chromosome inactivation for dosage compensation in females, the choice of which chromosome to inactivate is not always random. In marsupial mammals, which include the kangaroo, the koala, and the wombat, the X chromosome that is inactivated is always the one contributed by the father. The result is that

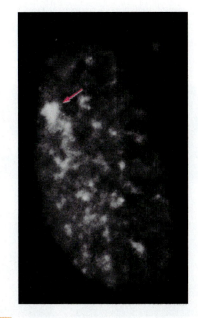

Figure 5.8 Fluorescence micrograph of a human cell, showing a Barr body (the bright spot at the upper left that is indicated by the arrow.) This cell is from a normal human female, and it has one Barr body. [Micrograph courtesy of A. J. R. de Jonge.]

female marsupials are not genetic mosaics of paternal and maternal X-linked genes. A female marsupial expresses the X-linked genes that she inherited from her mother.

■ The calico cat shows visible evidence of X-chromosome inactivation.

In some cases, the result of random X inactivation in females can be observed in the external phenotype. One example is the "calico" pattern of coat coloration in female cats. Two alleles affecting coat color are present in the X chromosome in cats. One allele results in an orange coat color, the other in a black coat color. Because a normal male has only one X chromosome, he has either the orange or the black allele. A female can be heterozygous for orange and black, and in this case the coat color is "calico"—a mosaic of orange and black patches mixed with patches of white. Figure 5.9 is a photograph of a female cat with the classic calico pattern. The orange and black patches result from X-chromosome inactivation. In cell lineages in which the X chromosome bearing the orange allele is inactivated, the X chromosome with the black allele is active and so the fur is black. In cell lineages in which the X chromosome with the black allele is inactivated, the orange allele in the active X chromosome results in orange fur.

The white patches have a completely different explanation than the orange and black patches. The white patches are due to an autosomal gene S for white spotting, which prevents pigment formation in the cell lineages in which it is expressed. Why the S gene is expressed in some cell lineages and not others is not known. Homozygous S/S cats have more white than heterozygous S/s cats. Ginger, the female shown in Figure 5.9, is homozygous S/S.

Figure 5.9 Ginger, a female cat heterozygous for the orange and black coat color alleles. She shows the classic "calico" pattern of patches of orange, black, and white.

■ Some genes in the X chromosome are also present in the Y chromosome.

The silencing of genes in the inactive X chromosome evolved gradually as the X and Y chromosomes progressively diverged from their ancestral chromosomes and the Y chromosome began to lose the function of most of its genes. The gene inactivation in the inactive X chromosome therefore affects individual genes and blocks of genes, and some genes in the inactive X are not silenced. Some of the genes that escape X inactivation have functional homologs in the Y chromosome, whereas others do not.

Two continuous regions that escape X inactivation, each several megabases in length, are found at the tips of the long and short arms. These are regions in which the Y chromosome does retain homologous genes that are functional. These regions of homology enable the X and Y chromosomes to synapse in spermatogenesis, and normally a crossover takes place that holds the chromosomes together to ensure their proper segregation during anaphase I. The regions of shared X–Y homology define the **pseudoautosomal regions:** *PARp* at the terminus of the short arms and *PARq* at the terminus of the long arms. Because crossing-over regularly takes place in the short pseudoautosomal regions, the rate of recombination per nucleotide pair is at least 20-fold greater in the pseudoautosomal regions than in the autosomes.

The pedigree patterns of inheritance of genes in the pseudoautosomal regions are indistinguishable from patterns characteristic of autosomal inheritance. The reason is that a mutant allele in a pseudoautosomal region is neither completely X-linked nor completely Y-linked but can move back and forth between the X and Y chromosomes because of recombination in the pseudoautosomal region. A gene that shows an autosome-like pattern of inheritance, but that is known from molecular studies to reside in a pseudoautosomal region, is said to show *pseudoautosomal inheritance.*

■ The pseudoautosomal region of the X and Y chromosomes has gotten progressively shorter in evolutionary time.

Comparative cytogenetic and molecular studies suggest that the X and Y chromosomes began their existence as a pair of ordinary autosomes in the common ancestor of modern mammals and birds. They started to diverge in DNA sequence and gene content at about the same time that the evolutionary lineage of mammals diverged from that of birds, some 300–350 million years ago (MY). One must assume that prior to this time, recombination took place at normal levels throughout the entire proto-

X and proto-Y chromosomes and that their gene contents were identical.

In the short arm of the Y chromosome, near the pseudoautosomal region but not within it, the Y chromosome contains the master sex-determining gene *SRY* at position Yp11.3. *SRY* codes for a protein transcription factor, the **testis-determining factor (TDF),** that triggers male embryonic development by inducing the undifferentiated embryonic genital ridge, the precursor of the gonad, to develop as a testis. (A *transcription factor* is a protein that stimulates transcription of its target genes.)

Once *SRY* had evolved as a sex-determining mechanism, the Y chromosome began to diverge in DNA sequence from the X chromosome, and the region of possible X–Y recombination became progressively restricted to the telomeric regions. In regions with no X–Y recombination, there is a steady selection pressure for genes in the Y chromosome to undergo mutational degeneration into nonfunctional states. This results from the forced heterozygosity of the Y chromosome, which allows multiple deleterious mutations to accumulate through time because there is no opportunity for

recombination to regenerate Y chromosomes that are free of deleterious mutations. Hence any Y-linked gene whose function is nonessential will tend to degenerate gradually because of the accumulation of mutations, and at the same time there will be selection pressure for dosage compensation of the homologous gene in the X chromosome. Eventually, only the dosage-compensated X-linked gene will remain functional.

Apparently blocks of genes were removed from the region of X–Y recombination in large chunks. Molecular evidence for this conclusion is summarized in Figure 5.10. Shown at the left are the locations of some protein-coding sequences in the short arm of the modern X chromosome from band Xp11 to the telomere. All of these genes have homologous sequences in the modern Y chromosome, as shown at the right. The amount of sequence divergence between the X and Y homologs shows a remarkable pattern. In the positions of the codons where a nucleotide substitution can occur without changing the encoded amino acid, the proportion of nucleotide differences between the X and Y homologs for the genes *GYG2–AMELX* is 0.07–0.11,

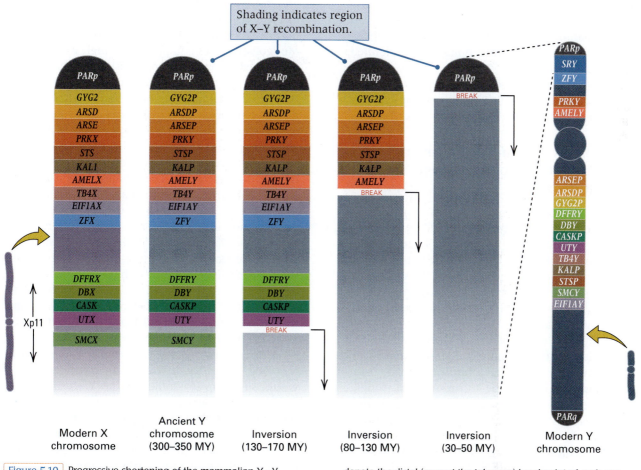

Figure 5.10 Progressive shortening of the mammalian X–Y pseudoautosomal region through time due to inversions in the Y chromosome, inferred from DNA sequence data. The arrows denote the distal (nearest the telomere) breakpoint of each successive inversion interrupting the pseudoautosomal region. [Data from B. T. Lahn and D. C. Page. 1999. *Science* 286: 964.]

for *TB4X–UTX* it is 0.23–0.36, for *SMCX* it is 0.52, and for other genes outside the region shown it is > 0.94. Because the evolutionary rate of nucleotide substitutions at synonymous sites in mammalian genes is known, we can say that these levels of divergence correspond to divergence times of 30–50 million years (MY) for *GYG2–AMELX*, 80–130 MY for *TB4X–UTX*, 130–170 MY for *SMCX*, and 300–350 MY for other genes outside the region shown. The simplest explanation is that these were the times when successive blocks of genes were removed from the region of X–Y recombination by chromosome rearrangements—namely, inversions in which the Y chromosome was broken at two points and the middle segment inverted before rejoining of the broken ends. For each inversion, one breakpoint must have occurred outside the region shown, but the location of the other breakpoint can be pinpointed to within the region between the blocks of genes. Hence, as shown in the evolutionary reconstruction in Figure 5.10, the 130–170 MY inversion breakpoint was adjacent to *UTY*, the 80–130 MY inversion breakpoint adjacent to *AMELY*, and the 30–50 MY year breakpoint adjacent to the present-day pseudoautosomal region on the short arm. As each of these inversions was fixed in the evolving Y chromosome, it removed the corresponding block of genes from the region of X–Y recombination, so that the rate of sequence divergence between the X and Y homologs accelerated. Other rearrangements in the Y chromosome, which it is not possible to trace from these data, led to some additional scrambling of the gene order in the Y chromosome.

■ The history of human populations can be traced through studies of the Y chromosome.

Because the Y chromosome does not undergo recombination along most of its length, genetic markers in the Y are completely linked and so remain together as the chromosome is transmitted from generation to generation. Therefore, the genetic relation between Y chromosomes can be traced, because chromosomes that are closely related will share more alleles along their length than will more distantly related chromosomes. The set of alleles at two or more loci present in a particular chromosome is called a **haplotype.** For many genealogical studies of the Y chromosome, simple-sequence repeat (SSR) polymorphisms are convenient because of their relatively high rate of mutation and the large number of alleles. The logic is that Y chromosomes with haplotypes that share alleles at each of 20–30 SSRs across the chromosome must have descended from the same ancestral Y chromosome in the very recent past. For haplo-

types differing at a single locus the genetic relationship is less close, for those differing at two loci it is still less close, and so forth. This simple logic is the basis of tracing population history through Y-chromosome polymorphisms. Haplotypes that share many alleles have a more recent common ancestral Y chromosome than haplotypes that share fewer alleles. Furthermore, because the rate of SSR mutation can be estimated, the time at which the ancestral chromosome existed can be deduced. This reasoning forms the basis of the estimate that the most recent common ancestor of all extant human Y chromosomes existed 50,000 years ago. Such estimates are not highly precise, and there are many assumptions that must be made. Other studies using different markers yield estimates of 150,000 years. The reasons for the discrepancy are still unclear. Nevertheless, much can be learned about human population history through studies of the Y chromosome. The following discussion highlights three specific examples.

A Legacy of Genghis Khan At its maximum extent stretching from China to Russia through to the Middle East and then into Eastern Europe, the Mongol Empire of the thirteenth century comprised the largest land empire that history has known. The founder was a man originally called Temujin, born in 1162. As a young man he organized a confederation of tribes, who around 1200 took to their small Mongolian ponies equipped with high wooden saddles and stirrups, and armed with bows and arrows began to conquer their neighbors. Soon thereafter, Temujin adopted the name Genghis Khan, which means Universal Ruler. He was often merciless, exterminating the men and boys of rebellious cities and kidnapping the women and girls. In answer to a question about the source of happiness, he is reputed to have said, "The greatest happiness is to vanquish your enemies, to chase them before you, to rob them of their wealth, to see those dear to them bathed in tears, to clasp to your bosom their wives and daughters." Through their multiple wives, concubines, and innumerable unrecorded sexual conquests, Genghis Khan and his descendants were very prolific. His eldest son Tushi had 40 acknowledged sons, and his grandson Kubilai Khan (under whom the Mongol Empire reached its maximum extent) had 22 acknowledged sons.

Although the legacy of Genghis Khan is well recorded in history, it was hardly expected that it would show up in studies of the Y chromosome. But the genotypes of 32 markers along the Y chromosome of 2123 men sampled from throughout a large region of Asia yielded the remarkable result in Figure 5.11 . Each circle represents a population sample, with its area proportion to the sample size. The red sectors denote the relative frequency of a group of nearly identical Y-chromosome haplo-

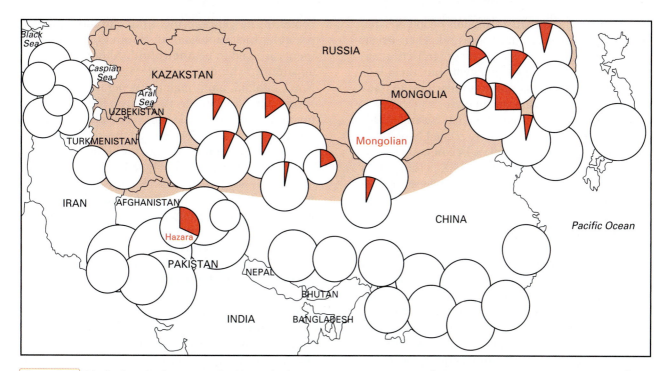

Figure 5.11 Distribution of Y-chromosome haplotypes (red), presumed to have descended from Genghis Khan or his close male relatives, among populations near and bordering the ancient Mongol Empire. [After T. Zerjal et al. 2003. *American Journal of Human Genetics* 72: 717.]

types, whereas the white sectors represent the relatively frequency of other haplotypes that are genetically much more diverse. The most recent common ancestor of the closely related haplotypes is estimated as existing 1000 ± 300 years ago. Furthermore, the geographical region in which the closely related haplotypes cluster is included largely within the Mongol Empire (shading). The sole exception is composed of the ethnic Haraza of Pakistan. This provides a clue to the origin of the closely related Y chromosomes, because the Hazara consider themselves to be of Mongol origin, and many claim to be direct male-line descendants of Genghis Khan. Whatever their origin, the closely related Y chromosomes are found in about 8 percent of the males throughout a large region of Asia. Direct proof of the connection with Genghis Khan could, in principle, be obtained by determining the haplotype of the Y chromosome in material recovered from his grave. He died in 1227 from injuries sustained in a fall from a horse, but his burial place is unknown.

A Legacy of the Cohanim The Lemba are a group of about 50,000 Bantu-speaking people living predominantly in South Africa and Zimbabwe. They drew attention about 100 years ago because of their vaguely Jewish customs including dietary restrictions and male circumcision, and especially because of their oral history of their ancestors arriving by boat from a city called Sena, variously placed in Yemen, Judea, Egypt, or Ethiopia. Studies of 12

polymorphic Y-chromosome markers among 136 Lemba males from six clans has shed some light on the situation. The Y chromosomes from the Lemba derive from one of two lineages. One is closely related to the Bantu and the other is clearly Semitic. About 50 percent of the Y chromosomes of one Lemba clan (the Buba) have haplotypes closely related to a haplotype of Judaic origin called the *Cohen modal haplotype*, because it occurs primarily in the Cohanim (the plural of Cohen), the priestly lineage said to be descended from Moses's brother Aaron. Although the Cohen model haplotype has a frequency at least 50 percent in the Cohanim, it is rare in other Semitic groups. This finding affords some support for the Lemba's oral history, and the estimated time for the most recent common ancestor of the Lemba and Cohanim Y chromosomes is roughly 3000–5000 years. The earliest of these dates would be consistent with the time when the Assyrian King Shalmaneser V sent the 10 tribes of Israel into exile. Sometimes known as the "black Jews of south Africa," the Lemba are technically not Jewish. Judaism is transmitted through the maternal lineage, and Lemba tradition holds that only men survived the perilous voyage from Sena.

Origin of European Gypsies Arriving in Eastern Europe about 1000 years ago, the Roma (Gypsies) were persecuted for centuries. They were held and bartered as slaves until the 1860s, and they were the only ethnic group besides Jews to be singled out for extermination in the Nazi death camps. Today they number more than 12 million people located in many countries around the world. Their origin has been disputed. The term "Gypsy" reflects a legend that they originated in Egypt, but their language (Romanes) has some similarities to languages of the Indian subcontinent.

Studies of the Y chromosome have clarified this situation, too. A group of closely related haplotypes was found among men in all of 14 Romani populations studied and accounted for 44.8 percent of all the Romani Y chromosomes. Elsewhere in the world, this haplotype is frequent only in the Indian subcontinent. In this study, mitochondrial DNA haplotypes were also examined. Mitochondrial DNA is also convenient for tracing population history because it does not undergo recombination and is transmitted through the female (see Chapter 14). A particular group of mitochondrial DNA haplotypes was found in 26.5 percent of the female lineages among the Romani populations. This haplotype, too, derives from the Indian subcontinent. The origin of the Y-chromosomal and mitochondrial DNA haplotypes, and the relatively high frequency of a small number of haplotypes among the Roma, are consistent with a small group of founders originating in the Indian subcontinent. Given the time of their appearance in Eastern Europe, it has been suggested that their migration was actually a flight from the armies of Mahmud of Ghazni invading from what is now Afghanistan about a thousand years ago.

5.2

Chromosome abnormalities are frequent in spontaneous abortions.

Approximately 15 percent of all recognized pregnancies in human beings terminate in spontaneous abortion, and in about half of all spontaneous abortions, the fetus has a major chromosome abnormality. **Table 5.1** summarizes the average rates of chromosome abnormality found per 100,000 recognized pregnancies in several studies. The term **trisomic** refers to an otherwise diploid organism that has an extra copy of an individual chromosome. Many of the spontaneously aborted fetuses have trisomy of one of the autosomes. Triploids, which have three sets of chromosomes (total count 69), and tetraploids, which have four sets of chromosomes (total count 92), are also common in spontaneous abortions. Triploids and tetraploids are examples of **euploid** conditions, because they have the same relative gene dosage as found in the diploid. In contrast, relative gene dosage is upset in a trisomic, because three copies of the genes located in the trisomic chromosome are present, whereas two

copies of the genes in the other chromosomes are present. Such unbalanced chromosome complements are said to be **aneuploid.** Although it is not apparent in the data in Table 5.1, in most organisms, euploid abnormalities generally have less severe phenotypic effects than aneuploid abnormalities. For example, in *Drosophila,* triploid females are viable, fertile, and nearly normal in morphology, whereas trisomy for either of the two large autosomes is invariably lethal (the larvae die at an early stage). In Table 5.1 the term *balanced translocation* refers to a euploid condition in which nonhomologous chromosomes have an interchange of parts, but all of the parts are present; the term *unbalanced translocation* refers to an aneuploid condition in which some part of the genome is missing. The much greater survivorship of the balanced translocation indicates that a euploid chromosomal abnormality is generally less harmful than an aneuploid chromosome abnormality.

When an otherwise diploid organism has a missing copy of an individual chromosome, the condition is known as **monosomy.** In most organisms, chromosome loss (resulting in monosomy) is a more frequent event than chromosome gain (resulting in trisomy). However, monosomies are

Table 5.1		
Chromosome abnormalities per 100,000 recognized human pregnancies		
Chromosome constitution	Number among spontaneously aborted fetuses	Number among live births
Normal	7,500	84,450
Trisomy		
13	128	17
18	223	13
21	350	113
Other autosomes	3,176	0
Sex chromosomes		
47,XYY	4	46
47,XXY	4	44
45,X	1,350	8
47,XXX	21	44
Translocations		
Balanced (euploid)	14	164
Unbalanced (aneuploid)	225	52
Polyploid		
Triploid	1,275	0
Tetraploid	450	0
Others (mosaics, etc.)	280	49
Total	15,000	85,000

conspicuously absent in the data on spontaneous abortions in Table 5.1. Their absence is undoubtedly due to another feature of monosomy:

key concept

A missing copy of a chromosome (monosomy) usually results in more harmful effects than an extra copy of the same chromosome (trisomy).

In human fertilizations, monosomic zygotes are probably created in even greater numbers than the trisomic zygotes, but monosomy is not found among aborted fetuses because the abortions take place so early in development that the pregnancy goes unrecognized. Taking these into account, it seems likely that 10–25 percent of all human zygotes have some form of aneuploidy. The spontaneous abortions summarized in Table 5.1, although they represent a huge fetal loss, serve the important biological function of eliminating many fetuses that are grossly abnormal in their development because of major chromosome abnormalities.

■ Down syndrome results from three copies of chromosome 21.

Table 5.1 demonstrates that monosomy or trisomy of most human autosomes is incompatible with life. There are three exceptions: trisomy 13, trisomy 18, and trisomy 21. The first two are rare conditions associated with major developmental abnormalities, and the affected infants can survive for only a few days or weeks.

Trisomy 21 is **Down syndrome** (or *Down's syndrome*), which occurs in about 1 in 750 live-born children. Its major symptom is mental retardation, but there can also be multiple physical abnormalities, such as heart defects. Affected children are small in stature because of delayed maturation of the skeletal system; their muscle tone is poor, resulting in a characteristic facial appearance; and they have a shortened life span of usually less than 50 years. Nevertheless, for a major chromosomal abnormality, the symptoms are relatively mild, and most children with Down syndrome can relate well to other people.

Children with Down syndrome usually take great pleasure in their surroundings, their families, their toys, their playmates. Happiness comes easily, and throughout life they usually maintain a childlike good humor. They are not burdened with the grown-up cares that come to most people with adolescence and adulthood. Life is simpler and less complex. The emotions that others feel seem to be less intense for them. They are sometimes sad, happy, angry, or irritable, like everyone else, but their moods are generally not so profound and they blow away more

quickly. . . . Children with Down syndrome, though slow, are still very responsive to their environment, to those around them, and to the affection and encouragement they receive from others. [Quoted from D. W. Smith and A. A. Wilson. *The Child with Down's Syndrome.* (Philadelphia: Saunders, 1973.)]

Most cases of Down syndrome are caused by nondisjunction, which means the failure of homologous chromosomes to separate in meiosis, as explained in Chapter 3. The result of chromosome-21 nondisjunction is one gamete that contains two copies of chromosome 21 and one that contains none. If the gamete with two copies participates in fertilization, then a zygote with trisomy 21 is produced. The gamete with no copy may also participate in fertilization, but zygotes with monosomy 21 do not survive even through the first few days or weeks of pregnancy. About three-fourths of the trisomy-21 fetuses also undergo spontaneous abortion (Table 5.1). If this were not the case, and all trisomy-21 fetuses survived to birth, the incidence of Down syndrome would rise to 1 in 250, approximately a threefold increase over the incidence actually observed.

Chromosome 21 is a small chromosome and therefore is somewhat less likely to undergo meiotic crossing-over than a longer one. Noncrossover bivalents sometimes have difficulty aligning at the metaphase plate because they lack a chiasma to hold them together, so there is an increased risk of nondisjunction. Among the events of nondisjunction that result in Down syndrome, about 40 percent are derived from such nonexchange bivalents.

For unknown reasons, nondisjunction of chromosome 21 is more likely to happen in oogenesis than in spermatogenesis, and so the abnormal

DOWN SYNDROME child.

© Stock Connection Distribution/Alamy Images.

Catch-21

Jerome Lejeune, Marthe Gautier, and Raymond Turpin 1959
National Center for Scientific Research, Paris, France
Study of the Somatic Chromosomes of Nine Down Syndrome Children (original in French)

Prior to this study, Down syndrome was one of the greatest mysteries in human genetics. One of the most common forms of mental retardation, the syndrome did not follow any pattern of Mendelian inheritance. Yet some families had two or more children with Down syndrome. (Many of these cases are now known to be due to a translocation involving chromosome 21.) This paper marked a turning point in human genetics by demonstrating that Down syndrome actually results from the presence of an extra chromosome. It was the first chromosomal disorder to be identified. The excerpt uses the term *telocentric*, which means a chromosome that has its centromere very near one end. In the human genome, the smallest chromosomes are three very small telocentric chromosomes. These are chromosomes 21 and 22 and the Y chromosome. A normal male has five small telocentrics (21, 21, 22, 22, and Y); a normal female has four (21, 21, 22, and 22). (The X is a medium-sized chromosome with its centromere somewhat off center.) In the accompanying table, note the variation in chromosome counts in the "doubtful" cells. Counting chromosomes was very difficult then, and investigators made many errors, either by counting two nearby chromosomes as one or by including in the count of one nucleus a chromosome that actually belonged to a nearby nucleus. Lejeune and collaborators wisely chose to ignore these doubtful counts and based their conclusion only on the "perfect" cells. Sometimes good science is a matter of knowing which data to ignore.

....

The culture of fibroblast cells from nine Down syndrome children reveals the presence of 47 chromosomes, the supernumerary chromosome being a small telocentric one. The hypothesis of the chromosomal determination of Down syndrome is considered. . . . The observations made in these nine cases (five boys and four girls) are recorded in the [accompanying] table.

> **There exists in Down syndrome children a small supernumerary telocentric chromosome.**

The number of cells counted in each case may seem relatively small. This is due to the fact that only the pictures [of the spread chromosomes] that claim a minimum of interpretation have been retained in this table. The apparent variation in the chromosome number in the "doubtful" cells, that is to say, cells in which each chromosome cannot be noted individually with certainty, has been pointed out by several authors. It does not seem to us that this phenomenon represents a cytological reality, but merely reflects the difficulties of a delicate technique. It therefore seems logical to prefer a small number of absolutely certain counts ("perfect" cells in the table) to a mass of doubtful observations, the statistical variance of which rests solely on the lack of precision of the observations. Analysis of the chromosome set of the "perfect" cells reveals the presence in Down syndrome boys of 6 small telocentric chromosomes (instead of 5 in the normal man) and 5 small telocentric ones in Down syndrome girls (instead of 4 in the normal woman). . . . It therefore seems legitimate to conclude that there exists in Down syndrome children a small supernumerary telocentric chromosome, accounting for the abnormal figure of 47. To explain these observations, the hypothesis of nondisjunction of a pair of small telocentric chromosomes at the time of meiosis can be considered. . . . It is, however, not possible to say that the supernumerary small telocentric chromosome is indeed a normal chromosome and at the present time the possibility cannot be discarded that a fragment resulting from another type of aberration is involved.

Source: Comptes rendus des séances de l'Académie des Sciences 248: 1721–1722.

		Number of chromosomes				
		"Doubtful" cells			"Perfect" cells	
		46	47	48	46	47	48
Boys	1	6	10	2	–	11	–
	2	–	2	1	–	9	–
	3	–	1	1	–	7	–
	4	–	3	–	–	1	–
	5	–	–	–	–	8	–
Girls	1	1	6	1	–	5	–
	2	1	2	–	–	8	–
	3	1	2	1	–	4	–
	4	1	1	2	–	4	–

gamete in Down syndrome is usually the egg. Furthermore, the risk of nondisjunction of chromosome 21 increases dramatically with the age of the mother, resulting in a risk of Down syndrome that reaches 6 percent in mothers age 45 and older (Figure 5.12). For this reason, many physicians recommend that older women who are pregnant have cells from the fetus tested in order to detect Down syndrome prenatally. This can be done 15–16 weeks after fertilization by *amniocentesis,* in which cells of the developing fetus are obtained by insertion of a fine needle through the wall of the uterus and into the sac of fluid (the *amnion*) that contains the fetus, or it can be done at 10–11 weeks after fertilization by a procedure called *chorionic villus sampling (CVS),* which uses cells from a zygote-derived embryonic membrane (the *chorion*) associated with the placenta. Early diagnosis is desirable, but CVS has about a threefold higher risk of inducing a miscarriage than does amniocentesis.

About 3 percent of all cases of Down syndrome are due not to simple nondisjunction but to an abnormality in chromosome structure. In these cases the risk of recurrence of the syndrome in subsequent children is very high—up to 20 percent of births. The high risk is caused by a chromosomal translocation in one of the parents, in which chromosome 21 has been broken and become attached to another chromosome. This situation is considered in Section 5.3.

■ Trisomic chromosomes undergo abnormal segregation.

In a trisomic organism, the segregation of chromosomes in meiosis is upset because the trisomic chromosome has two pairing partners instead of one. The behavior of the chromosomes in meiosis depends on the manner in which the homologous chromosome arms pair and on the chiasmata formed between them. In some cells, the three

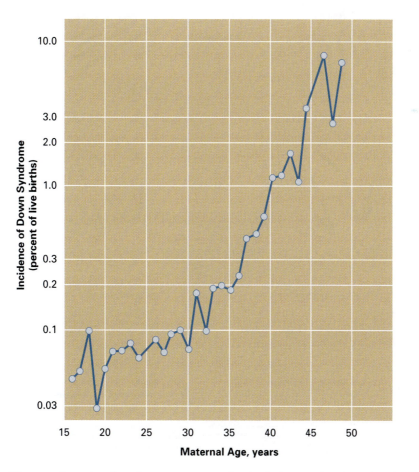

Figure 5.12 Frequency of Down syndrome (number of cases per 100 live births) related to age of mother. The graph is based on 438 Down syndrome births (among 330,859 total births) in Sweden in the period 1968 to 1970. [Data from E. B. Hook and A. Lindsjö. 1978. *Am. J. Human Genet.* 30: 19.]

chromosomes form a **trivalent** in which distinct parts of one chromosome are paired with homologous parts of each of the others (Figure 5.13 , part A). In metaphase, the trivalent is usually oriented with two centromeres pointing toward one pole and the other centromere pointing toward the other pole. The result is that at the end of both meiotic divisions, one pair of gametes contains two copies of the trisomic chromosome, and the other pair of gametes contains only a single copy. Alternatively, the trisomic chromosome can form one normal bivalent and one **univalent,** or unpaired chromosome, as shown in Figure 5.13, part B. In anaphase I, the bivalent disjoins normally and the univalent usually proceeds randomly to one pole or the other. Again, the end result is the formation of two prod-

ucts of meiosis that contain two copies of the trisomic chromosome and two products of meiosis that contain one copy. To state the matter in another way, a trisomic organism with three copies of a chromosome (say, C C C) will produce gametes half of which contain two copies (C C) and half of which contain one copy (C). When mated with a chromosomally normal individual, a trisomic is therefore expected to produce trisomic and normal progeny in a ratio of 1 : 1. This theoretical expectation is borne out in experimental organisms by matings of trisomics and in human beings by the finding that, in the few known cases of a person with Down syndrome reproducing, the frequency of Down syndrome in the offspring is approximately 50 percent.

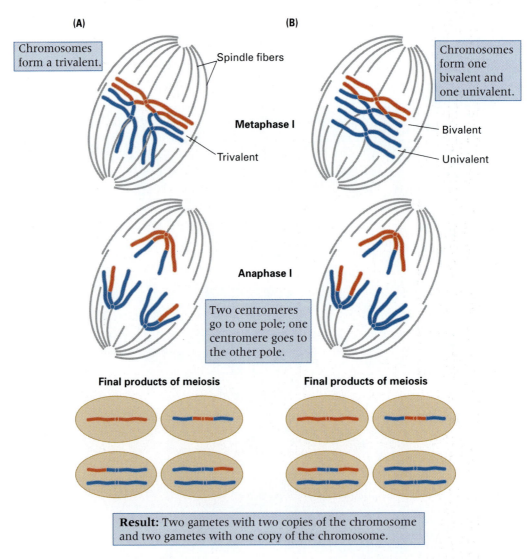

Figure 5.13 Meiotic synapsis in a trisomic. (A) Formation of a trivalent. (B) Formation of a bivalent and a univalent. Both types of synapsis result in one pair of gametes containing two copies of the trisomic chromosome and the other pair of gametes containing one copy of the trisomic chromosome.

■ An extra X or Y chromosome usually has a relatively mild effect.

Many types of sex-chromosome trisomies, as a group, are even more frequent among newborns than is trisomy 21 (Table 5.1). There are two reasons why extra sex chromosomes have phenotypic effects that are relatively mild compared with those of autosomal trisomies. First, the single-active-X principle results in the silencing of most X-linked genes in all but one X chromosome in each somatic cell. Second, the Y chromosome contains relatively few functional genes.

The four most common types of sex-chromosome abnormalities are described below. The karyotypes are given in the conventional fashion, with the total number of chromosomes listed first, followed by the sex chromosomes that are present. For example, in the designation 47,XXX the number 47 refers to the total number of chromosomes, and XXX indicates that the person has three X chromosomes.

- **47,XXX** This condition is often called **trisomy-X syndrome.** People with the karyotype 47,XXX are female. Many are phenotypically normal or nearly normal, though the frequency of mild mental disability is somewhat greater than it is among 46,XX females.

- **47,XYY** This condition is often called **double-Y syndrome.** These people are male and tend to be tall, but they are otherwise phenotypically normal. At one time it was thought that 47,XYY males developed severe personality disorders and were at a high risk of committing crimes of violence—a belief based on an elevated incidence of 47,XYY among violent criminals. Further study indicated that most 47,XYY males have slightly impaired mental function and that, although their rate of criminality is higher than that of normal males, the crimes are mainly nonviolent petty crimes such as theft. The majority of 47,XYY males are phenotypically and psychologically normal, have mental capabilities in the normal range, and have no criminal convictions.

- **47,XXY** This condition is called **Klinefelter syndrome.** Affected persons are male. They tend to be tall, do not undergo normal sexual maturation, are sterile, and in some cases have enlargement of the breasts. Mild mental impairment is common.

- **45,X** Monosomy of the X chromosome in females is called **Turner syndrome.** Affected persons are phenotypically female but are short in stature and do not exhibit sexual maturation. Mental abilities are typically within the normal range.

■ The rate of nondisjunction can be increased by chemicals in the environment.

Because a large fraction of aneuploid zygotes terminate in miscarriage or result in congenital defects or mental retardation, the identification of environmental hazards that may increase the incidence of meiotic errors is of great importance. Environmental risk factors that have been suggested include: radiation, smoking, alcohol consumption, oral contraceptives, fertility drugs, environmental pollutants, pesticides, among others. When significant effects have been found, they are usually small and not always reproducible, due in part to confounding effects of other factors such as maternal age. In view of the maternal-age effect, the female sex hormone estrogen and molecules resembling estrogen have long been under suspicion. With this background in mind, it was a great surprise to learn that modest concentrations of a common estrogen mimic known as bisphenol A [technical name 2,2-(4,4-dihydroxy-diphenol)propane] caused about an eightfold increase in the incidence of aneuploidy in mice.

Bisphenol A is the basic subunit of polycarbonate plastic products widely used as a can liner in the food and beverage industry. In its polymerized form it may be completely harmless, but the monomers can leach out of plastic products under certain conditions. The effect of bisphenol A was discovered by chance in studies of aneuploidy in the mouse. In the course of microscropic studies, the researchers noted a sudden increase in the incidence of meiotic cells in which one or more chromosomes failed to align properly on the metaphase II plate. At the same time the incidence of aneuploid oocytes increased from 1.4 percent to 11.6 percent.

The investigators succeeded in tracing the cause to the inadvertent use of a harsh alkaline detergent to wash the polycarbonate cages and water bottles. They suspected that the detergent was causing bisphenol A monomers to leach out of the plastic, and follow-up experiments with bisphenol A monomers demonstrated its effect on aneuploidy directly. Although the studies were carried out in female mice, the similarity of the meiotic process in female humans is a matter of concern. More generally, the study raises a warning flag about environmental agents that may increase the rate of aneuploidy, particularly agents that have estrogen-like activity.

5.3

Chromosome rearrangements can have important genetic effects.

This section deals with abnormalities in chromosome structure. There are several principal types of structural aberrations, each of which has characteristic genetic effects. Chromosome aberrations were initially discovered through their genetic effects, which, though confusing at first, were eventually understood as resulting from abnormal chromosome structure. This was later confirmed directly by microscopic observations.

■ A chromosome with a deletion has genes missing.

A chromosome sometimes arises in which a segment is missing. Such a chromosome is said to have a **deletion** or a **deficiency.** Deletions are generally harmful to the organism, and the usual rule is the larger the deletion, the greater the harm. Very large deletions are usually lethal, even when heterozygous with a normal chromosome. Small deletions are often viable when they are heterozygous with a structurally normal homolog, because the normal homolog supplies gene products that are necessary for survival. However, even small deletions are usually homozygous-lethal (when both members of a pair of homologous chromosomes carry the deletion).

Deletions can be detected genetically by making use of the fact that a chromosome with a deletion no longer carries the wildtype alleles of the genes that have been eliminated. For example, in *Drosophila*, many *Notch* deletions are large enough to remove the nearby wildtype allele of *white* also. When these deleted chromosomes are heterozygous with a structurally normal chromosome carrying the recessive *w* allele, the fly has white eyes because the wildtype w^+ allele is no longer present in the deleted *Notch* chromosome. This "uncovering" of the recessive allele implies that the corresponding wildtype allele of *white* has also been deleted. Once a deletion has been identified, its size can be assessed genetically by determining which recessive mutations in the region are uncovered by the deletion. This method is illustrated in Figure 5.14.

■ Rearrangements are apparent in giant polytene chromosomes.

In the nuclei of cells in the larval salivary glands and certain other tissues of *Drosophila* and other two-winged (dipteran) flies, there are giant chromosomes, called **polytene chromosomes,** that contain about 1000 DNA molecules laterally aligned (Figure 5.15). Each of these chromosomes

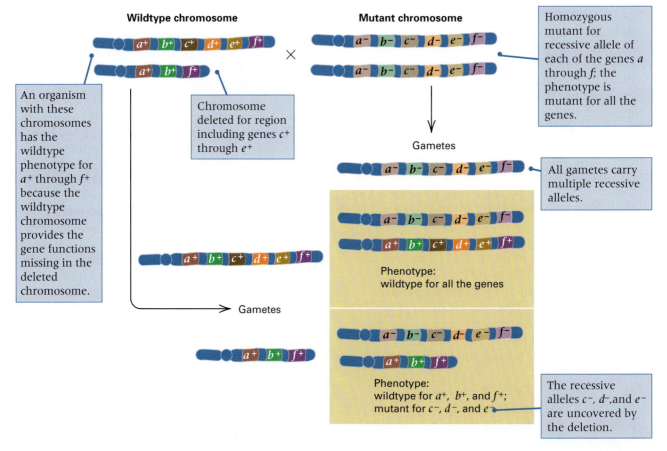

Figure 5.14 Mapping of a deletion by testcrosses. The F$_1$ heterozygotes with the deletion express the recessive phenotype of all deleted genes. The expressed recessive alleles are said to be "uncovered" by the deletion.

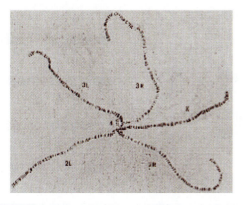

Figure 5.15 Polytene chromosomes from a larval salivary gland cell of *Drosophila melanogaster.* The extensive heterochromatic regions around each centromere remain unreplicated and join the base of each chromosome arm to a common *chromocenter* in the middle. [Courtesy of George Lefevre.]

has a volume many times greater than that of the corresponding chromosome at mitotic metaphase in ordinary somatic cells, as well as a constant and distinctive pattern of transverse banding. The polytene structures are formed by repeated replication of the DNA in a closely synapsed pair of homologous chromosomes without separation of the replicated chromatin strands or of the two chromosomes. Polytene chromosomes are atypical chromosomes and are formed in "terminal" cells; that is, the larval cells containing them do not divide fur-

ther during the development of the fly and are later eliminated in the formation of the pupa. However, the polytene chromosomes have been especially valuable in the genetics of *Drosophila* and are ideal for the study of chromosome rearrangements.

About 5000 darkly staining transverse bands have been identified in the polytene chromosomes of *D. melanogaster* (Figure 5.15). The linear array of bands, which has a pattern that is constant and characteristic for each species, provides a finely detailed **cytological map** of the chromosomes. The banding pattern is such that short regions in any of the chromosomes can be identified. Because of their large size and finely detailed morphology, polytene chromosomes are exceedingly useful for the study of deletions and other chromosome aberrations. For example, all the *Notch* deletions cause particular bands to be missing in the salivary chromosomes. Physical mapping of deletions also allows particular genes, otherwise known only from genetic studies, to be assigned to specific bands or regions in the salivary chromosomes.

Physical mapping of genes in part of the *Drosophila* X chromosome is illustrated in Figure 5.16. The banded chromosome is shown, and beneath it are the designations of the individual bands. On the average, each band contains about 20 kb of DNA, but there is considerable variation in DNA content from band to band. The mutant X chromosomes labeled I through VI in the figure have deletions.

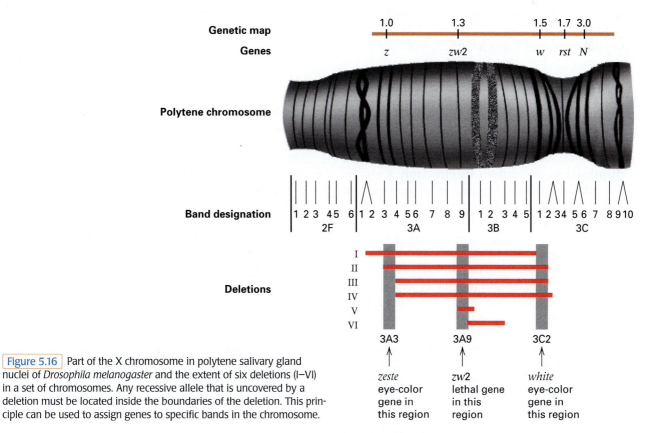

Figure 5.16 Part of the X chromosome in polytene salivary gland nuclei of *Drosophila melanogaster* and the extent of six deletions (I–VI) in a set of chromosomes. Any recessive allele that is uncovered by a deletion must be located inside the boundaries of the deletion. This principle can be used to assign genes to specific bands in the chromosome.

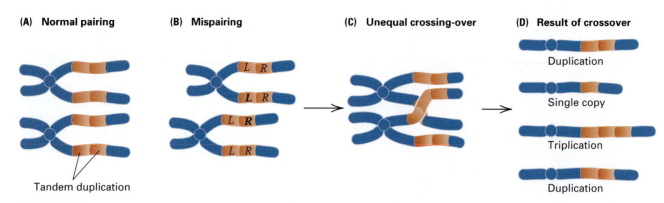

(A) **Normal pairing** (B) **Mispairing** (C) **Unequal crossing-over** (D) **Result of crossover**

Duplication

Single copy

Triplication

Duplication

Tandem duplication

Figure 5.17 An increase in the number of copies of a chromosome segment resulting from unequal crossing-over of tandem duplications (brown). (A) Normal synapsis of chromosomes with a tandem duplication. (B) Mispairing. The right-hand element of the lower chromosome is paired with the left-hand element of the upper chromosome. (C) Crossing-over within the mispaired duplication, which is called unequal crossing-over. (D) The outcome of unequal crossing-over.

The deleted part of each chromosome is shown in red. These deletions define regions along the chromosome, some of which correspond to specific bands. For example, the deleted region in both chromosomes I and II that is present in all the other chromosomes consists of band 3A3. In crosses, only deletions I and II uncover the mutation *zeste (z)*, so the *z* gene must be in band 3A3, as indicated at the top. Similarly, the recessive-lethal mutation *zw2* is uncovered by all deletions except VI; therefore, the *zw2* gene must be in band 3A9. As a final example, the *w* mutation is uncovered only by deletions II, III, and IV; thus the *w* gene must be in band 3C2. The *rst* (rough eye texture) and *N* (notched wing margin) genes are not uncovered by any of the deletions. These genes were localized by a similar analysis of overlapping deletions in regions 3C5 to 3C10.

■ A chromosome with a duplication has extra genes.

Some abnormal chromosomes have a region that is present twice. These chromosomes are said to have a **duplication.** A **tandem duplication** is one in which the duplicated segment is present in the same orientation immediately adjacent to the normal region in the chromosome. Tandem duplications are able to produce even more copies of the duplicated region by means of a process called **unequal crossing-over.** Part A of Figure 5.17 illustrates the chromosomes in meiosis of an organism that is homozygous for a tandem duplication (brown region). When they undergo synapsis, these chromosomes can mispair with each other, as illustrated in part B. A crossover within the mispaired part of the duplication (part C) will thereby produce a chromatid carrying three copies of the region, as well as a reciprocal product containing a single copy (part D).

■ Human color-blindness mutations result from unequal crossing-over.

Human color vision is mediated by three light-sensitive protein pigments present in the cone cells of the retina. Each of the pigments is related to *rhodopsin,* the pigment found in the rod cells that mediates vision in dim light. The light sensitivities of the cone pigments are toward blue, red, and green. These are our primary colors. We perceive all other colors as mixtures of these primaries. The gene for the blue-sensitive pigment is in chromosome 7, whereas the genes for the red and green pigments are in the X chromosome near the tip of the long arm, separated by less than 5 cM (roughly 5 Mb of DNA). Because the red and green pigments arose from the duplication of a single ancestral pigment gene and are still 96 percent identical in amino acid sequence, the genes are similar enough that they can pair and undergo unequal crossing-over. The process of unequal crossing-over is the genetic basis of red–green color blindness.

Almost everyone is familiar with **red–green color blindness;** it is one of the most common inherited conditions in human beings (Figure 5.18). Approximately 5 percent of males have some form of red–green color blindness. The preponderance of affected males immediately suggests X-linked inheritance, which is confirmed by pedigree studies. Affected males have normal sons and carrier daughters, and the carrier daughters have 50 percent affected sons and 50 percent carrier daughters.

There are several distinct varieties of red–green color blindness. Defects in red vision go by the names of *protanopia,* an inability to perceive red, and *protanomaly,* an impaired ability to perceive red. The comparable defects in green perception are called *deuteranopia* and *deuteranomaly,* respectively. Isolation of the red-pigment and green-pigment genes and study of their organization in people with normal and defective color vision have indi-

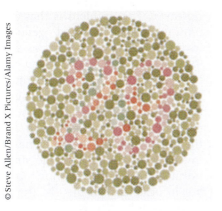

Figure 5.18 A standard color chart used in initial testing for color blindness. The pattern tests for an inability to distinguish red from green. Those with red–green color blindness will not be able to distinguish the red dots from the green and therefore will not see the red number.

cated quite clearly how the "-opias" and "-omalies" differ; they have also explained why the frequency of color blindness is so relatively high.

The organization of the red-pigment and green-pigment genes in men with normal vision is illustrated in part A of Figure 5.19. Unexpectedly, a significant proportion of normal X chromosomes contain two or three green-pigment genes. How these arise by unequal crossing-over is shown in part B. The red-pigment and green-pigment genes pair, and the crossover takes place in the region of homology between the genes. The result is a duplication of the green-pigment gene in one chromosome and a deletion of the green-pigment gene in the other.

The recombinational origin of the defects in color vision are illustrated in Figure 5.20. The top chromosome in part A is the result of deletion of the green-pigment gene shown in part B of Figure 5.19. Males with such an X chromosome have deuteranopia, or "green blindness." Other types of abnormal pigments result when crossing-over takes place within mispaired red-pigment and green-pigment genes. Crossing-over between the genes yields a **chimeric gene,** which is a composite gene, part of one joined with part of the other. The chimeric gene in part A of Figure 5.20 joins the 5′ end of the green-pigment gene with the 3′ end of the red-pigment gene. If the crossover point is toward the 5′ end of the gene, the resulting chimeric gene is mostly "red" in sequence, and hence the chromosome will cause deuteranopia or "green blindness." However, if the crossover point is near the 3′ end of the gene, most of the green-pigment gene remains intact, and the chromosome will cause deuteranomaly.

Chromosomes associated with defects in red vision are illustrated in part B of Figure 5.20. The chimeric genes are the reciprocal products of the

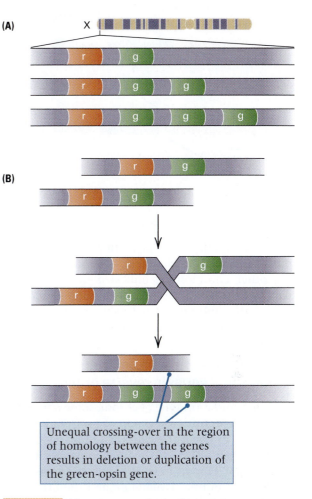

Unequal crossing-over in the region of homology between the genes results in deletion or duplication of the green-opsin gene.

Figure 5.19 (A) Organization of red-pigment and green-pigment genes in three wildtype X chromosomes. (B) Origin of multiple green-pigment genes by unequal crossing-over.

unequal crossovers that yield defects in green vision. In this case, the chimeric gene consists of the red-pigment gene at the 5′ end and the green-pigment gene at the 3′ end. If the crossover point is near the 5′ end, most of the red-pigment gene is

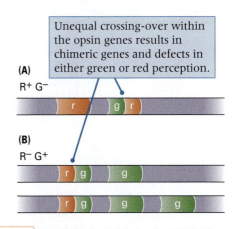

Unequal crossing-over within the opsin genes results in chimeric genes and defects in either green or red perception.

Figure 5.20 Genetic basis of absent or impaired red–green color vision. (A) Defects in green vision. (B) Defects in red vision.

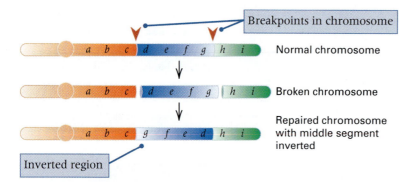

Breakpoints in chromosome

a b c d e f g h i — Normal chromosome

a b c d e f g h i — Broken chromosome

a b c g f e d h i — Repaired chromosome with middle segment inverted

Inverted region

Figure 5.21 Origin of an inversion by reversal of the region between two chromosomal breakpoints.

replaced with the green-pigment gene. The result is protanopia, or "red blindness." The same is true of the other chromosome indicated in Figure 5.20, part B. However, if the crossover point is near the 3' end, then most of the red-pigment gene remains intact and the result is protanomaly.

■ A chromosome with an inversion has some genes in reverse order.

Another important type of chromosome abnormality is an **inversion,** a segment of a chromosome in which the order of the genes is the reverse of the normal order (Figure 5.21). In an organism that is heterozygous for an inversion, one chromosome is structurally normal (wildtype), and the other carries an inversion. These chromosomes pass through mitosis without difficulty, because each chromosome duplicates and its chromatids are separated into the daughter cells without regard to the other chromosome. There is a problem in meiosis,

however. The problem is that the chromosomes are attracted gene-for-gene in the process of synapsis, as shown in Figure 5.22. In an inversion heterozygote, in order for gene-for-gene pairing to take place everywhere along the length of the chromosome, one or the other of the chromosomes must twist into a loop in the region in which the gene order is inverted. In Figure 5.22 the structurally normal chromosome is shown as looped, but in other cells it may be the inverted chromosome that is looped. In either case, the loop is called an **inversion loop.**

The loop itself does not create a problem. The looping apparently takes place without difficulty and can be observed through the microscope. As long as there is no crossing over within the inversion, the homologous chromosomes can separate normally at anaphase I. When there is crossing-over within the inversion loop, the chromatids involved in the crossover become physically joined, and the result is the formation of chromosomes containing large duplications and deletions. The products of the crossover can be deduced from Figure 5.23 by tracing along the chromatids in part A. The outer chromatids are the ones that do not participate in the crossover. One of these contains the inverted sequence and the other the normal sequence, as shown in part B. Because of the crossover, the inner chromatids, which did participate in the crossover, are connected. If the centromere is not included in the inversion loop, as is the case here, the result is a dicentric chromosome. The reciprocal product of the crossover is an acentric chromosome. Neither the dicentric chromosome nor the acentric chromosome can be included in a normal gamete. The acentric chromosome is usually lost because it lacks a centromere and, in any case, has a deletion of the a region and a duplication of the d region. The dicentric chromosome is also often lost because it is held on the meiotic spindle by the chromatid bridging between the centromeres; in any case, this chromosome is deleted for the d region and duplicated for the a region. Hence, when there is a crossover in the inversion loop,

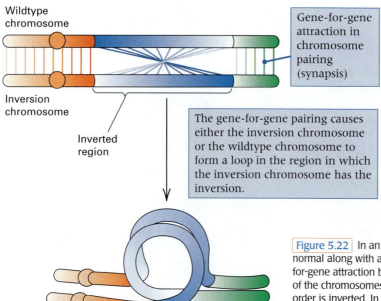

Wildtype chromosome

Inversion chromosome

Inverted region

Gene-for-gene attraction in chromosome pairing (synapsis)

The gene-for-gene pairing causes either the inversion chromosome or the wildtype chromosome to form a loop in the region in which the inversion chromosome has the inversion.

Figure 5.22 In an organism that carries a chromosome that is structurally normal along with a homologous chromosome with an inversion, the gene-for-gene attraction between the chromosomes during synapsis causes one of the chromosomes to form into a loop in the region in which the gene order is inverted. In this example, the structurally normal chromosome forms the loop. Only two of the four chromatids are shown.

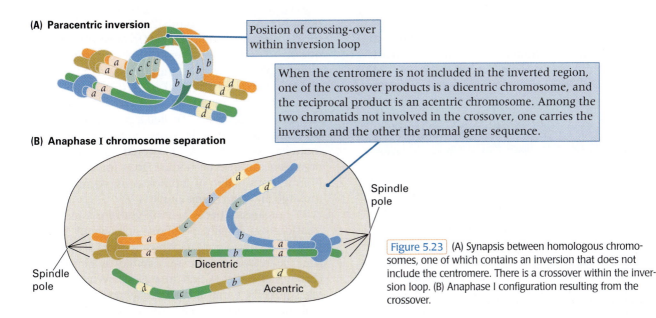

(A) Paracentric inversion

Position of crossing-over within inversion loop

When the centromere is not included in the inverted region, one of the crossover products is a dicentric chromosome, and the reciprocal product is an acentric chromosome. Among the two chromatids not involved in the crossover, one carries the inversion and the other the normal gene sequence.

(B) Anaphase I chromosome separation

Spindle pole

Dicentric

Acentric

Spindle pole

Figure 5.23 (A) Synapsis between homologous chromosomes, one of which contains an inversion that does not include the centromere. There is a crossover within the inversion loop. (B) Anaphase I configuration resulting from the crossover.

the only chromatids that can be recovered in the gametes are the chromatids that did not participate in the crossover. One of these carries the inversion and the other does not.

The inversion in Figure 5.23, in which the centromere is not included in the inverted region, is known as a **paracentric inversion,** which means inverted "beside" (*para-*) the centromere. As seen in the figure, the products of crossing-over include a dicentric and an acentric chromosome.

When the inversion does include the centromere, it is called a **pericentric inversion,** which means "around" (*peri-*) the centromere. Chromatids with duplications and deficiencies are

also created by crossing-over within the inversion loop of a pericentric inversion, but in this case the crossover products are monocentric. The situation is illustrated in Figure 5.24, part A. The diagram is identical to that in Figure 5.23 except for the position of the centromere. The products of crossing-over can again be deduced by tracing the chromatids. In this case, both products of the crossover are monocentric, but one chromatid carries a duplication of *a* and a deletion of *d,* and the other carries a duplication of *d* and a deletion of *a* (Figure 5.24, part B). Although either of these chromosomes could be included in a gamete, the duplication and deficiency usually cause inviability. Thus, as with

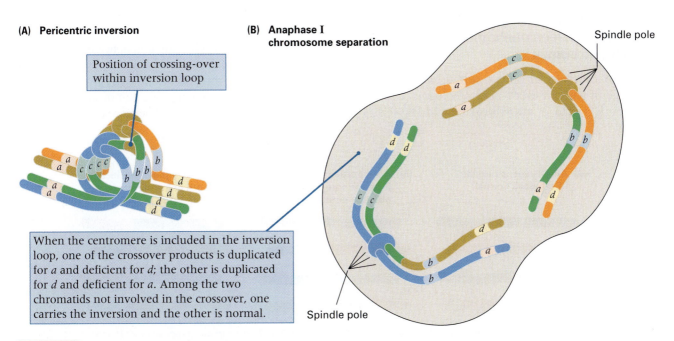

(A) Pericentric inversion

Position of crossing-over within inversion loop

When the centromere is included in the inversion loop, one of the crossover products is duplicated for *a* and deficient for *d*; the other is duplicated for *d* and deficient for *a*. Among the two chromatids not involved in the crossover, one carries the inversion and the other is normal.

(B) Anaphase I chromosome separation

Spindle pole

Spindle pole

Figure 5.24 (A) Synapsis between homologous chromosomes, one of which carries an inversion that includes the centromere. A crossover within the inversion loop is shown. (B) Anaphase I configuration resulting from the crossover.

the paracentric inversion, the products of recombination are not recovered, but for a different reason. Among the chromatids that do not participate in the crossing-over in part A of Figure 5.24, one carries the pericentric inversion and the other has the normal sequence.

Q A Moment to Think

Problem: The accompanying illustration is a simplified version of pairing between a chromosome with an inversion and its normal homolog, which is extremely useful when the question of interest concerns the consequences of crossing-over within the inversion loop. In the simplified version, the inverted region alone is shown paired and the normal regions unpaired. In the case of a paracentric inversion, this means that the centromeres are on opposite sides. (In living cells, the inverted region forms a loop.) In this diagram it is straightforward to work out the consequences of double crossovers within an inversion loop, which can happen if the inverted region is sufficiently long. Use the illustration to deduce the consequences of all four types of double crossover that are possible when the first takes place between the genes B and C and the second takes place between the genes C and D. For each type of double crossover, specify the alleles present in the resulting chromosomes and indicate whether the chromosome is acentric, monocentric, or dicentric. (The answer can be found on page 192.)

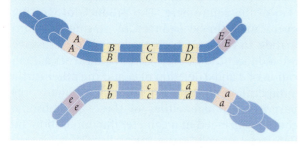

■ **Reciprocal translocations interchange parts between nonhomologous chromosomes.**

A chromosomal aberration resulting from the interchange of parts between nonhomologous chromosomes is called a **translocation.** In Figure 5.25 ,

organism A is homozygous for two pairs of structurally normal chromosomes. Organism B contains one structurally normal pair of chromosomes and another pair of chromosomes that have undergone an interchange of terminal parts. This organism is said to be *heterozygous* for the translocation. The translocation is properly called a **reciprocal translocation** because it consists of two reciprocally interchanged parts. As indicated in part C, an organism can also be homozygous for a translocation if both pairs of homologous chromosomes undergo an interchange of parts.

An organism that is heterozygous for a reciprocal translocation usually produces only about half as many offspring as normal, which is called **semisterility.** The reason for the semisterility is difficulty in chromosome segregation in meiosis. When meiosis takes place in a translocation heterozygote, the normal and translocated chromosomes must undergo synapsis as shown in Figure 5.26 . Ordinarily, there would also be chiasmata between nonsister chromatids in the arms of the homologous chromosomes, but these are not shown, as if the translocation were present in an organism with no crossing-over, such as a male *Drosophila*. Segregation from this configuration can take place in any of three ways. In the list that follows, the notation $1 + 2 \longleftrightarrow 3 + 4$ means that at the first meiotic anaphase, the chromosomes in Figure 5.26 labeled 1 and 2 go to one pole and those labeled 3 and 4 go to the opposite pole. The red numbers 1 and 4 indicate the two parts of the reciprocal translocation. The three types of segregation are

- $1 + 2 \longleftrightarrow 3 + 4$ This mode is called **adjacent-1 segregation.** Homologous centromeres go to opposite poles, but each normal chromosome goes with one part of the reciprocal translocation. All gametes formed from adjacent-1 segregation have a large duplication and deficiency for the distal part of the translocated chromosomes. (The *distal* part of a chromosome is the part farthest from the centromere.) The pair of gametes that originate from the $1 + 2$ pole are duplicated for

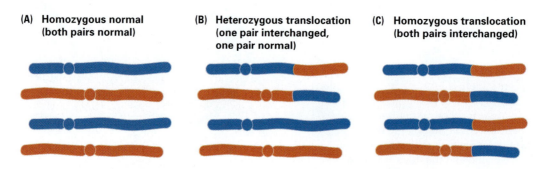

(A) Homozygous normal (both pairs normal) **(B) Heterozygous translocation (one pair interchanged, one pair normal)** **(C) Homozygous translocation (both pairs interchanged)**

Figure 5.25 (A) Two pairs of nonhomologous chromosomes in a diploid organism. (B) Heterozygous reciprocal translocation, in which two nonhomologous chromosomes (the two at the top) have interchanged terminal segments. (C) Homozygous reciprocal translocation.

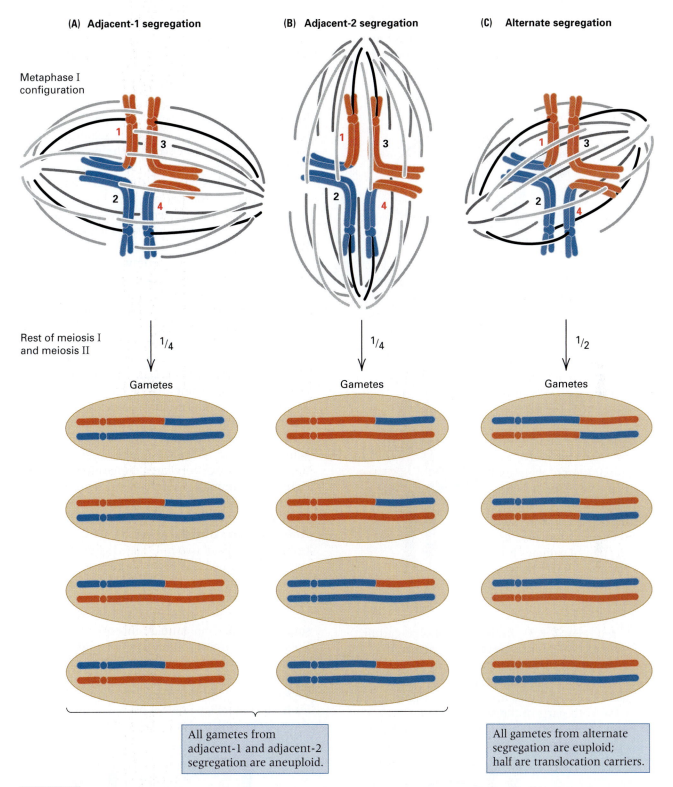

(A) Adjacent-1 segregation

(B) Adjacent-2 segregation

(C) Alternate segregation

Metaphase I configuration

Rest of meiosis I and meiosis II

$1/4$

$1/4$

$1/2$

Gametes

Gametes

Gametes

All gametes from adjacent-1 and adjacent-2 segregation are aneuploid.

All gametes from alternate segregation are euploid; half are translocation carriers.

Figure 5.26 A quadrivalent formed in the synapsis of a heterozygous reciprocal translocation and their expected frequencies. The translocated chromosomes are numbered in red, their normal homologs in black. No chiasmata are shown. (A) Adjacent-1 segregation: homologous centromeres separate at anaphase I; all of the resulting gametes have a duplication of one terminal segment and a deficiency of the other. (B) Adjacent-2 segregation: homologous centromeres go together at anaphase I; all of the resulting gametes have a duplication of one basal segment and a deficiency of the other. (C) Alternate segregation: half of the gametes receive both parts of the reciprocal translocation and the other half receive both normal chromosomes.

the distal part of the blue chromosome and deficient for the distal part of the red chromosome; the pair of gametes from the 3 + 4 pole have the reciprocal deficiency and duplication.

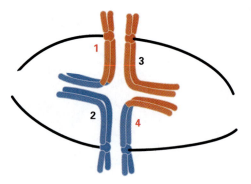

- **1 + 3 ⟷ 2 + 4** This mode is **adjacent-2 segregation,** in which homologous centromeres go to the same pole at anaphase I. In this case, all gametes have a large duplication and deficiency of the proximal part of the translocated chromosome. (The *proximal* part of a chromosome is the part closest to the centromere.) The pair of gametes from the 1 + 3 pole have a duplication of the proximal part of the red chromosome and a deficiency of the proximal part of the blue chromosome; the pair of gametes from the 2 + 4 have the complementary deficiency and duplication.

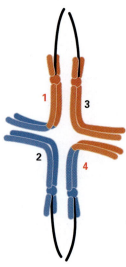

- **1 + 4 ⟷ 2 + 3** In this type of segregation, which is called **alternate segregation,** the gametes are all balanced (euploid), which means that none has a duplication or a deficiency. The gametes from the 1 + 4 pole have both parts of the reciprocal translocation; those from the 2 + 3 have both normal chromosomes.

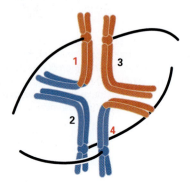

The semisterility of genotypes that are heterozygous for a reciprocal translocation results from lethality due to the duplication and deficiency gametes produced by adjacent-1 and adjacent-2 segregation. Although the expected frequencies of adjacent-1 : adjacent-2 : alternate segregation are approximately 1/4 : 1/4 : 1/2, in practice the frequency with which these types of segregation take place is strongly influenced by the position of the translocation breakpoints, by the number and distribution of chiasmata in the interstitial region between the centromere and each breakpoint, and by whether the quadrivalent tends to open out into a ring-shaped structure on the metaphase plate. Adjacent-1 segregation is usually quite frequent, which means that semisterility is to be expected from virtually all translocation heterozygotes.

Translocation semisterility is manifested in different life-history stages in plants and animals. Plants have an elaborate gametophyte phase of the life cycle—a haploid phase in which complex metabolic and developmental processes are necessary. In plants, large duplications and deficiencies are usually lethal in the gametophyte stage. Because the gametophyte produces the gametes, in higher plants the semisterility is manifested as pollen or seed lethality. In animals, by contrast, minimal gene activity is necessary in the gametes, which function in spite of very large duplications and deficiencies. In animals, therefore, the semisterility is usually manifested as zygotic lethality.

A special type of *non*reciprocal translocation is a **Robertsonian translocation,** in which the centromeric regions of two nonhomologous acrocentric chromosomes become fused to form a single centromere (Figure 5.27). This kind of fusion happened in human evolution to create the present metacentric chromosome 2 from two acrocentric chromosomes in an ancient human ancestor (Figure 5.6). Robertsonian translocations are an important risk factor to be considered in Down syndrome. When chromosome 21 is one of the acrocentrics in a Robertsonian translocation, the rearrangement leads to a familial type of Down syndrome. An example in which chromosome 21 is joined with

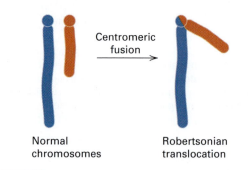

Normal chromosomes — Centromeric fusion → Robertsonian translocation

Figure 5.27 Formation of a Robertsonian translocation by fusion of two acrocentric chromosomes in the centromeric region.

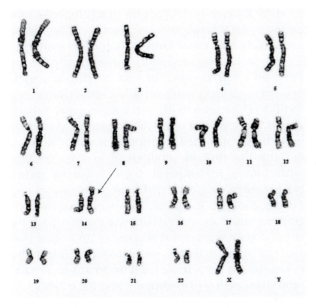

Figure 5.28 A karyotype of a child with Down syndrome, carrying a Robertsonian translocation of chromosomes 14 and 21 (arrow). [Courtesy Ms. Viola Freeman, Faculty of Health Sciences, Department of Pathology and Molecular Medicine, McMaster University.]

chromosome 14 is shown in Figure 5.28 (arrow). The heterozygous carrier is phenotypically normal, but a high risk of Down syndrome results from aberrant segregation in meiosis. Approximately 3 percent of children with Down syndrome are found to have one parent with such a translocation.

5.4

Polyploid species have multiple sets of chromosomes.

The genus *Chrysanthemum* illustrates **polyploidy,** an important phenomenon found frequently in higher plants. In polyploidy, a species has a genome composed of multiple complete sets of chromosomes. One *Chrysanthemum* species, a diploid species, has 18 chromosomes. A closely related species has 36 chromosomes. However, comparison of chromosome morphology indicates that the 36-chromosome species has two complete sets of the chromosomes found in the 18-chromosome species (Figure 5.29). The basic chromosome set in the group, from which all the other genomes are formed, is called the **monoploid** chromosome set. In *Chrysanthemum,* the monoploid chromosome number is 9. The diploid species has two complete copies of the monoploid set, or 18 chromosomes altogether. The 36-chromosome species has four copies of the monoploid set ($4 \times 9 = 36$) and is a **tetraploid.** The horticul-

Monoploid chromosome set

Diploid (18)

Tetraploid (36)

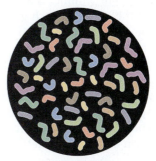

Hexaploid (54)

Octaploid (72)

Decaploid (90)

Figure 5.29
Chromosome numbers in diploid and polyploid species of *Chrysanthemum.* Each set of homologous chromosomes is depicted in a different color.

turalist's *Chrysanthemum* has 54 chromosomes (6×9, constituting the **hexaploid**). Other species have 72 chromosomes (8×9, the *octoploid*), and 90 chromosomes (10×9, the *decaploid*).

In meiosis, the chromosomes of all *Chrysanthemum* species synapse normally in pairs to form bivalents. The 18-chromosome species forms 9 bivalents, the 36-chromosome species forms 18 bivalents, the 54-chromosome species forms 27 bivalents, and so forth. Gametes receive one chromosome from each bivalent, so the number of chromosomes in the gametes of any species is exactly half the number of chromosomes in its somatic cells. The chromosomes present in the gametes of a species constitute the **haploid** set of chromosomes. In the species of *Chrysanthemum* with 54 chromosomes, for example, the haploid chromosome number is 27; in meiosis, 27 bivalents are formed, and so each gamete contains 27 chromosomes. When two such gametes come together in fertilization, the complete set of 54 chromosomes in the species is restored. Thus the gametes of a polyploid organism are not always monoploid, as they are in a diploid organism; for example, a tetraploid organism has diploid gametes.

The distinction between the term *monoploid* and the term *haploid* is subtle:

- The *monoploid* chromosome set is the basic set of chromosomes that is multiplied in a polyploid series of species, such as *Chrysanthemum*.

- The *haploid* chromosome set is the set of chromosomes present in a gamete, irrespective of the chromosome number in the species.

Confusion can arise because of diploid organisms, in which the monoploid chromosome set and the haploid chromosome set are the same. It helps to clarify the difference by considering the tetraploid, which contains four monoploid chromosome sets and in which the haploid gametes are diploid.

Polyploidy is widespread in certain plant groups. Among various groups of flowering plants, 30–80 percent of existing species are thought to have originated as some form of polyploid. Valuable agricultural crops that are polyploid include wheat, oats, cotton, potatoes, bananas, coffee, sugar cane, peanuts, and apples. Polyploidy often leads to an increase in the size of individual cells, and polyploid plants are often larger and more vigorous than their diploid ancestors; however, there are many exceptions to these generalizations. Polyploidy is rare in vertebrate animals, but it is found in a few groups of invertebrates. One reason why polyploidy is rare in animals is the difficulty in regular segregation of the sex chromosomes. For example, a tetraploid animal with XXXX females and XXYY males would produce XX eggs and XY sperm (if all chromosomes paired to form bivalents), so the progeny would be exclusively XXXY and unlike either of the parents.

Polyploid plants found in nature nearly always have an even number of sets of chromosomes, because organisms with an odd number have low fertility. Organisms with three monoploid sets of chromosomes are known as **triploids.** As far as growth is concerned, a triploid is quite normal, because the triploid condition does not interfere with mitosis; in mitosis in triploids (or any other type of polyploid), each chromosome replicates and divides just as in a diploid. However, because each chromosome has more than one pairing partner, chromosome segregation is severely upset in meiosis, and most gametes are defective. Unless the organism can perpetuate itself by means of asexual reproduction, it will eventually become extinct.

The infertility of triploids is sometimes of commercial benefit. For example, the seeds of "seedless" watermelons are small and edible because the plant is triploid and most of the seeds fail to develop to full size. In oysters, triploids are produced by treating fertilized diploid eggs with a chemical that causes the second polar body of the egg to be retained. The triploid oysters are sterile and do not spawn. In Florida and in certain other states, weed control in waterways is aided by the release of weed-eating fish (the grass carp), which do not become overpopulated, and hence a problem themselves, because the released fish are sterile triploids.

■ Polyploids can arise from genome duplications occurring before or after fertilization.

Polyploid organisms can be produced in two principal ways, which are illustrated in Figure 5.30 for the example of tetraploidy. In the mechanism known as **sexual polyploidization,** the increase in chromosome number takes place in *meiosis* through the formation of *unreduced gametes* that have double the normal complement of chromosomes. Unreduced gametes are formed in many species at frequencies of 1–40 percent, and the frequency can be under genetic control. For example, in the potato, a single recessive mutation that acts during pollen formation causes the first-division and second-division meiotic spindles to be oriented in the same direction (rather than being at right angles to each other as in nonmutant cells), with the result that a pollen nucleus forms around each of the two adjacent groups of telophase II chromosomes, yielding unreduced gametes. Also in the potato, a different recessive mutation acts to eliminate the second meiotic division during the formation of female gametes, again resulting in unreduced gametes. Part A of Figure 5.30 shows two unreduced $2n$ gametes yielding a $4n$ tetraploid, but there are many other possibilities. For example, union of an unreduced $2n$ gamete with a normal n gamete yields a $3n$ triploid.

The other principal mechanism of polyploid formation is **asexual polyploidization** (Figure 5.30, part B), in which the doubling of the chromosome number takes place in *mitosis*. Chromosome doubling through an abortive mitotic division is called **endoreduplication.** In a plant species that can undergo self-fertilization, endoreduplication creates a new, genetically stable species, because if the chromosomes in the tetraploid can pair two-by-two in meiosis, they can segregate regularly and yield gametes with a full complement of chromosomes. Self-fertilization of such a tetraploid restores the chromosome number, so the tetraploid condition can be perpetuated.

The genetics of tetraploid species and that of other polyploids is more complex than that of diploid species, because the organism carries more than two alleles of any gene. With two alleles in a diploid, only three genotypes are possible: *AA, Aa,* and *aa.* In a tetraploid, by contrast, five genotypes are possible: *AAAA, AAAa, AAaa, Aaaa,* and *aaaa.* Among these genotypes, the middle three represent different types of tetraploid heterozygotes.

An octoploid species (eight sets of chromosomes) can be generated by sexual or asexual polyploidization of a tetraploid. Again, if only bivalents form in meiosis, then an octoploid organism can be perpetuated sexually by self-fertilization or through crosses with other octoploids. Furthermore, cross-fertilization between an octoploid and a tetraploid results in a hexaploid (six sets of chromosomes). Repeated episodes of polyploidization and cross-fertilization may ultimately produce an entire polyploid series of closely related organisms that differ in chromosome number, as exemplified in Chrysanthemum.

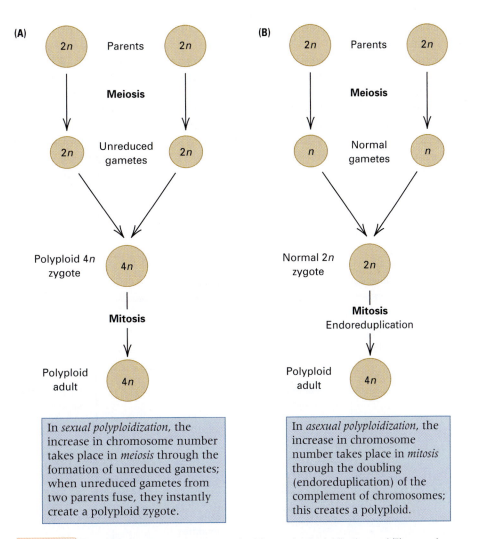

In *sexual polyploidization*, the increase in chromosome number takes place in *meiosis* through the formation of unreduced gametes; when unreduced gametes from two parents fuse, they instantly create a polyploid zygote.

In *asexual polyploidization*, the increase in chromosome number takes place in *mitosis* through the doubling (endoreduplication) of the complement of chromosomes; this creates a polyploid.

Figure 5.30 Formation of a tetraploid organism by (A) sexual polyploidization and (B) asexual polyploidization. The symbol n stands for the monoploid chromosome number.

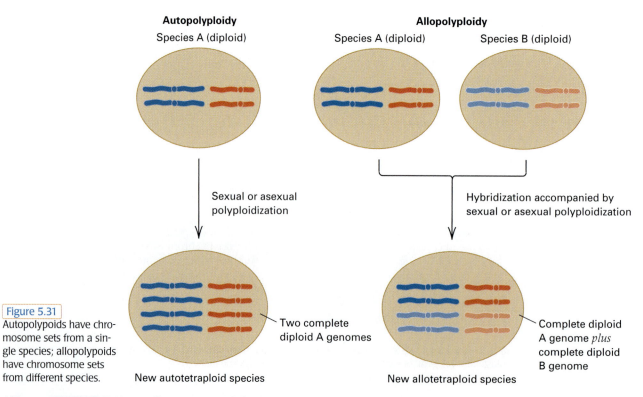

A Moment to Think

Answer to Problem: A two-strand double crossover results in two monocentric chromosomes ($A\,B\,C\,D\,E$ and $a\,d\,c\,b\,e$), which are the chromatids not involved in either crossover, plus two monocentric chromosomes ($A\,B\,c\,D\,E$ and $a\,d\,C\,b\,e$), which are the chromatids involved in the crossover. One type of three-strand double crossover results in a monocentric chromosome ($a\,d\,c\,b\,e$), not involved in either crossover, plus a dicentric ($a\,d\,C\,B\,A$), an acentric ($e\,b\,C\,D\,E$), and another monocentric ($A\,B\,c\,D\,E$). The other type of three-strand double crossover results in a monocentric chromosome ($A\,B\,C\,D\,E$), not involved in either crossover, plus a dicentric ($A\,B\,c\,d\,a$), an acentric ($e\,b\,c\,D\,E$), and another monocentric ($a\,d\,C\,b\,e$). A four-strand double crossover results in two dicentric chromosomes ($A\,B\,C\,d\,a$ and $A\,B\,c\,d\,a$) and two acentric chromosomes ($e\,b\,C\,D\,E$ and $e\,b\,c\,D\,E$). Note that genes can be exchanged between a normal and an inverted chromosome if there is double crossing-over.

■ Polyploids can include genomes from different species.

Chrysanthemum represents a type of polyploidy, known as **autopolyploidy,** in which all chromosomes in the polyploid species derive from a single diploid ancestral species (Figure 5.31, part A). In many cases of polyploidy, the polyploid species have complete sets of chromosomes from two or more *different* ancestral species. Such polyploids are known as **allopolyploids** (Figure 5.31, part B). They derive from occasional hybridization between different diploid species when pollen from one species germinates on the stigma of another species and sexually fertilizes the ovule, followed by endoreduplication in the zygote to yield a hybrid plant in which each chromosome has a pairing partner in meiosis. The pollen may be carried to the wrong flower by wind, insects, or other pollinators. Part B of Figure 5.31 illustrates hybridization between species A and B in which endoreduplication leads to the formation of an allopolyploid (in this case, an *allotetraploid*), which carries a complete diploid genome from each of its two ancestral species. The formation of allopolyploids through hybridization and endoreduplication is an extremely important process in plant evolution and plant breeding. At least half of all naturally occurring polyploids are allopolyploids. Cultivated wheat provides an excellent example of allopolyploidy. Cultivated bread wheat is a hexaploid with 42 chromosomes constituting a complete diploid genome of 14 chromosomes from each of three ancestral species. The 42-chromosome allopolyploid is thought to have originated by the series of hybridizations and endoreduplications outlined in Figure 5.32.

The ancestral origin of the chromosome sets in an allopolyploid can often be revealed by the technique of chromosome painting, in which chromosomes are "painted" different colors via hybridization with DNA strands labeled with fluorescent dyes. DNA from each of the putative ancestral species is isolated, denatured, and labeled with a different fluorescent dye. Then the labeled single strands are spread on a microscope slide and allowed to renature with homologous strands present in the chromosomes of the allopolyploid species.

Figure 5.31
Autopolypoids have chromosome sets from a single species; allopolypoids have chromosome sets from different species.

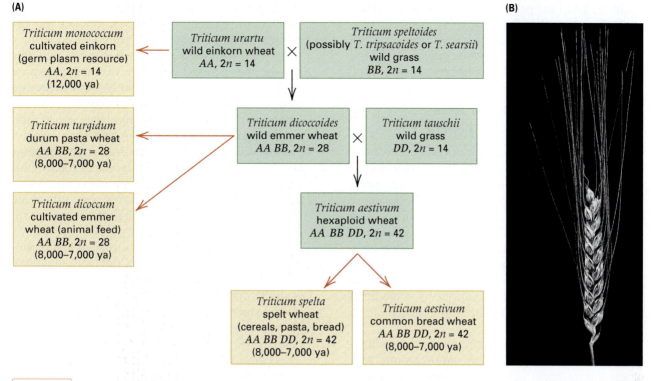

(A)

Triticum monococcum
cultivated einkorn
(germ plasm resource)
AA, $2n = 14$
(12,000 ya)

Triticum urartu
wild einkorn wheat
AA, $2n = 14$

$\times$

Triticum speltoides
(possibly T. tripsacoides or T. searsii)
wild grass
BB, $2n = 14$

Triticum turgidum
durum pasta wheat
$AA\ BB$, $2n = 28$
(8,000–7,000 ya)

Triticum dicoccoides
wild emmer wheat
$AA\ BB$, $2n = 28$

$\times$

Triticum tauschii
wild grass
DD, $2n = 14$

Triticum dicoccum
cultivated emmer
wheat (animal feed)
$AA\ BB$, $2n = 28$
(8,000–7,000 ya)

Triticum aestivum
hexaploid wheat
$AA\ BB\ DD$, $2n = 42$

Triticum spelta
spelt wheat
(cereals, pasta, bread)
$AA\ BB\ DD$, $2n = 42$
(8,000–7,000 ya)

Triticum aestivum
common bread wheat
$AA\ BB\ DD$, $2n = 42$
(8,000–7,000 ya)

(B)

Figure 5.32 Repeated hybridization and polyploidization in the origin of wheat. (A) Each of the A, B, and D genomes has 7 chromosomes, and $2n$ is the total chromosome number for each species. Wild species are in green boxes, and domesticated species are in yellow boxes along with the approximate time of domestication (ya = years ago). (B) The spike of *T. turgidum,* one of the earliest cultivated wheats. [B courtesy of Gordon Kimber.]

An example of chromosome painting to detect allopolyploidy is shown in Figure 5.33. The flower is from a variety of crocus called Golden Yellow. Its genome contains seven pairs of chromosomes, shown painted in yellow and green. Golden Yellow was thought to be an allopolyploid formed by hybridization of two closely related species, followed by endoreduplication of the chromosomes in the hybrid. The putative ancestral species are *Crocus flavus,* which has four pairs of chromosomes, and *Crocus angustifolius,* which has three pairs of chromosomes. To paint the chromosomes of Golden Yellow, DNA from *C. flavus* was isolated and labeled with a fluorescent green dye, and that from *C. angustifolius* was isolated and labeled with a fluorescent yellow dye. The result of the chromosome painting is very clear: Three pairs of chromosomes hybridize with the green-labeled DNA from

Figure 5.33 Flower of the crocus, variety Golden Yellow, and chromosome painting that reveals its origin as an allopolyploid. Its seven pairs of chromosomes are shown at the right. The chromosomes in green hybridized with DNA from *C. angustifolius,* which has three pairs of chromosomes, and those in yellow hybridized with DNA from *C. flavus,* which has four pairs of chromosomes. [Courtesy of J. S. Heslop-Harrison, University of Leicester, UK. With permission of the *Annals of Botany.*]

C. flavus, and four pairs of chromosomes hybridize with the yellow-labeled DNA from *C. angustifolius.* This pattern of hybridization strongly supports the hypothesis of the allopolyploid origin of Golden Yellow.

■ Plant cells with a single set of chromosomes can be cultured.

An organism is monoploid if it develops from a cell with a single monoploid set of chromosomes. Meiosis cannot take place normally in the germ cells of a monoploid, because each chromosome lacks a pairing partner, and hence monoploids are usually sterile. Monoploid organisms are quite rare, but they occur naturally in certain insect species (ants, bees) in which males are derived from unfertilized eggs. These monoploid males are fertile because the gametes are produced by a modified meiosis in which chromosomes do not separate in meiosis I.

Monoploids are important in plant breeding, because in the selection of diploid organisms with desired properties, favorable recessive alleles may be masked by heterozygosity. This problem can be avoided by studying monoploids, provided that their sterility can be overcome. In many plants, the production of monoploids capable of reproducing can be stimulated by conditions that yield aberrant cell divisions. Two techniques make this possible.

With some diploid plants, monoploids can be derived from cells in the anthers (the pollen-bearing structures). Extreme chilling of the anthers causes some of the haploid cells destined to become pollen grains to begin to divide. These cells are monoploid as well as haploid. If the cold-shocked cells are placed on an agar surface containing suitable nutrients and certain plant hormones, a small dividing mass of cells called an *embryoid* forms. A subsequent change of plant hormones in the growth medium causes the embryoid to form a small plant with roots and leaves that can be potted in soil and allowed to grow normally. Because monoploid cells have only a single set of chromosomes, their genotypes can be identified without regard to the dominance or recessiveness of individual alleles. A plant breeder can then select a monoploid plant with the desired traits. In some cases, the desired genes are present in the original diploid plant and are merely sorted out and selected in the monoploids. In other cases, the anthers are treated with mutagenic agents in the hope of producing the desired traits.

When a desired mutation is isolated in a monoploid, it is necessary to convert the monoploid into a homozygous diploid because the monoploid plant is sterile and does not produce seeds. The monoploid is converted into a diploid by treatment of the meristematic tissue (the growing point of a stem or branch) with the substance **colchicine.** This chemical is an inhibitor of the formation of the mitotic spindle. When the treated cells in the monoploid meristem begin mitosis, the chromosomes replicate normally; however, the colchicine blocks metaphase and anaphase, so the result is endoreduplication (doubling of each chromosome in a given cell). Many of the cells are killed by colchicine, but fortunately for the plant breeder, a few of the monoploid cells are converted into the diploid state (Figure 5.34). The colchicine is removed to allow continued cell multiplication, and many of the now-diploid cells multiply to form a small sector of tissue that can be recognized microscopically. If placed on a nutrient-agar surface, this tissue will develop into a complete plant. Such plants, which are completely homozygous, are fertile and produce normal seeds.

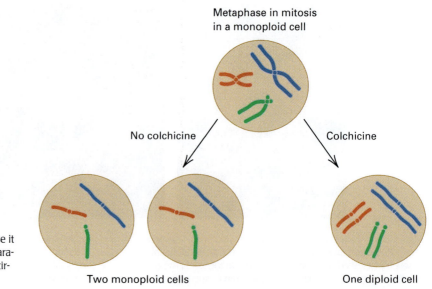

Figure 5.34 Production of a diploid from a monoploid by treatment with colchicine. The colchicine results in endoreduplication because it disrupts the spindle and thereby prevents separation of the chromatids after the centromeres (circles) divide.

Metaphase in mitosis in a monoploid cell

No colchicine

Colchicine

Two monoploid cells

One diploid cell

The grass family illustrates the importance of polyploidy and chromosome rearrangements in genome evolution.

The cereal grasses are our most important crop plants. They include rice, wheat, maize, millet, sugar cane, sorghum, and other cereals. The genomes of grass species vary enormously in size. The smallest, at 400 Mb, is found in rice; the largest, at 16,000 Mb, is found in wheat. Although some of the difference in genome size results from the fact that wheat is an allohexaploid whereas rice is a diploid, a far more important factor is the large variation from one species to the next in types and amount of repetitive DNA sequences present. Each chromosome in wheat contains approximately 25 times as much DNA as each chromosome in rice.

For comparison, maize has a genome size of 2500 Mb; it is intermediate in size among the grasses and approximately the same size as the human genome.

In spite of the large variation in chromosome number and genome size in the grass family, there are a number of genetic and physical linkages between single-copy genes that are remarkably conserved amid a background of very rapidly evolving repetitive DNA sequences. In particular, each of the conserved regions can be identified in all the grasses and referred to a similar region in the rice genome. The situation is as depicted in Figure 5.35. The rice chromosome pairs are numbered R1 through R12, and the conserved regions within each chromosome are indicated by lowercase letters—for example, R1a and R1b. In each of the other species, each chromosome pair is diagrammed according to the arrangement of segments of the rice genome that contain single-copy DNA sequences homologous to those in the corre-

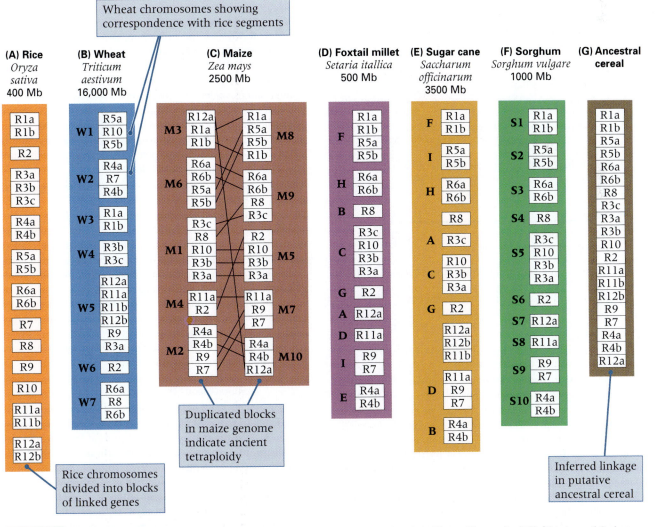

Figure 5.35 Conserved linkages (synteny groups) between the rice genome (A) and that of other grass species: wheat (B), maize (C), foxtail millet (D), sugar cane (E), and sorghum (F). Genome sizes are given in millions of base pairs (Mb). [Courtesy of Graham Moore. From G. Moore, K. M. Devos, Z. Wang, and M. D. Gale. 1995. *Current Opinion Genet. Devel.* 5: 737.]

sponding region of the chromosome of the species in question. For example, the wheat monoploid chromosome set is designated W1 through W7. One region of W1 contains single-copy sequences that are homologous to those in rice segment R5a, another contains single-copy sequences that are homologous to those in rice segment R10, and still another contains single-copy sequences that are homologous to those in rice segment R5b. The genomes of the other grass species can be aligned with those of rice as shown. Each of such conserved

genetic and physical linkages is called a **synteny group.**

Synteny groups are found in other species comparisons as well. For example, the human and mouse genomes share about 180 synteny groups owing to about an equal number of chromosome rearrangements that took place in the approximately 80 million years since the species last shared a common ancestor. These synteny groups are often useful in identifying the mouse homolog of a human gene.

chapter summary

5.1 Human beings have 46 chromosomes in 23 pairs.

- The standard human karyotype consists of 22 pairs of autosomes and two sex chromosomes.
- Chromosomes with no centromere, or with two centromeres, are genetically unstable.
- Dosage compensation adjusts the activity of X-linked genes in females.
- The calico cat shows visible evidence of X-chromosome inactivation.
- Some genes in the X chromosome are also present in the Y chromosome.
- The pseudoautosomal region of the Y chromosome has gotten progressively shorter in evolutionary time.
- The history of human populations can be traced through studies of the Y chromosome.

5.2 Chromosome abnormalities are frequent in spontaneous abortions.

- Down syndrome results from three copies of chromosome 21.
- Trisomic chromosomes undergo abnormal segregation.
- An extra X or Y chromosome has a relatively mild effect.
- The rate of nondisjunction can be increased by chemicals in the environment.

Fetuses that contain an abnormal number of autosomes usually either fail to complete normal embryonic development or die shortly after birth, though people with Down syndrome (trisomy 21) sometimes survive for several decades. Persons with excess sex chromosomes survive, because the Y chromosome contains relatively few genes other than the master sex-controller *SRY*, and because only one X chromosome is genetically active in the cells of females (dosage compensation through the single-active-X principle). The mosaic orange and black pattern of the female "calico" cat results from X inactivation, because these alternative coat color alleles are X-linked in cats.

5.3 Chromosome rearrangements can have important genetic effects.

- A chromosome with a deletion has genes missing.
- Rearrangements are apparent in giant polytene chromosomes.
- A chromosome with a duplication has extra genes.
- Human color-blindness mutations result from unequal crossing-over.

Most structural abnormalities in chromosomes are duplications, deletions, inversions, or translocations. In a duplication, there are two copies of a chromosomal segment. In a deletion, a chromosomal segment is missing. An organism can often tolerate an imbalance of gene dosage resulting from small duplications or deletions, but large duplications or deletions are nearly always harmful.

- A chromosome with an inversion has some genes in reverse order.

A chromosome that contains an inversion has a group of adjacent genes in reverse of the normal order. Expression of the genes is usually unaltered, so inversions rarely affect viability. However, crossing-over between an inverted chromosome and its noninverted homolog in meiosis yields abnormal chromatids. Crossing-over within a heterozygous paracentric inversion yields an acentric chromosome and a dicentric chromosome, both of which also have a duplication and a deficiency. Crossing-over within a heterozygous pericentric inversion yields monocentric chromosomes, but both have a duplication and a deficiency.

- Reciprocal translocations interchange parts between nonhomologous chromosomes.

Two nonhomologous chromosomes that have undergone an exchange of parts constitute a reciprocal translocation. Organisms that contain a reciprocal translocation, as well as the normal homologous chromosomes of the translocation, produce fewer offspring (this is called semisterility) because of abnormal segregation of the chromosomes in meiosis. The semisterility is caused by the aneuploid gametes produced in adjacent-1 and adjacent-2 segregation. Alternate segregation yields equal numbers of normal and translocation-bearing gametes. In genetic crosses, the semisterility of a heterozygous translocation behaves like a dominant genetic marker that can be mapped like any other gene; however, what is actually mapped is the breakpoint of the translocation.

A translocation may also be nonreciprocal. A Robertsonian translocation is a type of nonreciprocal translocation in which the long arms of two acrocentric chromosomes are attached to a common centromere. In human beings, Robertsonian translocations that include chromosome 21 account for about 3 percent of all cases of Down syndrome, and the parents have a high risk of recurrence of Down syndrome in a subsequent child.

5.4 Polyploid species have multiple sets of chromosomes.

- Polyploids can arise from genome duplications occurring before or after fertilization.
- Polyploids can include genomes from different species.
- Plant cells with a single set of chromosomes can be cultured.

5.5 The grass family illustrates the importance of polyploidy and chromosome rearrangements in genome evolution.

issues & ideas

- Although autosomal trisomy is common among human fetuses that undergo spontaneous abortion, autosomal monosomies are almost unknown. How can this observation be explained?

- Why do most chromosome rearrangements pass through mitosis without upsetting the process?

- What are the four major classes of abormality in chromosome structure?

- What types of chromosomal abnormalities in meiosis are associated with an inversion? How are these related to the position of the centromere? To crossing-over?

- What types of chromosomally abnormal gametes are associated with a translocation?

- Why do inversions and translocations cause reproductive abnormalities only when they are heterozygous?

- How does the ability to form bivalents in meiosis contribute to the production of gametes that have the same number and types of chromosomes?

- Why would the presence of sex chromosomes be a hinderance to the evolution of a series of related species with different levels of polyploidy?

- Why do most naturally occurring polyploid species have an even-number multiple of the monoploid chromosome set?

- Distinguish between sexual and asexual polyploidization. In which type does the chromosome number double in meiosis? In mitosis?

key terms & concepts

acentric chromosome	dicentric chromosome	monoploid	tandem duplication
adjacent-1 segregation	dosage compensation	monosomy	testis-determining factor
adjacent-2 segregation	double-Y syndrome	paracentric inversion	(TDF)
allopolyploid	Down syndrome	pericentric inversion	tetraploid
alternate segregation	duplication	polyploidy	translocation
aneuploid	endoreduplication	polytene chromosome	triploid
asexual polyploidization	euploid	pseudoautosomal region	trisomic
autopolyploidy	haploid	reciprocal translocation	trisomy-X syndrome
Barr body	haplotype	red-green color blindness	trivalent
chimeric gene	hexaploid	Robertsonian translocation	Turner syndrome
chromosome painting	inversion	semisterility	unequal crossing-over
colchicine	inversion loop	sexual polyploidization	univalent
cytological map	karyotype	single-active-X principle	viral oncogene
deficiency	Klinefelter syndrome	synteny group	X inactivation
deletion			

1. _____ The Y-linked gene *SRY* encodes this key protein that acts in switching the human developmental pathway from female to male.

2. _____ Syndrome associated with 47, XXY.

3. _____ Small region of the X and Y chromosome in which genes are inherited in a fashion indistinguishable from autosomal inheritance.

4. _____ In mammals, this process underlies the single-active-X principle that results in dosage compensation.

5. _____ Procedure for differentially labeling each chromosome in a karyotype by means of hybridization with chromosome-specific DNA linked to various fluorescent tags.

6. _____ Two or more genes descended from a common ancestral gene that are adjacent along a chromosome.

7. _____ Process producing the chimeric genes associated with various forms of red-green color blindness.

8. _____ A single crossover in the loop of this type of inversion results in acentric and dicentric products.

9. _____ Type of segregation of a heterozygous reciprocal translocation that results in euploid gametes.

10. _____ A consequence of aneuploid gametes produced by adjacent-1 and adjacent-2 segregation from a heterozygous reciprocal translocation.

11. _____ Formation of a polyploid by the union of unreduced gametes.

12. _____ A group of genes that remains physically linked together along a chromosome.

solutions: step by step

Problem 1

Indicate what types of gametes would be expected, and in what proportions, from an organism that is:
(a) Trisomic for a single chromosome.
(b) Monosomic for a single chromosome.

■ **Solution** (a) In a trisomic organism, either two of the chromosomes will form a pair and the other remain unpaired, or all three chromosomes will come together, each chromosome pairing along part of its length with both of the others. These situations are shown in part A of the illustration shown below. The anaphase I segregation determines the chromosomes included in the gametes. In either type of pairing, two gametes receive two copies of the chromosome and two gametes receive one copy. Therefore, trisomic segregation leads to disomic and monosomic gametes in an expected ratio of 1 : 1.

(b) In a monosomic organism, the monosomic chromosome must remain unpaired. At anaphase I, it goes to one pole or the other. The end result is two gametes containing no copies of the chromosome and two gametes containing one copy. Therefore, monosomic segregation leads to nullisomic (missing the chromosome) gametes and monosomic gametes in an expected ratio of 1 : 1 [illustration part B].

Problem 2

A certain chromosome in maize, when heterozygous with a standard chromosome carrying multiple genetic markers, fails to yield recombinants between genetic markers in a region of the short arm. Cytological investigation reveals many meiotic anaphases in which there is a chromosome bridge connecting the centromeres of a dicentric chromosome as the centromeres are pulled to opposite poles.
(a) What chromosomal abnormality can account for the suppression of recombination and the anaphase bridges?
(b) Are the anaphase bridges seen at anaphase I or at anaphase II?

■ **Solution** (a) Suppression of recombination in one segment of a chromosome is typically associated with a heterozygous inversion. This is a reasonable hypothesis in the present case, because the suppression occurs in heterozygotes with a standard, structurally normal chromosome. The appearance of chromosome bridges indicates that the inversion is a paracentric inversion. (Pericentric inversions also suppress recombination but do not result in chromosome bridges.)
(b) If the correct explanation is a paracentric inversion, the bridges would be observed in anaphase I.

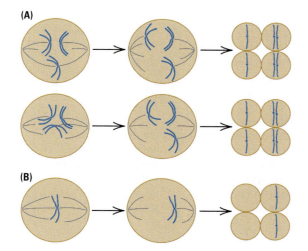
(A)

(B)

Problem 3

Asiatic wild cotton (Gossypium arboreum) has 13 pairs of chromosomes, and an American wild species (G. thurberi) also has 13 pairs of chromosomes. Interspecific crosses between the species are sterile because of highly irregular chromosome pairing in meiosis. The American cultivated cotton (G. hirsutum) has 26 pairs of chromosomes and is fully fertile. Crosses between G. arboreum and G. hirsutum and crosses between G. thurberi and G. hirsutum produce plants that, in meiosis, exhibit 13 pairs of chromosomes (bivalents) and 13 chromosomes with no pairing partner (univalents). What do these cytological data tell us about the genetic origin of present-day American cultivated cotton?

■ Solution It is helpful to draw a diagram showing the 13 paired chromosomes in G. arboreum and G. thurberi. The fact that the cross between these species yields a hybrid in which the chromosomes do not pair normally indicates that the chromosomes of the two parental species are quite different genetically. One possibility for the origin of G. hirsutum is that it arose by endoreduplication in such a hybrid, which would account for the 26 chromosome pairs in G. hirsutum. The backcross results are also consistent with this hypothesis, because each yields a hybrid with 13 bivalents and 13 univalents. These results are illustrated in the accompanying

diagram. The data therefore suggest that present-day American cultivated cotton originally arose as an allo-tetraploid from a hybrid between a wild species related to G. arboreum and another wild species related to G. thurberi.

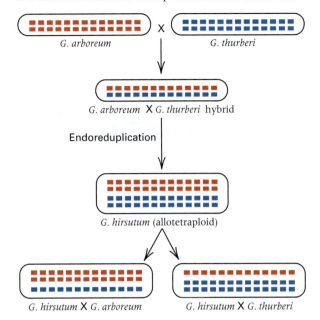

concepts in action: problems for solution

5.1 Explain why the typical calico cat is a female with patches of orange fur and patches of black fur.

5.2 What are the most likely karyotypes of a newborn baby with 47 chromosomes? With 45 chromosomes?

5.3 A recessive mutation in the human genome results in a condition called anhidrotic ectodermal dysplasia, which is associated with an absence of sweat glands. The condition can be detected by studies of the electrical conductivity of the skin, because skin without sweat glands has a lower electrical conductivity (higher resistance) than normal skin. In kinships in which the recessive allele is segregating, affected males are found to show low conductance uniformly across their skin surface, as do affected females. However, many females show a mosaic pattern with normal conductance in some patches of skin and low conductance in others. The pattern of tissue lacking sweat glands is different for each mosaic female examined. How could this pattern of gene expression be explained?

5.4 The vast majority of progeny from either 47,XXX females or 47,XYY males are karyotypically normal 46,XX or 46,XY. Is this finding expected?

5.5 How many Barr bodies are present in each of the following human conditions?
(a) Klinefelter syndrome
(b) Turner syndrome
(c) Down syndrome
(d) Double-Y syndrome
(e) Trisomy-X syndrome

5.6 A spontaneously aborted human fetus was found to have the karyotype 92,XXYY. What happened to the chromosomes in the zygote to result in this karyotype?

5.7 Inversions are often called "suppressors" of crossing-over. Is this term literally true? If not, what does it actually mean?

5.8 A chromosome has the gene sequence $A\ B\ C\ D\ E\ F\ G$. What is the sequence following an inversion of genes C through E? After a deletion of genes C through E? Two chromosomes with the sequences $A\ B\ C\ D\ E\ F\ G$ and $M\ N\ O\ P$ $Q\ R\ S\ T\ U\ V$ undergo a reciprocal translocation after breaks in $E-F$ and $S-T$. What are the possible products? Which products are genetically stable?

5.9 A female cat with orange fur mates with a male with black fur. The resulting litter includes a male calico kitten that, when mature, proves to be sterile. Suggest a likely explanation.

5.10 Recessive genes a, b, c, d, e, and f are closely linked in a chromosome, but their order is unknown. Three deletions in the region are examined. One deletion uncovers a, d, and e; another uncovers c, d, and f; and the third uncovers b and c. What is the order of the genes?

5.11 Genes a, b, c, d, e, and f are closely linked in a chromosome, but their order is unknown. Three deletions in the region are found to uncover recessive alleles of the genes as follows: (1) deletion 1 uncovers a, b, and d; (2) deletion 2 uncovers a, d, c, and e; (3) deletion 3 uncovers e and f. What is the order of the genes? In this problem, you will see that there is enough information to order most, but not all, of the genes. Suggest what experiments you might carry out to complete the ordering.

5.12 Six bands in a salivary gland chromosome of *Drosophila* are shown in the accompanying figure, along with the extent of five deletions (Del1−Del5).

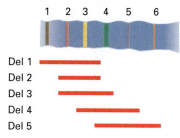

Recessive alleles *a, b, c, d, e,* and *f* are known to be in the region, but their order is unknown. When the deletions are heterozygous with each allele, the following results are obtained:

	a	*b*	*c*	*d*	*e*	*f*
Del 1	−	−	−	+	+	+
Del 2	−	+	−	+	+	+
Del 3	−	+	−	+	−	+
Del 4	+	+	−	−	−	+
Del 5	+	+	+	−	−	−

In this table, the − means that the deletion is missing the corresponding wildtype allele (the deletion uncovers the recessive allele) and + means that the corresponding wild-type allele is still present. Use these data to infer the position of each gene relative to the salivary gland chromosome bands.

5.13 A phenotypically normal woman has a child with Down syndrome. The woman is found to have 45 chromosomes. What kind of chromosome abnormality can account for these observations? How many chromosomes does the affected child have? How does this differ from the usual chromosome number and karyotype of a child with Down syndrome?

5.14 Curly wings (*Cy*) is a dominant mutation in the second chromosome of *Drosophila*. A heterozygous *Cy* male was irradiated with x-rays and crossed with homozygous wildtype females, and the *Cy* sons were mated individually with homozygous wildtype females. From one cross, the progeny were

curly males	146	curly females	0
wildtype males	0	wildtype females	163

What abnormality in chromosome structure is the most likely explanation for these results? (Remember that crossing-over does not take place in male *Drosophila*.)

5.15 *Drosophila virilis* has 6 pairs of chromosomes in somatic cells, consisting of 5 acrocentric chromosome pairs and 1 small "dot" chromosome pair. The closely related species *D. texana* has 5 pairs of chromosomes, consisting of 4

acrocentric pairs, 1 metacentric pair, and the small "dot" pair. Hybrids between these species have meiotic cells in which there are 4 bivalents (including the "dot") and 1 trivalent. The trivalent consists of the metacentric chromosome of *D. texana* paired with two of the acrocentrics of *D. virilis*, oriented such that the three centromeres are close together. Suggest an explanation.

5.16 In *Drosophila melanogaster*, the genes for *brown eyes* (*bw*) and *humpy thorax* (*hy*) are about 12 map units distant on the same arm of chromosome 2. A paracentric inversion spans about one-third of this region but does not include the genes mentioned. Explain what recombinant frequency between *bw* and *hy* you would expect in females that are:
(a) Homozygous for the inversion.
(b) Heterozygous for the inversion.

5.17 Semisterile tomato plants heterozygous for a reciprocal translocation between chromosomes 5 and 11 were crossed with chromosomally normal plants homozygous for the recessive mutation *broad leaf* on chromosome 11. When semisterile F$_1$ plants were crossed with the plants of *broad-leaf* parental type, the following phenotypes were found in the backcross progeny:

semisterile broad-leaf	38
fertile broad-leaf	242
semisterile normal-leaf	282
fertile normal-leaf	33

(a) What is the recombination frequency between the *broad-leaf* gene and the translocation breakpoint in chromosome 11?
(b) What ratio of phenotypes in the backcross progeny would have been expected if the *broad-leaf* gene had not been on the chromosome involved in the translocation?

5.18 The herb genus *Tragopogon* shows a great deal of interspecific hybridization leading to allopolyploids as new species. Explain how three different species of *Tragopogon* (each with $2n = 12$) could hybridize and produce a new species with a chromosome number of 36.

5.19 Several different species of rhododendrons exist, and they have the following numbers of somatic chromosomes: 26, 39, 52, 78, 104, and 156.
(a) What process might produce this series of species with varying chromosome numbers?
(b) What would you predict to be the basic monoploid number of chromosomes?
(c) How many sets of chromosomes are present in the species with 156 chromosomes?

5.20 A genetically wildtype natural isolate of *Neurospora crassa* was crossed with a laboratory strain carrying a recessive allele, *a*, known to be 10 map units from the centromere. The resulting asci showed only 2 percent second-division segregation, and many of the asci contained inviable ascospores. One of the *a*-bearing ascospores from a second-division segregation ascus was germinated and mated with the original wildtype isolate. In this case, the resulting asci showed 20 percent second-division segregation. How can you account for these results?

geNETics on the web

GeNETics on the Web will introduce you to some of the most important sites for finding genetic information on the Internet. To explore these sites, visit the Jones and Bartlett home page at

http://www.jbpub.com/genetics

For the book *Essential Genetics: A Genomics Perspective,* choose the link that says **Enter GeNETics on the Web**. You will be presented with a chapter-by-chapter list of highlighted keywords. Select any highlighted keyword and you will be linked to a Web site containing genetic information related to the keyword.

• **Down syndrome** affects 350,000 people in the United States, and there are approximately 5,000 new cases every year. Since the syndrome was first described by the physician John Langdon Down in 1866, many myths about the condition have come into being, such as that the condition is very rare, that all affected children are alike, and that all affected children have severe learning disabilities. A discussion of the myths and truths about Down syndrome can be found at this keyword site.

• Amniocentesis for the prenatal diagnosis of chromosome disorders is usually performed between 15 and 18 weeks after a woman's last menstrual period. A somewhat different procedure called chorionic villus sampling (**CVS**) can be carried out earlier in pregnancy, usually between 10 and 12 weeks after a woman's last menstrual period. At this keyword site you can find a nontechnical description of CVS and a discussion of its uses and potential dangers.

further readings

Bickmore, W. A., and A. T. Sumner. 1989. Mammalian chromosome banding: An expression of genome organization. *Trends in Genetics* 5: 144.

Cicchetti, D., and M. Beeghly, eds. 1990. *Children with Down Syndrome: A Developmental Perspective.* New York: Cambridge University Press.

Epstein, C. J. 1988. Mechanisms of the effects of aneuploidy in mammals. *Annual Review of Genetics* 22: 51.

Gresham, D., et al. 2001. Origins and divergence of the Roma (Gypsies). *American Journal of Human Genetics* 69: 1314.

Hernandez, D., and E. M. C. Fisher. 1999. Mouse autosomal trisomy: Two's company, three's a crowd. *Trends in Genetics* 15: 241.

Hsu, T. H. 1979. *Human and Mammalian Cytogenetics.* New York: Springer-Verlag.

Hunt, P. A., et al. 2003. Bisphenol A exposure causes meiotic aneuploidy in the female mouse. *Current Biology* 13: 546.

Hurst, L. D., and J. P. Randerson. 1999. An exceptional chromosome. *Trends in Genetics* 15: 383.

Kimber, G., and M. Feldman. 1987. *Wild Wheat: An Introduction.* Columbia: University of Missouri Press.

Manning, C. H., and H. O. Goodman. 1981. Parental origin of chromosomes in Down's syndrome. *Human Genetics* 59: 101.

Miller, O. J., and E. Therman. 2000. *Human Chromosomes.* 4th ed. New York: Springer-Verlag.

Nathans, J. 1989. The genes for color vision. *Scientific American,* February.

Ronald, P. C. 1998. Making rice disease-resistant. *Scientific American,* November.

Stebbins, G. L. 1971. *Chromosome Evolution in Higher Plants.* Reading, MA: Addison-Wesley.

Stewart, G. D., T. J. Hassold, and D. M. Kurnit. 1988. Trisomy 21: Molecular and cytogenetic studies of nondisjunction. *Advances in Human Genetics* 17: 99.

Thomas, M. G., et al. 2000. Y chromosomes traveling south: The Cohen modal haplotype and the origins of the Lemba—the "Black Jews of southern Africa." *American Journal of Human Genetics* 66: 674.

Wagner, R. P., M. P. Maguire, and R. L. Stallings. 1993. *Chromosomes.* New York: Wiley-Liss.

White, M. J. D. 1977. *Animal Cytology and Evolution.* London: Cambridge University Press.

Zerjal, T., et al. 2003. The genetic legacy of the Mongols. *American Journal of Human Genetics* 72: 717.

The spiny pufferfish, *Diodon holacanthus*, has a genome size of 800 million base pairs. The genome of its close relative, the Japanese pufferfish, *Takifugu rubripes*, is only 400 million base pairs—a mere 13 percent of the size of the human genome. These fish and their relatives are also famous because some of their tissues contain tetrodotoxin, a potent neurotoxin that blocks voltage-gated sodium channels on the surface of nerve membranes. Eating a piece of contaminated fish as small as the size of a quarter results in rapid onset of dizziness, headache, nausea, vomiting, progressive paralysis, and death within 4-6 hours. [© Ablestock]

key concepts

- Prokaryotes generally have smaller genomes (less DNA) than higher eukaryotes.
- Among eukaryotes, there is no consistent relationship between genome size and organismic complexity.
- A DNA strand is a polymer of A, T, G, and C deoxyribonucleotides joined 3′ to 5′ by phosphodiester bonds.
- Hydrogen bonding between the A−T and G−C base pairs helps hold the two DNA strands in a duplex together.
- In DNA replication, each parental strand serves as a template for a daughter strand that is synthesized in the 5′ → 3′ direction (successive nucleotides are added only at the 3′ end).

- Each type of restriction endonuclease enzyme cleaves double-stranded DNA at a particular sequence of bases, usually 4 or 6 nucleotides in length.
- In the polymerase chain reaction, short oligonucleotide primers are used in successive cycles of DNA replication to amplify selectively a particular region of a DNA duplex.
- The DNA fragments produced by a restriction enzyme can be separated by electrophoresis, isolated, sequenced, and manipulated in other ways.

(A)

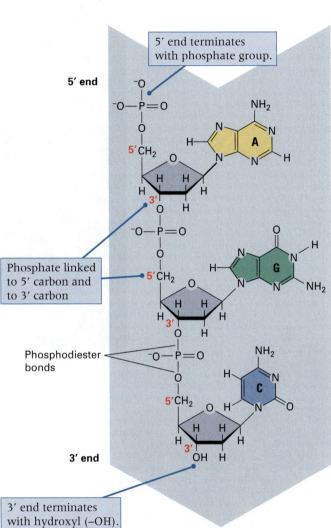

5′ end terminates with phosphate group.

5′ end

Phosphate linked to 5′ carbon and to 3′ carbon

Phosphodiester bonds

3′ end

3′ end terminates with hydroxyl (–OH).

(B)

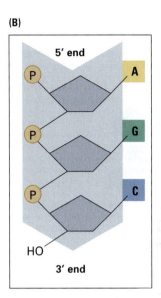

5′ end

3′ end

Figure 6.4 Three nucleotides at the 5′ end of a single poly-nucleotide strand. (A) The chemical structure of the sugar–phosphate linkages, showing the 5′-to-3′ orientation of the strand (the red numbers are those assigned to the carbon atoms). (B) A common schematic way to depict a polynucleotide strand.

6.3

Duplex DNA is a double helix in which the bases form hydrogen bonds.

Figure 6.5 shows several representations of double-stranded DNA. The duplex molecule of DNA consists of two polynucleotide chains twisted around one another to form a right-handed helix in which adenine and thymine are paired, as are guanine and cytosine (Figure 6.5). Each chain makes one complete turn every 34 Å. The bases are spaced at 3.4 Å, so there are ten bases per helical turn in each strand, or ten base pairs per turn of the double helix. Each base is paired to its partner base in the other strand by a hydrogen bond. A **hydrogen bond** is a weak bond in which two negatively charged atoms share a hydrogen atom. Hydrogen bonds contribute to holding the strands together, as does the stacking of the base pairs on top of one another so as to exclude water mole-

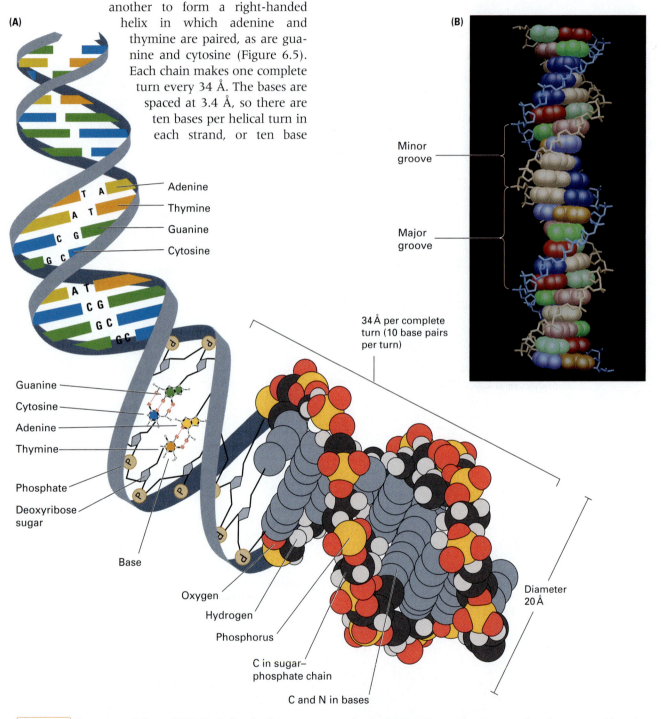

Figure 6.5 Two representations of DNA illustrating the three-dimensional structure of the double helix. (A) In a "ribbon diagram," the sugar–phosphate backbones are depicted as bands, with horizontal lines used to represent the base pairs. (B) A computer model of the standard form of DNA. The stick figures are the sugar–phosphate chains winding around outside the stacked base pairs, forming a major groove and a minor groove. The color coding for the base pairs is A, red or pink; T, dark green or light green; G, dark brown or beige; C, dark blue or light blue. The bases depicted in dark colors are those attached to the blue sugar–phosphate backbone; the bases depicted in light colors are attached to the beige backbone. [B, Courtesy of Antony M. Dean.]

cules. The paired bases are planar, parallel to one another, and perpendicular to the long axis of the double helix.

For encoding genetic information, the central feature of DNA structure is the A−T and G−C pairing between the bases:

The principles of A−T and G−C base pairing explain two generalizations about the relative amounts of the bases found in all double-stranded DNA:

- Number of adenine bases [A] equals number of thymine bases [T], so [A] = [T].
- Number of guanine bases [G] equals number of cytosine bases [C], so [G] = [C].

Although [A] = [T] and [G] = [C] in double-stranded DNA, the proportion of bases that are either G or C (called the *percent G + C*) varies among species but is constant in all cells of an organism. For example, human DNA has 39 percent G + C on the average, but there can be large variations in base composition along the chromosomes. The regional variation can be observed microscopically because regions relatively poor in G + C content give rise to dark bands when the chromosomes are stained with Giemsa (Figure 5.2 on page 165).

The adenine–thymine base pair and the guanine–cytosine base pair are illustrated in Figure 6.6. Note that an A−T pair has two hydrogen bonds and a G−C pair has three hydrogen bonds. This means that the hydrogen bonding between G and C is stronger in the sense that it requires more energy to break. The specificity of base pairing means that the sequence of bases along one polynucleotide strand of the DNA is matched (complementary) with the base sequence in the other strand. However, the base pairs along a DNA duplex can be arranged in any order, and the sequence of bases differs from one part of the molecule to another and from species to species. Because there is no restriction on the base sequence, DNA has a virtually unlimited capability to code for a variety of different protein molecules.

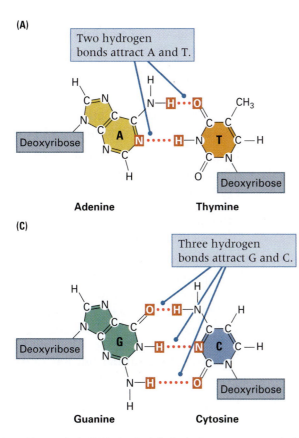

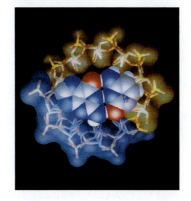

Figure 6.6 Normal base pairs in DNA. On the left, the hydrogen bonds (dotted lines) and the joined atoms are shown in red. (A, B) An A−T base pair. (C, D) A G−C base pair. In the space-filling models (B and D), the colors are C, gray; N, blue; O, red; and H (shown in the bases only), white. Each hydrogen bond is depicted as a white disk squeezed between the atoms that share the hydrogen. The stick figures on the outside represent the backbones winding around the stacked base pairs. [B and D, Courtesy of Antony M. Dean.]

The backbone of each polynucleotide strand in the double helix in Figure 6.5 consists of deoxyribose sugars alternating with phosphate groups that link the 3' carbon atom of one sugar to the 5' carbon of the next in line. The two polynucleotide strands of the double helix run in opposite directions, as can be seen from the orientation of the deoxyribose sugars in Figure 6.7 . The paired strands are said to be **antiparallel.** Figure 6.5 also shows that there are two grooves spiraling along outside of the double helix. These grooves are not symmetrical in size. The large one is called the *major groove,* the smaller one the *minor groove.* Proteins that interact with double-stranded DNA often have regions that make contact with the base pairs by fitting into the major groove, the minor groove, or both.

The diagrams of the DNA duplexes in parts A and B of Figure 6.5 are static and so somewhat misleading. DNA is in fact a very dynamic molecule that is constantly in motion. In some regions, the strands can separate briefly and then come together again. Furthermore, although the right-handed double helix in Figure 6.5 is the standard helix, DNA can form more than 20 slightly different variants of right-handed helices, and some regions can even form helices in which the strands twist to the left. If there are complementary stretches of nucleotides in the same strand, a single strand, separated from its partner, can fold back upon itself like a hairpin. Even triple helices, consisting of three strands, can form in regions of DNA that contain suitable base sequences.

6.4

Replication uses each DNA strand as a template for a new one.

The process of replication, in which each strand of the double helix serves as a **template** for the synthesis of a new strand, is simple in principle (Figure 6.8). It requires only that the hydrogen bonds joining the bases break to allow separation of the chains and that appropriate free nucleotides of the four types pair with the newly accessible bases in each strand. In practice, however, replication is a complex of geometric processes that require a variety of enzymes and other proteins. These processes are examined in this section.

■ Nucleotides are added one at a time to the growing end of a DNA strand.

The primary function of any mode of DNA replication is to reproduce the base sequence of the parent molecule. The specificity of base pairing—adenine with thymine and guanine with cytosine—provides the mechanism used by all genetic replication systems. Furthermore,

- Nucleotide monomers are added one by one to the end of a growing strand by an enzyme called a *DNA polymerase.*

- The sequence of bases in each newly replicated strand, or **daughter strand,** is complementary to the base sequence in the old strand, or **parental strand,** being replicated. For example, wherever an adenine nucleotide is present in the parental strand, a thymine nucleotide will be added to the growing end of the daughter strand.

The following section explains how the two strands of a daughter molecule are physically related to the two strands of the parental molecule.

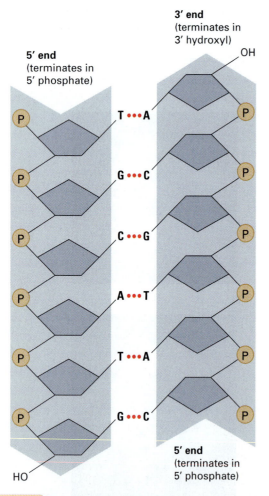

Figure 6.7 A segment of a DNA molecule showing the antiparallel orientation of the complementary strands. The arrows indicate the 5'-to-3' direction of each strand. The phosphate groups (P) join the 3' carbon atom of one deoxyribose to the 5' carbon atom of the adjacent deoxyribose.

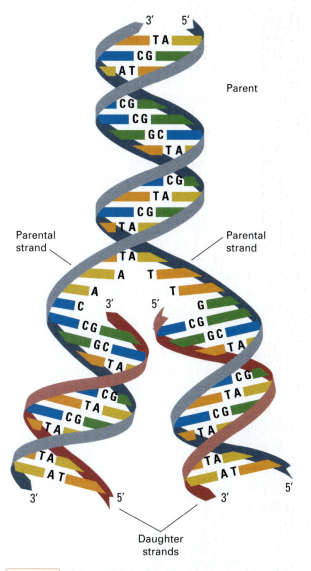

IN THEIR FIRST PAPER on the structure of DNA, Watson and Crick remarked that "it has not escaped our notice that the specific base pairing we have postulated immediately suggests a copying mechanism." The mechanism they proposed is shown in Figure 6.8.

Figure 6.8 Watson–Crick model of DNA replication. The newly synthesized strands are in red. Each of the new strands is elongated *only at the 3' end.*

◾ DNA replication is semiconservative: The parental strands remain intact.

The mode of replication diagrammed in Figure 6.8 is called **semiconservative replication** because each parental DNA strand serves as a template for a new strand. In the semiconservative mode of replication, each parental DNA strand serves as a template for one new strand, and as each new strand is formed, it is hydrogen-bonded to its parental template. As replication proceeds, the parental double helix unwinds and then rewinds again into two new double helices, each of which contains one originally parental strand and one newly formed daughter strand.

In theory, DNA could be replicated by a number of mechanisms other than the semiconservative

mode. However, the reality of semiconservative replication was demonstrated experimentally by Matthew Meselson and Franklin Stahl in 1958. Their experiment made use of a newly developed high-speed centrifuge (an *ultracentrifuge*) that could spin a solution so fast that molecules differing only slightly in density could be separated. In their experiment, the heavy ^{15}N isotope of nitrogen was used for physical separation of parental and daughter DNA molecules. DNA isolated from the bacterium *E. coli* grown in a medium containing ^{15}N as the only available source of nitrogen is denser than DNA from bacteria grown in media with the normal ^{14}N isotope. These DNA molecules can be separated in an ultracentrifuge because they have about the same density as a very concentrated solution of cesium chloride (CsCl).

When a CsCl solution containing DNA is centrifuged at high speed, the Cs^+ ions gradually sediment toward the bottom of the centrifuge tube. This movement is counteracted by diffusion (the random movement of molecules), which prevents complete sedimentation. At equilibrium, a linear gradient of increasing CsCl concentration—and of density—is present from the top to the bottom of the centrifuge tube. The DNA also moves upward or downward in the tube to a position in the gradient at which the density of the solution is equal to its own density. At equilibrium, a mixture of ^{14}N-containing ("light") and ^{15}N-containing ("heavy") *E. coli* DNA will separate into two distinct zones in a density gradient even though they differ only slightly in density. It is for this reason that the separation technique is called *equilibrium density-gradient centrifugation.*

The Meselson–Stahl experiment is a textbook example of hypothesis-driven science. In other words, they had a hypothesis for the mechanisms of DNA replication (the Watson–Crick model), derived predictions of this model that would distinguish it from other alternatives, and then carried out an experiment to learn whether the predictions would be verified or falsified. The predictions they derived are illustrated in Figure 6.9. They imagined a situation in which bacteria were grown for many generations in a ^{15}N-containing medium so that all parental DNA strands would be "heavy." At one point, the cells are transferred to a ^{14}N-containing medium so that newly synthesized DNA strands will be "light." What would happen if duplex DNA were isolated from samples of cells taken from the culture at intervals, and equilibrium density-gradient centrifugation carried out to determine the density of the molecules? With semiconservative replication, the expected result of the experiment is as shown in Figure 6.9. After one round of replication, each duplex should consist of one heavy and one light strand, so all daughter molecules have

intermediate density. After two rounds of replication, the duplexes containing an original parental strand would again be intermediate in density, but now there is an equal number of duplexes consisting of two light strands, so two bands differing in density are expected. After a third round of replication, DNA duplexes of light and intermediate density would again be expected, but in this generation their expected ratio of abundances are 3 : 1, as shown by the ribbon diagrams.

The actual result of the Meselson–Stahl experiment is shown in Figure 6.10. Each photograph shows the image of a centrifuge tube taken in ultraviolet light of wavelength 260 nm (nanometers), which is absorbed by DNA in solution. The positions of the DNA molecules in the density gradient are therefore indicated by the dark bands that absorb the light. Each photograph is oriented such that the bottom of the tube is at the right and the top is at the left. To the right of each photograph is a graph showing the absorbance of the ultraviolet light from the top of the centrifuge tube to the bottom. In each trace, the peaks correspond to the

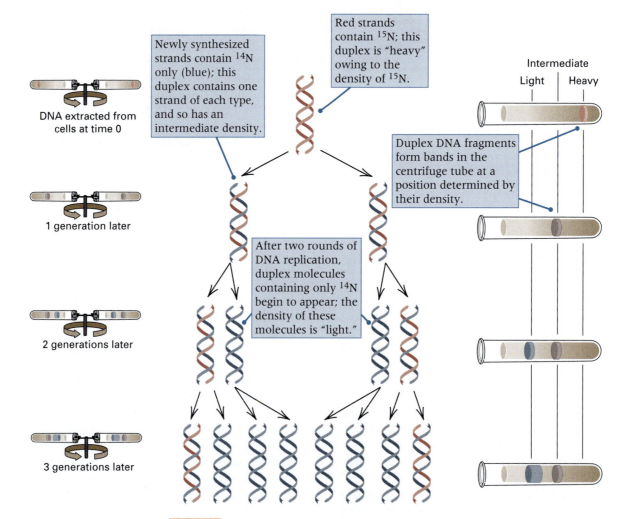

Figure 6.9 Predictions of semiconservative DNA replication.

positions of the bands in the photographs, but the height and width of each peak allow the amount of DNA in each band to be quantified.

At the start of the experiment (generation 0), all of the DNA is heavy (^{15}N). After the transfer to ^{14}N

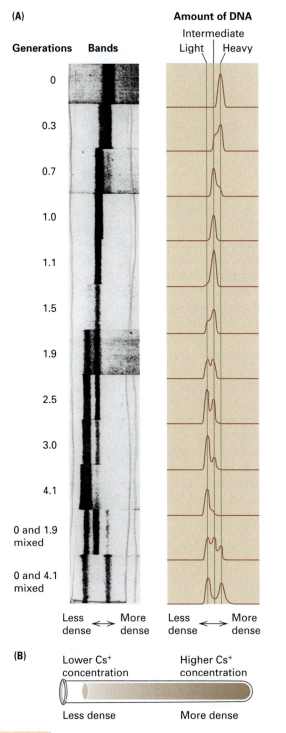

(A)

Generations   Bands

Amount of DNA

Intermediate
Light | Heavy

0
0.3
0.7
1.0
1.1
1.5
1.9
2.5
3.0
4.1
0 and 1.9 mixed
0 and 4.1 mixed

Less dense ←→ More dense   Less dense ←→ More dense

(B)   Lower Cs$^+$ concentration   Higher Cs$^+$ concentration

Less dense   More dense

Figure 6.10 The Meselson–Stahl experiment on DNA replication. (A) Photographs of the centrifuge tubes taken with ultraviolet light, with the centrifuge tubes oriented as shown in part B. The smooth curves in part B show quantitatively the amount of absorption of the ultraviolet light across each tube. [Photograph courtesy of Matthew Meselson. From M. Meselson and F. Stahl. 1958. *Proc. Natl. Acad. Sci. USA* 44: 671.]

medium, a band of lighter density begins to appear and gradually becomes more prominent as the cells replicate their DNA and divide. After 1.0 generations of growth (one round of replication of the DNA molecules and a doubling of the number of cells), all of the DNA had a "hybrid" density exactly intermediate between the densities of ^{15}N-DNA and ^{14}N-DNA. The finding of molecules with a hybrid density indicates that the replicated molecules contain equal amounts of the two nitrogen isotopes. After 1.9 generations of replication in the ^{14}N medium, approximately half of the DNA had the density of DNA with ^{14}N in both strands ("light" DNA), and the other half had the hybrid density. After 3.0 generations, the ratio of light to hybrid DNA was approximately 3 : 1, and after 4.1 generations, it was approximately 7 : 1. This distribution of ^{15}N atoms is precisely the result predicted from semiconservative replication of the Watson–Crick structure, as illustrated in Figure 6.9. Similar experiments with replicating DNA from numerous viruses, bacteria, and higher organisms have also demonstrated semiconservative replication.

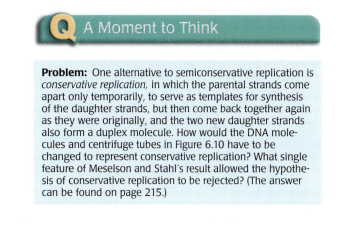

Q A Moment to Think

Problem: One alternative to semiconservative replication is *conservative replication,* in which the parental strands come apart only temporarily, to serve as templates for synthesis of the daughter strands, but then come back together again as they were originally, and the two new daughter strands also form a duplex molecule. How would the DNA molecules and centrifuge tubes in Figure 6.10 have to be changed to represent conservative replication? What single feature of Meselson and Stahl's result allowed the hypothesis of conservative replication to be rejected? (The answer can be found on page 215.)

In the Meselson–Stahl experiment, the DNA was extensively fragmented when isolated, so the form of the molecule was unknown. Later, the isolation of unbroken molecules and their examination by other techniques showed that the DNA in *E. coli* cells is actually circular.

■ **DNA strands must unwind to be replicated.**

The first proof that *E. coli* DNA replicates as a circle came from an experiment in which cells were grown in a medium containing radioactive thymine (^{3}H-thymine) so that all DNA synthesized would be radioactive. The DNA was isolated without fragmentation and placed on photographic film. Each radioactive decay caused a tiny black spot to appear in the film, and after several months there were enough spots to visualize the DNA with a microscope; the pattern of black spots on the film showed the location of the molecule. One of the now-

Figure 6.11 Autoradiogram of the intact replicating chromosome of an *E. coli* cell that has grown in a medium containing ^{3}H-thymine for slightly less than two generations. The continuous lines of dark grains were produced by electrons emitted by decaying ^{3}H atoms in the DNA molecule. The pattern is seen by light microscopy. [From J. Cairns. 1963. *Cold Spring Harbor Symp. Quant. Biol.* 28: 44.]

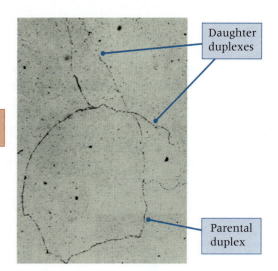

Actual length 1.6 mm (4.6×10^6 base pairs)

Daughter duplexes

Parental duplex

famous images from this experiment appears in Figure 6.11. The actual length of the *E. coli* chromosome is 1.6 mm (4.6 million base pairs). The pattern of black grains on the film traces the path of the circular, replicating molecule. A replicating circle is schematically like the Greek letter θ (theta), so this mode of replication is usually called θ **replication.**

The position along a molecule at which DNA replication begins is called a **replication origin,** and the region in which parental strands are separating and new strands are being synthesized is called a **replication fork.** The process of generating a new replication fork is **initiation.** In most bacteria, bacteriophage, and viruses, DNA replication is initiated at a unique origin of replication. Furthermore, with only a few exceptions, two replication forks move in opposite directions from the origin (Figure 6.12), which means that DNA nearly always replicates bidirectionally.

Some circular DNA molecules, including those of a number of bacterial and eukaryotic viruses, replicate by a process that does not include a θ-shaped intermediate. This replication mode is called **rolling-circle replication.** In this process, replication starts with a single-strand cleavage at a specific sugar–phosphate bond in a double-stranded circle (Figure 6.13). This cleavage produces two chemically distinct ends: a 3′ end (at which the nucleotide has a free 3′-OH group) and a 5′ end (at which the nucleotide has a free 5′-P group). The DNA is synthesized by the addition of successive deoxynucleotides to the 3′ end with simultaneous displacement of the 5′ end from the circle. As replication proceeds around the circle, the 5′ end rolls out as a tail of increasing length.

In most cases, as the tail is extended, a complementary chain is synthesized, which results in a double-stranded DNA tail. Because the displaced

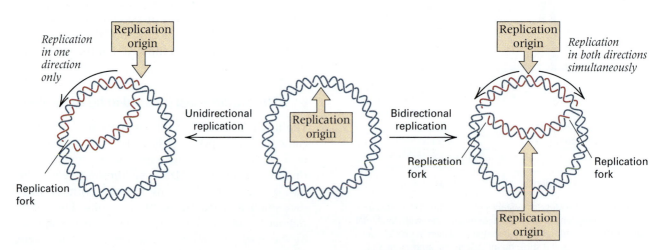

Replication in one direction only

Replication origin

Unidirectional replication

Replication origin

Replication fork

Bidirectional replication

Replication origin

Replication in both directions simultaneously

Replication fork

Replication fork

Replication origin

Figure 6.12 The distinction between unidirectional and bidirectional DNA replication. In unidirectional replication, there is only one replication fork; bidirectional replication requires two replication forks. The curved arrows indicate the direction of movement of the forks. Most DNA replicates bidirectionally.

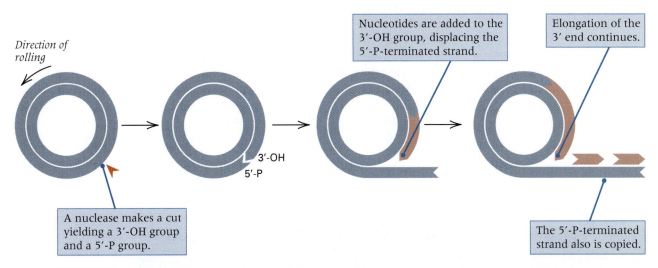

Direction of rolling

A nuclease makes a cut yielding a 3'-OH group and a 5'-P group.

3'-OH
5'-P

Nucleotides are added to the 3'-OH group, displacing the 5'-P-terminated strand.

Elongation of the 3' end continues.

The 5'-P-terminated strand also is copied.

Figure 6.13 Rolling-circle replication. Newly synthesized DNA is in red. The displaced strand forming the "tail" is replicated in short fragments.

A A Moment to Think

Answer to Problem: The accompanying diagram shows the predicted result of conservative replication. Note that a hybrid molecule of one light and one heavy strand is never formed. The finding of molecules of hybrid density allowed the hypothesis of conservative replication to be rejected.

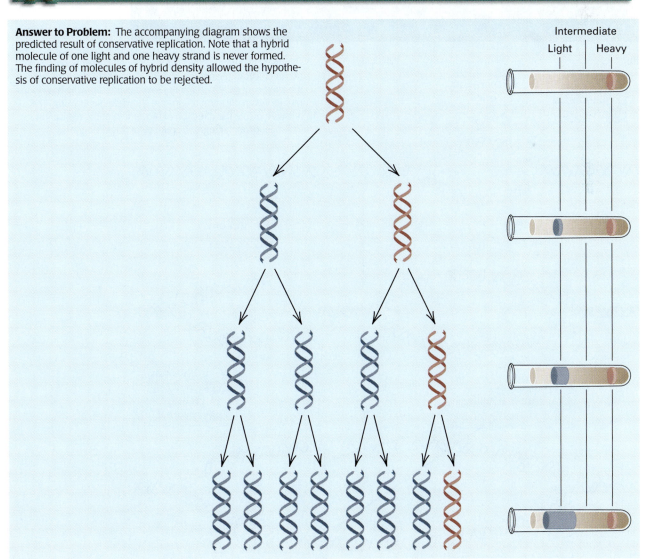

strand is chemically linked to the newly synthesized DNA in the circle, replication does not terminate, and extension proceeds without interruption, forming a tail that may be many times longer than the circumference of the circle. Rolling-circle replication is a common feature in late stages of replication of double-stranded DNA phages that have circular intermediates. An important example of rolling-circle replication will be examined in Chapter 7, where matings between donor and recipient *E. coli* cells are described.

■ Eukaryotic DNA molecules contain multiple origins of replication.

Although the DNA duplex in a eukaryotic chromosome is linear, it also replicates bidirectionally. Replication is initiated almost simultaneously at many sites along the DNA. The structures resulting from the numerous origins are seen in electron micrographs as multiple loops along the DNA molecule (part A of Figure 6.14). Multiple initiation is a means of reducing the total replication time of a

(A)

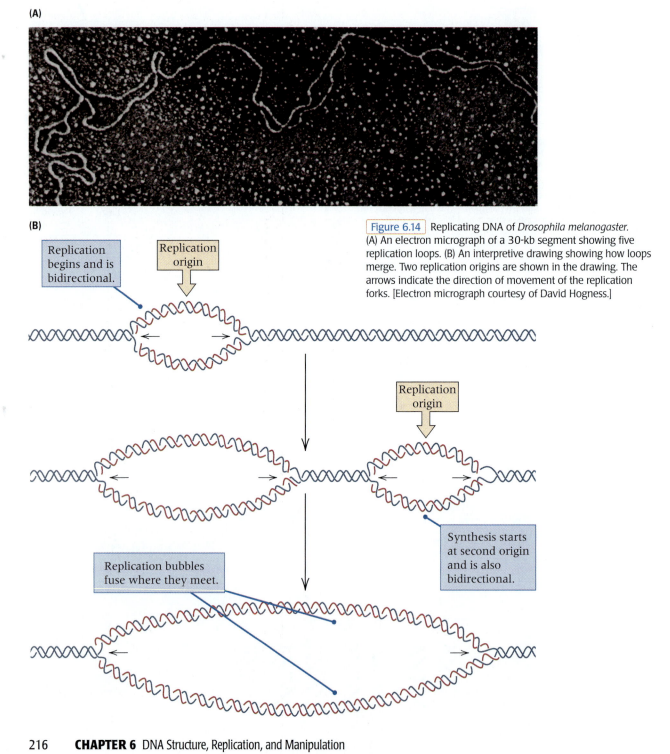

(B)

Replication begins and is bidirectional.

Replication origin

Replication origin

Synthesis starts at second origin and is also bidirectional.

Replication bubbles fuse where they meet.

Figure 6.14 Replicating DNA of *Drosophila melanogaster*. (A) An electron micrograph of a 30-kb segment showing five replication loops. (B) An interpretive drawing showing how loops merge. Two replication origins are shown in the drawing. The arrows indicate the direction of movement of the replication forks. [Electron micrograph courtesy of David Hogness.]

large molecule. In eukaryotic cells, movement of each replication fork proceeds at a rate of approximately 10 to 100 nucleotide pairs per second. For example, in *D. melanogaster,* the rate of replication is about 50 nucleotide pairs per second at 25°C. Because the DNA molecule in the largest chromosome in *Drosophila* contains about 7×10^7 nucleotide pairs, replication from a single bidirectional origin of replication would take about 8 days. Developing *Drosophila* embryos actually use about 8500 replication origins per chromosome, which reduces the replication time to a few minutes. In a typical eukaryotic cell, origins are spaced about 40,000 nucleotide pairs apart, which allows each chromosome to be replicated in 15 to 30 minutes. Because not all chromosomes replicate simultaneously, complete replication of all chromosomes in eukaryotes usually takes from 5 to 10 hours.

So far, we have considered only certain geometrical features of DNA replication. In the next section, the enzymes and other proteins used in DNA replication are described.

6.5

Many proteins participate in DNA replication.

Some of the main molecular players in DNA replication are illustrated in Figure 6.15. Unwinding the double helix to separate the parental strands

requires a **helicase** protein that hydrolyzes ATP to drive the unwinding reaction. Most cells have several helicases specialized for different roles, such as replication, recombination, or repair. Once unwound, the strands of the double helix would tend to come together again spontaneously, so they must be stabilized as single strands to serve as templates for DNA synthesis. This stabilization is a function of a **single-stranded DNA binding protein (SSB).** The SSB binds single-stranded DNA tightly and cooperatively, and it has an affinity for single-stranded DNA at least 1000-fold greater than that for double-stranded DNA. It is this strong tendency for SSB to bind with single strands that stabilizes the templates for replication. In *E. coli,* apparently the same SSB is used in DNA replication, recombination, and repair.

Because the two strands of a replicating helix must make a full rotation to unwind each of the turns, some kind of swivel mechanism must exist to avoid the buildup of so much stress farther along the helix that strand separation would be brought to a halt. In *E. coli,* for example, only about 10 percent of the genome could be replicated before the torsional stress caused by unwinding became too great to continue. The swivel that relieves this stress is an enzyme called **gyrase,** which cleaves both strands of a DNA duplex, swivels the ends of the broken strands to relieve the torsional stress, and then rejoins the strands

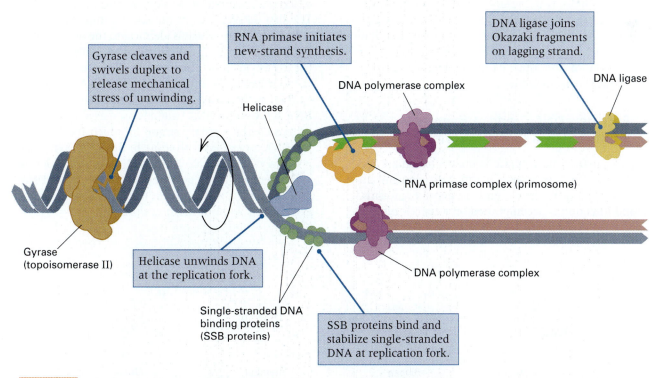

Figure 6.15 Role of some of the key proteins in DNA replication. The DNA polymerase complex and the primase complex are both composed of multiple different polypeptide subunits. The DNA polymerase that joins precursor fragments where they meet, is not shown.

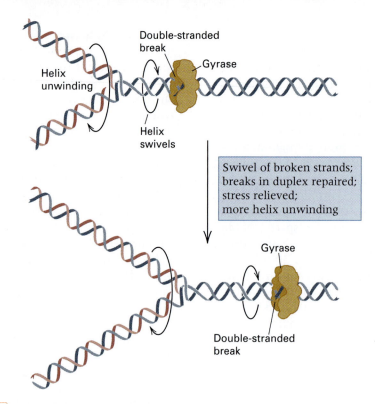

Double-stranded break

Gyrase

Helix unwinding

Helix swivels

Swivel of broken strands; breaks in duplex repaired; stress relieved; more helix unwinding

Gyrase

Double-stranded break

Figure 6.16 | DNA gyrase introduces a double-stranded break ahead of the replication fork and swivels the cleaved ends around the central axis to relieve the stress of helix unwinding.

(Figure 6.16). Enzymes capable of catalyzing breakage and rejoining of DNA strands are known as *topoisomerases*. Gyrase is called a topoisomerase II because it makes a double-stranded break.

As the helix is being unwound by the helicase, the template strands stabilized by SSB, and the torsional stress relieved by the gyrase, the first few nucleotides are synthesized to serve as a *primer* for elongation of the new daughter strands. Primer synthesis is considered next.

■ Each new DNA strand or fragment is initiated by a short RNA primer.

The major DNA polymerase is unable to initiate the synthesis of a new strand; it can only elongate an existing strand at the 3′ end. In most organisms, strand initiation is accomplished by a special type of RNA polymerase. RNA is usually a single-stranded nucleic acid consisting of four types of nucleotides joined together by $3' \rightarrow 5'$ phosphodiester bonds (the same chemical bonds as those in DNA). Two chemical differences distinguish RNA from DNA (Figure 6.17). The first difference is in the sugar component. RNA contains **ribose,** which is identical to the deoxyribose of DNA except for the presence of an $-OH$ group on the 2′ carbon atom. The second difference is in one of the four bases: The thymine found in DNA is replaced by the closely related pyrimidine *uracil* (U) in RNA. In RNA synthesis, a DNA strand is used as a template to form a complementary strand in which the bases in the DNA are paired with those in the RNA. Synthesis is catalyzed by an enzyme called an **RNA polymerase.** RNA polymerases differ from DNA polymerases in that they can initiate the synthesis of RNA chains without needing a primer.

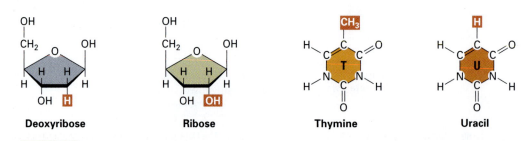

Deoxyribose   Ribose   Thymine   Uracil

Figure 6.17 | Differences between DNA and RNA. The chemical groups in red are the distinguishing features of deoxyribose and ribose and of thymine and uracil.

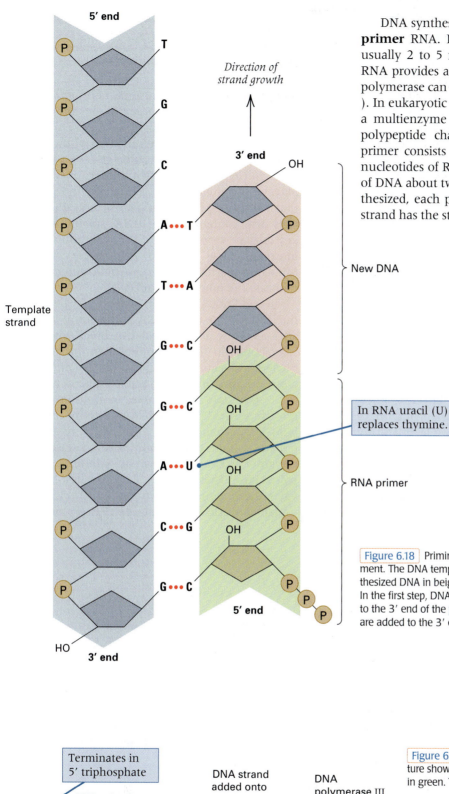

DNA synthesis is initiated by a short stretch of **primer** RNA. Bacterial primers are very short, usually 2 to 5 nucleotides. This short stretch of RNA provides a free 3'-OH onto which the DNA polymerase can add deoxynucleotides (Figure 6.18). In eukaryotic cells, the primer is synthesized by a multienzyme complex composed of 15 to 20 polypeptide chains called a **primosome.** The primer consists of an initial stretch of about 12 nucleotides of RNA to which is attached a stretch of DNA about twice as long. While it is being synthesized, each precursor fragment in the lagging strand has the structure shown in Figure 6.19.

Figure 6.18 Priming of DNA synthesis with an RNA segment. The DNA template strand is shown in blue, newly synthesized DNA in beige. The RNA segment is shown in green. In the first step, DNA polymerase adds a deoxyribonucleotide to the 3' end of the primer. Successive deoxyribonucleotides are added to the 3' end of the growing chain.

Figure 6.19 Each new DNA strand has the structure shown here. The short stretch of RNA is shown in green. The primer is later removed.

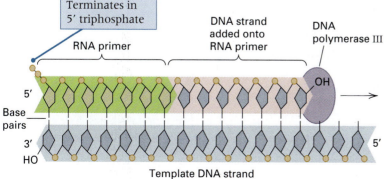

■ DNA polymerase has a proofreading function that corrects errors in replication.

The enzyme **DNA polymerase** forms the sugar–phosphate bond (the phosphodiester bond) between adjacent nucleotides in a new DNA acid chain. The reaction catalyzed by a DNA polymerase is the formation of a phosphodiester bond between the free 3'-OH group of the chain being extended and the innermost phosphorus atom of the nucleoside triphosphate being incorporated at the 3' end (Figure 6.20). What happens is that the 3' hydroxyl group at the 3' terminus of the growing strand attacks the innermost phosphate of the incoming nucleotide and forms a phosphodiester bond, releasing the two outermost phosphates. The result is as follows:

key concept

DNA synthesis proceeds by the elongation of primer chains, *always in the 5' → 3' direction.*

Recognition of the appropriate incoming nucleoside triphosphate in replication depends on base pairing with the opposite nucleotide in the template chain. DNA polymerase usually catalyzes the polymerization reaction that incorporates the new nucleotide at the primer terminus only when the correct base pair is present. The same DNA polymerase is used to add each of the four deoxynucleo-side phosphates to the 3'-OH terminus of the growing strand.

Two DNA polymerases are needed for DNA replication in *E. coli*—DNA polymerase I (abbreviated Pol I) and DNA polymerase III (Pol III). Polymerase III is the major replication enzyme. Pol III exists in the cell as a large protein complex that is responsible not only for the elongation of DNA molecules but also for the initiation of the replication fork at origins of replication and the addition of deoxynucleotides to the RNA primers. Polymerase I plays an essential, but secondary, role in replication that will be described in a later section. Eukaryotic cells also contain several DNA polymerases. The key enzyme responsible for the replication of chromosomal DNA is called polymerase delta (δ). Mitochondria have their own DNA polymerase to replicate the mitochondrial DNA.

In addition to their ability to polymerize nucleotides, the major DNA polymerases also have an *exonuclease* activity that can break phosphodiester bonds in the sugar–phosphate backbones of nucleic acid chains. DNA polymerases I and III of *E. coli* have an exonuclease activity that acts only at the 3' terminus and removes the nucleotide added most recently. This exonuclease activity provides a built-in mechanism for correcting rare errors in polymerization. Occasionally, a polymerase adds to the growing chain a nucleotide with an incorrectly paired base. The presence of an unpaired base activates the exonuclease activity, which cleaves the unpaired nucleotide from the 3'-OH end of the

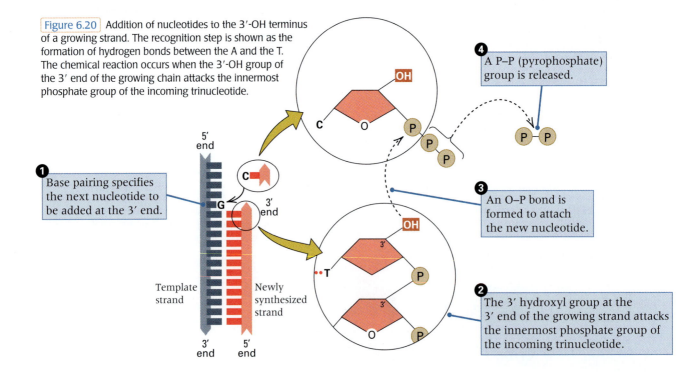

Figure 6.20 Addition of nucleotides to the 3'-OH terminus of a growing strand. The recognition step is shown as the formation of hydrogen bonds between the A and the T. The chemical reaction occurs when the 3'-OH group of the 3' end of the growing chain attacks the innermost phosphate group of the incoming trinucleotide.

❶ Base pairing specifies the next nucleotide to be added at the 3' end.

❷ The 3' hydroxyl group at the 3' end of the growing strand attacks the innermost phosphate group of the incoming trinucleotide.

❸ An O–P bond is formed to attach the new nucleotide.

❹ A P–P (pyrophosphate) group is released.

Sickle-Cell Anemia: The First "Molecular Disease"

Vernon M. Ingram 1957
Cavendish Laboratory, University of Cambridge, England
Gene Mutations in Human Hemoglobin: The Chemical Difference Between Normal and Sickle-Cell Hemoglobin

The mutation in sickle-cell anemia results in a change in the molecular structure of hemoglobin, but what is the nature of this change? Ingram studied various peptide fragments of hemoglobin and found that the only difference resided in a peptide fragment of eight amino acids ("peptide number 4"). To study this fragment further, he used a method of "fingerprinting," in which digests of peptide 4 containing still smaller fragments were resolved into spots on a sheet of filter paper, first by separating the fragments on the basis of charge along one edge of the paper (electrophoresis) and then by separating on the basis of solubility (chromatography) in the other direction. The complete sequence of peptide 4 was deduced after determining the amino acid sequence of each of the short peptides in the fingerprints. In this case, the normal peptide number 4 has the amino acid sequence

Val–His–Leu–Thr–Pro–<u>Glu</u>–Glu–Lys

(V–H–L–T–P–<u>E</u>–E–K in the single-letter codes), whereas that from sickle-cell hemoglobin has the sequence

Val–His–Leu–Thr–Pro–<u>Val</u>–Glu–Lys

(V–H–L–T–P–<u>V</u>–E–K). The only difference is in the underlined amino acid. This was the first evidence that genes may code for polypeptides in a relatively simple manner, in which successive bits of DNA sequence encode successive amino acids in the polypeptide chain. (There were a few minor errors in Ingram's peptide sequences; they have been corrected here.)

• •

I reported [the previous year] that the globins of normal and sickle-cell anemia differed only in a small portion of their polypeptide chains. I have now found that out of nearly 300 amino acids in the two proteins, only one is different; one of the glumatic acid residues of normal hemoglobin is replaced by a valine residue in sickle-cell anemia hemoglobin. [These results] show, for the first time, that the effect of a single gene mutation is a change in one amino acid of the hemoglobin polypeptide. . . . Tryptic digests of the two proteins . . . were separated on a sheet of paper, using electrophoresis in one direction and chromatography in the other. . . . All peptides had identical chromatographic properties, except for one spot, peptide number 4. . . . Partial hydrolysis of this peptide . . . followed by "fingerprinting" gave the products indicated in the accompanying figure.

> The difference consists in a replacement of only one of nearly 300 amino acids— a very small change indeed.

The only difference found between the two peptides is that the first glutamic acid residue (E) in normal hemoglobin peptide is replaced by valine (V) in the hemoglobin S peptide. . . . Sickle-cell anemia is an example of a "molecular disease" that is due to an alteration in the structure of a large protein molecule. . . . The difference consists in a replacement of only one of nearly 300 amino acids—a very small change indeed. . . . The results presented are certainly what one would expect on the basis of the widely accepted hypothesis of gene action; the sequence of base pairs along the chain of nucleic acid provides the information which determines the sequence of amino acids in the polypeptide chain for which the particular gene is responsible. A substitution in the nucleic acid leads to a substitution in the polypeptide.

Source: Nature 180: 326–328.

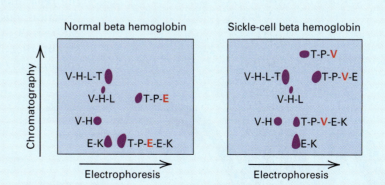

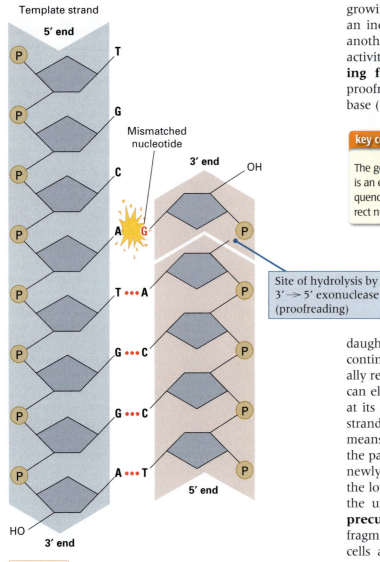

Template strand

5' end

Mismatched nucleotide

3' end   OH

A ⚡ G

Site of hydrolysis by 3' → 5' exonuclease (proofreading)

T ••• A

G ••• C

G ••• C

A ••• T

5' end

HO

3' end

Figure 6.21 The 3'-to-5' exonuclease activity of the proofreading function. The growing strand is cleaved to release a nucleotide containing the base G, which does not pair with the base A in the template strand.

growing chain (Figure 6.21). Because it cleaves off an incorrect nucleotide and gives the polymerase another chance to get it right, the exonuclease activity of DNA polymerase is also called the **editing function** or **proofreading function.** The proofreading function can "look back" only one base (the one added last). Nevertheless,

key concept

The genetic significance of the proofreading function is that it is an error-correcting mechanism that serves to reduce the frequency of mutation resulting from the incorporation of incorrect nucleotides in DNA replication.

■ One strand of replicating DNA is synthesized in pieces.

In the model of replication suggested by Watson and Crick (see Figure 6.8), both daughter strands were supposed to be replicated as continuous units. No known DNA molecule actually replicates in this way. Because DNA polymerase can elongate a newly synthesized DNA strand only at its 3' end, within a single replication fork both strands grow in the 5' → 3' orientation, which means that they grow in opposite directions along the parental strands (Figure 6.22). One strand of the newly made DNA is synthesized continuously (in the lower fork in Figure 6.22). The other strand (in the upper fork in Figure 6.22) is made in small **precursor fragments.** The size of the precursor fragments is 1000–2000 base pairs in prokaryotic cells and 100–200 base pairs in eukaryotic cells. Because synthesis of the discontinuous strand is initiated only at intervals, there is always at least one single-stranded region of the parental strand present on one side of the replication fork. Figure 6.22 also makes it clear that the 3'-OH terminus of the continuously replicated strand is always closer to the replication fork than the 5'-P terminus of the discontinuously replicated strand; this is the physical basis of the terms **leading strand** and **lagging strand** that are used for the continuously and discontinuously replicating strands, respectively.

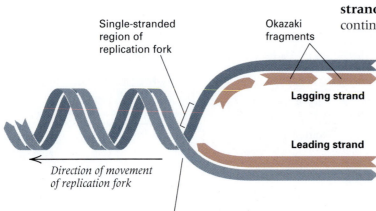

Single-stranded region of replication fork

Okazaki fragments

Lagging strand

Leading strand

Direction of movement of replication fork

Replication fork

Figure 6.22 Short fragments in the replication fork. For each tract of base pairs, the lagging strand is synthesized later than the leading strand.

■ Precursor fragments are joined together when they meet.

The precursor fragments are ultimately joined to yield a continuous strand of DNA. This strand contains no RNA sequences, so the final stitching together of the lagging strand must require

- Removal of the RNA primer
- Replacement with a DNA sequence
- Joining where adjacent DNA fragments come into contact

Primer removal and replacement in *E. coli* is accomplished by a special DNA polymerase (Pol I) that removes one ribonucleotide at a time through its exonuclease activity and replaces it with a deoxyribonucleotide through its polymerase activity. In eukaryotes, the primer RNA is removed as an intact unit (Figure 6.23). When the polymerase complex meets the RNA of the next precursor fragment in line (part A), a protein called RPA (replication protein A) joins the complex. RPA is a single-stranded DNA binding protein that unwinds the RNA and a short segment of DNA from the double helix and

Figure 6.23 Sequence of events in the joining of adjacent precursor fragments in eukaryotes.

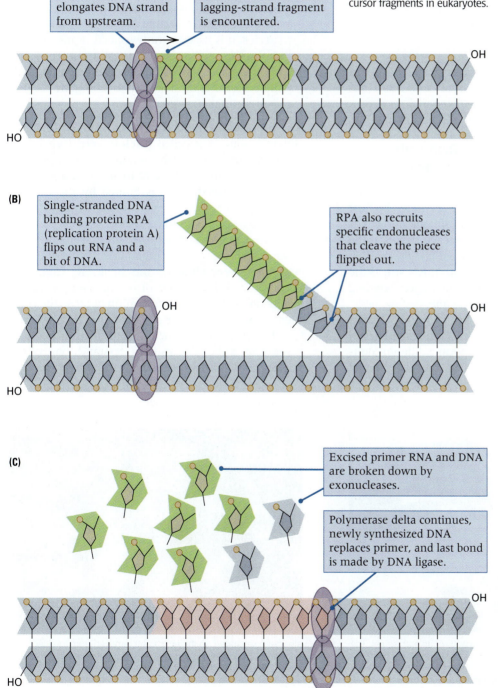

(A)
DNA polymerase delta elongates DNA strand from upstream.

RNA primer for previous lagging-strand fragment is encountered.

(B)
Single-stranded DNA binding protein RPA (replication protein A) flips out RNA and a bit of DNA.

RPA also recruits specific endonucleases that cleave the piece flipped out.

(C)
Excised primer RNA and DNA are broken down by exonucleases.

Polymerase delta continues, newly synthesized DNA replaces primer, and last bond is made by DNA ligase.

stabilizes the unwound single strand by binding to it (Figure 6.22, part B). RPA also recruits endonucleases that cleave the unwound single strand from the double helix, and these also cleave the bond connecting the RNA and DNA stretches in the excised segment. The polymerase complex then replaces the excised segment with DNA nucleotides, and the enzyme **DNA ligase** catalyzes the formation of the final bond connecting the two precursor fragments (part C). As this is happening the RNA and DNA components of the excised segment are broken down by enzymes and the nucleotides are recycled.

6.6

Knowledge of DNA structure makes possible the manipulation of DNA molecules.

This and the following sections show how our knowledge of DNA structure and replication has been put to practical use in the development of procedures for the isolation and manipulation of DNA.

■ Single strands of DNA or RNA with complementary sequences can hybridize.

One of the most important features of DNA is that the two strands of a duplex can be separated by heat without breaking any of the phosphodiester bonds that join successive nucleotides in each strand. If the temperature is maintained sufficiently high, random molecular motion will keep the strands apart. If the temperature is lowered so that hydrogen bonding between complementary base sequences is stable, then under the proper conditions, two single strands that are complementary or nearly complementary in sequence can come together to form a different double helix. The separation of DNA strands is called **denaturation,** and the coming together **renaturation.** The practical applications of denaturation and renaturation are many:

- A small part of a DNA fragment can be renatured with a much larger DNA fragment. This principle is used in identifying specific DNA fragments in a complex mixture.

- A DNA fragment from one gene can be renatured with similar fragments from other genes in the same genome; this principle is used to identify genes that are similar, but not identical, in sequence and that have related functions.

- A DNA fragment from one species can be renatured with similar sequences from other species. This allows the isolation of genes that have the same or related functions in multiple species. It is used to study aspects of molecular evolution, such as how differences in sequence are correlated with differences in function, and the patterns and rates of change in gene sequences as they evolve.

The process of renaturating DNA strands from two different sources is called **nucleic acid hybridization** because the double-stranded molecules are "hybrid." The initial phase of hybridization is a slow process because the rate is limited by the random chance that a region of two complementary strands will come together to form a short sequence of correct base pairs. This initial pairing step is followed by a rapid pairing of the remaining complementary bases and rewinding of the helix. Rewinding is accomplished in a matter of seconds, and its rate is independent of DNA concentration because the complementary strands have already found each other.

The example of nucleic acid hybridization in Figure 6.24 will enable us to understand some of the molecular details and also to see how hybridization

Courtesy of Dr. S. Korsmeyer/National Cancer Institute.

FRAGMENTS OF DNA being placed in the wells of an ararose gel. The solution contains a purple dye that moves faster than any of the DNA fragments. When the dye has migrated toward the far edge of the gel, it indicates that the DNA fragments behind the dye have

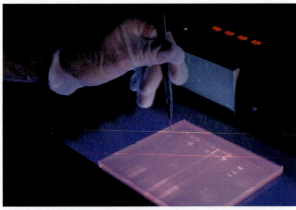

Courtesy of James Gathany/CDC.

also been separated, so the gel can be removed from the electrophoresis apparatus in preparation for Southern blotting or other procedures. In the gel at the right a fluorescent dye has also been added to make the DNA bands visible with ultraviolet light.

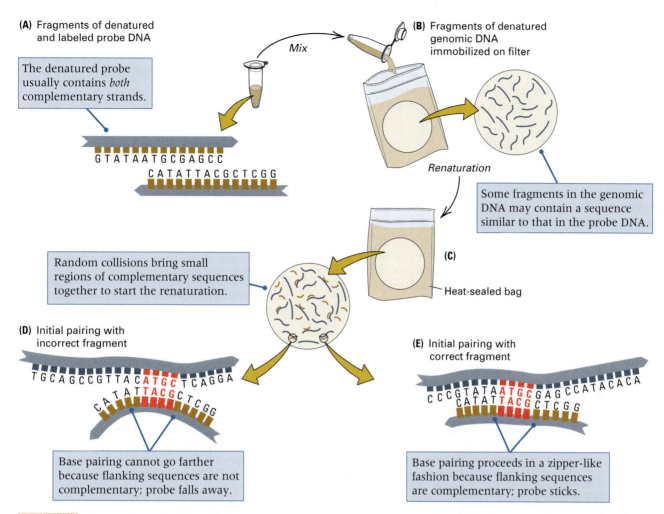

(A) Fragments of denatured and labeled probe DNA

The denatured probe usually contains *both* complementary strands.

G T A T A A T G C G A G C C
C A T A T T A C G C T C G G

Mix

(B) Fragments of denatured genomic DNA immobilized on filter

Renaturation

Some fragments in the genomic DNA may contain a sequence similar to that in the probe DNA.

(C)

Heat-sealed bag

Random collisions bring small regions of complementary sequences together to start the renaturation.

(D) Initial pairing with incorrect fragment

T G C A G C C G T T A C A T G C T C A G G A
C A T A T T A C G C T C G G

Base pairing cannot go farther because flanking sequences are not complementary; probe falls away.

(E) Initial pairing with correct fragment

C C C G T A T A A T G C G A G C C A T A C A C A
C A T A T T A C G C T C G G

Base pairing proceeds in a zipper-like fashion because flanking sequences are complementary; probe sticks.

Figure 6.24 Nucleic acid hybridization. (A) Duplex molecules of probe DNA (obtained from a clone) are denatured and (B) placed in contact with a filter to which is attached denatured strands of genomic DNA. (C) Under the proper conditions of salt concentration and temperature, short complementary stretches come together by random collision. (D) If the sequences flanking the paired region are not complementary, then the pairing is unstable and the strands come apart again. (E) If the sequences flanking the paired region are complementary, then further base pairing stabilizes the renatured duplex.

is used to "tag" and identify a particular DNA fragment. Shown in part A is a solution of denatured DNA, called the **probe**, in which each molecule has been labeled with either radioactive atoms or light-emitting molecules. Probe DNA usually contains denatured forms of both strands present in the original duplex molecule. Part B in Figure 6.24 is a diagram of genomic DNA fragments that have been immobilized on a nitrocellulose filter. When the probe is mixed with the genomic fragments (part C), random collisions bring short, complementary stretches together. If the region of complementary sequence is short (part D), then random collision cannot initiate renaturation because the flanking sequences cannot pair; in this case the probe falls off almost immediately. If, however, a collision brings short sequences together in the correct register (part E), then this initiates renaturation, because the pairing proceeds zipper-like from the initial contact. The main point is that DNA fragments are able to hybridize only if the length of the region in which

they can pair is sufficiently long. Some mismatches in the paired region can be tolerated. How many mismatches are allowed is determined by the conditions of the experiment: The lower the temperature at which the hybridization is carried out, and the higher the salt concentration, the greater the proportion of mismatches that are tolerated.

■ **Restriction enzymes cleave duplex DNA at particular nucleotide sequences.**

One of the problems with breaking large DNA molecules into smaller fragments by random shearing is that the fragments containing a particular gene, or part of a gene, will all be of different sizes. With random shearing, because of the random length of each fragment, it is not possible to isolate and identify a *particular* DNA fragment. However, there is an important enzymatic technique, described in this section, that can be used for cleaving DNA molecules at specific sites.

As we saw in Chapter 4, members of a class of enzymes knows as **restriction enzymes** or, more specifically, as *restriction endonucleases*, are able to cleave DNA molecules at the positions at which particular, short sequences of bases are present. For example, the enzyme *Bam*HI recognizes the double-stranded sequence

$$5'-GGATCC-3'$$
$$3'-CCTAGG-5'$$

and cleaves each strand between the G-bearing nucleotides shown in red. Figure 6.25 shows the recognition sequence for *Bam*HI and the cleavage reaction that takes place.

Table 6.2 lists six of the several hundred restriction enzymes that are known. Most restriction enzymes are isolated from bacteria, and they are named after the species in which they were found. *Bam*HI, for example, was isolated from *Bacillus amyloliquefaciens* strain H, and it is the first (I) restriction enzyme isolated from this organism. Most restriction enzymes recognize only one short base sequence, usually four or six nucleotide pairs. The enzyme binds with the DNA at these sites and makes a break in each strand of the DNA molecule, producing 3'-OH and 5'-P groups at each position (Figure 6.25). The nucleotide sequence recognized for cleavage by a restriction enzyme is called the **restriction site** of the enzyme. The restriction enzymes in Table 6.2 all cleave their restriction site asymmetrically (at a different site on the two DNA strands), but some restriction enzymes cleave symmetrically (at the same site in both strands). The former leave *sticky ends* because each end of the cleaved site has a small, single-stranded overhang that is complementary in base sequence to the other end. In contrast, enzymes that have symmetrical cleavage sites yield DNA fragments that have *blunt ends*.

In virtually all cases, the restriction site of a restriction enzyme reads the same on both strands, provided that the opposite polarity of the strands is taken into account; for example, each strand in the restriction site of *Bam*HI reads 5'-GGATCC-3'. A DNA sequence with this type of symmetry is called a *palindrome*. (In ordinary English, a palindrome is a word or phrase that reads the same forward and backward; an example "madam.")

Restriction enzymes have the following important characteristics:

- Most restriction enzymes recognize a single restriction site.
- The restriction site is recognized without regard to the source of the DNA.
- Because most restriction enzymes recognize a unique restriction-site sequence, the number of cuts in the DNA from a particular organism is determined by the number of restriction sites that are present.

The DNA fragment produced by a pair of adjacent cuts in a DNA molecule is called a **restriction fragment.** A large DNA molecule will typically be cut into many restriction fragments of different sizes. For example, an *E. coli* DNA molecule, which contains 4.6×10^6 base pairs, is cut into several hundred to several thousand fragments, and mammalian nuclear DNA is cut into more than a million fragments.

Because of the sequence specificity, *a particular restriction enzyme produces a unique set of fragments for a particular DNA molecule.* Another enzyme will produce a different set of fragments from the same DNA molecule. Part A of Figure 6.26 shows the sites of cutting of *E. coli* phage λ DNA by the enzymes *Eco*RI and *Bam*HI. A map showing the unique sites of cutting of the DNA of a particular organism by a single enzyme is called a **restriction map.** The family of fragments

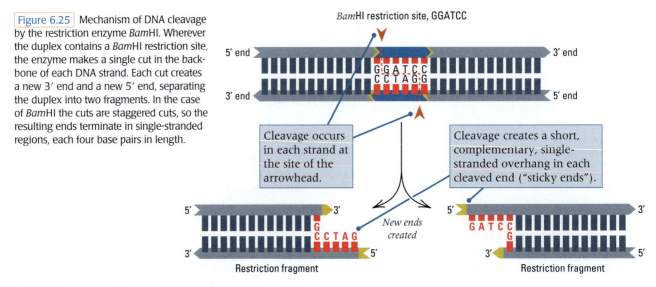

Figure 6.25 Mechanism of DNA cleavage by the restriction enzyme *Bam*HI. Wherever the duplex contains a *Bam*HI restriction site, the enzyme makes a single cut in the backbone of each DNA strand. Each cut creates a new 3' end and a new 5' end, separating the duplex into two fragments. In the case of *Bam*HI the cuts are staggered cuts, so the resulting ends terminate in single-stranded regions, each four base pairs in length.

Table 6.2

Some restriction endonucleases, their sources, and their cleavage sites

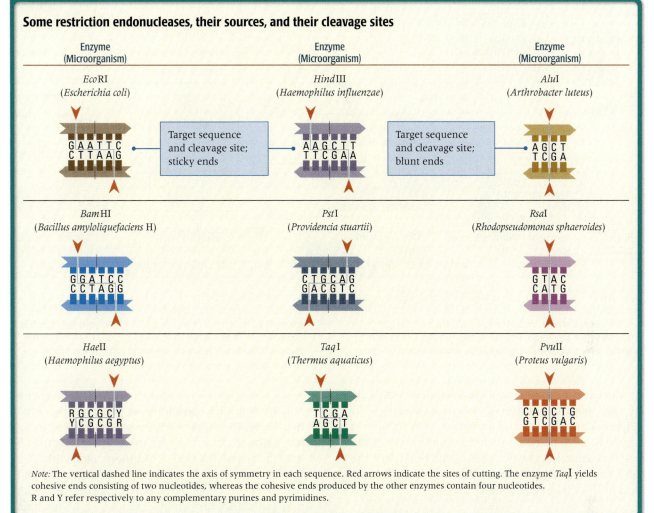

| Enzyme (Microorganism) | | Enzyme (Microorganism) | | Enzyme (Microorganism) |

*Eco*RI (*Escherichia coli*) — G A A T T C / C T T A A G

*Hind*III (*Haemophilus influenzae*) — A A G C T T / T T C G A A

*Alu*I (*Arthrobacter luteus*) — A G C T / T C G A

Target sequence and cleavage site; sticky ends

Target sequence and cleavage site; blunt ends

*Bam*HI (*Bacillus amyloliquefaciens* H) — G G A T C C / C C T A G G

*Pst*I (*Providencia stuartii*) — C T G C A G / G A C G T C

*Rsa*I (*Rhodopseudomonas sphaeroides*) — G T A C / C A T G

*Hae*II (*Haemophilus aegyptus*) — R G C G C Y / Y C G C G R

*Taq*I (*Thermus aquaticus*) — T C G A / A G C T

*Pvu*II (*Proteus vulgaris*) — C A G C T G / G T C G A C

Note: The vertical dashed line indicates the axis of symmetry in each sequence. Red arrows indicate the sites of cutting. The enzyme *Taq*I yields cohesive ends consisting of two nucleotides, whereas the cohesive ends produced by the other enzymes contain four nucleotides. R and Y refer respectively to any complementary purines and pyrimidines.

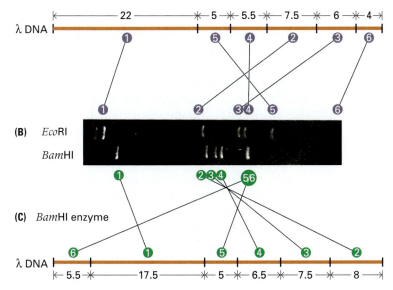

(A) *Eco*RI enzyme

λ DNA — |← 22 →|← 5 →|← 5.5 →|← 7.5 →|← 6 →|← 4 →|

(B) *Eco*RI / *Bam*HI

(C) *Bam*HI enzyme

λ DNA — |← 5.5 →|← 17.5 →|← 5 →|← 6.5 →|← 7.5 →|← 8 →|

Figure 6.26 (A, C) Restriction maps of λ DNA for two restriction enzymes, *Eco*RI (A) and *Bam*HI (C). The vertical bars indicate the sites of cutting. The numbers indicate the approximate length of each fragment in kilobase pairs. (B) An electrophoresis gel of *Bam*HI and *Eco*RI enzyme digests of λ DNA. Numbers indicate fragments in order from largest (1) to smallest (6); the circled numbers on the maps correspond to the numbers beside the gel. The DNA has not undergone electrophoresis long enough to separate bands 5 and 6 of the *Bam*HI digest.

6.6 Knowledge of DNA structure makes possible the manipulation of DNA molecules   227

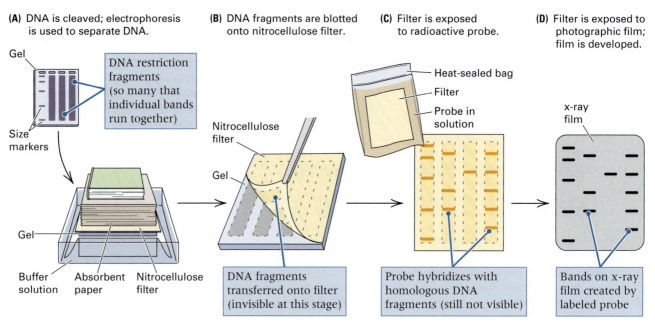

(A) DNA is cleaved; electrophoresis is used to separate DNA.

(B) DNA fragments are blotted onto nitrocellulose filter.

(C) Filter is exposed to radioactive probe.

(D) Filter is exposed to photographic film; film is developed.

Gel

DNA restriction fragments (so many that individual bands run together)

Size markers

Nitrocellulose filter

Gel

Gel

Buffer solution / Absorbent paper / Nitrocellulose filter

Heat-sealed bag
Filter
Probe in solution

x-ray film

DNA fragments transferred onto filter (invisible at this stage)

Probe hybridizes with homologous DNA fragments (still not visible)

Bands on x-ray film created by labeled probe

Figure 6.27 Southern blot. (A, B) DNA restriction fragments are separated by electrophoresis, blotted from the gel onto a nitrocellulose or nylon filter, and chemically attached by the use of ultraviolet light. (C) The strands are denatured and mixed with radioactive or light-sensitive probe DNA, which binds with complementary sequences present on the filter. The bound probe remains, whereas unbound probe washes off. (D) Bound probe is revealed by darkening of photographic film placed over the filter. The positions of the bands indicate which restriction fragments contain DNA sequences homologous to those in the probe.

produced by a single enzyme can be detected easily by gel electrophoresis of enzyme-treated DNA (Figure 6.26, part B), and particular DNA fragments can be isolated by cutting out the small region of the gel that contains the fragment and removing the DNA from the gel. Gel electrophoresis for the separation of DNA fragments is described next.

■ Specific DNA fragments are identified by hybridization with a probe.

Several techniques enable a researcher to locate a particular DNA fragment in a gel. One of the most generally applicable procedures is the **Southern blot** (Figure 6.27). In this procedure, a gel in which DNA molecules have been separated by electrophoresis is placed in contact with a sheet of nitrocellulose or other suitable membrane in such a way that the DNA is transferred with the relative positions of the DNA bands maintained (part B). Then the DNA is treated to denature into single strands. The nitrocellulose, to which the single-stranded DNA binds tightly, is then exposed to RNA or DNA (the probe) that has been labeled with either radioactive or light-emitting molecules. The mixing is performed under conditions that lead complementary strands to hybridize to form duplex molecules (part C). The radioactive or light-emitting probe becomes stably bound (resistant to removal by washing) to the DNA only at positions at which base sequences complementary to the probe are present, so that duplex molecules can form. The label is located by placing the paper in contact with

x-ray film; after development of the film, blackened regions indicate positions at which the probe hybridized to complementary sequences. For example, a cloned DNA fragment from one species may be used as probe DNA in a Southern blot with DNA from another species; the probe will hybridize only with restriction fragments containing DNA sequences that are sufficiently complementary to allow stable duplexes to form.

6.7

The polymerase chain reaction makes possible the amplification of a particular DNA fragment.

It is also possible to obtain large quantities of a particular DNA sequence merely by selective replication. The method for selective replication is called the **polymerase chain reaction (PCR),** and it uses DNA polymerase and a pair of short, synthetic oligonucleotides, usually about 20 to 30 nucleotides in length, that are complementary in sequence to the ends of the DNA sequence to be amplified and so can serve as primers for strand elongation. Starting with a mixture containing as little as one molecule of the fragment of interest, repeated rounds of DNA replication increase the number of molecules exponentially. For example, starting with a single molecule, 25 rounds of DNA replication will result in $2^{25} = 3.4 \times 10^7$ molecules. This number of molecules of the amplified fragment is so much greater than that of the other unamplified molecules in the original mixture that the amplified DNA can often be used without further purification. For example, a single fragment of 3000 base pairs in *E. coli* accounts for only 0.06 percent of the total DNA in this organism. However, if this single fragment were replicated through 25 rounds of replication, 99.995 percent of the resulting mixture would consist of the amplified sequence.

An outline of the polymerase chain reaction is shown in Figure 6.28. The DNA sequence to be amplified and the oligonucleotide sequences are shown in contrasting colors. The oligonucleotides act as primers for DNA replication because they

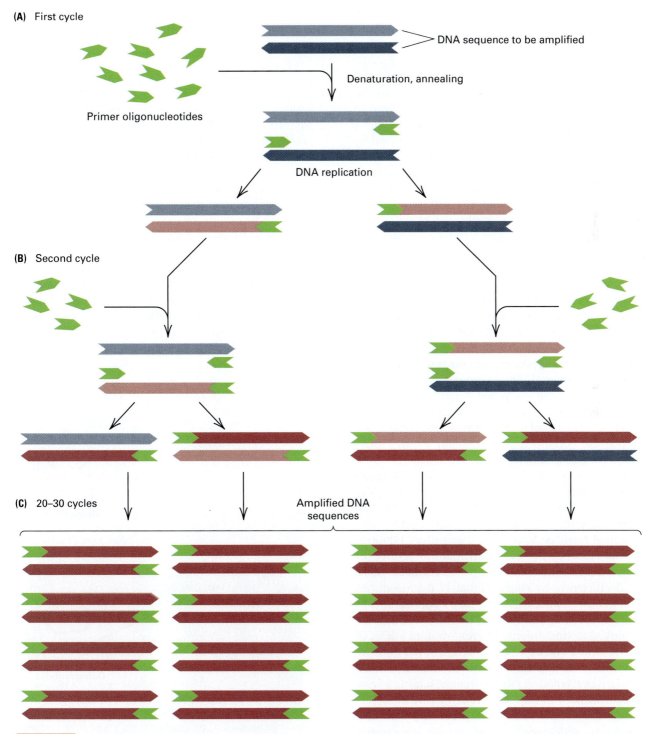

Figure 6.28 Polymerase chain reaction (PCR) for amplification of particular DNA sequences. Only the region to be amplified is shown. Oligonucleotide primers (green) that are complementary to the ends of the target sequence (blue) are used in repeated rounds of denaturation, annealing, and DNA replication. Newly replicated DNA is shown in pink. The number of copies of the target sequence doubles in each round of replication, eventually overwhelming any other sequences that may be present.

anneal to the ends of the sequence to be amplified and become the substrates for chain elongation by DNA polymerase. In the first cycle of PCR amplification, the DNA is denatured to separate the strands. The denaturation temperature is usually around 95°C. Then the temperature is decreased to allow annealing in the presence of a vast excess of the primer oligonucleotides. The annealing temperature is typically in the range of 50°C to 60°C, depending largely on the G + C content of the oligonucleotide primers. The temperature is raised slightly, to about 70°C, for the elongation of each primer. The first cycle in PCR produces two copies of each molecule containing sequences complementary to the primers. The second cycle of PCR is similar to the first. The DNA is denatured and then renatured in the presence of an excess of primer oligonucleotides, whereupon the primers are elongated by DNA polymerase; after this cycle there are four copies of each molecule present in the original mixture. The steps of denaturation, renaturation, and replication are repeated from 20 to 30 times, and in each cycle, the number of molecules of the amplified sequence is doubled. The theoretical result of 25 rounds of amplification is 2^{25} copies of each template molecule present in the original mixture.

Implementation of PCR with conventional DNA polymerases is not practical, because at the high temperature necessary for denaturation, the polymerase is itself irreversibly unfolded and becomes inactive. However, DNA polymerase isolated from certain bacteria is heat-stable because the organisms normally live in hot springs at temperatures well above 90°C, such as are found in Yellowstone National Park. These organisms are said to be *thermophiles*. The most widely used heat-stable DNA polymerase is called *Taq* polymerase because it was originally isolated from the thermophilic bacterium *Thermus aquaticus.*

PCR amplification is very useful for generating large quantities of a specific DNA sequence. The principal limitation of the technique is that the DNA sequences at the ends of the region to be amplified must be known so that primer oligonucleotides can be synthesized. In addition, sequences longer than about 5000 base pairs cannot be replicated efficiently by conventional PCR procedures. On the other hand, there are many applications in which PCR amplification is useful. PCR can be employed to study many different mutant alleles of a gene whose wildtype sequence is known in order to identify the molecular basis of the mutations. Similarly, variation in DNA sequence among alleles present in natural populations can easily be determined using PCR. The PCR procedure has also come into widespread use in clinical laboratories for diagnosis. To take just one very important example, the presence of the human immunodeficiency virus (HIV), which causes acquired immune deficiency syndrome (AIDS), can be detected in trace quantities in blood banks by means of PCR using primers complementary to sequences in the viral genetic material. These and other applications of PCR are facilitated by the fact that the procedure lends itself to automation—for example, the use of mechanical robots to set up the reactions.

6.8

Chemical terminators of DNA synthesis are used to determine the base sequence.

A great deal of information about gene structure and gene expression can be obtained by direct determination of the sequence of bases in a DNA molecule. The most widely used method for base sequencing is described in this section. No technique can determine the sequence of bases in an entire chromosome in a single experiment, and so chromosomes are first cut into fragments of a size that can be sequenced easily. To obtain the sequence of a long stretch of DNA, a set of overlapping fragments must be prepared, the sequence of each is determined, and all sequences are then combined.

The **dideoxy sequencing method** employs DNA synthesis in the presence of small amounts of nucleotides that contain the sugar **dideoxyribose** instead of deoxyribose (see Figure 6.29). Dideoxy-

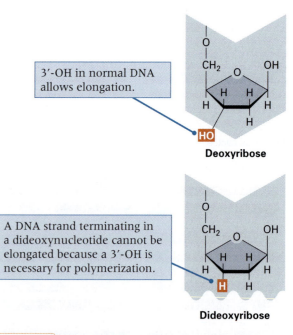

Figure 6.29 Structures of normal deoxyribose and the dideoxyribose sugar used in DNA sequencing. The dideoxyribose has a hydrogen atom (red) attached to the 3' carbon, in contrast with the hydroxyl group (red) at this position in deoxyribose. Because the 3' hydroxyl group is essential for the attachment of the next nucleotide in line in a growing DNA strand, the incorporation of a dideoxynucleotide immediately terminates synthesis.

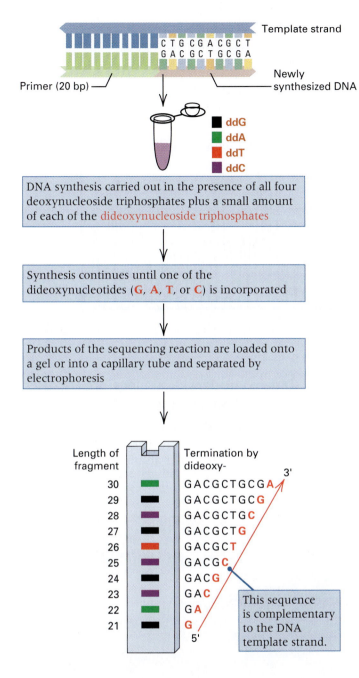

Template strand
C T G C G A C G C T
G A C G C T G C G A

Primer (20 bp)

Newly synthesized DNA

- ddG (black)
- ddA (green)
- ddT (red)
- ddC (purple)

DNA synthesis carried out in the presence of all four deoxynucleoside triphosphates plus a small amount of each of the dideoxynucleoside triphosphates

Synthesis continues until one of the dideoxynucleotides (G, A, T, or C) is incorporated

Products of the sequencing reaction are loaded onto a gel or into a capillary tube and separated by electrophoresis

Length of fragment | Termination by dideoxy-
30 — GACGCTGCGA 3'
29 — GACGCTGCG
28 — GACGCTGC
27 — GACGCTG
26 — GACGCT
25 — GACGC
24 — GACG
23 — GAC
22 — GA
21 — G 5'

This sequence is complementary to the DNA template strand.

Figure 6.30 Dideoxy method of DNA sequencing. Lengths of the terminated DNA fragments are shown at the left of the gel. The sequence of the daughter strand is read from the bottom of the gel according to the color of each band.

lengths of the fragments are determined by the positions in the daughter strand at which a particular dideoxynucleotide was incorporated. The fragments are separated by size using electrophoresis in a gel or capillary tube, and the dideoxy terminator is identified by its fluorescence. The base sequence is determined by the following rule:

key concept

If a fragment containing n nucleotides is generated in the reaction containing a particular dideoxynucleotide (determined by the color of the fluorescent band), then position n in the *daughter strand* is occupied by the base present in the dideoxynucleotide. The numbering is from the 5' nucleotide of the primer.

For example, if a 400-base fragment was terminated by the dideoxy form of dATP, then the 400th base in the daughter strand produced by DNA synthesis must be an adenine (A). Because most native duplex DNA molecules consist of complementary strands, it does not matter whether the sequence of the template strand or the daughter strand is determined. The sequence of the template strand can be deduced from the daughter strand because their nucleotide sequences are complementary. In practice, however, both strands of a molecule are usually sequenced independently and compared in order to minimize errors.

■ **The incorporation of a dideoxynucleotide terminates strand elongation.**

The procedure for sequencing a DNA fragment is diagrammed in Figure 6.30. The sequencing reaction is carried out in the presence of a small amount of fluorescently labeled dideoxynucleotides (G, black; A, green; T, red; C, purple). The products of DNA synthesis are then separated by electrophoresis. In principle, the sequence can be read directly from the gel. Starting at the bottom, the sequence of the newly synthesized strand reads

5'-GACGCTGCGA-3'

However, a substantial improvement in efficiency is accomplished by continuing the electrophoresis until each band, in turn, drops off the bottom of the gel. As each band comes off the bottom of the gel, the fluorescent dye that it contains is excited by laser light, and the color of the fluorescence is read automatically by a photocell and recorded in a

ribose lacks the 3'-OH group, which is essential for attachment of the next nucleotide in a growing DNA strand, so incorporation of a dideoxynucleotide instead of a deoxynucleotide immediately terminates further synthesis of the strand. To sequence a DNA strand, a DNA synthesis reaction is carried out. The reaction mixture contains the single-stranded DNA template to be sequenced, a single oligonucleotide primer complementary to a stretch of the template strand, all four deoxyribonucleoside triphosphates, and a small amount of *each* of the dideoxynucleoside triphosphates, each labeled with a different fluorescent constituent. The reaction produces a set of fragments that terminate at the point at which a dideoxynucleotide was randomly incorporated in place of the normal deoxynucleotide. Therefore, the

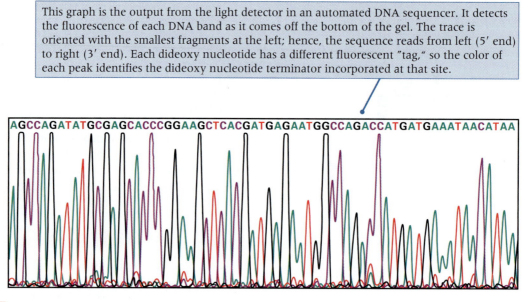

This graph is the output from the light detector in an automated DNA sequencer. It detects the fluorescence of each DNA band as it comes off the bottom of the gel. The trace is oriented with the smallest fragments at the left; hence, the sequence reads from left (5' end) to right (3' end). Each dideoxy nucleotide has a different fluorescent "tag," so the color of each peak identifies the dideoxy nucleotide terminator incorporated at that site.

AGCCAGATATGCGAGCACCCGGAAGCTCACGATGAGAATGGCCAGACCATGATGAAATAACATAA

Figure 6.31 Trace of the fluorescence pattern obtained from a DNA sequencing gel by automated detection of the fluorescence of each band as it comes off the bottom of the gel during continued electrophoresis.

computer. Figure 6.31 is a trace of the fluorescence pattern that might emerge at the bottom of a gel after continued electrophoresis. The nucleotide sequence is read directly from the colors of the alternating peaks along the trace.

■ **Dideoxynucleoside analogs are also used in the treatment of diseases.**

Our knowledge of DNA structure and replication has applications not only in procedures for the manipulation of DNA but also in the development

Courtesy of Monte Later/Yellowstone National Park/NPS.

HOT SPRINGS at Yellowstone National Park in Wyoming, a source of the bacteria *Thermus aquaticus* and the Taq polymerase used in the polymerase chain reaction.

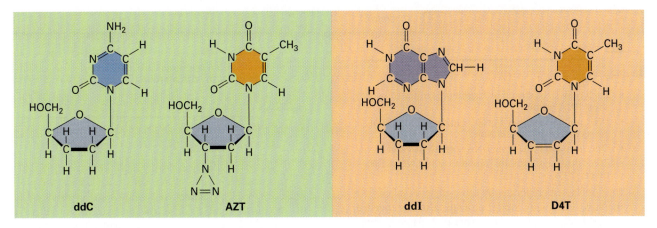

ddC   AZT   ddI   D4T

Figure 6.32 A few of the drugs that have been found to be effective in the treatment of AIDS by interfering with the replication of HIV virus. The technical names of the sub-

stances are as follows: ddC is 2′,3′-dideoxycytidine; AZT is 3′-azido-2′,3′-dideoxythymidine; D4T is 2′,3′-didehydro-2′,3′-dideoxythymidine; and ddI is 2′,3′-dideoxyinosine.

of drugs for clinical use. One approach to the treatment of AIDS offers an example. A number of dideoxynucleoside analogs are effective in inhibiting replication of the viral genetic material. A few of these are illustrated in Figure 6.32. Recall that a *nucleoside* is a base attached to a sugar without a phosphate. A nucleoside *analog* is a molecule similar, but not identical, in structure to a nucleoside. In Figure 6.32, ddC is the normal dideoxyribocytidine nucleoside. It is effective against AIDS, as are the dideoxynucleoside analogs AZT, D4T, and ddI (and

other such analogs). The nucleoside, rather than the nucleotide, is used in therapy because the nucleotide, having a highly charged phosphate group, cannot cross the cell membrane as easily. The drugs that have emerged from our basic knowledge of DNA structure and replication demonstrate that "pure" science may have many unforeseen practical applications. The basic experiments on DNA were carried out long before the recognition of AIDS as a distinct infectious disease and the discovery that HIV virus was the causative agent.

chapter summary

6.1 Genome size can differ tremendously, even among closely related organisms.

DNA content varies widely among species of organisms. Among small viruses the genome size is only a few thousand nucleotides, prokaryotes have genomes on the order of a few million nucleotides, and eukaryotes have genome sizes ranging from 10 to 100,000 megabases. The C-value paradox refers to the fact that among eukaryotes, there is no consistent relationship between genome size and organismic complexity.

6.2 DNA is a linear polymer of four deoxyribonucleotides.

6.3 Duplex DNA is a double helix in which the bases form hydrogen bonds.

DNA is a double-stranded polymer consisting of deoxyribonucleotides. A nucleotide has three components: a base, a sugar (deoxyribose in DNA, ribose in RNA), and a phosphate. Sugars and phosphates alternate in forming a single polynucleotide chain with one terminal 3′-OH group and one termi-

nal 5′-P group. In double-stranded (duplex) DNA, the two strands are antiparallel: Each end of the double helix carries a terminal 3′-OH group in one strand and a terminal 5′-P group in the other strand. Four bases are found in DNA: adenine (A) and guanine (G), which are purines, and cytosine (C) and thymine (T), which are pyrimidines. Equal numbers of purines and pyrimidines are found in double-stranded DNA, because the bases are paired as A−T pairs and G−C pairs. This pairing holds the two polynucleotide strands together in a double helix. The base composition of DNA varies from one organism to the next. The information content of a DNA molecule resides in the sequence of bases along the chain, and each gene consists of a unique sequence.

6.4 Replication uses each DNA strand as a template for a new one.

- Nucleotides are added one at a time to the growing end of a DNA strand.
- DNA replication is semiconservative: The parental strands remain intact.
- DNA strands must unwind to be replicated.
- Eukaryotic DNA molecules contain multiple origins of replication.

chapter summary *continued*

6.5 Many proteins participate in DNA replication.

- Each new DNA strand or fragment is initiated by a short RNA primer.
- DNA polymerase has a proofreading function that corrects errors in replication.
- One strand of replicating DNA is synthesized in pieces.
- Precursor fragments are joined together when they meet.

The double helix replicates by using enzymes called DNA polymerases, but many other proteins also are needed. Replication is semiconservative in that each parental single strand, called a template strand, is found in one of the double-stranded progeny molecules. Semiconservative replication was first demonstrated in the Meselson–Stahl experiment, which used equilibrium density-gradient centrifugation to separate DNA molecules containing two ^{15}N-labeled strands, two ^{14}N-labeled strands, or one of each. Replication proceeds by a DNA polymerase (1) bringing in a nucleoside triphosphate with a base capable of hydrogen-bonding with the corresponding base in the template strand and (2) joining the 5'-P group of the nucleotide to the free 3'-OH group of the growing strand. (The terminal P—P from the nucleoside triphosphate is cleaved off and released.) Because double-stranded DNA is antiparallel, only one strand (the leading strand) grows in the direction of movement of the replication fork. The other strand (the lagging strand) is synthesized in the opposite direction as short fragments that are subsequently joined together. DNA polymerases cannot initiate synthesis, so a primer is always needed. The primer is an RNA fragment made by an RNA polymerase enzyme; the RNA primer is removed at later stages of replication. DNA molecules of prokaryotes usually have a single replication origin; eukaryotic DNA molecules usually have many origins.

6.6 Knowledge of DNA structure makes possible the manipulation of DNA molecules.

- Single strands of DNA or RNA with complementary sequences can hybridize.
- Restriction enzymes cleave duplex DNA at particular nucleotide sequences.

- Specific DNA fragments are identified by hybridization with a probe.

6.7 The polymerase chain reaction makes possible the amplification of a particular DNA fragment.

Restriction enzymes cleave DNA molecules at the positions of specific sequences (restriction sites) of usually four or six nucleotides. Each restriction enzyme produces a unique set of fragments for any particular DNA molecule. These fragments can be separated by electrophoresis and used for purposes such as DNA sequencing. The positions of particular restriction fragments in a gel can be visualized by means of a Southern blot, in which radioactive probe DNA is mixed with denatured DNA made up of single-stranded restriction fragments that have been transferred to a filter membrane after electrophoresis. The probe DNA will form stable duplexes (anneal or renature) with whatever fragments contain sufficiently complementary base sequences, and the positions of these duplexes can be determined. Particular DNA sequences can also be amplified by means of the polymerase chain reaction (PCR), in which short synthetic oligonucleotides are used as primers to replicate and amplify, repeatedly, the sequence between them.

6.8 Chemical terminators of DNA synthesis are used to determine the base sequence.

- The incorporation of a dideoxynucleotide terminates strand elongation.
- Dideoxynucleoside analogs are also used in the treatment of diseases.

The base sequence of a DNA molecule can be determined by dideoxynucleotide sequencing. In this method, the DNA is isolated in discrete fragments. Complementary strands of each fragment are sequenced, and the sequences of overlapping fragments are combined to yield the complete sequence. The dideoxy sequencing method uses dideoxynucleotides to terminate daughter-strand synthesis and reveal the identity of the base present in the daughter strand at the site of termination.

issues & ideas

- What are the four bases commonly found in DNA? Which form base pairs?
- What is the relationship between the amount of DNA in a somatic cell and the amount in a gamete?
- What chemical feature at the 3' end of a DNA strand in the process of being synthesized is essential for elongation? Can the strand also be elongated at the 5' end?
- What does it mean to say that the two strands in duplex DNA are antiparallel?

- If the paired strands in duplex DNA were parallel rather than antiparallel, would replication still involve a leading strand and a lagging strand? Explain.
- Why is the polymerase chain reaction so extremely specific in amplifying a single region of DNA? Why is the technique so extremely powerful in multiplying the sequence?
- What feature of DNA replication guarantees that the incorporation of a dideoxynucleotide will terminate strand elongation?

key terms & concepts

antiparallel
C-value paradox
daughter strand
denaturation
dideoxyribose
dideoxy sequencing method
DNA ligase
DNA polymerase
editing function
5' end
genome
gyrase
helicase

hydrogen bond
initiation
kilobase (kb)
lagging strand
leading strand
megabase (Mb)
nucleic acid hybridization
nucleoside
nucleotide
parental strand
phosphodiester bond
polymerase chain reaction
    (PCR)

polynucleotide chain
precursor fragment
primer
primosome
probe
proofreading function
purine
pyrimidine
renaturation
replication fork
replication origin
restriction enzyme
restriction fragment

restriction map
restriction site
ribose
RNA polymerase
rolling-circle replication
semiconservative replication
single-stranded DNA binding
    protein (SSB)
Southern blot
template
θ replication
3' end

1. _____ The observation that, among eukaryotes, the genome size of a species has no relationship to its metabolic, developmental, or behavioral complexity.

2. _____ Chemical unit in a nucleic acid that consists of a base attached to a sugar without any phosphate groups.

3. _____ What the famous Meselson-Stahl experiment showed.

4. _____ Mode of replication of a circular DNA molecule in which the 5' end of a nicked strand peels off while the 3' end is elongated.

5. _____ Any place along a DNA molecule where replication can be initiated.

6. _____ Short stretch of RNA needed to start the replication of a strand of DNA; also a short stretch of DNA used to initiate synthesis in the polymerase chain reaction.

7. _____ In DNA replication, the strand synthesized in long, continuous stretches.

8. _____ Opposite of the leading strand.

9. _____ Another term for the editing function of DNA polymerase.

10. _____ The ability of separated complementary nucleic acid strands to form duplexes; another term for nucleic acid hybridization.

11. _____ Nucleic acid hybridization in which separated restriction fragments are transferred to a solid supporting material, such as a filter, and then hybridized with a labeled probe to identify the locations of complementary sequences.

12. _____ Method of DNA sequencing based on terminating incomplete strands.

solutions: step by step

Problem 1

A single-stranded DNA molecule is replicated *in vitro* by extending a primer oligonucleotide in the presence of nucleoside triphosphates whose two outermost phosphate groups (called the γ and β phosphates) are labeled with radioactive ^{32}P but whose innermost phosphate (the α phosphate) is not labeled. Is the resulting double-stranded DNA labeled or unlabeled? Explain.

■ Solution Recall that the addition of each new nucleotide in strand elongation requires a nucleoside triphosphate, but in the reaction itself the covalent bond connecting the α and β phosphates is cleaved, releasing the still-connected β and γ phosphates into the medium. This means that none of the ^{32}P from each labeled triphosphate will be incorporated into the phosphodiester bond of the growing DNA strand, so the daughter duplex will be unlabeled.

Problem 2

The polymerase chain reaction is used to amplify a region of human DNA of length 3.2 kb from a DNA solution prepared from nuclei of human cells. The human genome has a size of 3,200 Mb, or 3.2×10^9 base pairs, per haploid genome. (a) Prior to amplification, what proportion of the DNA in the solution consists of the 3.2-kb target sequence? Assume that the target sequence is present in one copy per haploid genome. (b) Each round of amplification doubles the number of target molecules. How many rounds of replication would be required to reach a stage in which the amplified sequence constitutes more than 99.9 percent of all the DNA in the solution?

■ Solution (a) The original DNA solution contains one 3.2-kb target sequence per 3200 Mb haploid genome. The proportion of DNA consisting of the target sequence is therefore

$$\frac{3.2 \times 10^3}{3.2 \times 10^9} = 1 \times 10^{-6} = 0.0001 \text{ percent}$$

(b) Because each round of amplification doubles the number of target molecules, after n rounds of replication there will be 2^n target molecules for each haploid human genome present in the original solution. Each of these has a length of 3200 bp, so the total amount of amplified target DNA will be $2^n \times 3200$ bp. This DNA is newly created and therefore increases the total amount of DNA in the solution. After n rounds of replication, the amount of DNA present per haploid genome is $2^n \times 3200$ bp (the newly created material) + 3.2×10^9 bp (the original material). The question asks for the value of n for which the fraction of newly created DNA constitutes 99.9 percent of the total DNA in solution. The inequality to be solved is

$$\frac{2^n \times 3200 \text{ bp}}{2^n \times 3200 \text{ bp} + 3.2 \times 10^9 \text{ bp}} \geq 0.999$$

from which we obtain

$$n \geq \frac{1}{\log(2)} \times \log\left[\frac{(3.2 \times 10^9)(0.999)}{(3200)(1 - 0.999)}\right] = 29.9$$

This means that 30 rounds of amplification increase the percentage of target DNA in the solution by a factor of almost 10^6.

Problem 3

A solution containing single-stranded DNA with the sequence

```
5'-ATGGTGCACCTGACTCCTGAGGAGAAGTCTNNNNNNNN-3'
```

undergoes DNA replication *in vitro* in the presence of all four nucleoside triphosphates plus an amount of dideoxyadenosine triphosphate sufficient to compete for incorporation with deoxyadenosine triphosphate. The run of N's represents the nucleotides that bind with the oligonucleotide primer. What DNA fragments are expected?

■ **Solution** Replication will proceed normally for all A, G, and C nucleotides in the template strand, but it will terminate at a T wherever a dideoxyadenosine was incorporated instead of deoxyadenosine. The resulting fragments will be as shown, where XXXXXXXX represents the nucleotides in the oligonucleotide primer.

```
5'-XXXXXXXXA-3'
5'-XXXXXXXXAGA-3'
5'-XXXXXXXXAGACTTCTCCTCA-3'
5'-XXXXXXXXAGACTTCTCCTCAGGA-3'
5'-XXXXXXXXAGACTTCTCCTCAGGAGTCA-3'
5'-XXXXXXXXAGACTTCTCCTCAGGAGTCATTCA-3'
5'-XXXXXXXXAGACTTCTCCTCAGGAGTCATTCACA-3'
5'-XXXXXXXXAGACTTCTCCTCAGGAGTCATTCACACCA-3'
5'-XXXXXXXXAGACTTCTCCTCAGGAGTCATTCACACCAT-3'
```

concepts in action: problems for solution

6.1 To what chemical group in a DNA chain is an incoming nucleotide added? What chemical group in the incoming nucleotide reacts with the DNA terminus?

6.2 Consider a culture of *E. coli* cells grown for many generations in a ^{15}N-containing medium. The cells are washed and transferred to a ^{14}N-containing medium. After exactly two chromosome replications in the second medium, the DNA is extracted without any breakage whatsoever. What density classes would result and in what proportions?

6.3 What is meant by the statement that the DNA replication fork is asymmetrical?

6.4 The haploid genome of the wall cress, *Arabidopsis thaliana*, contains 100 Mb and has 10 chromosomes. If a particular chromosome contains 10 percent of the DNA in the haploid genome, what is the approximate length of its DNA molecule in micrometers? (There are 10^{-4} micrometers per angstrom unit.)

6.5 For the chromosome of *A. thaliana* in the previous problem, estimate the time of replication, assuming that there is only one origin of replication (exactly in the middle), that replication is bidirectional, and that the rate of DNA synthesis is:
(a) 1500 nucleotide pairs per second (typical of bacterial cells).
(b) 50 nucleotide pairs per second (typical of eukaryotic cells).

6.6 The double-stranded DNA molecule of a newly discovered virus was found by electron microscopy to have a length of 68 micrometers (68×10^4 Å).
(a) How many nucleotide pairs are present in one of these molecules?
(b) How many complete turns of the two polynucleotide chains are present in such a double helix?

6.7 Which of the following sequences are palindromes and which are not? Explain your answer. Symbols such as (A/T) mean that the site may be occupied by (in this case) either A or T, and N stands for any nucleotide.
(a) 5'-AATT-3'
(b) 5'-AAAA-3'
(c) 5'-AANTT-3'
(d) 5'-AA(A/T)AA-3'
(e) 5'-AA(G/C)TT-3'

6.8 Consider the restriction enzymes *Bam*HI (cleavage site 5'-G↓GATCC-3') and *Sau*3A (cleavage site 5'-↓GATC-3'), where the downward arrow denotes the site of cleavage in each strand. Is every *Bam*HI site a *Sau*3A site? Is every *Sau*3A site a *Bam*HI site? Explain your answer.

6.9 Specify the function or enzymatic activity of the following enzymes or enzyme complexes that participate in DNA replication: primosome, gyrase, DNA ligase, polymerase I (Pol I), and polymerase III (Pol III).

6.10 In early studies of the properties of the precursor fragments in DNA synthesis, it was shown that these fragments could hybridize to both strands of *E. coli* DNA. This was taken as evidence that they are synthesized on both branches of the replication fork. However, one feature of the replication of *E. coli* DNA that was not known at the time invalidates this conclusion. What is this characteristic?

6.11 A friend brings you three samples of nucleic acid and asks you to determine each sample's chemical identity (whether DNA or RNA) and whether the molecules are double-stranded or single-stranded. You use powerful nucleases to degrade each sample completely to its constituent nucleotides and then determine the approximate relative proportions of nucleotides. The results of your assay follow. What can you tell your friend about the nature of these samples?

Sample 1: dGMP 13%   dCMP 14%   dAMP 36%   dTMP 37%
Sample 2: dGMP 12%   dCMP 36%   dAMP 47%   dTMP 5%
Sample 3:   GMP 22%   CMP 47%   AMP 17%   UMP 14%

6.12 A new technique is used for determining the base composition of double-stranded DNA. Rather than giving the relative amounts of each of the four bases, A, T, G, and C, the procedure yields the value of the ratio of A to C. If this ratio is 1/3, what are the relative amounts of the four bases?

6.13 A 3.1-kilobase linear fragment of DNA was digested with *Pst*I and produced a 2.0-kb fragment and a 1.1-kb fragment. When the same 3.1-kb fragment was cut with *Hin*dIII, it yielded a 1.5-kb fragment, a 1.3-kb fragment, and a 0.3-kb fragment. When the 3.1-kb molecule was cut with a mixture of the two enzymes, fragments of 1.5, 0.8, 0.5, and 0.3 kb resulted. Draw a map of the original 3.1-kb fragment, and label the restriction sites and the distances between these sites.

6.14 The *dusky* mutation is an X-linked recessive in *Drosophila* that causes small, dark wings. In a stock of wildtype flies, you find a single male that has the dusky phenotype. In wildtype flies, the *dusky* gene is contained within an 8-kb *Xho*I restriction fragment. When you digest genomic DNA from the mutant male with *Xho*I and probe with the 8-kb restriction fragment on a Southern blot, you find that the size of the labeled fragment is 10 kb. You clone the 10-kb fragment and use it as a probe for a polytene chromosome *in situ* hybridization in a number of different wildtype strains, and you notice that this fragment hybridizes to multiple locations along the polytene chromosomes. Each wildtype strain has a different pattern of hybridization. What do these data suggest about the origin of the *dusky* mutation that you isolated?

6.15 The following sequence of bases is present along one chain of a DNA duplex that has opened up at a replication fork, and synthesis of an RNA primer on this template begins by copying the base shown in red.

5'-TCTGATATCAGTACG-3'

If the RNA primer consists of eight nucleotides, what is its base sequence?

6.16 For the replication bubble illustrated here, indicate the leading strand and the lagging strand at each replication fork, and identify the ends as 3' or 5'.

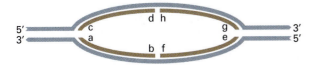

6.17 Identify the chemical group that is at the indicated terminus of the daughter strand of the extended branch of the rolling circle in the accompanying diagram.

6.18 For the fluorescent color coding A = green, T = red, G = black, and C = purple, read the DNA sequence indicated in the accompanying gel diagram.

6.19 For the fluorescent color coding A = green, T = red, G = black, and C = purple, read the DNA sequence in each of the accompanying gel diagrams. How does the sequence in gel A differ from that in gel B?

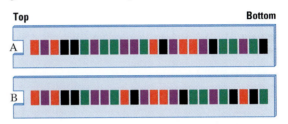

6.20 The first evidence for semiconservative replication of DNA in eukaryotes made use of an artificial nucleotide of thymidine called bromodeoxyuridine (BUdR), in which the methyl group in thymine is replaced with an atom of bromine. When chromatids whose DNA contains BudR are stained with certain fluorescent dyes, the chromatids with one strand labeled and one unlabeled fluoresce very brightly (light blue in the accompanying illustration), whereas those with both strands labeled fluoresce dully (dark blue). The illustration depicts the fluorescence patterns of chromosomes in mitotic metaphase after one and two rounds of DNA replication in the presence of BUdR, and the dotted lines represent the DNA strands in the DNA duplex present in each chromatid. Depict the BUdR labeling of each chromatid by (1) making the line solid if the strand is fully labeled with BUdR or (2) leaving it dashed if it is half labeled with BUdR.

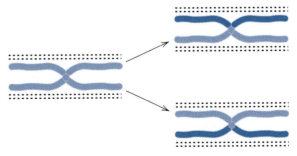

further readings

Bauer, W. R., F. H. C. Crick, and J. H. White. 1980. Supercoiled DNA. *Scientific American,* July.

Blow, J. J., ed. 1996. *Eukaryotic DNA Replication.* New York: IRL Press.

Cairns, J. 1966. The bacterial chromosome. *Scientific American,* January.

Danna, K., and D. Nathans. 1971. Specific cleavage of Simian Virus 40 DNA by restriction endonuclease of *Hemophilus influenzae. Proceedings of the National Academy of Sciences,* USA 68: 2913.

Davies, J. 1995. Vicious circles: Looking back on resistance plasmids. *Genetics* 139: 1465.

DePamphilis, M. L., ed. 2000. *Concepts in Eukaryotic DNA Replication.* Cold Spring Harbor, NY: Cold Spring Harbor Laboratory Press.

Donovan, S., and J. F. X. Diffley. 1996. Replication origins in eukaroytes. *Current Opinion in Genetics & Development* 6: 203.

Grimaldi, D. A. 1996. Captured in amber. *Scientific American,* April.

Grunstein, M. 1992. Histones as regulators of genes. *Scientific American,* October.

Hubscher, U., and J. M. Sogo. 1997. The eukaryotic DNA replication fork. *News in Physiological Sciences* 12: 125.

Kelley, T., ed. 1988. *Eukaryotic DNA Replication.* Cold Spring Harbor, NY: Cold Spring Harbor Laboratory Press.

Kornberg, A. 1995. *DNA Replication.* 2d ed. New York: Freeman.

Kornberg, R. D., and A. Klug. 1981. The nucleosome. *Scientific American,* February.

Mullis, K. B. 1990. The unusual origin of the polymerase chain reaction. *Scientific American,* April.

Neidhardt, F. C., R. Curtiss III, J. L. Ingraham, et al, eds. 1996. Escherichia coli *and* Salmonella typhimurium: *Cellular and Molecular Biology* (2 volumes). 2d ed. Washington, DC: American Society for Microbiology.

Olby, R. C. 1994. *The Path to the Double Helix: The Discovery of DNA.* New York: Dover.

Singer, M., and P. Berg. 1991. *Genes & Genomes.* Mill Valley, CA: University Science Books.

Watson, J. D. 1968. *The Double Helix.* New York: Atheneum.

geNETics on the web

GeNETics on the Web will introduce you to some of the most important sites for finding genetic information on the Internet. To explore these sites, visit the Jones and Bartlett home page at

http://www.jbpub.com/genetics

For the book *Essential Genetics: A Genomics Perspective,* choose the link that says Enter **GeNETics on the Web**. You will be presented with a chapter-by-chapter list of highlighted keywords. Select any highlighted keyword and you will be linked to a Web site containing genetic information related to the keyword.

- The concept of the polymerase chain reaction (**PCR**) occurred to Kary Mullis one night while he was traveling on Route 128 from San Francisco to Mendocino. He immediately realized that this approach would be unique in its ability to amplify, at an exponential rate, a specific nucleotide sequence present in a vanishingly small quantity amid a much larger background of total nucleic acid. Once its feasibility was demonstrated, PCR was quickly recognized as a major technical advance in molecular biology. The new technique earned Mullis the 1993 Nobel Prize in chemistry, and today it is the basis for a large number of experimental and diagnostic procedures. At this keyword site you can learn more about the development of the PCR from Mullis's original conception.

- Many people think that **birth defects** are rare and happen only to other people. But birth defects affect more than 150,000 newborns each year and are the leading cause of infant death and disability. Some 3000 to 5000 different birth defects have been described. The most common are listed at this keyword site. The list includes only those conditions that can be diagnosed immediately at birth; it does not include such conditions as cystic fibrosis, Tay–Sachs disease, and sickle-cell anemia, which become apparent in the first weeks or months after birth. Note that defects of the heart and circulatory system affect more infants than any other type of birth defect.

In spite of the scrupulous cleanliness of health professionals, some of the most troublesome bacterial infections originate in hospitals. They are difficult to treat because the bacteria are resistant to the antibiotics commonly used in hospitals. Why do you suppose this is the case? [© Photos.com]

key concepts

- Bacteria take advantage of several mechanisms by which DNA sequences can move from one DNA molecule to another, from one cell to another, or even from one bacterial species to another; these mechanisms have led to the evolution of multiple-antibiotic-resistant bacteria.

- Some bacteria are capable of DNA transfer and genetic recombination.

- In *E. coli,* the F (fertility) plasmid can mobilize the chromosome for transfer to another cell in the process of conjugation.

- Some types of bacteriophages can incorporate bacterial genes and transfer them into new host cells in the process of transduction.

- DNA molecules from related bacteriophages that are present in the same host cell can undergo genetic recombination.

- Some bacteriophages are able to integrate their DNA into that of the host cell, where it replicates along with the host DNA and is transmitted to progeny cells.

7

The Genetics of Bacteria and Their Viruses

chapter organization

7.1 Many DNA sequences in bacteria are mobile and can be transferred between individuals and among species.

7.2 Mutations that affect a cell's ability to form colonies are often used in bacterial genetics.

7.3 Transformation results from the uptake of DNA and recombination.

7.4 In bacterial mating, DNA transfer is unidirectional.

7.5 Some phages can transfer small pieces of bacterial DNA.

7.6 Bacteriophage DNA molecules in the same cell can recombine.

7.7 Lysogenic bacteriophages do not necessarily kill the host.

the human connection
One Gene, One Enzyme

Chapter Summary
Issues & Ideas
Key Terms & Concepts
Solutions: Step by Step
Concepts in Action: Problems for Solution
Further Readings

geNETics on **the web**

Bacteria and their viruses (bacteriophage) have unique and diverse reproductive systems with multiple and novel mechanisms of genetic exchange. Some bacterial DNA sequences can become mobile by any of a variety of mechanisms. This feature enables them to become widely disseminated within a bacterial population and even to spread between species. In this chapter we discuss the genetic systems of bacteria and bacteriophage. We begin by examining **mobile DNA:** sequences that can be transferred between DNA molecules and from one cell to another. The ability to share genes in this manner, even among different bacterial species, is a unique feature of bacterial genetic systems.

7.1

Many DNA sequences in bacteria are mobile and can be transferred between individuals and among species.

A high percentage of bacteria isolated from clinical infections are resistant to one or more antibiotics. Most of them are resistant to multiple antibiotics. Some are resistant to all antibiotics in routine use. The problem has become so severe that many of the antibiotics that were at one time most effective and had the fewest side effects are now virtually useless. The widespread antibiotic-resistance genes almost never originate from new mutations in the bacterial genome. They are acquired, usually several at a time, in various forms of mobile DNA.

■ A plasmid is an accessory DNA molecule, often a circle.

Plasmids are nonessential DNA molecules that exist inside bacterial cells. They replicate independently of the bacterial genome and segregate to the progeny when a bacterial cell divides, so they can be maintained indefinitely in a bacterial lineage. Many plasmids are circular DNA molecules, but others are linear. The number of copies of a particular plasmid in a cell varies depending on the mechanism by which replication is regulated. High-copy-number plasmids are found in as many as 50 copies per host cell, whereas low-copy-number plasmids are present in 1 to 2 copies per cell. Plasmids range in size from a few kilobases to a few hundred kilobases (Figure 7.1) and are found in most bacterial species that have been studied. In *E. coli,* most plasmids are either quite small (up to about 10 kb) or quite large (greater than 40 kb). A typical *E. coli* isolate contains three different small plasmids, each present in multiple copies per cell, and one large plasmid present in a single copy per cell. The presence of plasmids can be detected physically by electron microscopy, as in Figure 7.1, or by gel electrophoresis of DNA samples. Some plasmids can be detected because of phenotypic characteristics that they confer on the host cell. The phenotype most commonly studied is antibiotic resistance. For example, a plasmid containing a tetracycline-resistance gene (*tet-r*) will enable the host bacterial cell to form colonies on medium containing tetracycline.

Plasmids rely on the DNA-replication enzymes of the host cell for their reproduction, but the initi-

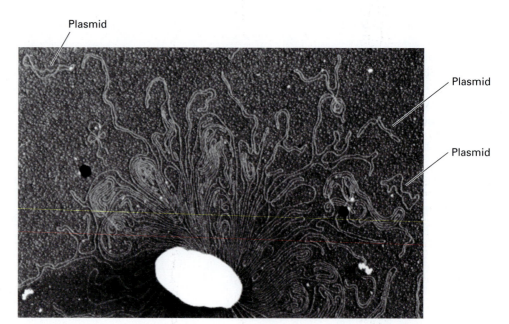

Plasmid

Plasmid

Plasmid

Plasmid

Figure 7.1 Electron micrograph of a ruptured *E. coli* cell, showing released chromosomal DNA and several plasmid molecules. [Courtesy of David Dressler and Huntington Potter.]

ation of replication is controlled by plasmid genes. In high-copy-number plasmids, replication is initiated multiple times during replication of the host genome, but in low-copy-number plasmids, replication is initiated only once, or little more than once, per round of replication of the host genome. All types of plasmids contain sequences that promote their segregation into both daughter cells produced by fission of the host cell, so spontaneous loss of plasmids is uncommon.

■ The F plasmid is a conjugative plasmid.

Many large plasmids contain genes that enable the plasmid DNA to be transferred between cells. The transfer is mediated by a tube-like structure called a pilus (plural pili), formed between the cells, through which the plasmid DNA passes (Figure 7.2). The joining of bacterial cells in the transfer process is called **conjugation,** and the plasmids that can be transferred in this manner are called **conjugative plasmids.** Not all plasmids are conjugative plasmids. Most small plasmids are nonconjugative: They can be maintained in a bacterial lineage as the cells divide, but they do not contain the approximately

20 genes necessary for pilus assembly or those for DNA transfer. Hence they are unable to be transferred on their own. As we shall see later, however, they are able to employ the genetic trickery of recombination in order to tag along with conjugative plasmids, and in this way nonconjugative plasmids can be mobilized for cell-to-cell transfer.

The pilus between the *E. coli* cells in Figure 7.2 is an *F pilus* whose synthesis results from the presence of a conjugative plasmid called the **F factor** (the F stands for *fertility*). Cells that contain the F plasmid are donors and are designated the **F+ cells** ("F plus"); those lacking F are recipients and are designated the **F− cells** ("F minus"). The F plasmid is a low-copy-number plasmid, present in 1 to 2 copies per cell. It replicates once per cell cycle and segregates to both daughter cells in cell division. The F factor is approximately 100 kb in length and contains many genes that govern its maintenance in the cell and its transmission between cells.

Conjugation begins with physical contact between a donor cell and a recipient cell, as in Figure 7.2. Once the pilus contacts the F− cell, the pilus retracts and the cell membranes of the donor and recipient are brought face-to-face. Then the donor DNA moves through a pore in the membrane from the donor to the recipient. The transfer is always accompanied by replication of the plasmid. Contact between an F+ and an F− cell initiates rolling-circle replication (explained in Section 6.4), which results in the transfer of a single-stranded

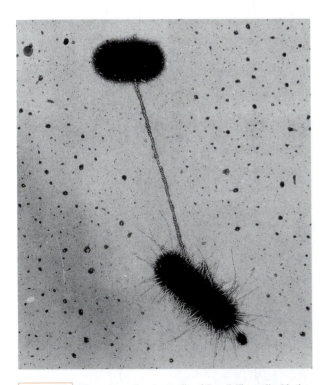

Figure 7.2 Pilus connecting two cells of *E. coli*. The cell with the multiple appendages (used for colonizing the intestine) contains an F plasmid that encodes the proteins necessary to produce the pilus. Prior to transfer of the plasmid DNA, the pilus shortens somewhat and draws the cells closer together. For ease of visualization, this pilus is coated with a bacteriophage that attaches to the F pilus. [Courtesy of C. C. Brinton, Jr., and J. Carnahan.]

Ouch! The abrasion needs to be cleaned and bandaged. Nowadays there may be little danger, but in the pre-antibiotic era, serious infections could result.

linear branch of the rolling circle to the recipient cell. During transfer, DNA is synthesized in both donor and recipient (Figure 7.3). Leading-strand synthesis in the donor replaces the transferred single strand, and lagging-strand synthesis in the recipient converts the transferred single strand into double-stranded DNA. When transfer is complete, the linear F strand becomes circular again in the recipient cell. Note that because one replica remains in the donor while the other is transferred

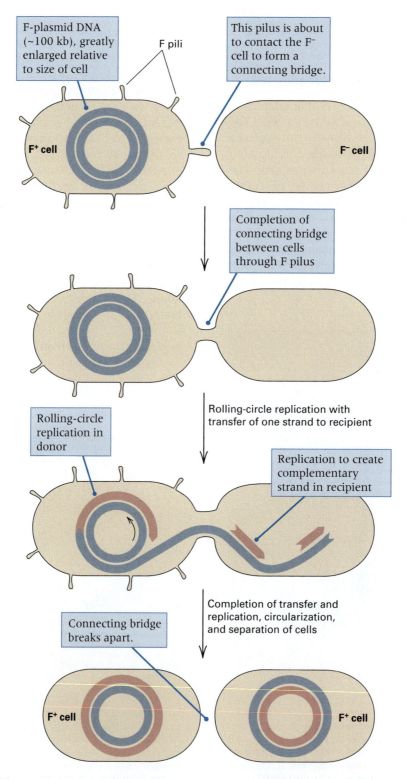

Figure 7.3 Transfer of F from an F+ to an F− cell. Pairing of the cells triggers rolling-circle replication. Pink represents DNA synthesized during pairing. For clarity, the bacterial chromosome is not shown, and the plasmid is drawn overly large; the plasmid is in fact much smaller than a bacterial chromosome.

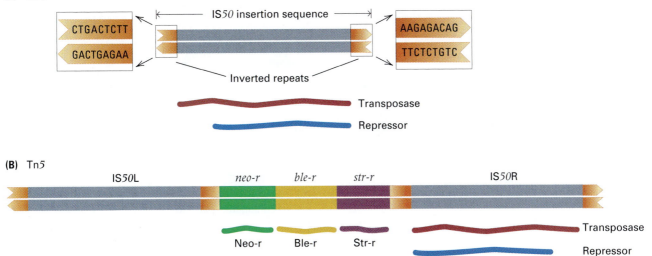

(A) IS*50*

IS*50* insertion sequence

CTGACTCTT
GACTGAGAA

AAGAGACAG
TTCTCTGTC

Inverted repeats

Transposase

Repressor

(B) Tn*5*

IS*50*L   *neo-r*   *ble-r*   *str-r*   IS*50*R

Neo-r   Ble-r   Str-r

Transposase

Repressor

Figure 7.4 Transposable elements in bacteria. (A) Insertion sequence IS*50*. The element is terminated by short, nearly perfect inverted-repeat sequences, the terminal nine base pairs of which are indicated. IS*50* contains a region that codes for the transposase and for a repressor of transposition. The coding regions are identical in the region of overlap, but the repressor is somewhat shorter because it begins at a different place. (B) Composite transposon Tn*5*. The central sequence contains genes for resistance to neomycin, *neo-r*; bleomycin, *ble-r*; and streptomycin, *str-r*. It is flanked by two copies of IS*50* in inverted orientation. The left-hand element (IS*50*L) contains mutations and is nonfunctional, so the transposase and repressor are made by the right-hand element (IS*50*R).

to the recipient, after transfer both cells contain F and can function as donors. The F⁻ cell has been converted into an F⁺ cell.

The transfer of the F plasmid requires only a few minutes. In laboratory cultures, if a small number of donor cells are mixed with an excess of recipient cells, F spreads throughout the population in a few hours, and all cells ultimately become F⁺. Transfer is not so efficient under natural conditions, and only about 10 percent of naturally occurring *E. coli* cells contain the F factor. Conjugation normally takes place only between F⁺ and F⁻ cells, because the F plasmid contains two genes for *surface exclusion*, which prevents an F⁺ cell from conjugating with any other cell containing the same or a closely related plasmid. Most conjugative plasmids have analogous exclusion mechanisms.

■ Insertion sequences and transposons play a key role in bacterial populations.

Transposable elements are DNA sequences that can jump from one position to another or from one DNA molecule to another. Bacteria contain a wide variety of transposable elements. The smallest and simplest are **insertion sequences,** or *IS elements,* which are typically 1–3 kb in length and usually encode only the *transposase* protein required for transposition and one or more additional proteins that regulate the rate of transposition. Like many transposable elements in eukaryotes, they possess inverted-repeat sequences at their termini, which

are used by the transposase for recognizing and mobilizing the IS element. Upon insertion, they create a short, direct duplication of the target sequence at each end of the inserted element. The DNA organization of the insertion sequence IS*50* is diagrammed in part A of Figure 7.4.

Other transposable elements in bacteria contain one or more genes unrelated to transposition that can be mobilized along with the transposable element; this type of element is called a **transposon.** The length of a typical transposon is several kilobases, but a few are much longer. Much of the widespread antibiotic resistance among bacteria is due to the spread of transposons that include one or more (usually multiple) antibiotic-resistance genes. When a transposon mobilizes and inserts into a conjugative plasmid, it can be widely disseminated among different bacterial hosts by means of conjugation.

Some transposons have composite structures with antibiotic resistance sandwiched between insertion sequences, as is the case with the Tn*5* element illustrated in part B of Figure 7.4, which terminates in two IS*50* elements in inverted orientation. Transposons are usually designated by the abbreviation Tn followed by an italicized number (for example, Tn*5*). When it is necessary to refer to genes carried in such an element, the usual designations for the genes are used. For example, Tn*5* (*neo-r ble-r str-r*) contains genes for resistance to three different antibiotics: neomycin, bleomycin, and streptomycin.

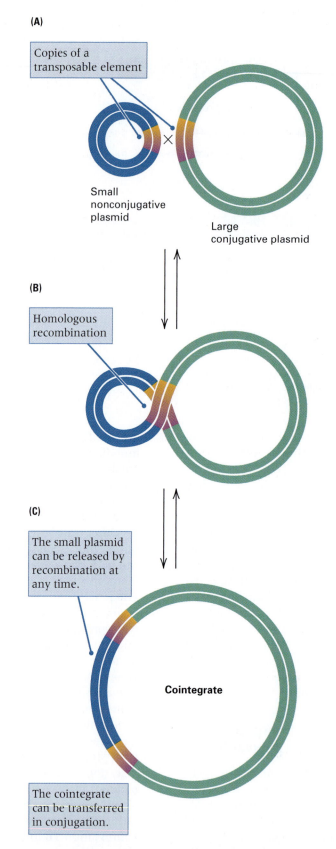

(A)

Copies of a transposable element

Small nonconjugative plasmid

Large conjugative plasmid

(B)

Homologous recombination

(C)

The small plasmid can be released by recombination at any time.

Cointegrate

The cointegrate can be transferred in conjugation.

Figure 7.5 Cointegrate formed between two plasmids by recombination between homologous sequences (for example, copies of a transposable element) present in both plasmids.

■ Nonconjugative plasmids can be mobilized by cointegration into conjugative plasmids.

Nonconjugative and conjugative plasmids typically coexist in the same cell along with host genomic DNA, and when a transposable element is mobilized, all of the DNA molecules present are potential targets for insertion. In time, many of the plasmids in a bacterial lineage can acquire copies of transposable elements present in the host DNA, and the host DNA can acquire copies of transposable elements present in the plasmids. In this manner, transposable elements become disseminated among independently replicating DNA molecules. The result is that most bacteria contain multiple copies of different types of transposable elements, some in the host genome, some in plasmids, and some in both. In *E. coli*, for example, natural isolates contain an average of 1 to 6 genomic copies of each of six naturally occurring IS elements, and among the cells that contain a particular IS element, 20 to 60 percent also contain copies in one or more plasmids.

Thus it happens that many nonconjugative and conjugative plasmids present in a bacterial cell come to carry one or more copies of the same transposable element. Because these copies are homologous DNA sequences, they can serve as substrates for recombination. When two plasmids undergo recombination in a region of homology, the result is as shown in Figure 7.5. The recombination forms a composite plasmid called a **cointegrate.** If one of the participating plasmids is nonconjugative and the other is conjugative, then the cointegrate is also a conjugative plasmid and so can be transferred in conjugation. After conjugation, the nonconjugative plasmid can become free of the cointegrate by recombination between the same sequences that created it. By the mechanism of cointegrate formation, therefore, nonconjugative plasmids can temporarily ride along with conjugative plasmids and be transferred from cell to cell.

■ Integrons have special site-specific recombinases for acquiring antibiotic-resistance cassettes.

In the evolution of multiple antibiotic resistance, bacteria have also made liberal use of a set of enzymes known as *site-specific recombinases,* which were present in bacterial populations and functioned in the evolution of other traits long before the antibiotic era. Each type of **site-specific recombinase** binds with a specific nucleotide sequence in duplex DNA. When the site is present in each of two duplex DNA molecules, the recombinase brings the sites together and catalyzes a reciprocal exchange between the duplexes. An example is shown in Figure 7.6, where the site-specific recombinase joins a circular DNA molecule with a linear DNA molecule. Note that the reaction can

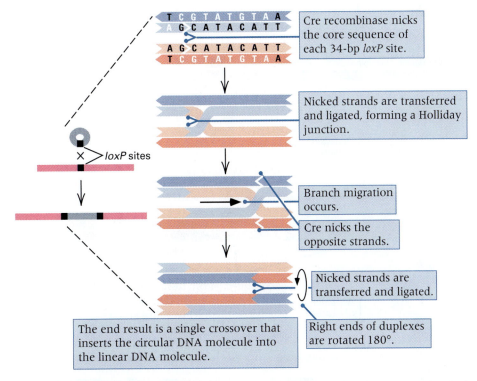

Cre recombinase nicks the core sequence of each 34-bp *loxP* site.

Nicked strands are transferred and ligated, forming a Holliday junction.

Branch migration occurs.

Cre nicks the opposite strands.

Nicked strands are transferred and ligated.

Right ends of duplexes are rotated 180°.

The end result is a single crossover that inserts the circular DNA molecule into the linear DNA molecule.

Figure 7.6 A site-specific recombinase catalyzes a reciprocal exchange between two specific sequences. No other sequences can serve as substrates. The recognition site for the Cre recombinase is *loxP*.

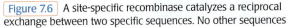

proceed in the reverse direction, too, and free the circle from the cointegrate.

An example of a site-specific recombinase is an enzyme called the *Cre recombinase*, which is encoded in a gene in the *E. coli* bacteriophage P1. The Cre recognition sequence is called *loxP*; it is 34 bp in length and contains the central asymmetrical core sequence shown in Figure 7.6. Recombination between two *loxP* sites preserves the *loxP* sequences because the participating sites are identical, and hence the recombination reaction is reversible. Some site-specific recombinases favor the reaction that brings two molecules together into a cointegrate. Others (including Cre) favor the reaction that splits a cointegrate into two separate molecules. Some site-specific recombinases bring together and recombine sites that are similar but not identical; in these cases the recombination does not preserve the recognition sites, and so the reaction is not necessarily reversible.

Site-specific recombinases are used in the assembly of multiple-antibiotic-resistance units called *integrons*. An **integron** is a DNA element that encodes a site-specific recombinase as well as a recognition region that allows other sequences with similar recognition regions to be incorporated into the integron by recombination. The elements that integrons acquire are known as *cassettes*. In the context of integrons, a **cassette** is a circular antibiotic-resistance-coding region flanked by a recognition region for an integron. Because the site-specific recombinase integrates cassettes, the integron recombinase is usually called an **integrase.**

Several different types of integrons have been characterized. The best known of these are the Class 1 integrons, which include a site-specific recombinase denoted Int1 and, invariably, a coding region (*sul1*) that confers resistance to sulphonamide antibiotics. The molecular structure of a Class 1 integron is shown in part A of Figure 7.7 . Also

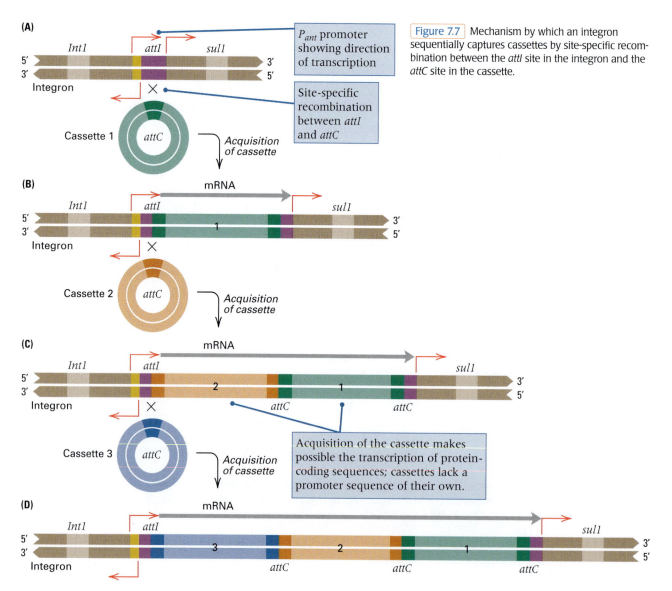

Figure 7.7 Mechanism by which an integron sequentially captures cassettes by site-specific recombination between the *attI* site in the integron and the *attC* site in the cassette.

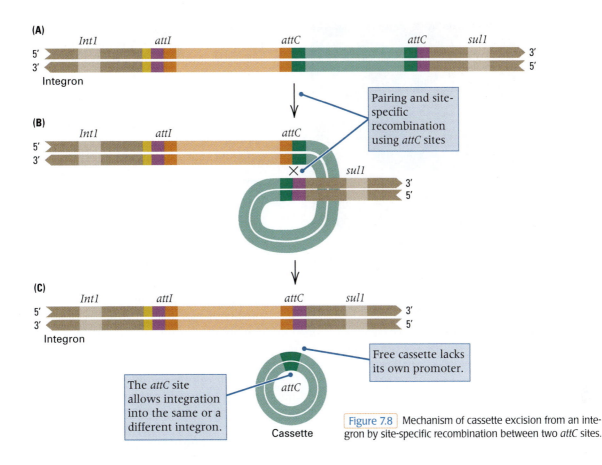

(A)

(B)

Pairing and site-specific recombination using *attC* sites

(C)

Free cassette lacks its own promoter.

The *attC* site allows integration into the same or a different integron.

Cassette

Figure 7.8 Mechanism of cassette excision from an integron by site-specific recombination between two *attC* sites.

shown is the mechanism by which antibiotic-resistance cassettes are sequentially acquired. The Int1 integrase catalyzes a site-specific recombination between a sequence denoted *attI* present in the integron and a similar sequence denoted *attC* (also called the *59-base element*) in the cassette. All *attC* regions are similar, but no two are identical.

Figure 7.7, part A, shows how a cassette is captured by site-specific recombination between *attI* and *attC*. In general, antibiotic-resistance cassettes contain protein-coding regions but lack the *promoter* sequences needed to initiate transcription. They can be transcribed only by read-through transcription from an adjacent promoter. The integron provides the needed promoter, called P_{ant}, at a position upstream from the *attI* site, so that when a cassette is captured, the coding sequence can be expressed. More than 40 different promoterless cassettes have been identified that encode proteins for resistance to antibiotics including β-lactams, aminoglycosides, chloramphenicol, trimethoprim, and streptothricin.

Once one cassette is in place, as shown in part B of Figure 7.7, a second can be captured using the same *attI* site and the *attC* present in the new cassette. Note that the new cassette is integrated immediately adjacent to the *attI* site and that the mRNA produced from the P_{ant} promoter includes the coding sequences for both cassettes. In part C of Figure 7.7 a third cassette is added to the integron,

and the mRNA from P_{ant} becomes even longer. When there are multiple cassettes, as shown here, all of them are cotranscribed from P_{ant}, but the downstream coding sequences are transcribed less frequently because there is a greater chance that transcription will terminate before reaching them. This constraint sets a practical limit on the number of cassettes that can be transcribed efficiently, but integrons with up to 10 antibiotic-resistance cassettes have been found.

The Int1 integrase can also catalyze the reverse of the cassette-capture reaction, though at a much lower level. This reaction generally uses two *attC* sites. An example is shown in Figure 7.8. Site-specific recombination between adjacent *attC* sites releases a circular cassette containing a promoterless coding sequence. The cassette cannot replicate because it lacks an origin of replication, but the capture reaction is efficient enough that the cassette will often be recaptured by the same integron (which repositions the cassette immediately adjacent to *attI*) or by a different integron in the same cell (adding to the repertoire of cassettes it already contains).

Although integrons cannot mobilize themselves, they are present in transposons, conjugative plasmids, and nonconjugative plasmids, as well as in bacterial chromosomes. The integrons that are parts of mobile DNA elements are particularly important in the evolution of antibiotic resistance, because

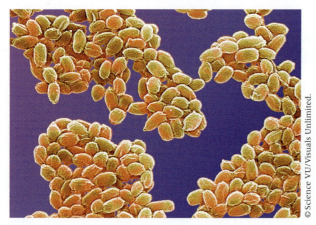

Anthrax is an agent that can be used as a biological weapon. It is easy to grow, but harder to "weaponize" (disseminate widely in deadly forms). Nevertheless, unweaponized anthrax spores sent through the mail have caused multiple deaths and great disruption in the United States.

they can capture antibiotic-resistance cassettes and thereby make possible not only the transcription of the antibiotic-resistance coding sequences but also their mobilization.

■ Bacteria with resistance to multiple antibiotics are an increasing problem in public health.

In nature, a conjugative plasmid can, through time, accumulate different transposons containing multiple independent antibiotic-resistance genes, or transposons containing integrons that have acquired multiple-antibiotic-resistance cassettes, with the result that the plasmid confers resistance to a large number of completely unrelated antibiotics. These multiple-resistance plasmids are called **R plasmids.** Some R plasmids are closely related to the F plasmid and clearly evolved from the F factor. The evolution of R plasmids is promoted by the use (and overuse) of antibiotics, which selects for resistant cells because, in the presence of antibiotics, resistant cells

have a growth advantage over sensitive cells. The presence of multiple antibiotics in the environment selects for multiple-drug resistance. Serious clinical complications result when plasmids resistant to multiple drugs are transferred to bacterial pathogens, or agents of disease. Infections with some pathogens that contain R factors are extremely difficult to treat, because the pathogen may be resistant to most or all antibiotics currently in use.

7.2

Mutations that affect a cell's ability to form colonies are often used in bacterial genetics.

Bacteria can be grown both in liquid medium and on the surface of a semisolid growth medium hardened with agar. Bacteria used in genetic analysis are usually grown on an agar surface in plastic petri dishes (called *plates*). A single bacterial cell placed on a solid medium will grow and divide many times, forming a visible cluster of cells called a *colony* (Figure 7.9). The number of bacterial cells in a suspension can be determined by spreading a known volume of the suspension on an agar surface and counting the colonies that form. Typical *E. coli* cultures contain as many as 10^9 cells/ml. The appearance of colonies, or the ability or inability to form colonies, on a particular medium can sometimes be used to identify the genotype of the cell that produced the colony.

As in other organisms, genetic analysis in bacteria requires mutants. In bacteria, mutations that affect metabolic pathways or antibiotic resistance are particularly useful. There are three principal types of mutations.

- **Antibiotic-resistant mutants** These mutants are able to grow in the presence of an antibiotic, such as streptomycin (Str) or tetracycline (Tet). For example, streptomycin-sensitive (Str-s) cells have the wildtype phenotype and fail to form colonies on medium containing streptomycin, but streptomycin-resistant (Str-r) mutants can form colonies.

- **Nutritional mutants** Wildtype bacteria can synthesize most of the complex nutrients

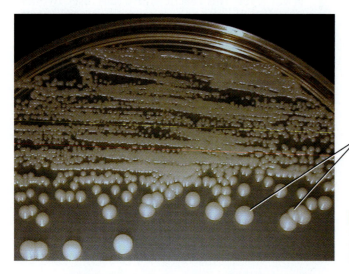

Colonies

Figure 7.9 A petri dish with bacterial colonies that have formed on a solid medium. The heavy streaks of growth result from colonies so densely packed that there is no space between them. [Courtesy of M. A. Lachance.]

they need from simple molecules present in the growth medium. The wildtype cells are said to be **prototrophs.** The ability to grow in simple medium can be lost by mutations that disable the enzymes used in synthesizing the complex nutrients. Mutant cells are unable to synthesize an essential nutrient and thus cannot grow unless the required nutrient is supplied in the medium. Such a mutant bacterium is said to be an **auxotroph** for the particular nutrient. For example, a methionine auxotroph cannot grow on a **minimal medium** containing only inorganic salts and a source of energy and carbon atoms (such as glucose), but such Met^- cells can grow if the minimal medium is supplemented with methionine.

- **Carbon-source mutants** These mutants cannot utilize particular substances as sources of energy or carbon atoms. For example, Lac^- mutants cannot utilize the sugar lactose for growth and are unable to form colonies on minimal medium containing lactose as the only carbon source.

A medium on which all wildtype cells form colonies is called a **nonselective medium.** Mutants and wildtype cells may or may not be distinguishable by growth on a nonselective medium. If the medium allows growth of only one type of cell (either wildtype or mutant), it is said to be a **selective medium.** For example, a medium containing streptomycin is selective for the Str-r (resistant) phenotype and selective against the Str-s (sensitive) phenotype, and minimal medium containing lactose as the sole carbon source is selective for Lac^+ cells and against Lac^- cells.

In bacterial genetics, phenotype and genotype are designated in the following way. A phenotype is designated by three letters, the first of which is capitalized; a superscript + or − denotes the presence or absence of the designated character; and s or r denotes sensitivity or resistance, respectively. A genotype is designated by lowercase italicized letters. Thus, a cell unable to grow without a supplement of leucine (a leucine auxotroph) has a Leu^- phenotype, and this would usually result from a leu^- mutation in one of the genes required for leucine biosynthesis. Often the − superscript is omitted, but we will use it consistently to avoid ambiguity.

7.3

Transformation results from the uptake of DNA and recombination.

As we saw in Chapter 1, important evidence that DNA is the genetic material came from experiments in which DNA from a heat-killed virulent strain of a pneumonia-causing bacterium was able to convert genetically cells of another strain from nonvirulent into virulent. The process of genetic alteration by pure DNA is **transformation,** and we know much more about it now than was known in 1944 when the experiments were carried out.

In transformation, recipient cells acquire genes from free DNA molecules in the surrounding medium. In laboratory experiments, DNA isolated from donor cells is added to a suspension of recipient cells. In nature, DNA can become available by spontaneous breakage (lysis) of donor cells. Either way, transformation begins with uptake of a DNA fragment from the surrounding medium by a recipient cell and terminates with *one strand* of donor DNA replacing the homologous segment in the recipient DNA. Most bacterial species are probably capable of the recombination step, but the ability of most bacteria to take up DNA efficiently is limited. Even in a species capable of transformation, DNA is able to penetrate only some of the cells in a growing population. However, appropriate chemical treatment of cells of these species yields a population of cells that are competent to take up DNA.

Transformation affords a convenient technique for gene mapping. DNA that is isolated from a donor bacterium is invariably broken into small fragments. With suitable recipient cells and excess DNA, transformation takes place at a frequency of about one transformed cell per 10^3 cells. If two genes, a and b, are so widely separated in the donor chromosome that they are always contained in two different DNA fragments, then the probability of simultaneous transformation (**cotransformation**) of an $a^- b^-$ recipient into wildtype $a^+ b^+$ is the product of the probabilities of transformation of each marker separately, or roughly $10^{-3} \times 10^{-3}$, which equals one wildtype transformant per 10^6 recipient cells. However, if the two genes are so near one another that they are often present in a single donor fragment, then the frequency of cotransformation is nearly the same as the frequency of single-gene transformation, or one wildtype transformant per 10^3 recipients. The general principle is as follows:

> **key concept**
>
> Cotransformation of two genes at a frequency substantially greater than the product of the single-gene transformation frequencies implies that the two genes are close together in the bacterial chromosome.

Studies of the ability of various pairs of genes to be cotransformed also yield gene order. For example, if genes a and b can be cotransformed, and genes b and c can be cotransformed, but genes a

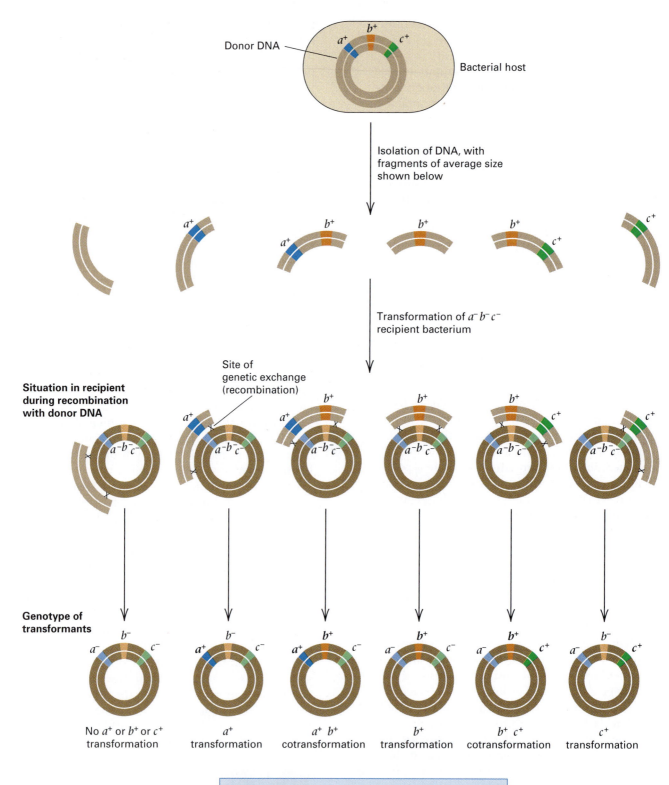

Donor DNA

Bacterial host

Isolation of DNA, with fragments of average size shown below

a^+ | b^+ | a^+ b^+ | b^+ | b^+ c^+ | c^+

Transformation of $a^- b^- c^-$ recipient bacterium

Situation in recipient during recombination with donor DNA

Site of genetic exchange (recombination)

Genotype of transformants

No a^+ or b^+ or c^+ transformation

a^+ transformation

a^+ b^+ cotransformation

b^+ transformation

b^+ c^+ cotransformation

c^+ transformation

No $a^+ b^+ c^+$ or $a^+ b^- c^+$ cotransformation occurs because the distance between a and c is too great.

Figure 7.10 Cotransformation of linked markers. Markers a and b are near enough to each other that they are often present on the same donor fragment, as are markers b and c. Markers a and c are not near enough to undergo cotransformation. The gene order must therefore be a b c. The size of the transforming DNA, relative to that of the bacterial chromosome, is greatly exaggerated.

and c cannot, then the gene order must be $a\ b\ c$ (Figure 7.10).

7.4

In bacterial mating, DNA transfer is unidirectional.

Conjugation is a process in which DNA is transferred from a bacterial donor cell to a recipient cell by cell-to-cell contact. We have already examined this process in the context of conjugative transmission of plasmids. In this section we shall see how the same process can transfer genes present in the bacterial chromosome.

■ **The F plasmid can integrate into the bacterial chromosome.**

Transfer of chromosomal genes between *E. coli* cells was first observed by Joshua Lederberg in 1951. Although it was not known at the time, the exchange took place because the donor cells were F+, and in a few cells the F factor had become integrated into the bacterial chromosome (Figure 7.11). These are known as **Hfr cells.** Hfr stands for *high frequency of recombination,* which

refers to the relatively high frequency with which donor genes are transferred to the recipient. The integration process is essentially the same as the formation of a cointegrate between two plasmids illustrated in Figure 7.5. Insertion sequences (Section 7.1) are key players in the origin of Hfr bacteria from F+ cells, because the F plasmid normally integrates through genetic exchange between an IS element present in F and a homologous copy that has transposed to an essentially random site in the bacterial chromosome. Because the F factor can exist either separate from the chromosome or incorporated into it, it qualifies as an **episome:** a genetic element that can exist free in the cell or as a segment of DNA integrated into the chromosome.

In an Hfr cell (Figure 7.11), the bacterial chromosome remains circular, though enlarged about 2 percent by the integrated F-factor DNA. Integration of F is an infrequent event, but single cells containing integrated F can be isolated and cultured. When an Hfr cell undergoes conjugation, the process of transfer of the F factor is initiated in the same manner as in an F+ cell. However, because the F factor is part of the bacterial chromosome, transfer from an Hfr cell also includes DNA from the bacterial chromosome.

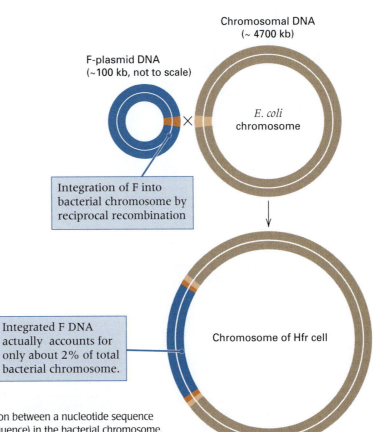

Figure 7.11   Integration of F (blue circle) by recombination between a nucleotide sequence in F and a homologous sequence (usually an insertion sequence) in the bacterial chromosome. The F plasmid DNA is shown greatly enlarged relative to the size of the bacterial chromosome.

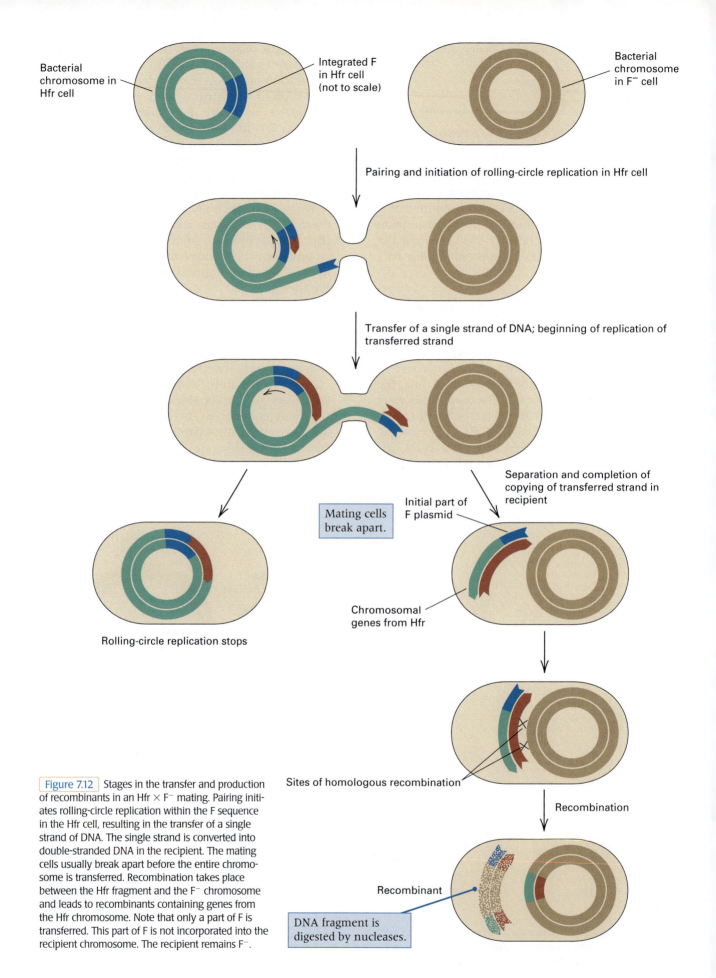

Bacterial chromosome in Hfr cell

Integrated F in Hfr cell (not to scale)

Bacterial chromosome in F⁻ cell

Pairing and initiation of rolling-circle replication in Hfr cell

Transfer of a single strand of DNA; beginning of replication of transferred strand

Mating cells break apart.

Separation and completion of copying of transferred strand in recipient

Initial part of F plasmid

Chromosomal genes from Hfr

Rolling-circle replication stops

Sites of homologous recombination

Recombination

Recombinant

DNA fragment is digested by nucleases.

Figure 7.12 Stages in the transfer and production of recombinants in an Hfr × F⁻ mating. Pairing initiates rolling-circle replication within the F sequence in the Hfr cell, resulting in the transfer of a single strand of DNA. The single strand is converted into double-stranded DNA in the recipient. The mating cells usually break apart before the entire chromosome is transferred. Recombination takes place between the Hfr fragment and the F⁻ chromosome and leads to recombinants containing genes from the Hfr chromosome. Note that only a part of F is transferred. This part of F is not incorporated into the recipient chromosome. The recipient remains F⁻.

■ Chromosome transfer begins at F and proceeds in one direction.

The Hfr $\times$ F$^-$ conjugation process is illustrated in Figure 7.12. The stages of transfer are much like those by which F is transferred to F$^-$ cells: coming together of donor and recipient cells, rolling-circle replication in the donor cell, and conversion of the transferred single-stranded DNA into double-stranded DNA by lagging-strand synthesis in the recipient. However, in the case of Hfr matings, the transferred DNA does not become circular and is not capable of further replication in the recipient because the transferred F factor is not complete. The replication and associated transfer of the chromosomal DNA are controlled by the integrated F; they are initiated in the Hfr chromosome at the same point in F at which replication and transfer begin within an unintegrated F plasmid. A part of F is the first DNA transferred, chromosomal genes are transferred next, and the remaining part of F is the last DNA to enter the recipient. Because the conjugating cells usually break apart long before the entire bacterial chromosome is transferred, the final segment of F is almost never transferred into the recipient.

Several differences between F transfer and Hfr transfer are notable.

- It takes 100 minutes under the usual conditions for an entire bacterial chromosome to be transferred—in contrast to about 2 minutes for the transfer of F. The difference in time is a result of the relative sizes of F and the chromosome (100 kb versus 4600 kb).

- In the transfer of Hfr DNA into a recipient cell, the mating pair usually breaks apart before the entire chromosome is transferred. Under typical experimental conditions, several hundred genes are transferred before the cells separate.

- In a mating between Hfr and F$^-$ cells, the F$^-$ recipient cell remains F$^-$ because cell separation usually takes place before the final segment of F is transferred.

- In Hfr transfer, some regions in the transferred DNA fragment become incorporated into the recipient chromosome. The incorporated regions replace homologous regions in the recipient chromosome. The result is that some F$^-$ cells become recombinants containing one or more genes from the Hfr donor cell. For example, in a mating between Hfr leu^+ and F$^-$ leu^-, some cells arise that are F$^-$ leu^+. However, *the genotype of the donor Hfr cell remains unchanged.*

Genetic analysis requires that recombinant recipients be identified. Because the recombinants derive from recipient cells, a method is needed to eliminate the donor cells. The usual procedure is to use an F$^-$ recipient containing an allele that can be selected. The selected allele should be located at such a place in the chromosome that most mating cells will have broken apart before the selected gene is transferred, and the selected allele must not be present in the Hfr cell. The selective agent can then be used to select the F$^-$ cells and eliminate the Hfr donors. Genes that confer antibiotic resistance are especially useful for this purpose. For instance, after a mating between Hfr leu^+ str-s and F$^-$ leu^- str-r cells, the Hfr Str-s cells can be selectively killed by plating the mating mixture on medium containing streptomycin. A selective medium that lacks leucine can then be used to distinguish between the nonrecombinant and the recombinant recipients. The F$^-$ leu^- parent cannot grow in medium lacking leucine, but recombinant F$^-$ leu^+ cells can grow because they possess a leu^+ gene. Thus, only recombinant recipients—that is, cells having the genotype leu^+ str-r—form colonies on a selective medium that contains streptomycin and lacks leucine.

When a mating is done in this way, the transferred marker that is selected by the growth conditions (leu^+ in this case) is called a **selected marker,** and the marker used to prevent growth of the donor (str-s in this case) is called the **counterselected marker.** Selection and counterselection are necessary in bacterial matings because recombinants constitute only a small proportion of the entire population of cells (in spite of the name *high frequency of recombination*).

■ The unit of distance in the *E. coli* genetic map is the length of chromosomal DNA transferred in one minute.

Genes in the bacterial chromosome can be mapped by Hfr $\times$ F$^-$ matings. However, the genetic map is quite different from linkage maps in eukaryotes in that it is based not on meiotic recombination but on transfer order. It is obtained by deliberate interruption of DNA transfer in the course of mating—for example, by violent agitation of the suspension of mating cells in a kitchen blender. The time at which a particular gene is transferred can be determined by breaking the mating cells apart at various times and noting the earliest time at which breakage no longer prevents recombinants from appearing. This procedure is called the **interrupted-mating technique.** When it is performed with Hfr $\times$ F$^-$ mat-

Table 7.1

Data showing the production of Leu$^+$ Str-r recombinants in a cross between Hfr *leu$^+$ str-s* and F$^-$ *leu$^-$ str-r* cells when mating is interrupted at various times

Minutes after mating	Number of Leu$^+$ Str-r recombinants per 100 Hfr cells
0	0
3	0
6	6
9	15
12	24
15	33
18	42
21	43
24	43
27	43

Note: Minutes after mating means minutes after the Hfr and F$^-$ cell suspensions are mixed. Extrapolation of the recombination data to a value of zero recombinants indicates that the earliest time of entry of the *leu$^+$* marker is 4 minutes.

ings, the number of recombinants of any particular allele increases with the time during which the cells are in contact. This phenomenon is illustrated in **Table 7.1**. The reason for the increase in the number of recombinants is that different Hfr $\times$ F$^-$ pairs initiate conjugation and chromosome transfer at slightly different times.

A greater understanding of the transfer process can be obtained by observing the results of a mating with several genetic markers. For example, consider the mating

Hfr a^+ b^+ c^+ d^+ e^+ *str-s* $\times$ F$^-$ a^- b^- c^- d^- e^- *str-r*

in which a^- cells require nutrient A, b^- cells require nutrient B, and so forth. At various times after mixing of the cells, samples are agitated violently and then plated on a series of media containing streptomycin and different combinations of the five substances A through E (in each medium, one of the five is left out). Colonies that form on the medium lacking A are a^+ *str-r*, those growing without B are b^+ *str-r*, and so forth. All of these data can be plotted on a single graph to give a set of curves, as shown

in part A of Figure 7.13. Four features of this set of curves are notable.

1. The number of recombinants in each curve increases with length of time of mating.

2. For each marker, there is a time (the **time of entry**) before which no recombinants are detected.

3. Each curve has a linear region that can be extrapolated back to the time axis, defining the time of entry of each gene a^+, b^+, . . . , e^+.

4. The number of recombinants of each type reaches a maximum, the value of which decreases with successive times of entry.

The explanation for the time-of-entry phenomenon is as follows. Not all donor cells start transferring DNA at the same time, so the number of recombinants increases with time. Transfer begins at a particular point in the Hfr chromosome (the replication origin of F). Genes are transferred in linear order to the recipient, and the time of entry of a gene is the time at which that gene first enters a recipient in the population. Separation of a mating pair prevents further transfer and limits the number of recombinants seen at a particular time.

The times of entry of the genes used in the mating just described can be placed on a map, as shown in Figure 7.13, part B. The numbers on this map are genetic distances between the markers, *measured as minutes between their times of entry.* Mating with another F$^-$ with genotype b^- e^- f^- g^- h^- *str-r* could be used to locate the three genes f, g, and h. Data for the second recipient would yield a map such as that shown in Figure 7.13, part C. Because genes b and e are common to both maps, the two maps can be combined to form a more complete map, as shown in Figure 7.13, part D.

Studies with different Hfr strains (Figure 7.13, part E) also are informative. It is usually found that different Hfr strains are distinguishable by their origins and directions of transfer, indicating that F can integrate at numerous sites in the chromosome and in either of two different orientations. Combining the maps obtained with different Hfr strains yields a composite map that is *circular*, as illustrated in Figure 7.13, part F. The circularity of the map is a result of the circularity of the *E. coli* chromosome in F$^-$ cells and the multiple points of integration of the F plasmid; if F could integrate at only one site and in one orientation, the map would be linear.

A great many such mapping experiments have been carried out, and the data have been combined to provide an accurate map of approximately 2000 genes throughout the *E. coli* chromo-

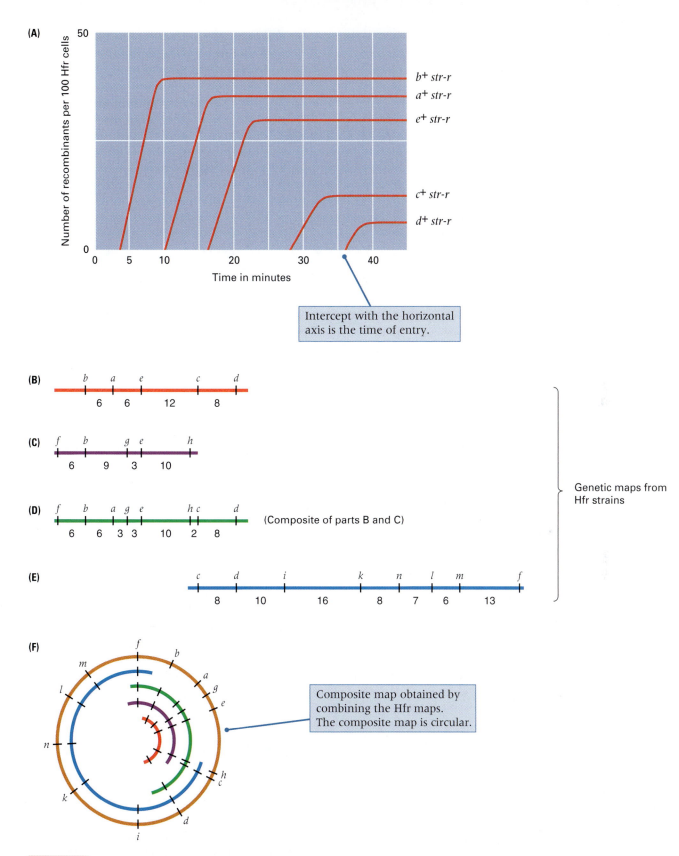

(A)

Number of recombinants per 100 Hfr cells (y-axis, 0 to 50)

Time in minutes (x-axis, 0 to 40)

Curves labeled: b^+ str-r, a^+ str-r, e^+ str-r, c^+ str-r, d^+ str-r

Intercept with the horizontal axis is the time of entry.

(B)

b — a — e — c — d

6   6   12   8

(C)

f — b — g — e — h

6   9   3   10

(D)

f — b — a — g — e — h — c — d

6   6   3   3   10   2   8

(Composite of parts B and C)

(E)

c — d — i — k — n — l — m — f

8   10   16   8   7   6   13

Genetic maps from Hfr strains

(F)

Composite map obtained by combining the Hfr maps. The composite map is circular.

Figure 7.13 Time-of-entry mapping. (A) Time-of-entry curves for one Hfr strain. (B) The linear map derived from the data in part A. (C) A linear map obtained with the same Hfr but with a different F⁻ strain containing the alleles b^- e^- f^- g^- h^-. (D) A composite map formed from the maps in parts B and C. (E) A linear map from another Hfr strain. (F) The circular map (gold) obtained by combining the two (green and blue) maps of parts D and E.

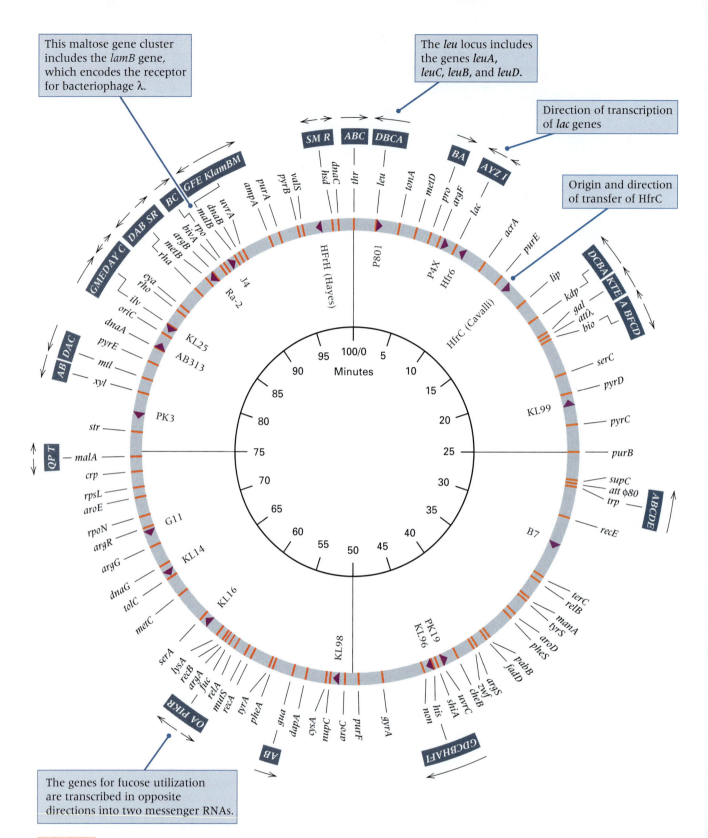

This maltose gene cluster includes the *lamB* gene, which encodes the receptor for bacteriophage λ.

The *leu* locus includes the genes *leuA*, *leuC*, *leuB*, and *leuD*.

Direction of transcription of *lac* genes

Origin and direction of transfer of HfrC

The genes for fucose utilization are transcribed in opposite directions into two messenger RNAs.

Figure 7.14 Circular genetic map of *E. coli*. Map distances are given in minutes; the total map length is 100 minutes. For some of the loci that encode functionally related gene products, the map order of the clustered genes is shown, along with the direction of transcription and length of transcript (black arrows). The purple arrowheads show the origin and direction of transfer of a number of Hfr strains. For example, HfrH transfers *thr* very early, followed by *leu* and other genes in a clockwise direction.

some. [Figure 7.14] is a map of the chromosome of *E. coli* containing a sample of the mapped genes. Both the DNA molecule and the genetic map are circular. The entire chromosome requires 100 minutes to be transferred (it usually breaks first), so the total map length is 100 minutes. In the outer circle, the arrows indicate the direction of transcription and the coding region included in each transcript. The purple arrowheads show the origin and direction of transfer of a number of Hfr strains. Transfer from HfrC, for example, goes counterclockwise starting with *purE acrA lac.*

■ Some F plasmids carry bacterial genes.

Occasionally, F is excised from Hfr DNA by an exchange between the same sequences used in the integration event. In some cases, however, the excision process is not a precise reversal of integration.

Instead, breakage and reunion take place between the nonhomologous regions at a boundary of F and nearby chromosomal DNA ([Figure 7.15]). Aberrant excision creates a plasmid that contains a fragment of chromosomal DNA, which is called an **F' plasmid** ("F prime"). By the use of Hfr strains that have different origins of transfer, F' plasmids carrying chromosomal segments from many regions of the chromosome have been isolated. These elements are extremely useful because they render any recipient diploid for the region of the chromosome carried by the plasmid. These diploid regions allow dominance tests and gene-dosage tests (studies of the effects on gene expression of increasing the number of copies of a gene). Because only a part of the genome is diploid, cells that contain an F' plasmid are **partial diploids,** also called *merodiploids.* Examples of genetic analysis using F' plasmids are given in Chapter 9 in a discussion of the *E. coli lac* genes.

7.5

Some phages can transfer small pieces of bacterial DNA.

In the process of **transduction,** bacterial DNA is transferred from one bacterial cell to another by a phage particle containing the DNA. Such a particle is called a **transducing phage.** Two types of transducing phages are known. A **generalized transducing phage** produces some particles that contain only DNA obtained from the host bacterium, rather than phage DNA; the bacterial DNA fragment can be derived from *any* part of the bacterial chromosome. A **specialized transducing phage** produces particles that contain both phage and bacterial genes linked in a single DNA molecule, but the bacterial genes are obtained from a *particular* region of the bacterial chromosome. In this section, we consider *E. coli* phage P1, a well-studied generalized transducing phage. Specialized transducing particles are discussed in Section 7.7.

During infection by P1, the phage makes a nuclease that cuts the bacterial DNA into fragments. Single fragments of bacterial DNA comparable in size to P1 DNA are occasionally packaged into phage particles in place of P1 DNA. The positions of the nuclease cuts in the host chromosome are random, so a transducing particle may contain a fragment derived from any region of the host DNA. A large population of P1 phages will contain a few particles carrying any bacterial gene. On the average, any particular gene is present in roughly one transducing particle per 10^6

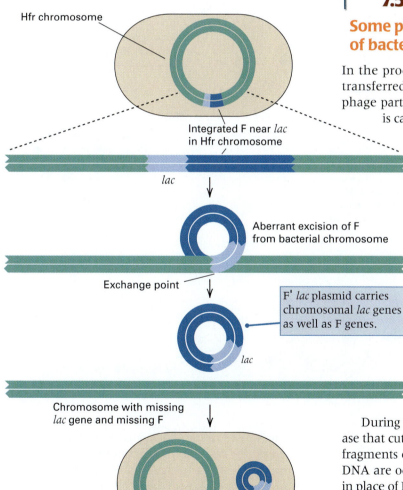

Figure 7.15 Formation of an F' *lac* plasmid by aberrant excision of F from an Hfr chromosome. Breakage and reunion are between nonhomologous regions.

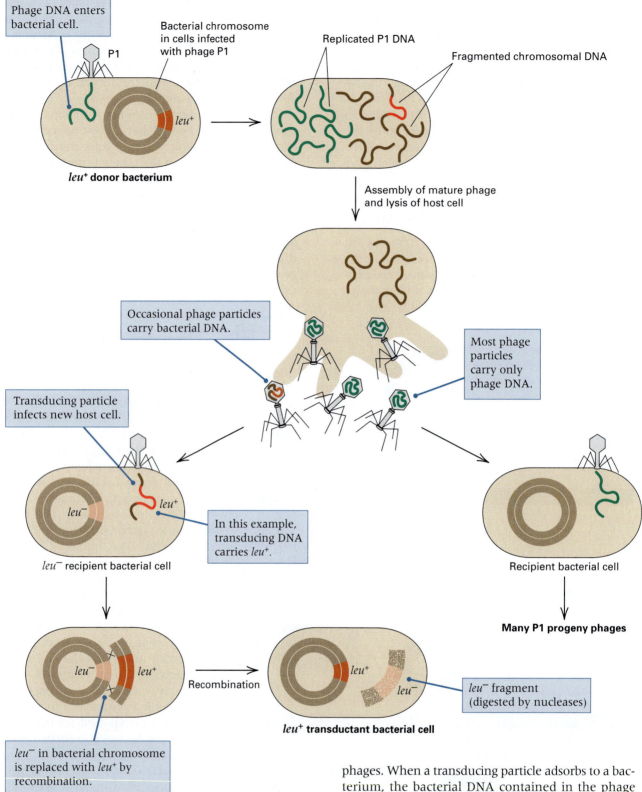

Phage DNA enters bacterial cell.

Bacterial chromosome in cells infected with phage P1

Replicated P1 DNA

Fragmented chromosomal DNA

leu^+ donor bacterium

Assembly of mature phage and lysis of host cell

Occasional phage particles carry bacterial DNA.

Most phage particles carry only phage DNA.

Transducing particle infects new host cell.

In this example, transducing DNA carries leu^+.

leu^- recipient bacterial cell

Recipient bacterial cell

Many P1 progeny phages

Recombination

leu^+ transductant bacterial cell

leu^- fragment (digested by nucleases)

leu^- in bacterial chromosome is replaced with leu^+ by recombination.

Figure 7.16 Transduction. Phage P1 infects a leu^+ donor, yielding predominantly normal P1 phages with an occasional one carrying bacterial DNA instead of phage DNA. If the phage population infects a bacterial culture, then the normal phages produce progeny phages, whereas the transducing particle yields a transductant. Note that the recombination step requires two crossovers. For clarity, double-stranded phage DNA is drawn as a single line.

phages. When a transducing particle adsorbs to a bacterium, the bacterial DNA contained in the phage head is injected into the cell and becomes available for recombination with the homologous region of the host chromosome. A typical P1 transducing particle contains from 100 to 115 kb of bacterial DNA.

Let us now examine the events that follow infection of a bacterium by a generalized transducing particle obtained, for example, by growth of P1 on wildtype $E. coli$ containing a leu^+ gene (Figure 7.16).

the human connection

One Gene, One Enzyme

George W. Beadle and Edward L. Tatum
1941
Stanford University, Stanford, California
Genetic Control of Biochemical Reactions in Neurospora

How do genes control metabolic processes? The suggestion that genes control enzymes was made very early in the history of genetics, most notably by the British physician Archibald Garrod in his 1908 book *Inborn Errors of Metabolism.* But the precise relationship between genes and enzymes was still uncertain. Perhaps each enzyme is controlled by more than one gene, or perhaps each gene contributes to the control of several enzymes. The classic experiments of Beadle and Tatum showed that the relationship is usually remarkably simple: One gene codes for one enzyme. The pioneering experiments united genetics and biochemistry, and for the "one gene, one enzyme" concept, Beadle and Tatum were awarded a Nobel Prize in 1958 (Joshua Lederberg shared the prize for his contributions to microbial genetics). Because we now know that some enzymes contain polypeptide chains encoded by two (or occasionally more) different genes, a more accurate statement of the principle is "one gene, one polypeptide." Beadle and Tatum's experiments also demonstrate the importance of choosing the right organism. *Neurospora* had been introduced as a genetic organism only a few years earlier, and Beadle and Tatum realized that they could take advantage of the ability of this organism to grow on a simple medium composed of known substances.

From the standpoint of physiological genetics the development and functioning of an organism consist essentially of an integrated system of chemical reactions controlled in some manner by genes. . . . In investigating the roles of genes, the physiological geneticist usually attempts to determine the physiological and biochemical bases of already known hereditary traits. . . . There are, however, a number of limitations inherent in this approach. Perhaps the most serious of these is that the investigator must in general confine himself to the study of non-lethal heritable characters. Such characters are likely to involve more or less non-essential so-called "terminal" reactions. . . . A second difficulty is that the standard approach to the problem implies the use of characters with visible manifestations. Many such characters involve morphological variations, and these are likely to be based on systems of biochemical reactions so complex as to make analysis exceedingly difficult. . . . Considerations such as those just outlined have led us to investigate the general problem of the genetic control of development and metabolic reactions by reversing the ordinary procedure and, instead of attempting to work out the chemical bases of known genetic characters, to set out to determine if and how genes control known biochemical reactions. The ascomycete *Neurospora* offers many advantages for such an approach and is well suited to genetic studies. Accordingly, our program has been built around this organism. The procedure is based on the assumption that x-ray treat-

ment will induce mutations in genes concerned with the control of known specific chemical reactions. If the organism must be able to carry out a certain chemical reaction to survive on a given medium, a mutant unable to do this will obviously be lethal on this medium. Such a mutant can be maintained and studied, however, if it will grow on a medium to which has been added the essential product of the genetically blocked reaction. . . . Among approximately 2000 strains [derived from single cells after x-ray treatment], three mutants have been found that grow essentially normally on the complete medium and scarcely at all on the minimal medium. One of these strains proved to be unable to synthesize vitamin B6 (pyridoxine). A second strain turned out to

> **These preliminary results appear to us to indicate that the approach may offer considerable promise as a method of learning more about how genes regulate development and function.**

be unable to synthesize vitamin B1 (thiamine). A third strain has been found to be unable to synthesize para-aminobenzoic acid. . . . These preliminary results appear to us to indicate that the approach may offer considerable promise as a method of learning more about how genes regulate development and function. For example, it should be possible, by finding a number of mutants unable to carry out a particular step in a given synthesis, to determine whether only one gene is ordinarily concerned with the immediate regulation of a given specific chemical reaction.

Source: Proceedings of the National Academy of Sciences of the USA 27: 499–506.

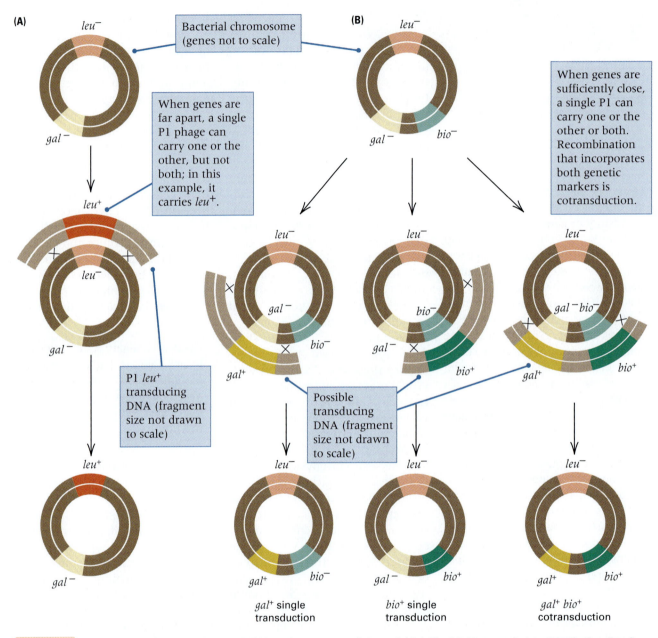

(A)

leu⁻

Bacterial chromosome (genes not to scale)

(B)

leu⁻

gal⁻

gal⁻   *bio⁻*

When genes are far apart, a single P1 phage can carry one or the other, but not both; in this example, it carries *leu⁺*.

When genes are sufficiently close, a single P1 can carry one or the other or both. Recombination that incorporates both genetic markers is cotransduction.

leu⁺

leu⁻

leu⁻

leu⁻

gal⁻

gal⁻   *bio⁻*

bio⁻

gal⁻   *bio⁻*

P1 *leu⁺* transducing DNA (fragment size not drawn to scale)

gal⁺

gal⁻   *bio⁺*

gal⁺   *bio⁺*

Possible transducing DNA (fragment size not drawn to scale)

leu⁺

leu⁻

leu⁻

leu⁻

gal⁻

gal⁺   *bio⁻*

gal⁻   *bio⁺*

gal⁺   *bio⁺*

gal⁺ single transduction

bio⁺ single transduction

gal⁺ bio⁺ cotransduction

Figure 7.17   Demonstration of linkage of the *gal* and *bio* genes by cotransduction. (A) A P1 transducing particle carrying the *leu⁺* allele can convert a *leu⁻ gal⁻* cell into a *leu⁺ gal⁻* genotype (but cannot produce a *leu⁺ gal⁺* genotype). (B) The transductants that could be formed by three possible types of transducing particles—one carrying *gal⁺*, one carrying *bio⁺*, and one carrying the linked alleles *gal⁺ bio⁺*. The third type results in cotransduction. For clarity, the distance between *gal* and *bio,* relative to that between *leu* and *gal,* is greatly exaggerated, and the size of the DNA fragment in a transducing particle, relative to the size of the bacterial chromosome, is not drawn to scale.

262   **CHAPTER 7** The Genetics of Bacteria and Their Viruses

If such a particle adsorbs to a bacterial cell of leu^- genotype and injects the DNA that it contains into the cell, the cell survives because the phage head contained only bacterial genes and no phage genes. A recombination event exchanging the leu^+ allele carried by the phage for the leu^- allele carried by the host converts the genotype of the host cell from leu^- into leu^+. In such an experiment, typically about one leu^- cell in 10^6 becomes leu^+. Such frequencies are easily detected on selective growth medium. For example, if the infected cell is placed on solid medium lacking leucine, it is able to multiply and a leu^+ colony forms. A colony does not form unless recombination inserted the leu^+ allele.

The small fragment of bacterial DNA contained in a transducing particle includes about 50 genes, so transduction provides a valuable tool for genetic linkage studies of short regions of the bacterial genome. Consider a population of P1 prepared from a leu^+ gal^+ bio^+ bacterium. This population contains particles able to transfer any of these alleles to another cell; that is, a leu^+ particle can transduce a leu^- cell to leu^+, or a gal^+ particle can transduce a gal^- cell to gal^+. Furthermore, if a leu^- gal^- culture is infected by phage, both leu^+ gal^- and leu^- gal^+ bacteria are produced. However, leu^+ gal^+ colonies do not arise, because the leu and gal genes are too far apart to be included in the same DNA fragment (part A of Figure 7.17).

The situation is quite different for a recipient cell with genotype gal^- bio^-, because the gal and bio genes are so closely linked that both genes are sometimes present in a single DNA fragment carried in a transducing particle—namely, a gal-bio particle (Figure 7.17, part B). However, not all gal^+ transducing particles also include bio^+, nor do all bio^+ particles include gal^+. The probability of both markers being in a single particle, and hence the probability of simultaneous transduction of both markers (**cotransduction**), depends on how close to each other the genes are. The closer they are, the greater the frequency of cotransduction. Cotransduction of the gal^+-bio^+ pair can be detected by plating infected cells on the appropriate growth medium. If bio^+ transductants are selected by spreading the infected cells on a glucose-containing medium that lacks biotin, both gal^+ bio^+ and gal^- bio^+ colonies grow. If these colonies are tested for the gal marker, 42 percent are found to be gal^+ bio^+ and the rest gal^- bio^+; similarly, if gal^+ transductants are selected, 42 percent are found to be gal^+ bio^+. In other words, the **frequency of cotransduction** of gal and bio is 42 percent, which means that 42 percent of all transducing particles that contain one gene also include the other.

Studies of cotransduction can be used to map closely linked genetic markers by means of three-factor crosses. Suppose, for example, that P1 is grown on wildtype bacteria and used to transduce cells carrying a mutation of each of three closely linked genes. Cotransductants that contain various pairs of wildtype alleles are examined. The gene located in the middle can be identified because its wildtype allele is nearly always cotransduced with the wildtype alleles of the genes that flank it. For example, in part B of Figure 7.17, a genetic marker located between gal^+ and bio^+ would almost always be present in gal^+ bio^+ transductants.

7.6

Bacteriophage DNA molecules in the same cell can recombine.

The reproductive cycle of a phage is called the **lytic cycle.** In the lytic cycle, phage DNA enters a cell and replicates repeatedly, bacterial ribosomes are used to produce phage protein components, the newly synthesized phage DNA molecules are packaged into protein shells to form progeny phages, and the bacterium is split open (**lysis**), releasing the progeny phages from the cell. Phage progeny from a bacterium infected by one phage have the parental genotype, except for new mutations. However, if two phage particles that have *different* genotypes infect a single bacterial cell, new genotypes can arise by genetic recombination. This process differs significantly from genetic recombination in eukaryotes in two ways: (1) The number of participating DNA molecules varies from one cell to the next, and (2) reciprocal recombinants are not always recovered in equal frequencies from a single infected cell. Recombination in bacteriophage is the subject of this section.

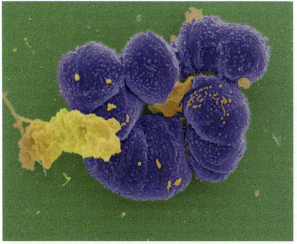

Clumps of *Streptococcus pyrogenes* bacteria (yellow) attacking human cells. These organisms are often responsible for infections of the skin, middle ear, or tonsils. The most dangerous strains are the life-threatening "flesh-eating" forms that cause a rapidly progressing destruction of the skin and underlying tissues.

■ Bacteriophages form plaques on a lawn of bacteria.

Phages are easily detected because in a lytic cycle, an infected cell breaks open and releases phage particles into the growth medium. The test is performed as outlined in part A of Figure 7.18 . A large number of bacteria (about 10^8) are placed on a solid medium. After a period of growth, a continuous turbid layer of bacteria results. If phages are present at the time the bacteria are placed on the medium, each phage adsorbs to a cell, and shortly afterward, the infected cell lyses and releases many progeny phages. Each of the progeny phages adsorbs to a nearby bacterium, and after another lytic cycle, these bacteria in turn release progeny phages that can infect still other bacteria in the vicinity. These cycles of infection continue, and after several hours, the descendants of each phage that was originally present destroy all of the bacteria in a localized area, giving rise to a clear, trans-

parent region—a **plaque**—in the otherwise turbid layer of confluent bacterial growth. Phages can multiply only in growing bacterial cells, so exhaustion of nutrients in the growth medium limits phage multiplication and the size of the plaque. Because a plaque is a result of an initial infection by one phage particle, the number of phage particles originally present on the medium can be counted.

The genotypes of phage mutants can be determined by studying the plaques. In some cases, the appearance of the plaque is sufficient. For example, phage mutations that decrease the number of phage progeny from infected cells often yield smaller plaques. Large plaques can be produced by mutants that cause premature lysis of infected cells, so that each round of infection proceeds more quickly (Figure 7.18, part B). Another type of phage mutation can be identified by the ability or inability of the phage to form plaques on a particular bacterial strain.

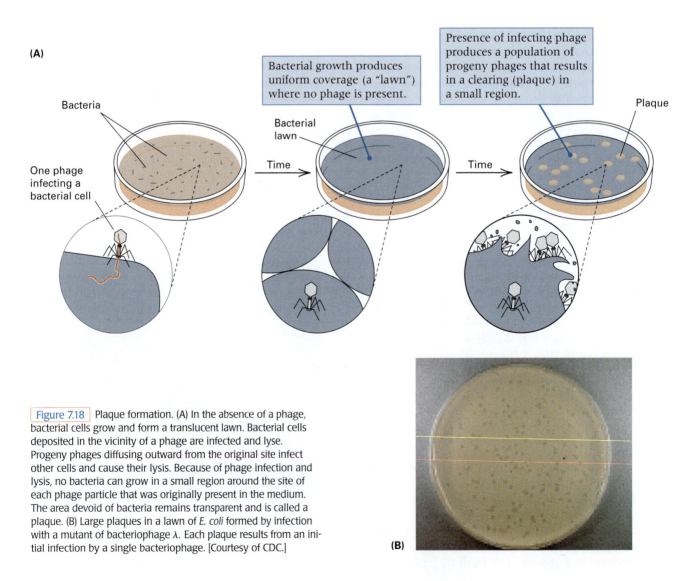

Figure 7.18 Plaque formation. (A) In the absence of a phage, bacterial cells grow and form a translucent lawn. Bacterial cells deposited in the vicinity of a phage are infected and lyse. Progeny phages diffusing outward from the original site infect other cells and cause their lysis. Because of phage infection and lysis, no bacteria can grow in a small region around the site of each phage particle that was originally present in the medium. The area devoid of bacteria remains transparent and is called a plaque. (B) Large plaques in a lawn of E. coli formed by infection with a mutant of bacteriophage λ. Each plaque results from an initial infection by a single bacteriophage. [Courtesy of CDC.]

Infection with two mutant bacteriophages yields recombinant progeny.

If two phage particles with different genotypes infect a single bacterium, some phage progeny are genetically recombinant. Figure 7.19 shows plaques resulting from the progeny of a mixed infection with *E. coli* phage T4 mutants. The r^- allele results in large plaques, and the h^- allele results in clear plaques. The cross is

$$r^- h^+ \quad \times \quad r^+ h^-$$
(large turbid plaque)   (small clear plaque)

Four plaque types can be seen in Figure 8.19. Two—the large turbid plaque and the small clear plaque—correspond to the phenotypes of the parental phages. The other two phenotypes—the large clear plaque and the small turbid plaque—are recombinants that correspond to the genotypes $r^- h^-$ and $r^+ h^+$, respectively. When many bacteria are infected, approximately equal numbers of reciprocal recombinant types are usually found among the progeny phage. In an experiment like that shown in Figure 7.19, in which each of the four genotypes yields a different phenotype of plaque morphology, the numbers of the genotypes can be counted by examining each of the plaques that is formed. The recombination frequency is the percentage of progeny phage that have recombinant genotypes.

Recombination occurs within genes.

The *rII* gene of phage T4 was used by Seymour Benzer to study the genetic fine structure of genes through recombination mapping. Using novel tech-

niques that reduced the number of required crosses from more than half a million to a mere several thousand, Benzer succeeded in mapping 2400 independent mutations in the *rII* gene.

Wildtype T4 bacteriophage is able to multiply in *E. coli* strains B and K12(λ) and gives small ragged plaques. (The strain K12(λ) has a phage λ genome integrated into the bacterial chromosome, which happens by means of a process discussed in Section 7.7). Mutations in the *rII* gene of T4 result in large round plaques on strain B but prevent propagation in strain K12(λ). If *E. coli* cells of strain B are infected with two different *rII* mutants, then recombination between the mutants can be detected, even though the frequency is very low, by taking advantage of the inability of *rII* mutants to grow on K12(λ). Plating the progeny phages on K12(λ) selects for growth of the rII^+ recombinant progeny, because only these recombinants *can* grow. Furthermore, because very large numbers of progeny phage can be examined (numbers of 10^{10} bacteriophages/ml are not unusual), even very low frequencies of recombination can be detected.

Some mutations failed to recombine with several mutations, each of which recombined with the others. These were interpreted to be deletion mutations, because they prevented recombination with two or more "point" mutations known to be at different sites in the gene. Each deletion eliminated a part of the bacteriophage genome, including a region of the *rII* gene. The use of deletions greatly simplified the ordering and mapping of thousands of mutations.

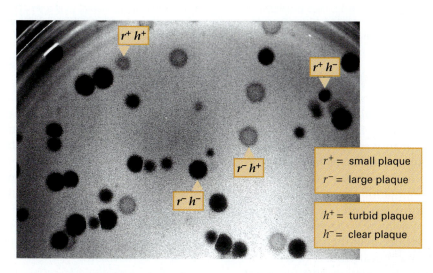

$r^+ h^+$

$r^+ h^-$

$r^- h^+$

$r^- h^-$

r^+ = small plaque
r^- = large plaque

h^+ = turbid plaque
h^- = clear plaque

Figure 7.19 A phage cross is performed by infecting host cells with both parental types of phage simultaneously. This example shows the progeny of a cross between T4 phage of genotypes r^- h^+ and r^+ h^- when both parental phage infect cells of *E. coli*. The $r^+ h^+$ and $r^- h^-$ genotypes are recombinants. [Courtesy of Leslie Smith and John W. Drake.]

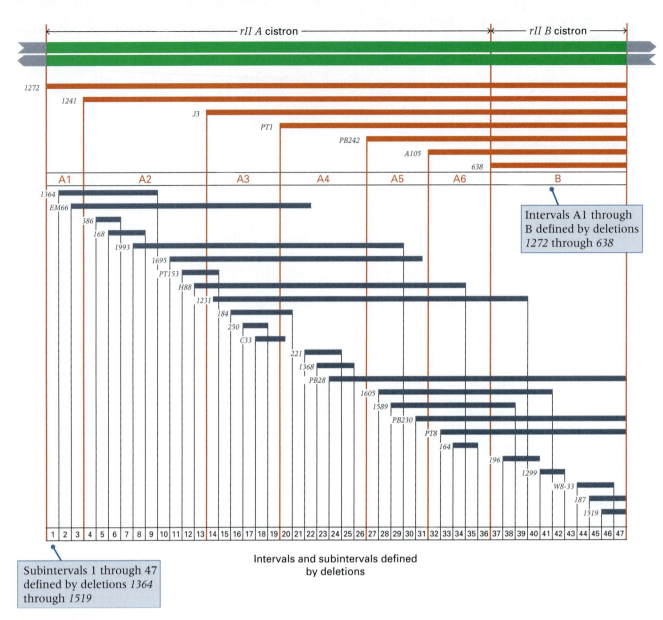

Intervals A1 through B defined by deletions *1272* through *638*

Subintervals 1 through 47 defined by deletions *1364* through *1519*

Intervals and subintervals defined by deletions

Figure 7.20 The array of deletion mutations used to divide the *rII* locus of bacteriophage T4 into 7 regions and 47 smaller subregions. The extent of each deletion is indicated by a horizontal bar. Any deletion endpoint used to establish a boundary between regions or subregions is indicated with a vertical line. [After S. Benzer. 1961. *Proc. Natl. Acad Sci. USA* 47: 403.]

Figure 7.20 depicts the array of deletion mutations used for mapping. Deletion mapping is based on the presence or absence of recombinants; each cross yields a yes-or-no answer and so avoids many of the ambiguities of genetic maps based on frequencies of recombination. In any cross between an unknown "point" mutation (for example, a simple nucleotide substitution) and one of the deletions, the presence of wildtype progeny means that the point mutation is outside of the region missing in the deletion. (The reciprocal product of recombination, which carries the deletion plus the point mutation, is not detected in these experiments.) On the other hand, if the point mutation is present in the region missing in the deletion, then wildtype

recombinant progeny cannot be produced. Because each cross clearly reveals whether a particular mutation is within the region missing in the deletion, deletion mapping also substantially reduces the amount of work needed to map a large number of mutations. The genetic map generated for a large number of independent *rII* mutations is given in Figure 7.21 .

The *rII* mutation and mapping studies were important because they gave experimental support to these conclusions:

- Genetic exchange can take place within a gene and probably between any pair of adjacent nucleotides.

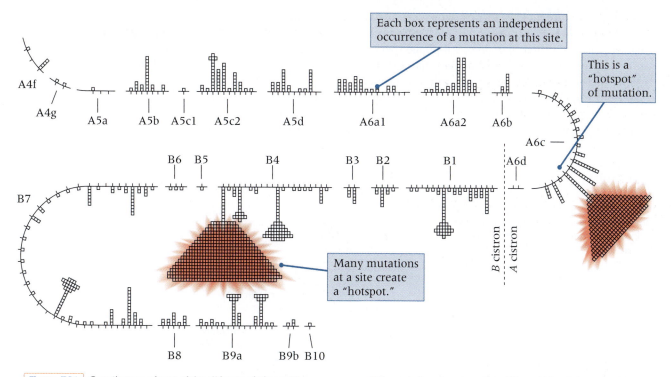

Each box represents an independent occurrence of a mutation at this site.

This is a "hotspot" of mutation.

Many mutations at a site create a "hotspot."

A4f A4g A5a A5b A5c1 A5c2 A5d A6a1 A6a2 A6b A6c A6d

B cistron A cistron

B7 B6 B5 B4 B3 B2 B1 B8 B9a B9b B10

Figure 7.21 Genetic map of part of the *rII* locus of phage T4. Each small square indicates a separate, independent occurrence of a mutation at the indicated site. The arrangement of sites within each A or B segment is arbitrary. [After S. Benzer. 1961. *Proc. Natl. Acad Sci. USA* 47: 403.]

- Mutations are not produced at equal frequencies at all sites within a gene. For example, the 2400 *rII* mutations were located at only 304 sites. One of these sites accounted for 474 mutations (Figure 7.21); a site that shows such a high frequency of mutation is called a **hotspot** of mutation. At sites other than hotspots, mutations were recovered only once or a few times.

The *rII* analysis was also important because it helped to distinguish experimentally between three distinct meanings of the word *gene*. Most commonly, the word *gene* refers to a unit of function. Physically, this corresponds to a protein-coding segment of DNA. Benzer assigned the term **cistron** to this unit of function, and that term is still occasionally used. The unit of function is normally defined experimentally by a complementation test (see Section 2.8), and indeed, two units of function, the *rIIA* cistron and the *rIIB* cistron, were defined by complementation. The limits of *rIIA* and *rIIB* are shown in Figures 7.20 and 7.21. The complementation between *rIIA* and *rIIB* is observed when two types of T4 phage, one with a mutation in *rIIA* and the other with a mutation in *rIIB*, are used simultaneously to infect *E. coli* strain K12(λ). The multiply infected cells produce normal numbers of phage progeny, most of which

carry either the parental *rIIA* mutation or the parental *rIIB* mutation. However, rare recombination between the mutant sites results in recombinant progeny phage that are *rII⁺*. In contrast, when K12(λ) is simultaneously infected with two phage that have different *rIIA* mutations or with two phage that have different *rIIB* mutations, no progeny phage are produced at all.

Besides meaning a unit of function, which is clarified by use of the term *cistron*, the term *gene* has two other distinct meanings: (1) the unit of genetic transmission that participates in recombination, and (2) the unit of genetic change or mutation. Physically, both the recombinational and the mutational units correspond to the individual nucleotides in a gene. Despite potential ambiguity, *gene* is still the most important word in genetics; in most cases, the shade of meaning intended is clear from the context.

7.7

Lysogenic bacteriophages do not necessarily kill the host.

The lytic cycle is one of two alternative phage life cycles. The alternative to the lytic cycle is called the **lysogenic cycle,** in which no progeny particles are produced, the infected bacterium survives, and a phage DNA molecule is transmitted to each bacterial progeny cell when the cell divides. All phage species can undergo a lytic cycle. Those phages that are also capable of the lysogenic cycle are called *temperate phage*, and those capable of only the lytic cycle are called *virulent phage*. In the lysogenic cycle, a replica of the infecting phage DNA becomes inserted, or *inte-*

grated, into the bacterial chromosome (Figure 7.22). The inserted DNA is called a **prophage**, and the surviving bacterial cell is called a **lysogen.** A lysogen is denoted by the designation of the bacterial strain followed by the name of the lysogenic phage in parentheses; for example, a clone of *E. coli* strain K12 that has become lysogenic for phage λ is denoted K12(λ). Many bacterial generations after a strain has become

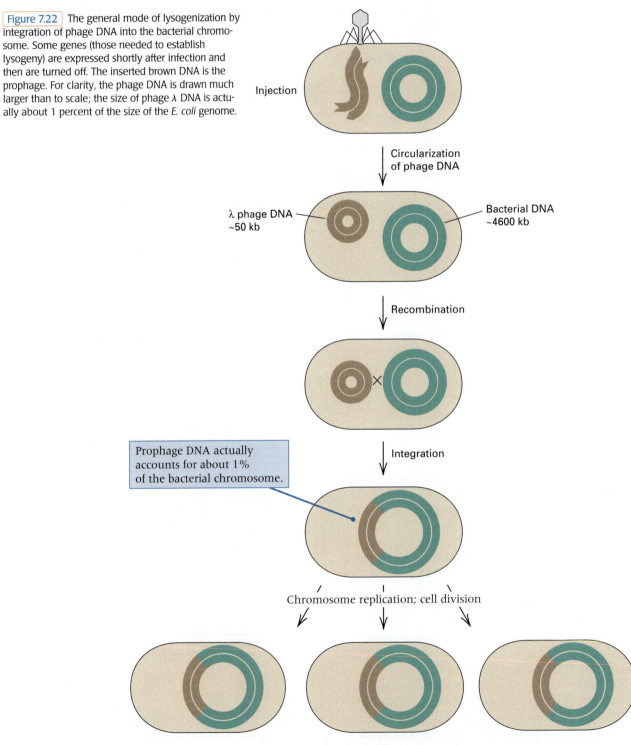

Figure 7.22 The general mode of lysogenization by integration of phage DNA into the bacterial chromosome. Some genes (those needed to establish lysogeny) are expressed shortly after infection and then are turned off. The inserted brown DNA is the prophage. For clarity, the phage DNA is drawn much larger than to scale; the size of phage λ DNA is actually about 1 percent of the size of the *E. coli* genome.

Injection

Circularization of phage DNA

λ phage DNA ~50 kb

Bacterial DNA ~4600 kb

Recombination

Prophage DNA actually accounts for about 1% of the bacterial chromosome.

Integration

Chromosome replication; cell division

Many lysogenic daughter cells

Problem: Although deletions were extremely important in Benzer's *rll* mapping experiments, they did not come labeled as "deletions" but had to be identified as deletions by experiments. Two principles were used in this identification: (1) deletions cannot undergo reverse mutation to wildtype, and (2) deletions do not map as "point mutations" in recombination experiments but fail to recombine with two or more mutations that do recombine with each other. This problem illustrates the second approach. The accompanying data are from some of Benzer's experiments showing the observed frequency of recombination between all combinations of 8 *rll* mutations. To avoid unnecessary complexity, the mutations have been renamed in sequential order, and the frequencies of recombination have been adjusted to make them perfectly additive. Identify which of the mutations are "point mutations" and which are deletions. Draw a genetic map showing the locations of the point mutations and the frequencies of recombination between adjacent mutations, and indicate the location and extent of each deletion. (The answer can be found on page 271.)

	r1	r2	r3	r4	r5	r6	r7	r8
r1	0	0	0.6	1.0	1.1	0	1.3	2.5
r2		0	0	0.1	0.2	0	0.4	1.6
r3			0	1.6	0.5	0	0.7	1.9
r4				0	2.1	0.5	2.3	3.5
r5					0	0	0.2	1.4
r6						0	0	0.9
r7							0	0
r8								0

lysogenic, the prophage can be activated and excised from the chromosome, and the lytic cycle can begin. Prophage activation results in a normal lytic cycle in which the host cell is killed and progeny phage are released.

A temperate phage, such as *E. coli* phage λ, when reproducing in its lytic cycle, undergoes general recombination, much as phage T4 does. A map of the phage λ genome is depicted in Figure 7.23; it is linear rather than circular. The DNA molecule in

the λ phage particle is also linear. Unlike the DNA molecules in T4 phage, however, every phage λ DNA molecule has identical ends. Indeed, the ends are single-stranded and complementary in sequence so that they can pair, forming a circular molecule. The single-stranded ends are called **cohesive ends** to indicate their ability to undergo base pairing.

The complete DNA sequence of the genome of λ phage has been determined, and many of the

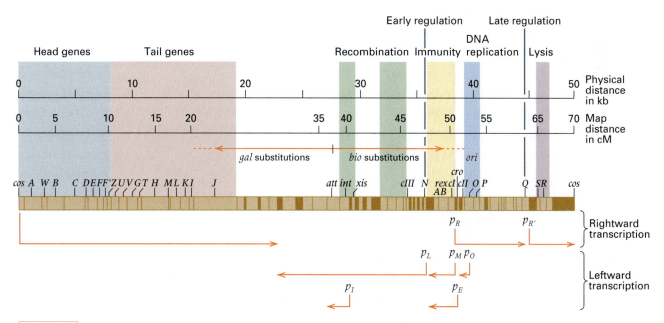

Figure 7.23 | Molecular and genetic maps of bacteriophage λ. The scale of the molecular map is length in kilobase pairs (kb); genetic distances are given in centimorgans (cM), equivalent in this case to percent recombination. Clusters of genes with related functions are indicated by the pastel-colored boxes. The positions of the regulatory genes *N* and *Q* are indicated by vertical lines at about 35 kb and 45 kb, respectively. Promoters are indicated by the letter *p*. The direction of transcription and length of transcript are indicated by arrows. Many of the genes known through mutant phenotypes are identified. Coding regions of unidentified function are indicated as unlabeled, light-colored rectangles; dark-colored regions indicate sequences unlikely to code for proteins. The λ *att* (attachment site) is the site of recombination for integration of the λ prophage. The origin of replication (*ori*) is the region denoted *O*. The extents of possible substitutions of λ DNA with *E. coli* DNA in the specialized transducing phages λ*dgal* and λ*dbio* are indicated by arrows just above the gene map; such transducing phages are discussed in the section on specialized transduction.

genes and gene products and their functions have been identified. The map of genes in Figure 7.23 shows where each gene is located along the DNA molecule, scaled in kilobase pairs (kb) rather than in terms of its position in the genetic map. The genes in λ, like those in most other bacteriophage, show extensive clustering by function. The left half of the map consists entirely of genes whose products (head and tail proteins) are required for assembly of the phage structure, and within this region the head genes and the tail genes themselves form subclusters. The right half of the λ genome also shows several gene clusters, which include genes for DNA replication, recombination, and lysis. The genes are clustered not only by function but also according to the time at which their products are synthesized. For example, the N gene acts early; genes O and P are active later; and genes Q, S, R, and the head-tail cluster are expressed last. The transcription patterns for mRNA synthesis are thus very simple and efficient. There are only two rightward transcripts, and all late genes except Q are transcribed into the same mRNA.

Two features of the lysogenic cycle of phage λ will be considered here: how the DNA is integrated and how it is excised. Several important observations emerged from interrupted-mating experiments and transductional analysis of *E. coli* cells lysogenic for phage λ.

- The bacterial *gal* and *bio* genes and the prophage are linked, with the gene order *gal* λ *bio*.
- The presence of the prophage increases the physical distance between *gal* and *bio*, as evidenced by the finding that the *gal* and *bio* genes can be cotransduced by P1 phage grown on a nonlysogen but not by P1 grown on a lysogen.

- Genetic mapping of prophage genes with respect to the *gal* and *bio* genes yields a prophage genetic map that is a permutation of the genetic map of the phage progeny obtained from standard phage crosses. That is, the genetic map from phage crosses is *A J att N R*, whereas the prophage map is *gal N R A J bio* (Figure 7.24).

These observations were a key source of insight into the biology of λ, because they implied that lysogeny was accompanied by circularization of the λ DNA and integration into the chromosome of the *E. coli* host. Furthermore,

1. In a λ lysogen, the λ prophage is *linearly* inserted between the *gal* and *bio* genes in the bacterial DNA.

2. Integration is the result of a recombination event that breaks and rejoins DNA at a particular site in λ and at a site in the bacterial chromosome between the *gal* and *bio* genes; the site of recombination is designated in Figures 7.23 and 7.24 as *att*.

These two features have been proved by studies of the structure of λ DNA and an analysis of the nature of the sites of exchange.

The DNA of λ is a linear molecule with cohesive ends, each twelve bases long. Each cohesive end (*cos*) is a single-stranded region at the end of the linear molecule (Figure 7.25). The *cos* sequences are complementary, so the termini can form base pairs. Pairing yields a circular molecule with two gaps that are joined by DNA ligase to create the closed circular molecule shown in Figure 7.25. Circularization takes place early in both the lytic and the lysogenic cycle and is necessary for both processes—for DNA replication in the lytic mode, for prophage integration in the lysogenic cycle.

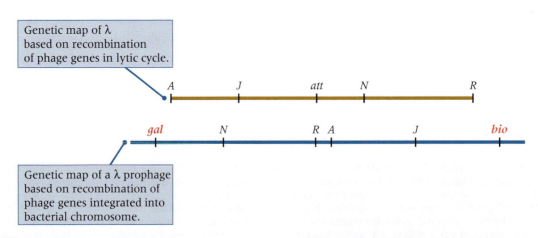

Genetic map of λ based on recombination of phage genes in lytic cycle.

Genetic map of a λ prophage based on recombination of phage genes integrated into bacterial chromosome.

Figure 7.24 The map order of genes in phage λ as determined by phage recombination (lytic cycle) and in the prophage (prophage order). The genes have been selected arbitrarily to provide reference points. Bacterial DNA is labeled in red letters.

Answer to Problem: The mutation *r2* fails to undergo recombination with *r1* and *r3,* but *r1* and *r3* do recombine with each other; hence *r2* is identified as a deletion. The mutation *r6* also fails to recombine with *r1* and *r3.* The mutation *r7* fails to recombine with *r6* and *r8,* but *r6* and *r8* do recombine with each other; evidently *r7* is a deletion that overlaps the *r6* deletion and includes the mutation *r8.* The genetic map for the point mutations is shown in gold in the accompanying illustration. The location and extent of each deletion are shown as a red bar beneath the genetic map. Each deletion is located precisely in the genetic map according to the frequency of recombination between either end of the deletion and the nearest genetic marker not included in the deletion. Note that there is insufficient information to specify the right-hand end of deletion *r7.*

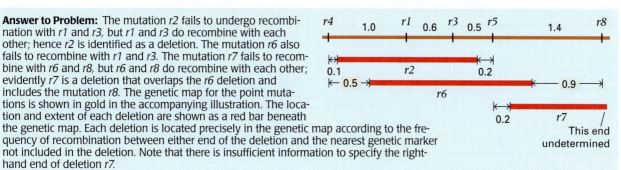

The sites of breakage and rejoining in the bacterial and phage DNA are called the *bacterial attachment site* and the *phage attachment site.* Each attachment site consists of three segments. The central segment has the same nucleotide sequence in both attachment sites and is the region in which the DNA molecules are broken and rejoined. The phage attachment site is denoted *POP'* (*P* for *phage*), and the bacterial

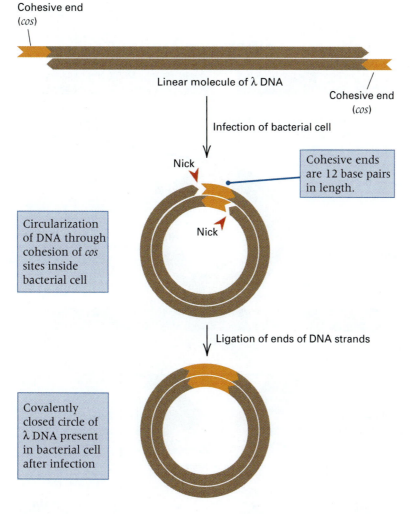

Figure 7.25 A diagram of a linear λ DNA molecule showing the cohesive ends (complementary single-stranded ends). Circularization by means of base pairing between the cohesive ends forms an open (nicked) circle, which is converted into a covalently closed (uninterrupted) circle by sealing (ligation) of the single-strand breaks. The length of the cohesive ends is 12 base pairs in a total molecule of approximately 50 kb.

attachment site is denoted BOB' (B for *bacteria*). A phage-specific *integrase* catalyzes a site-specific recombination event that results in integration of the λ DNA molecule into the bacterial DNA (Figure 7.26). The site-specific recombination between POP' and BOB' is very similar to the site-specific recombination between *loxP* sites illustrated in Figure 7.6 on page 247. The result is that the circular DNA of the phage becomes inserted into the circular DNA of the bacterial cell at the site of BOB'. The difference between the phage genetic map and the prophage genetic map is a consequence of the circularization of the phage DNA and the central location of POP'.

A lysogenic cell can replicate nearly indefinitely without the release of phage progeny. However, the prophage can sometimes become activated to undergo a lytic cycle in which the usual number of phage progeny are produced. This phenomenon is called **prophage induction,** and it is initiated by damage to the bacterial DNA. Prophage induction is often caused by some environmental agent that damages DNA, such as chemicals or radiation. The ability to be induced allows the phage to escape from a damaged cell. The biochemical mechanism of induction is complex and will not be described in this book, but the excision of the phage is straightforward.

Excision is another site-specific recombination event that reverses the integration process. Excision requires the phage enzyme integrase and a phage protein called **excisionase.** Genetic evidence and studies of physical binding of purified excisionase, integrase, and λ DNA indicate that excisionase binds to integrase and thereby enables the latter to recognize the prophage attachment sites BOP' and POB'. Once bound to these sites, integrase makes cuts in the O sequence and recreates the BOB' and POP' sites. This reverses the integration reaction, causing excision of the prophage (Figure 7.26).

When a cell is lysogenized, a block of phage genes becomes part of the bacterial chromosome, so the phenotype of the bacterium might be

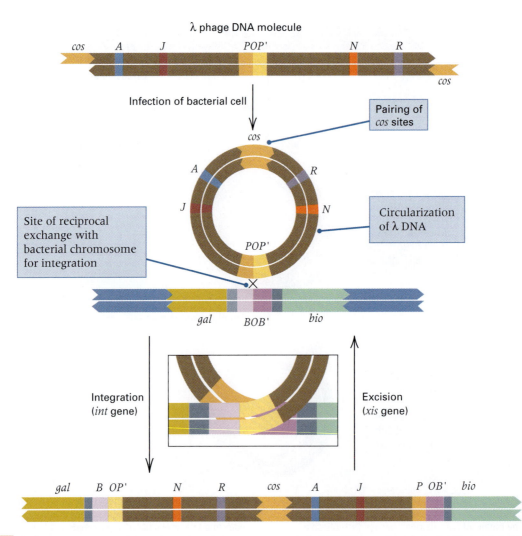

Figure 7.26 The geometry of integration and excision of phage λ. The phage attachment site is POP'. The bacterial attachment site is BOB'. The prophage is flanked by two hybrid attachment sites denoted BOP' and POB'.

expected to change—and indeed it does. Most phage genes in a prophage are kept in an inactive state by a **phage repressor** protein, the product of one of the phage genes. The repressor protein is synthesized initially by the infecting phage and then continually by the prophage. The gene that codes for the repressor is frequently the only prophage gene that is expressed in lysogens. If a lysogen is infected with a phage of the same type as the prophage—for example, λ infecting a λ lysogen—then the repressor present within the cell from the prophage prevents expression of the genes of the infecting phage. The resistance to infection by a phage identical with the prophage is called *immunity*. Thus λ does not form plaques on bacteria containing a λ prophage, because the infected cells are immune.

■ **Specialized transducing phages carry a restricted set of bacterial genes.**

When a bacterium lysogenic for phage λ is subjected to DNA damage that leads to induction, the prophage is usually excised from the chromosome precisely. However, once in every 10^6 or 10^7 cells, an excision error is made in which the sites of breakage and rejoining are displaced (Figure 7.27). The displaced sites of breakage and rejoining are not always located so as to produce a length of DNA that can fit in a λ phage head—the DNA may be too large or too small—but sometimes a molecule forms that can replicate and be packaged. In λ lysogens, the prophage lies between the *gal* and *bio* genes, and because the aberrant cut in the bacterial DNA can be either to the right or to the left of the

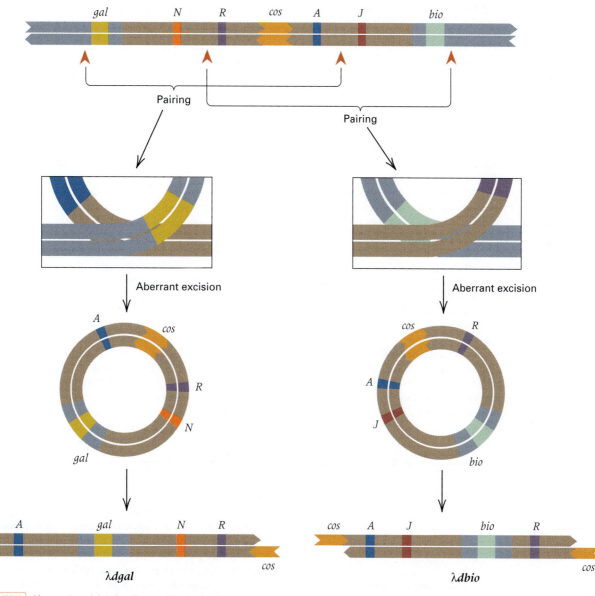

Figure 7.27 Aberrant excision leading to the production of specialized λ transducing phages. (A) Formation of a *gal* trans-ducing phage (λ*dgal*). (B) Formation of a *bio* transducing phage (λ*dbio*).

7.7 Lysogenic bacteriophages do not necessarily kill the host 273 prophage, aberrant phage particles can carry either the *bio* genes (cut at the right) or the *gal* genes (cut at the left). They are called λ*dbio* and λ*dgal* transducing particles (Figure 7.27). These are *specialized*

transducing phages because they can transduce only certain bacterial genes, in contrast to the P1-type generalized transducing particles, which can transduce any gene.

chapter summary

7.1 Many DNA sequences in bacteria are mobile and can be transferred between individuals and among species.

- A plasmid is an accessory DNA molecule, often a circle.
- The F plasmid is a conjugative plasmid.
- Insertion sequences and transposons play a key role in bacterial populations.
- Nonconjugative plasmids can be mobilized by cointegration into conjugative plasmids.
- Integrons have special site-specific recombinases for acquiring antibiotic-resistance cassettes.
- Bacteria with resistance to multiple antibiotics are an increasing problem in public health.

Bacterial cells often contain plasmids—nonessential DNA molecules that replicate and are transmitted to progeny cells. Some plasmids, such as the F (fertility) factor, are conjugative: They contain genes that make possible their transfer between cells. Other plasmids are nonconjugative, but through homologous recombination they can form a cointegrate with a conjugative plasmid and be transferred by a sort of hitchhiking. Most bacteria also possess transposable elements, which are capable of jumping from one position in a DNA molecule to another. Transposition does not require sequence homology. Bacterial insertion sequences (IS) are small transposable elements that code for their own transposase and that have inverted-repeat sequences at the ends. Some larger transposons have a composite structure consisting of a central region, often containing one or more antibiotic-resistance genes, flanked by two IS sequences in inverted orientation. DNA sequences called integrons encode a site-specific recombinase (integrase) and can capture circular cassettes of DNA by site-specific recombination between an *attI* site in the integron and an *attC* site in the cassette. Accumulation of antibiotic-resistance transposons and integrons in plasmids has given rise to multiple-drug-resistant R plasmids, which confer resistance to many chemically unrelated antibiotics.

7.2 Mutations that affect a cell's ability to form colonies are often used in bacterial genetics.

7.3 Transformation results from the uptake of DNA and recombination.

DNA can be transferred from one bacterium to another in three ways: transformation, transduction, and conjugation. In transformation, free DNA molecules obtained from

donor cells are taken up by recipient cells; by a recombinational mechanism, a single-stranded segment becomes integrated into the recipient chromosome, replacing a homologous segment.

7.4 In bacterial mating, DNA transfer is unidirectional.

- The F plasmid can integrate into the bacterial chromosome.
- Chromosome transfer begins at F and proceeds in one direction.
- The unit of distance in the *E. coli* genetic map is the length of chromosomal DNA transferred in one minute.
- Some F plasmids carry bacterial genes.

In conjugation, donor and recipient cells pair, and a single strand of DNA is transferred by rolling-circle replication from the donor cell to the recipient. Transfer is mediated by the transfer genes of the F plasmid. When F is present as a free plasmid, it becomes established in the recipient as an autonomously replicating plasmid; if F is integrated into the donor chromosome (that is, if the donor is an Hfr cell), only part of the donor chromosome is usually transferred, and this part can be maintained in the recipient only after an exchange event. About 100 minutes is required to transfer the entire *E. coli* chromosome, but the mating cells normally break apart before transfer is complete. DNA transfer starts at a particular point in the Hfr chromosome (the site of integration of F) and proceeds linearly. The times at which donor markers first enter recipient cells—the times of entry—can be arranged in order, yielding a map of the bacterial genome. We detect the map as circular because of the multiple sites at which an F plasmid integrates into the bacterial chromosome. Occasionally, F is excised from the chromosome in an Hfr cell; aberrant excision, in which one cut is made at the end of F and the other cut is made in the chromosome, gives rise to F' plasmids, which can transfer bacterial genes.

7.5 Some phages can transfer small pieces of bacterial DNA.

In transduction, a generalized transducing phage infects a donor cell, releases nuclease enzymes that fragment the host DNA, and occasionally packages fragments of host DNA into phage particles. The transducing particles contain no phage DNA but can inject donor DNA into a recipient bacterium; by genetic exchange, the transferred DNA can replace homologous DNA of the recipient, producing a recombinant bacterium called a transductant.

7.6 Bacteriophage DNA molecules in the same cell can recombine.

- Bacteriophages form plaques on a lawn of bacteria.
- Infection with two mutant bacteriophages yields recombinant progeny.
- Recombination occurs within genes.

When bacteria are infected with several phages, genetic exchange can take place between phage DNA molecules, generating recombinant phage progeny. Measurement of recombination frequency yields a genetic map of the phage. The fine-structure genetic analysis of the *rII* region led to the discovery of hotspots of mutation and to clarification of the term *gene* as meaning (1) the unit of function (cistron) defined by a complementation test, (2) the unit of genetic transmission that participates in recombination, or (3) the unit of genetic change or mutation.

7.7 Lysogenic bacteriophages do not necessarily kill the host.

- Specialized transducing phages carry a restricted set of bacterial genes.

In contrast to virulent phage, such as T4, the temperate phages possess mechanisms for recombining phage and bacterial DNA. A bacterium containing integrated phage DNA is called a lysogen, the integrated phage DNA is a prophage, and the overall phenomenon is lysogeny. Integration results from site-specific recombination between particular sequences, called attachment sites, in the bacterial and phage DNA. The bacterial and phage attachment sites are not identical but have a short common sequence in which the crossover takes place. Temperate phages circularize their DNA after infection. Because the phage has a single attachment site that is not terminally located, the order of the prophage genes is a permutation of the order of the genes in the phage particle. The integrated prophage DNA is stable, but if the bacterial DNA is damaged, prophage induction is initiated, the phage DNA is excised, and the lytic cycle of the phage ensues. Defective excision of a prophage yields a specialized transducing phage capable of transducing bacterial genes on either side of the phage attachment site.

issues & ideas

- How are antibiotic-resistance cassettes acquired by integrons? Once acquired, can they be lost?

- What role do antibiotic-resistance genes, transposable elements, and transmissible plasmids play in relation to certain pathogenic bacteria that are simultaneously resistant to multiple, chemically unrelated antibiotics?

- How could you distinguish between a bacterial strain that has the phenotype Lac$^+$ and one that has the phenotype Lac$^-$?

- How could you distinguish a Lac$^+$ Amp-r bacterial strain from one that is Lac$^-$ Amp-s? (Amp-r and Amp-s denote resistance and sensitivity, respectively, to the antibiotic ampicillin.)

- What is the physical basis of cotransformation? If two genes can be cotransformed, what does this observation

imply about the ability of each gene to be cotransformed with a genetic marker located between them?

- How does the physical state of the F factor differ between an F$^+$ bacterial cell and an Hfr bacterial cell?

- Is the F$^+$ state of a bacterial cell infectious? Is the Hfr state infectious? Explain why or why not.

- When an Hfr bacterial cell transfers its chromosomal DNA into an F$^-$ recipient cell, where in the chromosome does the transfer process begin? How long does it take to transfer the entire chromosome? Why does complete chromosome transfer happen only rarely?

- How does the process of transduction differ from that of transformation?

key terms & concepts

auxotroph	F$'$ plasmid	lysis	prophage induction
cassette	F$^+$ cells	lysogen	prototroph
cistron	F$^-$ cells	lysogenic cycle	R plasmid
cohesive ends	frequency of cotransduction	lytic cycle	selected marker
cointegrate	generalized transducing	minimal medium	selective medium
conjugation	phage	mobile DNA	site-specific recombinase
conjugative plasmid	Hfr cells	nonselective medium	specialized transducing phage
cotransduction	hotspot	partial diploid	time of entry
cotransformation	insertion sequence	phage repressor	transducing phage
counterselected marker	integrase	plaque	transduction
episome	integron	plasmid	transformation
excisionase	interrupted-mating technique	prophage	transposon
F factor			

key terms & concepts   *continued*

1. _____ An enzyme that carries out recombination between two specific DNA sequences.

2. _____ A DNA sequence that can change position in a DNA molecule or move from one molecule to another; insertion sequences and transposons are examples from bacteria.

3. _____ A plasmid that encodes proteins enabling it to be transferred from one bacterial cell to another.

4. _____ The formation of this type of molecule allows a nonconjugative plasmid to be transferred between cells by catching a ride on a conjugative plasmid.

5. _____ A type of conjugative plasmid that includes genes for resistance to one or more antibiotics.

6. _____ Bacterial cells containing an integrated F factor enabling bacterial genes to be transferred in the process of conjugation.

7. _____ In a cross between Hfr cells and F⁻ cells, the genetic marker used to eliminate the Hfr cells.

8. _____ Bacterial growth medium containing only those nutrients that are essential for growth of wildtype cells.

9. _____ A bacterial mutant that requires one or more nutrients for growth beyond those required by wildtype cells.

10. _____ Among cells showing transduction for one bacterial genetic marker, the frequency of cells in which a second genetic marker is also transduced.

11. _____ A small clearing in an otherwise continuous overgrowth of bacterial cells on a solid surface caused by successive generations of bacteriophage growth at that position.

12. _____ A protein encoded in temperate bacteriophage that keeps the prophage form from exiting the bacterial DNA and undergoing the lytic cycle.

solutions: step by step

Problem 1

Cotransduction experiments were carried out to determine the order of the closely linked genes *tolC, metC,* and *ebg* in the chromosome of *E. coli.* P1 phage grown on a strain of genotype

$$tolC^+ \; metC^+ \; ebg^+$$

were used to transduce a recipient strain of genotype

$$tolC^- \; metC^- \; ebg^-$$

The results were as shown in the accompanying table. What order of the genes is consistent with these results?

Selected marker	Genotypes of unselected markers	Observed percent
$tolC^+$	$metC^+ \; ebg^+$	2
	$metC^+ \; ebg^-$	12
	$metC^- \; ebg^+$	30
	$metC^- \; ebg^-$	56
$metC^+$	$tolC^+ \; ebg^+$	1
	$tolC^- \; ebg^+$	0
	$tolC^+ \; ebg^-$	34
	$tolC^- \; ebg^-$	65

■ **Solution** The gene order is specified if we can deduce the gene in the middle. One approach is to note that the wildtype allele of the gene in the middle will be cotransduced at high frequency (usually greater than 90 percent) when the flanking markers are both transduced. Consider the results when $tolC^+$ is the selected gene: Among the $tolC^+ \; metC^+$ transductants, 2/14 are ebg^+, which suggests that ebg is not in

the middle; among the $tolC^+ \; ebg^+$ transductants, 2/32 are $metC^+$, which suggests that $metC$ is not in the middle. These comparisons would suggest that $tolC$ is the gene in the middle, but it is important to see whether this hypothesis is consistent with the other results. Among the $metC^+ \; ebg^+$ transductants, 1/1 are $tolC^+$, but this is too little information for us to make an inference. However, among the $metC^+ \; tolC^+$ transductants, only 1/35 are ebg^+, which again implies that ebg is not in the middle. The data are therefore all consistent with the hypothesis that the gene order is $metC-tolC-ebg$ (or $ebg-tolC-metC$) and in any case are inconsistent with either of the other genes being in the middle.

Problem 2

Four independent integrations of the F factor into the chromosome of an unusual strain of *E. coli* yielded four different Hfr derivatives of the strain (HfrW, HfrX, HfrY, and HfrZ), each with a different origin and possibly a different direction of transfer of markers. These were examined in interrupted-mating experiments and were found to transfer chromosomal genes at the times shown in the accompanying table.

(a) Draw a circular genetic map, with position 0 (also 100) minutes at the top, showing the order of the chromosomal genes and the distance (in minutes) between adjacent genes. (The marker *leu* is near 2 minutes on the standard map.) Annotate the genetic map with arrows indicating the origin and direction of transfer of each Hfr and the distance (in minutes) from the origin of transfer to the first marker transferred.

(b) How does the genetic map of this particular strain compare with that of the standard *E. coli* strain in Figure 7.14? Suggest an explanation of any discrepancy between the genetic maps.

					Genetic marker				
Hfr	*his*	*lac*	*leu*	*lip*	*pheS*	*pro*	*pyrD*	*terC*	*tonA*	*trp*
W	−	20	−	11	−	−	3	−	−	−
X	−	−	18	−	−	−	31	−	20	25
Y	−	13	−	22	−	6	−	2	−	−
Z	19	−	−	−	12	4	−	8	−	−

■ **Solution** (a) Consider each Hfr in turn and, starting with the earliest-entering gene, write down the name of each gene as it enters. The difference in time of entry between adjacent genes is the distance in minutes between the genes. When a partial genetic map of gene transfer from each Hfr has been made, arrange the maps so that any shared markers between two or more Hfr strains coincide. This process yields the genetic maps shown here, where the arrows indicate direction of transfer and the numbers are times of transfer in minutes.

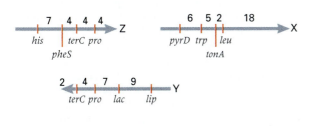

Now arrange the composite genetic map in the manner required, in the form of a circle of 100 minutes, with 0 minutes at the top and *leu* at about 2 minutes. The map is as shown in the illustration on the right.

(b) Comparison of this map with that in Figure 7.14 indicates that a group of markers, present in a contiguous segment of chromosome, is in the wrong order as compared with the standard map. The simplest explanation is that the strain in question has undergone an inversion of part of the chromosome, the breakpoints in the *tonA–pro* and *trp–terC* intervals in the standard map.

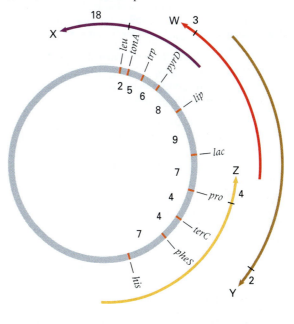

concepts in action: problems for solution

7.1 A Class 1 integron that carries two antibiotic-resistance cassettes traps a third cassette encoding geldanamycin resistance. Where in the integron is the new cassette inserted relative to the promoter at which transcription is initiated?

7.2 If 1×10^6 phage are mixed with 1×10^6 bacteria and all phage adsorb, what fraction of the bacteria remain uninfected?

7.3 Transfer of DNA in conjugation is always accompanied by replication. What type of replication takes place and where does it originate?

7.4 Is it possible for a plasmid without any genes to exist?

7.5 Numbers of phage or bacteria in a suspension are usually so large that single colonies or plaques would be impossible to observe without suitable dilution. The usual dilutions are 100-fold, in which 0.1 ml of the suspension is mixed with 9.9 ml of dilution buffer, or 10-fold, in which 1 ml of suspension is mixed with 9 ml of dilution buffer. Usually serial dilutions are necessary, in which the suspension is diluted once into dilution buffer, the resulting suspension mixed thoroughly and diluted a second time, the resulting suspension mixed thoroughly and diluted a third time, and so forth. The serial dilution factors multiply, so, for example, two 100-fold and two 10-fold dilutions yield

an overall dilution factor of $100 \times 100 \times 10 \times 10 = 10^6$. Using these principles, estimate the number of viable λ phage per milliliter (the phage titer) in a suspension from the following data. The original suspension was serially diluted through four dilutions of 100-fold each and one dilution of 10-fold. From the final dilution, a volume of 0.1 ml was mixed with a great excess of growing bacteria and then spread over nutrient agar in a petri dish and incubated overnight. The next day, 36 plaques were visible.

7.6 You are given a suspension of bacteria and told that it contains 5×10^7 viable cells per milliliter. What combination of 100-fold and 10-fold serial dilutions would you carry out so that 0.1 ml of the final dilution would contain approximately 50 viable cells?

7.7 A suspension of *E. coli* was serially diluted through two dilutions of 100-fold each and two dilutions of 10-fold each. From the final dilution, a volume of 0.1 ml was spread over nutrient agar in a petri dish and incubated overnight. The next day, 42 colonies were visible. Estimate the number of viable bacteria per milliliter in the original undiluted suspension.

7.8 Given that bacteriophage λ has a genome size of 50 kilobase pairs, what is the approximate genetic length of λ prophage in minutes? (*Hint:* The length of the *E. coli* genome is 4.6×10^6 base pairs.)

concepts in action: problems for solution *continued*

7.9 What is the difference between a selected marker and a counterselected marker? Why are both necessary in an Hfr $\times$ F$^-$ mating?

7.10 A cross of an Hfr strain of genotype h^+ and an F$^-$ strain of genotype h^- is carried out. When tetracycline sensitivity is used for counterselection, the number of recombinant colonies is 1000-fold lower than when streptomycin sensitivity is used for counterselection. In the former case the recombinants are h^+ *tet-r*, and in the latter case they are h^+ *str-r*. Suggest an explanation for the difference.

7.11 An Hfr strain transfers genes in alphabetical order, *a b c*. In an Hfr $a^+ b^+ c^+$ *str-s* $\times$ F$^-$ $a^- b^- c^-$ *str-r* mating, do all b^+ *str-r* recombinants receive the a^+ allele? Are all b^+ *str-r* recombinants also a^+? Why or why not?

7.12 A temperate bacteriophage has the gene order *a b c d e f g h*, whereas the prophage of the same phage has the gene order *g h a b c d e f*. What information does this permutation give you about the location of the phage attachment site?

7.13 Bacteriophage P1 transduction using a donor bacterial strain that has the wildtype alleles of the closely linked markers *a, b, c,* and *d* shows the following percentages of cotransduction:

$a-b$: 29	$a-c$: 2	$a-d$: 5
$b-c$: 0	$b-d$: 1	$c-d$: 50

What is the order of the genetic markers?

7.14 Bacterial cells of genotype *pur$^-$ pro$^+$ his$^+$* were transduced with P1 bacteriophage grown on bacteria of genotype *pur$^+$ pro$^-$ his$^-$*. Transductants containing *pur$^+$* were selected and tested for the unselected markers *pro* and *his*. The numbers of *pur$^+$* colonies with each of four genotypes were as follows:

pro$^+$	*his$^+$*	102
pro$^-$	*his$^+$*	25
pro$^+$	*his$^-$*	160
pro$^-$	*his$^-$*	1

What is the gene order?

7.15 You are studying a biochemical pathway in *E. coli* that leads to the production of substance A. You isolate a set of mutants, each of which is unable to grow on minimal medium unless it is supplemented with A. By performing appropriate matings, you group all the mutants into four complementation groups (genes) designated *a1, a2, a3,* and *a4*. You know beforehand that the biochemical pathway for the production of A includes four intermediates: B, C, D, and E. You test the nutritional requirements of the mutants by growing them on minimal medium supplemented with each of these intermediates in turn. The results are summarized in the following table, where the plus signs indicate growth and the minus signs indicate failure to grow.

	A	B	C	D	E
a1	+	+	+	−	+
a2	+	−	+	−	+
a3	+	−	−	−	+
a4	+	−	−	−	−

Determine the order in which the substances A, B, C, D, and E are most likely to participate in the biochemical pathway, and indicate the enzymatic steps by arrows. Label each arrow with the name of the gene that codes for the corresponding enzyme.

7.16 The genes *A, B, G, H, I,* and *T* were tested in all possible pairs for cotransduction with bacteriophage P1. Only the following pairs were found able to be cotransduced: *G* and *H, G* and *I, T* and *A, I* and *B, A* and *H*. What is the order of the genes along the chromosome? Explain your logic.

7.17 The order of the genes in the λ phage virus is

$$A\ B\ C\ D\ E\ att\ int\ xis\ N\ CI\ O\ P\ Q\ S\ R$$

(a) Given that the bacterial attachment site, *att*, is between *gal* and *bio* in the bacterial chromosome, what is the prophage gene order?
(b) A mutant phage is discovered that has the reverse gene order in the prophage as in the wildtype prophage. What does this say about the orientation of the *att* site in regard to the termini of the phage chromosome?
(c) A wildtype λ lysogen is infected with another λ phage carrying a genetic marker, *Z*, located between *E* and *att*. The superinfection gives rise to a rare, doubly lysogenic *E. coli* strain that carries both λ and λZ prophage. Assuming that the second phage also entered the chromosome at an *att* site, diagram two possible arrangements of the prophages in the bacterial chromosome and indicate the locations of the bacterial genes *gal* and *bio*.

7.18 An experiment was carried out in *E. coli* to map five genes around the chromosome using each of three different Hfr strains. The genetic markers were *bio, met, phe, his,* and *trp*. The Hfr strains were found to transfer the genetic markers at the times indicated here. Construct a genetic map of the *E. coli* chromosome that includes all five genetic markers, the genetic distances in minutes between adjacent gene pairs, and the origin and direction of transfer of each Hfr. Complete the missing entries in the table, which are indicated by question marks.

Hfr1 markers	*bio*	*met*	*phe*	*his*
Time of entry	26	44	66	?

Hfr2 markers	*phe*	*met*	*bio*	*trp*
Time of entry	?	26	44	75

Hfr3 markers	*phe*	*his*	?	*bio*
Time of entry	6	27	35	?

7.19 *Salmonella enterica* is closely related to *Escherichia coli*. It can be infected with the F plasmid, which can integrate into the chromosome to produce Hfr strains. These can be mated with F$^-$ *E. coli* to study the order and time of entry of genetic markers. The following data pertain to times of entry of four genetic markers in crosses of *E. coli* Hfr $\times$ *E. coli* F$^-$ and *S. enterica* Hfr $\times$ *E. coli* F$^-$.

	ile	*met*	*pro*	*arg*
E. coli Hfr $\times$ *E. coli* F$^-$	28	20	6	22
S. enterica Hfr $\times$ *E. coli* F$^-$	4	22	47	18

geNETics on the web

GeNETics on the Web will introduce you to some of the most important sites for finding genetic information on the Internet. To explore these sites, visit the Jones and Bartlett home page at

http://www.jbpub.com/genetics

For the book *Essential Genetics: A Genomics Perspective,* choose the link that says **Enter GeNETics on the Web**. You will be presented with a chapter-by-chapter list of highlighted keywords. Select any highlighted keyword and you will be linked to a Web site containing genetic information related to the keyword.

- With about 100 bacteriophage, 200 plasmid, and 50 bacterial genomes completely sequenced, merely finding the information that one seeks is not always as straightforward as it may seem. The National Center for Biotechnology Information (NCBI) maintains a site to facilitate the use of databases and software for analyzing data about genetics, molecular biology, and biochemistry. Browse the site map at the keyword **NCBI** and find the annotated maps of the *E. coli* and bacteriophage λ genomes.

- When bacteriophage were first discovered early in the twentieth century, they

opened up the possibility of **phage therapy**, or the use of specific kinds of bacteriophage to kill pathogenic bacteria. This avenue remained unexplored after the early 1940s when antibiotics were developed and began to be widely used in the treatment of bacterial infections. Today, with the spread of bacteria resistant to most or all available antibiotics causing increasing concern, phage therapy is the subject of renewed research. At this keyword site you can learn about the historical and ecological context of this research and find a realistic assessment of its prospects for success.

(a) How do the genetic maps of *E. coli* and *S. enterica* compare with respect to these genes?

(b) What are the origin and direction of transfer in each Hfr?

(c) Approximately how fast does the *S. enterica* Hfr transfer chromosomal DNA relative to the *E. coli* Hfr?

7.20 A time-of-entry experiment was carried out with the mating

$$\text{Hfr } a^+ \, b^+ \, c^+ \, d^+ \, str\text{-}s \times \text{F}^- \, a^- \, b^- \, c^- \, d^- \, str\text{-}r$$

The data in the accompanying table were obtained. Make a graph showing the number of recombinants per 100 Hfr (*y*-axis) against time of mating (*x*-axis) for each gene, and from this graph determine the time of entry of each gene.

Time of mating in minutes	Number of recombinants of indicated genotype per 100 Hfr			
	a^+ str-r	b^+ str-r	c^+ str-r	d^+ str-r
0	0.01	0.006	0.008	0.0001
10	5	0.1	0.01	0.0001
15	50	3	0.1	0.0005
20	95	35	2	0.001
25	97	80	20	0.001
30	98	82	43	0.01
40	98	80	40	8
50	99	80	40	12
60	98	81	42	16
70	99	80	41	16

further readings

Adelberg, E. A., ed. 1966. *Papers on Bacterial Genetics.* Boston: Little, Brown.

Campbell, A. 1976. How viruses insert their DNA into the DNA of the host cell. *Scientific American,* December.

Clowes, R. D. 1975. The molecules of infectious drug resistance. *Scientific American,* July.

Davies, J. 1995. Vicious circles: Looking back on resistance plasmids. *Genetics* 139: 1465.

Drlica, K., and M. Riley. 1990. *The Bacterial Chromosome.* Washington, DC: American Society for Microbiology.

Groth, A. C., and M. P. Calos. 2004. Phage integrases: Biology and applications. *Journal of Molecular Biology* 335: 667.

Hawkey, P. M., and C. J. Munday. 2004 Multiple resistance in gram-negative bacteria. *Reviews in Medical Microbiology* 15: 51.

Hopwood, D. A., and K. E. Chater, eds. 1989. *Genetics of Bacterial Diversity.* New York: Academic Press.

Komano, T. 1999. Shufflons: Multiple inversion systems and integrons. *Annual Review of Genetics* 33: 171.

Levy, S. B. 1998. The challenge of antibiotic resistance. *Scientific American,* March.

Losick, R., and D. Kaiser. 1997. Why and how bacteria communicate. *Scientific American,* February.

Low, K. B., and R. Porter. 1978. Modes of genetic transfer and recombination in bacteria. *Annual Review of Genetics* 12: 249.

Miller, R. V. 1998. Bacterial gene swapping in nature. *Scientific American,* January.

Neidhardt, F. C., R. Curtiss III, J. L. Ingraham, et al, eds. 1996. Escherichia coli *and* Salmonella typhimurium: *Cellular and Molecular Biology* (2 volumes). 2d ed. Washington, DC: American Society for Microbiology.

Novick, R. P. 1980. Plasmids. *Scientific American,* December.

Robertson, B. D., and T. F. Meyer. 1992. Genetic variation in pathogenic bacteria. *Trends in Genetics* 8: 422.

Snyder, L., and W. Champness. 1977. *Molecular Genetics of Bacteria.* Washington, DC: American Society for Microbiology.

Small and flexible biconcave disks, mammalian red blood cells are able to squeeze through tiny capillaries. In the process of their formation in the bone marrow, red blood cells lose their nucleus. All protein synthesis therefore relies on long-lived, premade messenger RNA. [© Chad Baker/Photodisc/Getty Images]

key concepts

- In gene expression, information in the base sequence of DNA is used to dictate the linear order of amino acids in a polypeptide by means of an RNA intermediate.

- Transcription of an RNA from one strand of the DNA is the first step in gene expression.

- In eukaryotes, the RNA transcript is modified and may undergo splicing to make the messenger RNA.

- The messenger RNA is translated on ribosomes in groups of three bases (codons), each specifying an amino acid through an interaction with molecules of transfer RNA.

8

The Molecular Genetics of Gene Expression

chapter organization

8.1 Polypeptide chains are linear polymers of amino acids.

8.2 The linear order of amino acids is encoded in a DNA base sequence.

8.3 The base sequence in DNA specifies the base sequence in an RNA transcript.

8.4 RNA processing converts the original RNA transcript into messenger RNA.

8.5 Translation into a polypeptide chain takes place on a ribosome.

8.6 The genetic code for amino acids is a triplet code.

8.7 Several ribosomes can move in tandem along a messenger RNA.

the human connection
Poly-U

Chapter Summary
Issues & Ideas
Key Terms & Concepts
Solutions: Step by Step
Concepts in Action: Problems for Solution
Further Readings

geNETics **on** the web

The term **gene expression** refers to the process by which information contained in genes is decoded to produce other molecules that determine the phenotypic traits of organisms. An overview of gene expression was presented in Chapter 1. The process is initiated when the information contained in the base sequence of DNA is copied into a molecule of RNA, and the process culminates when the molecule of RNA is used to determine the linear order of amino acids in a polypeptide chain. This chapter will increase your understanding of these events. The principal steps in gene expression are as follows:

1. RNA molecules are synthesized by an enzyme, *RNA polymerase,* which uses a segment of a single strand of DNA as a **template strand** to produce a strand of RNA complementary in base sequence to the template DNA. The overall process by which the segment corresponding to a particular gene is selected and an RNA molecule is made is called **transcription.**

2. In the nucleus of eukaryotic cells, the RNA usually undergoes chemical modification called **RNA processing.**

3. The processed RNA molecule is used to specify the order in which amino acids are joined together to form a polypeptide chain. In this manner, the amino acid sequence in a polypeptide is a direct consequence of the base sequence in the DNA. The production of an amino acid sequence from an RNA base sequence is called **translation,** and the protein made is called the **gene product.**

8.1

Polypeptide chains are linear polymers of amino acids.

Proteins are the molecules responsible for catalyzing most intracellular chemical reactions (enzymes), for regulating gene expression (regulatory proteins), and for determining many features of the structures of cells, tissues, and viruses (structural proteins). A protein is composed of one or more chains of linked amino acids called **polypeptide chains.** Twenty different amino acids are commonly found in polypeptides, and they can be joined in any number and in any order. Because the number of amino acids in a polypeptide usually ranges from a hundred to a thousand, an enormous number of polypeptide chains differing in amino acid sequence can be formed.

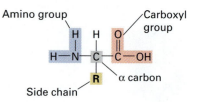

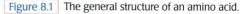
Figure 8.1 The general structure of an amino acid.

Each amino acid contains a carbon atom (the α carbon) to which is attached one carboxyl group ($-COOH$), one amino group ($-NH_2$), and a side chain commonly called an **R group.** In Figure 8.1, the α carbon is shown in gray, the carboxyl group in red, the amino group in blue, and the R group in gold. The R groups are generally chains or rings of carbon atoms bearing various chemical groups. The simplest R groups are those of glycine ($-H$) and of alanine ($-CH_3$). The chemical structures of all 20 amino acids are shown in Figure 8.2.

Polypeptide chains are formed when the carboxyl group of one amino acid joins with the amino group of a second amino acid to form a **peptide bond** (Figure 8.3, part A). In a polypeptide chain, the α-carbon atoms alternate with peptide groups to form a backbone that has an ordered array of side chains (Figure 8.3, part B). The opposite ends of a polypeptide molecule are chemically different. One end has a free $-NH_2$ group and is called the **amino terminus;** the other end has a free $-COOH$ group and is the **carboxyl terminus.** Polypeptides are synthesized by the addition of successive amino acids to the carboxyl end of the growing chain. Conventionally, the amino acids of a polypeptide chain are numbered starting at the amino terminus. Therefore, the amino acids are numbered in the order in which they are added to the chain during synthesis.

Owing to interactions between amino acids in the polypeptide chain, most polypeptide chains fold back on themselves in a convoluted manner into a unique three-dimensional shape, in some cases assisted by interactions with other proteins in the cell. About 70 to 75 percent of polypeptide chains fold correctly within milliseconds after release from the ribosome. Exceptionally long polypeptide chains, or ones with a slow or very complex folding pathway, are assisted in their folding by specialized proteins discussed in Section 8.5.

Many protein molecules consist of more than one polypeptide chain. When this is the case, the protein is said to contain *subunits.* The subunits may be identical or different. For example, hemoglobin, the oxygen carrier of blood, consists of four subunits—two of the α polypeptide chain and two of the β polypeptide chain.

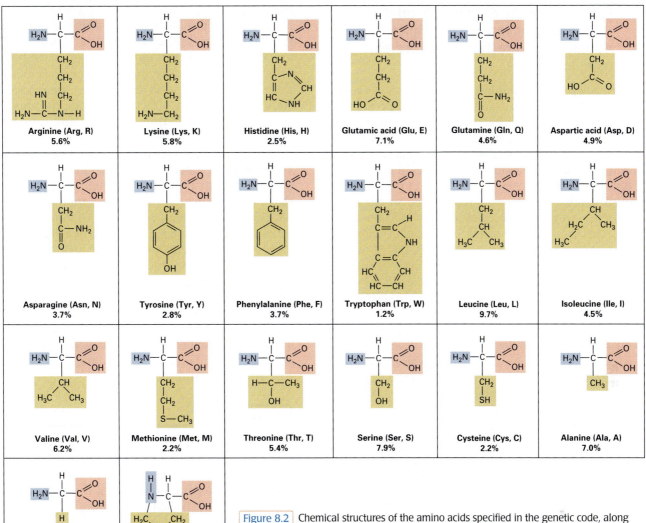

Arginine (Arg, R) 5.6%

Lysine (Lys, K) 5.8%

Histidine (His, H) 2.5%

Glutamic acid (Glu, E) 7.1%

Glutamine (Gln, Q) 4.6%

Aspartic acid (Asp, D) 4.9%

Asparagine (Asn, N) 3.7%

Tyrosine (Tyr, Y) 2.8%

Phenylalanine (Phe, F) 3.7%

Tryptophan (Trp, W) 1.2%

Leucine (Leu, L) 9.7%

Isoleucine (Ile, I) 4.5%

Valine (Val, V) 6.2%

Methionine (Met, M) 2.2%

Threonine (Thr, T) 5.4%

Serine (Ser, S) 7.9%

Cysteine (Cys, C) 2.2%

Alanine (Ala, A) 7.0%

Glycine (Gly, G) 6.7%

Proline (Pro, P) 6.1%

Figure 8.2 Chemical structures of the amino acids specified in the genetic code, along with their conventional three-letter and one-letter abbreviations. Note that proline does not have the same general structure as the rest because it lacks a free amino group. The percentage values give the relative abundance of each amino acid averaged over all human proteins.

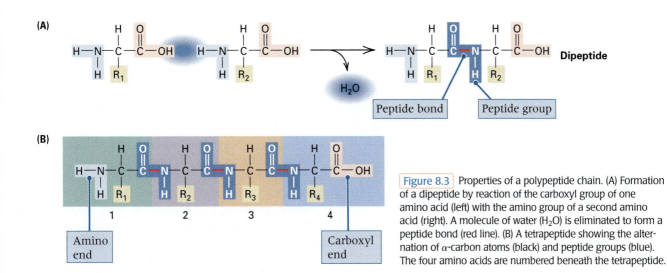

Figure 8.3 Properties of a polypeptide chain. (A) Formation of a dipeptide by reaction of the carboxyl group of one amino acid (left) with the amino group of a second amino acid (right). A molecule of water (H_2O) is eliminated to form a peptide bond (red line). (B) A tetrapeptide showing the alternation of α-carbon atoms (black) and peptide groups (blue). The four amino acids are numbered beneath the tetrapeptide.

8.1 Polypeptide chains are linear polymers of amino acids 283

■ **Human proteins, and those of other vertebrates, have a more complex domain structure than do the proteins of invertebrates.**

Most polypeptide chains include regions that can fold in upon themselves to acquire well-defined structures of their own, which interact with other structures formed in other regions of the molecule. Each of these relatively independent folding units is known as a **domain.** The domains in a protein molecule often have specialized functions, such as the binding of substrate molecules, cofactors needed for enzyme activity, or regulatory molecules that modulate activity. The individual domains in a protein usually have independent evolutionary origins, but through duplication of their coding regions and genomic rearrangements, they can come together in various combinations to create genes with novel functions of benefit to the organism. Just as the use of interchangeable parts facilitates airplane development and manufacture, so too does the use of interchangeable domains facilitate the evolution of new proteins.

Protein domains can be identified through computer analysis of the amino acid sequence. When these methods are applied to the human genome sequence, two interesting conclusions emerge:

- Only a minority (about 7 percent) of human proteins and protein domains are specific to vertebrates.
- Human proteins tend to have a more complex domain architecture (linear arrangement of domains) than proteins found in invertebrates. On average, human proteins contain about 1.8 times as many domain architectures as those of the worm or fly, and 5.8 times as many domain architectures as those of yeast.

These comparisons support the following principle:

> **key concept**
>
> Vertebrate genomes, including the human genome, have relatively few proteins or protein domains not found in other organisms. Their complexity arises in part from innovations in bringing together preexisting domains to create novel proteins that have more complex domain architectures than those found in other organisms.

8.2

The linear order of amino acids is encoded in a DNA base sequence.

Most genes contain the information for the synthesis of only one polypeptide chain. Furthermore, the linear order of nucleotides in a gene determines the linear order of amino acids in a polypeptide. This point was first proved by studies of the tryptophan synthase gene *trpA* in *E. coli*, a gene in which many mutations had been obtained and accurately mapped genetically. The effects of numerous mutations on the amino acid sequence of the enzyme were determined by directly analyzing the amino acid sequences of the wildtype and mutant enzymes. Each mutation was found to result in a single amino acid substituting for the wildtype amino acid in the enzyme. More important, the order of the mutations in the genetic map was the same as the order of the affected amino acids in the polypeptide chain (Figure 8.4). This attribute of genes and polypeptides is called **colinearity,** which means that the sequence of base pairs in DNA determines the sequence of amino acids in the polypeptide in a colinear, or point-to-point, manner. Colinearity is universally found in prokaryotes. However, we will see later that in eukaryotes, non-informational DNA sequences interrupt the continuity of most genes; in these genes, the order but not the spacing between the mutations correlates with amino acid substitution.

8.3

The base sequence in DNA specifies the base sequence in an RNA transcript.

The first step in gene expression is the synthesis of an RNA molecule copied from the segment of DNA that constitutes the gene. The basic features of the production of RNA are described in this section.

■ **The chemical synthesis of RNA is similar to that of DNA.**

Although the essential chemical characteristics of the enzymatic synthesis of RNA are generally similar to those of DNA, described in Chapter 6, there are also some important differences.

- Each RNA molecule produced in transcription derives from a single strand of DNA, because in any particular region of the DNA, usually only one strand serves as a template for RNA synthesis.
- The precursors in the synthesis of RNA are the four ribonucleoside 5'-triphosphates: adenosine triphosphate (ATP), guanosine triphosphate (GTP), cytidine triphosphate (CTP), and uridine triphosphate (UTP). They differ from the DNA precursors only in that the sugar is ribose rather than deoxyribose and the base uracil (U) replaces thymine (T) (Figure 8.5).
- The sequence of bases in an RNA molecule is determined by the sequence of bases in the

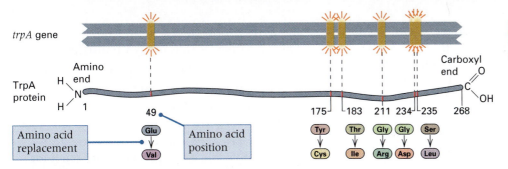

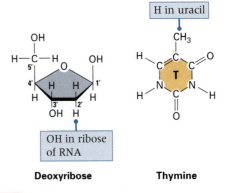

Figure 8.4 Colinearity of DNA and protein in the *trpA* gene of *E. coli*.

DNA template. Each base added to the growing end of the RNA chain is chosen for its ability to base-pair with the DNA template strand. Thus the bases C, T, G, and A in the DNA template cause G, A, C, and U, respectively, to be added to the growing end of the RNA molecule.

• In the synthesis of RNA, a sugar–phosphate bond is formed between the 3'-hydroxyl group of one nucleotide and the 5'-triphosphate of the next nucleotide in line (Figure 8.6, parts A and B). The chemical bond formed is the same as that in the synthesis of DNA, but the enzyme is different. The enzyme used in transcription is **RNA polymerase** rather than DNA polymerase.

Deoxyribose   **Thymine**

Figure 8.5 Differences between the structures of ribose and deoxyribose and between those of uracil and thymine.

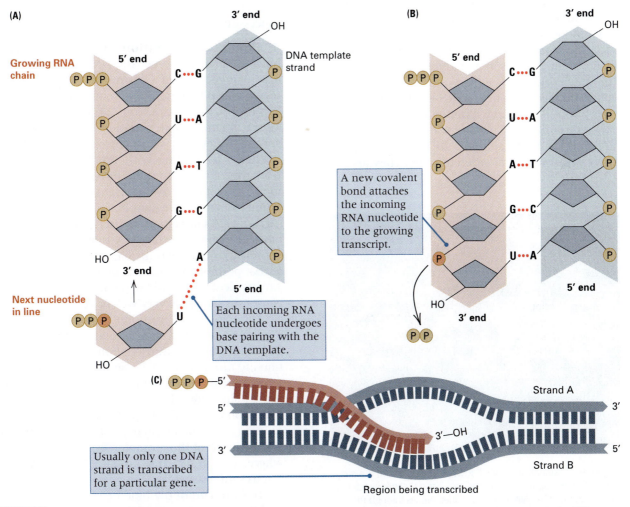

Figure 8.6 RNA synthesis. (A) Base pairing with the template strand. (B) The polymerization step. (C) Geometry of RNA synthesis.

RNA is copied from only one strand of a segment of a DNA molecule—in this example, strand B—without the need for a primer.

8.3 The base sequence in DNA specifies the base sequence in an RNA transcript   285

- Nucleotides are added only to the 3'-OH end of the growing chain; as a result, the 5' end of a growing RNA molecule bears a triphosphate group. The $5' \rightarrow 3'$ direction of chain growth is the same as that in DNA synthesis (Figure 8.6, part C).

- RNA polymerase (unlike DNA polymerase) is able to initiate chain growth without a primer (Figure 8.6, part C).

■ Eukaryotes have several types of RNA polymerase.

RNA polymerases are large, multisubunit complexes whose active form is called the **RNA polymerase holoenzyme.** Bacterial cells have only one RNA polymerase holoenzyme, which contains six polypeptide chains. At its widest dimension it is 150Å (Figure 8.7), about the same as a stretch of 45 nucleotides in duplex DNA. But in transcriptional initiation, the holoenzyme actually contacts 70 to 90 bp of DNA, which means that the DNA must wrap around the holoenzyme. Once transcription begins, the region of contact is reduced to about 35 nucleotides, centered on the nucleotide being added. The *processivity* of RNA polymerase (the number of nucleotides transcribed without dissociating from the template) is impressive: more than 10^4 nucleotides in prokaryotes and more than 10^6 nucleotides in eukaryotes. Processivity is important, because once the RNA polymerase separates from the template, it cannot resume synthesis.

Eukaryotic RNA polymerases are even larger and include more subunits in the holoenzyme. There are also several different types. They are de-

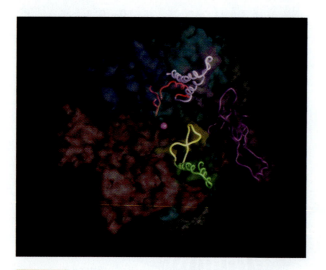

Figure 8.7 Subunit structure of RNA polymerase from the bacterium *Thermus aquaticus.* The complex has a U-shaped channel running through it. DNA to be transcribed passes through this channel. [Courtesy of Rachel Anne Mooney and Robert Landick. 1999. *Cell* 98: 687.]

noted RNA polymerase I, II, and III, and each makes a particular class of RNA transcript:

- *RNA polymerase I* is used exclusively in producing the transcript that becomes processed into ribosomal RNA.

- *RNA polymerase II* is the workhorse eukaryotic polymerase responsible for transcribing all protein-coding genes as well as the genes for a number of small nuclear RNAs (U1, U2, U3, and so forth) used in RNA processing.

- *RNA polymerase III* is used in transcribing all transfer RNA genes as well as the 5S component of ribosomal RNA.

■ Particular nucleotide sequences define the beginning and end of a gene.

How does RNA polymerase determine which strand of DNA should be transcribed? How does the enzyme recognize where transcription of the template strand should begin? How does the enzyme recognize where transcription should stop? These are critical features in the regulation of RNA synthesis that can be described in terms of four discrete processes:

1. Promoter recognition The RNA polymerase binds to DNA wherever the DNA has a particular base sequence called a **promoter.** Many promoter regions have had their base sequences determined. Typical promoters are from 20 to 200 bases long. There is substantial variation in base sequence among promoters, which correlates with different strengths in binding with the RNA polymerase. However, most promoters have certain sequence motifs in common. Two consensus sequences often found in promoter regions in *E. coli* are illustrated in Figure 8.8. A **consensus sequence** is a sequence of bases determined by majority rule: Each base in the consensus sequence is the base most often observed at that position among a set of observed sequences. Any particular observed sequence may resemble the consensus sequence very well or very poorly.

The consensus promoter regions in *E. coli* are TTGACA, centered approximately 35 base pairs upstream from the transcription start site (conventionally numbered the +1 site), and TATAAT, centered approximately 10 base pairs upstream from the +1 site. The −10 sequence, which is called the **TATA box,** is similar to sequences found at corresponding positions in many eukaryotic promoters. The positions of the promoter sequences determine where, and on which strand, the RNA polymerase begins synthesis.

The strength of the binding of RNA polymerase to different promoters varies greatly, which causes differences in the extent of expression from one

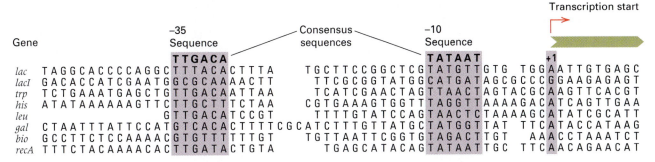

Gene		−35 Sequence		Consensus sequences		−10 Sequence		Transcription start +1
lac	TAGGCACCCCAGGC	TTTACA	CTTTA		TGCTTCCGGCTCG	TATGTT	GTG	TGGAATTGTGAGC
lacI	GACACCATCGAATG	GCGCAA	AACTT		TTCGCGGTATGG	CATGATA	GCGCCC	GGAAGAGAGT
trp	TCTGAAATGAGCTG	TTGACA	ATTAA		TCATCGAACTAGT	TAACT	AGTACGCAAG	TTCACGT
his	ATATAAAAAAGTTC	TTGCTT	TCTAA		CGTGAAAGTGGTT	TAGGT	AAAAGACAT	CAGTTGAA
leu		GTTGACA	TCCGT		TTTTGTATCCAGT	AACTCT	AAAAGCATAT	CGCATT
gal	CTAATTTATTCCATG	TCACA	CTTTTCGCAT		CTTTGTTATGC	TATGGT	TAT	TTCATACCATAAG
bio	GCCTTCTCCAAAAC	GTGTTT	TTTGT		TGTTAATTCGGTG	TAGACT	TGT	AAACCTAAATCT
recA	TTTCTACAAAACAC	TTGACA	CTGTA		TGAGCATACAGT	ATAAT	TGC	TTCAACAGAACAT

Figure 8.8 Base sequences in promoter regions of several genes in *E. coli*. The consensus sequences located 10 and 35 nucleotides upstream from the transcription start site (+1) are indicated. Promoters vary tremendously in their ability to promote transcription. Much of the variation in promoter strength results from differences between the promoter elements and the consensus sequences at −10 and −35.

gene to another. Most of the differences in promoter strength result from variations in the −35 and −10 promoter elements and in the number of bases between them. Promoter strength among *E. coli* genes differs by a factor of 10^4, and most of the variation can be attributed to the promoter sequences themselves. In general, the more closely the −35 and −10 promoter elements resemble the consensus sequences, the stronger the promoter. The situation is somewhat different in eukaryotes, where other types of DNA sequences (*enhancers*) interact with promoters to determine the level of transcription (Chapter 9).

2. Chain initiation After the initial binding step, RNA polymerase initiates RNA synthesis at a nearby transcription start site, labeled the +1 site in Figure 8.8. The first nucleoside triphosphate is placed at this site, the next nucleotide in line is attached to the 3′ carbon of the ribose, and so forth. Only one of the DNA strands serves as the template for transcription. Because RNA is synthesized in the 5′ → 3′ direction, the DNA template is traversed in the 3′ → 5′ direction. Therefore, relative to the orientation of the promoter sequences shown in Figure 8.8, the template DNA strand is the *partner* of the strand illustrated. Take the *lac* promoter as an example. Transcription begins on the opposite strand at the nucleotide labeled +1 and proceeds from left to right. Hence the base sequence of the RNA transcript is the same as that of the DNA strand illustrated (except that RNA contains U where T appears in DNA), and so the *lac* RNA sequence begins AAUUGUGAGC . . . .

3. Chain elongation After initiation at the +1 site, RNA polymerase moves along the DNA template strand, adding nucleotides to the growing RNA chain (part C of Figure 8.6). Each new nucleotide is added to the 3′ end of the chain, so RNA chains resemble DNA chains in growing in the 5′ → 3′ direction. Part C of Figure 8.6 also shows that transcription separates the partner strands of the DNA duplex only in a short region around the

point of chain elongation. As the RNA polymerase moves along the template strand, only about 17 base pairs of the DNA duplex (less than two turns of the double helix) are unwound at any time. Once the RNA polymerase has passed, the DNA strands are released and the duplex forms again, with the part of the RNA chain already synthesized trailing off as a separate polynucleotide strand.

4. Chain termination Special sequences also terminate RNA synthesis. When the RNA polymerase reaches a transcription termination sequence in

Lactose intolerance results from reduced expression of an enzyme necessary to metabolize the sugar lactose found in milk. The condition leads to painful digestive upsets after eating dairy products such as ice cream. The condition is infrequent in Caucasian populations, but very common in certain African and Asian groups.

8.3 The base sequence in DNA specifies the base sequence in an RNA transcript   287

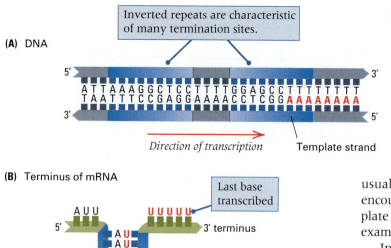

(A) DNA

Inverted repeats are characteristic of many termination sites.

```
5'  ATTAAAGGCTCCTTTTGGAGCCTTTTTTTT  3'
    TAATTTCCGAGGAAAACCTCGGAAAAAAAA
3'                                  5'
```

Direction of transcription → Template strand

(B) Terminus of mRNA

Last base transcribed

5' A U U ... U U U U U 3' terminus

The inverted repeats can undergo base pairing and enable the RNA transcript to form a hairpin structure.

Figure 8.9 (A) Base sequence of the transcription termination region for the set of tryptophan-synthesizing genes in *E. coli*. (B) The 3' terminus of the RNA transcript, folded to form a stem-and-loop structure. The sequence of Us found at the end of the transcript in this and many other prokaryotic genes is shown in red. The RNA polymerase, not shown here, terminates transcription when the loop forms in the transcript.

the DNA, the polymerase enzyme disassociates from the DNA, and the newly synthesized RNA molecule is released. Two kinds of termination events are known: (1) those that are self-terminating and depend only on the transcription termination sequence in the DNA, and (2) those that require the presence of a termination protein in addition to the transcription termination sequence. Self-termination is the usual case; it takes place when the polymerase encounters a particular sequence of bases in the template strand that causes the polymerase to stop. An example from *E. coli* is shown in Figure 8.9.

Initiation of a second round of transcription need not await completion of the first. By the time an RNA transcript reaches a size of 50 to 60 nucleotides, the RNA polymerase has moved along the DNA far enough from the promoter that the promoter becomes available for another RNA polymerase to initiate a new transcript. Such reinitiation can take place repeatedly, and a gene can be cloaked with numerous RNA molecules in various degrees of completion. The micrograph in Figure 8.10 shows a region of the DNA of the newt *Triturus* containing tandem repeats of a particular gene. Each gene is associated with many growing RNA molecules. The shortest ones are at the promoter end of the gene, the longest near the gene terminus.

Genetic experiments in *E. coli* yielded the first demonstration of the existence of promoters. A class of Lac⁻ mutations, denoted p^-, was isolated that was unusual in two respects:

- All p^- mutations were closely linked to the *lacZ* gene.
- Any p^- mutation eliminated activity of a wild-type *lacZ* gene *present in the same DNA molecule.*

The need for an adjacent genetic configuration to eliminate *lacZ* activity can be seen by examining a cell with two copies of the *lacZ* gene. Such cells can be produced through the use of F'*lacZ* plasmids, which contain a copy of *lacZ* in an F plasmid. Infection with an F'*lacZ* plasmid yields a cell with two copies of *lacZ*—one in the chromosome and another in the F'. Transcription of the *lacZ* gene enables the cell to synthesize the enzyme β-galactosidase. **Table 8.1** shows that a wildtype *lacZ* gene (*lacZ⁺*) is inactive when a p^- mutation is present in the same DNA molecule (either in the bacterial chromosome or in an F' plasmid). This result can be seen by comparing entries 4 and 5. Analysis of the RNA shows that in a cell with the genotype p^- *lacZ⁺*, the *lacZ⁺* gene is not transcribed. On the other hand, cells of genotype p^+*lacZ⁻* produce a mutant RNA. The p^- mutations are called *promoter mutations.*

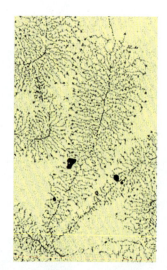

Figure 8.10 Electron micrograph of part of the DNA of the newt *Triturus viridescens* containing tandem repeats of genes being transcribed into ribosomal RNA. The thin strands forming each feather-like array are RNA molecules. A gradient of lengths can be seen for each rRNA gene. Regions in the DNA between the thin strands are spacer DNA sequences, which are not transcribed. [Courtesy of Oscar Miller and Barbara R. Bisatty.]

Table 8.1

Effect of promoter mutations on transcription of the *lacZ* gene

Genotype	Transcription of *lacZ*$^+$ gene
1. p^+lacZ^+	Yes
2. p^-lacZ^+	No
3. p^+lacZ^+/p^+lacZ^-	Yes
4. p^-lacZ^+/p^+lacZ^-	No
5. p^+lacZ^+/p^-lacZ^-	Yes

Note: lacZ$^+$ is the wildtype gene; the lacZ$^-$ mutant produces a nonfunctional enzyme.

Mutations have also been instrumental in defining the transcription termination region. For example, mutations have been isolated that create a new termination sequence upstream from the normal one. When such a mutation is present, an RNA molecule is made that is shorter than the wildtype RNA. Other mutations eliminate the terminator, resulting in a longer transcript.

■ Messenger RNA directs the synthesis of a polypeptide chain.

The RNA molecule produced from a DNA template is the **primary transcript.** Each gene has only one DNA strand that serves as the template strand, but which strand is the template strand can differ from gene to gene along a DNA molecule. Therefore, in an extended segment of a DNA molecule, primary transcripts would be seen growing in either of two directions (Figure 8.11), depending on which DNA strand functions as a template in a particular gene. In prokaryotes, the primary transcript serves directly as the **messenger RNA (mRNA)** used in polypeptide

synthesis. In eukaryotes, the primary transcript is generally processed before it becomes mRNA.

Not all base sequences in an mRNA molecule are translated into the amino acid sequences of polypeptides. For example, translation of an mRNA molecule rarely starts exactly at one end and proceeds to the other end; instead, initiation of polypeptide synthesis may begin many nucleotides downstream from the 5′ end of the RNA. The untranslated 5′ segment of RNA is called the **5′ untranslated region.** This is followed by an **open reading frame (ORF),** which specifies the polypeptide chain. A typical ORF in an mRNA molecule is between 500 and 3000 bases long (depending on the number of amino acids in the protein), but it may be much longer. The 3′ end of an mRNA molecule following the ORF also is not translated; it is called the **3′ untranslated region.**

In prokaryotes, most mRNA molecules are degraded within a few minutes after synthesis. In eukaryotes, a typical lifetime is several hours, although some last only minutes whereas others persist for days. In both kinds of organisms, the degradation enables cells to dispose of molecules that are no longer needed and to recycle the nucleotides in synthesizing new RNAs. The short lifetime of prokaryotic mRNA is an important factor in regulating gene activity (Chapter 9).

8.4

RNA processing converts the original RNA transcript into messenger RNA.

Although the process of transcription is very similar in prokaryotes and eukaryotes, there are major differences in the relationship between the transcript and the mRNA used for polypeptide synthesis. In prokaryotes, the immediate product of transcription (the primary transcript) is mRNA; in contrast, *the primary transcript in eukaryotes must be converted*

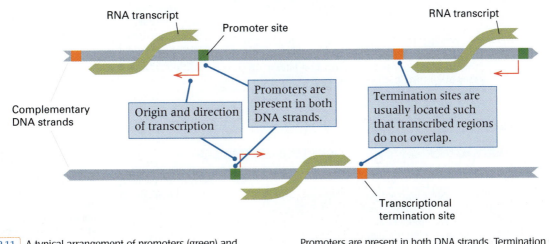

Figure 8.11 A typical arrangement of promoters (green) and termination sites (red) in a segment of a DNA molecule.

Promoters are present in both DNA strands. Termination sites are usually located such that transcribed regions do not overlap.

into mRNA. The conversion of the original transcript into mRNA is called *RNA processing.* It usually consists of three types of events:

- The 5′ end is altered by the addition of a modified guanosine in an uncommon 5′−5′ linkage (instead of the typical 3′−5′ linkage); this terminal group is called the **cap;** the 5′ cap is necessary for the ribosome to bind with the mRNA to begin protein synthesis.

- The 3′ end is usually modified by the addition of a sequence called the **poly-A tail,** which can consist of as many as 200 consecutive A-bearing nucleotides; the poly-A tail is thought to help regulate mRNA stability.

- Certain regions internal to the transcript (*introns*) are removed by splicing. This process is described next. The segments that are excised from the primary transcript are called **introns** or *intervening sequences.* Accompanying the excision of introns is a rejoining of the coding segments (**exons**) to form the mRNA molecule. The excision of the introns and the joining of

the exons to form the final mRNA molecule is called **RNA splicing.**

The events that constitute RNA processing begin even while transcription is still in progress, and the events are *coupled processes,* which means that occurrence of one event initiates the next. Some of the interconnections are shown in Figure 8.12 . A key player in the coupling is the carboxyl terminal domain of the large subunit of RNA polymerase II, which contains a series of nearly identical repeats of a sequence of seven amino acids. When key amino acids in this domain become phosphorylated, the RNA polymerase recruits the capping machinery, and when they are dephosphorylated the capping machinery is released. Phosphorylation of other amino acids in the domain helps recruit the machinery for splicing and polyadenylation.

The effect of the coupled processes is to greatly increase the speed and specificity of RNA processing. Without such coupling, RNA processing would be dependent on diffusion, and many mistakes would be made, especially in splicing the often large introns that separate relatively small exons. The recruitment of the splicing machinery while transcription is still taking place greatly facilitates correct splicing. Introns are not necessarily spliced in exactly the same order in which they are transcribed, however. The order in which splicing takes place depends on the size and nucleotide composition of the introns, as well as on the overall rate of transcription.

(A) Transcription initiation and elongation

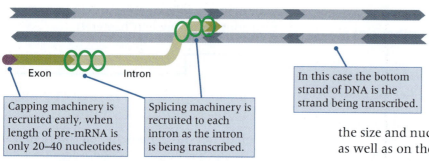

Exon Intron

Capping machinery is recruited early, when length of pre-mRNA is only 20–40 nucleotides.

Splicing machinery is recruited to each intron as the intron is being transcribed.

In this case the bottom strand of DNA is the strand being transcribed.

(B) Termination

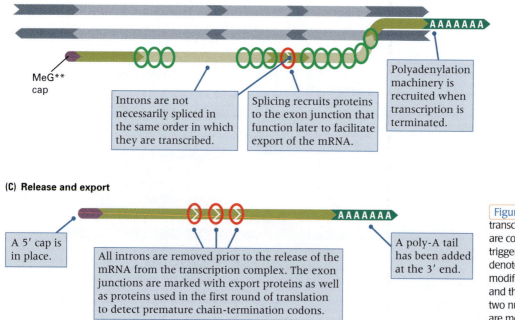

MeG** cap

AAAAAAA

Introns are not necessarily spliced in the same order in which they are transcribed.

Splicing recruits proteins to the exon junction that function later to facilitate export of the mRNA.

Polyadenylation machinery is recruited when transcription is terminated.

(C) Release and export

AAAAAAA

A 5′ cap is in place.

All introns are removed prior to the release of the mRNA from the transcription complex. The exon junctions are marked with export proteins as well as proteins used in the first round of translation to detect premature chain-termination codons.

A poly-A tail has been added at the 3′ end.

Figure 8.12 In eukaryotes, transcription and RNA processing are coupled. Each step (A, B, C) triggers the next in line. MeG denotes 7-methylguanosine (a modified form of guanosine), and the two asterisks indicate two nucleotides whose riboses are methylated.

Numerous interconnecting links couple the various steps in transcription with those in RNA processing. For example, proteins that bind with RNA polymerase to promote elongation also help recruit the splicing machinery, and the splicing machinery in turn stimulates elongation so that genes containing introns are more efficiently transcribed. The splicing machinery also helps recruit the polyadenylation machinery. The principal steps in RNA processing are all completed prior to release of the mRNA from the transcription complex. As each intron is spliced, proteins bind to the junction between the exons (Figure 8.12, parts B and C). Some of these function after release of the mRNA to facilitate its export from the nucleus to the cytoplasm. Other of these proteins function in the first round of translation to identify defective mRNA molecules that are subsequently destroyed.

■ Splicing removes introns from the RNA transcript.

RNA splicing takes place in nuclear particles known as **spliceosomes.** These abundant particles are composed of protein and several specialized small RNA molecules, which are present in the cell as small nuclear ribonucleoprotein particles; the underlined letters give the acronym for these particles: **snRNPs.** The specificity of splicing comes from the five small snRNP RNAs denoted U1, U2, U4, U5, and U6, which contain sequences complementary to the splice junctions, to the branchpoint region of the intron, and/or to one another; as many as 100 spliceosome proteins may also be required for splicing. The ends of the intron are brought together by U1 RNA, which forms base pairs with nucleotides in the intron at both the 5′ and the 3′ ends. U2 RNA binds to the branchpoint region. U2 RNA interacts with a paired complex of U4/U6 RNAs, resulting in a complex in which U2 RNA ends up paired with U6 RNA and the intron of the transcript (Figure 8.13). All of these dynamic interactions bring the branchpoint region near to the donor splice site and allow the A in the branch-

Figure 8.13 | Dynamic interactions between some small nuclear RNAs present in snRNPs that are involved in splicing. (A) U6 snRNA is usually found complexed with U4 snRNA. (B) U2 snRNA forms a stable foldback structure on its own. (C) Essential to the splicing reaction is destabilization of the U4−U6 structure and formation of a U2−U6 structure in which U2 is base-paired with part of the intron. An A in the paired region attacks the G at the 5′ splice junction, initiating the splicing reaction. The nucleotides in bold are critical to the structures, judging by their having been conserved in very diverse species. Note that G−U base pairs are allowed in double-stranded RNA. [After H. D. Madhani and C. Guthrie. 1994. *Annual Reviews of Genetics* 28: 1.]

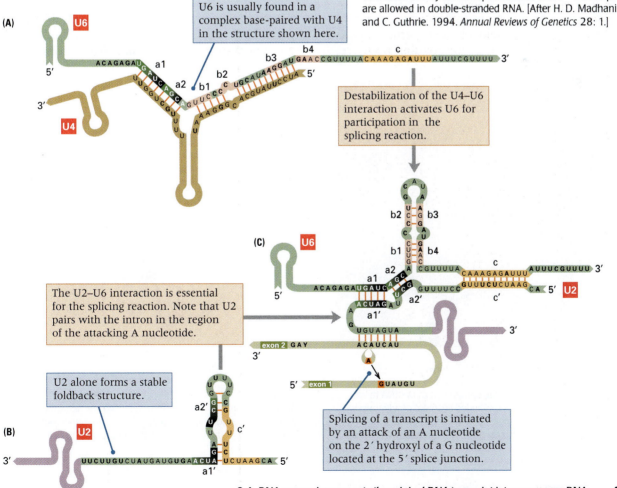

point to attack the G of the donor splice site, freeing the upstream exon and forming the looped intermediate (Figure 8.13). U5 RNA helps line up the two exons and somehow facilitates the final step in splicing, which results in scission of the intron from the downstream exon and in ligation of the upstream and downstream exons.

Introns are also present in some genes in organelles, such as mitochondria, but the mechanisms of their excision differ from those of introns in nuclear genes because organelles do not contain spliceosomes. In one class of organelle introns, the intron contains a sequence coding for a protein that participates in removing the intron that codes for it. The situation is even more remarkable in the splicing of a ribosomal RNA precursor in the ciliate *Tetrahymena*. In this case, the splicing reaction is intrinsic to the folding of the precursor; that is, the RNA precursor is *self-splicing* because the folded precursor RNA creates its own RNA-splicing activity. The self-splicing *Tetrahymena* RNA was the first example found of an RNA molecule that could function as an enzyme in catalyzing a chemical reaction; such enzymatic RNA molecules are usually called **ribozymes.**

The existence and the positions of introns in a particular primary transcript are readily demonstrated by renaturing the transcribed DNA with the fully processed mRNA molecule. The DNA–RNA hybrid can then be examined by electron microscopy. An example of adenovirus mRNA (fully processed) and the corresponding DNA are

shown in Figure 8.14 . The DNA copies of the introns appear as single-stranded loops in the hybrid molecule, because no corresponding RNA sequence is available for hybridization.

The number of introns per RNA molecule varies considerably from one gene to the next. One of the major genes for inherited breast cancer in women (*BRCA1*) contains 21 introns spread across more than 100,000 bases. More than 90 percent of the primary transcript is excised in processing, yielding a processed mRNA of about 7800 bases, which codes for a polypeptide chain of 1863 amino acids. Among human genes with a simpler intron–exon structure is that for α-globin, which contains two introns. Introns vary greatly in size as well as in number. In human beings and other mammals, most introns range in size from 100 to 10,000 bases, and in the processing of a typical primary transcript, the amount of discarded RNA ranges from about 50 percent to more than 90 percent. In lower eukaryotes, such as yeast, nematodes, and fruit flies, genes generally have fewer introns than do genes in mammals, and the introns tend to be much smaller.

■ Human genes tend to be very long even though they encode proteins of modest size.

Table 8.2 summarizes features of the "typical" human gene. Both the median and the mean values are given because many of the size distributions have a

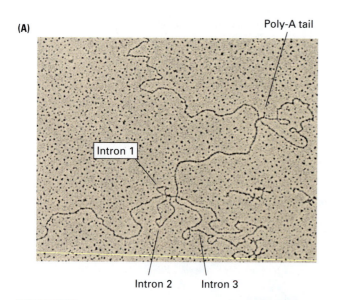

(A)

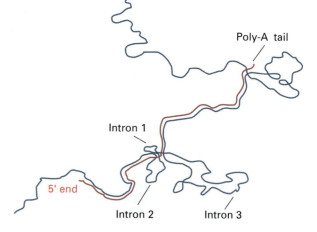

(B)

Figure 8.14 (A) An electron micrograph of a DNA–RNA hybrid obtained by annealing a single-stranded segment of adenovirus DNA with one of its mRNA molecules. The loops are single-stranded DNA. (B) An interpretive drawing. RNA and DNA strands are shown in red and black, respectively. Four regions do not anneal, creating three single-stranded DNA segments that correspond to the introns and the poly-A tail of the mRNA molecule. [Electron micrograph courtesy of Tom Broker and Louise Chow.]

Table 8.2

Characteristics of human genes

Gene feature	Median	Mean
Size of internal exon	122 bp	145 bp
Number of exons	7	8.8
Size of introns	1023 bp	3356 bp
5' untranslated region	240 bp	300 bp
3' untranslated region	400 bp	770 bp
Length of coding sequence	1101 bp	1341 bp
Number of amino acids (aa)	367	447
Extent of genome occupied	14 kb	27 kb

Source: Data from E. S. Lander et al. 2001. *Nature* 409: 860.

very long tail, rendering the mean potentially misleading. For example, whereas the mean number of exons is 8.8, this average is unduly influenced by some genes that have a very large number of exons, such as the gene for the muscle protein titin, which includes 178 exons (the largest number for any human gene). Similarly, the distribution of intron sizes is strongly skewed. The most common intron length peaks at 87 nucleotides, but the tail of the distribution is so stretched out that the mean is 3365 nucleotides. The median is the value that splits the distribution in the middle: Half the values are above the median and half below.

One noteworthy feature of Table 8.2 is that human genes tend to be spread over a larger region of the genome than those in worms or flies. Most human genes consist of small exons separated by long introns, and many genes are over 100 kb in length. The average human gene occupies 27 kb of genomic DNA, yet only 1.3 kb (about 5 percent) is used to encode amino acids. The picture is not much different for the medians. The median gene length is 14 kb, of which only 1.1 kb (about 8 percent) is used to encode amino acids. Most of the added length is due to the long introns in human genes. The longest human gene is that for the muscle protein dystrophin, which is 2.4 Mb in length.

■ Many exons code for distinct protein-folding domains.

The existence of an elaborate splicing mechanism shared among all eukaryotes implies that introns must be very ancient. Introns may play a role in gene evolution by serving as the boundaries of exons encoding amino acid sequences that are more or less independent in their folding character-

istics. For example, the central exon of the β-globin gene codes for a domain that folds around an iron-containing molecule of heme. The correlation between exons and domains found in some genes suggests that the genes were originally assembled from smaller pieces. In some cases, the ancestry of the exons can be traced. For example, the human gene for the low-density lipoprotein receptor that participates in cholesterol regulation shares exons with certain blood-clotting factors and epidermal growth factors. The model of protein evolution through the combination of different exons is called the **exon shuffle** model. The mechanism for combining exons from different genes is not known, but we have already seen that the proteins of human beings and other vertebrates tend to have more complex domain architectures than do proteins found in other organisms.

8.5

Translation into a polypeptide chain takes place on a ribosome.

The synthesis of every protein molecule in a cell is directed by an mRNA originally copied from DNA. Protein production includes two kinds of processes: (1) information-transfer processes, in which the RNA base sequence determines an amino acid sequence, and (2) chemical processes, in which the amino acids are linked together. The complete series of events is called translation.

The translation system consists of five major components:

1. **Messenger RNA** Messenger RNA is needed to bring the ribosomal subunits together (described below) and to provide the coding sequence of bases that determines the amino acid sequence in the resulting polypeptide chain.

2. **Ribosomes** These components are particles on which protein synthesis takes place. They move along an mRNA molecule and align successive transfer RNA molecules; the amino acids are attached one by one to the growing polypeptide chain by means of peptide bonds. Ribosomes consist of two separate RNA–protein particles (the small subunit and the large subunit), which come together in polypeptide synthesis to form a mature ribosome.

3. **Transfer RNA, or tRNA** The sequence of amino acids in a polypeptide is determined by the base sequence in the mRNA by means of a set of adaptor molecules, the tRNA molecules, each of which is attached to a particular amino acid. Each successive group of three adjacent

bases in the mRNA forms a **codon** that binds to a particular group of three adjacent bases in the tRNA (an anticodon), bringing the attached amino acid into line for elongation of the growing polypeptide chain.

4. **Aminoacyl-tRNA synthetases** Each enzyme in this set of molecules catalyzes the attachment of a particular amino acid to its corresponding tRNA molecule. A tRNA attached to its amino acid is called an **aminoacylated tRNA** or a **charged tRNA.**

5. **Initiation, elongation, and termination factors** Polypeptide synthesis can be divided into three stages: initiation, elongation, and termination. Each stage requires specialized molecules.

In prokaryotes, all of the components for translation are present throughout the cell; in eukaryotes, they are located in the cytoplasm, as well as in mitochondria and chloroplasts.

■ In eukaryotes, initiation takes place by scanning the mRNA for an initiation codon.

In overview, the process of translation begins with an mRNA molecule binding to a ribosome. The aminoacylated tRNAs are brought along sequentially, one by one, to the ribosome that is translating the mRNA molecule. Peptide bonds are made between successive amino acids. At each step, the carboxyl end of the growing chain is attached to the amino group of the amino acid on the incoming tRNA. The growing chain is thereby handed off from tRNA to tRNA until translation is completed and the finished polypeptide chain is released from the ribosome.

We will examine the processes of translation as they occur in eukaryotes, pointing out differences in the prokaryotic mechanism that are significant. In the predominant mode of translation **initiation** in eukaryotes, the 5′ cap on the mRNA is instrumental (Figure 8.15). The elongation factor eIF4F first binds to the cap and then recruits eIF4A and eIF4B (part A). This creates a binding site for the other components of the initiation complex, which consist of a charged $tRNA^{Met}$ (that serves as an initiator tRNA), bound with elongation factor eIF2, and a small 40S ribosomal subunit together with elongation factors eIF3 and eIF5. These components all come together at the 5′ cap and form the 48S initiation complex (part B).

Once the initiation complex has formed, it moves along the mRNA in the 3′ direction, scanning for the first occurrence of the nucleotide

sequence AUG, the **start codon** that signals the start of polypeptide synthesis. When this motif is encountered, the AUG is recognized as the initial methionine codon, and polypeptide synthesis begins. At this point eIF5 causes the release of all the initiation factors and the recruitment of a large 60S ribosomal subunit (part C). This subunit includes three binding sites for tRNA molecules. These sites are called the **E (exit) site,** the **P (peptidyl) site,** and the **A (aminoacyl) site.** Note that at the beginning of polypeptide synthesis, the initiator methionine tRNA is located in the P site and that the A site is the next site in line to be occupied. The tRNA binding is accomplished by hydrogen bonding between bases in the AUG codon in the mRNA and the three-base **anticodon** in the tRNA.

■ Elongation takes place codon by codon.

Recruitment of other elongation factors into the initiation complex begins the **elongation** phase of polypeptide synthesis. Elongation consists of three processes executed iteratively:

1. Bringing each new aminoacylated tRNA into line.

2. Forming the new peptide bond to elongate the polypeptide.

3. Moving the ribosome to the next codon along the mRNA.

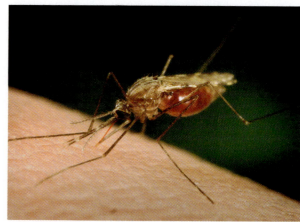

Courtesy of James Gathany and Dr. Frank Collins, University of Notre Dame/CDC.

About 40 percent of the world's population is at risk of malaria. Every year, more than one million children, most of them in Africa, die from the deadliest form of the disease caused by the protozoan parasite *Plasmodium falciparum.* The death toll is equivalent to a Boeing 747, fully loaded with children under the age of 5, crashing every four hours. The parasite is transmitted through mosquitoes. This mosquito is *Anopheles funestus,* which is second only to *Anopheles gambiae* as a major transmitter of the disease in Africa.

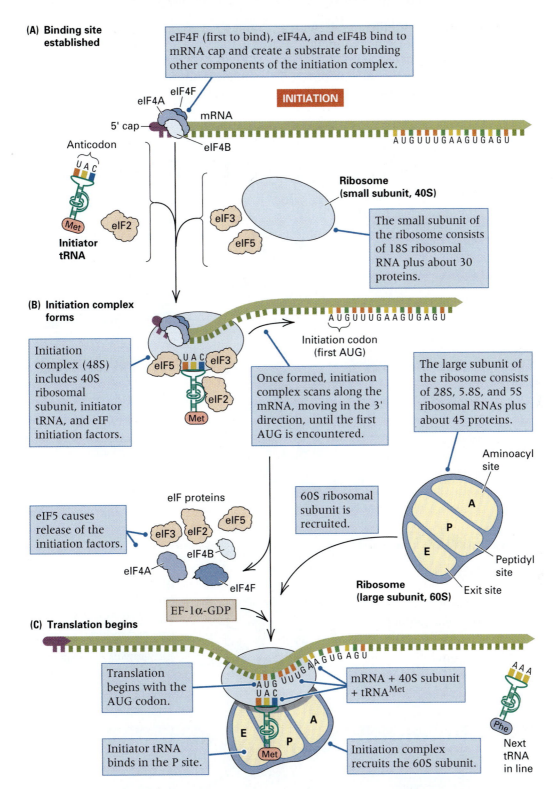

(A) Binding site established

eIF4F (first to bind), eIF4A, and eIF4B bind to mRNA cap and create a substrate for binding other components of the initiation complex.

eIF4F

eIF4A

5' cap

mRNA

eIF4B

INITIATION

AUGUUUGAAGUGAGU

Anticodon

UAC

Met

Initiator tRNA

eIF2

eIF3

eIF5

Ribosome (small subunit, 40S)

The small subunit of the ribosome consists of 18S ribosomal RNA plus about 30 proteins.

(B) Initiation complex forms

AUGUUUGAAGUGAGU

Initiation codon (first AUG)

Initiation complex (48S) includes 40S ribosomal subunit, initiator tRNA, and eIF initiation factors.

eIF5 UAC eIF3

eIF2

Met

Once formed, initiation complex scans along the mRNA, moving in the 3' direction, until the first AUG is encountered.

The large subunit of the ribosome consists of 28S, 5.8S, and 5S ribosomal RNAs plus about 45 proteins.

Aminoacyl site

A

P

E

Peptidyl site

Exit site

Ribosome (large subunit, 60S)

eIF proteins

eIF5 causes release of the initiation factors.

eIF3 eIF2 eIF5

eIF4B

eIF4A

eIF4F

60S ribosomal subunit is recruited.

EF-1α-GDP

(C) Translation begins

UUUGAAGUGAGU

Translation begins with the AUG codon.

AUG UAC

mRNA + 40S subunit + tRNA^{Met}

E P A

Met

Initiator tRNA binds in the P site.

Initiation complex recruits the 60S subunit.

AAA

Phe

Next tRNA in line

Figure 8.15 Initiation of protein synthesis. (A) The initiation complex forms at the 5' end of the mRNA. (B) This consists of one 40S ribosomal subunit, the initiator tRNA^{Met}, and the eIF initiation factors. (C) The initiation complex recruits a 60S ribosomal subunit in which the tRNA^{Met} occupies the P (peptidyl) site of the ribosome. This complex travels along the mRNA until the first AUG is encountered, at which codon translation begins.

The process of elongation is illustrated in Figure 8.16. The key players in providing the energy for translation are the elongation factors EF-2 and EF-1α, which alternately occupy the same ribosomal binding site. In their active forms (EF-2-GTP and EF-1α-GTP) the molecules are bound with guanosine triphosphate (GTP). Hydrolysis of the GTP to GDP releases the energy to move the ribosomal subunits along the messenger RNA as well as to carry out the reactions needed to grow the

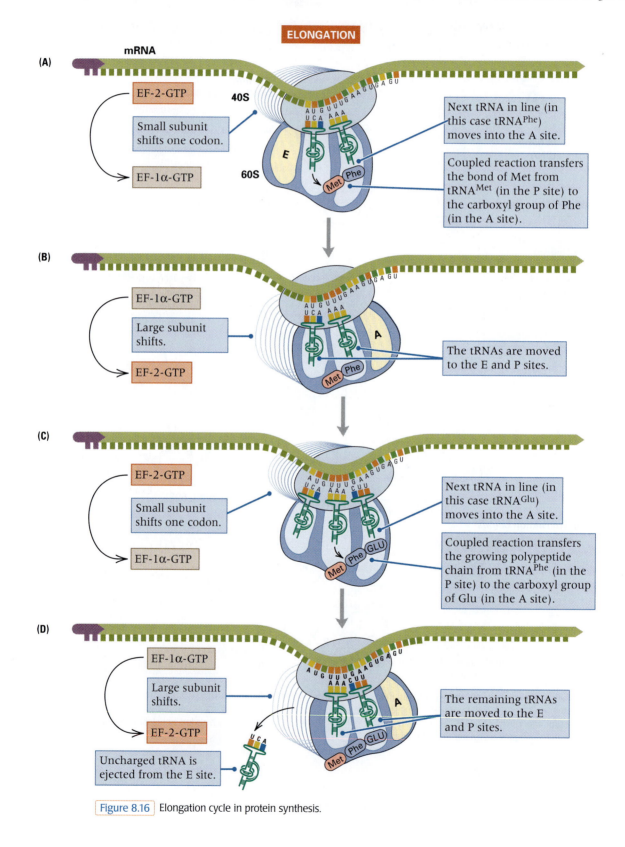

Figure 8.16 Elongation cycle in protein synthesis.

polypeptide chain. Conversion of either elongation factor from its GTP-bound form into its GDP-bound form lowers its affinity for the ribosome, and the GDP-bound form diffuses away and is replaced by the GTP-bound form of the alternate elongation factor.

In the first step of elongation, the 40S ribosomal subunit moves one codon farther along the messenger RNA, and the charged tRNA corresponding to the new codon (in this case, $tRNA^{Phe}$) is brought into the A site on the 60S subunit (Figure 8.16, part A). Once the A site is filled, a **peptidyl transferase** activity catalyzes a coupled reaction in which the bond connecting the methionine to the $tRNA^{Met}$ is transferred to the amino group of the phenylalanine, forming the first peptide bond. Peptidyl transferase activity is not due to a single molecule but requires multiple components of the 60S subunit, including several proteins and the 28S ribosomal RNA in the 60S subunit. Some evidence indicates that the actual catalysis is carried out by the 28S RNA, which suggests that 28S is an example of a ribozyme at work.

In the next step in chain elongation (part B), the 60S subunit swings forward to catch up with the 40S subunit, and at the same time the tRNAs in the P and A sites of the large subunit are shifted to the E and P sites, respectively.

One cycle of elongation is now completed, and the entire procedure is repeated for the next codon (part C). The 40S subunit shifts one codon to the right, the next aminoacylated tRNA (in this case, $tRNA^{Glu}$) is brought into the A site, and a new peptide bond is formed between the carboxyl group of Phe and the amino group of Glu. As shown in part D, the large subunit swings forward while at the same time the tRNAs in the P and A sites are shifted into the E and P sites. At this point, the tRNA that formerly occupied the E site is ejected from the ribosome.

Polypeptide elongation consists of the steps C → D → C → D carried out repeatedly until a termination codon is encountered. The elongation cycle happens relatively rapidly. Under optimal conditions, eukaryotes synthesize a polypeptide chain at the rate of about 15 amino acids per second. Elongation in prokaryotes is a little faster (about 20 amino acids per second), but the essential processes are very similar. In *E. coli*, the sizes of the ribosomal subunits are 30S (small) and 50S (large), and the complete ribosome is 70S. Figure 8.17 shows a large ribosomal subunit from *E. coli*, reconstructed from the x-ray diffraction structure, depicting the locations of the tRNA molecules in their binding pockets: E in gold, P in blue, and A in green. Above the tRNAs is a channel through which the mRNA is moved along as translation progresses. The small subunit fits on top of the large subunit, leaving enough space for the tRNA molecules to bind. In prokaryotes the source of energy for elongation is also GTP hydrolysis. The *E. coli* analogs of EF-1α and EF-2 are EF-Tu and EF-G, respectively.

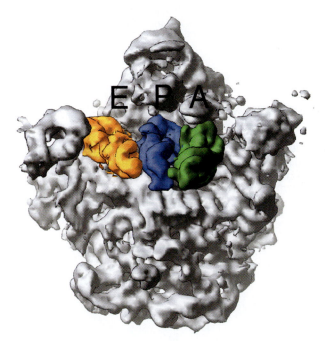

Figure 8.17 Cutaway view of a bacterial ribosome, showing the groove in which the tRNAs are bound and the positions of tRNAs when present in the E (exit) site in gold, the P (peptidyl) site in blue, and the A (aminoacyl) site in green. [Courtesy of Jamie H. Cate, M. M. Yusupova, G. Zh. Yusupova, T. N. Earnest, and H. F. Noller. 1999. *Science* 285: 2095. © 1999 AAAS.]

■ A termination codon signals release of the finished polypeptide chain

Compared to initiation and elongation, the termination of polypeptide synthesis—the **release phase**—is simple (Figure 8.18). When a stop codon is encountered, the tRNA holding the polypeptide remains in the P site, and a *release factor (RF)* binds with the ribosome. GTP hydrolysis provides the energy to cleave the polypeptide from the tRNA to which it is attached, as well as to eject the release factor and dissociate the 80S ribosome from the mRNA. At this point the 40S and 60S subunits are recycled to initiate translation of another mRNA. Eukaryotes have only one release factor that recognizes all three **stop codons**: UAA, UAG, and UGA.

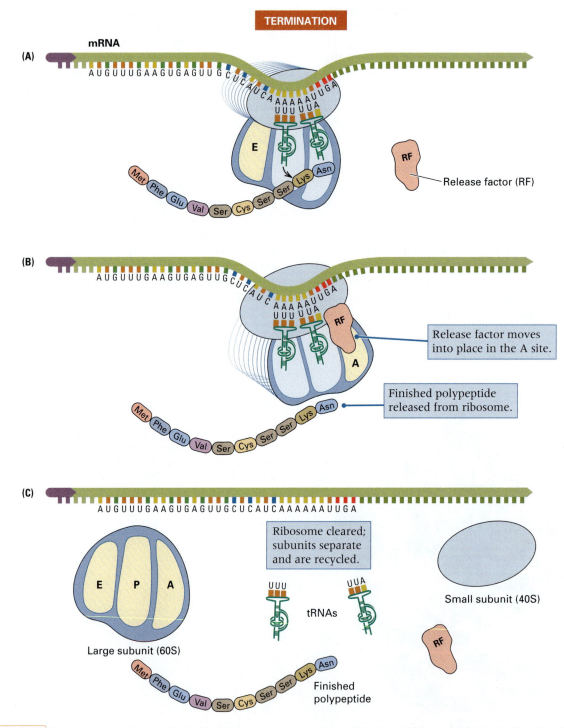

Figure 8.18 Termination of protein synthesis. When a stop codon is reached (A), no tRNA can bind to that site (B), which causes the release of the newly formed polypeptide and the remaining bound tRNA (C).

The situation differs in prokaryotes. In *E. coli*, the release factor RF-1 recognizes the stop codons UAA and UAG, whereas release factor RF-2 recognizes UAA and UGA. A third release factor, RF-3, is also required, but its function is uncertain.

■ Most polypeptide chains fold correctly as they exit the ribosome.

Each polypeptide chain tends to fold into a unique three-dimensional shape determined primarily by its sequence of amino acids. Generally speaking, polypeptide molecules fold so that amino acids with charged, hydrophilic side chains tend to be on the surface of the protein (in contact with water) and those with uncharged, hydrophobic side chains tend to be internal (hidden from water). Specific folded configurations also result from hydrogen bonding between peptide groups. Two fundamental polypeptide structures are the alpha (α) helix and the beta (β) sheet (Figure 8.19). An α helix is formed by hydrogen bonding between peptide groups that are close together in the polypeptide backbone. In an α helix, often represented as a coiled ribbon, the backbone is twisted so that the N−H in each peptide group is hydrogen-bonded with the C=O in the peptide group located four amino acids farther along the helix. The helical twist may be right-handed or left-handed, but

right-handed α helices are more common. Both α helices in Figure 8.19 are right-handed. In contrast, a β sheet is formed by hydrogen bonding between peptide groups in distant parts of the polypeptide chain, or even in different polypeptide chains. In a β sheet, often represented as parallel "flat" ribbons, the backbones of the interacting polypeptide chains are held flat and relatively rigid (forming a "sheet"), because alternate N−H groups in one backbone are hydrogen-bonded with alternate C=O groups in the backbone of the adjacent chain. In each polypeptide chain, alternate C=O and N−H groups are free to form hydrogen bonds with their counterparts in a different chain on the opposite side, so a β sheet can consist of multiple aligned segments in the same or different polypeptides. The orientation of the backbones in a β sheet may be antiparallel (adjacent backbones reversed in orientation relative to their amino and carboxyl ends) or parallel, but antiparallel is more common. In Figure 8.19, both β sheets are antiparallel. The rules of folding are so complex that, except for the simplest proteins, the final shape of a protein cannot usually be predicted from the amino acid sequence alone.

As polypeptides are being synthesized, they pass through a tunnel in the large ribosomal subunit that is long enough to include about 35 amino acids. The diameter of this tunnel is wide enough to accommodate an α helix but not so wide as to allow more

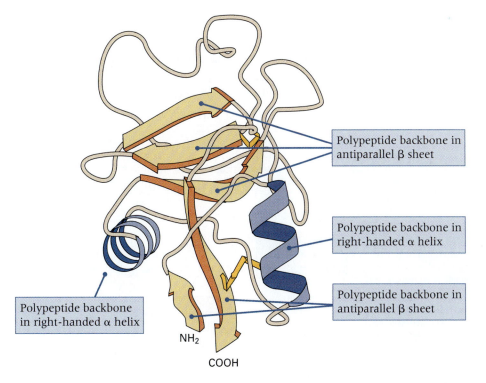

Figure 8.19 A "ribbon" diagram of the path of the backbone of a polypeptide, in this example a mannose-binding protein, showing the way in which the polypeptide is folded. The flat arrows represent β sheets, each of which is held to its neighboring sheet by hydrogen bonds. Helical regions are shown as coiled ribbons. [Adapted from William I. Weis, Kurt Drickamer, and Wayne A. Hendrickson. 1992. *Nature* 360: 127.]

complex structures to form. As the polypeptide emerges from the tunnel, it enters into a sort of cradle formed by a protein associated with the ribosome, which in prokaryotes is known as *trigger factor.* This cradle provides a protected space where the emerging polypeptide is able to undergo its folding process. About 70 to 75 percent of polypeptide chains fold properly as they emerge from the ribosomal tunnel into this protected space (Figure 8.20, part A).

But some polypeptide chains need additional help to fold properly. These tend to be large polypeptide chains composed of multiple folding domains that fold slowly, so that hydrophobic residues are exposed to the high concentration of macromolecules in the cytoplasm. Under such crowded conditions the exposed hydrophobic groups often attract each other and bind together,

forming inactive protein aggregates (part B). The proper folding of more complex polypeptides is aided by proteins called **chaperones** (part C). These proteins bind to hydrophobic groups and unstructured regions to shield them from aggregation, and by repeated cycles of binding and release they give the polypeptide time to find its proper folding pathway. The most complex proteins with very slow and inefficient folding pathways are shielded by a special class of proteins known as *chaperonins.* These form large, hollow cylindrical structures that trap the unstable intermediates inside and allow them to fold in a protected environment (part D). In eukaryotes, the most abundant polypeptides that make use of the chaperonin cylinders for folding are the cytoskeletal proteins actin and tubulin.

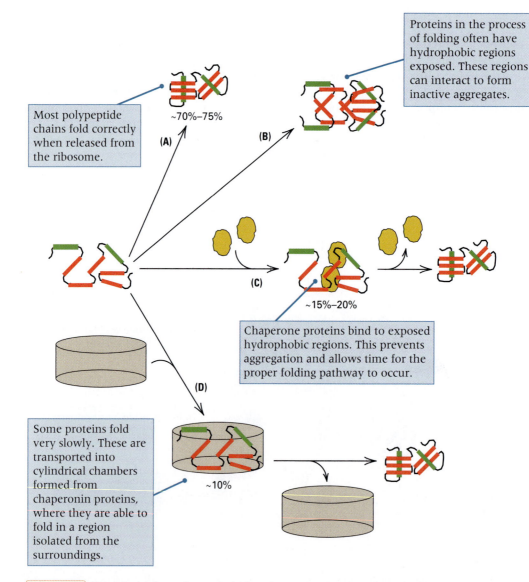

Most polypeptide chains fold correctly when released from the ribosome.

Proteins in the process of folding often have hydrophobic regions exposed. These regions can interact to form inactive aggregates.

Chaperone proteins bind to exposed hydrophobic regions. This prevents aggregation and allows time for the proper folding pathway to occur.

Some proteins fold very slowly. These are transported into cylindrical chambers formed from chaperonin proteins, where they are able to fold in a region isolated from the surroundings.

Figure 8.20 Alternative pathways in protein folding. The green regions represent α helices and the red regions β sheets.

■ Prokaryotes often encode multiple polypeptide chains in a single mRNA.

In prokaryotes, mRNA molecules have no cap, and there is no scanning mechanism to locate the first AUG. In *E. coli*, for example, translation is initiated when two initiation factors (IF-1 and IF-3) interact with the 30S subunit at the same time that another initiation factor (IF-2) binds with a special initiator tRNA charged with formylmethionine, symbolized tRNA^{fMet}. These components come together and combine with an mRNA, but not at the end. The attachment occurs by hydrogen bonding between the 3' end of the 16S RNA present in the 30S subunit and a special sequence, the **ribosome-binding site,** in the mRNA (also called the *Shine–Dalgarno sequence*). Together, the 30S + tRNA^{fMet} + mRNA complex recruits a 50S subunit, in which the tRNA^{fMet} is positioned in the P site and aligned with the AUG initiation codon, just as in part C of Figure 8.15. In the assembly of the completed ribosome, the initiation factors dissociate from the complex.

The major difference between translational initiation in prokaryotes and that in eukaryotes has an important implication. In eukaryotes, because of the scanning mechanism of initiation, a single mRNA can usually encode only one polypeptide chain. In prokaryotic mRNA, by contrast, the ribosome-binding site can be present anywhere near an AUG, so polypeptide synthesis can begin at any AUG that is closely preceded by a ribosome-binding site. Prokaryotes put this feature to good use. In prokaryotes, mRNA molecules commonly contain information for the amino acid sequences of several different polypeptide chains; such a molecule is called a **polycistronic mRNA.** (*Cistron* is a term often used to mean a base sequence that encodes a single polypeptide chain.) In a polycistronic mRNA, each polypeptide coding region is preceded by its own ribosome-binding site and AUG initiation codon. After the synthesis of one polypeptide is finished, the next along the way is translated (Figure 8.21). The genes contained in a polycistronic mRNA molecule often encode the different proteins of a metabolic pathway. For example, in *E. coli*, the ten enzymes needed to synthesize histidine are encoded by one polycistronic mRNA molecule. The use of polycistronic mRNA is an economical way for a cell to regulate the synthesis of related proteins in a coordinated manner.

In all organisms, an important feature of translation is that it proceeds in a particular direction along the mRNA and the polypeptide.

> **key concept**
>
> The mRNA is translated from an initiation codon to a stop codon in the 5'-to-3' direction. The polypeptide is synthesized from the amino end toward the carboxyl end by the addition of amino acids, one by one, to the carboxyl end.

For example, a polypeptide with the sequence $NH_2-Met-Pro-\cdots-Gly-Ser-COOH$ would start with methionine as the first amino acid in the chain and end with serine as the last amino acid added to

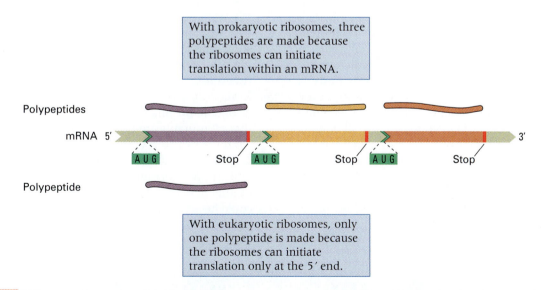

Figure 8.21 | Different products are translated from a three-cistron mRNA molecule by the ribosomes of prokaryotes and eukaryotes. The prokaryotic ribosome translates all of the genes, but the eukaryotic ribosome translates only the gene nearest the 5' terminus of the mRNA. Translated sequences are shown in purple, yellow, and orange, stop codons in red, the ribosome binding sites in green, and the spacer sequences in light green.

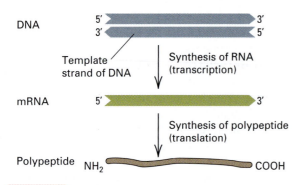

Figure 8.22 Direction of synthesis of RNA with respect to the coding strand of DNA, and of synthesis of protein with respect to mRNA.

the chain. The directions of synthesis are illustrated schematically in Figure 8.22.

By convention, in writing nucleotide sequences, we place the 5′ end at the left, and in writing amino acid sequences, we place the amino end at the left. Polynucleotides are generally written so that both synthesis and translation proceed from left to right, and polypeptides are written so that synthesis proceeds from left to right. This convention is used in all of our subsequent discussions of the genetic code.

8.6

The genetic code for amino acids is a triplet code.

Only four bases in DNA are needed to specify the 20 amino acids in proteins because a combination of three adjacent bases is used for each amino acid, as well as for the signals that start and stop protein synthesis. Each sequence of three adjacent bases in mRNA is a codon that specifies a particular amino acid (or chain termination). The **genetic code** is the list of all codons and the amino acids that they encode. Before the genetic code was determined experimentally, it was assumed that if all codons had the same number of bases, then each codon would have to contain at least three bases. Codons consisting of pairs of bases would be insufficient, because four bases can form only $4^2 = 16$ pairs; triplets of bases would suffice because four bases can form $4^3 = 64$ triplets. In fact, the genetic code is

a **triplet code,** and all 64 possible codons carry information of some sort. Most amino acids are encoded by more than one codon. Furthermore, in the translation of mRNA molecules, the codons do not overlap but are used sequentially (Figure 8.23).

■ Genetic evidence for a triplet code came from three-base insertions and deletions.

Although theoretical considerations suggested that each codon must contain at least three letters, codons having more than three letters could not be ruled out. The first widely accepted proof for a triplet code came from genetic experiments using *rII* mutants of bacteriophage T4 that had been induced by replication in the presence of the chemical *proflavin*. These experiments were carried out in 1961 by Francis Crick and collaborators. Proflavin-induced mutations typically resulted in total loss of function, which the investigators suspected were due to single-base insertions or deletions. Analysis of the properties of these mutations led directly to the deduction that the code is read three nucleotides at a time from a fixed point; in other words, there is a **reading frame** to each mRNA. Mutations that delete or add a base pair shift the reading frame and are called **frameshift mutations.** Figure 8.24 illustrates the profound effect of a frameshift mutation on the amino acid sequence of the polypeptide produced from the mRNA of the mutant gene.

The genetic analysis of the structure of the code began with an *rII* mutation called *FC0*, which was arbitrarily designated (+), as though it had an inserted base pair. (This was a lucky guess; when *FC0* was sequenced, it did turn out to have a single-base insertion.) If *FC0* has a (+) insertion, then it should be possible to revert the *FC0* allele to "wild-type" by deletion of a nearby base. Selection for r^+ revertants was carried out by isolating plaques formed on a lawn of an *E. coli* strain K12 that was lysogenic for phage λ. The basis of the selection is that *rII* mutants are unable to propagate in K12(λ). Analysis of the revertants revealed that each still carried the original *FC0* mutation, along with a second (suppressor) mutation that reversed the effects of the *FC0* mutation. The suppressor mutations could be separated by recombination from the original mutation by crossing each revertant to wildtype;

Direction of reading of codons in translation

Figure 8.23 Bases in an RNA molecule are read sequentially in the 5′ → 3′ direction, in groups of three.

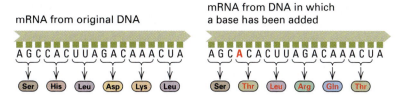

mRNA from original DNA

mRNA from DNA in which a base has been added

A G C C A C U U A G A C A A A C U A

A G C **A** C A C U U A G A C A A A C U A

Ser His Leu Asp Lys Leu

Ser Thr Leu Arg Gln Thr

Figure 8.24 The change in the amino acid sequence of a protein caused by the addition of an extra base, which shifts the reading frame. A deleted base also shifts the reading frame.

each suppressor mutation proved to be an *rII* mutation that, by itself, would cause the *r* (rapid lysis) phenotype. If *FC0* had an inserted base, then the suppressors should all result in deletion of a base pair; hence each suppressor of *FC0* was designated (−). The consequences of three such revertants for the translational reading frame are illustrated using ordinary three-letter words in Figure 8.25 . The (−) mutations are designated (−)₁, (−)₂, and (−)₃, and those parts of the mRNA translated in the correct reading frame are indicated in green.

In the *rII* experiments, all of the individual (−) suppressor mutations were used, in turn, to select other "wildtype" revertants, with the expectation that these revertants would carry new suppressor mutations of the (+) variety, because the (−)(+) combination should yield a phage able to form plaques on K12(λ).

Various double-mutant combinations were made by recombination. Usually any (+) (−) combination, or any (−)(+) combination, resulted in a wildtype phenotype, whereas (+)(+) and (−)(−) double-mutant combinations always resulted in the mutant phenotype. The most revealing result came when triple mutants were made. Usually, the (+)(+)(+) and (−)(−)(−) triple mutants yielded the wildtype phenotype!

The phenotypes of the various (+) and (−) combinations were interpreted in terms of a reading frame. The initial *FC0* mutation, a +1 insertion, shifts the reading frame, resulting in incorrect amino acid sequence from that point on and thus a nonfunctional protein (Figure 8.25). Deletion of a base pair nearby will restore the reading frame, although the amino acid sequence encoded between the two mutations will be different and

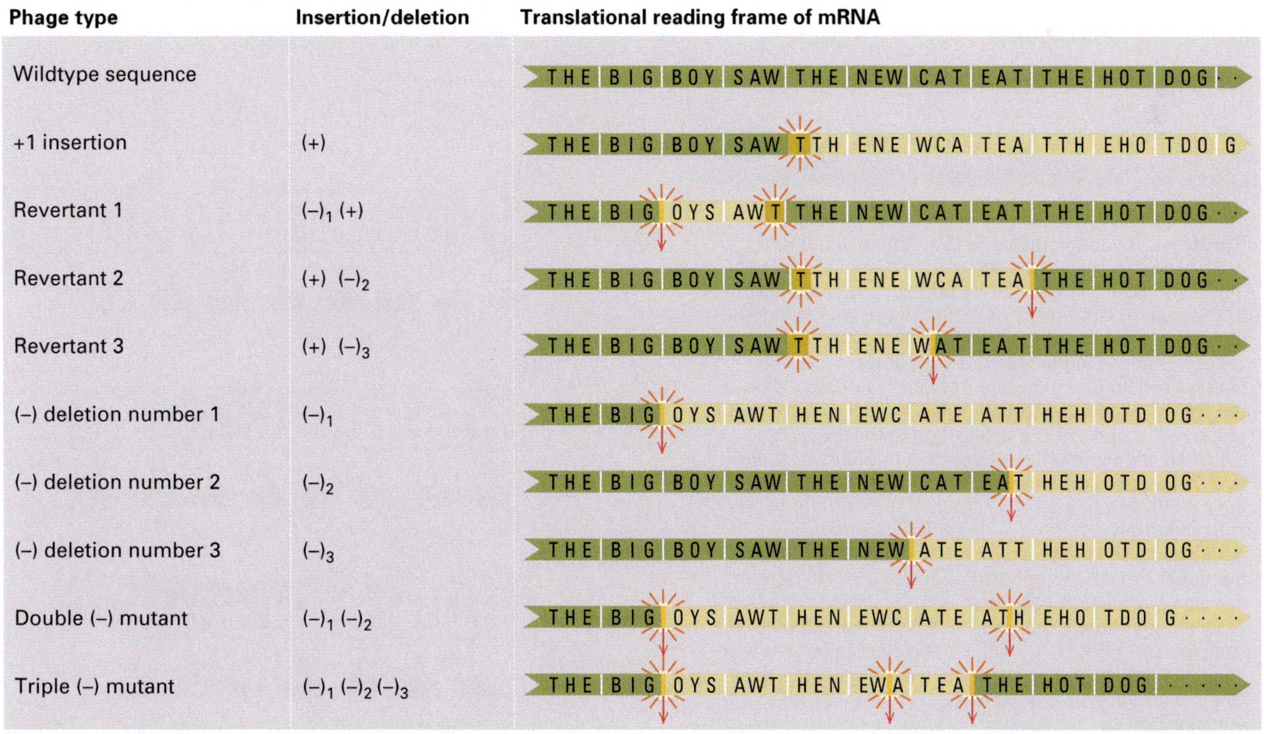

Phage type	Insertion/deletion	Translational reading frame of mRNA
Wildtype sequence		THE BIG BOY SAW THE NEW CAT EAT THE HOT DOG
+1 insertion	(+)	THE BIG BOY SAW TTH ENE WCA TEA TTH EHO TDO G
Revertant 1	(−)₁ (+)	THE BIG OYS AWT THE NEW CAT EAT THE HOT DOG
Revertant 2	(+) (−)₂	THE BIG BOY SAW TTH ENE WCA TEA THE HOT DOG
Revertant 3	(+) (−)₃	THE BIG BOY SAW TTH ENE WAT EAT THE HOT DOG
(−) deletion number 1	(−)₁	THE BIG OYS AWT HEN EWC ATE ATT HEH OTD OG
(−) deletion number 2	(−)₂	THE BIG BOY SAW THE NEW CAT EAT HEH OTD OG
(−) deletion number 3	(−)₃	THE BIG BOY SAW THE NEW ATE ATT HEH OTD OG
Double (−) mutant	(−)₁ (−)₂	THE BIG OYS AWT HEN EWC ATE ATH EHO TDO G
Triple (−) mutant	(−)₁ (−)₂ (−)₃	THE BIG OYS AWT HEN EWA TEA THE HOT DOG

Figure 8.25 Interpretation of the *rII* frameshift mutations showing that combinations of appropriately positioned single-base insertions (+) and single-base deletions (−) can restore the correct reading frame (green). The key finding was that a combination of three single-base deletions, as shown in the bottom line, also restores the correct reading frame. Two single-base deletions do not restore the reading frame. These classic experiments gave strong genetic evidence that the genetic code is a triplet code.

8.6 The genetic code for amino acids is a triplet code   303

incorrect. In $(+)(+)$ and $(-)(-)$ double mutants, the reading frame is shifted by two bases; the protein made is still nonfunctional. However, in the $(+)(+)(+)$ and $(-)(-)(-)$ triple mutants, the reading frame is restored, even though all amino acids encoded within the region bracketed by the outside mutations are incorrect; the protein made is one amino acid longer for $(+)(+)(+)$ and one amino acid shorter for $(-)(-)(-)$ (Figure 8.25).

The genetic analysis of the $(+)$ and $(-)$ mutations strongly supported the following conclusions:

- Translation of an mRNA starts from a fixed point.
- There is a single reading frame maintained throughout the process of translation.
- Each codon consists of three nucleotides.

Crick and his colleagues also drew other inferences from these experiments. First, in the genetic code, most codons must function in the specification of an amino acid. Second, each amino acid must be specified by more than one codon. They reasoned that if each amino acid had only one codon, then only 20 of the 64 possible codons could be used for coding amino acids. In this case, most frameshift mutations should have affected one of the remaining 44 "noncoding" codons in the reading frame, and hence a nearby frameshift of the opposite polarity mutation should not have suppressed the original mutation. Consequently, the code was deduced to be one in which more than one codon can specify a particular amino acid.

■ Most of the codons were determined from *in vitro* polypeptide synthesis.

Polypeptide synthesis can be carried out in cell extracts containing ribosomes, tRNA molecules, mRNA molecules, and the various protein factors needed for translation. If radioactive amino acids are added to the extract, radioactive polypeptides are made. Synthesis continues for only a few minutes because mRNA is gradually degraded by various nucleases in the mixture. The elucidation of the genetic code began with the observation that when the degradation of mRNA was allowed to go to completion and the synthetic polynucleotide polyuridylic acid (poly-U) was added to the mixture as an mRNA molecule, a polypeptide consisting only of phenylalanine (Phe—Phe—Phe—···) was synthesized. From this simple result and knowledge that the code is a triplet code, it was concluded that UUU must be a codon for the amino acid phenylalanine. Variations on this basic experiment identified other codons. For example, when a long sequence of guanines was added at the terminus of the poly-U, the polyphenylalanine was terminated by a sequence of glycines, indicating that GGG is a glycine codon (Figure 8.26). A trace of leucine or tryptophan was also present in the glycine-terminated polyphenylalanine. Incorporation of these amino acids was directed by the codons UUG and UGG at the transition point between U and G. When a single guanine was added to the terminus

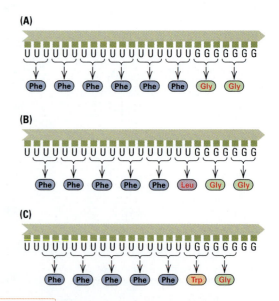

(A)

U U U U U U U U U U U U U U U U U U U G G G G G G

Phe Phe Phe Phe Phe Phe Gly Gly

(B)

U U U U U U U U U U U U U U U U U U U G G G G G G

Phe Phe Phe Phe Phe Leu Gly Gly

(C)

U U U U U U U U U U U U U U U U U U U G G G G G G

Phe Phe Phe Phe Phe Trp Gly

Figure 8.26 Polypeptide synthesis using 5'-UUUU . . . UUGGGGGGG-3' as an mRNA in three different reading frames, showing the reasons for the incorporation of glycine, leucine, and tryptophan.

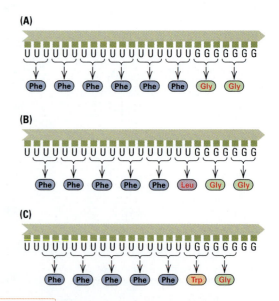

A Moment to Think

Problem: The nontemplate DNA strand in an amino-acid–coding portion of a gene is shown here, along with the amino acid sequence of the polypeptide chain encoded in this region. This particular region of the polypeptide is quite tolerant of amino acid replacements, so many missense mutations in the region do not destroy function. Four frameshift mutations are isolated: $-A$ is a deletion of the red A, $+A$ is a duplication of the red A, $-G$ is a deletion of the red G, and $+G$ is a duplication of the red G. As expected, all four mutations lead to a nonfunctional protein. Recombination is used to combine the mutations, in the expectation that a "$-$" deletion of one nucleotide will be compensated for by a "$+$" addition slightly farther along, because the second mutation will shift the reading frame back to the normal reading frame, and changes in the amino acids between the mutations should not destroy function. An unexpected result was obtained. Although the combination $+A$ with $-G$ did, indeed, restore protein function, the combination $-A$ with $+G$ was still mutant. How can you account for this result? (The answer can be found on page 309.)

```
GCTACAAAAATGACTGGCACTGCA
AlaThrLysMetThrGlyThrAla
```

304 **CHAPTER 8** The Molecular Genetics of Gene Expression

the human connection

Poly-U

Marshall W. Nirenberg and J. Heinrich Matthaei 1961
National Institutes of Health, Bethesda, Maryland
The Dependence of Cell-Free Protein Synthesis in E. coli *upon Naturally Occurring or Synthetic Polyribonucleotides*

In the years following the discovery of DNA structure by Watson and Crick in 1953, the biological implications of the discovery were largely ignored. A principal reason was that most biochemists still held strongly to the conviction that DNA had nothing to do with protein synthesis. The prevailing view was that proteins were made from small preexisting peptides by enzymes that joined the peptides together step by step in a specific order. It had been suggested that proteins might be made by amino acids being laid down in sequence upon an RNA template, but this hypothesis also was largely ignored. Not until this important paper appeared in 1961 was it shown that proteins are made by stepwise joining of individual amino acids in a sequence specified by a molecule of template RNA. The key finding was that in a cell-free mixture capable of supporting protein synthesis, the artificial polynucleotide poly-U (polyuridylic acid) resulted in the synthesis of a protein consisting only of the amino acid phenylalanine. The requirements for protein synthesis also included ribosomes (necessary for translation) and small RNA molecules (which include the charged transfer RNAs). After this paper appeared, the race was on to decipher the genetic code by which RNA specifies the amino acids in a protein.

A stable cell-free system has been obtained from *E. coli* which incorporates [radioactive] valine into protein at a rapid rate. . . . The present communication describes a requirement for template RNA, needed for amino acid incorporation even in the presence of soluble [small] RNA molecules and ribosomes. The amino acid incorporation stimulated by the addition of template RNA has many properties expected of protein synthesis. Naturally occurring RNA as well as synthetic polynucleotides were active. The synthetic polynucleotide appears to contain a code for the synthesis of a "protein" containing only one amino acid. . . . [Specifically,] the addition of polyuridylic acid resulted in a remarkable stimulation of [radioactive] phenylalanine incorporation. Phenylalanine incorporation was almost completely dependent upon the addition of polyuridylic acid, and incorporation proceeded at a linear rate for approximately 30 minutes. No other polynucleotide tested could replace polyuridylic acid. . . . The product of the reaction had the same apparent solubility as authentic polyphenylalanine...[and contained] phenylalanine and no other amino acids. . . . The results indicate the polyuridylic acid contains the

> **The results indicate the polyuridylic acid contains the information for the synthesis of a protein having the characteristics of polyphenylalanine.**

information for the synthesis of a protein having the characteristics of polyphenylalanine. . . . One or more uridylic acid residues therefore appears to be the code for phenylalanine. Whether the code is of the singlet, triplet, etc., type has not yet been determined. Polyuridylic acid seemingly functions as a synthetic template or messenger RNA, and this stable, cell-free system may well synthesize any protein corresponding to meaningful information contained in added RNA.

Source: Proceedings of the National Academy of Sciences USA 47: 1588–1602.

of a poly-U chain, the polyphenylalanine was terminated by leucine. Thus UUG is a leucine codon, and UGG must be a codon for tryptophan. Similar experiments were carried out with poly-A, which yielded polylysine, and with poly-C, which produced polyproline.

Other experiments led to a complete elucidation of the code. Three codons,

$$UAA \quad UAG \quad UGA$$

were found to be stop signals for translation.

■ **Redundancy and near-universality are principal features of the genetic code.**

The *in vitro* translation experiments with components isolated from the bacterium *E. coli* have been repeated with components obtained from many species of bacteria, yeast, plants, and animals. The standard genetic code deduced from these experiments is considered to be almost universal, because the same codon assignments can be made for nuclear genes in nearly all organisms that have been examined. However, some minor differences in

Table 8.3

The standard genetic code

First position (5' end)	Second position				Third position (3' end)
	U	**C**	**A**	**G**	
U	UUU Phe ⎫ F / UUC Phe ⎬ / UUA Leu ⎫ L / UUG Leu ⎭	UCU Ser ⎫ / UCC Ser ⎬ S / UCA Ser / UCG Ser ⎭	UAU Tyr ⎫ Y / UAC Tyr ⎬ / **UAA Stop** / **UAG Stop**	UGU Cys ⎫ C / UGC Cys ⎬ / **UGA Stop** / UGG Trp W	U / C / A / G
C	CUU Leu ⎫ / CUC Leu ⎬ L / CUA Leu / CUG Leu ⎭	CCU Pro ⎫ / CCC Pro ⎬ P / CCA Pro / CCG Pro ⎭	CAU His ⎫ H / CAC His ⎬ / CAA Gln ⎫ Q / CAG Gln ⎭	CGU Arg ⎫ / CGC Arg ⎬ R / CGA Arg / CGG Arg ⎭	U / C / A / G
A	AUU Ile ⎫ / AUC Ile ⎬ I / AUA Ile / **AUG Met** M	ACU Thr ⎫ / ACC Thr ⎬ T / ACA Thr / ACG Thr ⎭	AAU Asn ⎫ N / AAC Asn ⎬ / AAA Lys ⎫ K / AAG Lys ⎭	AGU Ser ⎫ S / AGC Ser ⎬ / AGA Arg ⎫ R / AGG Arg ⎭	U / C / A / G
G	GUU Val ⎫ / GUC Val ⎬ V / GUA Val / GUG Val ⎭	GCU Ala ⎫ / GCC Ala ⎬ A / GCA Ala / GCG Ala ⎭	GAU Asp ⎫ D / GAC Asp ⎬ / GAA Glu ⎫ E / GAG Glu ⎭	GGU Gly ⎫ / GGC Gly ⎬ G / GGA Gly / GGG Gly ⎭	U / C / A / G

Note: Each amino acid is given its conventional abbreviation in both the single-letter and three-letter format. The codon AUG, which codes for methionine (green), is generally used for initiation. The codons are conventionally written with the 5' base on the left and the 3' base on the right.

codon assignments are found in certain protozoa and in the genetic codes of organelles.

The standard genetic code is shown in **Table 8.3**. Note that four codons—the three stop codons and the start codon—are signals. Altogether, 61 codons specify amino acids. In many cases several codons direct the insertion of the same amino acid into a polypeptide chain. This feature confirms the inference from the *rII* frameshift mutations that the genetic code is *redundant* (also called *degenerate*). In a redundant genetic code, some amino acids are encoded by two or more different codons. In the actual genetic code, all amino acids except tryptophan and methionine are specified by more than one codon. This **redundancy** is not random. For example, with the exception of serine, leucine, and arginine, all codons that correspond to the same amino acid are in the same box of Table 8.3; that is, *synonymous codons usually differ only in the third base.* For example, GGU, GGC, GGA, and GGG all code for glycine. Moreover, in all cases in which two codons code for the same amino acid, the third base is either A or G (both purines) or T or C (both pyrimidines).

The codon assignments shown in Table 8.3 are completely consistent with all chemical observations and with the amino acid sequences of wild-type and mutant proteins. In virtually every case in which a mutant protein differs by a single amino acid from the wildtype form, the amino acid replacement can be accounted for by a single base change between the codons corresponding to the two different amino acids. For example, substitution of glutamic acid by valine, which occurs in sickle-cell hemoglobin, results from a change from GAG to GUG in codon six of the β-globin mRNA.

■ An aminoacyl-tRNA synthetase attaches an amino acid to its tRNA.

The decoding operation by which the base sequence within an mRNA molecule becomes translated into the amino acid sequence of a protein is accomplished by charged tRNA molecules, each of which is linked to the correct amino acid by an aminoacyl-tRNA synthetase.

The tRNA molecules are small single-stranded nucleic acids ranging in size from about 70 to 90 nucleotides. Like all RNA molecules, they have a 3'-OH terminus, but the opposite end terminates with a 5'-monophosphate, rather than a 5'-triphosphate, because tRNA molecules are

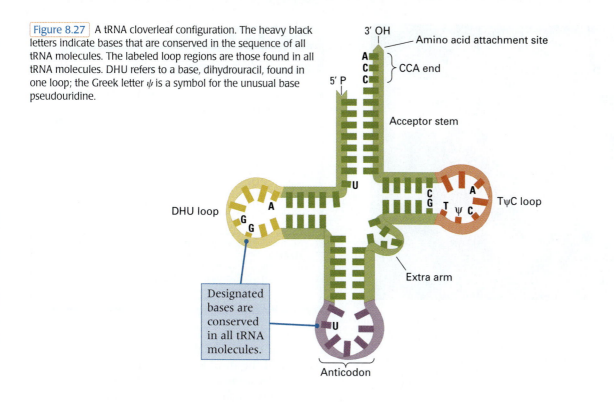

Figure 8.27 A tRNA cloverleaf configuration. The heavy black letters indicate bases that are conserved in the sequence of all tRNA molecules. The labeled loop regions are those found in all tRNA molecules. DHU refers to a base, dihydrouracil, found in one loop; the Greek letter ψ is a symbol for the unusual base pseudouridine.

cleaved from a larger primary transcript. Internal complementary base sequences form short double-stranded regions, causing the molecule to fold into a structure in which open loops are connected to one another by double-stranded stems (Figure 8.27). In two dimensions, a tRNA molecule is drawn as a planar cloverleaf. Its three-dimensional structure is more complex, as is shown in Figure 8.28, where part A shows a skeletal model of a yeast tRNA molecule for phenylalanine and part B is an interpretive drawing. All tRNA molecules have similar structures.

(A)

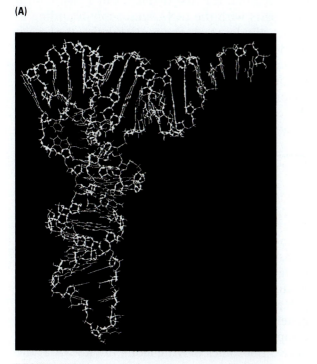

(B)

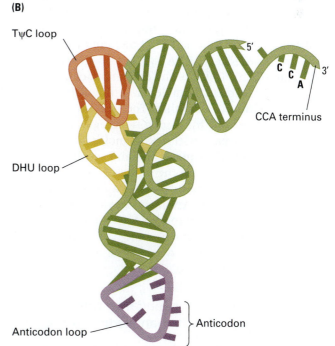

Figure 8.28 Yeast phenylalanine tRNA (called tRNA^{Phe}). (A) A skeletal model. (B) A schematic diagram of the three-dimensional structure of yeast tRNA^{Phe}. [Courtesy of Sung-Hou Kim.]

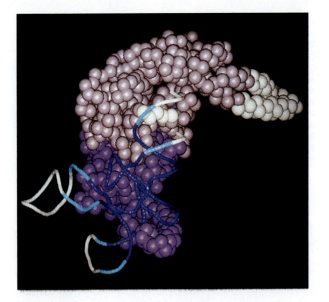

Figure 8.29 Three-dimensional structure of seryl-tRNA synthetase (solid spheres) complexed with its tRNA. Note that there are many points of contact between the enzyme and the tRNA. The molecules are from *Thermus thermophilus*. [Courtesy of Stephen Cusack. Reprinted with permission from S. Cusack, V. Biou, A. Yaremchuk, and M. Tukalo, *Science* 263: 1404. © 1996 AAAS.]

Particular regions of each tRNA molecule are used in the decoding operation. One region is the *anticodon* sequence, which consists of three bases that can form base pairs with a codon sequence in the mRNA. No normal tRNA molecule has an anticodon complementary to any of the stop codons UAG, UAA, and UGA. A second critical site, which all tRNAs share, is the CCA terminus at the 3' end where the amino acid is attached. A specific aminoacyl-tRNA synthetase transfers the amino acid onto the A residue. At least one (and usually only one) aminoacyl-tRNA synthetase exists for each amino acid. To make the correct attachment, the synthetase must be able to distinguish one tRNA molecule from another. The necessary distinction is provided by recognition regions that encompass many parts of the tRNA molecule.

The different tRNA molecules and synthetases are designated by stating the name of the amino acid that is linked to a particular tRNA molecule by a specific synthetase; for example, seryl-tRNA synthetase attaches serine to tRNA^{Ser} (Figure 8.29). When an amino acid has become attached to a tRNA molecule, the tRNA is said to be *charged*. An uncharged tRNA lacks an amino acid.

■ Much of the code's redundancy comes from wobble in codon–anticodon pairing.

Several features of the genetic code and of the decoding system suggest that base pairing between the codon and the anticodon has special features.

Table 8.4

Wobble rules for tRNAs of *E. coli* and *S. cerevisiae*

First base in anticodon (5' position)	Allowed base in third codon position (3' position)	
	E. coli	*S. cerevisiae*
A	U	—
C	G	G
U	A or G	A
G	C or U	C or U
I	A, C, or U	C or U

Notes: In *S. cerevisiae*, an A at the 5' position in the anticodon is always modified to I, which indicates inosine; inosine is structurally similar to adenosine except that the $-NH_2$ is replaced with $-OH$. Likewise, a U at the first anticodon position is often modified in this organism.

First, the code is highly redundant. Second, the identity of the third base of a codon is often unimportant. In some cases, any nucleotide will do; in others, any purine or any pyrimidine serves the same function. Third, the number of distinct tRNA molecules present in an organism is less than the number of codons; because all codons are used, the anticodons of some tRNA molecules must be able to pair with more than one codon. Experiments with several purified tRNA molecules showed this to be the case.

To account for these observations, the **wobble** concept was advanced in 1966 by Francis Crick. He proposed that the first two bases in a codon form base pairs with the tRNA anticodon according to the usual rules (A−U and G−C) but that the base at the 5' end of the anticodon is less spatially constrained than the first two and can form hydrogen bonds with more than one base at the 3' end of the codon. His suggestion was essentially correct, but the allowed base pairs differ somewhat among organisms (**Table 8.4**).

8.7

Several ribosomes can move in tandem along a messenger RNA.

In most prokaryotes and eukaryotes, the unit of translation is almost never simply one ribosome traversing an mRNA molecule. After about 25 amino acids have been joined together in a polypeptide chain, an AUG initiation codon is completely free of the ribosome, and a second initiation complex can form. The overall configuration is that of two ribosomes moving along the mRNA at the same speed. When the second ribosome has moved along a distance similar to that traversed by the

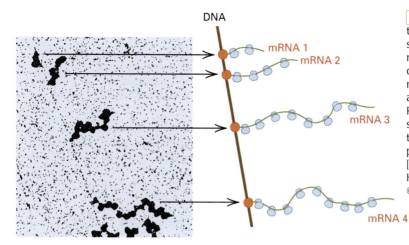

DNA

mRNA 1
mRNA 2
mRNA 3
mRNA 4

first, a third ribosome can attach to the initiation site. The process of movement and reinitiation continues until the mRNA is covered with ribosomes at a density of about one ribosome per 80 nucleotides. This large translation unit is called a **polysome,** and this is the usual form of the translation unit in both prokaryotes and eukaryotes. Figure 8.30 is an electron micrograph and interpretive drawing of polysomes that have formed along several mRNA molecules in *E. coli.*

Figure 8.30 also illustrates a feature of gene expression that is found only in prokaryotes. Because prokaryotes lack a nuclear envelope separating the location of DNA from that of the ribosomes, transcription of an mRNA and translation of the mRNA can take place in rapid succession. The 5′ end of an mRNA molecule is synthesized first. This end includes the ribosome-binding site, followed, in order, by the initiating AUG codon and the rest of the coding sequence. Because

translation takes place in the 5′ → 3′ direction, the first part of the mRNA becomes available for translation even before the rest of the transcript is finished. The absence of a nucleus therefore makes possible the simultaneous execution, or **coupling,** of transcription and translation. In Figure 8.30, the DNA molecule is actively being transcribed into a number of mRNA molecules (four are shown). Each of the mRNA molecules, in turn, is actively being translated by polysomes. Transcription of DNA begins in the upper left part of the micrograph. The lengths of the polysomes increase with distance from the transcription initiation site because the mRNA is farther from that site and hence longer. Coupled transcription and translation cannot take place in eukaryotes because the mRNA is synthesized and processed in the nucleus and is only later transported through the nuclear envelope to the cytoplasm, where the ribosomes are located.

A A Moment to Think

Answer to Problem: In this kind of problem, the best strategy is first to deduce the amino acid sequence of each mutant polypeptide chain. The answer often becomes clear immediately. In this case, the mutant amino acid sequences for all four mutants and the two types of double mutants are shown. Note that the −A mutation results in a termination codon (UGA), which terminates translation at that codon (indicated by the raised dot). The −A with +G combination also has this termination codon, so it remains mutant; even though the frameshift is corrected by the +G, the correction is ineffectual because translation is already terminated prior to this point. The +A with −G combination is nonmutant, because the frameshifts do compensate, and there is no termination codon in the region between them.

```
−A   GCUACAAAAUGACUGGCACUGCA        +A with −G
     AlaThrLys·                          GCUACAAAAAAUGACUGCACUGCA
                                         AlaThrLysAsnAspCysThrAla
+A   GCUACAAAAAAUGACUGGCACUGCA
     AlaThrLysAsnAspTrpHisCys       −A with +G

−G   GCUACAAAAAUGACUGCACUGCA            GCUACAAAAUGACUGGGCACUGCA
     AlaThrLysMetThrAlaLeu               AlaThrLys·

+G   GCUACAAAAAUGACUGGGCACUGCA
     AlaThrLysMetThrGlyHisCys
```

chapter summary

8.1 Polypeptide chains are linear polymers of amino acids.

- Human proteins, and those of other vertebrates, have a more complex domain structure than do the proteins of invertebrates.

8.2 The linear order of amino acids is encoded in a DNA base sequence.

The flow of information from a gene to its product is from DNA to RNA to protein. A sequence of bases in a DNA molecule is first used as a template to direct the synthesis of a molecule of RNA with a complementary sequence of bases. This RNA either is the mRNA (in prokaryotes) or is processed into the mRNA (in eukaryotes), and then the sequence of bases in the mRNA is used to specify the order of amino acids in a polypeptide chain. The characteristics of different proteins are determined by the sequence of amino acids of the polypeptide chain and by the way in which the chain is folded. Each gene is usually responsible for the synthesis of a single polypeptide.

8.3 The base sequence in DNA specifies the base sequence in an RNA transcript.

- The chemical synthesis of RNA is similar to that of DNA.
- Eukaryotes have several types of RNA polymerase.
- Particular nucleotide sequences define the beginning and end of a gene.
- Messenger RNA directs the synthesis of a polypeptide chain.

8.4 RNA processing converts the original RNA transcript into messenger RNA.

- Splicing removes introns from the RNA transcript.
- Human genes tend to be very long even though they encode proteins of modest size.
- Many exons code for distinct protein-folding domains.

The process in which a molecule of RNA is made that is complementary in base sequence to a template strand of DNA is called transcription. The RNA transcript is produced by the enzyme RNA polymerase, which joins ribonucleoside triphosphates by the same chemical reaction used in DNA synthesis. RNA polymerase differs from DNA polymerase in that a primer is not needed to initiate synthesis. Transcription is initiated when RNA polymerase binds to a promoter sequence. Transcription begins at the transcription start site downstream from the promoter and continues until a termination site is reached. Only one strand of duplex DNA is transcribed. Transcription along the DNA template proceeds in the $3' \rightarrow 5'$ direction, and the RNA is produced in the $5' \rightarrow 3'$ direction. (That is, successive ribonucleotides are added to the growing transcript at the $3'$ end.) In prokaryotes, the transcript is used directly as mRNA (messenger RNA) in polypeptide synthesis. In eukaryotes, the RNA primary transcript is processed even as transcription is taking place: Noncoding sequences called introns are removed, the exons are spliced together, and the termini are modified by the formation of a cap at the $5'$ end and the addition of a poly-A tail at the $3'$ end.

8.5 Translation into a polypeptide chain takes place on a ribosome.

- In eukaryotes, initiation takes place by scanning the mRNA for an initiation codon.
- Elongation takes place codon by codon.
- A termination codon signals release of the finished polypeptide chain.
- Most polypeptide chains fold correctly as they exit the ribosome.
- Prokaryotes often encode multiple polypeptide chains in a single mRNA.

After mRNA is formed, polypeptide chains are synthesized by translation of the mRNA molecule. Translation includes initiation, elongation, and release of the polypeptide chain. In eukaryotes, the initiation complex scans from the $5'$ cap to locate the first AUG; in prokaryotes, the initiation complex forms at any ribosome-binding site in the mRNA. In elongation, the process of translocation ratchets the ribosomal subunits along the mRNA as each new codon is encountered and each new peptide bond is formed. Elongation makes possible the successive reading of the nucleotide sequence of an mRNA molecule in groups of three nucleotides (the codons). When a stop codon is encountered, one or more release factors bind with the ribosome and cause release of the finished polypeptide and dissociation of the ribosome from the mRNA.

8.6 The genetic code for amino acids is a triplet code.

- Genetic evidence for a triplet code came from three-base insertions and deletions.
- Most of the codons were determined from *in vitro* polypeptide synthesis.
- Redundancy and near-universality are principal features of the genetic code.
- An aminoacyl-tRNA synthetase attaches an amino acid to its tRNA.
- Much of the code's redundancy comes from wobble in codon–anticodon pairing.

8.7 Several ribosomes can move in tandem along a messenger RNA.

Several ribosomes can translate an mRNA molecule simultaneously, forming a polysome. In prokaryotes, translation often begins before synthesis of mRNA is completed; in eukaryotes, this does not occur because mRNA is made in the nucleus, whereas the ribosomes are located in the cytoplasm. Prokaryotic mRNA molecules are often polycistronic, encoding several different polypeptides. This is not possible in eukaryotes, because only the AUG site nearest the $5'$ terminus of the mRNA can be used to initiate polypeptide synthesis; thus eukaryotic mRNA is monocistronic.

issues & ideas

- What is meant by the term *gene expression*? Would you make a distinction between gene *expression* and gene *regulation*? Why or why not?

- Would you regard an original text and its translation into another language as "colinear"? Explain your answer.

- In a eukaryotic cell, four general types of RNA molecules are used in gene expression. What are these types of RNA called? Which is not involved in gene expression in prokaryotic cells, and why not?

- Give an example of a genetic system that does not use the standard genetic code.

- What does it mean to say that the standard genetic code is redundant? Which (if any) amino acids are encoded by one codon? By two? By three? By four? By five? By six?

- What is a frameshift mutation? Explain how *rII* recombinants containing multiple, single-nucleotide frameshift mutations were used to show that the messenger RNA is translated in consecutive groups of three nucleotides.

- Suppose that a duplex DNA molecule undergoes two double-stranded breaks that tightly flank the promoter of a gene and that the promoter region is inverted before the backbones are rejoined by repair enzymes. Would you expect the inverted promoter to be able to recruit the transcription complex? What, if anything, would be wrong with the transcript of the gene?

key terms & concepts

A (aminoacyl) site
amino terminus
aminoacylated tRNA
aminoacyl-tRNA synthetase
anticodon
cap
carboxyl terminus
chain elongation
chain initiation
chain termination
chaperone
charged tRNA
codon
colinearity
consensus sequence
coupled transcription–
  translation

domain
E (exit) site
elongation factor
exon
exon shuffle
5′ untranslated region
frameshift mutation
gene expression
gene product
genetic code
initiation factor
inosine (I)
intron
messenger RNA (mRNA)
open reading frame (ORF)
P (peptidyl) site
peptide bond

peptidyl transferase
poly-A tail
polycistronic mRNA
polypeptide chain
polysome
primary transcript
promoter
promoter recognition
R group
reading frame
redundancy
release phase
ribosome
ribosome-binding site
ribozyme
RNA polymerase
RNA polymerase holoenzyme

RNA processing
RNA splicing
scanning
snRNP
spliceosome
start codon
stop codon
TATA box
template strand
termination factor
3′ untranslated region
transcription
transfer RNA (tRNA)
translation
triplet code
wobble

1. _____ Triplet of bases that specifies an amino acid.

2. _____ Process in which introns are removed from an RNA transcript.

3. _____ The tRNA-binding site on the ribosome that attracts the incoming charged tRNA.

4. _____ Type of enzyme that attaches an amino acid to its corresponding tRNA.

5. _____ Unique feature of the 5′ end of a eukaryotic messenger RNA not found in prokaryotic messenger RNA.

6. _____ Process by which eukaryotic ribosomes find the correct AUG initiation codon.

7. _____ The end of the polypeptide chain synthesized first.

8. _____ Table of correspondences of the full set of codons with the full set of amino acids.

9. _____ This feature of codon-anticodon base pairing accounts for the observation that certain bases at the 5′ end of an anticodon can pair with any of two or more bases at the 3′ end of the codon.

10. _____ An RNA molecule that functions as a catalyst to promote a biochemical reaction.

11. _____ Region in a polypeptide chain that undergoes folding relatively independently of other regions in the same chain.

12. _____ Type of protein that helps other proteins to fold properly.

Problem 1

The International Union of Biochemistry and Molecular Biology (IUBMB) has designated a single-letter code for abbreviating the nucleotide bases that allows for ambiguous assignments. The code is shown in the accompanying dia-gram. The same code is used for DNA as for RNA. For ambiguous nucleotides, T and U are regarded as equivalent. Assuming standard Watson–Crick pairing between the two nucleotide strands shown, complete the sequence of the bottom strand, using the appropriate symbol from the standard ambiguity code.

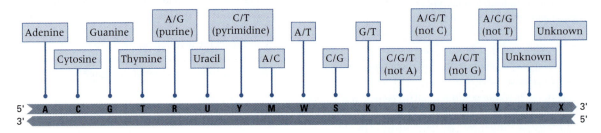

Solution The ambiguity codes are very useful not only for designating uncertain nucleotides in DNA sequences but also for summarizing the redundancies in the genetic code (see Step-by-Step Problem 2). The pairing relationships are straightforward for A, T, G, C, and U, but for ambiguous nucleotides one has to enumerate the possibilities and then select the symbol that expresses these ambiguities. One peculiar feature is that some symbols pair with themselves.

For example, W (A or T) in one strand must also have a W (T or A) in the other strand, where the convention is that the paired nucleotides, though ambiguous, must obey the Watson–Crick pairing rules. All of the pairings can be worked out in this way, and the results are shown in the accompanying diagram. There are two symbols—namely N and X—in use for "any nucleotide," so these can be paired however it is convenient.

5'	A	C	G	T	R	U	Y	M	W	S	K	B	D	H	V	N	X	3'
3'	T	G	C	A	Y	A	R	K	W	S	M	V	H	D	B	N/X	X/N	5'

Problem 2

Rewrite the genetic code table using as many as possible of the single-letter codes for ambiguous bases established by the International Union of Biochemistry and Molecular Biology (IUBMB), as shown in Step-by-Step Problem 1.

Solution This problem requires that you examine the standard genetic code and select the proper symbol for ambiguous nucleotides. The version of the genetic code that results is shown here. It has considerably fewer entries than the standard format, and it shows the general structure of the code at a glance.

			Second nucleotide in codon			
		T	C	A	G
First nucleotide in codon	T	TTY Phe **F** TTR Leu **L**	TCN Ser **S**	TAY Tyr **Y** TAR Stop	TGY Cys **C** TGA Stop TGG Trp **W**
	C	CTN Leu **L**	CCN Pro **P**	CAY His **H** CAR Gln **Q**	CGN Arg **R**
	A	ATH Ile **I** ATG Met **M**	ACN Thr **T**	AAY Asn **N** AAR Lys **K**	AGY Ser **S** AGR Arg **R**
	G	GTN Val **V**	GCN Ala **A**	GAY Asp **D** GAR Glu **E**	GGN Gly **G**

8.1 Which of the following characteristics are shared by polypeptide chains and nucleotide chains?
(a) Both have a sugar–phosphate backbone.
(b) Both are chains consisting of 20 types of repeating units.
(c) Both types of molecules contain unambiguous and interconvertible triplet codes.
(d) Both types of polymers are linear and unbranched.
(e) None of the above is correct.

8.2 The concept that a strand of DNA serves as a template for transcription of an RNA, which is translated into a polypeptide, is known as the "central dogma" of gene expression. All three types of molecules have a polarity. In the DNA template and the RNA transcript, the polarity is determined by the free 3' or 5' group at opposite ends of the polynucleotide chains; in a polypeptide chain, the polarity is determined by the free amino group (N terminal) or carboxyl group (C terminal) at opposite ends. Each of the following statements describes one possible polarity of the DNA template, the RNA transcript, and the polypeptide chain, respectively, in temporal order of use as a template or in synthesis. Which statement is correct?

(a) 5′ to 3′ DNA; 3′ to 5′ RNA; N terminal to C terminal.
(b) 3′ to 5′ DNA; 3′ to 5′ RNA; N terminal to C terminal.
(c) 3′ to 5′ DNA; 3′ to 5′ RNA; C terminal to N terminal.
(d) 3′ to 5′ DNA; 5′ to 3′ RNA; N terminal to C terminal.
(e) 5′ to 3′ DNA; 5′ to 3′ RNA; C terminal to N terminal.

8.3 What are the translation initiation and stop codons in the genetic code? In a random sequence of four ribonucleotides, all with equal frequency, what is the probability that any three adjacent nucleotides will be a start codon? A stop codon? In an mRNA molecule of random sequence, what is the average distance between stop codons?

8.4 A part of the *template* strand of a DNA molecule that codes for the 5′ end of an mRNA has the sequence 3′-TTTTACGGGAATTAGAGTCGCAGGATG-5′. What is the amino acid sequence of the polypeptide encoded by this region, assuming that the normal start codon is needed for initiation of polypeptide synthesis?

8.5 What codons could pair with the anticodon 5′-IAU-3′, given that I (inosine) can pair with "H" (A or U or C). What amino acid would be incorporated?

8.6 How many different sequences of nine ribonucleotides would code for each of the following amino acids?
(a) Met−His−Thr
(b) Met−Arg−Thr
Write the sequences using the symbols Y for any pyrimidine, R for any purine, and N for any nucleotide.

8.7 If DNA consisted of only two nucleotides (say, A and T) in any sequence, what is the minimum number of adjacent nucleotides that would be needed to specify uniquely each of the 20 amino acids?

8.8 An RNA molecule of sequence 5′-UUUUUUUUU-3′ (poly-U) codes for PhePhePhe (polyphenylalanine). If a G is added to the 5′ end of the molecule, the polyphenylalanine has a different amino acid at the amino terminus, and if a G is added to the 3′ end, there is a different amino acid at the carboxyl terminus. What are the amino acids?

8.9 What polypeptide products are made when the alternating polymer GUGU . . . is used in an *in vitro* protein-synthesizing system that does not need a start codon?

8.10 Some codons in the genetic code were determined experimentally by the translation of random polymers. If a ribonucleotide polymer is synthesized that contains 3/4 A and 1/4 C in random order, which amino acids will the resulting polypeptide contain, and in what frequencies?

8.11 The amino acid methionine (Met) is the unique "start" codon for polypeptide synthesis, and any mRNA with an open reading frame that lacks a methionine codon to initiate translation remains untranslated and is eventually degraded. Yet many polypeptides found in mature functional proteins do not, in fact, have methionine at their amino end. Suggest an explanation.

8.12 Suppose a primitive living organism is discovered on Mars. It has a genetic system similar to our own in that the sequence of subunits in the genetic material (nucleic acid) is used as a code to specify the linear sequence of subunits of a different type of molecule (protein). In this organism, the nucleic acids are made up of four kinds of nucleotides, but the proteins contain only five kinds of amino acids. The organism produces 100 different proteins, each 10 amino acids in length.
(a) What is the minimum allowable number of bases in a codon in this organism?
(b) In order to make the 10 proteins, what is the minimum number of nucleotides in the genome of the organism?

8.13 Make a sketch of a mature eukaryotic messenger RNA molecule hybridized to the transcribed strand of DNA of a gene that contains two introns, oriented with the promoter region of the DNA at the left. Clearly label the DNA and the mRNA. Use the letters that follow to label the location and/or boundaries of each segment. Some letters may be used several times, as appropriate; and some, which are not applicable, may not be used at all.
(a) 5′ end   (b) 3′ end
(c) Promoter region   (d) Attenuator
(e) Intron   (f) Exon
(g) Polyadenylation signal (h) Leader region
(i) Ribosome-binding site (j) Translation start codon
(k) Translation stop codon (l) 5′ cap
(m) Poly-A tail

8.14 Two *E. coli* genes, *A* and *B*, are known from mapping experiments to be very close to each other. A deletion mutation is isolated that eliminates the activity of both *A* and *B*. Neither the A nor the B protein can be found in the mutant, but a novel protein is isolated in which the amino-terminal 30 amino acids are identical to those of the *B* gene product and the carboxyl-terminal 30 amino acids are identical to those of the *A* gene product.
(a) With regard to the 5′ → 3′ orientation of the nontranscribed DNA strand, is the order of the genes *A B* or *B A*?
(b) Can you make any inference about the number of bases deleted?

8.15 The table at the bottom of this page shows matching regions of the DNA, mRNA, tRNA, and amino acids encoded in a particular gene. The mRNA is shown with its 5′ end at the left, and the tRNA anticodon is shown with its 3′ end at the left. The vertical lines define the reading frame.
(a) Complete the nucleic acid sequences, assuming normal Watson–Crick pairing between each codon and anticodon.
(b) Is the DNA strand that is transcribed the top strand or the bottom strand?
(c) Translate the mRNA in all three reading frames.
(d) Specify the nucleic acid strand(s) whose sequence could be used as a probe in a Southern blot hybridization, in which the hybridization is carried out against genomic DNA.
(e) Specify the nucleic acid strand(s) whose sequence could be used as a probe in a Northern blot hybridization, in which the hybridization is carried out against mRNA.

Problem 8.15

DNA double helix	– C – – – A T – C – – – – – – – – – – – A T – – G T
	T – – – – – – T – – – – – – – – – C A – – – – –
mRNA	– – A A C – – – – – – – – – – G C – – – G C – –
tRNA anticodon	– – – – – – – – – G C A – – – – – – – – – –
Amino acids	| | | | | Trp | | | |

8.16 A double mutant produced by recombination contains two single nucleotide frameshifts separated by about 20 base pairs. The first is an insertion, the second a deletion. The amino acid sequences of the wildtype and mutant polypeptide in this part of the protein are as follows:
Wildtype:

```
Lys—Lys—Tyr—His—Gln—Trp—Thr—Cys—Asn
```

Double Mutant:

```
Lys—Gln—Ile—Pro—Pro—Val—Asp—Met—Asn
```

What are the original and the double-mutant mRNA sequences? Which nucleotide in the wildtype sequence is the frameshift addition? Which nucleotide in the double-mutant sequence is the frameshift deletion? (In working this problem, you will find it convenient to use the conventional symbols Y for unknown pyrimidine, R for unknown purine, N for unknown nucleotide, and H for A or C or T.)

8.17 What polypeptide products are made when the alternating polymer GUCGUC . . . is used in an *in vitro* protein-synthesizing system that does not need a start codon?

8.18 Prior to the demonstration that messenger RNA is translated in consecutive, nonoverlapping groups of three nucleotides (codons), the possibility of an *overlapping code* had to be considered. Such overlapping codes could be rejected because they impose constraints on which consecutive pairs of amino acids could be found in proteins. To understand why, suppose that the standard triplet codons were translated with an overlap of two. (In other words, the last two nucleotides of any codon are also the first two

nucleotides of the next codon in line.) Which amino acids could follow:
(a) Lys?
(b) Met?
(c) Tyr?
(d) Trp?

8.19 Protein synthesis occurs with high fidelity. In prokaryotes, incorrect amino acids are inserted at the rate of approximately 10^{-3} (that is, one incorrect amino acid per 1000 translated). What is the probability that a polypeptide of 300 amino acids has *exactly* the amino acid sequence specified in the mRNA? For an active enzyme consisting of 4 subunits (a tetramer), each 1000 amino acids in length, what is the probability that every amino acid in every subunit was translated without error?

8.20 Consider a hypothetical prokaryote with the unusual property that some of its mRNA molecules are circular, but they are translated according to the standard genetic code using any 5′-AUG- 3′ for initiation. The circular mRNA shown here has its 5′ → 3′ orientation in the clockwise direction. Deduce the amino acid sequence.

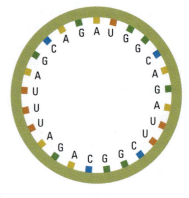

geNETics on the web

GeNETics on the Web will introduce you to some of the most important sites for finding genetic information on the Internet. To explore these sites, visit the Jones and Bartlett home page at

http://www.jbpub.com/genetics

For the book *Essential Genetics: A Genomics Perspective,* choose the link that says **Enter GeNETics on the Web**. You will be presented with a chapter-by-chapter list of highlighted keywords. Select any highlighted keyword and you will be linked to a Web site containing genetic information related to the keyword.

- A **sequence logo** is a graphical representation of a set of nucleotide sequences that are binding sites for proteins or other macromolecules. The logo uses letters of differing size to display the relative frequencies of alternative bases at each position in the sequence, so that information about variation in the sequence motif is not lost as it is in a conventional consensus sequence. This keyword site contains logos for many sequence motifs that are important in gene expression. Check out the logos for yeast TATA boxes as well as the splice donor and splice acceptor sites in human RNA processing to judge how strongly the "consensus sequences" are conserved.

- **Cystic fibrosis** is among most common serious inherited human disorders. It results from any of a number of mutations that cause a defect in the cystic fibrosis transmembrane conductance regulator (CFTR), one of a family of membrane chloride channels. The gene is designated *Cftr.* The mutant allele of *Cftr* most frequently encountered in patients is ΔF508, a deletion of three nucleotides that deletes a phenylalanine at position 508 in the CFTR polypeptide chain. Although the ΔF508 deletion accounts for about two-thirds of the defective Cftr alleles in the population, more than 700 mutant alleles have been described. Most of these are listed at this keyword site.

further readings

Barrell, B. G., A. T. Bankier, and J. Drouin. 1979. A different genetic code in human mitochondria. *Nature* 282: 189.

Blumenthal, T. 1995. *Trans*-splicing and polycistronic transcription in *Caenorhabditis elegans. Trends in Genetics* 11: 132.

Crick, F. H. C. 1979. Split genes and RNA splicing. *Science* 204: 264.

Garrett, R. A., S. R. Douthwaite, A. Liljas, et al. 2000. *The Ribosome: Structure, Function, Antibiotics, and Cellular Interaction.* Washington, DC: American Society for Microbiology.

Hanke, J., D. Brett, I. Zastrow, et al. 1999. Alternative splicing of human genes: More the rule than the exception? *Trends in Genetics* 15: 389.

Hartl, F. U., and M. Hayer-Hartl. 2002. Protein folding: Molecular chaperones in the cytosol: from nascent chain to folded protein. *Science* 295: 1852.

Haseltine, W. A. 1997. Discovering genes for new medicines. *Scientific American,* March.

Henkin, T. M. 1996. Control of transcription termination in prokaryotes. *Annual Review of Genetics* 30: 35.

Hill, W. E., and A. Dahlberg, eds. 1990. *The Ribosome: Structure, Function, and Evolution.* Washington, DC: American Society for Microbiology.

Jackson, R . J., and M. Wickens. 1997. Translational controls impinging on the $5'$-untranslated region and initiation factor proteins. *Current Opinion in Genetics & Development* 7: 233.

Kim, J. L., D. B. Nikolov, and S. K. Burley. 1993. Co-crystal structure of TBP recognizing the minor groove of a TATA element. *Nature* 3656: 520.

Kim, Y., J. H. Geiger, S. Hahn, and P. B. Sigler. 1993. Crystal structure of a yeast TBP/TATA-box complex. *Nature* 365: 512.

Lee, M. S., and P. A. Silver. 1997. RNA movement between the nucleus and the cytoplasm. *Current Opinion in Genetics & Development* 7: 212.

Maniatis, T., and R. Reed. 2002. An extensive network of coupling among gene expression machines. *Nature* 416: 499.

Neidhardt, F. C., R. Curtiss III, J. L. Ingraham, et al, eds. 1996. Escherichia coli *and* Salmonella typhimurium: *Cellular and Molecular Biology* (2 volumes). 2d ed. Washington, DC: American Society for Microbiology.

Nirenberg, M. 1963. The genetic code. *Scientific American,* March.

Ross, J. 1996. Control of messenger RNA stability in higher eukaryotes. *Trends in Genetics* 12: 171.

Taylor, J. H., ed. 1965. *Selected Papers on Molecular Genetics.* New York: Academic Press.

Wickens, M., P. Anderson, and R. J. Jackson. 1997. Life and death in the cytoplasm: Messages from the $3'$ end. *Current Opinion in Genetics & Development* 7: 220.

Chocolate was introduced into Spain by Hernán Cortéz, who in 1519 was served a chocolate drink called *xocoatle* (the *xo* is pronounced *ch*) by the Aztec ruler Montezuma. The Spanish managed to keep the ingredients secret for over a hundred years, but the recipe finally leaked to France and, soon thereafter, England. Today, the average American consumes about 10 pounds of chocolate candy per year. [© Photodisc]

key concepts

- Genes can be regulated at any level, including transcription, RNA processing, translation, and posttranslation.
- Control of transcription is an important mechanism of gene regulation.
- Transcriptional control can be negative ("on unless turned off") or positive ("off unless turned on"); many genes include regulatory regions for both types of regulation.
- Most genes have multiple, overlapping regulatory mechanisms that operate at more than one level, from transcription through posttranslation.
- In prokaryotes, the genes coding for related functions are often clustered in the genome and controlled jointly by a regulatory protein that binds with an "operator" region at one end of the cluster. This type of gene organization is known as an operon.

- In eukaryotes, genes are not organized into operons. Genes at dispersed locations in the genome are coordinately controlled by one or more "enhancer" DNA sequences located near each gene that interact with transcriptional activator proteins to allow gene expression.
- The transcription complex in eukaryotes consists of numerous protein components that are recruited to the promoter of a gene whose chromatin has been suitably reconfigured.

9

Molecular Mechanisms of Gene Regulation

chapter organization

9.1 Regulation of transcription is a common mechanism in prokaryotes.

9.2 In prokaryotes, groups of adjacent genes are often transcribed as a single unit.

9.3 Gene activity can be regulated through transcriptional termination.

9.4 Eukaryotes regulate transcription through transcriptional activator proteins, enhancers, and silencers.

9.5 Gene expression can be affected by heritable chemical modifications in the DNA.

9.6 Regulation also takes place at the levels of RNA processing and decay.

9.7 Regulation can also take place at the level of translation.

9.8 Some developmental processes are controlled by programmed DNA rearrangements.

the human connection
X-ing Out Gene Activity

Chapter Summary
Issues & Ideas
Key Terms & Concepts
Solutions: Step by Step
Concepts in Action: Problems for Solution
Further Readings

 geNETics on the web

A vertebrate animal contains approximately 200 different cell types with specialized functions. Yet with very few exceptions, all cells in an organism have the same genome. The cell types differ in which genes are active. For example, the genes for hemoglobin are expressed at a high level only in the precursors of red blood cells. The subject of **gene regulation** encompasses the mechanisms that determine the types of cells in which a gene will be transcribed, when it will be transcribed, where the transcript will start along the DNA, where it will terminate, how the transcript will be spliced, when the mRNA will be exported to the cytoplasm, when and how often the mRNA will be translated, and the duration of time before the mRNA is degraded.

9.1

Regulation of transcription is a common mechanism in prokaryotes.

In bacteria and bacteriophages, on–off gene activity is often controlled through transcription. Under conditions when a gene product is needed, transcription of the gene is turned "on"; under other conditions, transcription is turned "off." The term *off* should not be taken literally. In bacteria, few examples are known of a system being switched off completely. When transcription is in the "off" state, a basal level of gene expression nearly always remains, often averaging one transcriptional event or fewer per cell generation; hence "off" really means that there is very little synthesis of the gene product. Extremely low levels of expression are also found in certain classes of genes in eukaryotes. Regulatory mechanisms other than the on–off type also are known in both prokaryotes and eukaryotes; in these examples, the level of expression of a gene may be modulated in gradations from low to high according to conditions in the cell.

In bacterial systems, when several enzymes act in sequence in a single metabolic pathway, usually either all or none of the enzymes are produced. This **coordinate regulation** results from control of the synthesis of one or more mRNA molecules that are *polycistronic*; these mRNAs encode all of the gene products that function in the same metabolic pathway. This type of regulation is not found in eukaryotes because, as we saw in Chapter 8, eukaryotic mRNA is monocistronic.

■ In negative regulation, the default state of transcription is "on."

The molecular mechanisms of regulation usually fall into either of two broad categories: *negative regulation* and *positive regulation*. In a system subject to **negative regulation** (Figure 9.1, part A), the default state is "on," and transcription takes place

(A) Negative regulation of transcription

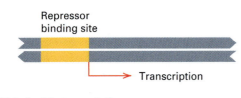

(B) Inducible transcription

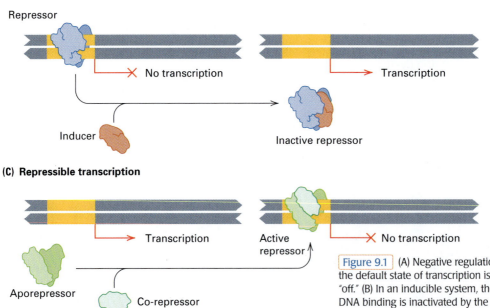

(C) Repressible transcription

Figure 9.1 (A) Negative regulation of transcription means that the default state of transcription is "on" unless a repressor turns it "off." (B) In an inducible system, the repressor is a protein whose DNA binding is inactivated by the inducer. (C) In a repressible system, the repressor is formed by the interaction between an aporepressor protein and a co-repressor.

until it is turned "off" by a **repressor** protein that binds to the DNA upstream from the transcriptional start site. A negatively regulated system may be either *inducible* (part B) or *repressible* (part C), depending on how the active repressor is formed. In **inducible transcription,** a repressor DNA-binding protein normally keeps transcription in the "off" state. In the presence of a small molecule called the **inducer,** the repressor binds preferentially with the inducer and loses its DNA-binding capability, allowing transcription to occur. Many degradative (catabolic) pathways are inducible and use the initial substrate of the degradative pathway as the inducer. In this way, the enzymes used for degradation are not synthesized unless the substrate is present in the cell.

In **repressible transcription** (part C), the default state is "on" until an active repressor is formed to turn it "off." In this case the regulatory protein is called an **aporepressor,** and it has no DNA-binding activity on its own. The active repressor is formed by the combination of the aporepressor and a small molecule known as the **co-repressor.** Presence of the co-repressor thereby results in the cessation of transcription. Repressible regulation is often found in the control of the synthesis of enzymes that participate in biosynthetic (anabolic) pathways; in these cases the final product of the pathway is frequently the co-repressor. In this way, the enzymes of the biosynthetic pathway are not synthesized until the concentration of the final product becomes too low to cause repression.

■ In positive regulation, the default state of transcription is "off."

In a positively regulated system (), the default state of transcription is "off," and binding with a regulatory protein is necessary to turn it "on." The protein that turns transcription on is a **transcriptional activator protein.** Negative and positive regulation are not mutually exclusive, and many systems are both positively and negatively regulated, utilizing two regulators to respond to different conditions in the cell. Negative regulation is more common in prokaryotes, positive regulation in eukaryotes.

Some genes exhibit **autoregulation,** which means that the protein product of a gene regulates its own transcription. In negative autoregulation, the protein inhibits transcription, and high concentrations of the protein result in less transcription of the mRNA. This mechanism automatically adjusts the steady-state level of the protein in the cell. In positive autoregulation, the protein stimulates transcription: As more protein is made, transcription increases to the maximum rate. Positive autoregulation is a common way for weak induc-

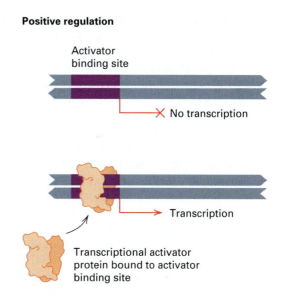

Positive regulation

Figure 9.2 In positive regulation, the default state of transcription is "off." Transcription is stimulated by the binding of a transcriptional activator protein.

tion to be amplified. Only a weak signal is necessary to get production of the protein started, but then the positive autoregulation takes over and stimulates further production to the maximum level.

Next we examine two classical systems of regulation found in the bacterium *Escherichia coli.* These serve as specific examples of the general concepts introduced in Figures 9.1 and 9.2. We shall see that in the real world, most genes have overlapping mechanisms of control that include both positive and negative regulatory elements.

9.2

In prokaryotes, groups of adjacent genes are often transcribed as a single unit.

Analysis of gene regulation was first carried out in detail for the genes responsible for degradation of the sugar lactose in *E. coli.* Much of the terminology used to describe regulation came from this genetic analysis.

■ The first regulatory mutations that were discovered affected lactose metabolism.

In *E. coli,* two proteins are necessary for the metabolism of lactose: the enzyme **β-galactosidase,** which cleaves lactose (a β-galactoside sugar) to yield galactose and glucose, and a transporter molecule, **lactose permease,** which is required for the entry of lactose into the cell. The existence of two different proteins in the lactose-utilization system was first shown by a combination of genetic experiments and biochemical analysis.

First, hundreds of mutants unable to use lactose as a carbon source, designated Lac$^-$ mutants, were isolated. Some of the mutations were in the *E. coli* chromosome, and others were in an F′ *lac*, a plasmid carrying the genes for lactose utilization. By performing F′ × F$^-$ matings, investigators constructed partial diploids with the genotypes F′ *lac$^-$* / *lac$^+$* or F′ *lac$^+$* / *lac$^-$*. (The genotype of the plasmid is given to the left of the slash and that of the chromosome to the right.) It was observed that all of these partial diploids always had a Lac$^+$ phenotype (that is, they made both β-galactosidase and permease). Other partial diploids were then constructed in which both the F′ *lac* plasmid and the chromosome carried a *lac$^-$* allele. When these were tested for the Lac$^+$ phenotype, it was found that all of the mutants initially isolated could be placed into two complementation groups, called *lacZ* and *lacY*, a result that implies that the *lac* system consists of at least two genes. Complementation is indicated by the observation that the partial diploids F′ *lacY$^-$ lacZ$^+$* / *lacY$^+$ lacZ$^-$* and F′ *lacY$^+$ lacZ$^-$* / *lacY$^-$ lacZ$^+$* had a Lac$^+$ phenotype, producing both β-galactosidase and permease. However, the genotypes F′ *lacY$^-$ lacZ$^+$* / *lacY$^-$ lacZ$^+$* and F′ *lacY$^+$ lacZ$^-$* / *lacY$^+$ lacZ$^-$* had the Lac$^-$ phenotype; they were unable to synthesize permease and β-galactosidase, respectively. Hence the *lacZ* gene codes for β-galactosidase and the *lacY* gene for permease. (A third gene that participates in lactose metabolism was discovered later; it was not included among the early mutants because it is not essential for growth on lactose.) Close physical proximity of the *lacZ* and *lacY* genes was deduced from a high frequency of cotransduction observed in genetic mapping experiments. In fact, *lacZ* and *lacY* are adjacent in the chromosome.

■ Lactose-utilizing enzymes can be inducible (regulated) or constitutive.

The on–off nature of the genes responsible for lactose utilization is evident in the following observations:

- If a culture of Lac$^+$ *E. coli* is grown in a medium lacking lactose or any other β-galactoside, the intracellular concentrations of β-galactosidase and permease are exceedingly low—roughly one or two molecules per bacterial cell. However, if lactose is present in the growth medium, the number of each of these molecules is about 10^3-fold higher.

- If lactose is added to a Lac$^+$ culture growing in a lactose-free medium (also lacking glucose, a point we will discuss shortly), both β-galactosidase and permease are synthesized nearly simultaneously, as shown in Figure 9.3. Analysis of the total mRNA present in the cells

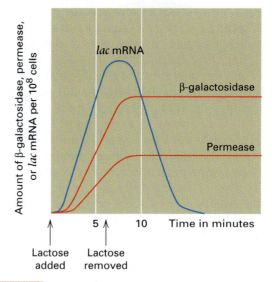

Figure 9.3 The "on–off" nature of the *lac* system. The *lac* mRNA appears soon after lactose or another inducer is added; β-galactosidase and permease appear at nearly the same time but are delayed with respect to mRNA synthesis because of the time required for translation. When lactose is removed, no more *lac* mRNA is made, and the amount of *lac* mRNA decreases because of the degradation of mRNA already present. Both β-galactosidase and permease are stable proteins: Their amounts remain constant even when synthesis ceases. However, their concentration per cell gradually decreases as a result of repeated cell divisions.

before and after the addition of lactose shows that almost no *lac* mRNA (the polycistronic mRNA that codes for β-galactosidase and permease) is present before lactose is added and that the addition of lactose triggers synthesis of the *lac* mRNA.

These two observations led to the view that transcription of the lactose genes is *inducible transcription* and that lactose is an *inducer* of transcription. Some analogs of lactose are also inducers, such as a sulfur-containing analog denoted IPTG (isopropylthiogalactoside), which is convenient for experiments because it induces, but is not cleaved by, β-galactosidase. The inducer IPTG is taken up by the cells and maintained at a constant level, whether or not the β-galactosidase enzyme is present.

Mutants were also isolated in which *lac* mRNA was synthesized, and the enzymes produced, in the *absence* of an inducer as well as in its presence. Because of their constant synthesis, with or without inducer, the mutants were called **constitutive.** They provided the key to understanding induction. Mutants were also obtained that failed to produce *lac* mRNA and the enzymes even when the inducer was present. These uninducible mutants fell into two classes, *lacI^s* and *lacP$^-$*. The characteristics of the mutants are shown in **Table 9.1** and discussed in the following sections.

Table 9.1

Characteristics of partial diploids containing several combinations of *lacI*, *lacO*, and *lacP* alleles

Genotype	Synthesis of *lac* mRNA	Lac phenotype
1. F′ *lacO^c lacZ^+/lacO^+ lacZ^+*	Constitutive	+
2. F′ *lacO^+ lacZ^+/lacO^c lacZ^+*	Constitutive	+
3. F′ *lacI^- lacZ^+/lacI^+ lacZ^+*	Inducible	+
4. F′ *lacI^+ lacZ^+/lacI^- lacZ^+*	Inducible	+
5. F′ *lacO^c lacZ^-/lacO^+ lacZ^+*	Inducible	+
6. F′ *lacO^c lacZ^+/lacO^+ lacZ^-*	Constitutive	+
7. F′ *lacI^s lacZ^+/lacI^+ lacZ^+*	Uninducible	−
8. F′ *lacI^+ lacZ^+/lacI^s lacZ^+*	Uninducible	−
9. F′ *lacP^- lacZ^+/lacP^+ lacZ^+*	Inducible	+
10. F′ *lacP^+ lacZ^+/lacP^- lacZ^+*	Inducible	+
11. F′ *lacP^+ lacZ^-/lacP^- lacZ^+*	Uninducible	−
12. F′ *lacP^+ lacZ^+/lacP^- lacZ^-*	Inducible	+

Repressor shuts off messenger RNA synthesis.

In Table 9.1, genotypes 3 and 4 show that $lacI^-$ mutations are recessive. In the absence of inducer, a $lacI^+$ cell does not make *lac* mRNA, whereas the mRNA is made in a $lacI^-$ mutant. These results suggest that

> **key concept**
>
> The *lacI* gene is a regulatory gene whose product is the repressor protein that keeps the system turned off. Because the repressor is necessary to shut off mRNA synthesis, regulation by the repressor is negative regulation.

A $lacI^-$ mutant lacks the repressor and hence is constitutive. Wildtype copies of the repressor are present in a $lacI^+$ / $lacI^-$ partial diploid, so transcription is repressed. It is important to note that the single $lacI^+$ gene prevents synthesis of *lac* mRNA from both the F′ plasmid and the chromosome. Therefore, the repressor protein must be diffusible within the cell to be able to shut off mRNA synthesis from both DNA molecules present in a partial diploid.

On the other hand, genotypes 7 and 8 indicate that the $lacI^s$ mutations are dominant and act to shut off mRNA synthesis from both the F′ plasmid and the chromosome, whether or not the inducer is present (the superscript in $lacI^s$ signifies *super-repressor*.) The $lacI^s$ mutations result in repressor molecules that fail to recognize and bind the inducer and thus permanently shut off *lac* mRNA

synthesis. Genetic mapping experiments placed the *lacI* gene nearly adjacent to the *lacZ* gene and established the gene order *lacI lacZ lacY*. How the *lacI* repressor prevents synthesis of *lac* mRNA will be explained shortly.

The lactose operator is an essential site for repression.

Entries 1 and 2 in Table 9.1 show that $lacO^c$ mutants are dominant. However, the dominance is evident only in certain combinations of *lac* mutations, as can be seen by examining the partial diploids shown in entries 5 and 6. Both combinations are Lac$^+$ because a functional *lacZ* gene is present. However, in the combination shown in entry 5, synthesis of β-galactosidase is inducible even though a $lacO^c$ mutation is present. The difference between the two combinations in entries 5 and 6 is that in entry 5, the $lacO^c$ mutation is present in the same DNA molecule as the $lacZ^-$ mutation, whereas in entry 6, $lacO^c$ is contained in the same DNA molecule as $lacZ^+$. The key feature of these results is that

> **key concept**
>
> A $lacO^c$ mutation causes constitutive synthesis of β-galactosidase only when the $lacO^c$ and $lacZ^+$ alleles are contained in the same DNA molecule.

The $lacO^c$ mutation is said to be ***cis*-dominant,** because only genes in the *cis* configuration (in the same DNA molecule as that containing the mutation) are expressed in dominant fashion. Confirmation of this conclusion comes from an important biochemical observation: The mutant enzyme from the $lacZ^-$ allele is synthesized constitutively in a $lacO^c lacZ^-$ / $lacO^+ lacZ^+$ partial diploid (entry 5), whereas the wildtype enzyme from the $lacZ^+$ allele is synthesized only if an inducer is added. All $lacO^c$ mutations are located between the *lacI* and *lacZ* genes; hence the gene order of the four genetic elements of the *lac* system is

$$lacI \quad lacO \quad lacZ \quad lacY$$

An important feature of all $lacO^c$ mutations is that they cannot be complemented (a characteristic feature of all *cis*-dominant mutations); that is, a $lacO^+$ allele cannot alter the constitutive activity of a $lacO^c$ mutation. This observation implies that the *lacO* region does not encode a diffusible product and must instead define a site in the DNA that determines whether synthesis of the product of the adjacent *lacZ* gene is inducible or constitutive. The *lacO* region is called the **operator.** In a subsequent section, we will see that the operator is in fact a *binding site* in the DNA for the repressor protein.

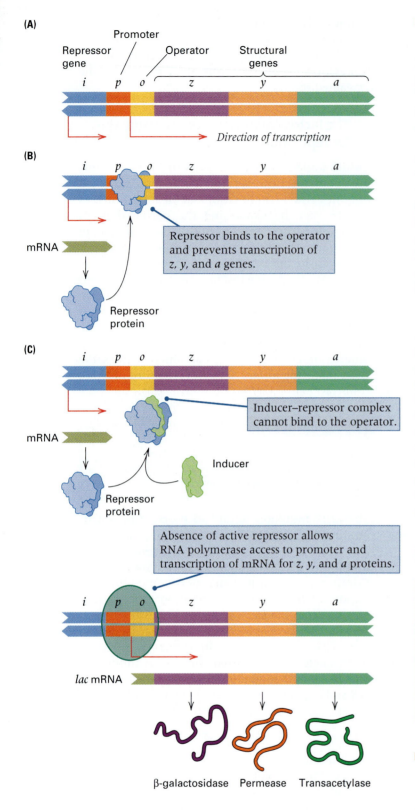

(A)

Repressor gene · Promoter · Operator · Structural genes

i · p · o · z · y · a

Direction of transcription

(B)

i · p · o · z · y · a

mRNA

Repressor binds to the operator and prevents transcription of z, y, and a genes.

Repressor protein

(C)

i · p · o · z · y · a

mRNA

Inducer–repressor complex cannot bind to the operator.

Inducer

Repressor protein

Absence of active repressor allows RNA polymerase access to promoter and transcription of mRNA for z, y, and a proteins.

i · p · o · z · y · a

lac mRNA

β-galactosidase · Permease · Transacetylase

Figure 9.4 (A) Organization of the *lac* operon, not drawn to scale; the p and o sites are actually much smaller than the other elements and together comprise only 83 base pairs. (B) A diagram of the *lac* operon in the repressed state. (C) A diagram of the *lac* operon in the induced state. The inducer alters the shape of the repressor so that the repressor can no longer bind to the operator. The common abbreviations i, p, o, z, y, and a are used instead of *lacI*, *lacO*, and so forth. The *lacA* gene is not essential for lactose utilization.

■ The lactose promoter is an essential site for transcription.

Entries 11 and 12 in Table 9.1 show that $lacP^-$ mutations, like $lacO^c$ mutations, are *cis*-dominant. The *cis*-dominance can be seen in the partial diploid in entry 11. The genotype in entry 11 is uninducible, in contrast to the partial diploid of entry 12, which is inducible. The difference between the two genotypes is that in entry 11, the $lacP^-$ mutation is in the same DNA molecule with $lacZ^+$, whereas in entry 12, the $lacP^-$ mutation is combined with $lacZ^-$. This observation means that a wildtype $lacZ^+$ remains inexpressible in the presence of $lacP^-$; no *lac* mRNA is transcribed from that DNA molecule. The $lacP^-$ mutations map between *lacI* and *lacO*, and the order of the five genetic elements of the *lac* system is

$$lacI \quad lacP \quad lacO \quad lacZ \quad lacY$$

As expected because of the *cis*-dominance of $lacP^-$ mutations, they cannot be complemented; that is, a $lacP^+$ allele on another DNA molecule cannot supply the missing function to a DNA molecule carrying a $lacP^-$ mutation. Thus *lacP*, like *lacO*, must define a site that determines whether synthesis of *lac* mRNA will take place. Because synthesis does not occur if the site is defective or missing, *lacP* defines an essential site for mRNA synthesis. The *lacP* region is called the **promoter.** It is a site at which RNA polymerase binding takes place to allow initiation of transcription.

Q A Moment to Think

Problem: The LacY permease is able to transport lactobionic acid as well as lactose, but LacZ cleaves lactobionic acid very inefficiently. Certain mutant forms of LacZ have an altered substrate specificity that allows the mutant enzyme to cleave lactobionic acid. These mutants are able to grow in medium containing lactobionic acid as the sole source of carbon and energy. However, they cannot grow unless they also have a $lacO^c$ mutation or, when the operator is wildtype, unless the *lac* inducer molecule IPTG (isopropylthiogalactoside) is present in the medium. How would you explain these results? (The answer can be found on page 324.)

■ The lactose operon contains linked structural genes and regulatory sequences.

The genetic regulatory mechanism of the *lac* system was first explained by the *operon model* of François Jacob and Jacques Monod, which is illustrated in Figure 9.4 . (The figure uses the alternative abbreviations i, o, p, z, y, and a for *lacI*, *lacO*, *lacP*, *lacZ*, *lacY*, and *lacA*.) The operon model has the following features:

1. The lactose-utilization system consists of two kinds of components: *structural genes* (*lacZ* and

X-ing Out Gene Activity

Mary F. Lyon 1961
Medical Research Council, Harwell, England
Gene Action in the X Chromosome of the Mouse (Mus musculus L.)

How do organisms solve the problem that females have two X chromosomes whereas males have only one? As we have seen in Chapter 5, unless there were some type of correction (called dosage compensation), the unequal number would mean that for all the genes in the X chromosome, cells in females would have twice as much gene product as cells in males. It would be difficult for the developing organism to cope with such a large difference in dosage for so many genes. The problem of dosage compensation has been solved by different organisms in different ways. The hypothesis put forward in this paper is that in the mouse (and by inference in other mammals), the mechanism is very simple: One of the X chromosomes, chosen at random in each cell lineage early in development, becomes inactivated and remains inactivated in all descendant cells in the lineage. In certain cells the inactive X chromosome becomes visible in interphase as a deeply staining "sex-chromatin body." We now know that X inactivation takes place sequentially from an "X-inactivation center," and that about 15 percent of X-linked genes escape inactivation to some degree. We also know that in marsupial mammals, such as the kangaroo, it is always the paternal X chromosome that is inactivated.

• •

It has been suggested that the so-called sex chromatin body is composed of one heteropyknotic [that is, deeply staining during interphase] X chromosome. . . . The present communication suggests that evidence of mouse genetics indicates: (1) that the heteropyknotic X chromosome can be either paternal or maternal in origin, in different cells of the same animal; (2) that it is genetically inactivated. The evidence has two main parts. First, the normal phenotype of XO females in the mouse shows that only one active X chromosome is necessary for normal development, including sexual development. The second piece of evidence concerns the mosaic phenotype of female mice heterozygous for some sex-linked [X-linked] mutants. All sex-linked mutants so far known affecting coat colour cause a "mottled" or "dappled" phenotype, with patches of normal and mutant colour. . . . It is here suggested that this mosaic phenotype is due to the inactivation of one or other X chromosome early in embryonic development. If this is true, pigment cells descended from the cells in which the chromosome carrying the mutant gene was inactivated will give rise to a normal-coloured patch and those in which the chromosome carrying the normal gene was inactivated will give rise to a mutant-coloured patch. . . . Thus this hypothesis predicts that for all sex-linked genes of the mouse in which the phenotype is due to localized gene action the heterozygote will have a mosaic appearance. . . . The genetic evidence does not indicate at what early stage of embryonic development the inactivation of the one X chromosome occurs. . . . The sex-chromatin body is thought to be formed from one X chromosome in the rat and in the opossum. If this should prove to be the case in all mammals, then all female mam-

> **It is here suggested that this [X-linked] mosaic phenotype is due to the inactivation of one or other X chromosome early in embryonic development.**

mals heterozygous for sex-linked mutant genes would be expected to show the same phenomena as those in the mouse. The coat of the tortoiseshell cat, being a mosaic of the black and yellow colours of the two homozygous genotypes, fulfills this expectation.

Source: Nature 190: 372.

---

lacY), which encode proteins needed for the transport and metabolism of lactose, and *regulatory elements* (the repressor gene *lacI*, the promoter *lacP*, and the operator *lacO*).

2. The products of the *lacZ* and *lacY* genes are coded by a single polycistronic mRNA molecule. The linked structural genes, together with *lacP* and *lacO*, constitute the *lac* **operon.** (The third protein, encoded by *lacA*, is also translated from the polycistronic mRNA. This protein is the enzyme β-galactoside transacetylase; it is used in the metabolism of certain β-galactosides other than lactose and will not concern us here.)

3. The promoter mutations (*lacP⁻*) eliminate the ability to synthesize *lac* mRNA.

4. The product of the *lacI* gene is a repressor, which binds to a unique sequence of DNA bases constituting the operator.

5. When the repressor is bound to the operator, initiation of transcription of *lac* mRNA by RNA polymerase is prevented.

6. Inducers stimulate mRNA synthesis by binding to and inactivating the repressor. In the presence of an inducer, the operator is not bound with the repressor, and the promoter is available for the initiation of mRNA synthesis.

Note that regulation of the operon requires that the *lacO* operator either overlap or be adjacent to the promoter of the structural genes, because binding with the repressor prevents transcription. Proximity of *lacI* to *lacO* is not strictly necessary, because the *lacI* repressor is a soluble protein and is therefore diffusible throughout the cell. The presence of inducer has a profound effect on the DNA-binding properties of the repressor; the inducer–repressor complex has an affinity for the operator that is approximately 10^3 smaller than that of the repressor alone.

When the operon is induced, the numbers of protein molecules of β-galactosidase, permease, and transacetylase are in the ratio 1.0 : 0.5 : 0.2. These differences are partly due to the order of the genes in the mRNA. Downstream cistrons are less likely to be translated because of failure of reinitiation when an upstream cistron has finished translation.

The operon model is supported by a wealth of experimental data and explains many of the features of the *lac* system, as well as numerous other negatively regulated genetic systems in prokaryotes. One aspect of the regulation of the *lac* operon—the effect of glucose—has not yet been discussed. Examination of this feature indicates that the *lac* operon is also subject to positive regulation, as we will see in the next section.

A Moment to Think

Answer to Problem: In order to grow on lactobionic acid, even the mutant LacZ strain must have both *lacZ* and *lacY* expressed. Evidently lactobionic acid is not an inducer of the *lac* operon, and therefore a *lacO*⁺ mutant cannot grow in lactobionic acid unless IPTG, an inducer of the *lac* operon, is also present to induce transcription of the operon. The finding with *lacO*ᶜ mutants is consistent with this explanation, because when LacY and the mutant LacZ are produced constitutively, the strains can grow in lactobionic acid.

■ The lactose operon is also subject to positive regulation.

The function of β-galactosidase in lactose metabolism is to form glucose by cleaving lactose. (The other cleavage product, galactose, also is ultimately converted into a glucose derivative by the enzymes of the galactose operon.) If both glucose and lactose

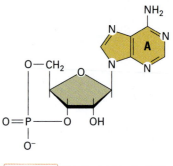

Figure 9.5 Structure of cyclic AMP.

are present in the growth medium, transcription of the *lac* operon is shut down until virtually all of the glucose in the medium has been consumed. The observation that no *lac* mRNA is made in the presence of glucose implies that another element, in addition to an inducer, is needed for initiating *lac* mRNA synthesis.

The inhibitory effect of glucose on expression of the *lac* operon is indirect. The small molecule *cyclic adenosine monophosphate* (cAMP), shown in Figure 9.5, is widely distributed in animal tissues and in multicellular eukaryotic organisms, in which it is important in mediating the action of many hormones. It is also present in *E. coli* and many other bacteria, where it has a different function. Cyclic AMP is synthesized by the enzyme *adenylyl cyclase,* and the concentration of cAMP is regulated indirectly by glucose metabolism. When bacteria are growing in a medium that contains glucose, the cAMP concentration in the cells is quite low. In a medium containing glycerol or any carbon source that requires aerobic metabolism for degradation, or when the bacteria are otherwise starved of an energy source, the cAMP concentration is high (**Table 9.2**). Glucose levels help regulate the cAMP concentration in the cell, and *cAMP regulates the activity of the lac operon* (as well as that of several other operons that control degradative metabolic pathways).

Table 9.2

Concentration of cyclic AMP in cells growing in media with the indicated carbon sources

Carbon source	cAMP concentration
Glucose	Low
Glycerol	High
Lactose	High
Lactose + glucose	Low
Lactose + glycerol	High

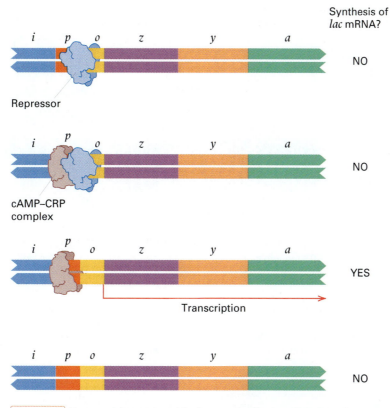

Synthesis of *lac* mRNA?

Repressor			NO
cAMP–CRP complex			NO
Transcription			YES
			NO

Figure 9.6 | Four regulatory states of the *lac* operon: The *lac* mRNA is synthesized only when cAMP–CRP is present and repressor is absent.

E. coli and many other bacterial species contain a protein called the *cyclic AMP receptor protein* (CRP), which is encoded by a gene called *crp*. Mutations of either the *crp* or the adenyl cyclase gene prevent synthesis of *lac* mRNA, which indicates that both CRP function and cAMP are required for *lac* mRNA synthesis. CRP and cAMP bind to one another, forming a complex denoted **cAMP–CRP.** The presence of cAMP–CRP is necessary for induction, because *crp* and adenyl cyclase mutants are unable to make *lac* mRNA even when a *lacI⁻* or a *lacOᶜ* mutation is present. The reason for the requirement is that transcription cannot occur unless the cAMP–CRP complex is bound to a specific DNA sequence in the promoter region (Figure 9.6). Unlike the repressor, which is a *negative* regulator, the cAMP–CRP complex is a *positive* regulator. The positive and negative regulatory systems of the *lac* operon are independent of each other.

Experiments carried out *in vitro* with purified *lac* DNA, *lac* repressor, cAMP–CRP, and RNA polymerase have established two further points:

1. In the absence of the cAMP–CRP complex, RNA polymerase binds only weakly to the promoter, but its binding is stimulated when cAMP–CRP is also bound to the DNA. The weak binding rarely leads to initiation of transcription,

because the correct interaction between RNA polymerase and the promoter does not take place.

2. If the repressor is bound to the operator, then RNA polymerase cannot stably bind to the promoter.

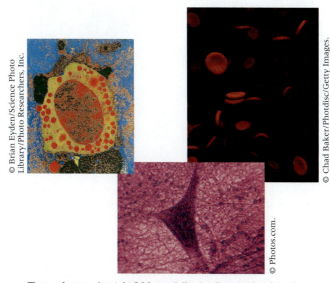

Three of approximately 200 specialized cell types that form in complex eukaryotic organisms: a white blood cell (top left), a neuron (center), and some red blood cells (top right).

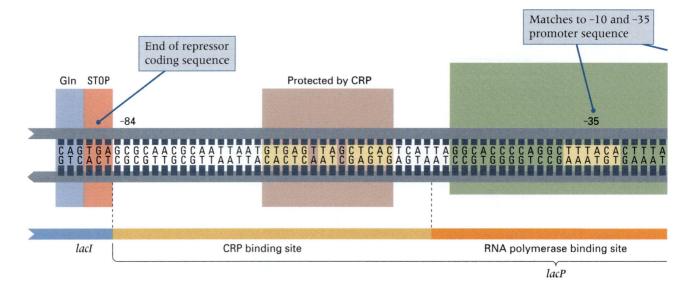

Matches to –10 and –35 promoter sequence

End of repressor coding sequence

Gln STOP

Protected by CRP

–84

–35

```
CAGTGAGCGCAACGCAATTAATGTGAGTTAGCTCACTCATTAGGCACCCCAGGCTTTACACTTTA
GTCACTCGCGTTGCGTTAATTACACTCAATCGAGTGAGTAATCCGTGGGGTCCGAAATGTGAAAT
```

lacI | CRP binding site | RNA polymerase binding site

lacP

Figure 9.7 (above and facing page) The base sequence of the control region of the *lac* operon. Sequences protected from DNase digestion by binding of the key proteins are indicated in the upper part. The end of the *lacI* gene is shown at the extreme left; the ribosome binding site is the site at which the ribosome binds to the *lac* mRNA. The consensus sites for CRP binding and for RNA polymerase promoter binding are indicated along the bottom.

Figure 9.8 Structure of the *lac* operon repression loop. The *lac* repressor, shown in violet, binds to two DNA regions (red) consisting of the symmetrical operator region indicated in Figure 10.7 and a second region immediately upstream from the CRP binding site. Within the loop is the CRP binding site (medium blue), shown bound with CRP protein (dark blue). The –10 and –35 promoter regions are in green. [Courtesy of Mitchell Lewis. Reprinted with permission from M. Lewis, G. Chang, N. C. Horton, M. A. Kercher, H. C. Pace, M. A. Schumacher, R. G. Brennan, and P. Lu. 1996. *Science* 271: 1247. © AAAS]

These results explain how lactose and glucose function together to regulate transcription of the *lac* operon. The relationship of these elements to one another, to the start of transcription, and to the base sequence in the region is depicted in Figure 9.7.

A great deal is also known about the three-dimensional structure of the regulatory states of the *lac* operon. Figure 9.8 shows that there is actually a 93-base-pair loop of DNA that forms in the operator region when it is in contact with the repressor. This loop corresponds to the *lac* operon region −82 to +11, numbered as in Figure 9.7. The DNA region in red corresponds, on the right-hand side, to the operator region centered at +11 and, on the left-hand side, to a second repressor binding site immediately upstream and adjacent to the CRP binding site. The *lac* repressor tetramer (violet) is shown bound to these sites. The DNA loop is formed by the region between the repressor binding sites and includes, in medium blue, the CRP binding site, to which the CRP protein (dark blue) is shown bound. The DNA regions in green are the −10 and −35 sites in the *lacP* promoter indicated in Figure 9.7. In this configuration, the *lac* operon is not transcribed. Removal of the repressor opens up the loop and allows transcription to occur.

■ **Tryptophan biosynthesis is regulated by the tryptophan operon.**

The tryptophan (*trp*) operon of *E. coli* contains structural genes for enzymes that synthesize the amino acid tryptophan. This operon is regulated in such a

326 **CHAPTER 9** Molecular Mechanisms of Gene Regulation

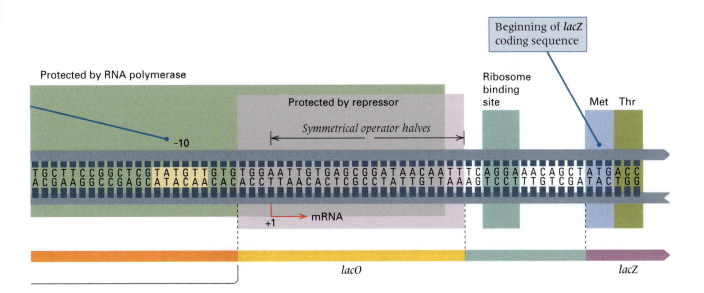

way that when adequate tryptophan is present in the growth medium, transcription of the operon is repressed. However, when the supply of tryptophan is insufficient, transcription takes place. Regulation in the *trp* operon is similar to that in the *lac* operon because mRNA synthesis is regulated negatively by a repressor. However, it differs from regulation of *lac* in that tryptophan acts as a co-repressor, which stimulates binding of the repressor to the *trp* operator to shut off synthesis. The *trp* operon is a *repressible* rather than an *inducible* operon. Furthermore, because the *trp* operon codes for a set of biosynthetic enzymes rather than degradative enzymes, neither glucose nor cAMP–CRP functions in regulation of the *trp* operon.

A simple on–off system, as in the *lac* operon, is not optimal for a biosynthetic pathway. For example, a situation may arise in which some tryptophan is present in the growth medium, but the amount is not enough to sustain optimal growth. Under these conditions, it is advantageous to synthesize tryptophan, but at less than the maximum

possible rate. Cells adjust to this situation by means of a regulatory mechanism in which the amount of transcription in the derepressed state is determined by the concentration of tryptophan in the cell. This regulatory mechanism is found in many operons responsible for amino acid biosynthesis.

Tryptophan is synthesized in five steps, each requiring a particular enzyme. The genes encoding these enzymes are adjacent in the *E. coli* chromosome and are in the same linear order as the order in which the enzymes function in the biosynthetic pathway. The genes are called *trpE, trpD, trpC, trpB,* and *trpA,* and the enzymes are translated from a single polycistronic mRNA molecule. The *trpE* coding region is the first one translated. Upstream (on the 5' side) of *trpE* are the promoter, the operator, and two regions called the *leader* and the *attenuator,* which are designated *trpL* and *trpa* (not *trpA*), respectively (Figure 9.9). The repressor gene, *trpR,* is located quite far from this operon.

The regulatory protein of the *trp* operon is the product of the *trpR* gene. Mutations in either this

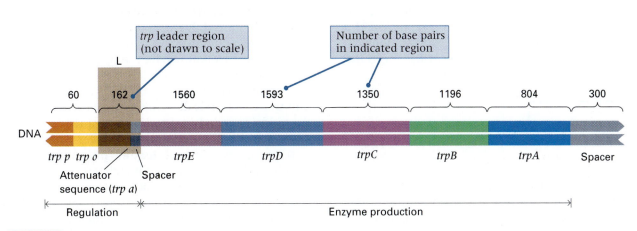

Figure 9.9 The *trp* operon in *E. coli.* For clarity, the regulatory region is enlarged with respect to the coding region. The actual size of each region is indicated by the number of base pairs. Region L is the leader.

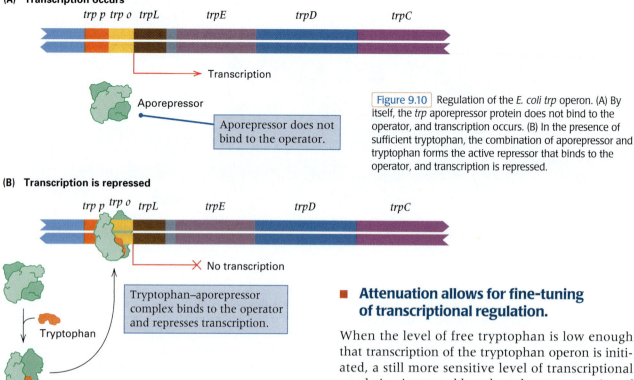

(A) Transcription occurs

trp p trp o trpL   trpE   trpD   trpC

Transcription

Aporepressor

Aporepressor does not bind to the operator.

Figure 9.10 Regulation of the *E. coli trp* operon. (A) By itself, the *trp* aporepressor protein does not bind to the operator, and transcription occurs. (B) In the presence of sufficient tryptophan, the combination of aporepressor and tryptophan forms the active repressor that binds to the operator, and transcription is repressed.

(B) Transcription is repressed

trp p trp o trpL   trpE   trpD   trpC

✕ No transcription

Tryptophan–aporepressor complex binds to the operator and represses transcription.

Tryptophan

Active aporepressor

gene or the operator cause constitutive initiation of transcription of *trp* mRNA. The *trpR* gene product is the *trp* aporepressor. It does not bind to the operator unless it is first bound to tryptophan; that is, the aporepressor and the tryptophan molecule join together to form the active *trp* repressor, which binds to the operator. The reaction scheme is outlined in Figure 9.10 . When there is insufficient tryptophan, the aporepressor adopts a conformation unable to bind with the *trp* operator, and the operon is transcribed (Figure 9.10, part A). When tryptophan is present at a sufficiently high concentration, some molecules bind with the aporepressor and cause it to change conformation into the active repressor. The active repressor binds with the *trp* operator and prevents transcription (Figure 9.10, part B). This is the basic on–off regulatory mechanism.

9.3

Gene activity can be regulated through transcriptional termination.

The lactose and tryptophan operons show how gene activity can be regulated through the initiation of transcription. There are also mechanisms for gene regulation through the termination of transcription. In this section we consider two examples.

■ Attenuation allows for fine-tuning of transcriptional regulation.

When the level of free tryptophan is low enough that transcription of the tryptophan operon is initiated, a still more sensitive level of transcriptional regulation is exerted based on the concentration of charged tryptophan tRNA. This type of regulation is called **attenuation,** and it uses translation to control transcription. If translation of the leader region of the mRNA takes place, it causes termination of transcription even before the first structural gene of the operon is transcribed.

Attenuation results from interactions between DNA sequences present in the leader region of the *trp* transcript. In wildtype cells, transcription of the *trp* operon is often initiated. However, in the presence of even small amounts of tryptophan, most of the mRNA molecules terminate in a specific 28-base region within the leader sequence. The result of termination is an RNA molecule containing only 140 nucleotides that stops short of the genes coding for the *trp* enzymes. The 28-base region in which termination takes place is called the **attenuator.** The base sequence of this region (Figure 9.11) contains the usual features of a termination site, including a potential stem-and-loop configuration in the mRNA followed by a sequence of eight uridylates.

In the tryptophan operon in *E. coli,* termination of transcription is determined by whether a small peptide encoded in the leader sequence can be translated. This coding sequence, shown in Figure 9.12 , specifies a **leader polypeptide** 14 amino acids in length, and it includes two adjacent tryptophan codons at positions 10 and 11. When there is sufficient charged tryptophan tRNA to allow translation of these codons, the nascent transcript adopts a conformation in which the attenuator is exposed, and transcription is terminated. On the other hand, when there is insufficient charged tryptophan tRNA to allow translation of the leader polypeptide, the ribosome stalls at the tryptophan codons; in this case the attenuator is hidden, and transcription continues through the entire operon.

The mechanism of attenuation is diagrammed in Figure 9.13. Part A shows the leader RNA molecule, including the two tryptophan codons in the leader polypeptide. Region 2 has a nucleotide sequence that enables it to pair either with region 1 or with region 3. In the purified RNA, region 1 pairs with region 2, and region 3 pairs with region 4. Part B shows the configuration in a cell in which there is sufficient tRNA^{Trp} to allow transla-

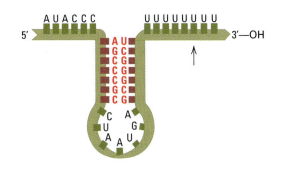

Figure 9.11 The terminal region of the *trp* attenuator sequence. The arrow indicates the final uridylate in attenuated RNA. Nonattenuated RNA continues past this point. The nucleotides in red letters form a stem by base-pairing within the RNA.

Figure 9.12 The sequence of bases in the *trp* leader mRNA, showing the leader polypeptide, the two tryptophan codons (red letters), and the beginning of the TrpE protein. The numbers 23 and 91 are the numbers of bases in the sequence that, for clarity, are not shown.

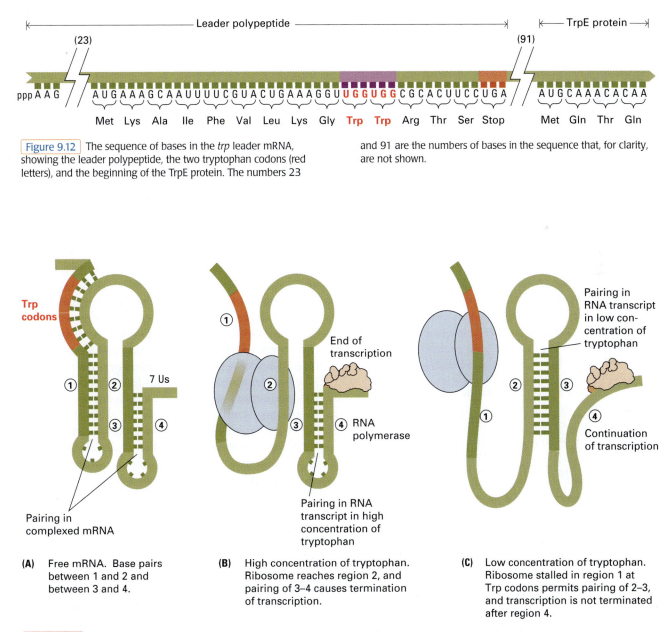

(A) Free mRNA. Base pairs between 1 and 2 and between 3 and 4.

(B) High concentration of tryptophan. Ribosome reaches region 2, and pairing of 3–4 causes termination of transcription.

(C) Low concentration of tryptophan. Ribosome stalled in region 1 at Trp codons permits pairing of 2–3, and transcription is not terminated after region 4.

Figure 9.13 The mechanism of attenuation in the *E. coli trp* operon. The tryptophan codons are highlighted in red.

tion of the leader polypeptide. The ribosome translocates beyond the Trp codons and blocks region 2, so the pairing that forms is between region 3 and region 4; this creates the transcriptional terminator, with termination occurring at the run of uridylates that follows region 4. Part C shows what happens when the ribosome stalls at the Trp codons as a result of insufficient $tRNA^{Trp}$. In this case, region 2 preferentially pairs with region 3, which disrupts the conformation of the terminator, allowing transcription to continue through the rest of the operon. The fine-tuning of this system takes place at intermediate concentrations of tryptophan, when the fraction of nascent transcripts that are completed depends on how frequently translation is stalled, which in turn depends on the intracellular concentration of charged tryptophan tRNA.

In summary, attenuation is a fine-tuning mechanism of regulation superimposed on the basic negative control of the *trp* operon:

> **key concept**
>
> When charged tryptophan tRNA is present in amounts that support translation of the leader polypeptide, transcription is terminated, and the *trp* enzymes are not synthesized. When the level of charged tryptophan tRNA is too low, transcription is not terminated, and the *trp* enzymes are made. At intermediate concentrations, the fraction of transcription initiation events that result in completion of *trp* mRNA depends on how frequently translation is stalled, which in turn depends on the intracellular concentration of charged tryptophan tRNA.

Many operons responsible for amino acid biosynthesis (for example, the leucine, isoleucine, phenylalanine, and histidine operons) are regulated by attenuators that function by forming alternative paired regions in the transcript. In the histidine operon, the coding region for the leader polypeptide contains seven adjacent histidine codons, and in the phenylalanine operon, the coding region for the leader polypeptide contains seven phenylalanine codons. This pattern is characteristic of operons in which attenuation is coupled with translation. Through translation of these leader polypeptides, the cell monitors the level of aminoacylated tRNA charged with the amino acid that is the end product of each amino acid biosynthetic pathway. Note that:

> **key concept**
>
> Attenuation cannot take place in eukaryotes because transcription and translation are uncoupled; transcription takes place in the nucleus and translation in the cytoplasm.

■ **Riboswitches combine with small molecules to control transcriptional termination.**

Transcription termination can also be triggered by direct binding of a small molecule to a 5′ untranslated leader mRNA. The mechanism is that the 5′ leader is able to adopt either of two conformations according to whether it binds with the small molecule. In the *antiterminator* conformation, transcription of the gene continues past the leader and through the remaining part of the gene. In the *terminator* conformation, which is triggered by binding with the small molecule, transcription is terminated. An RNA leader sequence able to switch between an antiterminator conformation and a terminator conformation is known as a **riboswitch.** Comparison of genomic sequences indicates that riboswitches are present in archaea, eubacteria, and eukarya.

Riboswitches have been described that bind with flavin mononucleotide, thiamine pyrophosphate, vitamin B_{12}, guanine, adenine, uncharged tRNA, lysine, or S-adenosylmethionine to regulate synthesis or transport of the corresponding metabolites. As a specific example, Figure 9.14, part A depicts the leader mRNA of the *yitJ* gene, which is involved in methionine biosynthesis in *Bacillus subtilis*. Nucleotides shown in red and blue can undergo two pairing configurations. In one, they form stem-and-loop structures of an anti-antiterminator (AAT) and a terminator (T), the latter followed by a string of uridylate residues. In this conformation, the RNA leader terminates transcription. The red and blue nucleotides can also pair with each other to form an antiterminator (AT) as shown at the upper right. In this conformation, transcription continues through the *yitJ* gene. The presence of S-adenosylmethionine (SAM), a modified form of methionine, results in conversion of the leader RNA from the terminator form (AAT and T loops) to the read-through form (AT loop) (part B). Sequences similar to the 5′ leader that binds S-adenosylmethionine are found upstream of 11 transcriptional units that encode 26 proteins for sulfur metabolism in *B. subtilis*.

9.4

Eukaryotes regulate transcription through transcriptional activator proteins, enhancers, and silencers.

Many eukaryotic genes are **housekeeping genes** that encode essential metabolic enzymes or cellular components and are expressed constitutively at relatively low levels in all cells. Other genes differ in

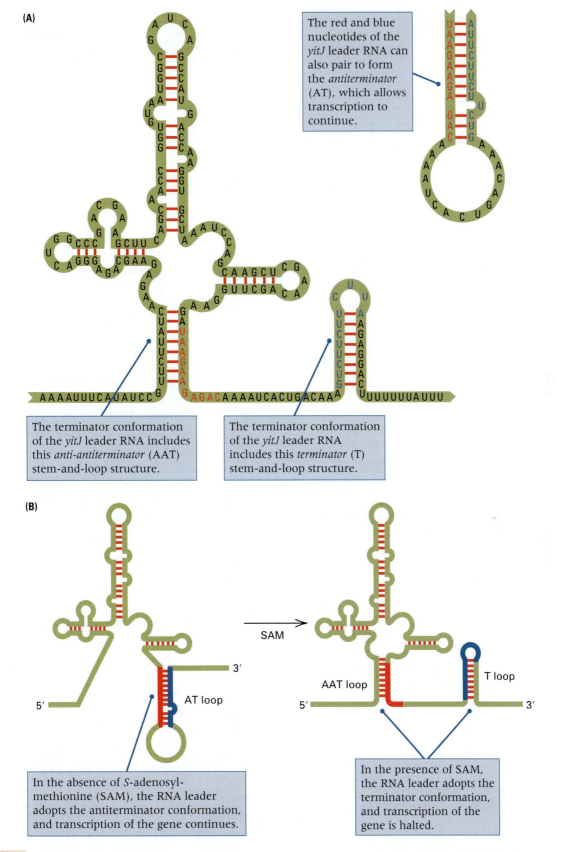

(A)

The red and blue nucleotides of the *yitJ* leader RNA can also pair to form the *antiterminator* (AT), which allows transcription to continue.

The terminator conformation of the *yitJ* leader RNA includes this *anti-antiterminator* (AAT) stem-and-loop structure.

The terminator conformation of the *yitJ* leader RNA includes this *terminator* (T) stem-and-loop structure.

(B)

SAM

AT loop

AAT loop

T loop

5'

3'

In the absence of *S*-adenosyl-methionine (SAM), the RNA leader adopts the antiterminator conformation, and transcription of the gene continues.

In the presence of SAM, the RNA leader adopts the terminator conformation, and transcription of the gene is halted.

Figure 9.14 Riboswitch regulation of transcription termination by the *yitJ* leader RNA in *Bacillus subtilis.* (A) Structural model of the untranslated leader RNA of the *yitJ* gene is depicted in the terminator (T) conformation, which includes an anti-antiterminator (AAT) stem-and-loop structure. The red and blue nucleotides in the AAT and T stems can also pair to form an antiterminator (AT).

(B) The presence of S-adenosylmethionine (SAM) results in conversion from the read-through form to the terminator form. [After B. A. M. McDaniel, F. J. Grundy, I. Artsimovitch, and T. M. Henkin. 2003. *Proceedings of the National Academy of Sciences USA* 100: 3083. © 2003 National Academy of Sciences USA]

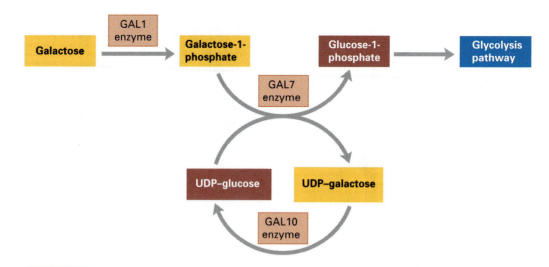

Figure 9.15 Metabolic pathway by which galactose is converted to glucose-1-phosphate in the yeast *Saccharomyces cerevisiae*.

their expression according to cell type or stage of the cell cycle. These genes are often regulated at the level of transcription. Typically, levels of expression of eukaryotic genes may differ 2- to 10-fold between the uninduced and induced levels. This contrasts with the more dramatic differences seen in prokaryotes, in which the ratio between the uninduced and induced levels may be as great as 1000-fold.

■ Galactose metabolism in yeast illustrates transcriptional regulation.

To introduce transcriptional regulation in eukaryotes, we first examine the control of galactose metabolism in yeast and compare it with the *lac* operon in *E. coli*. The first steps in the biochemical pathway for galactose degradation are illustrated in Figure 9.15 . Three enzymes, encoded by the genes *GAL1*, *GAL7*, and *GAL10*, are required for conversion of galactose into glucose-1-phosphate. These three structural genes are tightly linked genetically, as shown in Figure 9.16 . Despite the tight linkage of

the three genes, the genes are not part of an operon. The mRNAs are monocistronic. The *GAL1* and *GAL10* mRNAs are synthesized from divergent promoters lying between the genes, and *GAL7* mRNA is synthesized from its own promoter. On the other hand, the genes are inducible because the mRNAs are synthesized only when galactose is present.

Constitutive and uninducible mutants have been observed. In two types of mutants, *gal80* and *gal81ᶜ*, the mutants synthesize *GAL1*, *GAL7*, and *GAL10* mRNAs constitutively. Another type of mutant, *gal4*, is uninducible: It does not synthesize the mRNAs whether or not galactose is present. The characteristics of the mutants are shown in **Table 9.3**. This table uses the convention in yeast genetics that mutant alleles are denoted with the gene symbol written in small letters and wildtype alleles with the gene symbol in capital letters;

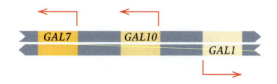

Figure 9.16 The linked *GAL* genes of *Saccharomyces cerevisiae*. Arrows indicate the transcripts produced. The *GAL1* and *GAL10* transcripts come from divergent promoters, *GAL7* from its own promoter.

Table	9.3

Characteristics of diploids containing various combinations of *gal80*, *gal4*, and *gal81ᶜ* alleles

Genotype	Synthesis of *GAL1*, *GAL7*, and *GAL10* mRNAs	Gal phenotype
1. *gal80 GAL1/GAL80 GAL1*	Inducible	+
2. *gal4 GAL1/GAL4 GAL1*	Inducible	+
3. *gal81ᶜ GAL1/GAL81 GAL1*	Constitutive	+

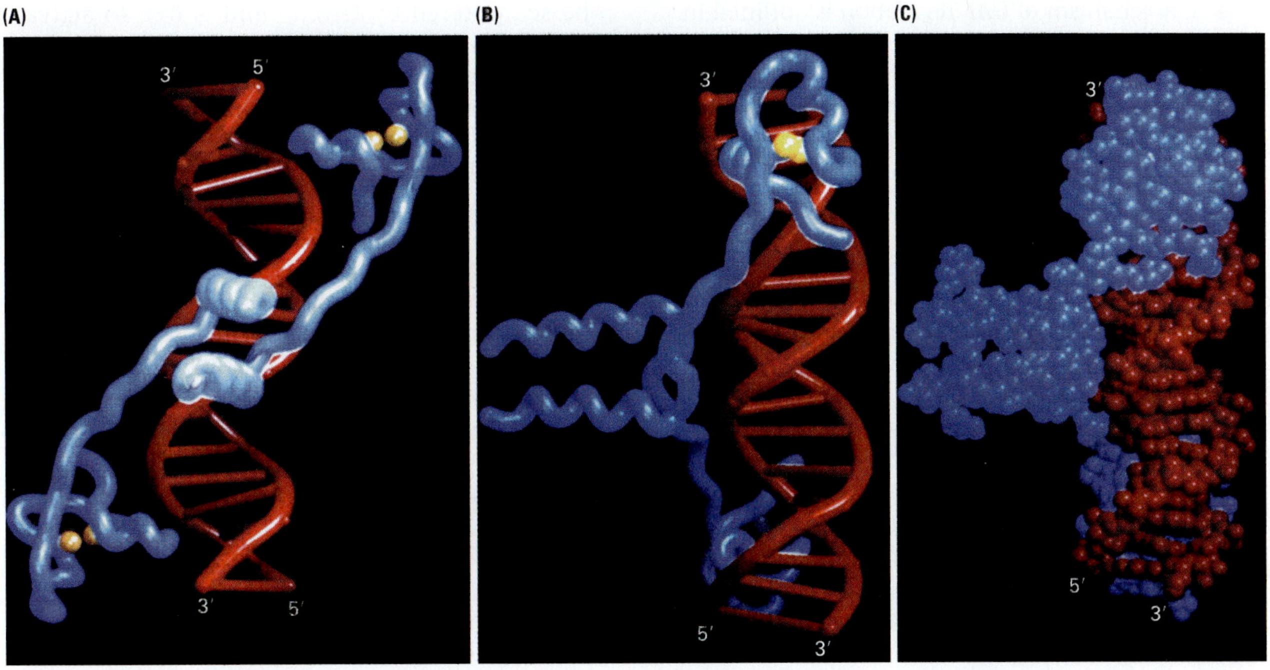

(A) **(B)** **(C)**

Figure 9.17 Three-dimensional structure of the GAL4 protein (blue) bound to DNA (red). The protein is composed of two polypeptide subunits held together by the coiled regions in the middle. The DNA-binding domains are at the extreme ends, and each physically contacts three base pairs in the major groove of the DNA. The zinc ions in the DNA-binding domains are shown in yellow. The views in (A) and (B) are at right angles; (C) is a space-filling model. [Courtesy of Dr. Stephen C. Harrison. Reprinted by permission from *Nature* 356: 408–414, S. C. Harrison, R. Marmorstein, M. Carey, and M. Ptashre. Copyright 1992 Macmillan Magazines Ltd.]

hence *gal80* and *gal4* are mutant alleles, whereas *GAL80* and *GAL4* are their wildtype counterparts.

The *gal80* mutation is recessive (entry 1 of Table 9.3). Thus, superficially, it behaves like a *lacI⁻* mutation. The wildtype *GAL80* allele does indeed encode a protein that negatively regulates transcription, and although the GAL80 protein is called a "repressor," it acts not by binding to an operator but by binding to a transcriptional activator protein and rendering it inactive. The activator that GAL80 binds to is the product of the *GAL4* gene, denoted GAL4.

The GAL4 protein is required for transcription of all three *GAL* genes. It is a positive regulatory protein that activates transcription of the three *GAL* genes individually, starting at a different site upstream from each gene. The GAL4 protein bound with its target site in the DNA is shown in Figure 9.17, in which the GAL4 protein (a dimer) is shown in blue and the DNA molecule in red. The small yellow spheres represent ions of zinc, which are essential components in the DNA binding.

© Photodisc

J. Scott Applewhite / Associated Press

Selective breeding has dramatically changed the size, shape, and color of cultivated varieties of wild crop plants. Wild tomatoes, for example, are about the size of the cultivated cherry tomato. Application of modern methods of genetic engineering to produce genetically modified organisms has led, in a few cases, to considerable controversy.

9.4 Eukaryotes regulate transcription through transcriptional activator proteins, enhancers, and silencers   **333**

The mechanism of *GAL* regulation is outlined in Figure 9.18. Although GAL80 superficially resembles a "repressor" protein in that *gal80/gal80* homozygotes produce the GAL enzymes constitutively, the GAL80 protein does not bind with DNA. Rather it has two binding sites, one for GAL4 (the transcriptional activator) and one for galactose (the substrate of the *GAL* pathway). In the absence of galactose (part A), GAL80 protein binds with the GAL4 protein and sequesters it, so that GAL4 is not free to activate transcription. In the presence of galactose (part B), GAL80 binds with the galactose. This alters the conformation of GAL80 and causes it to release GAL4, which then activates transcription. The mysterious constitutive mutation, *gal81^c*, is dominant and results in constitutive synthesis of all three mRNAs. It behaves as might be expected of an operator mutation, but it cannot be comparable to *lacO^c* because it does not map near *GAL1*, *GAL7*, or *GAL10*. The explanation is that "*gal81^c*" is actually a mutation in the *GAL4* gene that renders the GAL4 protein unable to bind with GAL80 protein (part C). Hence, the mutant GAL4 protein cannot be sequestered by GAL80 and is free to activate transcription in the absence of galactose, whether or not wildtype GAL4 protein is also present.

The main point of these comparisons is that the superficial similarity between the constitutive and uninducible mutations in the *lac* operon of *E. coli* and the *GAL* genes of yeast is not indicative of similar molecular regulatory mechanisms. Some physiological similarities remain in that, in both prokaryotes and eukaryotes, the genes for a particular metabolic (or developmental) pathway are expressed in a coordinated manner in response to a signal. The principle at work is that alternative molecular mechanisms are often employed to achieve similar ends.

■ Transcription is stimulated by transcriptional activator proteins.

The GAL4 protein is an example of a *transcriptional activator protein,* which must bind with an upstream DNA sequence in order to prepare a gene for transcription. Some transcriptional activator proteins work by direct interaction with one or more com-

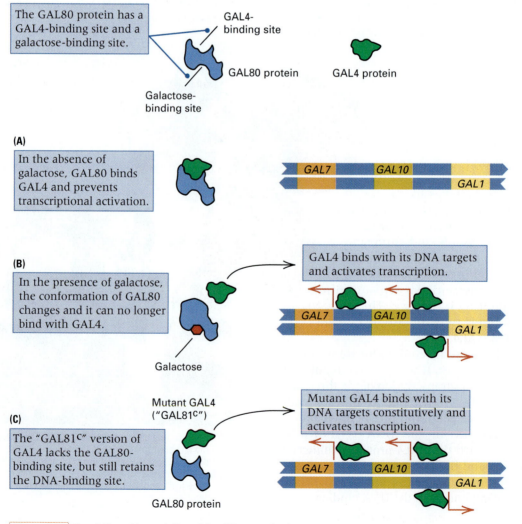

Figure 9.18 Regulation of transcription of the *GAL* genes by the GAL80 and GAL4 proteins.

ponents in the transcription complex, and in this way they recruit the transcription complex to the promoter of the gene to be activated. Other transcriptional activator proteins may initiate transcription by an already assembled transcription complex. In either case, the activator proteins are essential for the transcription of genes that are positively regulated.

Many transcriptional activator proteins can be grouped into categories on the basis of characteristics shared by their amino acid sequences. For example, one category has a *helix–turn–helix* motif, which consists of a sequence of amino acids forming a pair of α-helices separated by a bend; the helices are so situated that they can fit neatly into the grooves of a double-stranded DNA molecule. The helix–turn–helix motif is the basis of the DNA-binding ability, although the sequence specificity of the binding results from other parts of the protein.

A second large category of transcriptional activator proteins includes a DNA-binding motif called a *zinc finger* because the folded structure incorporates a zinc ion. An already familiar example is the GAL4 transcriptional activator protein in yeast (Figure 9.17), in which the zinc ions at the extreme ends are shown in yellow. The DNA sequence recognized by the GAL4 protein is a symmetrical sequence, 17 base pairs in length, which includes a CCG triplet at each end that makes direct contact with the zinc-containing domains.

■ Enhancers increase transcription; silencers decrease transcription.

Some transcriptional activator proteins bind with particular DNA sequences known as **enhancers.** Enhancer sequences are typically rather short (usually fewer than 20 base pairs) and are found at a variety of locations around the gene they regulate. Most enhancers are upstream of the transcriptional start site (sometimes many kilobases away), others are in introns within the coding region, and a few are even located at the 3' end of the gene. They are able to function as enhancers irrespective of their orientation; hence, an enhancer sequence can be in either the transcribed strand or the nontranscribed strand.

One of the most thoroughly studied enhancers is in the mouse mammary tumor virus and determines transcriptional activation by the glucocorticoid steroid hormone. The enhancer binds to a specific sequence of eight base pairs that is present at five different sites in the viral genome (Figure 9.19), providing five binding sites for the hormone–receptor complex that activates transcription.

Enhancers are essential components of gene organization in eukaryotes because they enable genes to be transcribed only when proper transcriptional activators are present. Some enhancers respond to molecules outside the cell—for example, steroid hormones that form receptor–hormone complexes. Other enhancers respond to molecules that are produced inside the cell (for example, during development), and these enhancers enable the genes under their control to participate in cellular differentiation or to be expressed in a tissue-specific manner. Many genes are under the control of several different enhancers, so they can respond to a variety of different molecular signals, both external and internal.

Some genes are also subjected to regulation by transcriptional **silencers,** which are short nucleotide sequences that are targets for DNA-binding proteins that, once recruited to the site, promote the assembly of large protein complexes that prevent transcription of the silenced genes. Examples of such silencing complexes include the set of *Drosophila* proteins called the PcG (Polycomb group) proteins, which silence certain genes during development.

■ The eukaryotic transcription complex includes numerous protein factors.

The eukaryotic **transcription complex** is an aggregate of protein factors that combines with the promoter region of a gene to initiate transcription. The factors necessary for transcription include a transcriptional activator protein that interacts with at least one protein subunit of the transcription complex to recruit the transcription complex to the gene. Many enhancers activate transcription by means of **DNA looping,** which refers to physical interactions

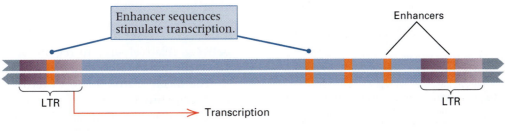

Figure 9.19 Positions, in the mouse mammary tumor virus, of enhancers (orange) that allow transcription of the viral sequence to be induced by glucocorticoid steroid hormone. LTR stands for the long terminal repeated sequences found at the extreme ends of the virus.

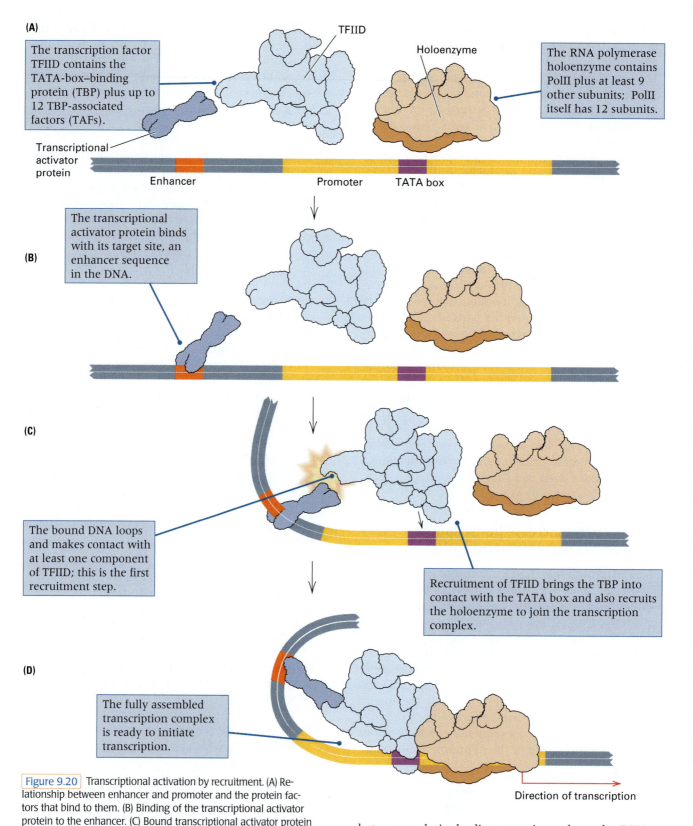

(A)

The transcription factor TFIID contains the TATA-box–binding protein (TBP) plus up to 12 TBP-associated factors (TAFs).

TFIID

Holoenzyme

The RNA polymerase holoenzyme contains PolII plus at least 9 other subunits; PolII itself has 12 subunits.

Transcriptional activator protein

Enhancer   Promoter   TATA box

(B)

The transcriptional activator protein binds with its target site, an enhancer sequence in the DNA.

(C)

The bound DNA loops and makes contact with at least one component of TFIID; this is the first recruitment step.

Recruitment of TFIID brings the TBP into contact with the TATA box and also recruits the holoenzyme to join the transcription complex.

(D)

The fully assembled transcription complex is ready to initiate transcription.

Direction of transcription

Figure 9.20 Transcriptional activation by recruitment. (A) Relationship between enhancer and promoter and the protein factors that bind to them. (B) Binding of the transcriptional activator protein to the enhancer. (C) Bound transcriptional activator protein makes physical contact with a subunit in the TFIID complex, which contains the TATA-box–binding protein, and attracts ("recruits") the complex to the promoter region. (D) The Pol II holoenzyme and any remaining general transcription factors are recruited by TFIID, and the transcription complex is fully assembled and ready for transcription. In the cell, not all of the Pol II is found in the holoenzyme, and not all of the TBP is found in TFIID. In this illustration, transcription factors other than those associated with TFIID and the holoenzyme are not shown.

between relatively distant regions along the DNA. The mechanism is illustrated in Figure 9.20.

The **basal transcription factors,** or **general transcription factors,** are proteins in the transcription complex that are used widely in the transcription of many different genes. The basal transcription factors in eukaryotes have been highly conserved in evolution. A minimal set necessary for accurate transcription *in vitro* includes TFIIB, TFIID, TFIIE, TFIIF, TFIIH, and Pol II. (TF in these designations stands for *transcription factor*.) These components can assemble *in vitro* in stepwise fashion on a promoter. The first step is recruitment of TFIID, itself a complex of proteins that includes a **TATA-box–binding protein (TBP),** which binds with the promoter in the region of the TATA box, and about 10 other proteins, called **TBP-associated factors (TAFs),** which are the components that respond specifically to activator proteins. The TBP binds to the DNA in the minor groove and then bends the DNA by about 80 degrees. The Pol II RNA polymerase is also found in a complex with multiple protein subunits called the **Pol II holoenzyme**.

It is not yet clear whether the transcription complex is recruited to the promoter and assembled stepwise, as *in vitro* studies suggest, or recruited in the form of one or more large, preassembled complexes, which the composition of the Pol II holoenzyme suggests is the case. For simplicity, Figure 9.20 shows recruitment of one preassembled complex that includes TFIID, which in turn recruits the preassembled Pol II holoenzyme. To activate transcription (part B), the transcriptional activator protein binds to an enhancer in the DNA and to one of the TAF subunits in the TFIID complex. This interaction attracts ("recruits") the TFIID complex to the region of the promoter (part C). Attraction of the TFIID to the promoter also recruits the Pol II holoenzyme (part D), as well as any remaining general transcription factors. Once these components are brought together, the transcriptional complex is ready for transcription to begin.

Experimental evidence for transcriptional activation by recruiting preassembled complexes has come from studies of a number of artificial proteins created by fusing a protein with a DNA-binding domain to one of the protein subunits in TFIID. Such fusion proteins act as transcriptional activa-

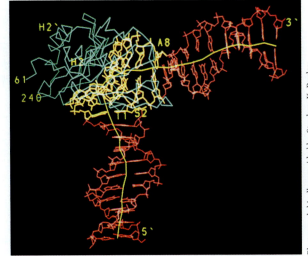

Model showing how TATA-box–binding protein binds to its target in duplex DNA. Note the pronounced kink in the DNA duplex resulting from interacting with the protein.

tors wherever they bind to DNA (provided that a promoter is nearby), because the TFIID is "tethered" to the DNA-binding domain and so the "recruitment" of TFIID is automatic. Similarly, fusion proteins that are tethered to a subunit of the holoenzyme can recruit the holoenzyme to the promoter. In this case, TFIID and the remaining general transcription factors are also attracted to the promoter, and the transcriptional complex is assembled. These experiments suggest that a transcriptional activator protein can activate transcription by interacting with subunits of either the TFIIA complex or the holoenzyme.

As Figure 9.20 suggests, the fully assembled transcription complex in eukaryotes is a very large structure. A real example, taken from early development in *Drosophila*, is shown in Figure 9.21 . In this case, the enhancers, located a considerable distance upstream from the gene to be activated, are bound by the transcriptional activator proteins BCD and HB, which are products of the genes *bicoid (bcd)* and *hunchback (hb)*, respectively; these transcriptional activators function in establishing the

Figure 9.21 An example of transcriptional activation during *Drosophila* development. The transcriptional activators in this example are bicoid protein (BCD) and hunchback protein (HB). The numbered subunits are TAFs (TBP-associated factors) that, together with TBP (TATA-box–binding protein), correspond to TFIID. BCD acts through a 110-kilodalton TAF, and HB through a 60-kilodalton TAF. The transcriptional activators act via enhancers to cause recruitment of the transcriptional apparatus. The fully assembled transcription complex includes TBP and TAFs, RNA polymerase II, and general transcription factors TFIIA, TFIIB, TFIIE, TFIIF, and TFIIH.

9.4 Eukaryotes regulate transcription through transcriptional activator proteins, enhancers, and silencers **337**

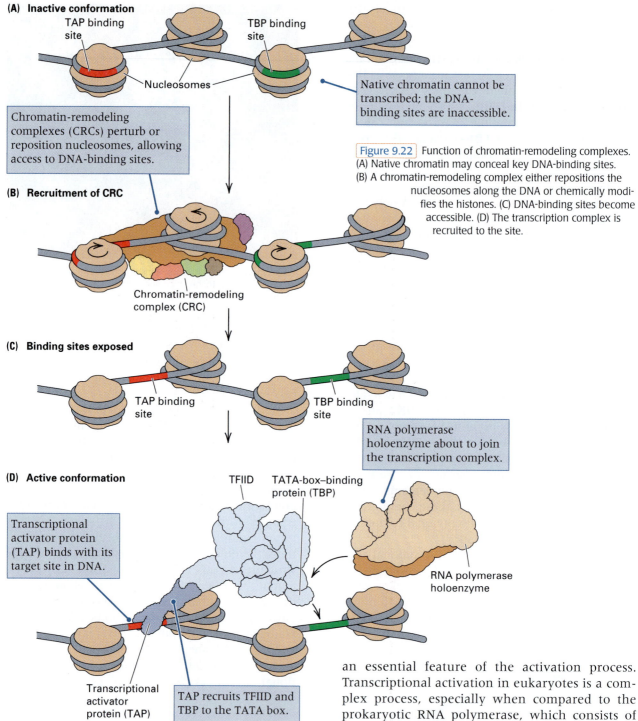

(A) Inactive conformation

TAP binding site

TBP binding site

Nucleosomes

Native chromatin cannot be transcribed; the DNA-binding sites are inaccessible.

Chromatin-remodeling complexes (CRCs) perturb or reposition nucleosomes, allowing access to DNA-binding sites.

Figure 9.22 Function of chromatin-remodeling complexes. (A) Native chromatin may conceal key DNA-binding sites. (B) A chromatin-remodeling complex either repositions the nucleosomes along the DNA or chemically modifies the histones. (C) DNA-binding sites become accessible. (D) The transcription complex is recruited to the site.

(B) Recruitment of CRC

Chromatin-remodeling complex (CRC)

(C) Binding sites exposed

TAP binding site

TBP binding site

RNA polymerase holoenzyme about to join the transcription complex.

(D) Active conformation

TFIID

TATA-box–binding protein (TBP)

Transcriptional activator protein (TAP) binds with its target site in DNA.

RNA polymerase holoenzyme

Transcriptional activator protein (TAP)

TAP recruits TFIID and TBP to the TATA box.

anterior–posterior axis in the embryo. (*Drosophila* development is discussed in Chapter 11.) Note the position of the TATA box in the promoter of the gene. The TATA box binding is the function of the TBP. The functions of a number of other components of the transcription complex have also been identified. For example, the TFIIH contains both helicase and kinase activity to separate the DNA strands and to phosphorylate RNA polymerase II.

Phosphorylation allows the polymerase to leave the promoter and elongate mRNA. The looping of the DNA effected by the transcriptional activators is an essential feature of the activation process. Transcriptional activation in eukaryotes is a complex process, especially when compared to the prokaryotic RNA polymerase, which consists of only six polypeptide chains.

■ **Chromatin-remodeling complexes prepare chromatin for transcription.**

Eukaryotic DNA is typically found in the form of chromatin packaged with nucleosomes (Chapter 3). Special mechanisms are required for transcriptional activator proteins and the transcription complex to acquire access to the DNA. The existence of such mechanisms is implied by the observation that the components of transcription sufficient to transcribe purified DNA *in vitro* are unable to initiate transcription of purified chromatin. The nucleosomes in chromatin must prevent the transcription complex from either binding to DNA or using it as a template.

Several different multiprotein complexes have been identified that can restructure chromatin and that enable it to be transcribed. These are known as **chromatin-remodeling complexes (CRCs).** All of these complexes use energy derived from ATP to restructure chromatin. The molecular mechanism of chromatin remodeling is unknown, and because there are several distinct types of CRCs, there may be several mechanisms. In one general class of models, the CRC disrupts nucleosome structure without displacing the nucleosomes, rendering the DNA accessible to transcriptional activator proteins, the TATA-box–binding protein, and other components of the transcription complex. In another general class of models, the CRC repositions the nucleosomes along the DNA, making key DNA-binding sites accessible. An example is illustrated in

Figure 9.22 . Part A shows a transcriptionally inactive chromatin conformation, with the DNA-binding sites for a transcriptional activator protein (TAP) and TATA-box–binding protein (TBP) sequestered in nucleosomes and unavailable. Recruitment of a CRC to the site results in repositioning of the nucleosomes (part B), which renders the binding sites accessible (part C). In this chromatin configuration, TAP and TBP can bind with the DNA and recruit the rest of the transcription complex.

■ **Some eukaryotic genes have alternative promoters.**

Some eukaryotic genes have two or more promoters that are active in different cell types. The different promoters result in different primary transcripts that contain the same protein-coding regions. An example from *Drosophila* is shown in Figure 9.23 . The gene code for alcohol dehydrogenase, and its

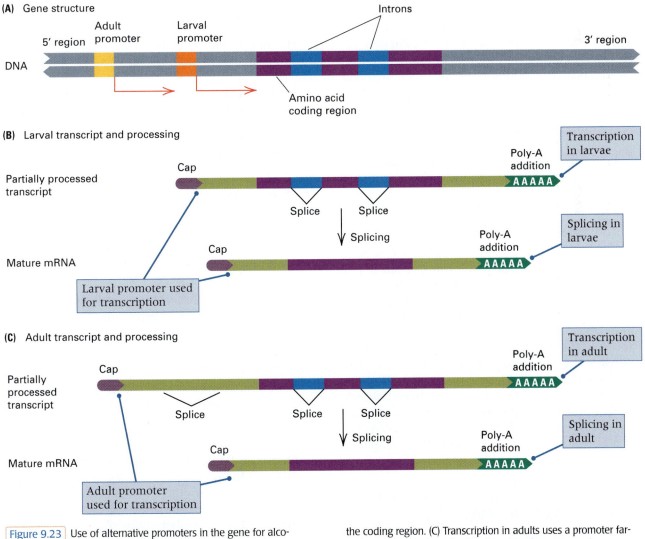

Figure 9.23 Use of alternative promoters in the gene for alcohol dehydrogenase in *Drosophila*. (A) The overall gene organization includes two introns within the amino acid coding region. (B) Transcription in larvae uses the promoter nearest the 5′ end of the coding region. (C) Transcription in adults uses a promoter farther upstream, and much of the larval leader sequence is removed by splicing.

organization in the genome, shown in part A, includes three protein-coding regions interrupted by two introns. Transcription in larvae (part B) uses a different promoter from that used in transcription in adults (part C). The adult transcript has a longer 5' leader sequence, but most of this sequence is eliminated in RNA splicing. Alternative promoters make possible the independent regulation of transcription in larvae and adults.

9.5

Gene expression can be affected by heritable chemical modifications in the DNA.

In this section we discuss some examples of *epigenetic* regulation of gene activity. The prefix *epi* means "besides" or "in addition to"; **epigenetic** therefore refers to heritable changes in gene expression that are due not to changes in the DNA sequence itself, but to something "in addition to" the DNA sequence, usually either chemical modification of the bases, or protein factors bound with the DNA. We shall see that there is a great deal yet to be learned about the molecular mechanisms by which epigenetic modifications are established and maintained.

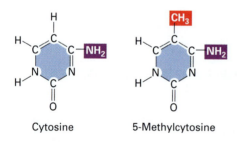

Cytosine   5-Methylcytosine

In most higher eukaryotes, a proportion of the cytosine bases are modified by the addition of a methyl (CH_3) group to the number-5 carbon atom. The cytosines are incorporated in their normal, unmodified form in the course of DNA replication, and then the methyl group is added by an enzyme called **DNA methylase.** In mammals, cytosines are modified preferentially in 5'-CG-3' dinucleotides. Many mammalian genes have CG-rich regions upstream of the coding region that provide multiple sites for methylation; these are called *CpG islands,* where the "p" represents the phosphate group in the polynucleotide backbone.

■ **Transcriptional inactivation is associated with heavy DNA methylation.**

A number of observations suggest that heavy methylation is associated with genes for which the rate of transcription is low. One example is the inactive X chromosome in mammalian cells, which is extensively methylated. In fact, in adult mammals, the majority of CpG dinucleotides in all chromosomes are methylated in somatic cells. The unmethylated CpGs are usually associated with the promoters of active housekeeping genes. The widespread methylation of inactive genes in adult somatic cells is thought to minimize accidental, low-level transcription from them.

Although there is a very strong correlation between heavy methylation and transcriptional silencing, heavy methylation may result from an earlier epigenetic signal that marks a gene for silencing and that recruits the methylase. If there is such an earlier signal, then it implies that methylation is the result of gene inactivity as well as its mechanism. In any case, treatment of cells with the cytosine analog *azacytidine* reverses methylation and can restore transcriptional activity. For example, in cell culture, some lineages of rat pituitary tumor cells express the gene for prolactin, whereas other related lineages do not. The gene is methylated in the nonproducing cells but is not methylated in the producers. Reversal of methylation in the nonproducing cells via azacytidine results in prolactin expression.

■ **In mammals, some genes are imprinted by methylation in the germ line.**

Mammals feature an unusual type of epigenetic silencing known as **genomic imprinting**, a process with the following characteristics:

- Imprinting occurs in the germ line.
- It affects at most a few hundred genes (many of them located in clusters).
- It is accompanied by heavy methylation (though the primary signal for imprinting is unknown).
- Imprinted genes are differentially methylated in the female and male germ lines.
- Once imprinted and methylated, a silenced gene remains transcriptionally inactive during embryogenesis.
- Imprints are erased early in germ-line development, then later reestablished according to sex-specific patterns.

Although mammalian gametes are extensively methylated, most of the DNA is demethylated in preimplantation development, except for imprinted genes that retain their sex-specific patterns of methylation. The embryonic DNA is remethylated beginning after implantation, gradually attaining the heavy methylation levels found in adult somatic cells. In the germ line, the original imprints are erased when the DNA is globally demethylated, and remethylation takes place later in germ-line development. All remethylated genes acquire identical patterns of methylation in the germ line of both sexes, except for those few that have sex-specific patterns of imprinting and differential methylation. The imprinted genes undergo methylation during oocyte growth prior to ovulation in females, and probably around the time of birth in males. Because the methylation associated with imprinting is retained throughout embryonic development, any gene that is imprinted in either the female or the male germ line has, effectively, only one active copy in the embryo.

The epigenetic, sex-specific gene silencing associated with imprinting is dramatically evident in a pair of syndromes characterized by neuromuscular defects, mental retardation, and other abnormalities. These are *Prader–Willi syndrome* and *Angelman syndrome*. Both conditions are associated with rare, spontaneous deletions that include chromosomal region *15q11*. If the deletion takes place in the father, the result is Prader–Willi syndrome, whereas if it takes place in the mother, the result is Angelman syndrome. The reason is that *15q11* includes at least three genes (*SNRPN, necdin,* and *UBE3A*) that are imprinted and differentially methylated in the gametes. Part A of Figure 9.24

shows the pattern of imprinting of these three genes in a normal embryo. *SNRPN* and *necdin* are imprinted in the egg, *UBE3A* in the sperm. In the embryo, therefore, *UBE3A* is transcriptionally active in the maternal chromosome, and *SNRPN* and *necdin* in the paternal chromosome. In the germ line of female and male embryos, shown in Figure 9.24, part B, the imprints are erased and reset according to sex: In the female both homologs have *SNRPN* and *necdin* imprinted, whereas in the male both homologs have *UBE3A* imprinted. If a normal, imprinted female gamete is fertilized by a sperm with a *15q11* deletion, the embryo has no transcriptionally active copy of either *SNRPN* or *necdin* and develops Prader–Willi syndrome. On the other hand, if a normal, imprinted male gamete fertilizes an egg with a *15q11* deletion, the embryo has no transcriptionally active copy of *UBE3A* and develops Angelman syndrome. These syndromes demonstrate not only the epigenetic control of gene expression by imprinting but also differential imprinting in the sexes and the clustering of imprinted genes in the genome.

Why is there imprinting? One suggestion is that it evolved in early mammals with polyandry (each female mating with a series of males). In such a situation, it is to a male's benefit to silence genes that

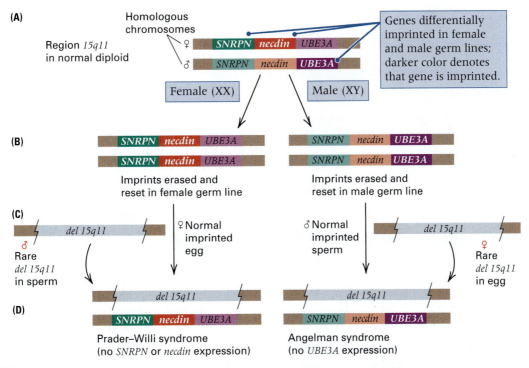

Figure 9.24 Imprinting of genes in chromosomal region *15q11* results in different neuromuscular syndromes, depending on which parent contributes a *15q11* deletion and which an imprinted chromosome. (A) Pattern of imprinting in a normal diploid. The maternal chromosome is at the top, the paternal chromosome at the bottom. Imprinted and transcriptionally inactive genes are indicated. (B) In the germ line, the imprints are erased and reset in either female-specific or male-specific patterns. (C) An individual who inherits a maternally imprinted chromosome along with a *15q11* deletion has Prader–Willi syndrome, whereas one who inherits a paternally imprinted chromosome along with a *15q11* deletion has Angelman syndrome. Other genes in the region, not shown, may also be imprinted.

conserve maternal resources at the expense of the fetus, because this strategy maximizes the father's immediate reproduction. But it is to a female's benefit to silence genes that allocate resources to the fetus at the expense of the mother, because this strategy maximizes the female's long-term reproduction. This hypothesis is supported by the fact that some imprinted genes do affect the allocation of resources between mother and fetus in the direction that would be predicted. On the other hand, many genes that are imprinted have no obvious connection to maternal–fetal conflict.

9.6

Regulation also takes place at the levels of RNA processing and decay.

Although transcriptional control of gene expression is of major importance, transcription is by no means the only level at which gene activity can be regulated. In this section we consider some mechanisms that act at the level of primary-transcript splicing or at the level of mRNA stability.

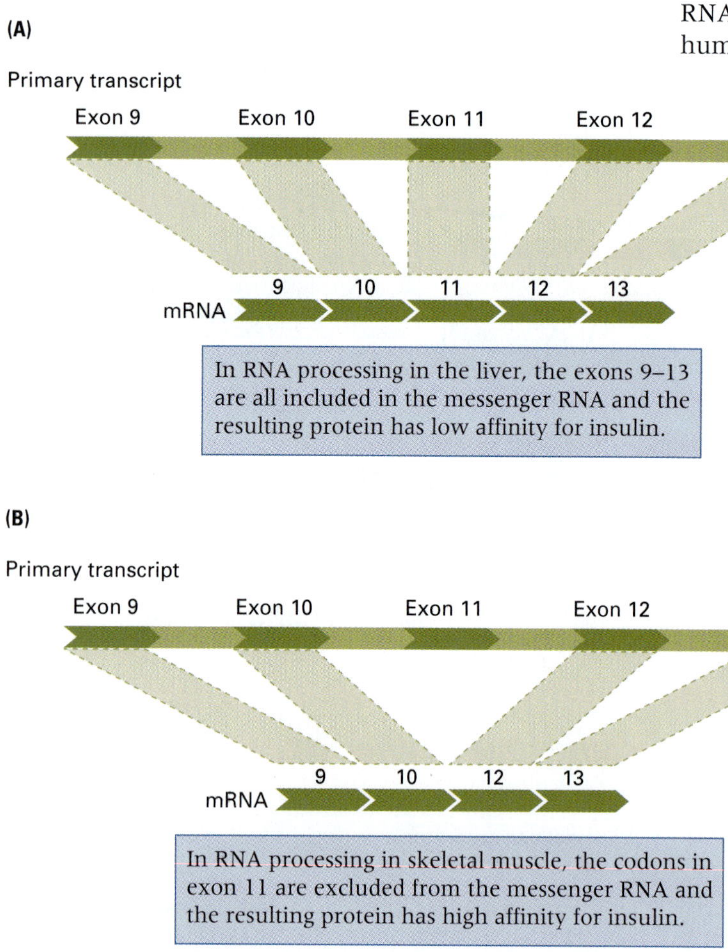

(A)

Primary transcript

Exon 9 | Exon 10 | Exon 11 | Exon 12 | Exon 13

9 | 10 | 11 | 12 | 13

mRNA

In RNA processing in the liver, the exons 9–13 are all included in the messenger RNA and the resulting protein has low affinity for insulin.

(B)

Primary transcript

Exon 9 | Exon 10 | Exon 11 | Exon 12 | Exon 13

9 | 10 | 12 | 13

mRNA

In RNA processing in skeletal muscle, the codons in exon 11 are excluded from the messenger RNA and the resulting protein has high affinity for insulin.

Figure 9.25 Alternative splicing of the primary transcript of the gene encoding the chain of the insulin receptor in humans and other mammals. (A) Splicing in the liver results in the low-affinity long form. (B) Splicing in skeletal muscle results in the high-affinity long form.

■ The primary transcripts of many genes are alternatively spliced to yield different products.

Even when the same promoter is used to transcribe a gene, different cell types can produce different quantities of the protein (or even different proteins) because of differences in the mRNA produced in processing. The reason is that the same transcript can be spliced differently from one cell type to the next. The different splicing patterns may include exactly the same protein-coding exons, in which case the protein is identical, but the rate of synthesis differs because the mRNA molecules are not translated with the same efficiency. In other cases, the protein-coding part of the transcript has a different splicing pattern in each cell type, and the resulting mRNA molecules code for proteins that are not identical even though they share certain exons. We have already noted (in Chapter 8) that transcripts in the human genome are frequently spliced in alternative ways; because of this, the approximately 30,000 human genes may encode 64,000 to 96,000 different proteins. Alternative RNA processing is one of the principal sources of human genetic complexity.

The insulin receptor gene in humans and other mammals provides an example of alternative splicing that results in the inclusion or exclusion of exon 11 in the messenger RNA. The resulting forms of the polypeptide chain differ in length by 12 amino acids. The relevant part of the primary transcript is shown in Figure 9.25. In the liver, all 20 exons are found in the mRNA for the long form of the receptor protein (part A), whereas in skeletal muscle exon 11 is eliminated along with the flanking introns and excluded from the mRNA for the short form (part B). The long form of the receptor shows low affinity for insulin and is expressed in tissues such as the liver that are exposed to relatively high concentrations of insulin. The short form of the protein has a high affinity for insulin and is expressed preferentially in tissues such as skeletal muscle that are normally exposed to lower levels of insulin. Alternative splicing thus offers the possibility of generating proteins with different properties from the same gene. The record estimated number of proteins derived from a single gene is held by the *Dscam* gene of *Drosophila*, which could give rise to 38,016 different proteins. How many are actually made is unknown, but it is likely to be a large number because 49 of the first 50 *Dscam* cDNAs isolated from embryos were different. Although the human genome is estimated to contain only 30,000 to 40,000 genes, the number of different proteins produced is several to many times greater because of alternative splicing.

The coding capacity of the human genome is enlarged by extensive alternative splicing.

Compared with genes in the worm or fly, human genes are spread over a larger region of the genome, and the primary transcripts are longer. Many human genes are alternatively spliced to yield multiple protein products. At least one-third of all human genes, and perhaps as many as two-thirds, are alternatively spliced. Among those that are alternatively spliced, the average number of distinct mRNAs produced from the primary transcript is in the range 2 to 7. The average number of different mRNAs per gene across the genome is in the range 2 to 3, which includes genes that produce a single mRNA as well as those that produce multiple different mRNAs. The alternative splicing greatly expands the number of protein products that can be encoded in a relatively small number of genes:

key concept

Alternative splicing is an important source of human genetic complexity. Although the number of human genes exceeds that in worms or flies by a factor of about two, the number of different human proteins may be greater than that in worms or flies by a factor of about five.

Different messenger RNAs can differ in their persistence in the cell.

A short-lived mRNA produces fewer protein molecules than a long-lived mRNA, so features that affect the rate of mRNA stability affect the level of gene expression. One route of degradation is the *deadenylation-dependent pathway,* which begins with enzymatic trimming of length of the poly-A tail on the mRNA. When the poly-A tail is trimmed to a length of 25 to 60 nucleotides, the mRNA becomes susceptible to a decapping enzyme that removes the 5' cap and renders the molecule unable to initiate translation; from this state the mRNA is rapidly degraded by exonucleases. An alternative pathway is the *deadenylation-independent pathway,* which is initiated either with decapping or with endonuclease cleavage of the mRNA, after which digestion goes to completion by exonuclease activity. The deadenylation-independent pathway is particularly active for mRNAs that contain early chain termination codons or unspliced introns, and it prevents the accumulation of truncated polypeptides in the cell.

RNA interference (RNAi) results in the destruction of RNA transcripts.

An important mechanism regulating the stability of RNA transcripts is a phenomenon known as **RNA interference (RNAi),** in which the introduction of a few hundred nucleotide pairs of *double-stranded RNA* triggers degradation of RNA transcripts containing homologous sequences. Originally discovered in *Caenorhabditis elegans,* RNAi has been shown to occur in a wide variety of organisms, including trypanosomes, planaria, *Drosophila,* amphibians, and mammals including humans. Although the silenced genes are apparently transcribed at the normal level, the transcripts are degraded before they leave the nucleus. The effect is highly specific and requires only a few double-stranded RNA molecules per cell to trigger the silencing. Moreover, the RNAi effect can be transmitted from cell to cell.

The molecular basis of RNA interference is an active field of investigation, but the outlines are already clear (Figure 9.26). The double-stranded RNA is cleaved by an enzyme known as *dicer* into short stretches of 21 to 22 nucleotide pairs (part A). These fragments are incorporated into a ribonucleoprotein complex called *RISC (RNA-induced silencing complex;* part B). Once the RISC is formed, the 21- or 22-mer oligonucleotides are used as templates to

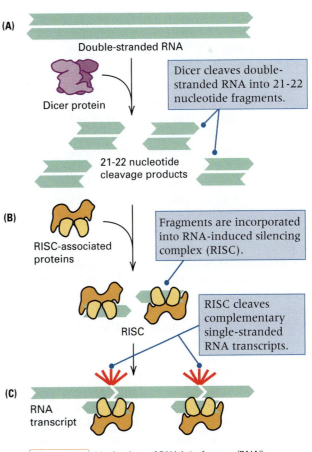

Figure 9.26 Mechanism of RNA interference (RNAi).

recognize RNA molecules containing complementary sequences, which are then enzymatically degraded and destroyed (part C). Double-stranded RNA from one cell can also be transported into neighboring cells by means of specific transmembrane proteins, and in this way the effects of RNAi can spread from cell to cell. The discovery of RNAi has created a great deal of excitement in genetics because of its potential in research as well as practical applications. In effect, RNAi affords a method of producing the equivalent of mutations that drastically reduce gene expression in organisms that do not have well-developed systems of mutagenesis and genetic manipulation.

9.7

Regulation can also take place at the level of translation.

Because transcription and translation are uncoupled in eukaryotes, gene expression can be regulated at the level of translation separately from transcription. The principal types of translational control are

- Inability of an mRNA molecule to be translated except under certain conditions
- Regulation of the overall rate of protein synthesis
- Inhibition or activation of translation by small regulatory RNAs that undergo base pairing with the mRNA

An important example of translational regulation is that of activating previously untranslated cytoplasmic mRNAs. This mechanism is prominent in early development, when newly fertilized eggs synthesize at a rapid rate many new proteins, virtually all of which derive from preexisting cytoplasmic mRNAs. In a few cases the molecular mechanism of mRNA activation is known. For example, in *Drosophila*, the mRNAs for the genes *bicoid, Toll,* and *torso* become activated because of the cytoplasmic elongation of their poly-A tail.

A dramatic example of translational control is the extension of the lifetime of silk fibroin mRNA in the silkworm. During cocoon formation, the silk gland synthesizes a single type of protein, silk fibroin, in very large amounts. The amount of fibroin is increased by three different mechanisms. First, the silk-gland cells become highly polyploid, accumulating thousands of copies of each chromosome. Second, transcription of the fibroin gene is initiated at a strong promoter, which results in the creation of about 10^4 fibroin mRNA molecules per gene copy in a period of a few days. Third, the fibroin mRNA molecule has a very long lifetime. In

contrast to a typical eukaryotic mRNA molecule, which has a lifetime of about 3 hours, fibroin mRNA survives for several days, during which each mRNA molecule is translated repeatedly to yield 10^5 fibroin molecules. Thus each fibroin gene copy yields about 10^9 protein molecules in the few days during which the cocoon is being created.

■ Small regulatory RNAs can control translation by base-pairing with the messenger RNA.

Small regulatory RNAs that control translation have been described in both prokaryotes and eukaryotes, and analyses of genome sequences suggest that there will be many more examples. Although only a handful of cases have been studied in detail, most seem to involve regulatory RNAs that are complementary in sequence to part of the mRNA whose translation they control. An RNA sequence complementary to an mRNA is called an **antisense RNA.** The antisense regulatory RNAs act by pairing with the mRNA to either inhibit or activate translation (Figure 9.27). Bacterial regulatory RNAs often control translation of several mRNAs and serve as global regulators of cellular processes.

Figure 9.27 shows an example of a small regulatory RNA that relieves oxidative stress in *E. coli.* One of the genes derepressed in the presence of hydrogen peroxide is *oxyS,* which encodes a regulatory RNA called OxyS. This RNA binds to several mRNAs. For any given interaction, only short stretches of OxyS are complementary to the target RNA and able to pair with it. For example, two separate regions of OxyS bind to the mRNA of the gene *flhA,* which encodes a transcriptional activator protein (part A). The complementary regions are very short, in this example only seven nucleotides. One of the complementary regions is near the AUG translational start, and the other is more than 40 nucleotides upstream. Base pairing between OxyS and the *flhA* mRNA conceals the ribosome-binding site and prevents translation. (Such a bipartite complex composed of a small regulatory RNA and an mRNA is called a *kissing complex.*)

Small regulatory RNAs can also activate translation. An example is the DsrA regulatory RNA from *E. coli* shown in part B. In this case the mRNA whose translation is controlled is from a gene *rpoS,* which encodes a sigma factor for RNA polymerase that allows transcription of a new set of RNAs from a special set of promoters at stationary phase in cell cultures when the cell density is high and the cells begin to slow their growth and division. The 5' end of the *rpoS* mRNA is self-complementary and can form a hairpin that hides the ribosome binding site and the translational start site. Two virtually con-

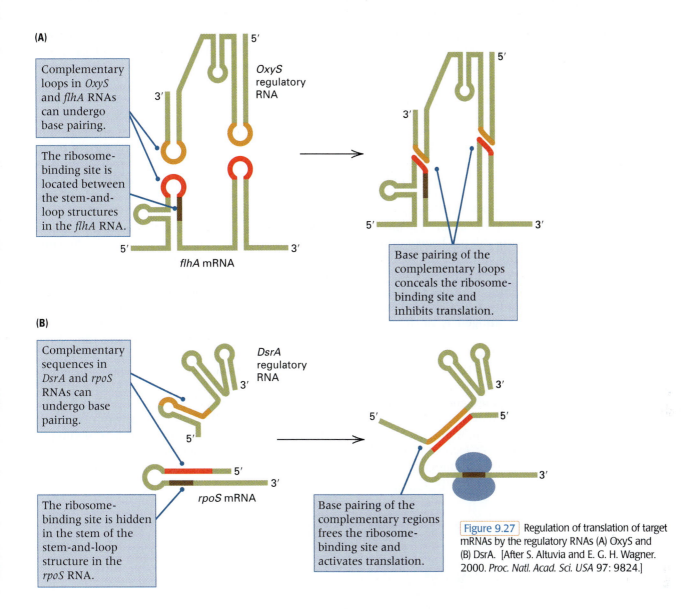

(A)

Complementary loops in *OxyS* and *flhA* RNAs can undergo base pairing.

OxyS regulatory RNA

The ribosome-binding site is located between the stem-and-loop structures in the *flhA* RNA.

flhA mRNA

Base pairing of the complementary loops conceals the ribosome-binding site and inhibits translation.

(B)

Complementary sequences in *DsrA* and *rpoS* RNAs can undergo base pairing.

DsrA regulatory RNA

The ribosome-binding site is hidden in the stem of the stem-and-loop structure in the *rpoS* RNA.

rpoS mRNA

Base pairing of the complementary regions frees the ribosome-binding site and activates translation.

Figure 9.27 Regulation of translation of target mRNAs by the regulatory RNAs (A) OxyS and (B) DsrA. [After S. Altuvia and E. G. H. Wagner. 2000. *Proc. Natl. Acad. Sci. USA* 97: 9824.]

tiguous regions of DsrA bind to the *rpoS* mRNA, and when binding occurs, the ribosome binding site becomes free and translation can occur.

Regulatory RNAs that control developmental timing in the nematode *C. elegans* were the first examples of small regulatory RNAs described in eukaryotes. The *lin4* regulatory RNA binds to the 3′ untranslated region of *lin14* mRNA, which results in degradation of the *lin14* mRNA and thereby prevents translation. Regulatory RNAs that govern development in plants have also been described.

9.8
Some developmental processes are controlled by programmed DNA rearrangements.

The introduction of changes in DNA sequence is an unusual mechanism of gene regulation, but there are a number of important examples. In some cases

the changes are reversible, but in others they are permanent. A permanent change in DNA sequence implies that in the cell lineage in which it takes place, the genotype becomes permanently altered. Such irreversible changes take place only in somatic cells, not in the germ line, so they are not genetically transmitted.

■ Programmed deletions take place in the development of the mammalian immune system.

Perhaps the most impressive examples of programmed DNA rearrangement take place in the bone-marrow–derived (B) cells and thymus-derived (T) cells that play key roles in the vertebrate immune system. B cells produce and secrete circulating antibodies that combine with foreign antigens and mark them for destruction, whereas T cells have specific receptors that recognize foreign antigens at the cell surface. Both antibodies and T-cell receptors have highly variable amino acid sequences from one B cell or T cell to the next, and much of this variability is generated by DNA rearrangements that assemble a relatively small number of coding sequences into a very large number of possible combinations, one in each B cell or T cell. We will examine the process of DNA splicing that generates diversity in antibodies. A similar process accounts for the variability of T-cell receptors.

A normal mammal is capable of producing more than 10^8 different antibodies, only a fraction of which are produced at any one time. Each B cell can produce a single type of antibody, but the antibody is not secreted until the cell has been stimulated by the appropriate antigen. Once stimulated, the B cell undergoes successive mitoses and eventually produces a lineage of genetically identical cells that secrete the antibody. Antibody secretion may continue even if the antigen is no longer present.

There are five distinct classes of antibodies known as IgG, IgM, IgA, IgD, and IgE (Ig stands for *immunoglobulin*). These classes perform specialized functions in the immune response and exhibit certain structural differences. However, each contains two types of polypeptide chains differing in size: a large subunit called the *heavy (H) chain* and a small subunit called the *light (L) chain*. Immunoglobulin G

(IgG) is the most abundant class of antibodies and has the simplest molecular structure (Figure 9.28). An IgG molecule consists of two heavy and two light chains held together by disulfide bridges (two covalently joined sulfur atoms) and has the overall shape of the letter Y. The sites for antibody specificity that combine with the antigen are located in the upper half of the arms above the fork of the Y. Each IgG molecule with a different specificity has a different amino acid sequence for the heavy and light chains in this part of the molecule. These specificity regions are called the **variable antibody regions** of the heavy and light chains. The remaining regions of the polypeptide are the **constant antibody regions,** which are called constant because they have virtually the same amino acid sequence in all IgG molecules.

Initial understanding of the genetic mechanisms responsible for variability in the amino acid sequences of antibody polypeptide chains came from cloning a gene for the light chain of IgG. The critical observation was made by comparing the nucleotide sequence of the gene in embryonic cells with that in mature antibody-producing B cells. In the genome of a B cell that was actively producing the antibody, the DNA segments corresponding to the constant and variable regions of the light chain were found to be very close together. However, in

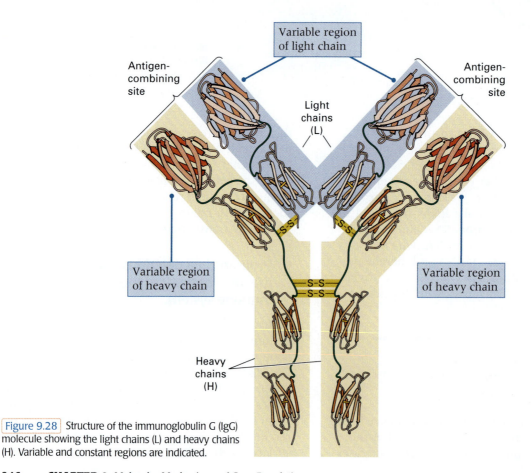

Figure 9.28 Structure of the immunoglobulin G (IgG) molecule showing the light chains (L) and heavy chains (H). Variable and constant regions are indicated.

embryonic cells, these same DNA sequences were located far apart. Similar results were obtained for the variable and constant regions of the heavy chains: Segments encoding these regions were close together in B cells but widely separated in embryonic cells.

Complete DNA sequencing of the genomic region that codes for antibody proteins revealed not only the reason for the different gene locations in B cells and embryonic cells but also the mechanism for the origin of antibody variability. Cells in the germ line contain a small number of genes corresponding to the constant region of the light chain, and these are close together along the DNA. Separated from them, but on the same chromosome, is another cluster consisting of a much larger number of genes that correspond to the variable region of the light chain. In the differentiation of a B cell, one gene for the constant region is spliced (cut and joined) to one gene for the variable region, and this DNA splicing produces a complete light-chain antibody gene. A similar DNA splicing mechanism yields the constant and variable regions of the heavy chains.

The formation of a finished antibody gene is somewhat more complicated than this overview implies, because light-chain genes consist of three parts and heavy-chain genes consist of four parts. DNA splicing in the origin of a light chain is illustrated in Figure 9.29 . For each of two parts of the variable region, the germ line contains multiple coding sequences called the **variable (V)** and **joining (J) antibody regions.** In the differentiation of a B cell, a deletion makes possible the joining of one of the V regions with one of the J regions. The DNA joining process is called **V–J joining,** and it can create many combinations of the V and J regions. Since the transcriptional enhancer lies in the region between the rightmost J region and the C region, the rearranged gene always contains the transcriptional enhancer. This also ensures that only the rearranged gene will be transcribed. When transcribed, this joined V–J sequence forms the 5' end of the light-chain RNA transcript. Transcription continues on through the DNA region coding for the **constant (C) antibody region** of the gene. RNA splicing subsequently joins the V–J region with the C region, creating the light-chain mRNA.

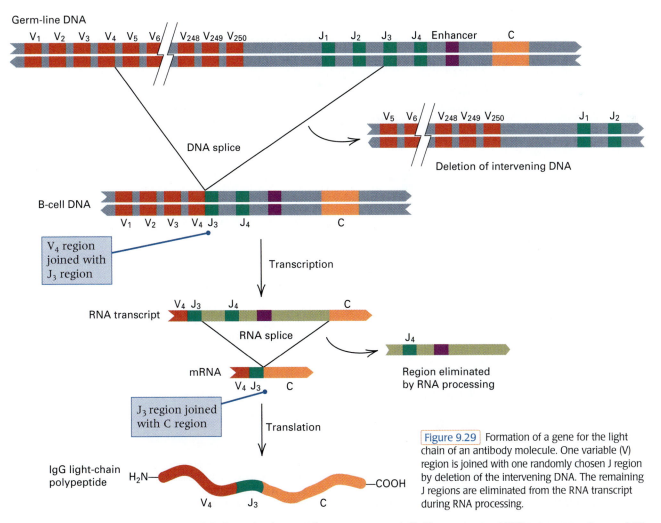

Figure 9.29 Formation of a gene for the light chain of an antibody molecule. One variable (V) region is joined with one randomly chosen J region by deletion of the intervening DNA. The remaining J regions are eliminated from the RNA transcript during RNA processing.

Combinatorial joining also takes place in the genes for the antibody heavy chains, but in this case the DNA splicing joins the heavy-chain counterparts of V and J with a third set of sequences, called D (for diversity), located between the V and J clusters.

■ Programmed transpositions take place in the regulation of yeast mating type.

Rearrangement of DNA sequences underlies the phenomenon of **mating-type interconversion** in budding yeast, *Saccharomyces cerevisiae*. This organism has two mating types, denoted **a** and α. Mating between haploid **a** and haploid α cells produces the **a**α diploid, which can undergo meiosis to produce four-spored asci containing haploid **a** and α spores in the ratio 2 : 2. If a single yeast spore of either the **a** or the α genotype is cultured in isolation from other spores, then mating between progeny cells would not be expected because the daughter cells would all have the same mating type. However, many strains of *S. cerevisiae* have a mating system called **homothallism,** in which some cells undergo a conversion into the opposite mating type. This allows matings between cells in what would otherwise be a pure culture.

The outlines of mating-type interconversion are shown in Figure 9.30. An original haploid spore (in this example, α) undergoes germination to produce two progeny cells. Both the mother cell (the original parent) and the daughter cell have mating type α, as expected from a normal mitotic division. However, in the next cell division, a switching (interconversion) of mating type takes place in both the mother cell and its *new* progeny cell, in which the original α mating type is replaced with the **a** mating type. After this second cell division is complete, the α and **a** cells are able to undergo mating because they now are of opposite mating types. Fusion of the nuclei produces the **a**α diploid, which undergoes mitotic divisions and later sporulation to produce **a** and α haploid spores again.

The genetic basis of mating-type interconversion is DNA rearrangement, as illustrated in Figure 9.30. The gene that controls mating type is the *MAT* gene in chromosome III, which can have either of two allelic forms, **a** or α. If the allele in a haploid cell is *MAT***a,** then the cell has mating type **a;** if the allele is *MAT*α, then the cell has mating type α. However, both genotypes normally contain both **a** and α genetic information in the form of unexpressed *cassettes* present in the same chromosome. The *HML*α

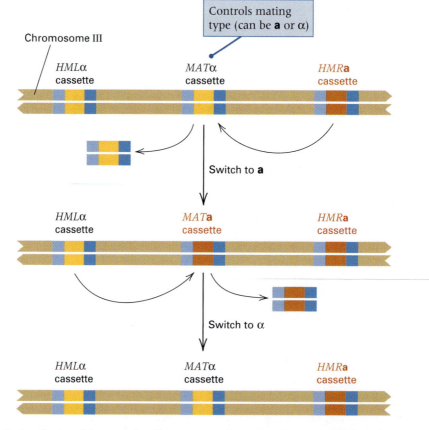

Figure 9.30 Genetic basis of mating-type interconversion. The mating type is determined by the DNA sequence present at the *MAT* locus. The *HML* and *HMR* loci are cassettes containing unexpressed mating-type genes, either α or **a.** In the interconversion from α to **a**, the α genetic information present at *MAT* is replaced with the **a** genetic information from *HMR***a**. In the switch from **a** to α, the **a** genetic information at *MAT* is replaced with the α genetic information from *HML*α.

cassette, about 200 kb distant from the *MAT* gene, contains the α DNA sequence; and the *HMR***a** cassette, about 150 kb distant from *MAT* on the other side, contains the **a** DNA sequence. When mating-type interconversion occurs, a specific endonuclease, encoded by the *HO* gene elsewhere in the genome, is produced and cuts both strands of the DNA in the *MAT* region. The double-stranded break initiates a process in which genetic information in the unexpressed cassette containing the opposite mating type becomes inserted into *MAT*. In this process, the DNA sequence in the donor cassette is duplicated, so the mating type becomes converted, but the same genetic information is retained in unexpressed form in the cassette. The terminal regions

of *HML, MAT,* and *HMR* are identical (they are shown in light blue and dark blue in Figure 9.30), and these regions are critical in pairing of the regions and interconversion.

Figure 9.30 illustrates two sequential mating-type interconversions. In the first, an α cell (containing the *MATα* allele) undergoes conversion into **a,** using the DNA sequence contained in the *HMR***a** cassette. The converted cell has the genotype *MAT***a.** In a later generation, a descendant **a** cell may become converted into mating type α, using the unexpressed DNA sequence contained in *HMLα*. This cell has the genotype *MATα*. Mating-type switches can occur repeatedly in the lineage of any particular cell.

chapter summary

9.1 Regulation of transcription is a common mechanism in prokaryotes.

- In negative regulation, the default state of transcription is "on."
- In positive regulation, the default state of transcription is "off."

Transcriptional regulation may be either negative ("on" unless specifically turned "off") or positive ("off" unless specifically turned "on"). A negative regulatory system may be inducible (when the repressor protein is inactivated by combining with the inducer) or repressible (when the repressor is formed by the combination of aporepressor protein and the co-repressor). Positively regulated systems are typically controlled by transcriptional activator proteins.

9.2 In prokaryotes, groups of adjacent genes are often transcribed as a single unit.

- The first regulatory mutations that were discovered affected lactose metabolism.
- Lactose-utilizing enzymes can be inducible (regulated) or constitutive.
- Repressor shuts off messenger RNA synthesis.
- The lactose operator is an essential site for repression.
- The lactose promoter is an essential site for transcription.
- The lactose operon contains linked structural genes and regulatory sequences.
- The lactose operon is also subject to positive regulation.
- Tryptophan biosynthesis is regulated by the tryptophan operon.

Lactose metabolism in *E. coli* is a classic example of an inducible system. When lactose is present, the genes for the enzymes necessary to metabolize lactose are transcribed; when lactose is absent, transcription is repressed. Two proteins needed to utilize lactose (permease, required for the entry of lactose into the cell, and β-galactosidase, the enzyme that cleaves lactose) are tightly linked and tran-

scribed into a single polycistronic mRNA molecule, *lac* mRNA. Immediately adjacent to the promoter for *lac* mRNA is a regulatory sequence of bases called the operator. A repressor protein, the product of the *lacI* gene, binds with the operator and prevents transcription. The inducer binds to the repressor and inactivates it. Therefore, in the absence of lactose, the permease and β-galactosidase genes are not transcribed, but in the presence of lactose, there is no active repressor, and the *lac* promoter becomes accessible to RNA polymerase. The operator, the promoter, and the structural genes are adjacent to one another and together constitute the *lac* operon. Repressor mutations have been isolated that inactivate the repressor protein, and operator mutations are known that prevent recognition of the operator by an active repressor; such mutations cause continuous production of *lac* mRNA and are said to be constitutive.

The lac operon and other sugar-utilization operons are subject to positive regulation through binding with the cAMP–CRP complex. (The cAMP stands for *cyclic AMP*, and CRP for *cyclic AMP receptor protein*.) When glucose is present in the growth medium, the level of cAMP is low, and not enough cAMP–CRP is formed to allow transcription of the operons. Only when glucose is absent is the concentration of cAMP sufficient to produce enough cAMP–CRP to make transcription possible.

Biosynthetic enzyme systems exemplify another type of transcriptional regulation in prokaryotes. In the tryptophan (*trp*) operon in *E. coli*, transcription of the operon is controlled by the concentration of tryptophan in the growth medium. When excess tryptophan is present, it binds with the *trp* aporepressor to form the active repressor that prevents transcription.

9.3 Gene activity can be regulated through transcriptional termination.

- Attenuation allows for fine-tuning of transcriptional regulation.
- Riboswitches combine with small molecules to control transcriptional termination.

Regulation of the *trp* operon is also subject to fine tuning by attenuation, in which transcription is initiated continually but the transcript forms a hairpin structure that results in premature termination. The frequency of termination of transcription is determined by the availability of charged tryptophan tRNA: With decreasing concentrations of tryptophan, termination occurs less often, thereby allowing transcription of the operon and production of the enzymes for tryptophan synthesis. Attenuators also regulate operons for the synthesis of other amino acids.

In some cases, a small molecule may bind directly with the leader region of an RNA transcript and induce a switch in conformation from one that allows transcription to a terminator conformation. The combination of the small molecule with the leader RNA constitutes a riboswitch.

9.4 Eukaryotes regulate transcription through transcriptional activator proteins, enhancers, and silencers.

- Galactose metabolism in yeast illustrates transcriptional regulation.
- Transcription is stimulated by transcriptional activator proteins.
- Enhancers increase transcription; silencers decrease transcription.
- The eukaryotic transcription complex includes numerous protein factors.
- Chromatin-remodeling complexes prepare chromatin for transcription.
- Some eukaryotic genes have alternative promoters.

Many genes in eukaryotes are regulated at the level of transcription. Transcriptional activators bind to DNA sequences known as enhancers, which are usually short sequences that may be present at a variety of positions around the genes they regulate. In the recruitment model of transcriptional activation, a transcriptional activator protein interacts directly with one or more protein components of the transcription complex, such as TFIID, a complex that includes the TATA-box–binding protein (TBP) and multiple TBP-associated factors (TAFs), or the Pol II holoenzyme. The fully assembled transcriptional apparatus in eukaryotes consists of a complex assemblage of TFIID, RNA polymerase II, and numerous other basal transcription factors. A prerequisite to formation of an active transcription complex in native chromatin is that the nucleosomes be modified or repositioned, a function of several chromatin-remodeling complexes. Some genes contain alternative promoters used in different tissues; other genes use a single promoter, but the transcripts are spliced in different ways. Alternative splicing can result in mRNA molecules that are translated with different efficiencies, or even in different proteins if there is alternative splicing of the protein-coding exons.

9.5 Gene expression can be affected by heritable chemical modifications in the DNA.

- Transcriptional inactivation is associated with heavy DNA methylation.
- In mammals, some genes are imprinted by methylation in the germ line.

In some organisms, epigenetic mechanisms of regulation also occur, in which the DNA is modified either chemically or by protein binding in such a way as to cause a genetically transmissible change in pattern of gene expression. Epigenetic transcriptional silencing of a gene is often associated with heavy cytosine methylation. Imprinting in mammals is an epigenetic, sex-specific modification of a limited number of genes that takes place in the female and male germ lines and that influences gene activity in embryogenesis. The imprints are erased early in germ-line development but then reestablished according to their sex-specific patterns.

9.6 Regulation also takes place at the levels of RNA processing and decay.

- The primary transcripts of many genes are alternatively spliced to yield different products.
- The coding capacity of the human genome is enlarged by extensive alternative splicing.
- Different messenger RNAs can differ in their persistence in the cell.
- RNA interference (RNAi) results in the destruction of RNA transcripts.

9.7 Regulation can also take place at the level of translation.

- Small regulatory RNAs can control translation by base-pairing with the messenger RNA.

9.8 Some developmental processes are controlled by programmed DNA rearrangements.

- Programmed deletions take place in the development of the mammalian immune system.
- Programmed transpositions take place in the regulation of yeast mating type.

Genetic regulatory mechanisms also occasionally include programmed DNA rearrangements. Examples include the generation of diversity in antibodies and T-cell receptors in mammals and the phenomenon of mating-type interconversion in yeast.

issues & ideas

- What is *positive regulation* of transcription? What is *negative regulation* of transcription? What is the role of the repressor in each case? Give an example of each type of regulation.

- What is autoregulation? Distinguish between positive and negative autoregulation. Which would be used to amplify a weak induction signal? Which would be used to prevent overproduction?

- What class of *lac* mutants demonstrated that the presence of lactose in the growth medium was not necessary for expression of the genes for lactose utilization?

- How does an operon result in coordinate control of the genes included? Are operons usually found in eukaryotic organisms?

- In what sense does attentuation provide a "fine-tuning" mechanism for operons that control amino acid biosynthesis?

- Explain how a small molecule can regulate transcription by means of a riboswitch.

- What is a transcriptional activator protein? A transcriptional enhancer? A chromatin-remodeling complex? What role do these elements play in eukaryotic gene regulation?

- How does the possibility of alternative splicing affect the generality of the statement that one gene encodes one polypeptide chain?

- What is meant by the term *epigenetic regulation?* Explain how epigenetic regulation can be mediated through cytosine methylation.

- What is the phenomenon of RNA interference (RNAi)? How is RNAi used in genetic analysis?

key terms & concepts

antisense RNA
aporepressor
attenuation
attenuator
autoregulation
basal transcription factors
β-galactosidase
cAMP–CRP complex
cassette
chromatin-remodeling complex (CRC)
cis-dominant
constant (C) antibody regions

constitutive mutant
coordinate regulation
co-repressor
DNA looping
DNA methylase
enhancer
epigenetic
general transcription factor
gene regulation
genomic imprinting
homothallism
housekeeping gene
inducer

inducible transcription
joining (J) antibody regions
lactose permease
leader polypeptide
mating-type interconversion
negative regulation
operator
operon
Pol II holoenzyme
positive regulation
promoter
repressible transcription

repressor
riboswitch
RNA interference (RNAi)
silencer
TATA-box–binding protein (TBP)
TBP-associated factors (TAFs)
transcriptional activator protein
transcription complex
variable (V) antibody regions
V–J joining

1. _____ General term for a gene whose product functions in general metabolic or maintenance activities that are shared among a wide variety of cell types.

2. _____ Type of transcriptional regulation in which the default state of transcription is "off" unless a protein or proteins turns it "on."

3. _____ Transcriptional regulatory protein that must bind with a small-molecule co-repressor in order to adopt a conformation that can bind with its target sequence in the DNA.

4. _____ In prokaryotes, a group of adjacent genes transcribed into a single transcript and controlled by an operator.

5. _____ Type of mutant in which transcription takes place continuously.

6. _____ Mechanism of fine-tuning the regulation of many operons for the synthesis of amino acids, in which an initial transcript is terminated if there is enough charged tRNA available to translate a small leader polypeptide containing two or more codons for the amino acid.

7. _____ Mechanism in which termination of transcription is controlled by whether or not the leader sequence of the transcript binds with a small molecule.

8. _____ Region of a gene containing the TATA box to which the transcription complex is recruited.

9. _____ A eukaryotic regulatory DNA sequence able to function in either orientation and at many locations in or near a gene, which acts to help recruit transcriptional activator proteins to the gene.

10. _____ Refers to heritable changes in gene expression resulting from DNA methylation, chromatin conformation, or other modifications that are not permanent.

11. _____ An RNA molecule capable of base pairing with all or part of an RNA transcript.

12. _____ Process in which small fragments produced by the degradation of double-stranded RNA are used to identify RNA transcripts to be destroyed by cleavage.

solutions: step by step

Problem 1

The permease of *E. coli* that transports the α-galactoside melibiose can also transport lactose, but it is temperature-sensitive: Lactose can be transported into the cell at 30°C but not at 37°C. In a strain that produces the melibiose permease constitutively, what are the Lac phenotypes of a *lacZ⁻* and a *lacY⁻* mutant at 30°C and at 37°C?

■ **Solution** The *lacZ⁻* mutant cannot grow on lactose even if it is able to transport lactose into the cell; hence *lacZ⁻* mutants are Lac⁻ at both 30°C and 37°C. The *lacY⁻* mutant can grow on lactose (because it is *lacZ⁺*), but only if it can transport lactose into the cell. In a strain that produces the melibiose permease constitutively, the *lacY⁻* mutant is Lac⁺ at 30°C because it can grow on lactose owing to the ability of the melibiose permease to transport lactose at this temperature. But the *lacY⁻* mutant is Lac⁻ at 37°C because at this temperature, the melibiose permease is nonfunctional.

Problem 2

The arabinose operon (*ara*) in *E. coli* encodes the enzymes necessary for degradation of the 5-carbon sugar arabinose. It is one of the few degradative operons in *E. coli* that are controlled primarily by positive regulation of transcription. The protein AraC, encoded in the gene *araC,* works as follows. In the absence of arabinose, AraC binds to the operator and an initiator region and, by causing formation of a DNA loop, prevents transcription. In the presence of arabinose, AraC has a high affinity only for the initiator, where it binds and induces transcription of the operon in a polycistronic mRNA for the degradative enzymes AraB, AraA, and AraD (collectively called AraBAD). The operon is also positively regulated by the cyclic AMP receptor protein (CRP), in much the way the *lac* operon is regulated by CRP. Fill each box in the table shown here with a + or a − sign. A + (or −) in the column for AraC or CRP means that under the specified growth condition, the AraC or CRP protein is (or is not) bound with the arabinose initiator; a + (or −) in the column for AraBAD means that under the specified growth condition, the AraBAD proteins are (or are not) induced.

	Initiator binding only		Enzyme induction
Growth medium	**AraC**	**CRP**	**AraBAD**
Lactose	☐	☐	☐
Arabinose	☐	☐	☐
Arabinose + glucose	☐	☐	☐

■ **Solution** According to the description of how the operon is regulated, AraC binds only to the initiator when arabinose is present in the medium. The binding of AraC will induce transcription of the AraBAD polycistronic mRNA, but AraC alone is not sufficient. As in the *lac* operon, transcription can occur only if the cAMP–CRP complex is also bound with the operator. The cAMP–CRP complex will form when cells have a high level of cyclic AMP, such as when growing in lactose or arabinose medium, but not when the level of cyclic AMP is low, such as under growth in glucose medium. The transcription of the AraBAD mRNA results in the synthesis of the AraBAD enzymes. These considerations indicate how the table should be completed.

	Initiator binding only		Enzyme induction
Growth medium	**AraC**	**CRP**	**AraBAD**
Lactose	−	+	−
Arabinose	+	+	+
Arabinose + glucose	+	−	−

concepts in action: problems for solution

9.1 Is it inevitable that once mRNA for a protein is made, the protein will be synthesized? What sort of mechanisms might control the translational process?

9.2 Is it necessary for the gene that codes for the repressor of a bacterial operon to be near the structural genes? Why or why not?

9.3 Why are mutations of the *lac* operator often called *cis*-dominant? Why are some constitutive mutations of the *lac* repressor (*lacI*) called *trans*-recessive? Can you think of a way in which a noninducible mutation in the *lacI* gene might be *trans*-dominant?

9.4 Why is the *lac* operon of *E. coli* not inducible in the presence of glucose?

9.5 A mutation imparting constitutive synthesis of an enzyme of arginine biosynthesis in *Citrobacteria* was found. The enzyme is normally repressible by arginine.
(a) What two kinds of regulatory mutations might cause this phenotype?
(b) What kind of mutation in an enzyme of arginine biosynthesis might cause this phenotype?

9.6 If a wildtype *E. coli* strain is grown in a medium without lactose or glucose, how many proteins are bound to the *lac* operon? How many are bound if the cells are grown in glucose?

9.7 Among mammals, the reticulocyte cells in the bone marrow lose their nuclei in the process of differentiation into red blood cells. Yet the reticulocytes and red blood cells continue to synthesize hemoglobin. Suggest a mechanism by which hemoglobin synthesis can continue for a long period of time in the absence of the hemoglobin genes.

9.8 Consider a eukaryotic transcriptional activator protein that binds to an enhancer sequence and promotes transcription. What change in regulation would you expect from a duplication in which several copies of the enhancer were present instead of just one?

9.9 In regard to mating-type switching in yeast, what phenotype would you expect from a type α cell that has a deletion of the *HMR**a*** cassette?

9.10 Cells of genotype *lacI⁻ lacO⁺ lacZ⁺ lacY⁺* Hfr are mated with cells of genotype *lacI⁺ lacO⁺ lacZ⁺ lacY⁺* F⁻. In the absence of any inducer in the medium, no β-galactosidase is made. However, when the *lacI⁺ lacO⁺ lacZ⁺ lacY⁺* Hfr strain is mated with a strain of genotype *lacI⁻ lacO⁺ lacZ⁻ lacY⁻* F⁻ under the same conditions, β-galactosidase is synthesized for a short time after the Hfr and F⁻ cells have been mixed. Explain this observation.

9.11 When glucose is present in an *E. coli* cell, is the concentration of cyclic AMP high or low? Can a mutant with either an inactive *adenyl cyclase* gene or an inactive *crp* gene synthesize β-galactosidase? Does the binding of cAMP–CRP to DNA affect the binding of a repressor?

9.12 Would you expect the regulation of a gene to be affected by an inversion of a promoter sequence? By an inversion of an enhancer sequence? Explain your answers.

9.13 The leader sequence of the *metI* transcript involved in methionine biosynthesis in *Bacillus subtilis* acts as a riboswitch that binds S-adenosylmethionine (SAM) in a manner analogous to the *yitJ* leader RNA. What phenotype would be expected of cells with a mutation in *metI* in which the leader was unable to bind with SAM?

9.14 Imagine a bacterial species in which the methionine operon is regulated only by an attenuator and there is no repressor. In its mode of operation, the methionine attenuator is exactly analogous to the *trp* attenuator of *E. coli*. The relevant portion of the attenuator sequence in the RNA is

5′ - AAA<u>A</u>UGAUGAUGAUGAUGAUGAUGAUGGACUAA - 3′

The translation start site is located upstream from this sequence, and the region shown is in the correct reading frame. What phenotype (constitutive, wildtype, or Met⁻) would you expect for each of the types of mutant RNA below? Explain your reasoning.
(a) The red A is deleted.
(b) Both the red A and the underlined A are deleted.
(c) The first three As in the sequence are deleted.

9.15 A frameshift mutation occurs near the end of an exon. Does it affect the reading frame of the next exon in the processed mRNA? Explain your answer.

9.16 Two genotypes of *E. coli* are grown and assayed for levels of the enzymes of the *lac* operon. Using the information provided in the accompanying table for these two genotypes, predict the enzyme levels for the other genotypes listed in parts (a) through (d). The levels of activity are expressed in arbitrary units relative to those observed under the induced conditions.

	Uninduced level		Induced level	
Genotype	**Z**	**Y**	**Z**	**Y**
I⁺ O⁺ Z⁺ Y⁺	0.1	0.1	100	100
I⁺ Oᶜ Z⁺ Y⁺	25	25	100	100

(a) *I⁻ O⁺ Z⁺ Y⁺*
(b) F′ *I⁺ O⁺ Z⁻ Y⁻* / *I⁻ O⁺ Z⁺ Y⁺*
(c) F′ *I⁺ O⁺ Z⁻ Y⁻* / *I⁺ Oᶜ Z⁺ Y⁺*
(d) F′ *I⁺ O⁺ Z⁻ Y⁻* / *I⁻ Oᶜ Z⁺ Y⁺*

9.17 A mutant of *E. coli* is isolated that has a defective ribosomal protein such that translation stalls briefly whenever a codon for tryptophan is encountered, irrespective of the level of charged tryotophan tRNA. How would this mutation be expected to affect attenuation of the tryptophan operon?

9.18 Temperature-sensitive mutations in the *lacI* gene of *E. coli* render the repressor nonfunctional (unable to bind the operator) at 42°C but leave it fully functional at 30°C. Would β-galactosidase be expected to be produced:
(a) In the presence of lactose at 30°C?
(b) In the presence of lactose at 42°C?
(c) In the absence of lactose at 30°C?
(d) In the absence of lactose at 42°C?

9.19 The accompanying illustration shows a primary RNA transcript containing 6 exons, indicated by the rectangles labeled A–F. How many different protein products could result from alternative splicing to produce mRNAs that contain four or more of the exons?

9.20 Shown here is a primary RNA transcript with three exons (A, B, and C), which is alternatively processed in two ways to yield either an mRNA of sequence A–B or an mRNA of sequence B–C. An organism is mutant unless it has a functional product from both mRNAs. The black lines represent loss-of-function mutations in the exons.

(a) Draw the complementation matrix for the six mutations, using + to indicate complementation (wildtype phenotype) and − to represent lack of complementation (mutant phenotype).
(b) Explain what is unusual about this complementation matrix.

further readings

Baylin, S. B., and J. G. Herman. 2000. DNA hypermethylation in tumorigenesis: Epigenetics joins genetics. *Trends in Genetics* 16: 168.

Cairns, B. R. 1998. Chromatin remodeling machines: Similar motors, ulterior motives. *Trends in Biochemical Sciences* 23: 20.

Cliften, P., P. Sudarsanam, A. Desikan, et al. 2003. Finding functional features in *Saccharomyces* genomes by phylogenetic footprinting. *Science* 301: 71.

Feil, R., and S. Khosla. 1999. Genomic imprinting in mammals: An interplay between chromatin and DNA methylation? *Trends in Genetics* 15: 431.

Fugmann, S. D., A. I. Lee, P. E. Shockett, I. J. Villey, and D. G. Schatz. 2000. The RAG proteins and V(D)J recombination: Complexes, ends, and transposition. *Annual Review of Immunology* 18: 495.

Gottesman, S. 2002. Stealth regulation: Biological circuits with small RNA switches. *Genes and Development* 16: 2829.

Gralla, J. D. 1996. Activation and repression of *E. coli* promoters. *Current Opinion in Genetics & Development* 6: 526.

Grunberg-Manago, M. 1999. Messenger RNA stability and its role in control of gene expression in bacteria and phages. *Annual Review of Genetics* 33: 193.

Haber, J. E. 1998. Mating-type gene switching in *Saccharomyces cerevisiae*. *Annual Review of Genetics* 32: 561.

Hampsey, M. 1998. Molecular genetics of the RNA polymerase II general transcriptional machinery. *Microbiology and Molecular Biology Reviews* 62: 465.

Kornberg, R. D. 1999. Eukaryotic transcriptional control. *Trends in Biochemical Sciences* 24: M46.

Mandal, M., B. Boese, J. E. Barrick, W. C. Winkler, and R. R. Breaker. 2003. Riboswitches control fundamental biochemical pathways in *Bacillus subtilis* and other bacteria. *Cell* 113: 577.

Neidhardt, F. C., R. Curtiss III, J. L. Ingraham, et al, eds. 1996. Escherichia coli *and* Salmonella typhimurium: *Cellular and Molecular Biology* (2 volumes). 2d ed. Washington, DC: American Society for Microbiology.

Novina, C. D., and A. L. Roy. 1996. Core promoters and transcriptional control. *Trends in Genetics* 12: 351.

Ptashne, M., and A. Gann. 1997. Transcriptional activation by recruitment. *Nature* 386: 569.

Struhl, K. 1995. Yeast transcriptional regulatory mechanisms. *Annual Review of Genetics* 29: 651.

Tijan, R. 1995. Molecular machines that control genes. *Scientific American,* February.

Tijsterman, M., R. F. Ketting, and R. H. A. Plasterk. 2002. The genetics of RNA silencing. *Annual Review Of Genetics* 36: 489.

geNETics on the web

GeNETics on the Web will introduce you to some of the most important sites for finding genetic information on the Internet. To explore these sites, visit the Jones and Bartlett home page at

http://www.jbpub.com/genetics

For the book *Essentials of Genetics: A Genomics Perspective,* choose the link that says **Enter GeNETics on the Web**. You will be presented with a chapter-by-chapter list of highlighted keywords. Select any highlighted keyword and you will be linked to a Web site containing genetic information related to the keyword.

- The goal of studying **gene regulation** is to be able to understand the DNA-protein and protein-protein interactions at the same level of detail as computer scientists understand the connections in microcircuits. Among the seminal ideas in regulation is the concept that DNA-binding proteins bind to DNA with positive cooperatively, which means that, as each subunit of a complex is bound with DNA, it facilitates the binding of the next subunit. Cooperative binding creates efficient "molecular switches" to control transcription. This keyword site includes an excellent nontechnical summary of some of the key advances in studies of gene regulation.

- This keyword database of **transcription factors** gives easy access to sequences and structural diagrams showing the key motifs of each of the numerous types and subtypes of DNA binding proteins involved in transcriptional regulation. View examples of each of the four major categories of transcription factors: those having basic domains (includes leucine zipper factors and helix-loop-helix factors), those having zinc-coordinating DNA-binding domains (includes all zinc-finger factors), those having helix-turn-helix motifs (includes all homeo domains), and those having beta-scaffold factors with minor groove contacts (includes the MADS box factor that figures prominently in plant development).

Biology as commodity. The tulip craze in Holland during the years 1633-1637 was a period of frenzied investment in which homes, estates, and factories were sold or mortgaged to purchase the rights to rare mutant tulip bulbs yielding plants with unusual colors or color patterns. The flower had been introduced into Holland from Turkey around 1550 and soon gained popularity, ultimately culminating in the tulip craze. The speculative bubble burst in 1637, throwing many investors into economic ruin. [© Photos.com]

key concepts

- In recombinant DNA (gene cloning), DNA fragments are isolated, inserted into suitable vector molecules, and introduced into host cells (usually bacteria or yeast), where they are replicated.

- Large-scale automated DNA sequencing has resulted in the complete sequence of the genomes of many species of bacteria, archaeons, and eukaryotes including the human genome.

- Functional genomics using DNA microarrays enables the level of gene expression of all genes in the genome to be assayed simultaneously, which allows global patterns and coordinated regulation of gene expression to be investigated.

- Two-hybrid analysis of proteins allows protein-protein interaction networks to be identified.

- Recombinant DNA is widely used in research, medical diagnostics, and the manufacture of drugs and other commercial products.

- Transgenic organisms carry DNA sequences that have been introduced by germ-line transformation or other methods.

10

Genomics, Proteomics, and Genetic Engineering

chapter organization

10.1 Cloning a DNA molecule takes place in several steps.

10.2 A genomic DNA sequence is like a book without an index, and identifying genes and their functions is a major challenge.

10.3 Genomics and proteomics reveal genome-wide patterns of gene expression and networks of protein interactions.

10.4 Reverse genetics creates an organism with a designed mutation.

10.5 Genetic engineering is applied in medicine, industry, agriculture, and research.

the human connection
The Human Genome Sequence

Chapter Summary
Issues & Ideas
Key Terms & Concepts
Solutions: Step by Step
Concepts in Action: Problems for Solution
Further Readings

geNETics on the web

Entering into the 21st century, many geneticists began to change their focus from genes to genomes. The change in perspective was made possible by large-scale DNA sequencing that revealed the complete genomic sequences of many organisms. The genomic sequencing was accompanied by advances in computational methods for recognizing genes and for comparing genomes. Soon methods were devised for identifying which genes in the genome were transcribed in particular tissue types, at specific times in development, or at different stages of the cell cycle. Going beyond transcription, methods were also devised for identifying which proteins in the cell interact physically with which other proteins. New terms were coined to emphasize the changes in perspective. The field of **genomics** deals with the DNA sequence, organization, function, and evolution of genomes. The counterpart at the level of proteins is **proteomics,** which aims to identify all the proteins in a cell or organism (including any posttranslationally modified forms), as well as their cellular localization, functions, and interactions.

Genomics was made possible by the invention of techniques originally devised for the manipulation of genes and the creation of genetically engineered organisms with novel genotypes and phenotypes. We refer to this approach as **recombinant DNA,** but it also goes by the names *gene cloning* or *genetic engineering*. The basic technique is quite simple: DNA is isolated and cut into fragments by one or more restriction enzymes; then the fragments are joined together in a new combination and introduced back into a cell or organism to change its genotype in a directed, predetermined way. Such genetically engineered organisms are called **transgenic organisms.** Transgenics are often created for experimental studies, but an important application is the development of improved varieties of domesticated animals and crop plants, in which case a transgenic organism is often called a *genetically modified organism (GMO)*. Specific examples of genetically modified organisms are considered later in this chapter.

10.1

Cloning a DNA molecule takes place in several steps.

In genetic engineering, the immediate goal of an experiment is usually to insert a *particular* fragment of chromosomal DNA into a plasmid or a viral DNA molecule. This is accomplished by techniques for breaking DNA molecules at specific sites and for isolating particular DNA fragments.

■ Restriction enzymes cleave DNA into fragments with defined ends.

DNA fragments are usually obtained by the treatment of DNA samples with restriction enzymes. *Restriction enzymes* are nucleases that cleave DNA wherever it contains a particular short sequence of nucleotides that matches the *restriction site* of the enzyme (see Section 6.6). Most restriction sites consist of four or six nucleotides, within which the restriction enzyme makes two single-strand breaks, one in each strand, generating 3'-OH and 5'-P groups at each position. About a thousand restriction enzymes, nearly all with different restriction site specificities, have been isolated from microorganisms.

Most restriction sites are symmetrical in the sense that the sequence is identical in both strands of the DNA duplex. For example, the restriction enzyme *Eco*RI, isolated from *E. coli,* has the restriction site 5'-GAATTC-3'; the sequence of the other strand is 3'-CTTAAG-5', which is identical but written with the 3' end at the left. *Eco*RI cuts each strand between the G and the A. The term *palindrome* is used to denote this type of symmetrical sequence.

Soon after restriction enzymes were discovered, observations with the electron microscope indicated that the fragments produced by many restriction enzymes could spontaneously form circles. The circles could be made linear again by heating. On the other hand, if the circles that formed spontaneously were treated with DNA ligase, which joins 3'-OH and 5'-P groups, then they could no longer be made linear with heat because the ends were covalently linked by the DNA ligase. This observation was the first evidence for three important features of restriction enzymes:

- Restriction enzymes cleave DNA molecules in palindromic sequences.
- The breaks need not be directly opposite one another in the two DNA strands.
- Enzymes that cleave the DNA strands asymmetrically generate DNA fragments with complementary ends.

These properties are illustrated for *Eco*RI in Figure 10.1 .

Most restriction enzymes are like *Eco*RI in that they make staggered cuts in the DNA strands, producing single-stranded ends called **sticky ends** that can adhere to each other because they contain complementary nucleotide sequences. Some restriction enzymes (such as *Eco*RI) leave a single-stranded overhang at the 5' end (Figure 10.2 , part A); others leave a 3' overhang. A number of restriction enzymes cleave both DNA strands at the center of symmetry, forming **blunt ends.** Part B of Figure 10.2 shows the blunt ends produced by the enzyme *Bal*I. Blunt ends also can be ligated by DNA ligase.

However, whereas ligation of sticky ends re-creates the original restriction site, any blunt end can join with any other blunt end and not necessarily create a restriction site.

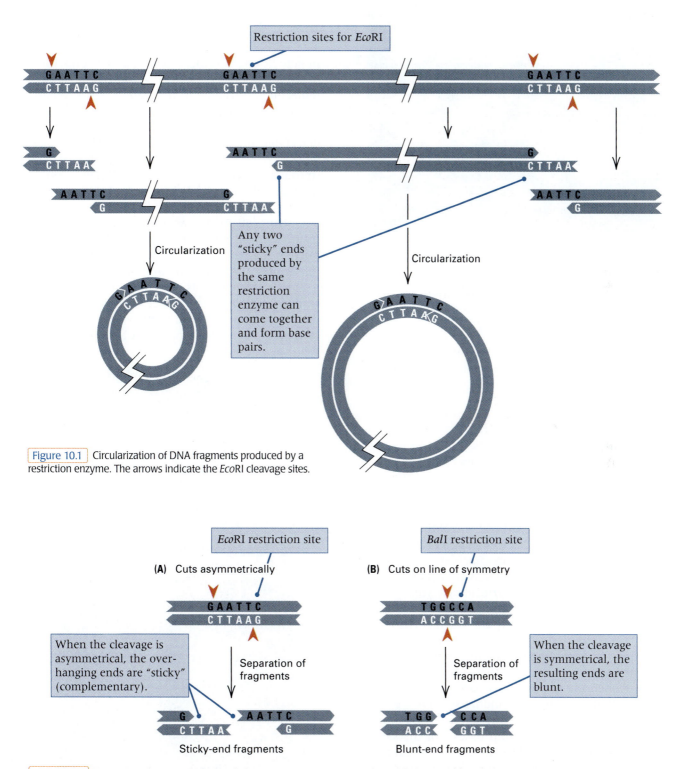

Figure 10.1 Circularization of DNA fragments produced by a restriction enzyme. The arrows indicate the *Eco*RI cleavage sites.

Figure 10.2 Two types of cuts made by restriction enzymes. The red arrowheads indicate the cleavage sites. (A) Cuts made in each strand at an equal distance from the center of symmetry of the restriction site. (B) Cuts made in each strand at the center of symmetry of the restriction site.

The seas of the world are teeming with microorganisms that cannot yet be cultured in the laboratory. Nevertheless, many of their genomic DNA sequences have been determined from bulk extractions of DNA from sea water. The genomic DNA sequences indicate a high diversity of very ancient organisms. Their inability to be cultured indicates highly specialized growth requirements.

Most restriction enzymes recognize their restriction sequence without regard to the source of the DNA. Thus,

key concept

Restriction fragments of DNA obtained from one organism have the same sticky ends as restriction fragments from another organism if they were produced by the same restriction enzyme.

This principle will be seen to be one of the foundations of recombinant DNA technology.

Because most restriction enzymes recognize a unique sequence, the number of cuts made in the DNA of an organism by a particular enzyme is limited. For example, an *E. coli* DNA molecule contains 4.6×10^6 base pairs, and any enzyme that cleaves a six-base restriction site will cut the molecule into about a thousand fragments. This number of fragments follows from the fact that any particular six-base sequence (including a six-base restriction site) is expected to occur in a random sequence every $4^6 = 4096$ base pairs, on the average, assuming equal frequencies of the four bases. For the same reason, mammalian nuclear DNA would be cut into about a million fragments. These large numbers are still small compared with the number that would be produced if breakage occurred at completely random sequences. Of special interest are the smaller DNA molecules, such as viral or plasmid

DNA, which may have from only one to ten sites of cutting (or even none) for particular enzymes. Plasmids that contain a single site for a particular enzyme are especially valuable, as we will see shortly.

■ Restriction fragments are joined end to end to produce recombinant DNA.

In genetic engineering, a particular DNA fragment of interest is joined to a *vector*, a relatively small DNA molecule that is able to replicate inside a cell and that usually contains one or more sequences able to confer antibiotic resistance (or some other detectable phenotype) on the cell. The simplest types of vectors are plasmids whose DNA is double-stranded and circular (Figure 10.3). When the DNA fragment of interest has been joined to the vector, the recombinant molecule is introduced into a cell by means of DNA transformation (Figure 10.4). Inside the cell, the recombinant molecule is replicated as the cell replicates its own DNA, and as the cell divides, the recombinant molecule is transmitted to the progeny cells. When a transformant containing the recombinant molecule has been isolated, the DNA fragment linked to the vector is said to be **cloned**. A **vector** is therefore a DNA molecule into

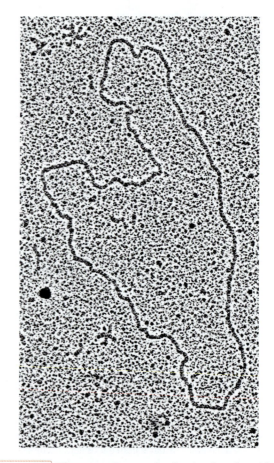

Figure 10.3 Electron micrograph of a circular plasmid used as a vector for cloning in *E. coli*.

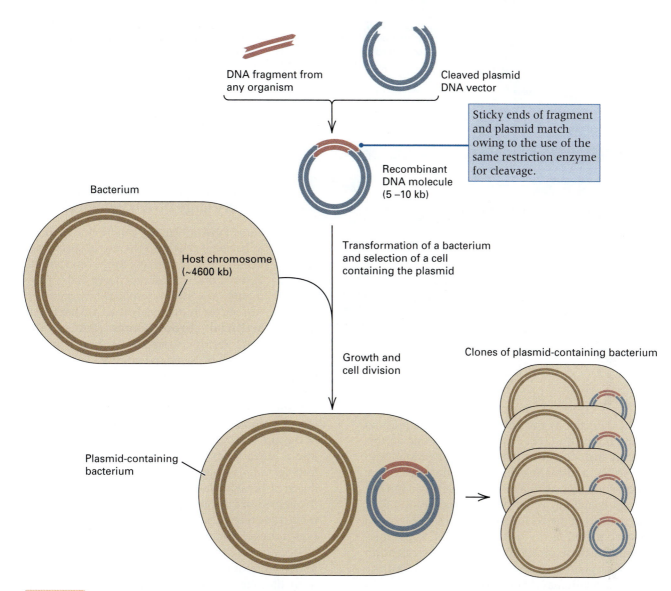

DNA fragment from any organism

Cleaved plasmid DNA vector

Sticky ends of fragment and plasmid match owing to the use of the same restriction enzyme for cleavage.

Recombinant DNA molecule (5–10 kb)

Bacterium

Host chromosome (~4600 kb)

Transformation of a bacterium and selection of a cell containing the plasmid

Growth and cell division

Clones of plasmid-containing bacterium

Plasmid-containing bacterium

Figure 10.4 An example of cloning. A fragment of DNA from any organism is joined to a cleaved plasmid. The recombinant plasmid is then used to transform a bacterial cell, where the recombinant plasmid is replicated and transmitted to the progeny bacteria. The bacterial host chromosome is not drawn to scale. It is typically about 1000 times larger than the plasmid.

which another DNA fragment can be cloned; it is a carrier for recombinant DNA. In the following sections, several types of vectors are described.

■ A vector is a carrier for recombinant DNA.

The most generally useful vectors have three properties:

1. The vector DNA can be introduced into a host cell relatively easily.

2. The vector contains a replication origin and so can replicate inside the host cell.

3. Cells containing the vector can usually be selected by a straightforward assay, most conveniently by allowing growth of the host cell on a solid selective medium.

The vectors most commonly used in *E. coli* are plasmids and derivatives of the bacteriophages λ and M13. Many other plasmids and viruses also have been developed for cloning into cells of animals, plants, and other bacteria. Recombinant DNA can be detected in host cells by means of genetic markers or phenotypic characteristics that are evident in the appearance of colonies or plaques. Plasmid and phage DNA can be introduced into cells by transformation, in which cells gain the ability to take up free DNA by exposure to a calcium chloride solution. Recombinant DNA can also be introduced into cells by a kind of electrophoretic procedure called *electroporation*. After introduction of the DNA, the cells that contain the recombinant DNA are plated on a solid medium. If the added DNA is a plasmid, colonies consisting of bacterial cells that contain the recombinant plasmid are formed, and the transformants can usu-

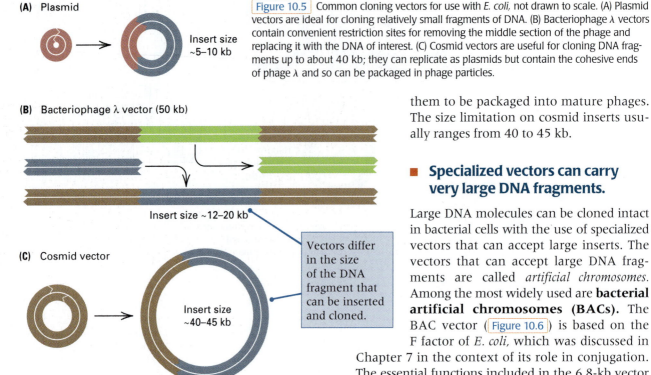

(A) Plasmid

Insert size ~5–10 kb

(B) Bacteriophage λ vector (50 kb)

Insert size ~12–20 kb

(C) Cosmid vector

Insert size ~40–45 kb

Vectors differ in the size of the DNA fragment that can be inserted and cloned.

Figure 10.5 Common cloning vectors for use with *E. coli,* not drawn to scale. (A) Plasmid vectors are ideal for cloning relatively small fragments of DNA. (B) Bacteriophage λ vectors contain convenient restriction sites for removing the middle section of the phage and replacing it with the DNA of interest. (C) Cosmid vectors are useful for cloning DNA fragments up to about 40 kb; they can replicate as plasmids but contain the cohesive ends of phage λ and so can be packaged in phage particles.

ally be detected by the phenotype that the plasmid confers on the host cell. For example, plasmid vectors typically include one or more genes for resistance to antibiotics, and plating the transformed cells on a selective medium with antibiotic prevents all but the plasmid-containing cells from growing (Section 8.2). Alternatively, if the vector is phage DNA, the infected cells are plated in the usual way to yield plaques. Variants of these procedures are used to transform animal or plant cells with suitable vectors, but the technical details may differ considerably.

Three types of vectors commonly used for cloning into *E. coli* are illustrated in Figure 10.5. Plasmids (part A) are most convenient for cloning relatively small DNA fragments (5 to 10 kb). Somewhat larger fragments can be cloned with bacteriophage λ (part B). The wildtype phage is approximately 50 kb in length, but the central portion of the genome is not essential for lytic growth and can be removed and replaced with donor DNA. After the donor DNA has been ligated in place, the recombinant DNA is packaged into mature phage *in vitro*, and the phage is used to infect bacterial cells. However, to be packaged into a phage head, the recombinant DNA must be neither too large nor too small, which means that the donor DNA must be roughly the same size as the portion of the λ genome that was removed. Most λ cloning vectors accept inserts ranging in size from 12 to 20 kb. Still larger DNA fragments can be inserted into cosmid vectors (part C). These vectors can exist as plasmids, but they also contain the complementary overhanging single-stranded ends of phage λ, which enables

them to be packaged into mature phages. The size limitation on cosmid inserts usually ranges from 40 to 45 kb.

■ Specialized vectors can carry very large DNA fragments.

Large DNA molecules can be cloned intact in bacterial cells with the use of specialized vectors that can accept large inserts. The vectors that can accept large DNA fragments are called *artificial chromosomes.* Among the most widely used are **bacterial artificial chromosomes (BACs).** The BAC vector (Figure 10.6) is based on the F factor of *E. coli,* which was discussed in Chapter 7 in the context of its role in conjugation. The essential functions included in the 6.8-kb vector are genes for replication (*repE* and *oriS*), for regulating copy number (*parA* and *parB*), and for resistance to the antibiotic chloramphenicol. BAC vectors with inserts greater than 300 kb can be maintained. DNA fragments in the appropriate size range can be produced by breaking larger molecules into fragments of the desired size by physical means, by treatment with restriction enzymes that have infrequent cleavage sites (for example, enzymes such as *Not*I and *Sfi*I), or by treatment with ordinary restriction enzymes under conditions in which only a fraction of the restriction sites are cleaved (*partial digestion*). Cloning the large molecules consists of mixing the large fragments of source DNA with the vector, ligation with

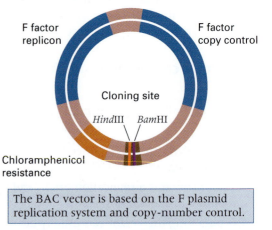

BAC, bacterial artificial chromosome

F factor replicon

F factor copy control

Cloning site

*Hind*III   *Bam*HI

Chloramphenicol resistance

The BAC vector is based on the F plasmid replication system and copy-number control.

Figure 10.6 Bacterial artificial chromosomes are based on a vector that contains the F factor as well as genes for regulating copy number.

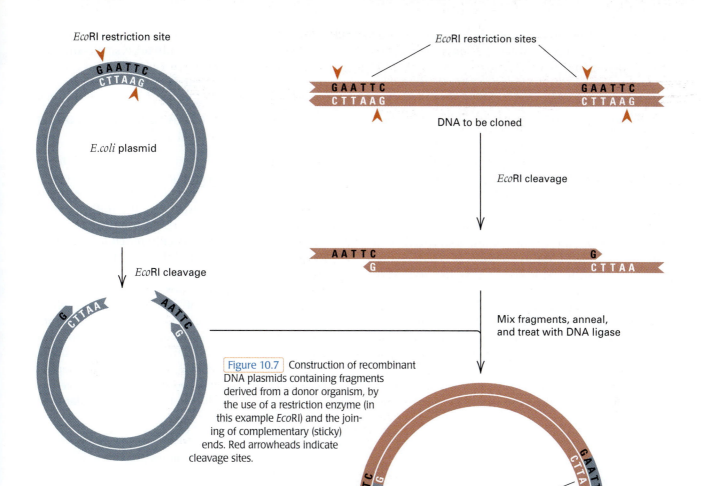

Figure 10.7 Construction of recombinant DNA plasmids containing fragments derived from a donor organism, by the use of a restriction enzyme (in this example EcoRI) and the joining of complementary (sticky) ends. Red arrowheads indicate cleavage sites.

DNA ligase, introduction of the recombinant molecules into cells of the host, and selection for the clones of interest. These methods are generally similar to those used for the production of recombinant molecules containing smaller inserts of cloned DNA.

BAC clones play a special role in genomic sequencing. Most genomes are sequenced in the form of many small single-stranded fragments, typically 500 to 1000 bp in length, cloned from random positions. This sequencing strategy is known as *shotgun sequencing,* and it needs to be carried out at high redundancy in order that the short sequence fragments can be recognized by their overlaps and assembled into a finished sequence. Each genomic region covered by overlapping clones is called a *contig*. Typically, a genomic sequence contains many gaps that prevent the contigs from being assembled. BAC clones are important because the sequences at the extreme ends of the cloned fragments give long-range information that allows adjacent contigs to be recognized and assembled in the correct orientation.

■ **Vector and target DNA fragments are joined with DNA ligase.**

The circularization of restriction fragments that have terminal single-stranded regions with complementary bases is illustrated in Figure 10.1. Because a particular restriction enzyme produces fragments with *identical* sticky ends, without regard for the source of the DNA, fragments from DNA molecules isolated from two different organisms can be joined, as shown in Figure 10.7. In this example, the restriction enzyme EcoRI is used to digest DNA from any organism of interest and to cleave a bacterial plasmid that contains only one EcoRI restriction site. The donor DNA is digested into many fragments (one of which is shown) and the plasmid into a single linear fragment. When the donor fragment and the linearized plasmid are mixed, recombinant molecules can form by base pairing between the complementary single-stranded ends. At this point, the DNA is treated with DNA ligase to seal the joints, and the donor fragment becomes permanently joined in a combination that may never have existed before. The ability to join a donor DNA fragment of interest to a vector is the basis of the recombinant DNA technology.

Joining sticky ends does not always produce a DNA sequence that has functional genes. For example, consider a linear DNA molecule that is cleaved into four fragments—A, B, C, and D—whose sequence in the original molecule was ABCD. Reassembly of the fragments can yield the original molecule, but because B and C have the same pair of sticky ends, molecules with the fragment arrangements ACBD and BADC can also form with the same probability as ABCD. Restriction fragments from the vector can also join together in the wrong order, but this potential problem can be eliminated by using a vector that has only one cleavage site for a particular restriction enzyme. Plasmids of this type are available (most have been created by genetic engineering). Many vectors contain unique sites for several different restriction enzymes, but generally only one enzyme is used at a time.

DNA molecules that lack sticky ends also can be joined. A direct method uses the DNA ligase made by *E. coli* phage T4. This enzyme differs from other DNA ligases in that it not only heals single-stranded breaks in double-stranded DNA but also can join molecules with blunt ends.

■ A recombinant cDNA contains the coding sequence of a eukaryotic gene.

Many genes in higher eukaryotes are very large. They can extend over hundreds of thousands of base pairs. Much of the length is made up of introns, which are excised from the mRNA in processing (Section 8.4). With such large genes, the length of the mRNA is usually much less than the length of the gene. Even if the large DNA sequence were cloned, expression of the gene product in bacterial cells would be impossible because bacterial cells are not capable of RNA splicing. Therefore, when a gene is so large that it is difficult to clone and express directly, it would be desirable to clone the coding sequence present in the mRNA to determine the base sequence and study the polypeptide gene product. The method illustrated in Figure 10.8 makes possible the direct cloning of any eukaryotic

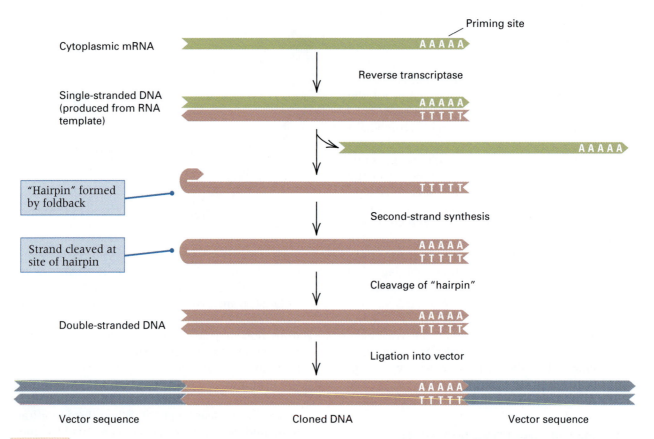

Figure 10.8 Reverse transcriptase produces a single-stranded DNA complementary in sequence to a template RNA. In this example, a cytoplasmic mRNA is copied. As indicated here, most eukaryotic mRNA molecules have a tract of consecutive A nucleotides at the 3' end, which serves as a convenient priming site. After the single-stranded DNA is produced, a foldback at the 3' end forms a hairpin that serves as a primer for second-strand synthesis. After the hairpin is cleaved, the resulting double-stranded DNA can be ligated into an appropriate vector either immediately or after PCR amplification. The resulting clone contains the entire coding region for the protein product of the gene.

coding sequence from cells in which the mRNA is present.

Cloning from mRNA molecules depends on an unusual polymerase, **reverse transcriptase,** which can use a single-stranded RNA molecule as a template and synthesize a complementary strand of DNA called **complementary DNA, or cDNA.** Like other DNA polymerases, reverse transcriptase requires a primer. The stretch of A nucleotides usually found at the 3′ end of eukaryotic mRNA serves as a convenient priming site, because the primer can be an oligonucleotide consisting of poly-T (Figure 10.8). Like any other single-stranded DNA molecule, the single strand of DNA produced from the RNA template can fold back upon itself at the extreme 3′ end to form a "hairpin" structure that includes a very short double-stranded region consisting of a few base pairs. The 3′ end of the hairpin serves as a primer for second-strand synthesis. The second strand can be synthesized either by DNA polymerase or by reverse transcriptase itself. Reverse transcriptase is the source of the second strand in RNA-based viruses that use reverse transcriptase, such as the human immunodeficiency virus (HIV). Conversion into a conventional double-stranded DNA molecule is achieved by cleavage of the hairpin by a nuclease.

In the reverse transcription of an mRNA molecule, the resulting full-length cDNA contains an uninterrupted coding sequence for the protein of interest. As we saw in Chapter 8, eukaryotic genes often contain DNA sequences, called *introns,* that are initially transcribed into RNA but are removed in the production of the mature mRNA. Because the introns are absent from the mRNA, the cDNA sequence is not identical with that in the genome of the original donor organism. However, if the purpose of forming the recombinant DNA molecule is to identify the coding sequence or to synthesize the gene product in a bacterial cell, then cDNA formed from processed mRNA is the material of choice for cloning. The joining of cDNA to a vector can be accomplished by available procedures for joining blunt-ended molecules (Figure 10.8).

Some specialized animal cells make only one protein, or a very small number of proteins, in large amounts. In these cells, the cytoplasm contains a great abundance of specific mRNA molecules, which constitute a large fraction of the total mRNA synthesized. An example is the mRNA for globin, which is highly abundant in reticulocytes while they are producing hemoglobin. The cDNA produced from purified mRNA from these cells is greatly enriched for the globin cDNA. Genes that are not highly expressed are represented by mRNA molecules whose abundance ranges from low to exceedingly low. The cDNA molecules produced from such rare RNAs will also be rare. The effi-

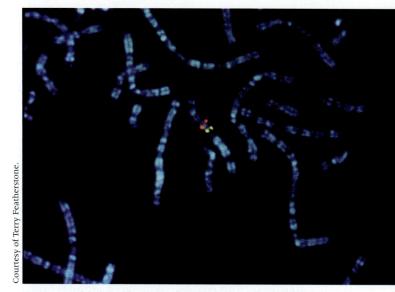

Courtesy of Terry Featherstone.

TWO CLONES, labeled with red or yellow fluorescent tags, hybridize to nearby sites on the same chromosome, but they clearly carry DNA from different parts of the chromosome.

ciency of cloning rare cDNA molecules can be markedly increased by PCR amplification prior to ligation into the vector. The only limitation on the procedure is the requirement that enough DNA sequence be known at both ends of the cDNA for appropriate oligonucleotide primers to be designed. PCR amplification of the cDNA produced by reverse transcriptase is called **reverse transcriptase PCR (RT-PCR).** The resulting amplified molecules contain the coding sequence of the gene of interest with very little contaminating DNA.

■ Loss of β-galactosidase activity is often used to detect recombinant vectors.

When a vector is cleaved by a restriction enzyme and renatured in the presence of many different restriction fragments from a particular organism, many types of molecules result, including such examples as a self-joined circular vector that has not acquired any fragments, a vector containing one or more fragments, and a molecule consisting only of many joined fragments. To facilitate the isolation of a vector containing a particular gene, some means is needed to ensure (1) that the vector does indeed possess an inserted DNA fragment, and (2) that the fragment is in fact the DNA segment of interest. This section describes several useful procedures for detecting the correct products.

In the use of transformation to introduce recombinant plasmids into bacterial cells, the initial goal is to isolate bacteria that contain the plasmid from a mixture of plasmid-free and plasmid-containing cells. A common procedure is to use a plasmid that possesses an antibiotic-resistance marker and to grow the transformed bacteria on a medium that contains the antibiotic: Only cells that contain plasmid can form a colony. An example of a cloning vector is the pBluescript plasmid illustrated in Figure 10.9 , part A. The entire plasmid is 2961 base pairs. Different regions contribute to its utility as a cloning vector.

- The plasmid origin of replication is derived from the *E. coli* plasmid ColE1. The ColE1 is a high-copy-number plasmid, and its origin of replication enables pBluescript and its recombinant derivatives to exist in approximately 300 copies per cell.

- The ampicillin-resistance gene allows for selection of transformed cells in medium containing ampicillin.

- The cloning site is called a *multiple cloning site* (MCS), or *polylinker,* because it contains unique cleavage sites for many different restriction enzymes and enables many types of restriction fragments to be inserted. In pBluescript, the MCS is a 108-bp sequence that contains cloning sites for 23 different restriction enzymes (Figure 10.9, part B).

- The detection of recombinant plasmids is by means of a region containing the *lacZ* gene from *E. coli,* shown in blue in Figure 10.9, part A. The basis of the selection is illustrated in Figure 10.10 . When the *lacZ* region is interrupted by a fragment of DNA inserted into the MCS, the recombinant plasmid yields Lac$^-$ cells because the interruption renders the lacZ region nonfunctional. Nonrecombinant plasmids do not contain a DNA fragment in the MCS and yield Lac$^+$ colonies. The Lac$^+$ and Lac$^-$ phenotypes can be distinguished by color when the cells are grown on a special β-galactoside compound called X-gal, which releases a deep blue dye when cleaved. On medium containing X-gal, Lac$^+$ colonies contain nonrecombinant plasmids and are a deep blue, whereas Lac$^-$ colonies contain recombinant plasmids and are white.

- The bacteriophage origin of replication is from the single-stranded DNA phage f1. When cells that contain a recombinant plasmid are infected with an f1 helper phage, the f1 origin enables a

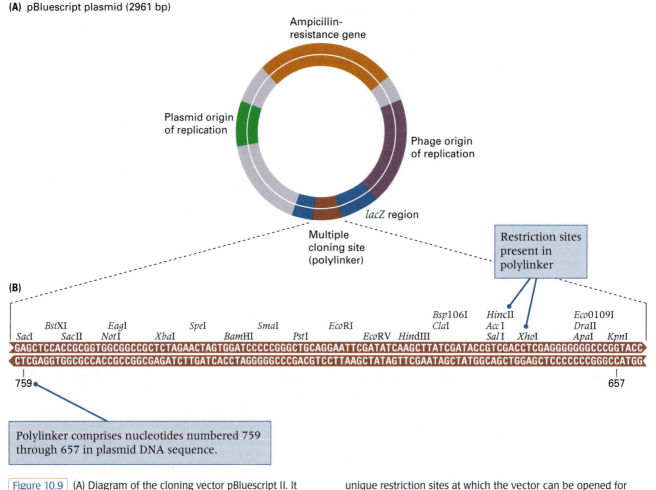

(A) pBluescript plasmid (2961 bp)

Ampicillin-resistance gene

Plasmid origin of replication

Phage origin of replication

lacZ region

Multiple cloning site (polylinker)

Restriction sites present in polylinker

(B)

Polylinker comprises nucleotides numbered 759 through 657 in plasmid DNA sequence.

Figure 10.9 (A) Diagram of the cloning vector pBluescript II. It contains a plasmid origin of replication, an ampicillin-resistance gene, a multiple cloning site (polylinker) within a fragment of the *lacZ* gene from *E. coli,* and a bacteriophage origin of replication. (B) Sequence of the multiple cloning site showing the unique restriction sites at which the vector can be opened for the insertion of DNA fragments. The numbers 657 and 759 refer to the position of the base pairs in the complete sequence of pBluescript. [Courtesy of Stratagene Cloning Systems, La Jolla, CA.]

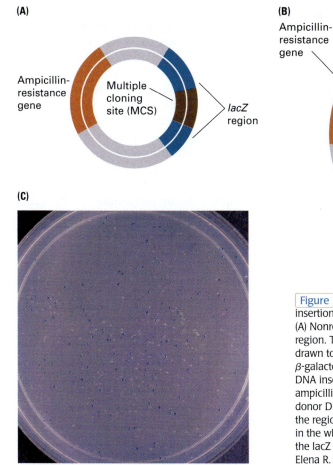

(A) Ampicillin-resistance gene — Multiple cloning site (MCS) — *lacZ* region

(B) Ampicillin-resistance gene — Inactive interrupted *lacZ* region + MCS — Inserted DNA

(C)

Figure 10.10 Detection of recombinant plasmids through insertional inactivation of a fragment of the *lacZ* gene from *E. coli.* (A) Nonrecombinant plasmid containing an uninterrupted *lacZ* region. The multiple cloning site (MCS) within the region (not drawn to scale) is sufficiently small that the plasmid still confers β-galactosidase activity. (B) Recombinant plasmid with donor DNA inserted into the multiple cloning site. This plasmid confers ampicillin resistance but not β-galactosidase activity, because the donor DNA interrupting the *lacZ* region is large enough to render the region nonfunctional. (C) Transformed bacterial colonies. Cells in the white colonies contain plasmids with inserts that disrupt the lacZ region; those in the blue colonies do not. [Courtesy of Elena R. Lozovsky.]

single strand of the inserted fragment, starting with *lacZ*, to be packaged in progeny phage. This feature is very convenient because it yields single-stranded DNA for sequencing. The plasmid shown in part A of Figure 10.9 is the SK(+) variety. There is also an SK(−) variety in which the f1 origin is in the opposite orientation and packages the complementary DNA strand.

All good cloning vectors have an efficient origin of replication, at least one unique cloning site for the insertion of DNA fragments, and a second gene whose interruption by inserted DNA yields a phenotype indicative of a recombinant plasmid. Once a **library,** or large set of clones, has been obtained in a particular vector, the next problem is how to identify the particular recombinant clones that contain the gene of interest.

Q A Moment to Think

Problem: Presence of a polylinker, or multiple cloning site (MCS), in a vector makes possible *directional cloning.* In this approach, the vector and the target sequence are both cleaved with the same two restriction enzymes, which are chosen so that their complementary sticky ends will ensure that the fragment of interest is inserted in a particular orientation in the vector. Consider, for example, the restriction sites in the vector MCS and the target sequence shown in the accompanying diagram. The vector sequence X to the left of the MCS is a promoter sequence, and the vector sequence to the right of the MCS is a transcriptional terminator. The target sequence is a protein-coding region that, in order to be expressed, must be oriented

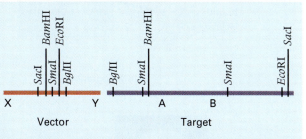

with A adjacent to, and to the right of, the promoter X. A geneticist therefore wants to create a recombinant molecule with the sequences oriented as sequence X−A−B−Y. The restriction enzymes *Sac*I, *Bam*HI, *Bgl*II, and *Eco*RI all produce sticky ends; *Sma*I produces blunt ends.
(a) If the vector and target were both digested with *Sma*I and the resulting fragments ligated in such a way that each vector molecule was ligated with one and only one target fragment, in what orientation would the A−B fragment be ligated into the polylinker?
(b) For directional cloning, what restriction enzyme (or enzymes) would you use to digest the vector and target so that after mixing of the fragments and ligation, the cloned DNA would have the sequence X−A−B−Y? (The answer can be found on page 370.)

10.1 Cloning a DNA molecule takes place in several steps   **367**

■ Recombinant clones are often identified by hybridization with labeled probe.

Genomic DNA or cDNA clones are often used to isolate additional clones containing DNA fragments with which they have sequences in common. This section discusses procedures that allow detection of the presence of any clone that contains a DNA fragment for which a complementary DNA or RNA sequence, labeled with radioactivity or by some

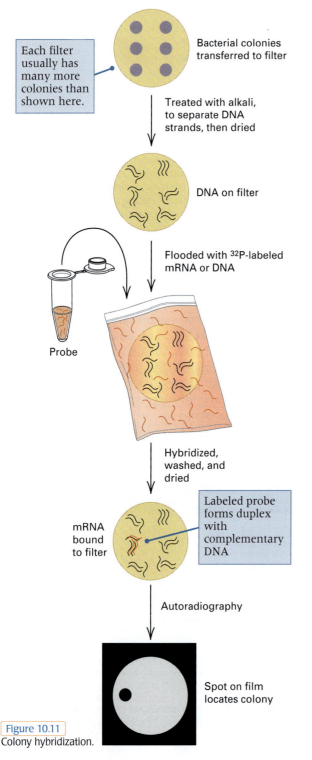

Each filter usually has many more colonies than shown here.

Bacterial colonies transferred to filter

Treated with alkali, to separate DNA strands, then dried

DNA on filter

Flooded with ^{32}P-labeled mRNA or DNA

Probe

Hybridized, washed, and dried

mRNA bound to filter

Labeled probe forms duplex with complementary DNA

Autoradiography

Spot on film locates colony

Figure 10.11
Colony hybridization.

other means, is available. The labeled nucleic acid used to detect the recombinant clones of interest is called the *probe*. A typical example is the use of a probe from a clone of cDNA to identify clones of genomic DNA that include parts of the coding sequence. Another typical example is the use of a probe from a cloned gene from one organism (for example, yeast or *Drosophila*) to identify clones that contain homologous DNA from another organism (for example, mouse or human being).

The procedure of **colony hybridization** makes it possible to detect the presence of any gene for which DNA or RNA labeled with radioactivity or some other means is available (Figure 10.11). Colonies to be tested are transferred (*lifted*) from a solid medium onto a nitrocellulose or nylon filter by gently pressing the filter onto the surface. A part of each colony remains on the agar medium, which constitutes the reference plate. The filter is treated with sodium hydroxide (NaOH), which simultaneously breaks open the cells and separates (*denatures*) the duplex DNA into single strands. The filter is then saturated with labeled probe complementary in base sequence to the gene being sought, and the DNA strands are allowed to form duplex molecules again (*renatured*). After washing to remove unbound probe, the positions of the bound probe identify the desired colonies. For example, with radioactively labeled probe, the desired colonies are located by means of autoradiography. A similar assay is done with phage vectors, but in this case plaques rather than colonies are lifted onto the filters.

If transformed cells can synthesize the protein product of a cloned gene or cDNA, then immunological techniques may also allow the protein-producing colony to be identified. In one method, the colonies are transferred as in colony hybridization, and the transferred copies are exposed to a labeled antibody directed against the particular protein. Colonies to which the antibody adheres are those that contain the gene of interest.

10.2

A genomic sequence is like a book without an index, and identifying genes and their functions is a major challenge.

Rapid automated DNA sequencing was instrumental in the success of the **Human Genome Project,** an international effort begun in 1990 to sequence the human genome and that of a number of model genetic organisms. The project achieved and surpassed its sequencing goals, as evidenced by the list shown in **Table 10.1**, which includes only a small sample of the organisms whose genomes have been completed. Genome sequences have also been

Table 10.1

Some examples of genomic sequencing

Domain	Organism	%(A + T)	Size (Mbp)	Comments/disease associations
Eubacteria				
Gamma (Enterobacteriaceae)	*Escherichia coli*	49	4.64	Normal intestinal flora
Gamma (Pasteurellaceae)	*Haemophilus influenzae*	62	1.83	Meningitis
Gamma (Pseudomonas group)	*Pseudomonas aeruginosa*	33	6.26	Respiratory infections
Gamma (Vibrionaceae)	*Vibrio cholerae*	53	4	Cholera
Epsilon subdivision	*Helicobacter pylori*	61	1.67	Duodenal ulcers
	Campylobacter jejuni	69	1.64	Gastroenteritis (food poisoning)
Alpha subdivision	*Rickettsia prowazekii*	71	1.11	Brill's disease and European epidemic typhus
Beta subdivision	*Neisseria meningitidis*	48	2.18	Meningitis
Bacillus/Clostridium group	*Bacillus subtilis*	56	4.20	Normal soil bacterium
	Mycoplasma genitalium	68	0.58	Smallest cellular genome
	Mycoplasma pneumoniae	60	0.82	Lung infections
	Ureaplasma urealyticum	75	0.75	Urethral infections
	Clostridium acetobutylicum	69	3.94	Food poisoning
	Streptococcus pyrogenes	61	1.85	Post-partum infections (puerperal fever)
Actinobacteria	*Mycobacterium tuberculosis*	34	4.41	Tuberculosis
Spirochaetales	*Treponema pallidum*	47	1.14	Syphilis
	Borrelia burgdorferi	71	0.91	Lyme disease
Chlamydia group	*Chlamydia pneumoniae*	59	1.23	Mild form of pneumonia
	Chlamydia trachomatis	59	1.04	Sexually transmitted
Aquificales	*Aquifex aeolicus*	57	1.55	Heat-tolerant, oxidizes hydrogen
Cyanobacteria	*Synechocystis* sp.	52	3.57	"Blue-green algae"
Deinococcus/Thermus	*Deinococcus radiodurans*	34	3.61	Radiation-resistant
Thermatogales	*Thermatoga maritima*	54	1.86	Anaerobic, hot marine mud
Archaea				
Euryarchaeota	*Methanococcus jannaschii*	69	1.66	Marine hydrothermal vents
	Archaeoglobus fulgidus	51	2.18	Marine hydrothermal sediment
	Pyrococcus abyssi	55	1.77	Extreme thermophile
	Pyrococcus furiosus	59	1.75	Extreme thermophile
	Pyrococcus horikoshii	58	1.74	Extreme thermophile
Crenarchaeota	*Aeropyrum pernix*	44	1.67	Marine sulfur vents
	Sulfolobus solfataricus	64	2.25	Yellowstone, USA, sulfur hot springs
Eukaryota				
Plant	*Arabidopsis thaliana*	65	125	Wall cress
	Oryza sativa	57	400	Rice
Fungi	*Saccharomyces cerevisiae*	62	12.07	Budding yeast
Protozoa	*Leishmania major*	37	33.6	Leishmaniasis
	Plasmodium falciparum	80	23	Malaria
Nematode	*Caenorhabditis elegans*	64	100	Soil organism
Insect	*Drosophila melanogaster*	57	165	Fermenting fruits
Rodent	*Mus musculus*	60	2500	Mouse
Primate	*Homo sapiens*	59	3000	Ubiquitous terrestrial biped

Source: Data courtesy of David Ussery, Peder Worning, Hans Henrik Stærfeldt, and Lars Juhl Jensen, Center for Biological Sequence Analysis, Department of Biotechnology, The Technical University of Denmark.

determined for numerous viruses, mitochondrial DNAs, and chloroplast DNAs, and a great deal of genomic sequencing activity is still underway.

On the other hand, genomic sequences are not self-explanatory. A genomic sequence has been called a book without an index. It is actually much worse. It is a book without spaces or punctuation, printed in an alphabet with only four letters. Especially for large, complex genomes in which much of the DNA does not code for proteins, and in which most protein-coding exons are relatively small and interrupted by large introns, it is a daunting challenge to parse a genomic sequence into its protein-coding exons, identify which protein-coding exons belong to the same gene, and recognize the upstream and downstream regulatory regions that control gene expression. The annotation of genomic sequences at this level is one aspect

Figure 10.12 Genes in the genome of *Mycoplasma genitalium* classified by function. [Data from C. M. Fraser, J. D. Gocayne, O. White, M. D. Adams, R. A. Clayton, R. D. Fleischmann, and 23 other authors. 1995. *Science* 270: 397.]

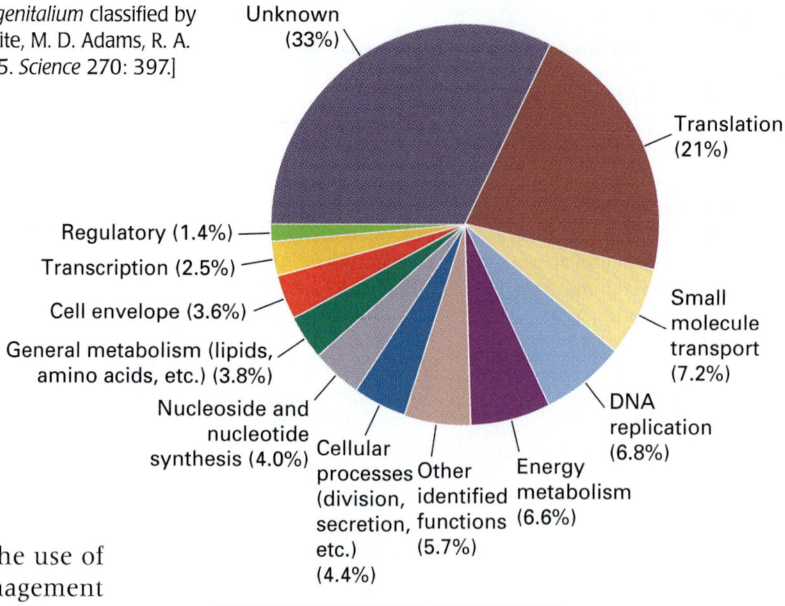

Unknown (33%)

Translation (21%)

Regulatory (1.4%)

Transcription (2.5%)

Cell envelope (3.6%)

General metabolism (lipids, amino acids, etc.) (3.8%)

Nucleoside and nucleotide synthesis (4.0%)

Cellular processes (division, secretion, etc.) (4.4%)

Other identified functions (5.7%)

Energy metabolism (6.6%)

DNA replication (6.8%)

Small molecule transport (7.2%)

Functions of 471 genes in the bacterium *Mycoplasma genitalium*

of **bioinformatics,** defined broadly as the use of computers in the interpretation and management of biological data.

Small, compact genomes are usually easier to annotate than more complex genomes. For example, the bacterium *Mycoplasma genitalium* is thought to have the smallest genome of any free-living organism—a circular DNA molecule 580 kb in length that includes only 471 genes. The organism is a parasite associated with ciliated epithelial cells of the genital and respiratory tracts of primates, including human beings. It belongs to a large group of bacteria (the mycoplasmas) that lack a cell wall and that parasitize a wide range of plant and animal hosts. Analysis of this genome enables us to identify what is probably a minimal set of genes necessary for a free-living cell.

The cellular processes in which the gene products of *M. genitalium* participate are summarized in Figure 10.12 . A substantial fraction of the genome is devoted to the synthesis of macromolecules such as DNA, RNA, and protein; another substantial fraction supports cellular processes and energy metabolism. There are very few genes for biosynthesis of small molecules. However, genes that encode proteins for salvaging and/or for transporting small molecules make up a significant fraction of the total, which underscores the fact that the bacterium is parasitic. The remaining genes are largely devoted to forming the cellular envelope and to helping the organism evade the immune system of the host.

Note in Figure 10.12 that one-third of the genes have no identified function. This finding is typical of genomic sequences. In many genomic sequences, the proportion of genes with no identified function exceeds 50%. Hence, even when the genes in a genome are correctly identified, there are numerous additional issues. Especially in multicellular eukaryotes, even for genes whose functions can be assigned, it is not usually known when dur-

ing the life cycle each gene is expressed, in which tissues it is expressed, or the presence, patterns, and tissue specificity of alternative spicing. Interactions among genes and gene products are also typically unknown. In some cases, useful information can be gained by identifying genes with similar sequences in other organisms, but then there is the problem of recognizing which sequences are sufficiently similar to suggest similar functions. Ultimately, the greatest challenge is to understand how the genes in the genome function and are coordinately regulated to control development, metabolism, reproduction, behavior, and response to the environment. Genomic sequencing should therefore be thought of as only a very early stage in the quest to understand the higher and more integrated levels of biological organization and function.

A A Moment to Think

Answer to Problem: **(a)** *Sma*I produces blunt ends, and any blunt end can be ligated onto any other blunt end. Hence the *Sma*I fragment A−B can be ligated into the polylinker in either of two orientations. The resulting clones are expected to be X−A−B−Y and X−B−A−Y in equal frequency. **(b)** A restriction enzyme produces fragments whose ends are identical, so either end can be ligated onto a complementary sticky end. To force the orientation X−A−B−Y, one needs to cleave the vector and the target with two restriction enzymes that produce different sticky ends. The site nearest X in the vector must match the site nearest A in the target, and the site nearest Y in the vector must match the site nearest B in the vector. In this case, if the vector and target are both cleaved with *Bam*HI and *Eco*RI, the resulting clone is expected to have the orientation X−A−B−Y. You should be able to convince yourself that no other combination of enzymes will work.

the human connection

The Human Genome Sequence

Eric S. Lander and 248 other investigators
2001
The Whitehead Institute for Biomedical
Research, Massachusetts Institute of
Technology, Boston, Massachusetts, and
23 other research institutions
*Initial Sequencing and Analysis of the Human
Genome*

Sequencing the human genome was no small challenge. At 3 billion base pairs, if the sequence were flashed on the display of an electronic clock at the rate of 10 bases per second, it would take almost 10 years merely to watch it! Actually, the genome was sequenced simultaneously by two groups, one an international public consortium whose report is excerpted here, and the other a private company, Celera Genomics, whose separate report in *Science* [291: 1304 (2001)] notes, "There are two fallacies to be avoided: determinism, the idea that all characteristics of the person are 'hard-wired' by the genome; and reductionism, the view that with complete knowledge of the human genome sequence, it is only a matter of time before our understanding of gene functions and interactions will provide a complete causal description of human variability." Shakespeare put the human experience and human biology in poetic perspective 400 years ago when he had Hamlet declare of humankind: "What a piece of work is a man! How noble in reason! In action how like an angel! In apprehension how like a god! The beauty of the world! The paragon of animals! And yet, what is this quintessence of dust?"

••

Here we report the results of a collaboration . . . to produce a draft sequence of the human genome. . . . The sequence was produced rapidly over a relatively short period . . . [with] . . . a continuous throughput exceeding 1,000 nucleotides per second, 24 hours per day, seven days per week. . . . The ultimate goal is to compile a complete list of all human genes and their encoded proteins, to serve as a "periodic table" for biomedical research. . . . Human genes tend to have small exons separated by long introns. . . . Many genes are over 100 kb long, the largest known example being the dystrophin (DMD) gene at 2.4 Mb. . . . The analysis allows us to estimate . . . the number of genes in the human genome . . . [as] . . . 30,000–35,000. . . . The human thus appears to have only about twice as many genes as worm or fly. However, human genes differ in important respects from those in worm and fly. They are spread out over much larger regions of genomic DNA, and they are used to construct more alternative transcripts. This may result in perhaps five times as many primary protein products in the human as in the worm or fly. . . . There also appears to be substantial innovation in the creation of new vertebrate proteins . . . evident at the level of domain architecture. . . . The Human Genome Project . . . provides a capstone for efforts in the past century to discover genetic information and a foundation for efforts in the coming century to understand it. . . . But the science is only a part of the challenge. . . . We must also involve society at

> **Understanding and wisdom will be required to ensure that these benefits are implemented broadly and equitably.**

large. . . . Understanding and wisdom will be required to ensure that these [scientific and medical] benefits are implemented broadly and equitably. To that end, serious attention must be paid to the many ethical, legal and social implications (ELSI) raised by the accelerated pace of genetic discovery. . . . The more we learn about the human genome, the more there is to explore.

Source: Nature 409: 860–920.

10.3

Genomics and proteomics reveal genome-wide patterns of gene expression and networks of protein interactions.

Genomic sequencing has made possible a new approach to genetics called **functional genomics,** which focuses on genome-wide patterns of gene expression and the mechanisms by which gene expression is coordinated. As changes take place in the cellular environment—for example, through development, aging, or changes in the external conditions—the patterns of gene expression change also. But genes are usually deployed in sets, not individually. As the level of expression of one coordinated set is decreased, the level of expression of a different coordinated set may be increased. How can one study tens of thousands of genes all at the same time?

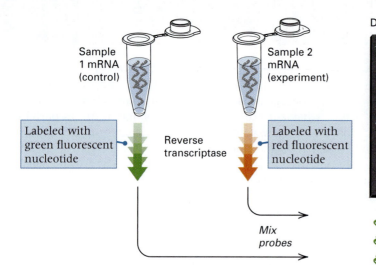

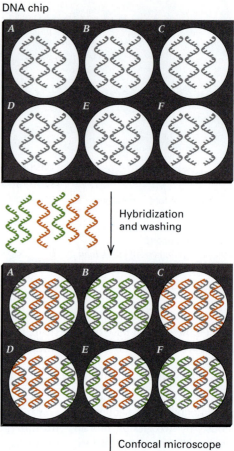

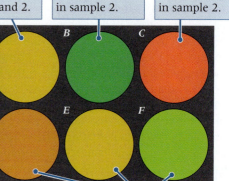

Figure 10.13 Principle of operation of one type of DNA microarray. At the top right are dried microdrops, each of which contains immobilized DNA strands from a different gene ($A-F$). These are hybridized with a mixture of fluorescence-labeled DNA samples obtained by reverse transcription of cellular mRNA. Competitive hybridization of red (experimental) and green (control) label is proportional to the relative abundance of each mRNA species in the samples. The relative levels of red and green fluorescence of each spot are assayed by microscopic scanning and displayed as a single color. Red or orange indicates overexpression in the experimental sample, green or yellow-green indicates underexpression in the experimental sample, and yellow indicates equal expression.

■ **DNA microarrays are used to estimate the relative level of gene expression of each gene in the genome.**

The study of genome-wide patterns of gene expression became feasible with the development of the **DNA microarray** (or *chip*), a flat surface about the size of a postage stamp on which 10,000 to 100,000 distinct spots are present, each containing a different immobilized DNA sequence suitable for hybridization with DNA or RNA isolated from cells growing under different conditions, from cells not exposed or cells exposed to a drug or toxic chemical, from different stages of development, or from different types or stages of a disease such as cancer. Two types of DNA chips are presently in use:

• A chip arrayed with oligonucleotides synthesized directly on the chip, one nucleotide at a time, by automated procedures; these chips typically have hundreds of thousands of spots per array.

• A chip arrayed with denatured, double-stranded DNA sequences of 500 to 5000 bp, in which the spots, each about a millionth of a drop in volume, are deposited by capillary action from miniaturized fountain-pen-like

devices mounted on the movable head of a flatbed robotic workstation; these chips typically have tens of thousands of spots per array.

Figure 10.13 shows one method by which DNA chips are used to assay the genome-wide levels of

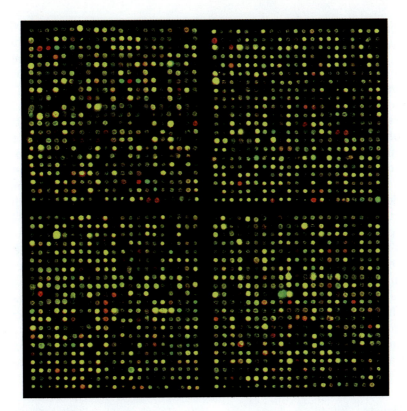

Figure 10.14 Small part of a yeast DNA chip showing 1764 spots, each specific for hybridization with a different mRNA sequence. The color of each spot indicates the relative level of gene expression in experimental and control samples. The complete chip for all yeast open reading frames includes over 6200 spots. [Courtesy of Jeffrey P. Townsend, Duccio Cavalieri, and the Harvard Center for Genomics Research.]

gene expression in an experimental sample relative to a control. At the upper right are shown six adjacent spots in the microarray, each of which contains a DNA sequence that serves as a probe for a different gene, *A* through *F*. At the left is shown the experimental protocol. Messenger RNA is first extracted from both the experimental and the control samples. This material is then subjected to one or more rounds of reverse transcription, as described in Section 10.1. In the experimental material (sample 2), the primer for reverse transcription includes a red fluorescent label; and in the control material (sample 1), the primer includes a green fluorescent label. When a sufficient quantity of labeled DNA strands have accumulated, the fluorescent samples are mixed and hybridized with the DNA chip.

The result of hybridization is shown in the middle part of Figure 10.13. Because the samples are mixed, the hybridization is competitive, and therefore the density of red or green strands bound to the DNA chip is proportional to the concentration of red or green molecules in the mixture. Genes that are overexpressed in sample 2 relative to sample 1 will have more red strands

hybridized to the spot, whereas those that are underexpressed in sample 2 relative to sample 1 will have more green strands hybridized to the spot.

After hybridization, the DNA chip is placed in a confocal fluorescence scanner that scans each *pixel* (the smallest discrete element in a visual image) first to record the intensity of one fluorescent label and then again to record the intensity of the other fluorescent label. These signals are synthesized to produce the signal value for each spot in the microarray. The signals indicate the relative levels of gene expression by color, as shown in Figure 10.14. A spot that is red or orange indicates high or moderate overexpression of the gene in the experimental sample; a spot that is green or yellow-green indicates high or moderate underexpression of the gene in the experimental sample; and a spot that is perfectly yellow indicates equal levels of gene expression in the samples. In this manner, DNA chips can assay the relative levels of any mRNA species whose abundance in the sample is more than one molecule per 10^5, and differences in expression as small as approximately twofold can be detected.

■ Microarrays reveal groups of genes that are coordinately expressed during development.

Gene-expression arrays have been used to identify groups of genes that are coordinately regulated in development. The example in Figure 10.15 shows expression profiles for 20 groups of genes in the early stages of development in *C. elegans*. In these experiments, time in development was measured in minutes relative to the four-cell stage. Relative levels of gene expression are plotted on a logarithmic scale and, hence, the changes in relative transcript abundance are often two or three orders of magnitude. Over the time period examined, the embryo undergoes a transition from control through maternal transcripts present in the egg to those transcribed in the embryo itself, and includes the times during which most of the major cell fates are specified. The microarrays used in these experiments allowed detection of transcripts from almost 9000 open reading frames, and the plots include traces for approximately 2500 genes, about 80% of all those that showed significant changes in transcript abundance over the time interval shown.

Up to the four-cell stage of development the patterns of transcription are all quite stable and then begin to change rapidly. Development in the earliest stages is supported largely by maternally derived transcripts. Many of these are cleared rapidly as development proceeds, for example, the transcripts plotted for clusters of 141, 244, and 568 genes in the lower-right panels. Production of transcripts from embryonic cells is clearly induced, as evidenced by the patterns for clusters of 431 and 153 genes in the panels at the upper right. The curves showing the disappearance of the maternal transcripts and appearance of the embryonic transcripts intersect at about the time of gastrulation, indicating a somewhat earlier (mid-blastula) transition from maternal to embryonic control of development. Many of the gene transcription patterns are very complex, with a transient peak of expression suggesting that the transcript (though not necessarily the protein prod-

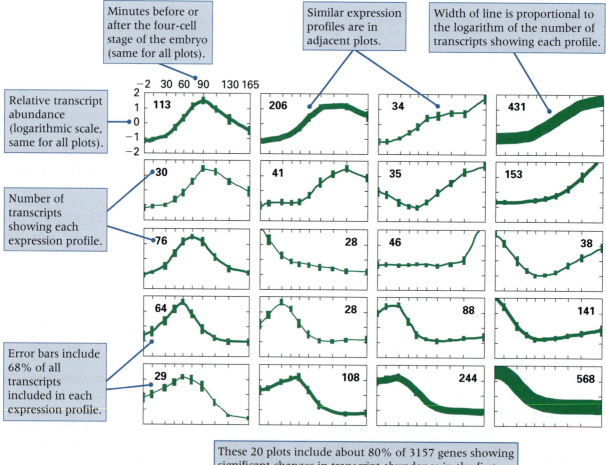

Figure 10.15 | Patterns of transcriptional regulation of about 2500 genes during the first approximately 2.75 hours of development in *C. elegans*. Complete development requires about 14 hours. [Courtesy of L. Ryan Baugh and Craig P. Hunter. From L. R. Baugh et al. 2003. *Development* 130: 889.]

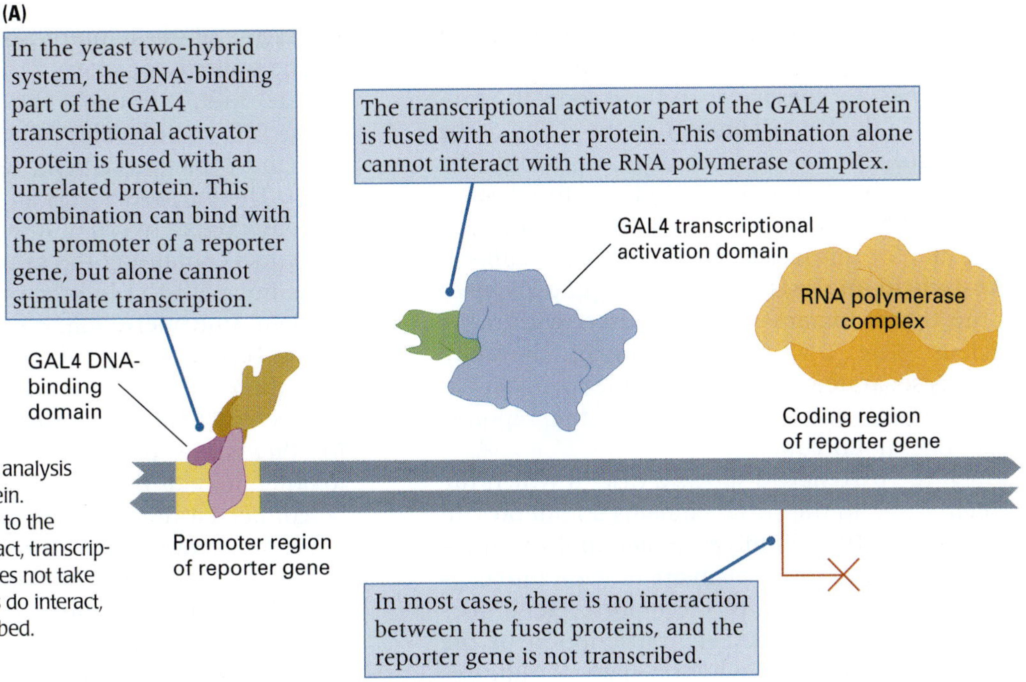

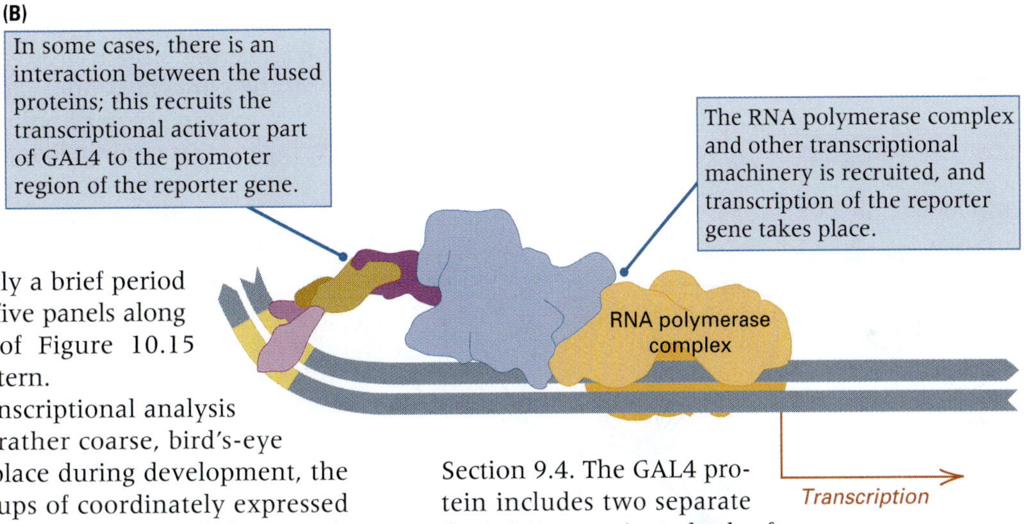

(A)

In the yeast two-hybrid system, the DNA-binding part of the GAL4 transcriptional activator protein is fused with an unrelated protein. This combination can bind with the promoter of a reporter gene, but alone cannot stimulate transcription.

The transcriptional activator part of the GAL4 protein is fused with another protein. This combination alone cannot interact with the RNA polymerase complex.

GAL4 transcriptional activation domain

RNA polymerase complex

GAL4 DNA-binding domain

Coding region of reporter gene

Figure 10.16 Two-hybrid analysis by means of the GAL4 protein. (A) When the proteins fused to the GAL4 domains do not interact, transcription of the reporter gene does not take place. (B) When the proteins do interact, the reporter gene is transcribed.

Promoter region of reporter gene

In most cases, there is no interaction between the fused proteins, and the reporter gene is not transcribed.

(B)

In some cases, there is an interaction between the fused proteins; this recruits the transcriptional activator part of GAL4 to the promoter region of the reporter gene.

The RNA polymerase complex and other transcriptional machinery is recruited, and transcription of the reporter gene takes place.

RNA polymerase complex

Transcription

uct) is needed for only a brief period in development. All five panels along the left-hand side of Figure 10.15 show this kind of pattern.

Although the transcriptional analysis in Figure 10.15 is a rather coarse, bird's-eye view of what takes place during development, the identification of groups of coordinately expressed genes is of considerable value in itself because it suggests that these genes may share common or overlapping *cis*-acting regulatory sequences that are controlled by common or overlapping sets of transcriptional activator proteins.

■ Yeast two-hybrid analysis reveals networks of protein interactions.

Biological processes can also be explored on a genomic scale at the level of protein–protein interactions. The rationale for studying such interactions is that proteins that participate in related cellular processes often interact with one another; hence, knowing which proteins interact can provide clues to the possible function of otherwise anonymous proteins.

One method for identifying protein-protein interactions makes use of the GAL4 transcriptional activator protein in budding yeast discussed in

Section 9.4. The GAL4 protein includes two separate domains or regions, both of which are necessary for transcriptional activation. One domain is the zinc-finger DNA-binding domain that binds with the target site in the promoter of the *GAL* genes that are activated, and the other domain is the transcriptional activation domain that makes contact with the transcriptional complex and actually triggers transcription. In the wildtype GAL4 protein, these domains are tethered together because they are parts of the same polypeptide chain.

The key to identifying protein–protein interactions through the use of GAL4 is that the coding regions for the separate domains can be taken apart and each fused to a coding region for a different protein. The strategy is shown in Figure 10.16 part A, where the GAL4 DNA-binding domain and the transcriptional activation domain are depicted as separate entities, each fused to a different polypeptide chain, shown in the vicinity of a *GAL* promoter.

The promoter is attached to a **reporter gene** whose transcription can be detected by means of, for example, a color change in the colony, the production of a fluorescent protein, or the ability of the cells to grow in the presence of an antibiotic. The fused DNA-binding domain and the fused transcriptional activation domain are both hybrid proteins, and for this reason the test system is called a **two-hybrid analysis.** In part A, the proteins fused to the GAL4 domains do not interact within the nucleus. The DNA-binding domain therefore remains separated from the transcriptional activation domain, and transcription of the reporter genes does not occur.

Figure 10.16 part B shows a case in which the protein fused to the GAL4 domains do interact. In this case the DNA-binding domain and the transcriptional activation domain are brought into contact, and transcription of the reporter gene does take place. In this manner, transcription of the reporter gene in the two-hybrid analysis indicates that the proteins fused to the GAL4 domains

undergo a physical interaction that brings the two hybrid proteins together.

An example of two-hybrid analysis is shown in Figure 10.17 , which depicts a network of 318 protein-protein interactions among 329 nuclear proteins in yeast. The purpose of this analysis was to compare the observed network of interactions with random networks containing the same number of interactions but with the interacting partners chosen at random. One interesting property of the network in Figure 10.17 as well as of other protein networks, is that there are fewer than expected interactions between proteins that are already highly connected. In other words, proteins that are highly connected to other proteins through many interactions tend to be connected not to other highly connected proteins, but to proteins with fewer connections. The systematic suppression of links between highly connected proteins has the effect of minimizing the extent to which random environmental or genetic perturbations in one part of the network spread to other parts of the network.

Two-hybrid analysis affords a powerful approach to discovering protein-protein interactions because it can be performed on a large scale, requires no protein purification, detects interactions that occur in living cells, and requires no information about the function of the proteins being tested. The method, however, does have some limitations. For example, the two-hybrid assay is qualitative, not quantitative, and so weak interactions cannot easily be distinguished from strong ones. The hybrid proteins are usually highly expressed to enhance the reliability of the assay, and so interactions can take place that would not take place at normal concentrations. The two-hybrid assay also requires that the protein–protein interactions take place in the nucleus, whereas some proteins may interact only in the environment of the cytoplasm. Finally, hybrid proteins may fold differently than native proteins, and the misfolded proteins may fail to interact when the native conformations do, or they may interact when the native conformations do not. The conclusion is that results from two-hybrid analyses need to be interpreted with care; nevertheless, the method has already yielded much valuable information.

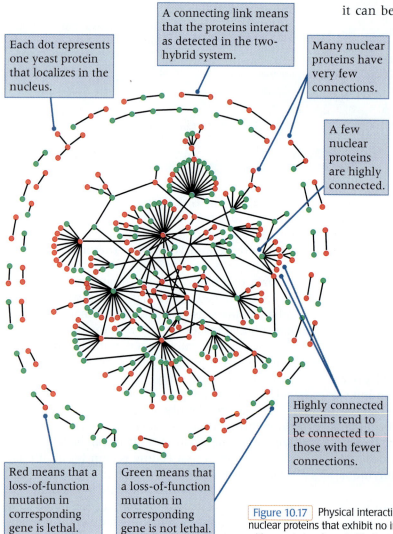

Each dot represents one yeast protein that localizes in the nucleus.

A connecting link means that the proteins interact as detected in the two-hybrid system.

Many nuclear proteins have very few connections.

A few nuclear proteins are highly connected.

Highly connected proteins tend to be connected to those with fewer connections.

Red means that a loss-of-function mutation in corresponding gene is lethal.

Green means that a loss-of-function mutation in corresponding gene is not lethal.

Figure 10.17 Physical interactions among nuclear proteins in yeast. Not shown are nuclear proteins that exhibit no interactions. [Courtesy of Sergei Maslov. Excerpted with permission from S. Maslov and K. Sneppen. 2002. *Science* 296: 910. © AAAS]

10.4

Reverse genetics creates an organism with a designed mutation.

Mutation has traditionally provided the raw material needed for genetic analysis. The customary procedure has been to use a mutant phenotype to recognize a mutant gene and then to identify the wildtype allele and its normal function. This approach has proved highly successful, as evidenced by numerous examples throughout this book. But the approach also has its limitations. For example, it may prove difficult or impossible to isolate mutations in genes that duplicate the functions of other genes or that are essential for the viability of the organism.

Recombinant DNA technology has made possible another approach in genetic analysis in which wildtype genes are cloned, intentionally mutated in specific ways, and introduced back into the organism to study the phenotypic effects of the mutations. Because the position and molecular nature of each mutation are precisely defined, a very fine level of resolution is possible in determining the functions of particular regions of nucleotide sequence. This type of analysis has been applied to defining the promoter and enhancer sequences that are necessary for transcription, the sequences necessary for normal RNA splicing, particular amino acids that are essential for protein function, and many other problems. The procedure is often called **reverse genetics** because it reverses the usual flow of study: Instead of starting with a mutant phenotype and trying to identify the wildtype gene, reverse genetics starts by making a mutant gene and studies the resulting phenotype. The following sections describe some techniques and applications of reverse genetics.

■ Recombinant DNA can be introduced into the germ line of animals.

Reverse genetics can be carried out in most organisms that have been extensively studied genetically, including the nematode *Caenorhabditis elegans, Drosophila*, the mouse, and many domesticated animals and plants. In nematodes, the basic procedure is to manipulate the DNA of interest in a plasmid that also contains a selectable genetic marker that will alter the phenotype of the transformed animal. The DNA is injected directly into the reproductive organs and sometimes spontaneously becomes incorporated into the chromosomes in the germ line. The result of transformation is observed and can be selected in the progeny of the injected animals because of the phenotype conferred by the selectable marker.

A somewhat more elaborate procedure is necessary for germ-line transformation in *Drosophila*. The usual method makes use of a 2.9-kb transposable element called the **P element,** which consists of a central region coding for transposase flanked by 31-base-pair inverted repeats (Figure 10.18, part A). A genetically engineered derivative of this *P* element, called *wings clipped,* can make functional transposase but cannot itself transpose because of deletions intro-

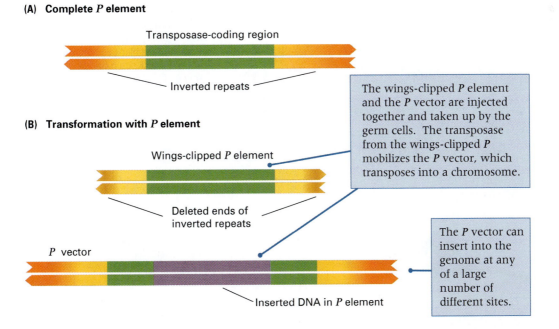

(A) **Complete *P* element**

Transposase-coding region

Inverted repeats

(B) **Transformation with *P* element**

Wings-clipped *P* element

The wings-clipped *P* element and the *P* vector are injected together and taken up by the germ cells. The transposase from the wings-clipped *P* mobilizes the *P* vector, which transposes into a chromosome.

Deleted ends of inverted repeats

P vector

The *P* vector can insert into the genome at any of a large number of different sites.

Inserted DNA in *P* element

Figure 10.18 | Transformation in *Drosophila* mediated by the transposable element *P*. (A) Complete *P* element containing inverted repeats at the ends and an internal transposase-coding region. (B) Two-component transformation system. The vector com-

ponent contains the DNA of interest flanked by the recognition sequences needed for transposition. The wings-clipped component is a modified *P* element that codes for transposase but cannot transpose itself because critical recognition sequences are deleted.

duced at the ends of the inverted repeats (Figure 10.18, part B). For germ-line transformation, the vector is a plasmid containing a P element that includes, within the inverted repeats, a selectable genetic marker (usually one affecting eye color), as well as a large internal deletion that removes much of the transposase-coding region. By itself, this P element cannot transpose because it makes no transposase, but it can be mobilized by the transposase produced by the wings-clipped or other intact P elements. In *Drosophila* transformation, any DNA fragment of interest is introduced between the ends of the deleted P element. The resulting plasmid and a different plasmid containing the wings-clipped element are injected into the region of the early embryo that contains the germ cells. The DNA is taken up by the germ cells, and the wings-clipped element produces functional transposase (Figure 10.18, part B). The transposase mobilizes the engineered P vector and results in its transposition into an essentially random location in the genome. Transformants are detected among the progeny of the injected flies because of the eye color or other genetic marker included in the P vector. Integration into the germ line is typically

very efficient: From 10 to 20 percent of the injected embryos that survive and are fertile yield one or more transformed progeny. However, the efficiency decreases with the size of the DNA fragment in the P element, and the effective upper limit is approximately 20 kb.

Transformation of the germ line in mammals can be carried out in several ways. The most direct is the injection of vector DNA into the nucleus of fertilized eggs, which are then transferred to the uterus of foster mothers for development. The vector is usually a modified retrovirus. **Retroviruses** have RNA as their genetic material and code for a reverse transcriptase that converts the retrovirus genome into double-stranded DNA that becomes inserted into the genome in infected cells. Genetically engineered retroviruses containing inserted genes undergo the same process. Animals that have had new genes inserted into the germ line in this or any other manner are called **transgenic** animals.

Another method of transforming mammals uses **embryonic stem cells** obtained from embryos a few days after fertilization (Figure 10.19). Although

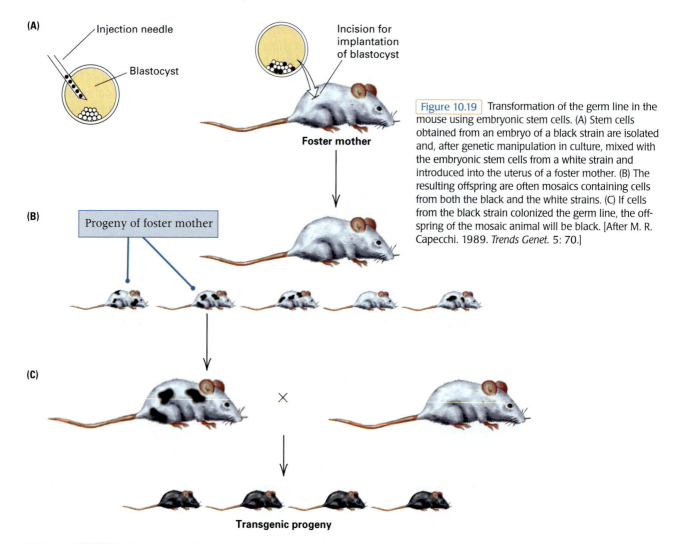

Figure 10.19 Transformation of the germ line in the mouse using embryonic stem cells. (A) Stem cells obtained from an embryo of a black strain are isolated and, after genetic manipulation in culture, mixed with the embryonic stem cells from a white strain and introduced into the uterus of a foster mother. (B) The resulting offspring are often mosaics containing cells from both the black and the white strains. (C) If cells from the black strain colonized the germ line, the offspring of the mosaic animal will be black. [After M. R. Capecchi. 1989. *Trends Genet.* 5: 70.]

378 **CHAPTER 10** Genomics, Proteomics, and Genetic Engineering

embryonic stem cells are not very hardy, they can be isolated and then grown and manipulated in culture; mutations in the stem cells can be selected or introduced using recombinant DNA vectors. The mutant stem cells are introduced into another developing embryo and transferred into the uterus of a foster mother (Figure 10.19, part A), where they become incorporated into various tissues of the embryo and often participate in forming the germ line. If the embryonic stem cells carry a genetic marker, such as a gene for black coat color, then mosaic animals can be identified by their spotted coats (part B). Some of these animals, when mated, produce black offspring (part C), which indicates that the embryonic stem cells had become incorporated into the germ line. In this way, mutations introduced into the embryonic stem cells while they were in culture may become incorporated into the germ line of living animals. Embryonic stem cells have been used to create strains of mice with mutations in genes associated with such human genetic diseases as cystic fibrosis. These strains serve as mouse models for studying the disease and for testing new drugs and therapeutic methods.

The procedure for introducing mutations into specific genes is called **gene targeting.** The specificity of gene targeting comes from the DNA sequence homology needed for homologous recombination. Two examples are illustrated in Figure 10.20, where the DNA sequences present in gene-targeting vectors are shown as looped configurations paired with homologous regions in the chromosome prior to recombination. The targeted gene is shown in pink. In part A of Figure 10.20, the vector contains the targeted gene interrupted by an insertion of a novel DNA sequence, and homologous recombination results in the novel sequence becoming inserted into the targeted gene in the genome. In part B of Figure 10.20, the vector contains only flanking sequences, not the targeted gene, so homologous recombination results in replacement of the targeted gene with an unrelated DNA sequence. In both cases, cells with targeted gene mutations can be selected by including an antibiotic-resistance gene, or other selectable genetic marker, in the sequences that are incorporated into the genome through homologous recombination.

■ Recombinant DNA can also be introduced into plant genomes.

A procedure for the transformation of plant cells makes use of a plasmid found in the soil bacterium *Agrobacterium tumefaciens* and related species. Infection of susceptible plants with this bacterium results in the growth of what are known as *crown gall tumors* at the entry site, which is usually a wound. Susceptible plants comprise about 160,000 species of flowering plants, known as the dicots, and include the great majority of the most common flowering plants.

The *Agrobacterium* contains a large plasmid of approximately 200 kb called the **Ti plasmid,** which includes a smaller (~25 kb) region known as the

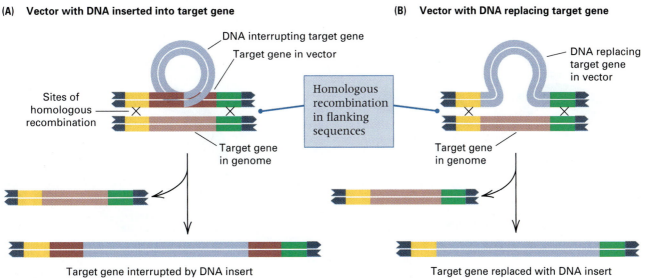

Figure 10.20 Gene targeting in embryonic stem cells. (A) The vector (top) contains the targeted sequence (red) interrupted by an insertion. Homologous recombination introduces the insertion into the genome. (B) The vector contains DNA sequences flanking the targeted gene. Homologous recombination results in replacement of the targeted gene with an unrelated DNA sequence. [After M. R. Capecchi. 1989. *Trends Genet.* 5: 70.]

T **DNA** flanked by 25-base-pair direct repeats (Figure 10.21, part A). The *Agrobacterium* causes a profound change in the metabolism of infected cells because of transfer of the *T* DNA into the plant genome. The *T* DNA contains genes coding for proteins that stimulate division of infected cells, hence causing the tumor, and also coding for enzymes that convert the amino acid arginine into an unusual derivative, generally *nopaline* or *octopine* (depending on the particular type of *Ti* plasmid), that the bacterium needs in order to grow. The transfer functions are present not in the *T* DNA itself but in another region of the plasmid called the *vir* (stands for *virulence*) region of about 40 kb that includes six genes necessary for transfer.

Transfer of *T* DNA into the host genome is similar in some key respects to bacterial conjugation, which we examined in Chapter 8. In infected cells, transfer begins with the formation of a nick that frees one end of the *T* DNA (part A), which peels off the plasmid and is replaced by rolling-circle replication (part B). The region of the plasmid that is transferred is delimited by a second nick at the other end of the *T* DNA, but the position of this nick is variable. The resulting single-stranded *T* DNA is bound with molecules of a single-stranded binding protein (SSBP) and is transferred into the plant cell and incorporated into the nucleus. There it is integrated into the chromosomal DNA by a mechanism that is still unclear (part C). Although the SSBP has certain similarities in amino acid sequence to the *recA* protein from *E. coli*, which plays a key role in homologous recombination, it is clear that integration of the *T* DNA does not require homology.

Use of *T* DNA in plant transformation is made possible by engineered plasmids in which the sequences normally present in *T* DNA are removed and replaced with those to be incorporated into the plant genome along with a selectable marker. A second plasmid contains the *vir* genes and permits mobilization of the engineered *T* DNA. In infected tissues, the *vir* functions mobilize the *T* DNA for transfer into the host cells and integration into the chromosome. Transformed cells are selected in culture by the use of the selectable marker and then grown into mature plants in accordance with the methods described in Section 5.4.

■ **Transformation rescue is used to determine experimentally the physical limits of a gene.**

One of the important applications of germ-line transformation is to define experimentally the limits of any particular gene along the DNA. Every gene includes upstream and downstream sequences that are necessary for its correct expression, as well as the coding sequences. As we saw in Chapter 9, these regulatory sequences control the time in development, the cell types, and the level at which transcription occurs. In defining the upstream and downstream limits of a gene, even knowing the complete nucleotide sequence of the coding region and the flanking DNA may be insufficient. The main reason is that there is no general method by which to identify regulatory sequences. Regulatory sequences are often composites of short, seemingly nondescript sequences, which are in fact the critical binding sites for regulatory proteins that control transcription.

To see how germ-line transformation is used to define the limits of a functional gene, consider the example of the *Drosophila* gene *white*, which, when mutated, results in flies with white eyes instead of

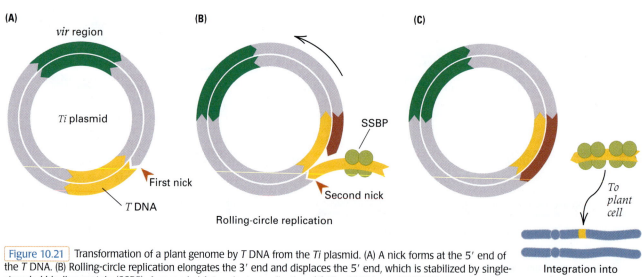

Figure 10.21 Transformation of a plant genome by *T* DNA from the *Ti* plasmid. (A) A nick forms at the 5′ end of the *T* DNA. (B) Rolling-circle replication elongates the 3′ end and displaces the 5′ end, which is stabilized by single-stranded binding protein (SSBP). A second nick terminates replication. (C) The SSBP-bound *T* DNA is transferred to a plant cell and inserts into the genome.

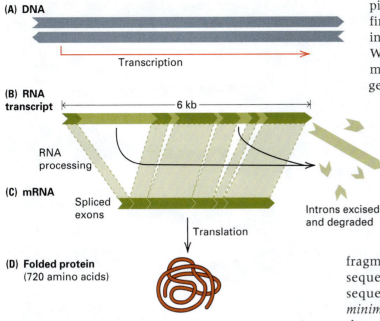

(A) DNA

Transcription

(B) RNA transcript

6 kb

RNA processing

(C) mRNA

Spliced exons

Introns excised and degraded

Translation

(D) Folded protein
(720 amino acids)

Figure 10.22 Genetic organization of the *Drosophila* gene *white*.

red. The genetic organization of *white* is illustrated in Figure 10.22. The primary RNA transcript (part B) is a little more than 6 kb in length and includes five introns that are excised and degraded during RNA processing. The resulting mRNA (part C) is translated into a protein of 720 amino acids (part D), which is a member of a large family of related ATP-dependent transmembrane proteins known as *ABC transporters* that regulate the traffic of their target molecules or ions across the cell membrane. The White transporter is located in the pigment-producing cells in the eye, and its target molecule is one of the key precursors of the eye-color pigments; hence flies with a mutant White transporter have white eyes.

The question addressed by germ-line transformation is this: How much DNA upstream and downstream of *white* is necessary for the fly to produce the wildtype transporter protein in the

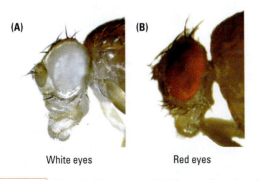

(A)   (B)

White eyes   Red eyes

Figure 10.23 Mutant white-eyed and wildtype red-eyed males of *Drosophila melanogaster*. [Courtesy of Elena R. Lozovsky.]

pigment cells? The experimental approach is first to clone a large fragment of DNA that includes the coding sequence for the wildtype White protein, then to use germ-line transformation to introduce this fragment into the genome of a fly that contains a *white* mutation. If the introduced DNA includes all the 5′ and 3′ regulatory sequences necessary for correct gene expression, then the phenotype of the resulting flies will be wildtype, because the wildtype gene is dominant to the *white* mutation. The ability of an introduced DNA fragment to correct a genetic defect in a mutant organism is called **transformation rescue** (Figure 10.23), and it means that the fragment contains all the essential regulatory sequences. The fragment may also include some sequences that are nonessential, so finding the *minimal* 5′ and 3′ flanking regions requires that the smaller pieces of the original fragment also be assayed for transformation rescue. In the case of *white*, the minimal DNA fragment is about 8.5 kb in length, starting about 2 kb upstream from the transcription start site.

10.5

Genetic engineering is applied in medicine, industry, agriculture, and research.

Recombinant DNA technology has revolutionized modern biology not only by opening up new approaches in basic research but also by making possible the creation of organisms with novel genotypes for practical use in agriculture and industry. In this section we examine a few of many applications of recombinant DNA.

■ Animal growth rate can be genetically engineered.

In many animals, the rate of growth is controlled by the amount of growth hormone produced. Transgenic animals with a growth-hormone gene under the control of a highly active promoter to drive transcription often grow larger than their normal counterparts. An example of a highly active promoter is found in the gene for *metallothionein*. The metallothioneins are proteins that bind heavy metals. They are ubiquitous in eukaryotic organisms and are encoded by members of a family of related genes. The human genome, for example, includes more than ten metallothionein genes that can be separated into two major groups according to their sequences. The promoter region of a metallothionein gene drives transcription of any gene to which it is attached, in response to heavy metals or steroid hormones. For example, when DNA constructs consisting of a rat growth-hormone gene under metallothionein control are used to produce transgenic mice, the resulting animals grow about twice as large as normal mice.

The effect of another growth-hormone construct is shown in Figure 10.24. The fish are coho salmon at 14 months of age. Those on the left are normal, whereas those on the right are transgenic animals that contain a salmon growth-hormone gene driven by a metallothionein regulatory region. Both the growth-hormone gene and the metallothionein gene were cloned from the sockeye salmon. As an indicator of size, the largest transgenic fish on the right has a length of about 42 cm. On average, the transgenic fish are 11 times heavier than their normal counterparts; the largest transgenic fish was 37 times the average weight of the nontransgenic animals. Not only do the transgenic salmon grow faster and become larger than normal salmon; they also mature faster.

Figure 10.24 Normal coho salmon (left) and genetically engineered coho salmon (right) containing a sockeye salmon growth-hormone gene driven by the regulatory region from a metallothionein gene. The transgenic salmon average 11 times the weight of the nontransgenic fish. The smallest fish on the left is about 4 inches long. [Courtesy of R. H. Devlin. From R. H. Devlin, T. Y. Tesaki, C. A. Biagi, E. M. Donaldson, P. Swanson, and W. K. Chan. 1994. *Nature* 371:209. © 1994 Macmillan Magazines Ltd.]

■ Crop plants with improved nutritional qualities can be created.

Beyond the manipulation of single genes, it is also possible to create transgenic organisms that have entire new metabolic pathways introduced. A remarkable example is in the creation of a genetically engineered rice that contains an introduced biochemical pathway for the synthesis of β-carotene, a precursor of vitamin A found primarily in yellow vegetables and greens. (Deficiency of vitamin A affects some 400 million people throughout the world, predisposing them to skin disorders and night blindness.) The β-carotene pathway includes four enzymes, which in the engineered rice are encoded in genes from different organisms (Figure 10.25). Two of the genes come from the common daffodil (*Narcissus pseudonarcissus*), whereas the other two come from the bacterium *Erwinia uredovora*. Each pair of genes was cloned into T DNA and transformed into rice using *Agrobacterium tumefaciens* by the mechanism outlined in Figure 10.21. Transgenic plants were then crossed to produce progeny containing all four enzymes. The engineered rice seeds contain enough β-carotene to provide the daily requirement of vitamin A in 300 grams of cooked rice; they even have a yellow tinge (Figure 10.25, part B).

People on high-rice diets are also prone to iron deficiency because rice contains a small phosphorus-storage molecule called *phytate*, which binds with iron and interferes with its absorption through the intestine. The transgenic β-carotene rice was also engineered to minimize this problem by introducing the fungal enzyme from *Aspergillus ficuum* that breaks down phytate, along with a gene encoding the iron-storage protein ferritin from the French bean, *Phaseolus vulgaris*, plus yet another gene from basmati rice that encodes a metallothionein-like gene that facilitates iron absorption in the human gut. Altogether, then, the transgenic rice strain rich in β-carotene and available iron contains six new genes taken from four unrelated species plus one gene from a totally different strain of rice!

■ The production of useful proteins is a primary impetus for recombinant DNA.

Among the most important applications of genetic engineering is the production of large quantities of particular proteins that are otherwise difficult to obtain (for example, proteins that are present in only a few molecules per cell or that are produced in only a small number of cells or only in human cells). The method is simple in principle. A DNA sequence cod-

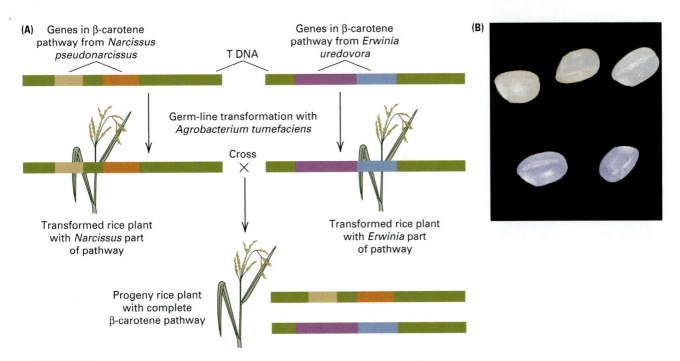

(A) Genes in β-carotene pathway from *Narcissus pseudonarcissus*

T DNA

Genes in β-carotene pathway from *Erwinia uredovora*

Germ-line transformation with *Agrobacterium tumefaciens*

Cross ×

Transformed rice plant with *Narcissus* part of pathway

Transformed rice plant with *Erwinia* part of pathway

Progeny rice plant with complete β-carotene pathway

(B)

Figure 10.25 Genetically engineered rice containing a biosynthetic pathway for β-carotene. (A) Enzymes in the pathway derive from genes in two different species. (B) Rice plants with both parts of the pathway produce grains with a yellowish cast (top) because of the β-carotene they contain, in contrast to the pure white grains (bottom) of normal plants. [B courtesy of Ingo Potrykus.]

ing for the desired protein is cloned in a vector adjacent to an appropriate regulatory sequence. This step is usually done with cDNA, because cDNA has all the coding sequences spliced together in the right order. Using a vector with a high copy number ensures that many copies of the coding sequence will be present in each bacterial cell, which can result in synthesis of the gene product at concentrations ranging from 1 to 5 percent of the total cellular protein. In practice, the production of large quantities of a protein in bacterial cells is straightforward, but there are often problems that must be overcome, because in the bacterial cell, which is a prokaryotic cell, the eukaryotic protein may be unstable, may not fold properly, or may fail to undergo necessary chemical modification. Many important proteins are currently produced in bacterial cells,

including human growth hormone, blood-clotting factors, and insulin. Patent offices in Europe and the United States have already issued tens of thousands of patents for the clinical use of the products of genetically engineered human genes. Figure 10.26 gives a breakdown of the numbers of patents issued for various clinical applications.

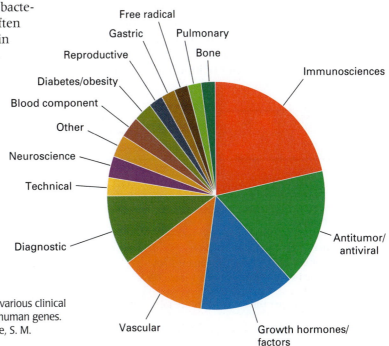

Figure 10.26 Relative numbers of patents issued for various clinical applications of the products of genetically engineered human genes. [Data from S. M. Thomas, A. R. W. Davies, N. J. Birtwistle, S. M. Crowther and J. F. Burke. 1996. *Nature* 380: 387.]

■ Animal viruses may prove useful vectors for gene therapy.

Genetic engineering in animal cells often exploits RNA retroviruses that use reverse transcriptase to make a double-stranded DNA copy of their RNA genome. The DNA copy then becomes inserted into the chromosomes of the cell. Ordinary transcription of DNA to RNA occurs only after the DNA copy is inserted. The infected host cell survives the infection, retaining the DNA copy of the retroviral RNA in its genome. These features of retroviruses make them convenient vectors for the genetic manipulation of animal cells, including those of birds, rodents, monkeys, and human beings.

Genetic engineering with retroviruses allows the possibility of altering the genotypes of animal cells. Because a wide variety of retroviruses are known, including many that infect human cells, genetic defects may be corrected by these procedures in the future. The recombinant DNA procedure employed with retroviruses consists of the *in vitro* synthesis of double-stranded DNA from the viral RNA by means of reverse transcriptase. The DNA is then cleaved with a restriction enzyme and, using techniques already described, any DNA fragment of interest is inserted. Transformation yields cells with the recombinant retroviral DNA permanently inserted into the genome. However, many retroviruses contain genes that result in uncontrolled proliferation of the infected cell, thereby causing a tumor. When retrovirus vectors are used for genetic engineering, the cancer-causing genes are first deleted. The deletion also provides the space needed for incorporation of the desired DNA fragment.

Attempts are currently under way to assess the potential use of retroviral vectors in **gene therapy,** or the correction of genetic defects in somatic cells by genetic engineering. Noteworthy successes so far include the correction of immunological deficiencies in patients with various kinds of inherited disorders.

However, a number of major problems stand in the way of gene therapy becoming widely used. At this time, there is no completely reliable way to ensure that a gene will be inserted only into the appropriate target cell or target tissue. For example, in one of the earliest clinical trials with a group of four patients treated with retroviral vectors for severe combined immunodeficiency disease, one of the patients had a retroviral insertion into a site that caused aberrant expression of a gene, *LMO-2,* associated with acute lymphoblastic leukemia.

A major breakthrough in disease prevention would come through the development of synthetic vaccines produced by recombinant DNA. Production of natural vaccines is often unacceptable because of the extreme hazards of working with large quantities of the active virus—for example, the human immunodeficiency virus (HIV) that causes acquired immune deficiency syndrome (AIDS). The danger can be minimized by cloning and producing viral antigens in a nonpathogenic organism. Vaccinia virus, the agent used in smallpox vaccination, has attracted much attention as a candidate for this application. Viral antigens are often on the surface of virus particles, and some of these antigens can be engineered into the coat of vaccinia. For example, engineered vaccinia virus with certain surface antigens of hepatitis B virus, influenza virus, and vesicular stomatitis virus (which kills cattle, horses, and pigs) has already proved effective in animal tests. One of the great challenges is malaria, which affects approximately 300 million people in Africa, Asia, and Latin America, causing approximately 1 to 1.5 million deaths per year. Control of the disease by drugs is increasingly compromised by the spread of resistance mutations. Successful vaccination of mice against malaria has been achieved using recombinant vaccinia virus expressing certain surface antigens of *Plasmodium berghei.* This result encourages hope that a vaccine against the human agent, *Plasmodium falciparum,* will eventually be developed.

chapter summary

10.1 Cloning a DNA molecule takes place in several steps.

- Restriction enzymes cleave DNA into fragments with defined ends.
- Restriction fragments are joined end to end to produce recombinant DNA.
- A vector is a carrier for recombinant DNA.
- Specialized vectors can carry very large DNA fragments.
- Vector and target DNA fragments are joined with DNA ligase.
- A recombinant cDNA contains the coding sequence of a eukaryotic gene.

- Loss of β-galactosidase activity is often used to detect recombinant vectors.
- Recombinant clones are often identified by hybridization with labeled probe.

Recombinant DNA technology makes it possible to modify the genotype of an organism in a directed, predetermined way by enabling different DNA molecules to be joined into novel genetic units, altered as desired, and reintroduced into the germ line. Restriction enzymes play a key role in the technique, because they can cleave DNA molecules within particular base sequences. Many restriction enzymes generate DNA fragments with complementary single-stranded ends, which can anneal and be ligated with similar fragments from other DNA molecules. The carrier DNA molecule that is used to propagate a desired DNA fragment is called a vector. The most common vectors are plasmids, phages, viruses, and bacterial artificial chromosomes (BACs). Transformation is an essential step in the propagation of recombinant molecules because it enables the recombinant DNA molecules to enter host cells, such as those of bacteria, yeast, or mammals. If the recombinant molecule has its own replication system or can use the host replication system, then it can replicate. Plasmid vectors become permanently established in the host cell; phage can multiply and produce a stable population of phage carrying source DNA; and retroviruses can be used to establish a gene in an animal cell.

10.2 A genomic sequence is like a book without an index, and identifying genes and their functions is a major challenge.

10.3 Genomics and proteomics reveal genome-wide patterns of gene expression and networks of protein interactions.

- DNA microarrays are used to estimate the relative level of expression of each gene in the genome.
- Microarrays reveal groups of genes that are coordinately expressed during development.
- Yeast two-hybrid analysis reveals networks of protein interactions.

Complete genomic DNA sequences have been determined for numerous mitochondria and chloroplasts, as well as for the genomes of many bacteria, archaeans, and eukaryotes including the human genome. Beyond analysis of the DNA sequences themselves, and beyond comparisons among organisms, genomics has spawned the field of functional genomics to describe and understand the genome-wide patterns of gene expression in cells under different conditions, including normal development and disease. Proteomics features large-scale studies to identify all proteins in each cell and their protein–protein interactions.

10.4 Reverse genetics creates an organism with a designed mutation.

- Recombinant DNA can be introduced into the germ line of animals.

- Recombinant DNA can also be introduced into plant genomes.
- Transformation rescue is used to determine experimentally the physical limits of a gene.

Recombinant DNA can be used to transform the germ line of animals or to genetically engineer plants. These techniques form the basis of reverse genetics, in which genes are deliberately mutated in specified ways and introduced back into the organism to determine the effects on phenotype. Reverse genetics is routine in genetic analysis in bacteria, yeast, nematodes, *Drosophila*, the mouse, and other organisms. In *Drosophila*, transformation employs a system of two vectors based on the transposable *P* element. One vector contains sequences that produce the *P* transposase; the other contains the DNA of interest between the inverted repeats of *P* and other sequences needed for mobilization by the transposase and insertion into the genome. Germ-line transformation in the mouse makes use of retrovirus vectors or embryonic stem cells. Dicotyledonous plants are transformed with T DNA derived from the *Ti* plasmid found in species of *Agrobacterium*, whose virulence genes promote a conjugation-like transfer of T DNA into the host plant cell, where it is integrated into the chromosomal DNA.

10.5 Genetic engineering is applied in medicine, industry, agriculture, and research.

- Animal growth rate can be genetically engineered.
- Crop plants with improved nutritional qualities can be created.
- The production of useful proteins is a primary impetus for recombinant DNA.
- Animal viruses may prove useful vectors for gene therapy.

Practical applications of recombinant DNA technology include the efficient production of useful proteins, the creation of novel genotypes for the synthesis of economically important molecules, the generation of DNA and RNA sequences for use in medical diagnosis, the manipulation of the genotype of domesticated animals and plants, the development of new types of vaccines, and the potential correction of genetic defects (gene therapy). Production of eukaryotic proteins in bacterial cells is sometimes hampered by protein instability, inability to fold properly, or failure to undergo necessary chemical modification. These problems are often eliminated by production of the protein in eukaryotic cells.

issues & ideas

- What does the term *recombinant DNA* mean? What are some of the practical uses of recombinant DNA?
- What features are essential in a bacterial cloning vector? How can a vector have more than one cloning site?
- What is the reaction catalyzed by the enzyme reverse transcriptase? How is this enzyme used in recombinant DNA technology?
- Explain the term *functional genomics*. What are DNA microarrays and how are they used in functional genomics? In the type of hybridization experiment described in the text, how would you interpret a spot on

a DNA microarray that is red or orange? yellow? green or yellow-green?

- Describe the two-hybrid system that makes use of the yeast GAL4 protein, and explain how the two-hybrid system detects interaction between proteins.
- What is a transgenic organism? What are some of the practical uses of transgenic organisms?
- Explain the role of recombination in gene targeting in embryonic stem cells. How can gene targeting be used to create a "knockout" mutation?

key terms & concepts

bacterial artificial
  chromosome (BAC)
bioinformatics
blunt ends
cloned DNA sequence
colony hybridization
complementary DNA
  (cDNA)
DNA microarray

embryonic stem cell
functional genomics
gene cloning
gene targeting
gene therapy
genetic engineering
genomics
Human Genome Project
library

proteomics
P transposable element
recombinant DNA
reporter gene
retrovirus
reverse genetics
reverse transcriptase
reverse transcriptase PCR
  (RT-PCR)

sticky ends
T DNA
Ti plasmid
transformation rescue
transgenic organism
two-hybrid analysis
vector

1. _____ Overhanging, complementary single-stranded ends of restriction fragments.

2. _____ A DNA molecule used as a vehicle for cloning a DNA fragment or gene.

3. _____ A usually large set of cloned DNA fragments.

4. _____ An enzyme that uses a single-stranded RNA template to produce a complementary double-stranded DNA copy.

5. _____ Term for the double-stranded DNA produced by reverse transcriptase.

6. _____ Use of computer methods to extract biological information from genomic sequences or other types of biological data.

7. _____ An ordered set of immobilized oligonucleotides or DNA fragments used in hybridization experiments to study genome-wide patterns of gene expression.

8. _____ Experimental method for identifying protein–protein interactions.

9. _____ In an experiment, a gene whose transcript or protein product is assayed to determine whether transcription takes place.

10. _____ Introduction of one or more mutations into a specific gene.

11. _____ Use of a transgenic organism to test for complementation between a cloned DNA fragment and a mutant gene.

12. _____ An organism that has been genetically modified using recombinant DNA and transformation.

solutions: step by step

Problem 1

What is the average distance between restriction sites for each of the following restriction enzymes? Assume that the DNA substrate has a random sequence with equal amounts of each base. The symbol N stands for any nucleotide, R for any purine (A or G), and Y for any pyrimidine (T or C).

(a) TCGA   *Taq*I
(b) GGTACC   *Kpn*I
(c) GTNAC   *Mae*III
(d) GGNNCC   *Nla*IV
(e) GRCGYC   *Acy*I

■ **Solution** (a) The average distance between restriction sites equals the reciprocal of the probability of occurrence of the restriction site. You must therefore calculate the probability of occurrence of each restriction site in a random DNA sequence. The probability of the sequence TCGA is

$$1/4 \times 1/4 \times 1/4 \times 1/4 = (1/4)^4 = 1/256$$

and so 256 bases is the average distance between *Taq*I sites.
(b) By the same reasoning, the probability of a *Kpn*I site is $(1/4)^6 = 1/4096$, so 4096 bases is the average distance between *Kpn*I sites.
(c) The probability of N (any nucleotide at a site) is 1, so the probability of the sequence GTNAC equals

$$1/4 \times 1/4 \times 1 \times 1/4 \times 1/4 = (1/4)^4 = 1/256$$

Therefore, 256 is the average distance between *Mae*III sites.
(d) The same reasoning yields the average distance between *Nla*IV sites as

$$(1/4 \times 1/4 \times 1 \times 1 \times 1/4 \times 1/4)^{-1} = 256 \text{ bases}$$

(e) The probability of an R (A or G) at a site is 1/2, and the probability of a Y (T or C) at a site is 1/2. Hence the probability of the sequence GRCGYC is

$$1/4 \times 1/2 \times 1/4 \times 1/4 \times 1/2 \times 1/4 = 1/1024$$

so there is an average of 1024 bases between *Acy*I sites.

Problem 2

A *sequence-tagged site* (*STS*) is a DNA sequence, present once per DNA molecule or per haploid genome, that can be amplified by the use of suitable oligonucleotide primers in the polymerase chain reaction in order to identify clones that contain the sequence. The table on the next page shows the presence or absence of seven sequence-tagged sites (STSs), numbered 1 through 7, in each of six BAC clones, designated A through F. In the table, + means that PCR primers specific to the STS are able to amplify the sequence from the clone; − means that the STS cannot be amplified from the clone. Use these data to make a map of the order of the BAC clones and of the STS markers located within the clones. In orienting the map, put the BAC clone A at the top.

	STS marker						
BAC clone	1	2	3	4	5	6	7
A	−	−	+	+	−	−	−
B	−	+	−	−	−	−	+
C	+	−	−	−	+	−	−
D	+	−	−	−	−	+	−
E	−	−	−	+	−	+	−
F	−	+	−	−	+	−	−

■ Solution The principle in ordering the BAC clones is that the DNA fragments present in any two (or more) clones that contain the same STS marker must overlap. The table contains no information bearing on the relative sizes of the BAC clones or on the distance between the STS markers. Thus each clone is drawn as though it had the same size, and the STS markers are spaced equally. Considering the

STS markers shared between clones, and starting with clone A, we obtain the map shown here. The mirror-image map with STS 3 on the right and STS 7 on the left is also a valid solution.

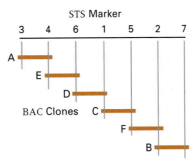

concepts in action: problems for solution

10.1 Reverse transcriptase, like most enzymes that make DNA, requires a primer. Explain why, when cDNA is to be made for the purpose of cloning a eukaryotic gene, a convenient primer is a short sequence of poly(dT). Why does this method not work with a prokaryotic messenger RNA?

10.2 A DNA molecule has 23 occurrences of the sequence 5′-AATT-3′ along one strand. How many times does the same sequence occur along the other strand?

10.3 A *kan-r tet-r* plasmid is treated with the restriction enzyme *Bgl*I, which cleaves the *kan* (kanamycin) gene. The DNA is annealed with and ligated to a *Bgl*I digest of *Neurospora* DNA and then used to transform *E. coli*.
(a) What antibiotic would you put in the growth medium to ensure that each colony has the plasmid?
(b) What antibiotic-resistance phenotypes would be found among the resulting colonies?
(c) Which phenotype is expected to contain *Neurospora* DNA inserts?

10.4 You want to introduce the human insulin gene into a bacterial host in hopes of producing a large amount of human insulin. Should you use the genomic DNA or the cDNA? Explain your reasoning.

10.5 After doing a restriction digest with the enzyme *Sse*I, which has the recognition site 5′-CCTGCA↓GG-3′ (the arrow indicates the position of the cleavage), you wish to separate the fragments in an agarose gel. In order to choose the proper concentration of agarose, you need to know the expected size of the fragments. Assuming equivalent amounts of each of the four nucleotides in the target DNA, what average fragment size would you expect?

10.6 You are given a plasmid containing part of a gene of *D. melanogaster*. The gene fragment is 303 base pairs long. You would like to amplify it using the polymerase chain reaction (PCR). You design oligonucleotide primers 19 nucleotides in length that are complementary to the plasmid sequences immediately adjacent to both ends of the cloning site. What would be the exact size of the resulting PCR product?

10.7 In cloning into bacterial vectors, why is it useful to insert DNA fragments to be cloned into a restriction site inside an antibiotic-resistance gene? Why is another gene for resistance to a second antibiotic also required?

10.8 A mutant allele is found to express the wildtype gene product, but at only about 20 percent of the wildtype level. The mutation is traced to an intron whose size has increased by 3.1 kb because of the presence of a DNA fragment with the restriction map shown here. The symbols *A, B, C, D, E, H, K, P, S,* and *X* represent cleavage sites for the restriction enzymes *Alu*I, *Bam*HI, *Cla*I, *Dde*I, *Eco*RI, *Hin*dIII, *Kpn*I, *Pst*I, *Sac*I, and *Xho*II, respectively. Does the restriction map of the insertion give any clues to what it is?

10.9 If the genomic and cDNA sequences of a gene are compared, what information does the cDNA sequence provide that is not obvious from the genomic sequence? What information does the genomic sequence contain that is not in the cDNA?

10.10 In studies of the operator region of an inducible operon in *E. coli*, the four constructs shown below were examined for level of transcription *in vitro*. The number associated with each construct is the relative level of transcription observed in the presence of the repressor protein. The symbols *E, B, H,* and *S* stand for the restriction sites *Eco*RI, *Bam*HI, *Hin*dIII, and *Sac*I. Construct (a) is the wildtype operator region, and in parts (b–d) the open boxes indicate restriction fragments that were deleted. What hypothesis about repressor–operator interactions can explain these results? How could this hypothesis be tested?

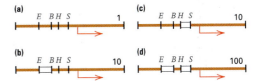

concepts in action: problems for solution   **387**

10.11 *Arabidopsis thaliana* has among the smallest genomes in higher plants, with a haploid genome size of about 100 Mb. If this genome is digested with *Not*I (an eight-base cutter), approximately how many DNA fragments would be produced? Assume equal and random frequencies of the four nucleotides.

10.12 How many DNA fragments would you expect if you digested the *E. coli* genome (containing 4.6 Mb) with the enzyme *Bam*HI, which has a six-base recognition sequence?

10.13 How frequently would the restriction enzymes *Taq*I (restriction site TCGA) and *Mae*III (restriction site GTNAC, in which N is any nucleotide) cleave double-stranded DNA molecules containing each of the following random sequences?
(a) 1/6 A, 1/6 T, 1/3 G, and 1/3 C
(b) 1/3 A, 1/3 T, 1/6 G, and 1/6 C

10.14 How many clones are needed to establish a library of DNA from a species of lemur with a diploid genome size of 6×10^9 base pairs if (1) fragments of average size 2×10^4 base pairs are used, and (2) one wants 99 percent of the genomic sequences to be in the library? (*Hint:* If the genome is cloned at random with x-fold coverage, the probability that a particular sequence will be missing is e^{-x}.)

10.15 Suppose that you digest the genomic DNA of a particular organism with *Sau*3A ($\downarrow$ GATC), where the arrow represents the cleavage site. Then you ligate the resulting fragments into a unique *Bam*HI (G $\downarrow$ GATCC) cloning site of a plasmid vector. Would it be possible to isolate the cloned fragments from the vector using *Bam*HI? From what proportion of clones would it be possible?

10.16 Digestion of a DNA molecule with *Hin*dIII yields two fragments of 2.2 kb and 2.8 kb. *Eco*RI cuts the molecule, creating 1.8-kb and 3.2-kb fragments. When treated with both enzymes, the same DNA molecule produces four fragments of 0.8 kb, 1.0 kb, 1.2 kb, and 2.0 kb. Draw a restriction map of this molecule.

10.17 A DNA microarray is hybridized with fluorescently labeled reverse-transcribed DNA as described in the text, where the control mRNA (C) is labeled with a green fluor and the experimental mRNA (E) with a red fluor. Indicate what you can conclude about the relative levels of expression of a spot in the microarray that fluoresces:
(a) Red
(b) Green
(c) Yellow
(d) Orange
(e) Lime green

10.18 Shown here is a restriction map of a 12-kb linear plasmid isolated from cells of *Borrelia burgdorferi*, a spirochete bacterium transmitted by the bite of *Ixodes* ticks that causes Lyme disease. The symbols *D, P, C, H, K, S,* and *A* represent cleavage sites for the restriction enzymes *Dde*I, *Pst*I, *Cla*I, *Hin*dIII, *Kpn*I, *Sac*I, and *Alu*I, respectively. In the accompanying gel diagram, show the positions at which bands would be found after digestion of the plasmid with the indicated restriction enzyme or enzymes.

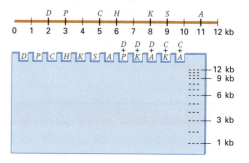

10.19 A functional genomics experiment is carried out using a DNA microarray to assay levels of gene expression in a species of bacteria. What genes would you expect to find overexpressed in cells grown in minimal medium compared to cells grown in complete medium?

10.20 A functional genomics experiment is carried out in *E. coli* to examine global levels of gene expression in various types of minimal growth medium. RNA extracted from the experimental culture is labeled with a molecule that fluoresces red, and RNA extracted from the control culture is labeled with a molecule that fluoresces green. The experimental and control samples are mixed prior to hybridization. Shown here are spots on the microarray corresponding to five genes: *trpE* (the first gene in the tryptophan biosynthetic operon), *lacI, lacZ, lacY,* and *crp* (which encodes the cyclic AMP receptor protein). Color the spots red, green, or yellow according to the relative levels of expression of each gene in the experimental and control cultures. (*Hint:* Before answering, think carefully about how the cyclic AMP receptor protein co-regulates the *lac* operon.)

Experimental minimal medium	Control minimal medium	Transcript				
		trpE	*lacI*	*lacZ*	*lacY*	*crp*
Glucose	Glucose	○	○	○	○	○
Glucose	Glycerol	○	○	○	○	○
Glycerol	Glucose	○	○	○	○	○
Lactose	Glucose	○	○	○	○	○
Glucose	Lactose	○	○	○	○	○
Lactose	Glycerol	○	○	○	○	○
Glycerol	Lactose	○	○	○	○	○

geNETics on the web

GeNETics on the Web will introduce you to some of the most important sites for finding genetic information on the Internet. To explore these sites, visit the Jones and Bartlett home page at

http://www.jbpub.com/genetics

For the book *Essentials of Genetics: A Genomics Perspective,* choose the link that says **Enter GeNETics on the Web**. You will be presented with a chapter-by-chapter list of highlighted keywords. Select any highlighted keyword and you will be linked to a Web site containing genetic information related to the keyword.

- This keyword site maintains an extensive list of **genome-sequencing projects** that are completed or are nearing completion. The list includes an astonishing number of eukaryotic genomes, prokaryotic genomes, genomes of organelles of numerous species, and genomes of viruses, bacteriophage, and plasmids. It is useful to have a biological dictionary or encyclopedia close at hand because the organisms are listed according to their official binomial Latin names—for example, *Oryza sativa*—rather than their common names, in this case rice.

- What are the **social implications** of advances in genomics? Some groups are worried because the application of genetic technologies poses ethical and legal issues of the foreknowledge of one's health, as well as issues of genetic privacy and insurability. Others are optimistic that the technologies will yield great benefits for medicine and society. This keyword site is devoted to the exploration of ethical, legal, and social issues related to sequencing the human genome, to the stimulation of public discussion, and to the formulation of public policies to prevent misuse of genetic information.

further readings

Azpirozleehan, R., and K. A. Feldmann. 1997. T-DNA insertion mutagenesis in *Arabidopsis*: Going back and forth. *Trends in Genetics* 13: 152.

Bishop, J. E., and M. Waldholz. 1990. *Genome.* New York: Simon and Schuster.

Blaese, R. M. 1997. Gene therapy for cancer. *Scientific American,* June.

Botstein, D., A. Chervitz, and J. M. Cherry. 1997. Yeast as a model organism. *Science* 277: 1259.

Capecchi, M. R. 1994. Targeted gene replacement. *Scientific American,* March.

Carlson, M. 2000. The awesome power of yeast biochemical genomics. *Trends in Genetics* 16: 49.

Chilton, M.-D. 1983. A vector for introducing new genes into plants. *Scientific American,* June.

Clark, M. S. 1999. Comparative genomics: The key to understanding the human genome project. *Bioessays* 21: 121.

Curtiss, R. 1976. Genetic manipulation of microorganisms: Potential benefits and hazards. *Annual Review of Microbiology* 30: 507.

Dujon, B. 1996. The yeast genome project: What did we learn? *Trends in Genetics* 12: 263.

Felgner, P. L. 1997. Nonviral strategies for gene therapy. *Scientific American,* June.

Friedmann, T. 1997. Overcoming the obstacles to gene therapy. *Scientific American,* June.

Gasser, C. S., and R. T. Fraley. 1992. Transgenic crops. *Scientific American,* June.

Gossen, J., and J. Vigg. 1993. Transgenic mice as model systems for studying gene mutations *in vivo. Trends in Genetics* 9: 27.

Granjeaud, S., F. Bertucci, and B. R. Jordan. 1999. Expression profiling: DNA arrays in many guises. *Bioessays* 21: 781.

Houdebine, L. M., ed. 1997. *Transgenic Animals: Generation and Use.* New York: Gordon and Breach.

Lockhart, D. J., and E. A. Winzeler. 2000. Genomics, gene expression and DNA arrays. *Nature* 405: 827.

Meisler, M. H. 1992. Insertional mutation of "classical" and novel genes in transgenic mice. *Trends in Genetics* 8: 341.

Nettelbeck, D. M., V. Jerome, and R. Muller. 2000. Gene therapy: Designer promoters for tumour targeting. *Trends in Genetics* 16: 174.

Pandey, A., and M. Mann. 2000. Proteomics to study genes and genomes. *Nature* 405: 837.

Sambrook, J., E. F. Fritsch, and T. Maniatis. 1989. *Molecular Cloning: A Laboratory Manual.* 2d ed. Cold Spring Harbor, NY: Cold Spring Harbor Laboratory.

Vukmirovic, O. G., and S. M. Tilghman. 2000. Exploring genome space. *Nature* 405: 820.

Watson, J. D. 1999. Interviews on genomics: Interview with Dr. James D. Watson. *Bioessays* 21: 175–178.

Butterflies use their wing color for mate recognition, predator avoidance, and other functions. Much remains to be learned about how the intricate patterns are genetically controlled. In the few cases that have been studied, however, the wing patterns are determined by employing genes used in other developmental processes. These genes are recruited for expression in the developing wing and are used for functions other than their original purposes. [© Photos.com.]

key concepts

- Transcription factors play a key role in the genetic control of development.
- Cell fate is determined by autonomous development and/or intercellular signaling.
- Developmental genes are often controlled by gradients of gene products, either within cells or across parts of the embryo.
- Transmembrane receptors often mediate signaling between cells.

- Cells can determine the fate of other cells through ligands that bind with their transmembrane receptors.
- Regulation of developmental genes is hierarchical: Genes expressed early in development regulate the activities of genes expressed later.
- Regulation of developmental genes is combinatorial: Each gene is controlled by a combination of other genes.

11

The Genetic Control of Development

chapter organization

11.1 Mating type in yeast illustrates transcriptional control of development.

11.2 The determination of cell fate in *C. elegans* development is largely autonomous.

11.3 Development in *Drosophila* illustrates progressive regionalization and specification of cell fate.

11.4 Floral development in *Arabidopsis* illustrates combinatorial control of gene expression.

the human connection
Maternal Impressions

Chapter Summary
Issues & Ideas
Key Terms & Concepts
Solutions: Step by Step
Concepts in Action: Problems for Solution
Further Readings

geNETics on the web

In the development of an organism, genes are expressed according to a prescribed program to ensure that as the fertilized egg divides repeatedly the resulting cells become specialized in an orderly way to give rise to the fully differentiated organism. Within what is usually a wide range of environments, the genotype determines not only the events that take place in development but also the temporal order in which the events unfold. The key process in development is **pattern formation,** which means the emergence of the spatially organized and specialized cells in the embryo from cell division and differentiation of the fertilized egg.

Genetic analyses of development often make use of mutations that alter developmental patterns. These mutations make it possible to identify genes that control development and to study the interactions among them. This chapter demonstrates how genetics is used in the study of development. To illustrate the principles, we focus on a specific example from each of four key model organisms whose genomes have been sequenced: *S. cerevisiae, C. elegans, D. melanogaster,* and *A. thaliana.*

11.1

Mating type in yeast illustrates transcriptional control of development.

Recall that budding yeast, *S. cerevisiae,* has two mating types denoted **a** and α. The specific mating type of a cell is controlled at the level of transcription. The alternative mating-type alleles *MAT***a** (mating type **a**) and *MAT*α (mating type α) both express a set of haploid-specific genes. They differ in that *MAT***a** also expresses a set of **a**-specific genes and *MAT*α also expresses a set of α-specific genes. The haploid-specific genes expressed in cells of both mating types include *HO,* which encodes the HO endonuclease used in mating-type interconversion, and *RME1,* which encodes a repressor of meiosis-specific genes. The functions of the expressed genes that differ in the mating types include (1) secretion of a mating peptide that arrests cells of the opposite mating type before DNA synthesis and prepares them for cell fusion, and (2) production of a receptor for the mating peptide secreted by the opposite mating type. Therefore, when **a** and α cells are in proximity, they prepare each other for mating and undergo fusion.

Regulation of mating type at the level of transcription takes place according to the regulatory interactions diagrammed in Figure 11.1 . These regulatory interactions were originally proposed on the basis of the phenotypes of various types of mutants, and most of the details have been confirmed by direct molecular studies. The symbols **a***sg,* α*sg,* and *hsg* represent the **a**-specific genes, the α-specific genes, and the haploid-specific genes, respectively; each set of genes is represented as a single segment, and lack of a "sunburst" indicates that transcription does not take place. In a cell of mating type **a** (part A), the *MAT***a** region is transcribed and produces a polypeptide called **a**1. By itself, **a**1 has no regulatory activity, and in the absence of any regulatory signal, **a***sg* and *hsg,* but not α*sg,* are transcribed. In a

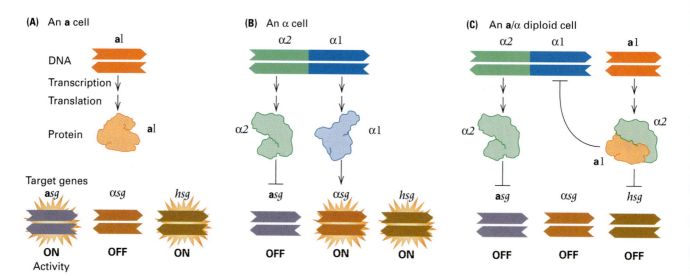

Figure 11.1 Transcriptional regulation of mating type in yeast. The symbols **a**sg, αsg, and *hsg* denote sets of **a**-specific genes, α-specific genes, and haploid-specific genes, respectively. Sets of genes represented with a "sunburst" are "on," and those unmarked are "off." (A) In an **a** cell, the **a**1 peptide is inactive, and the sets of genes manifest their default states of activity (**a**sg and *hsg* on and αsg off), so the cell is an **a** haploid. (B) In an α cell, the α2 peptide turns the **a**sg off and the α1 peptide turns the αsg on, so the cell is an α haploid. (C) In an **a**/α diploid, the α2 and **a**1 peptides form a complex that turns the *hsg* off, the α2 peptide turns the **a**sg off, and the αsg peptides manifest their default activity of off, so physiologically the cell is non-**a**, non-α, and non-haploid (that is, it is a normal diploid).

cell of mating type α (part B), the $MAT\alpha$ region is transcribed, and two regulatory proteins denoted $\alpha1$ and $\alpha2$ are produced: $\alpha1$ is a *positive regulator* of the α-specific genes, and $\alpha2$ is a *negative regulator* of the **a**-specific genes. The result is that αsg and hsg are transcribed, but transcription of **a**sg is turned off. Both $\alpha1$ and $\alpha2$ bind with particular DNA sequences upstream from the genes that they control.

In the diploid (Figure 11.1, part C), both MAT**a** and $MAT\alpha$ are transcribed, but the only polypeptides produced are **a**1 and $\alpha2$. The reason is that the **a**1 and $\alpha2$ polypeptides combine to form a negative regulatory protein that represses transcription of the $\alpha1$ gene in $MAT\alpha$ and of the haploid-specific genes. The $\alpha2$ polypeptide acting alone is a negative regulatory protein that turns off **a**sg. Because $\alpha1$ is not produced, transcription of αsg is not turned on. The overall result is that the αsg are not turned on because $\alpha1$ is absent, the **a**sg are turned off because $\alpha2$ is present, and the hsg are turned off by the **a**1/$\alpha2$ complex. This ensures that meiosis can occur (because expression of $RME1$ is turned off) and that mating-type switching ceases (because the HO endonuclease is absent). Thus the homothallic **a**α diploid is stable and can undergo meiosis. The result is that the **a**α diploid does not transcribe either the mating-type-specific set of genes or the haploid-specific genes.

The repression of transcription of the haploid-specific genes mediated by the **a**1/$\alpha2$ protein is an example of negative control of the type already familiar from the *lac* and *trp* systems in *E. coli* (Chapter 9). The interesting twist in the yeast example is that the $\alpha2$ protein has a regulatory role of its own in repressing transcription of the **a**-specific genes. Why does the $\alpha2$ protein, on its own, not repress the haploid-specific genes as well? The answer lies in the specificity of its DNA binding. By itself, the $\alpha2$ protein has low affinity for the target sequences in the haploid-specific genes. However, the **a**1/$\alpha2$ heterodimer has both high affinity and high specificity for the target DNA sequences in the haploid-specific genes.

11.2

The determination of cell fate in *C. elegans* development is largely autonomous.

The soil nematode *Caenorhabditis elegans* (Figure 11.2) is popular for genetic studies because it is small, easy to culture, and has a short generation time with a large number of offspring. The worms are grown on agar surfaces in petri dishes and feed on bacterial cells such as *E. coli*. Because they are microscopic in size, as many as 10^5 animals can be contained in a single petri dish. Sexually mature adults of *C. elegans* are capable of laying more than 300 eggs within a few days. At 20°C, it requires about 60 hours for the eggs to hatch, undergo four larval molts, and become sexually mature adults.

Nematodes are diploid organisms with two sexes. In *C. elegans*, the two sexes are the hermaphrodite and the male. The hermaphrodite contains two X chromosomes (XX), produces both functional eggs and functional sperm, and is capable of self-fertilization. The male produces only sperm and fertilizes the hermaphrodites. The sex-chromosome constitution of *C. elegans* consists of a single X chromosome; there is no Y chromosome, and the male karyotype is XO.

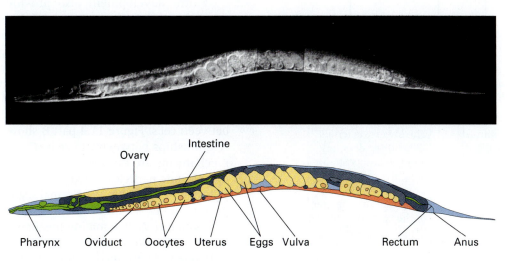

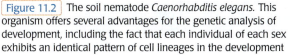

Figure 11.2 The soil nematode *Caenorhabditis elegans*. This organism offers several advantages for the genetic analysis of development, including the fact that each individual of each sex exhibits an identical pattern of cell lineages in the development of the somatic cells. DNA sequencing of the 100-megabase genome was the first eukaryote completed. [Photograph courtesy of Tim Schedl.]

■ Development in *C. elegans* exhibits a fixed pattern of cell divisions and cell lineages.

The transparent body wall of the worm has made it possible to study the division, migration, and death or differentiation of all cells present in the course of development. Nematode development is unusual in that the pattern of cell division and differentiation is virtually identical from one individual to the next. The result is that each sex shows the same geometry in the number and arrangement of somatic cells. The hermaphrodite contains exactly 959 somatic cells, and the male contains exactly 1031 somatic cells. The complete developmental history of each somatic cell is known.

The mechanisms that control early development can be studied genetically by isolating mutants with early developmental abnormalities and altered cell fates. In most organisms, it is difficult to trace the lineage of individual cells in development because the embryo is not transparent, the cells are small and numerous, and cell migrations are extensive. The *lineage* of a cell refers to the ancestor–descendant relationships among a group of cells. A cell lineage can be illustrated with a **lineage diagram,** a sort of cell pedigree that shows each cell division and indicates the terminal differentiated state of each cell. Figure 11.3 is a lineage diagram of a hypothetical cell A in which the cell

fate is either programmed cell death or one of the terminally differentiated cell types designated W, X, and Y. The letter symbols are the kind normally used for cells in nematodes, in which the name denotes the cell lineage according to ancestry and position in the embryo. For example, the cells A.a and A.p are, respectively, the anterior and posterior daughters of cell A, and A.aa and A.ap are the anterior and posterior daughters of cell A.a.

■ Cell fate is determined by autonomous development and/or intercellular signaling.

Two principal mechanisms progressively restrict the **cell fate,** or developmental outcome, of cells within a lineage.

- Developmental restriction may be **autonomous,** which means that it is determined by genetically programmed changes in the cells themselves.

- Cells may respond to **positional information,** which means that developmental restrictions are imposed by the position of cells within the embryo. Positional information may be mediated by signaling interactions between neighboring cells or by gradients in concentration of particular molecules.

Nematode development is largely autonomous, which means that in most cells, the developmental program unfolds automatically without the need for interactions with other cells. However, in the early embryo, some of the developmental fates are established by interactions among the cells. In later stages of development of these cells, the fates established early are reinforced by still other interactions between cells.

Worm development also provides important examples of the effects of intercellular signaling on determination. Figure 11.4A illustrates the first three cell divisions in the development of *C. elegans,* which result in eight embryonic cells that differ in genetic activity and developmental fate. The determination of cell fate in these early divisions is in part autonomous and in part results from interactions between cells. Figure 11.4 part B shows the lineage relationships between the cells. Cell-autonomous mechanisms are illustrated by the transmission of cytoplasmic particles called *polar granules* from the cells P0 to P1 to P2 to P3. Normal segregation of the polar granules is a function of microfilaments in the cytoskeleton. Cell-signaling mechanisms are illustrated by the effects of P2 on EMS and on ABp. The EMS fate is determined by the activity of the *mom-2* gene in P2. The P2 cell also produces a signaling molecule, APX-1, which determines the fate of ABp

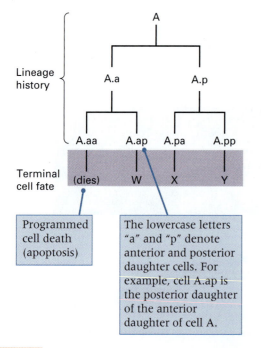

Figure 11.3 Hypothetical cell-lineage diagram. Different terminally differentiated cell fates are denoted W, X, and Y. One cell in the lineage (cell A.aa) undergoes programmed cell death.

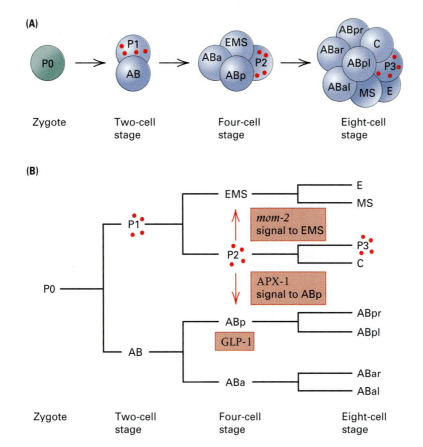

Figure 11.4 | Early cell divisions in *C. elegans* development. (A) Spatial organization of cells. (B) Lineage relationships of the cells. The transmission of the polar granules illustrates cell- autonomous development. The arrows denote cell-to-cell signaling mechanisms that determine developmental fate. [Reprinted with permission from Wade Roush. 1996. *Science* 272: 1871. © AAAS]

through the cell-surface receptor GLP-1. In contrast to *C. elegans*, in which many developmental decisions are cell-autonomous, in *Drosophila* and *Mus* (the mouse), regulation by cell-to-cell signaling is more the rule than the exception. The use of cell signaling to regulate development provides a sort of insurance that helps to overcome the death of individual cells in development that might happen by accident.

■ Developmental mutations often affect cell lineages.

Many mutations that affect cell lineages have been studied in nematodes, and they reveal several general features by which genes control development.

- The division pattern and fate of a cell are generally affected by more than one gene and can be disrupted by mutations in any of them.

- Most genes that affect development are active in more than one type of cell.

- Complex cell lineages often include simpler, genetically determined lineages within them; these components are called *sublineages* because

they are expressed as an integrated pattern of cell division and terminal differentiation.

- The lineage of a cell may be triggered autonomously within the cell itself or by signaling interactions with other cells.

- Regulation of development is controlled by genes that determine the different sublineages that cells can undergo and the individual steps within each sublineage.

The next section deals with some of the types of mutations that affect cell lineages and development.

■ Transmembrane receptors often mediate signaling between cells.

The controlling genes that cause cells to diverge in developmental fate are not always easy to recognize. For example, a mutant allele may identify a gene that is *necessary* for the expression of a particular developmental fate, but the gene may not be *sufficient* to determine the developmental fate of the cells in which it is expressed. This possibility complicates the search for genes that control major developmental decisions.

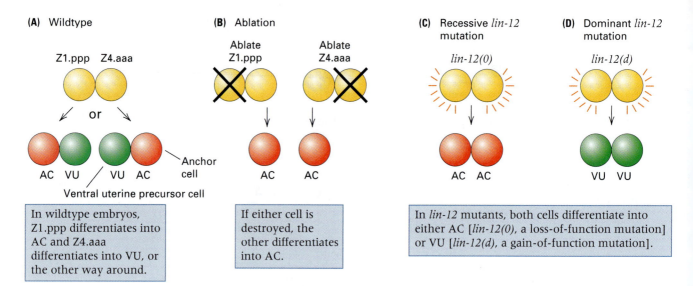

(A) Wildtype

Z1.ppp Z4.aaa

or

AC VU / VU AC — Anchor cell

Ventral uterine precursor cell

In wildtype embryos, Z1.ppp differentiates into AC and Z4.aaa differentiates into VU, or the other way around.

(B) Ablation

Ablate Z1.ppp   Ablate Z4.aaa

AC   AC

If either cell is destroyed, the other differentiates into AC.

(C) Recessive lin-12 mutation

lin-12(0)

AC AC

(D) Dominant lin-12 mutation

lin-12(d)

VU VU

In lin-12 mutants, both cells differentiate into either AC [lin-12(0), a loss-of-function mutation] or VU [lin-12(d), a gain-of-function mutation].

Figure 11.5 | Control of the fates of Z1.ppp and Z4.aaa in vulval development and genetic control of cell fate by the lin-12 gene. In recessive loss-of-function mutants [lin-12(0)], both cells become anchor cells; in dominant gain-of-function mutants [lin-12(d)], both cells become ventral uterine precursor cells.

Genes that control decisions about cell fate can sometimes be identified by the unusual characteristic that dominant or recessive mutations have opposite effects. That is, if alternative alleles of a gene result in opposite cell fates, then the product of the gene must be both necessary and sufficient for expression of the fate. Recessive mutations in genes that control development often result from **loss of function** in that the mRNA is not produced or the protein is inactive. Dominant mutations in developmental-control genes often result from **gain of function** in that the gene is overexpressed or is expressed at the wrong time.

In *C. elegans*, a relatively small number of genes have dominant and recessive alleles that affect the same cells in opposite ways. Among them is the *lin-12* gene, which controls developmental decisions in a number of cells. One example involves the cells denoted Z1.ppp and Z4.aaa in part A of Figure 11.5 . These cells lie side by side in the embryo, but they have quite different lineages. Normally, one of the cells differentiates into an *anchor cell* (AC), which participates in development of the vulva, and the other one differentiates into a *ventral uterine precursor cell* (VU). Z1.ppp and Z4.aaa are equally likely to become the anchor cell.

Direct cell–cell interaction between Z1.ppp and Z4.aaa controls the AC–VU decision. If either cell is burned away (ablated) by a laser microbeam, the remaining cell differentiates into an anchor cell (part B). This result implies that the preprogrammed fate of both Z1.ppp and Z4.aaa is that of an anchor cell. When either cell becomes committed to the anchor-cell fate, its contact with the other cell elicits the ventral-uterine-precursor-cell fate. As noted, recessive and dominant mutations of *lin-12* have opposite effects. Mutations in which *lin-12* activity is lacking or greatly reduced are denoted *lin-12(0)*. These mutations are recessive, and in the mutants both Z1.aaa and Z4.aaa become anchor cells (part C). In contrast, *lin-12(d)* mutations are those that cause *lin-12* activity to be overexpressed. These mutations are dominant or partly dominant, and in the mutants both Z1.aaa and Z4.ppp become ventral uterine precursor cells (part D).

The effects of *lin-12* mutations suggest that the wildtype gene product is a receptor of a developmental signal. The molecular structure of the *lin-12* gene product is typical of a **transmembrane receptor** protein containing regions that span the cell membrane. The LIN-12 protein shares domains with other proteins important in developmental control (Figure 11.6). The transmembrane region separates the LIN-12 protein into an extracellular part (the amino end) and an intracellular part (the carboxyl end). The extracellular part contains 13 repeats of a domain found in a mammalian peptide hormone, epidermal growth factor (EGF), as well as in the product of the *Notch* gene in *Drosophila*, which controls the decision between epidermal-cell and neural-cell fates. Nearer the transmembrane region, the amino end contains three repeats of a cysteine-rich domain also found in the *Notch* gene product. Inside the cell, the carboxyl part of the LIN-12 protein contains six repeats of a domain also found in the SWI6 proteins, which control cell division in yeast.

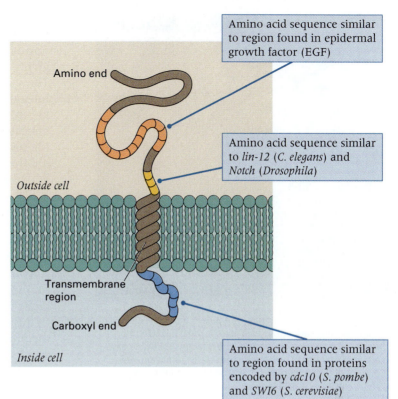

Figure 11.6 The structure of the LIN-12 protein is that of a receptor protein containing a transmembrane region and various types of repeated units that resemble those in epidermal growth factor (EGF) and other developmental control genes.

Amino acid sequence similar to region found in epidermal growth factor (EGF)

Amino acid sequence similar to *lin-12* (*C. elegans*) and *Notch* (*Drosophila*)

Amino end

Outside cell

Transmembrane region

Carboxyl end

Inside cell

Amino acid sequence similar to region found in proteins encoded by *cdc10* (*S. pombe*) and *SWI6* (*S. cerevisiae*)

■ Cells can determine the fate of other cells through ligands that bind with their transmembrane receptors.

The anchor cell expresses a signaling gene, called *lin-3*, that controls the fate of other cells in the development of the vulva. Figure 11.7 (below) illus-

trates five precursor cells, P4.p through P8.p, that participate in vulval development. Each precursor cell has the capability of differentiating into one of three fates, called the 1°, 2°, and 3° lineages, which differ according to whether descendant cells remain

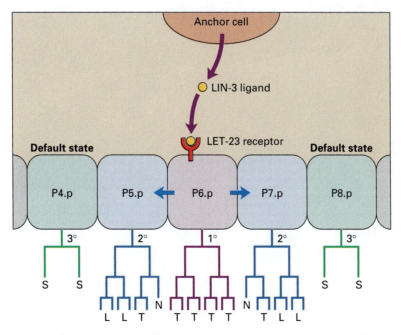

Figure 11.7 Determination of vulval differentiation by means of intercellular signaling. Cells P4.p through P8.p in the hermaphrodite give rise to lineages in the development of the vulva. The three types of lineages are designated 1°, 2°, and 3°. The 1° lineage is induced in P6.p by the ligand LIN-3 produced in the anchor cell (AC), which stimulates the LET-23 receptor tyrosine kinase in P6.p. The P6.p cell, in turn, produces a ligand that stimulates receptors in P5.p and P7.p to induce the 2° fate. On the other hand, the 3° fate is the default or baseline condition, which P4.p and P8.p adopt normally and all cells adopt in the absence of AC.

11.2 The determination of cell fate in *C. elegans* development is largely autonomous 397

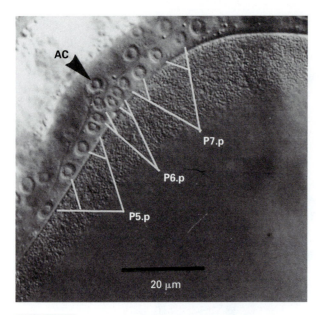

Figure 11.8 | Spatial organization of cells in the vulva, including the anchor cell (black arrowhead) and the daughter cells produced by the first two divisions of P5.p through P7.p (white tree diagrams). The length of the scale bar equals 20 micrometers. [Courtesy of G. D. Jongeward, T. R. Clandinin, and P. W. Sternberg. 1995. *Genetics* 139: 1553.]

in a syncytium (S) or divide longitudinally (L), transversely (T), or not at all (N). The precursor cells normally differentiate as shown in Figure 11.7, giving five lineages in the order 3°-2°-1°-2°-3°. The vulva itself is formed from the 1° and 2° cell lineages. The spatial arrangement of some of the key cells is shown in Figure 11.8. The black arrow indicates the anchor cell, and the white lines show the pedigrees of 12 cells. The four cells in the middle derive from P6.p, and the four on each side derive from P5.p and P7.p.

The important role of the *lin-3* gene product (LIN-3) is suggested by the opposite phenotypes of loss-of-function and gain-of-function alleles. Loss of LIN-3 results in the complete absence of vulval development, whereas overexpression of LIN-3 results in excess vulval induction. LIN-3 is a typical example of an interacting molecule, or **ligand,** that binds with an EGF-type transmembrane receptor. In this case the receptor is located in cell P6.p and is the product of the gene *let-23*. The LET-23 protein is a tyrosine-kinase receptor that, when bound with the LIN-3 ligand, stimulates a series of intracellular signaling events that ultimately results in the syn-

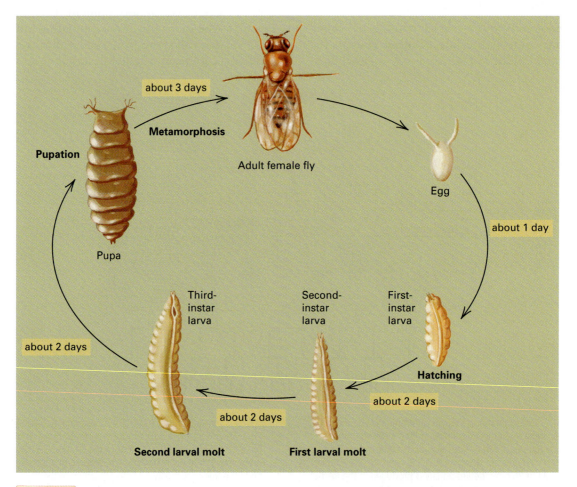

Figure 11.9 | Developmental program of *Drosophila melanogaster.* The durations of the stages are at 25°C.

398 **CHAPTER 11** The Genetic Control of Development

thesis of transcription factors that determine the 1° fate. Among the genes that are induced is a gene for yet another ligand, which binds with receptors on the cells P5.p and P7.p, causing these cells to adopt the 2° fate (horizontal arrows in Figure 11.7).

In vulval development, the adoption of the 3° lineages by the P4.p and P8.p cells is determined not by a positive signal but by the lack of a signal, because in the absence of the anchor cell, all of the cells P4.p through P8.p express the 3° lineage. Thus development of the 3° lineage is the uninduced or *default* state, which means that the 3° fate is preprogrammed into the cell and must be overridden by another signal if the cell's fate is to be altered.

11.3

Development in *Drosophila* illustrates progressive regionalization and specification of cell fate.

Many important insights into developmental processes have been gained from genetic analysis in *Drosophila*. The developmental cycle of *D. melanogaster*, summarized in Figure 11.9, includes

egg, larval, pupal, and adult stages. Early development includes a series of cell divisions, migrations, and infoldings that result in the *gastrula*. About 24 hours after fertilization, the first-stage larva, composed of about 10^4 cells, emerges from the egg. Each larval stage is called an *instar*. Two successive larval molts that give rise to the second- and third-instar larvae are followed by pupation and a complex metamorphosis that gives rise to the adult fly composed of more than 10^6 cells. In wildtype strains reared at 25°C, development requires from 10 to 12 days.

Early development in *Drosophila* takes place within the egg case (Figure 11.10, part A). The first nine mitotic divisions occur in rapid succession without division of the cytoplasm and produce a cluster of nuclei within the egg (part B). The nuclei migrate to the periphery, and the germ line is formed from about 10 **pole cells** set off at the posterior end (part C); the pole cells undergo two additional divisions and are reincorporated into the embryo by invagination. The nuclei within the embryo undergo four more mitotic divisions without division of the cytoplasm, forming the *syncytial blastoderm*, which contains about 6000 nuclei (part D). Cellularization of the blastoderm takes place from about 150 to 195 minutes after fertilization by the synthesis of membranes that separate the nuclei. The **blastoderm** formed by cellularization (part E) is a flattened hollow ball of cells that corresponds to the blastula in other animals.

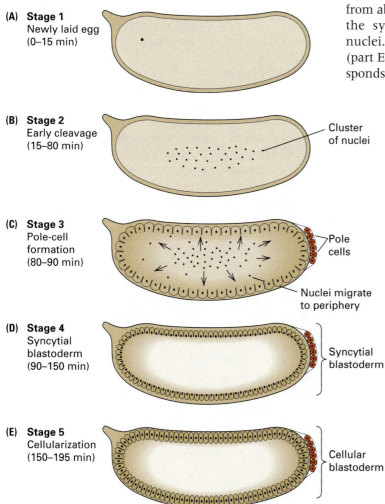

(A) **Stage 1**
Newly laid egg
(0–15 min)

(B) **Stage 2**
Early cleavage
(15–80 min)
Cluster of nuclei

(C) **Stage 3**
Pole-cell formation
(80–90 min)
Pole cells
Nuclei migrate to periphery

(D) **Stage 4**
Syncytial blastoderm
(90–150 min)
Syncytial blastoderm

(E) **Stage 5**
Cellularization
(150–195 min)
Cellular blastoderm

Figure 11.10 Early development in *Drosophila*. (A) The nucleus in the fertilized egg. (B) Mitotic divisions take place synchronously within a syncytium. (C) Some nuclei migrate to the periphery of the embryo, and at the posterior end, the pole cells (which form the germ line) become cellularized. (D) Additional mitotic divisions occur within the syncytial blastoderm. (E) Membranes are formed around the nuclei, giving rise to the cellular blastoderm.

the human connection

Maternal Impressions

David H. Skuse[1], Rowena S. James[2], Dorothy V. M. Bishop[3], Brian Coppin[4], Paola Dalton[2], Gina Aamodt-Leeper[1], Monique Barcarese-Hamilton[1], Catharine Creswell[1], Rhona McGurk[1], and Patricia A. Jacobs[2] 1997

[1]Institute of Child Health, London, England
[2]Salisbury District Hospital, Salisbury, Wiltshire, UK
[3]Medical Research Council Applied Psychology Unit, Cambridge, UK
[4]Princess Anne Hospital, Southampton, UK

Evidence from Turner's Syndrome of an Imprinted X-linked Locus Affecting Cognitive Function

Genetic imprinting is a difference in gene expression due to parental origin (Chapter 9). The mechanism of imprinting is thought to involve DNA methylation, in part because DNA methylation patterns can be inherited. Even though a relatively small number of genes are known to be imprinted in mammals, the existence of imprinting means that the maternal and paternal gametes do not contribute to embryonic development in completely equivalent ways. These authors realized that phenotypic differences due to X-chromosome imprinting might be revealed in Turner syndrome (45,X). In this chromosomal disorder, one X chromosome is missing as a result of nondisjunction. In 70 percent of Turner-syndrome females, the single X chromosome is maternal in origin ($45,X^m$); in the remaining 30 percent it is paternal ($45,X^p$). Although the 45,X chromosomal types are the same, the X^m and X^p chromosomes differ in parental origin. This controversial paper finds evidence that there may be significant behavioral differences (social interaction skills) in Turner syndrome patients, depending on the paternal origin of the chromosome. As the authors point out, this finding may have important implications for social development in chromosomally normal boys and girls.

....

Here we report a study of 80 females with Turner's syndrome, in 55 of which the X was maternally derived ($45,X^m$) and in 25 of which it was paternally derived ($45,X^p$). Members of the $45,X^p$ group were significantly better adjusted, with superior verbal and higher-order executive function skills, which mediate social interaction. Our observations suggest that there is a genetic locus for social cognition, which is imprinted and is not expressed from the maternally derived X chromosome. . . . If expressed only from the X chromosome of paternal origin, the existence of this locus could explain why 46,XY males (whose single X chromosome is maternal) are more vulnerable to developmental disorders of language and social cognition, such as autism, than are 46,XX females. . . . From a first-stage screening survey of parents and teachers using standardized tests, we discovered that 40% of $45,X^m$ girls of school age had received a statement of special educational needs, indicating academic failure, compared with 16% of $45,X^p$ subjects ($P < 0.05$); the figure in the general population is just 2%. We also found that clinically significant social difficulties affected 72.4% of the $45,X^m$ subjects over 11 years of age (21 of 29), compared with 28.6% of $45,X^p$ females (4 of 14) ($P < 0.02$). . . . Pilot interviews and observations showed that $45,X^m$ females in particular lacked flexibility and responsiveness in social interactions. We therefore devised a questionnaire relevant to social cognition. . . . The results for subjects from 6 to 18 years of age confirm there are significant differences between $45,X^m$ and $45,X^p$ in the predicted direction. . . . Normal boys also obtained significantly higher scores on the questionnaire than did normal girls, indicating significantly poorer social cognition. The magnitude and direction of this difference are compatible with the hypothesis that there is an imprinted locus on the X chromosome that influences

> **The existence of this locus could explain why 46,XY males . . . are more vulnerable to developmental disorders of language and social cognition than are 46,XX females.**

the development of social cognitive skills (although the finding is of course also compatible with other explanations of gender differences in behavior). . . . Males are substantially more vulnerable to a variety of developmental disorders of speech, language impairment and reading disability, as well as more severe conditions such as autism. Our findings are consistent with the hypothesis that the locus described, which we propose to be silent both in males and in $45,X^m$ females, acts synergistically with susceptibility loci elsewhere in the genome to increase the male-to-female prevalence ratio of such disorders.

Source: Nature 387: 705–708.

The experimental destruction of patches of cells within a *Drosophila* blastoderm results in localized defects in the larva and adult. This finding implies that cells in the blastoderm have predetermined developmental fates, with little ability to substitute in development for other, sometimes even adjacent, cells. Further evidence for this conclusion comes from experiments in which cells from a genetically marked blastoderm are implanted into host blastoderms. Blastoderm cells implanted into the equivalent regions of the host become part of the normal adult structures. However, blastoderm cells implanted into different regions develop autonomously and are not integrated into host structures.

Because of the relatively high degree of determination in the blastoderm, genetic analysis of *Drosophila* development has tended to focus on the early stages of development, when the basic body plan of the embryo is established and key regulatory processes become activated. The following sections summarize the genetic control of these early events.

■ **Mutations in a maternal-effect gene result in defective oocytes.**

Early development in *Drosophila* requires translation of maternal mRNA molecules present in the oocyte. Blockage of protein synthesis during this period arrests the early cleavage divisions. Expression of the zygote genome is also required, but the timing is different. Blockage of transcription of the zygote genome at any time after the ninth cleavage division prevents formation of the blastoderm.

Because the earliest stages of *Drosophila* development are programmed in the oocyte, mutations that affect oocyte composition or structure can upset development of the embryo. Genes that function in the mother that are needed for development of the embryo are called **maternal-effect genes,** and developmental genes that function in the embryo are called **zygotic genes.** The interplay between the two types of genes is as follows:

> **key concept**
>
> The zygotic genes interpret and respond to the positional information laid out in the egg by the maternal-effect genes.

Mutations in maternal-effect genes result in a phenotype in which homozygous females produce eggs unable to support normal embryonic development, whereas homozygous males produce normal sperm. Therefore, reciprocal crosses give dramatically different results. For example, a recessive maternal-effect mutation, *m,* will yield the following results in reciprocal crosses:

$$m/m\ ♀ \times +/+\ ♂ \rightarrow +/m\ \text{progeny}$$
(abnormal development)

$$+/+\ ♀ \times m/m\ ♂ \rightarrow +/m\ \text{progeny}$$
(normal development)

The $+/m$ progeny of the reciprocal crosses are genetically identical, but development is upset when the mother is homozygous m/m.

The reason why maternal-effect genes are needed in the mother is that the maternal-effect genes establish the polarity of the *Drosophila* oocyte even before fertilization takes place. They are active during the earliest stages of embryonic development, and they determine the basic body plan of the embryo. Maternal-effect mutations provide a valuable tool for investigating the genetic control of pattern formation and for identifying the molecules important in morphogenesis.

■ **Embryonic pattern formation is under genetic control.**

Some of the early stages in *Drosophila* development are shown in Figure 11.11. The larva that hatches from the egg features 14 superficially similar repeating units visible as a pattern of stripes along the main

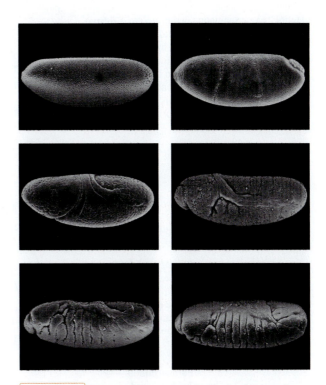

Figure 11.11 Representative stages of early development in *Drosophila* showing the pattern of segmentation that gives rise to the larval body plan. (Courtesy of Thomas C. Kaufman and Rudolph Turner.)

trunk (Figure 11.12). The stripes can be recognized externally by the bands of *denticles,* which are tiny, pigmented, tooth-like projections from the surface of the larva. The 14 stripes in the larva correspond to the segments that form from the embryo. Each **segment** is defined morphologically as the region between successive indentations formed by the sites of muscle attachment in the larval cuticle. The designations of the segments are indicated in Figure 11.12. There are three head segments (C1−C3), three thoracic segments (T1−T3), and eight abdominal segments (A1−A8.). In addition to the segments, another type of repeating unit is also important in development. These repeating units are called **parasegments;** each parasegment consists of the posterior region of one segment and the anterior region of the adjacent segment. Parasegments have a transient existence in embryonic development. Although they are not visible morphologically, they are important in gene expression because the boundaries of expression of many genes coincide with the boundaries of the parasegments rather than with those of the segments.

The early stages of pattern formation are determined by genes that are often called **segmentation genes** because they determine the origin and fate of the segments and parasegments. There are four classes of segmentation genes that differ in their times and patterns of expression in the embryo.

1. The *coordinate genes* determine the principal coordinate axes of the embryo: the anterior–posterior axis, which defines the front and rear; and the dorsal–ventral axis, which defines the top and bottom.

2. The *gap genes* are expressed in contiguous groups of segments along the embryo (Figure 11.13 , part A), and they establish the next level of spatial organization. Mutations in gap genes result in the absence of contiguous body segments, so gaps appear in the normal pattern of structures in the embryo.

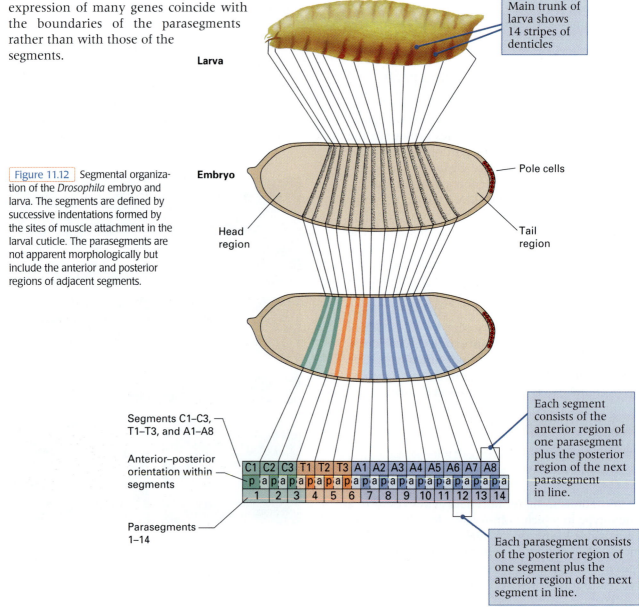

Figure 11.12 Segmental organization of the *Drosophila* embryo and larva. The segments are defined by successive indentations formed by the sites of muscle attachment in the larval cuticle. The parasegments are not apparent morphologically but include the anterior and posterior regions of adjacent segments.

Larva

Main trunk of larva shows 14 stripes of denticles

Embryo

Pole cells

Head region

Tail region

Segments C1–C3, T1–T3, and A1–A8

Anterior–posterior orientation within segments

Parasegments 1–14

Each segment consists of the anterior region of one parasegment plus the posterior region of the next parasegment in line.

Each parasegment consists of the posterior region of one segment plus the anterior region of the next segment in line.

| C1 | C2 | C3 | T1 | T2 | T3 | A1 | A2 | A3 | A4 | A5 | A6 | A7 | A8 |

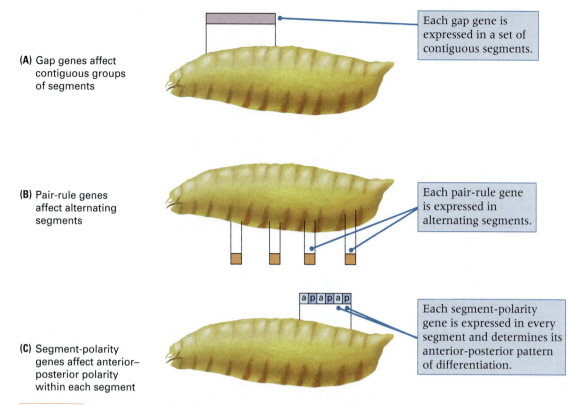

(A) Gap genes affect contiguous groups of segments

Each gap gene is expressed in a set of contiguous segments.

(B) Pair-rule genes affect alternating segments

Each pair-rule gene is expressed in alternating segments.

(C) Segment-polarity genes affect anterior–posterior polarity within each segment

a p a p a p

Each segment-polarity gene is expressed in every segment and determines its anterior-posterior pattern of differentiation.

Figure 11.13 | Patterns of expression of different types of segmentation genes.

3. The *pair-rule genes* determine the separation of the embryo into discrete segments (part B). Mutations in pair-rule genes result in missing pattern elements in alternate segments. The reason for the two-segment periodicity of pair-rule genes is that the genes are expressed in a zebra stripe pattern along the embryo.

4. The *segment-polarity genes* determine the pattern of anterior–posterior development within each segment of the embryo (part C). Mutations in segment-polarity genes affect all segments or parasegments in which the normal gene is active. Many segment-polarity mutants have the normal number of segments, but part of each segment is deleted and the remainder is duplicated in mirror-image symmetry.

Evidence for the existence of the four classes of segmentation genes—coordinate genes, gap genes, pair-rule genes, and segment-polarity genes—is presented in the following sections.

■ **Coordinate genes establish the main body axes.**

The **coordinate genes** are maternal-effect genes that establish early polarity through the presence of their products at defined positions within the oocyte or through gradients of concentration of their products. The genes that determine the anterior–posterior axis can be classified into three groups according to the effects of mutations in them, as illustrated in Figure 11.14 .

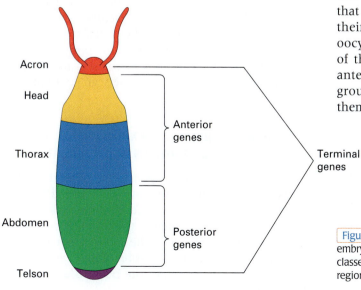

Acron
Head
Thorax
Abdomen
Telson

Anterior genes

Posterior genes

Terminal genes

Figure 11.14 | Regional differentiation of the early *Drosophila* embryo along the anterior–posterior axis. Mutations in any of the classes of genes shown result in elimination of the corresponding region of the embryo.

1. The first group of coordinate genes includes the *anterior genes*, which affect the head and thorax. The key gene in this class is *bicoid*. Mutations in *bicoid* produce embryos lacking the head and thorax that occasionally have abdominal segments in reverse polarity duplicated at the anterior end. The *bicoid* gene product is a transcription factor for genes determining anterior structures. Because the *bicoid* mRNA is localized in the anterior part of the early-cleavage embryo, these genes are activated primarily in the anterior region. The *bicoid* mRNA is produced in nurse cells (cells surrounding the oocyte) and exported to a localized region at the anterior pole of the oocyte. The protein product is less localized and, during the syncytial cleavages, forms an anterior–posterior concentration gradient with the maximum at the anterior tip of the embryo. The Bicoid protein is a transcriptional activator containing a helix–turn–helix motif for DNA binding. Genes affected by the Bicoid protein contain multiple upstream binding domains that consist of nine nucleotides resembling the consensus sequence 5'-TCTAATCCC-3'. Binding sites that differ by as many as two base pairs from the consensus sequence bind the Bicoid protein with high affinity, and sites that contain four mismatches bind with low affinity. The combination of high- and low-affinity binding sites determines the concentration of Bicoid protein needed for gene activation; genes with many high-affinity binding sites can be activated at low concentrations, but those with many low-affinity binding sites need higher concentrations. Such differences in binding affinity mean that the level of gene expression can differ from one regulated gene to the next along the Bicoid concentration gradient. It is the local concentration of the Bicoid protein that regulates the expression of critical gap genes along the embryo—for example, *hunchback*.

2. The second group of coordinate genes includes the *posterior genes*, which affect the abdominal segments (Figure 11.14). Some of the mutants also lack pole cells. One of the posterior mutations, *nanos*, yields embryos with defective abdominal segmentation but normal pole cells. The *nanos* mRNA is localized tightly to the posterior pole of the oocyte, and the gene product is a repressor of translation. Among the genes whose mRNA is not translated in the presence of Nanos protein is the gene *hunchback*. Hence *hunchback* expression is controlled jointly by the Bicoid and Nanos proteins, Bicoid protein activating transcription in an anterior–posterior gradient, and Nanos protein repressing translation in the posterior region.

3. The third group of coordinate genes includes the *terminal genes*, which simultaneously affect the most anterior structure (the acron) and the most posterior structure (the telson) (Figure 11.14). The key gene in this class is *torso*, which codes for a transmembrane receptor that is uniformly distributed throughout the embryo in the early developmental stages. The Torso receptor is activated by a signal released only at the poles of the egg by the nurse cells in that location.

Apart from the three sets of genes that determine the anterior-posterior axis of the embryo, a fourth set of genes determines the dorsal-ventral axis. The morphogen for dorsal/ventral determination is the product of the gene *dorsal,* which is present in a pronounced ventral-to-dorsal gradient in the late syncytial blastoderm.

■ Gap genes regulate other genes in broad anterior–posterior regions.

The main role of the coordinate genes is to regulate the expression of a small group of genes along the anterior–posterior axis. The genes are called **gap genes** because mutations in them result in the absence of pattern elements derived from a group of contiguous segments (Figure 11.13). Gap genes are zygotic genes. The gene *hunchback* serves as an example of the class because *hunchback* expression is controlled by offsetting effects of Bicoid and Nanos. Transcription of *hunchback* is stimulated in an anterior-to-posterior gradient by the Bicoid transcription factor, but posterior *hunchback* expression is prevented by translational repression owing to the posteriorly localized Nanos protein. In the early *Drosophila* embryo in part A of Figure 11.15 , the gradient of *hunchback* expression is indicated by the green fluorescence of an antibody specific to the hunchback gene product. The superimposed red fluorescence results from antibody specific to the product of *Krüppel,* another gap gene. The region of overlapping gene expression appears in yellow. The products of both *hunchback* and *Krüppel* are transcription factors of the zinc-finger type. Other gap genes also are transcription factors. Together, the gap genes have a pattern of regional specificity and partly overlapping domains of expression that enable them to act in combinatorial fashion to control the next set of genes in the segmentation hierarchy, the pair-rule genes.

■ Pair-rule genes are expressed in alternating segments or parasegments.

The coordinate and gap genes determine the polarity of the embryo and establish broad regions within which subsequent development takes place. As development proceeds, the progressively more refined organization of the embryo is correlated with the patterns of expression of the segmentation genes. Among these are the **pair-rule genes,** in

(A) (B) (C) (D)

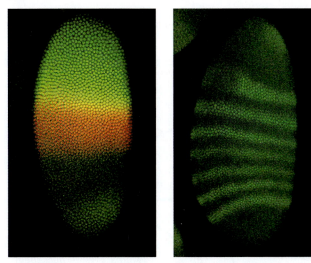
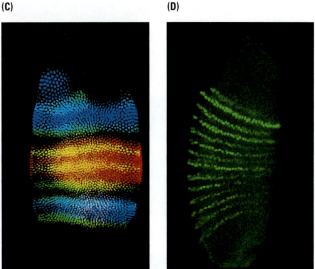

Figure 11.15 (A) An embryo of *Drosophila*, approximately 2.5 hours after fertilization, showing the regional localization of the *hunchback* gene product (green), the *Krüppel* gene product (red), and their overlap (yellow). (B) Characteristic seven stripes of expression of the gene *hairy* in a *Drosophila* embryo approximately 3 hours after fertilization. (C) Combined patterns of expression of *hairy* (green), *Krüppel* (red), and *giant* (blue) in a *Drosophila* embryo approximately 3 hours after fertilization.

Already there is considerable linear differentiation apparent in the patterns of gene expression. (D) Expression of the segment-polarity gene *engrailed* partitions the early *Drosophila* embryo into 14 regions. These eventually differentiate into three head segments, three thoracic segments, and eight abdominal segments. [Courtesy of James Langeland, Stephen Paddock, and Sean Carroll.]

which the mutant phenotype has alternating segments absent or malformed (Figure 11.13). For example, mutations of the pair-rule gene *even-skipped* affect even-numbered segments, and those of another pair-rule gene, *odd-skipped,* affect odd-numbered segments. The function of the pair-rule genes is to give the early *Drosophila* larva a segmented body pattern with both repetitiveness and individuality of segments. For example, there are eight abdominal segments that are repetitive in that they are regularly spaced and share several common features, but they differ in the details of their differentiation.

One of the earliest pair-rule genes expressed is *hairy,* whose pattern of expression is under both positive and negative regulation by the products of *hunchback, Krüppel,* and other gap genes. Expression of *hairy* occurs in seven stripes (Figure 11.15, part B). The striped pattern of pair-rule gene expression is typical, but the stripes of expression of one gene are usually slightly out of register with those of another. Together with the continued regional expression of the gap genes, the combinatorial patterns of gene expression in the embryo are already complex and linearly differentiated. Part C shows an embryo stained for the products of three genes—*hairy* (green), *Krüppel* (red), and *giant* (blue). The regions of overlapping expression appear as color mixtures—orange, yellow, light green, or purple. Even at this early stage in development, there is a unique combinatorial pattern of

gene expression in every segment and parasegment. The complexity of combinatorial control can be appreciated by considering that the expression of the *hairy* gene in stripe 7 depends on a promoter element smaller than 1.5 kb that contains a series of binding sites for the protein products of the genes *caudal, hunchback, knirps, Krüppel, tailless, huckebein, bicoid,* and perhaps still other proteins yet to be identified. The combinatorial patterns of gene expression of the pair-rule genes define the boundaries of expression of the segment-polarity genes, which function next in the hierarchy.

■ Segment-polarity genes govern differentiation within segments.

Whereas the pair-rule genes determine the body plan at the level of segments and parasegments, the **segment-polarity genes** create a spatial differentiation within each segment. The mutant phenotype has repetitive deletions of pattern along the embryo (Figure 11.13) and usually a mirror-image duplication of the part that remains. Among the earliest segment-polarity genes expressed is *engrailed,* whose stripes of expression approximately coincide with the boundaries of the parasegments and so divide each segment into anterior and posterior domains (Figure 11.15, part D).

Expression of the segment-polarity genes finally establishes the early polarity and linear differentiation of the embryo into segments and parasegments.

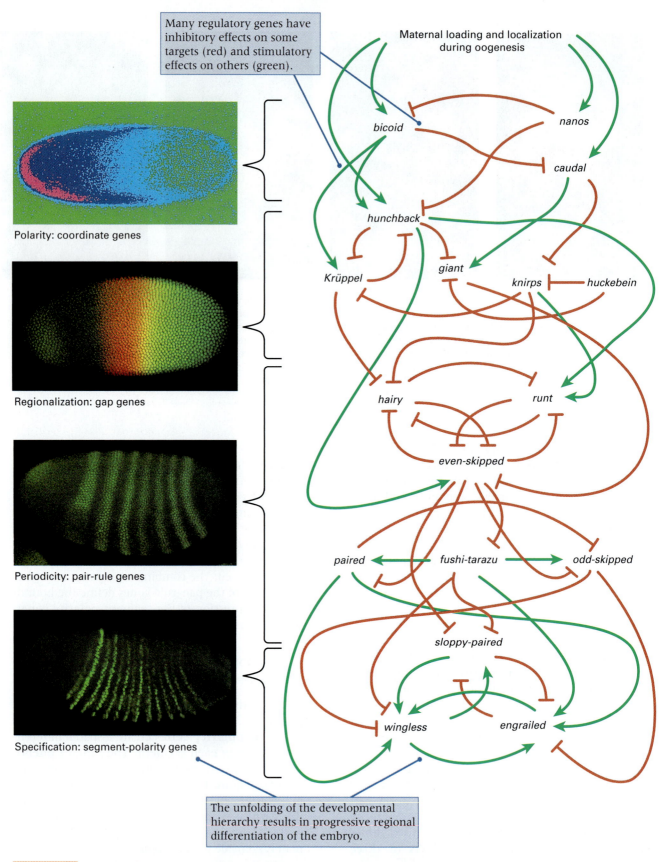

Many regulatory genes have inhibitory effects on some targets (red) and stimulatory effects on others (green).

Maternal loading and localization during oogenesis

Polarity: coordinate genes

Regionalization: gap genes

Periodicity: pair-rule genes

Specification: segment-polarity genes

bicoid

nanos

caudal

hunchback

Krüppel

giant

knirps

huckebein

hairy

runt

even-skipped

paired

fushi-tarazu

odd-skipped

sloppy-paired

wingless

engrailed

The unfolding of the developmental hierarchy results in progressive regional differentiation of the embryo.

Figure 11.16 | Hierarchy of regulatory interactions among genes controlling early development in *Drosophila*. [Diagram courtesy of George von Dassow; photos courtesy of James Langeland, Stephen Paddock, and Sean Carroll.]

Interactions among genes in the regulatory hierarchy ensure an orderly progression of developmental events.

Genes in the regulatory hierarchy are controlled by a complex set of interactions that ensure an orderly progression through the molecular events of development. Interactions among some of the coordinate genes, gap genes, pair-rule genes, and segment-polarity genes are shown in Figure 11.16. The green connectors indicate stimulatory effects, and the red connectors indicate inhibitory effects. Most of the genes are controlled by a complex set of stimulatory and inhibitory effects acting together. The coordinate genes act first to establish the polarity of the embryo, then the gap genes to differentiate large regions, after which the pair-rule genes establish the periodicity of the embryo indicated by the zebra stripes, and finally the segment-polarity genes act in the specification of the developmental identity and fate of each of the body segments. At each level in the regulatory hierarchy, the genes act to regulate other genes expressed at the same level, and also act to regulate the activity of genes that are expressed in the next downstream level in the hierarchy.

The segment-polarity genes also act to regulate downstream developmental genes that control the pathways of differentiation in each segment or parasegment, resulting ultimately in the morphology of the adult fly. The metamorphosis of the adult fly and how it emerges are discussed next.

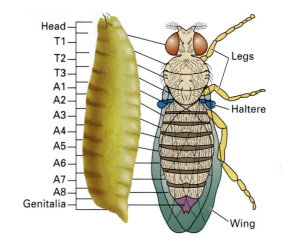

Figure 11.17 Relationship between larval and adult segmentation in *Drosophila*. Each of the three thoracic segments in the adult carries a pair of legs. The wings develop on the second thoracic segment (T2) and the halteres (flight balancers) on the third thoracic segment (T3).

Homeotic genes function in the specification of segment identity.

As with many other insects, the larvae and adults of *Drosophila* have a segmented body plan consisting of a head formed from segments C1–C3, a thorax formed from segments T1–T3, and an abdomen formed from segments A1–A8 (Figure 11.17). Metamorphosis makes use of about 20 structures called **imaginal disks** present inside the larvae (Figure 11.18). Formed early in development, the

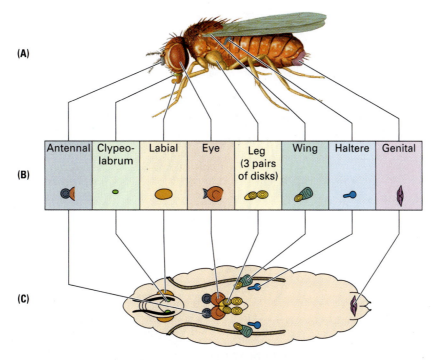

Figure 11.18 (A) Structures in the adult *Drosophila* correlated with the imaginal disks from which they arise. (B) General morphology of the disks late in larval development. (C) Larval locations of the imaginal disks.

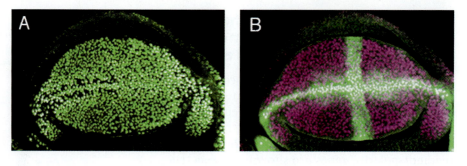

Figure 11.19 (A) Expression of the *vestigial* gene (green) in the developing wing imaginal disk. The approximately circular area of expression gives rise to the wing proper. (B) Visualization of the underlying boundary and quadrant patterns of *vestigial* expression in the same disk. [Courtesy of Stephen Paddock and Sean Carroll.]

imaginal disks ultimately give rise to the principal structures and tissues in the adult organism. Examples of imaginal disks include the pair of wing disks (one on each side of the body) that give rise to the wings and their attachments on thoracic segment T2, and the pair of haltere disks that give rise to the halteres (flight balancers) on thoracic segment T3 (Figure 11.17). During the pupal stage, when many larval tissues and organs break down, the imaginal disks progressively unfold and differentiate into adult structures. The morphogenic

events that take place in the pupa are initiated by the hormone *ecdysone*, secreted by the larval brain.

As in the early embryo, overlapping patterns of gene expression and combinatorial control guide later events in *Drosophila* development. The pattern of expression of a key gene in wing development, *vestigial*, in a wing disk is shown in part A of Figure 11.19 . The apparently uniform and approximately circular pattern of expression is actually the summation of *vestigial* response to two separate signaling pathways shown in Figure 11.20 , which

Figure 11.20 LThe uniform pattern of *vestigial* expression in the wing imaginal disk results from the superposition of two separate patterns. (A) The boundary expression pattern is determined by a dorsoventral signaling pathway. (B) The quadrant expression pattern is determined by the anteroposterior signaling pathway.

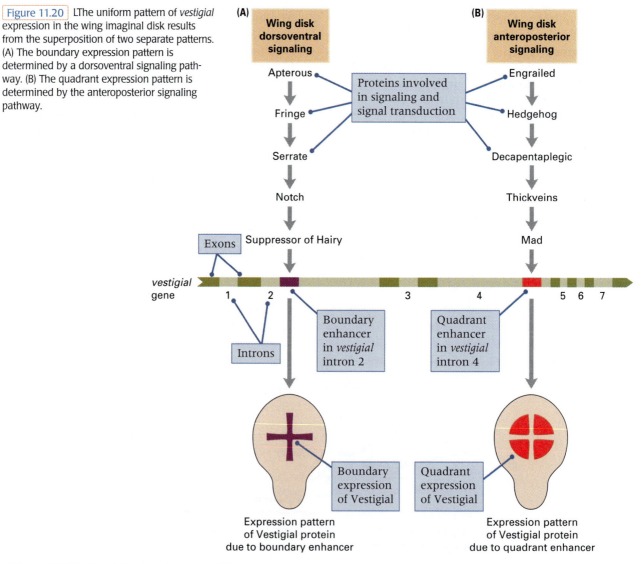

(A)

(B)

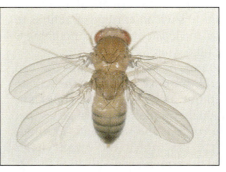

Haltere

Figure 11.21 (A) Wildtype *Drosophila* showing wings and halteres (the pair of knob-like structures protruding posterior to the wings). (B) A fly with four wings produced by mutations in the *bithorax* complex. The mutations convert the third thoracic segment into the second thoracic segment, and the halteres normally present on the third thoracic segment become converted into the posterior pair of wings. [Courtesy of E. B. Lewis.]

result in the cross-shaped and four-part patterns of expression shown at the bottom. Separate visualization of these patterns in the wing disk is shown in Figure 11.19, part B. The signaling pathway A in Figure 11.20 consists of the products of the genes *apterous, fringe, serrate,* and so forth; and pathway B consists of the products of the genes *engrailed, hedgehog,* and so forth. The Suppressor-of-Hairy protein binds to a *boundary enhancer* in the vestigial gene, which induces gene expression in the cross-shaped pattern. The Mad protein binds to a separate *quadrant enhancer,* which induces gene expression in the quadrant pattern. Such overlapping patterns of gene expression of *vestigial* and other genes in wing development ultimately yield the exquisitely fine level of cellular and morphological differentiation observed in the adult animal.

Among the genes that transform the periodicity of the *Drosophila* embryo into a body plan with linear differentiation are two small sets of **homeotic,** or ***HOX*, genes.** Homeotic mutations result in the transformation of one body segment into another, which is recognized by the misplaced development of structures that are normally present elsewhere in the embryo. One class of homeotic mutation is illustrated by *bithorax,* which causes transformation of the anterior part of the third thoracic segment into the anterior part of the second thoracic segment, with the result that the halteres normally formed from segment T3 are transformed into a pair of wings in addition to the pair normally formed from segment T2 (Figure 11.21, part B). The other class of homeotic mutation is illustrated by *Antennapedia,* which results in transformation of the antennae into legs. The *HOX* genes represented by *bithorax* and *Antennapedia* are in fact gene clusters. The cluster containing *bithorax* is designated BX-C (stands for *bithorax*-complex), and that containing *Antennapedia* is called ANT-C (stands for *Antennapedia*-complex). Both gene clusters were initially discovered through their homeotic effects in adults. Later they were shown to affect the identity of larval segments. The BX-C is primarily concerned with the development of larval segments T3 through A8 (Figure 11.17), with principal effects in T3 and A1. The ANT-C is primarily concerned with the development of the head (H) and thoracic segments T1 and T2.

The homeotic genes are transcriptional activators of other genes. Most *HOX* genes contain one or more copies of a characteristic sequence of about 180 nucleotides called a **homeobox,** which is also found in key genes concerned with the development of embryonic segmentation in organisms as diverse as segmented worms, frogs, chickens, mice, and human beings. Homeobox sequences are present in exons and code for a protein-folding domain that includes a helix–turn–helix DNA-binding motif.

■ *HOX* genes function at many levels in the regulatory hierarchy.

In *Drosophila,* a *HOX* gene called *Ultrabithorax (Ubx)* is normally active in segment T3, which gives rise to the halteres, and only in segment T3. Reduced *Ubx* expression makes the haltere more wing-like, and in the absence of *Ubx* expression, the haltere disk forms a structurally normal wing (Figure 11.21, part A). Given this information, it is tempting to say that *Ubx* "controls" the development of the haltere. But this inference is wrong, because it implies that *Ubx* expression is both necessary and sufficient for the development of the haltere. The fact is that *Ubx* homologs are expressed in T3 in probably all insects, but segment T3 carries halteres only in the dipteran insects. The ancestral condition in insects is similar to that in today's dragonfly, in which the hind wings are virtually identical to the forewings; the dragonfly *Ubx* homolog clearly does not make a haltere. In

11.3 Development in *Drosophila* illustrates progressive regionalization and specification of cell fate   **409**

the insect lineage leading to present-day lepidopterans, the hind wing became modified from the forewing, but not to such an extreme extent as in the dipterans; yet the lepidopteran *Ubx* homolog is expressed in the hind wing.

The resolution of the paradox of how *Ubx* can seemingly "control" the development of the haltere in dipterans without doing so in other insects is found in the gradual evolution of sets of genes that come under *Ubx* control. The situation is exemplified in Figure 11.22, in which the small circles denote genes and the arrows gene activation in segment T3. The condition in part A is assumed to be the ancestral state, in which the gene *c1* has evolved in such a way that it is activated by the *Ubx* homolog. In this organism, the active genes in T3 will consist of the genes activated by the *Ubx* homolog (*c1*), plus genes further along the regula-

tory hierarchy activated by these genes (*c2* and *c3*), plus other genes (not shown) whose expression in T3 is independent of the *Ubx* pathway. The complete set of *Ubx*-induced genes, direct and indirect, is shown at the right of the regulatory hierarchy. These are the genes that account for the slight modifications of the hind wing as compared with the forewing. In the absence of the *Ubx* homolog, the hind wing would lose these slight modifications and become even more like a forewing.

As evolution proceeds, mutations can occur that, by chance, either release a gene from *Ubx* control or bring a gene under *Ubx* control. Such mutations would typically include the loss or gain of *Ubx*-responsive enhancers. A newly evolved *Ubx* responsiveness, or loss of *Ubx* responsiveness, would not affect the expression of the same genes in other tissues, because *Ubx* is expressed spe-

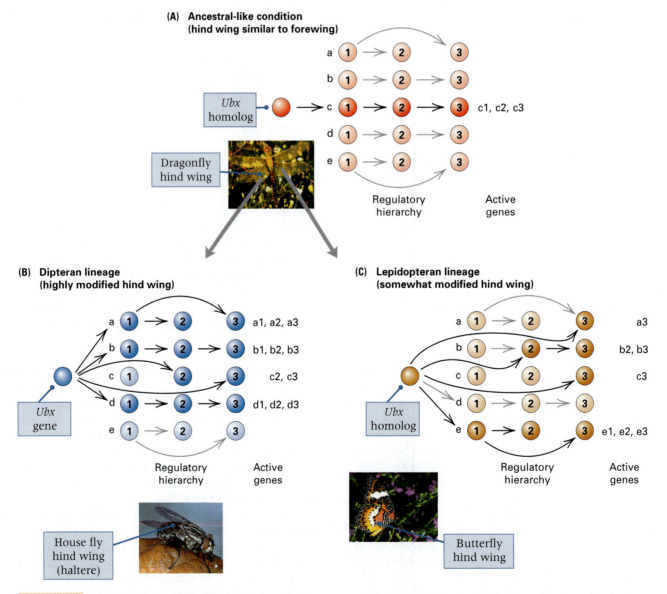

Figure 11.22 Evolutionary scheme by which the *Ubx* homolog expressed in segment T3 produces different developmental pathways for the hind wing in different insect lineages. As each line-age evolves, mutations gradually accumulate, through natural selection, that bring different sets of genes under the control of the *Ubx* homolog. [Based on ideas of Sean Carroll.]

cifically in T3. If the novel pattern of T3 gene expression is favored by natural selection, then the mutation will become incorporated into the species. After a long period of evolution, numerous genes will have evolved in this way, progressively modifying the hind-wing pattern in a manner favored by natural selection. In dipterans, for example, the selection was for extreme modification of size and structure to form the haltere. The result of the evolution is shown diagrammatically in Figure 11.22, part B by depicting $c1$ as having lost Ubx activation and $a1$ and $a3, b1, c2$ and $c3$, and $d1$ as having gained it. (Note that the Ubx gene can control genes anywhere in the regulatory hierarchy, not only at the beginning.) Similarly, in the lepidopteran lineage (part C), there was selection for less extreme modification, which is depicted as loss of $c1$ activation but gain of $a3, b2, c3$, and $e1$ and $e3$ activation. The point is that, independently in each lineage, the hind wing was modified gradually while a sequence of genes was brought under Ubx control or released from Ubx control. The result is that in dipterans the Ubx gene "controls" haltere development, whereas in lepidopterans it "controls" a less extreme modification of the forewing. In each lineage the Ubx homolog is critical in developmental control, but it regulates different genes. In each lineage also, mutations in Ubx will make the hind wing more similar to the forewing.

11.4

Floral development in *Arabidopsis* illustrates combinatorial control of gene expression.

As we have seen, most of the major developmental decisions in animals are made early in life, during embryogenesis. In higher plants, differentiation takes place almost continuously throughout life in regions of actively dividing cells called **meristems** in both the vegetative organs (root, stem, and leaves) and the floral organs (sepal, petal, stamen, and carpel). The shoot and root meristems are formed during embryogenesis and consist of cells that divide in distinctive geometric planes and at different rates to produce the basic morphological pattern of each organ system. The floral meristems are established by a reorganization of the shoot meristem after embryogenesis and eventually differentiate into floral structures characteristic of each particular species. One important difference between animal and plant development is that

> **key concept**
>
> In higher plants, as groups of cells leave the proliferating region of the meristem and undergo further differentiation into vegetative or floral tissue, their developmental fate is determined almost entirely by their position relative to neighboring cells.

The critical role of positional information in higher plant development stands in contrast to animal development, in which cell lineage often plays a key role in determining cell fate.

The plastic or "indeterminate" growth patterns of higher plants are the result of continuous production of both vegetative and floral organ systems. These patterns are conditioned largely by day length and the quality and intensity of light. The plasticity of plant development gives plants a remarkable ability to adjust to environmental insults. Figure 11.23 shows a tree that, over time, adjusts to the presence of a nearby fence by engulfing it into the trunk. Higher plants can also adjust remarkably well to a variety of genetic aberrations.

Figure 11.23 The ability of plant development to adjust to perturbations is illustrated by this tree. Encountering a fence, it eventually incorporates the fence into the trunk. [Courtesy of Robert Pruitt.]

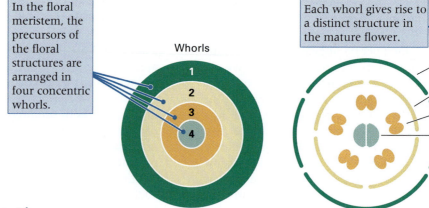

Figure 11.24 Origin of distinct floral structures from concentric whorls in the floral meristem.

In the floral meristem, the precursors of the floral structures are arranged in four concentric whorls.

Whorls
1
2
3
4

Each whorl gives rise to a distinct structure in the mature flower.

Sepal
Petal
Stamen
Carpel

(A)   (B)

■ Flower development in *Arabidopsis* is controlled by MADS box transcription factors.

Genetic analysis of *Arabidopsis thaliana,* a member of the mustard family, has revealed important principles in the genetic determination of floral structures. As is typical of flowering plants, the flowers of *Arabidopsis* are composed of four types of organs arranged in concentric rings, or whorls. Figure 11.24 illustrates the geometry, looking down at a flower from the top. From outermost to innermost, the whorls are designated 1, 2, 3, and 4 (part A). In the development of the flower, each whorl gives rise to a different floral organ (part B). Whorl 1 yields the sepals (the green, outermost floral leaves), whorl 2 the petals (the white, inner floral leaves), whorl 3 the stamens (the male organs, which form pollen), and whorl 4 the carpels (which fuse to form the ovary).

Mutations that affect floral development fall into three major classes, each with a characteristic phenotype (Figure 11.25). Compared with the wild-type flower (panel A), one class lacks sepals and petals (panel B), another class lacks petals and stamens (panel C), and the third class lacks stamens and carpels (panel D). On the basis of crosses between homozygous mutant organisms, these classes of mutants can be assigned to different complementation groups, each of which defines a different gene. The key genes and their mutant phenotypes are listed in **Table 11.1**.

- The phenotype lacking sepals and petals is caused by mutations in the gene *ap1 (apetala-1).*

- The phenotype lacking stamens and petals is caused by a mutation in either of two genes, *ap3 (apetala-3)* or *pi (pistillata).*

- The phenotype lacking stamens and carpels is caused by mutations in the gene *ag (agamous).*

These genes encode transcription factors that are members of the *MADS box* family of transcription factors. MADS box transcription factors include a common sequence motif consisting of 58 amino acids. They are involved frequently in transcriptional regulation in plants and to a lesser extent in animals.

Q A Moment to Think

Problem: Given the critical role of the Ap1, Ap3/Pi, and Ag transcription factors in floral determination, it might be speculated that triple mutants lacking all three types of transcription factors would have very strange flowers. Can you predict what the floral phenotype of an *ap1 pi ag* triple mutant would be? (The answer can be found on page 414.)

Table   11.1

Floral development in mutants of *Arabidopsis*

Genotype	Whorl			
	1	2	3	4
Wildtype	Sepals	Petals	Stamens	Carpels
ap1/ap1	Carpels	Stamens	Stamens	Carpels
ap3/ap3	Sepals	Sepals	Carpels	Carpels
pi/pi	Sepals	Sepals	Carpels	Carpels
ag/ag	Sepals	Petals	Petals	Sepals

(A) Wildtype

(B) *apetala-1 (ap1)*

(C) *pistillata (pi)*

(D) *agamous (ag)*

Figure 11.25 Phenotypes of the major classes of floral mutations in *Arabidopsis*. (A) The wildtype floral pattern consists of concentric whorls of sepals, petals, stamens, and carpels. (B) The homozygous mutation *ap1* (*apetala-1*) results in flowers missing sepals and petals. (C) Genotypes that are homozygous for either *ap3* (*apetala-3*) or *pi* (*pistillata*) yield flowers that have sepals and carpels but lack petals and stamens. (D) The homozygous mutation *ag* (*agamous*) yields flowers that have sepals and petals but lack stamens and carpels. [Courtesy of Elliot M. Meyerowitz and John Bowman. Part B from Elliot M. Meyerowitz. 1994. *Scientific American* 271: 56.]

■ **Flower development in *Arabidopsis* is controlled by the combination of genes expressed in each concentric whorl.**

The role of the *ap1, ap3, pi*, and *ag* transcription factors in the determination of floral organs can be inferred from the phenotypes of the mutations. The logic of the inference is based on the observation (see Table 11.1) that mutation in any of the genes eliminates two floral organs that arise from adjacent whorls. This pattern suggests that *ap1* is necessary for sepals and petals, *ap3* and *pi* are both necessary for petals and stamens, and *ag* is necessary for stamens and carpels. Because the mutant phenotypes are caused by loss-of-function alleles of the genes, it may be inferred that *ap1* is expressed in whorls 1 and 2, that *ap3* and *pi* are expressed in whorls 2 and 3, and that *ag* is expressed in whorls 3 and 4. The overlapping patterns of expression are shown in **Table 11.2**.

The model of gene expression in Table 11.2 suggests that floral development is controlled in combinatorial fashion by the four genes. Sepals develop from tissue in which only *ap1* is active; petals are evoked by a combination of *ap1, ap3*, and *pi*; stamens are determined by a combination of *ap3, pi*, and *ag*; and carpels derive from tissue in which only *ag* is expressed. This model is illustrated graphically

Table 11.2

Domains of expression of genes determining floral development

Whorl	Genes expressed	Determination
1	*ap1*	Sepal
2	*ap1* + *ap3* and *pi*	Petal
3	*ap3* and *pi* + *ag*	Stamen
4	*ag*	Carpel

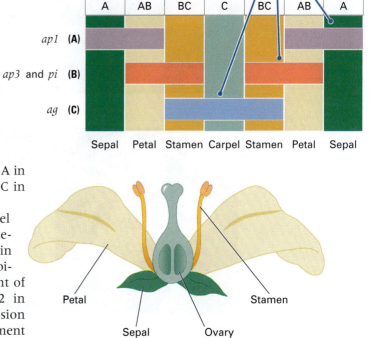

in Figure 11.26 . This model of floral determination is often called the **flower ABC model** because the wildtype activity of *ap1* was originally designated A, that of *ap3* and *pi* acting together as B, and that of *ag* as C. Therefore, the combination of activities present in each whorl would be represented as A in whorl 1, AB in whorl 2, BC in whorl 3, and C in whorl 4 (Figure 11.26).

You may have noted already that the model in Table 11.2 does not account for all of the phenotypic features of the *ap1* and *ag* mutations in Table 11.1. In particular, according to the combinatorial model in Table 11.2, the development of carpels and stamens from whorls 1 and 2 in homozygous *ap1* plants would require expression of *ag* in whorls 1 and 2. Similarly, the development of petals and sepals from whorls 3 and 4 in homozygous *ag* plants would require expression of *ap1* in whorls 3 and 4. This discrepancy can be explained if it is assumed that *ap1* expression and *ag* expression are mutually exclusive: In the presence of the AP1 transcription factor, *ag* is repressed; in the presence of the AG transcription factor, *ap1* is repressed. If this were the case, then in *ap1* mutants, *ag* expression would spread into whorls 1 and 2; in *ag* mutants, *ap1* expression would spread into whorls 3 and 4. This additional assumption enables us to explain the phenotypes of the single and even double mutants.

With the additional assumption we have made about *ap1* and *ag* interaction, the model in Table 11.2 fits the data. But is the model correct? For these genes, the patterns of gene expression, assayed by *in situ* hybridization of RNA in floral cells with labeled probes for each of the genes, fit the patterns in Table 11.2. In particular, *ap1* is expressed in whorls 1 and 2, *ap3* and *pi* in whorls 2 and 3, and *ag* in whorls 3 and 4. Furthermore, the seemingly arbitrary assumption about *ap1* and *ag* expression being mutually exclusive turns out to be true. In *ap1* mutants, *ag* is expressed in whorls 1 and 2; reciprocally, in *ag* mutants, *ap1* is expressed in whorls 3 and 4. It is also known how *ap3* and *pi* work together. The active transcription factor that corresponds to these genes is a dimeric protein composed of Ap3 and Pi polypeptides.

Each component polypeptide, in the absence of the other, remains inactive in the cytoplasm. Together, they form an active dimeric transcription factor that migrates into the nucleus.

A Moment to Think

Answer to Problem: The developing flower of the *ap1 pi ag* triple mutant still consists of four concentric whorls, but each whorl lacks any of the combinations of transcription factors necessary for the development of sepals, petals, stamens, or carpels. Each whorl therefore develops according to its default state, which is that of a leaf. Hence the predicted floral phenotype lacks all of the normal floral organs. The flowers consist merely of leaves arranged in concentric whorls. The accompanying photograph shows one flower from such a triple mutant. It is indeed very strange.

chapter summary

11.1 Mating type in yeast illustrates transcriptional control of development.

Mating type in budding yeast is controlled by transcription factors located in the *MAT* (mating type) locus. In a *MAT***a** cell, the haploid-specific genes (*hsg*), the **a**-specific genes (*asg*), and the α-specific genes (αsg) are all expressed at their default levels: The *hsg* are "on," the *asg* are "on," and the αsg are "off," yielding a cell that is an **a** haploid. In a *MAT*α cell, the $\alpha 1$ protein turns on transcription of the αsg, and the $\alpha 2$ protein shuts down transcription of the *asg*; the cell is therefore an α haploid. In a *MAT***a**/*MAT*α diploid, the **a**1 and $\alpha 2$ polypeptides acting together repress transcription of the $\alpha 1$ gene in *MAT*α and of the *hsg*, and the $\alpha 2$ polypeptide acting alone represses transcription of the *asg*; the result is a non-**a**, non-α nonhaploid (that is, a normal diploid).

11.2 The determination of cell fate in *C. elegans* development is largely autonomous.

- Development in *C. elegans* exhibits a fixed pattern of cell divisions and cell lineages.
- Cell fate is determined by autonomous development and/or intercellular signaling.
- Developmental mutations often affect cell lineages.
- Transmembrane receptors often mediate signaling between cells.
- Cells can determine the fate of other cells through ligands that bind with their transmembrane receptors.

The soil nematode *Caenorhabditis elegans* is used widely in studies of development, because many cell lineages in the organism undergo virtually autonomous development and the developmental program is identical from one organism to the next. Most lineages are affected by many genes, including genes that control the sublineages into which the lineage can differentiate.

Genes that control key points in development can often be identified by the unusual feature that recessive alleles (loss-of-function mutations) and dominant alleles (gain-of-function mutations) result in opposite effects on phenotype. For example, if loss of function results in failure to execute a developmental program in a particular anatomical position, then gain of function should result in execution of the program in an abnormal location. In this and other organisms, cell fate is often induced by specific proteins (ligands) that bind with their counterpart transmembrane receptors; the binding initiates a cascade of events within the cell that stimulates the expression of a specific developmental program.

11.3 Development in *Drosophila* illustrates progressive regionalization and specification of cell fate.

- Mutations in a maternal-effect gene result in defective oocytes.
- Embryonic pattern formation is under genetic control.
- Coordinate genes establish the main body axes.
- Gap genes regulate other genes in broad anterior–posterior regions.
- Pair-rule genes are expressed in alternating segments or parasegments.

- Segment-polarity genes govern differentiation within segments.
- Interactions among genes in the regulatory hierarchy ensure an orderly progression of developmental events.
- Homeotic genes function in the specification of segment identity.
- *HOX* genes function at many levels in the regulatory hierarchy.

Early development in *Drosophila* includes the formation of a syncytial blastoderm by early cleavage divisions without cytoplasmic division, the setting apart of pole cells that form the germ line at the posterior of the embryo, the migration of most nuclei to the periphery of the syncytial blastoderm, cellularization to form the cellular blastoderm, and determination of the blastoderm fate map at or before the cellular blastoderm stage. Metamorphosis into the adult fly makes use of about 20 imaginal disks present in the larva that contain developmentally committed cells that divide and develop into the adult structures. Most imaginal disks include several discrete groups of cells or compartments separated by boundaries that progeny cells do not cross.

Early development in *Drosophila* to the level of segments and parasegments requires four classes of segmentation genes: (1) coordinate genes that establish the basic anterior–posterior and dorsal–ventral aspect of the embryo, (2) gap genes for longitudinal separation of the embryo into regions, (3) pair-rule genes that establish an alternating on–off striped pattern of gene expression along the embryo, and (4) segment-polarity genes that refine the patterns of gene expression within the stripes and determine the basic layout of segments and parasegments. The segmentation genes can regulate themselves, other members of the same class, and genes of other classes further along the hierarchy. Together, the segmentation genes control the homeotic *(HOX)* genes that initiate the final stages of developmental specification.

Mutations in homeotic genes result in the transformation of one body segment into another. For example, *bithorax* causes transformation of the anterior part of the third thoracic segment T3 into the anterior part of the second thoracic segment T2. Homeotic genes act within developmental compartments to control other genes concerned with such characteristics as rates of cell division, orientation of mitotic spindles, and the capacity to differentiate bristles, legs, and other features.

11.4 Floral development in *Arabidopsis* illustrates combinatorial control of gene expression.

- Flower development in *Arabidopsis* is controlled by MADS box transcription factors.
- Flower development in *Arabidopsis* is controlled by the combination of genes expressed in each concentric whorl.

Developmental processes in higher plants differ significantly from those in animals in that developmental decisions continue throughout life in the meristem regions of the vegetative organs (root, stem, and leaves) and the floral organs (sepal, petal, stamen, and carpel). However, genetic control of plant development is mediated by transcription factors analogous to those in animals. Floral development in *Arabidopsis* is controlled through the combinatorial expression of a small set of MADS box transcription factors in each of a series of four concentric rings, or whorls, of cells that eventually form the sepals, petals, stamens, and carpels.

issues & ideas

- What is meant by *positional information* in regard to development? How can positional information affect cell fate?

- If a gene is both necessary and sufficient for determining a developmental pathway, why would loss-of-function mutants be expected to have a different phenotype than gain-of-function mutants?

- What is a transmembrane receptor? What is a ligand? What role do these types of molecules play in signaling between cells?

- Why was the study of maternal-effect lethal genes a key to deciphering the genetic control of early embryogenesis in *Drosophila?*

- Do plants have a germ line in the same sense as animals? What does the difference in germ-cell origin imply about the potential role of "somatic" mutations in the evolution of each type of organism?

- How does the genetic determination of floral development in *Arabidopsis* illustrate the principle of combinatorial control?

key terms & concepts

autonomous determination	gap gene	maternal-effect gene	positional information
blastoderm	homeobox	meristem	segment
cell fate	homeotic gene (*HOX* gene)	pair-rule gene	segmentation gene
cell lineage	imaginal disk	parasegment	segment-polarity gene
coordinate gene	ligand	pattern formation	transmembrane receptor
flower ABC model	lineage diagram	pole cell	zygotic gene
gain-of-function mutation	loss-of-function mutation		

1. _____ A developmental program that unfolds independently of external signals or position in the embryo.

2. _____ A pedigree of cells in a developing embryo.

3. _____ Molecule that binds with a receptor.

4. _____ The phenotypic effect of this type of mutation is usually expected to be dominant.

5. _____ In *Drosophila,* a type of cell from which the germ line develops.

6. _____ A gene required for females to be able to form functional oocytes.

7. _____ A developmental-control gene that, when mutated, results in the absence of contiguous segments along a *Drosophila* larva.

8. _____ Developmental-control gene that determines the anterior-posterior orientation of segments in the *Drosophila* embryo.

9. _____ A developmental-control gene that, when mutant, results in the replacement of one body part or structure with a different body part or structure.

10. _____ Any of about 20 structures in the *Drosophila* late larva from which discrete structures of the adult develop.

11. _____ Growing point in a higher plant.

12. _____ Theory for a developmental pattern that may be represented as A–AB–BC–C–BC–AB–A.

solutions: step by step

Problem 1

Flies of *Drosophila melanogaster* that are mutant for the gene *eyeless* are almost completely devoid of eye tissue in the head region normally occupied by the compound eyes. A deletion of *eyeless,* when homozygous, produces the eyeless phenotype. Transgenic flies that carry the wildtype *eyeless* gene under the control of a promoter that causes expression in abnormal body parts (called *ectopic expression*) develop misplaced eye tissue in these body parts. Some of the phenotypes are shown in the figure on the next page. Part A shows the result of ectopic expression in the entire head; eyes develop in the normal place, but eye tissue also develops instead of the antennae. Other results of ectopic expression are shown in parts B through D. Part B shows a wing with eye tissue growing out from it, part C a single antenna in which most of the third segment consists of eye tissue, and part D a middle leg with an eye outgrowth at the base of the tibia. What do these results imply about the role of the *eyeless* gene in eye development?

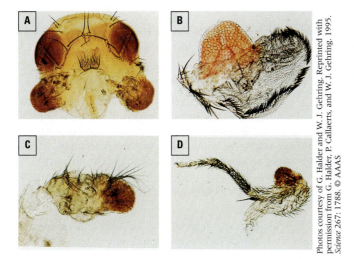

Photos courtesy of G. Halder and W. J. Gehring. Reprinted with permission from G. Halder, P. Callaerts, and W. J. Gehring. 1995. *Science* 267: 1788. © AAAS

results is that *eyeless* is a key regulatory gene in eye development. The expression of *eyeless* is *necessary* for eye tissue development and also *sufficient* for eye tissue development. (Interestingly, the human genome contains a homolog of *eyeless*, which also plays a critical role in the specification of eye development.)

Problem 2

Consider a hypothetical mutant protease that affects floral development in *Arabidopsis thaliana*. The protease has an altered substrate specificity that enables it to cleave and inactivate both Ap1 and Ag proteins (the products of *ap1* and *ag*, respectively). In view of the fact that tissue containing the Ap3/Pi dimeric protein, but neither Ap1 nor Ag alone, develops into floral organs intermediate between petals and stamens, what floral phenotype would be expected in the protease mutant?

■ **Solution** In whorl 1, Ap1 activity is missing, so this region will develop as a whorl of leaves. Likewise, in whorl 4, Ag activity is missing, so this region will develop as a whorl of leaves. In whorls 2 and 3, only Ap3/Pi is present, so these will develop as whorls of tissue intermediate between petals and stamens. Therefore, the flower phenotype will be leaves in whorls 1 and 4 and will be petal/stamen intermediates in whorls 2 and 3.

■ **Solution** The key feature of this situation is that loss-of-function mutants and gain-of-function mutants have opposite phenotypes. A deletion of *eyeless* is a loss-of-function mutation, and normal eye development is completely absent. The transgenic flies with ectopic expression are the equivalent of gain-of-function mutants, and they develop eyes in the ectopic locations. The inference from these

concepts in action: problems for solution

11.1 Why is transcription of the zygote nucleus dispensable in early *Drosophila* development but not in early development of the mouse?

11.2 Cell death (apoptosis) is responsible in part for shaping many organs and tissues in normal development. If a group of cells in the duck leg primordium that are destined to die are transplanted from their normal leg site to another part of the embryo just before they would normally die, they still die on schedule. The same operation performed a few hours earlier rescues the cells, and they do not die. How can you explain this observation?

11.3 Classify each of the following mutant alleles as a cell-lineage mutation, a homeotic mutation, or a pair-rule mutation.
(a) A mutant allele in *C. elegans* in which a cell that normally produces two daughter cells with different fates gives rise to two daughter cells with identical fates.
(b) A mutation in *Drosophila* that causes an antenna to appear at the normal site of a leg.
(c) A lethal mutation in *Drosophila* that is responsible for abnormal gene expression in alternating segments of the embryo.

11.4 How would you prove that a newly discovered mutant allele has a maternal effect?

11.5 Distinguish between a loss-of-function mutation and a gain-of-function mutation. Can the same gene undergo both types of mutations? Can the same allele have both types of effects?

11.6 The drug actinomycin D prevents RNA transcription but has little direct effect on protein synthesis. When ferti-

lized sea urchin eggs are immersed in a solution of the drug, development proceeds to the blastula stage, but gastrulation does not take place. How would you interpret this finding?

11.7 A mutant allele in the axolotl designated *o* is a maternal-effect lethal because embryos from *oo* females die at gastrulation, irrespective of their own genotype. However, the embryos can be rescued by injecting oocytes from *oo* females with an extract of nuclei from either $o^+ o^+$ or $o^+ o$ eggs. Injection of cytoplasm is not effective. Suggest an explanation for these results.

11.8 The same transmembrane receptor protein encoded by the *lin-12* gene is used in the determination of different developmental fates in different cell lineages. Suggest a mechanism by which the same receptor can determine different fates in different cell types.

11.9 Can the key genes involved in the specification of floral organ identity in *Arabidopsis* be regarded as homeotic genes? Explain your answer.

11.10 A particular gene is necessary, but not sufficient, for a certain developmental fate. What is the expected phenotype of a loss-of-function mutation in the gene? Is the allele expected to be dominant or recessive?

11.11 What floral phenotype would be expected of a mutation resulting in a loss-of-function of *ap3* and *pi*? A gain-of-function mutation in which both *ap3* and *pi* were expressed in whorls 1 and 4 as well as whorls 2 and 3? What do these results imply about Ap3/Pi being necessary or necessary and sufficient for the developmental fate of the four whorls?

11.12 What phenotype would be expected of a gain-of-function mutation in *Arabidopsis* that resulted in expression of *Ap3* and *Pi* in whorl 1?

11.13 The homeotic floral-identity genes *X, Y,* and *Z* in whirligigs act either separately or in pairs to control the identity of floral organs. The floral structure consists of four whorls, the outermost being whorl 1 and the innermost whorl 4. Gene product X is expressed in whorls 1 and 2, Y in 2 and 3, and Z in 3 and 4. The presence of X without Y induces sepals, X + Y together induce petals, Y + Z together induce stamens, and Z without Y induces carpels. In the absence of X, the domain of activity of Z expands to all four whorls. In the absence of Z, the domain of activity of X expands to all four whorls. Mutant alleles of *X, Y,* and *Z* eliminate characteristic floral organs from a specific whorl, and a different floral organ appears in its place. What would you expect for the phenotype of each of the following?
(a) A loss-of-function *X* allele
(b) A loss-of-function *Y* allele
(c) A loss-of-function *Z* allele

11.14 Using the information presented in the foregoing whirligig problem, deduce the expected floral phenotype of a double mutant homozygous for loss-of-function alleles of both *X* and *Y.*

11.15 Two classes of genes involved in segmentation of the *Drosophila* embryo are gap genes, which are expressed in one region of the developing embryo, and pair-rule genes, which are expressed in seven stripes. Homozygotes for mutations in gap genes lack a continuous block of larval segments; homozygotes for mutations in pair-rule genes lack alternating segments. You examine gene expression by mRNA *in situ* hybridization and find that (1) the embryonic expression pattern of gap genes is normal in all pair-rule mutant homozygotes and (2) the pair-rule gene expression pattern is abnormal in all gap gene mutant homozygotes. What do these observations tell you about the temporal hierarchy of gap genes and pair-rule genes in the developmental pathway of segmentation?

11.16 Edward B. Lewis, Christianne Nüsslein-Volhard, and Eric Wieschaus shared a 1995 Nobel Prize in physiology or medicine for their work on the developmental genetics of *Drosophila.* In their screen for developmental genes, Nüsslein-Volhard and Wieschaus initially identified 20 lines bearing maternal-effect mutations that produced embryos lacking anterior structures but having the posterior structures duplicated. When Nüsslein-Volhard mentioned this result to a colleague, the colleague was astounded to learn that mutations in 20 genes could give rise to this phenotype. Explain why his surprise was completely unfounded.

11.17 The nuclei of brain cells in the adult frog normally do not synthesize DNA or undergo mitosis. However, when transplanted into developing oocytes, the brain cell nuclei behave as follows: (a) In rapidly growing premeiosis oocytes, they synthesize RNA. (b) In more mature oocytes, they do not synthesize DNA or RNA, but their chromosomes condense and they begin meiosis. How would you explain these results?

11.18 Explain why, in accounting for the phenotypes of floral mutants in *Arabidopsis*, it was necessary to postulate that:
(a) In *agamous* mutants, the domain of expression of *apetala-1* expands to whorls 3 and 4.
(b) In *apetala-1* mutants, the domain of expression of *agamous* expands to whorls 1 and 2.

11.19 The autosomal gene *rosy* (*ry*) in *Drosophila* is the structural gene for the enzyme xanthine dehydrogenase (XDH), which is necessary for wildtype eye pigmentation. Flies of genotype *ry/ry* lack XDH activity and have rosy eyes. The X-linked gene *maroonlike* (*mal*) is also necessary for XDH activity, and *mal/mal*; ry^+/ry^+ females and *mal*/Y; ry^+/ry^+ males also lack XDH activity; they have maroonlike eyes. The cross mal^+/mal; *ry/ry* females × *mal*/Y; ry^+/ry^+ males produces *mal/mal*; ry^+/ry females and *mal*/Y; ry^+/ry males that have wildtype eye color even though their genotype would imply that they should have rosy eyes. Suggest an explanation.

11.20 Genetic analysis is being carried out on a specialized appendage of a marine invertebrate that is used in immobilizing prey. The wildtype appendage, shown at the top in the accompanying diagram, has three segments. The proximal segment, attached directly to the body, has a single row of sparse bristles; the medial segment has a single row of dense bristles; and the distal segment has two symmetrical rows of dense bristles. Three mutations that affect the appendage are isolated; they cause the phenotypes shown. The presence of a segment lacking bristles may be regarded as the default state of development of this appendage. Suggest a hypothesis to explain how the genes *a, b,* and *c* are involved in the developmental determination of this appendage.

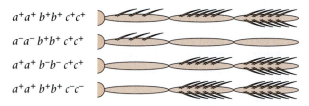

a^+a^+ b^+b^+ c^+c^+

a^-a^- b^+b^+ c^+c^+

a^+a^+ b^-b^- c^+c^+

a^+a^+ b^+b^+ c^-c^-

geNETics on the web

GeNETics on the Web will introduce you to some of the most important sites for finding genetic information on the Internet. To explore these sites, visit the Jones and Bartlett home page at

http://www.jbpub.com/genetics

For the book *Essential Genetics: A Genomics Perspective,* choose the link that says Enter **GeNETics on the Web**. You will be presented with a chapter-by-chapter list of highlighted keywords. Select any highlighted keyword and you will be linked to a Web site containing genetic information related to the keyword.

- A highly effective way to appreciate the complex, interactive hierarchy of gene activation and repression in *Drosophila* development is to explore the **interactive fly** at this keyword site. Created by Tom Brody and Judith Brody, this resource enables you to learn more about the stages and time course of developmental processes and much more about the genes that are involved—all richly illustrated with drawings and photographs.

- Check out this site to learn about other genes involved in **flower development** in *Arabidopsis* and to see pictures of the mutant phenotypes. You can also learn more about the family of over 30 *Arabidopsis* genes known as MADS-box genes, which encode transcription factors, many of which are important in flower development.

further readings

Capecchi, M. R., ed. 1989. *The Molecular Genetics of Early Drosophila and Mouse Development.* Cold Spring Harbor, NY: Cold Spring Harbor Laboratory.

Davis, G. K., and N. H. Patel. 1999. The origin and evolution of segmentation. *Trends in Biochemical Sciences* 24: M68.

De Robertis, E. M., G. Oliver, and C. V. E. Wright. 1990. Homeobox genes and the vertebrate body plan. *Scientific American,* July.

Gaul, U., and H. Jäckle. 1990. Role of gap genes in early *Drosophila* development. *Advances in Genetics* 27: 239.

Grunert, S., and D. St. Johnston. 1996. RNA localization and the development of asymmetry during *Drosophila* oogenesis. *Current Opinion in Genetics & Development* 6: 395.

Irish, V. 1987. Cracking the *Drosophila* egg. *Trends in Genetics* 3: 303.

Kaufman, T. C., M. A. Seeger, and G. Olsen. 1990. Molecular and genetic organization of the Antennapedia gene complex of *Drosophila melanogaster. Advances in Genetics* 27: 309.

Kornfeld, K. 1997. Vulval development in *Caenorhabditis elegans. Trends in Genetics* 13: 55.

Lawrence, P. A. 1992. *The Making of a Fly: The Genetics of Animal Design.* Oxford, England: Blackwell.

Ma, H. 1998. To be, or not to be, a flower: Control of floral meristem identity. *Trends in Genetics* 14: 26.

Meyerowitz, E. M. 1996. Plant development: Local control, global patterning. *Current Opinion in Genetics & Development* 6: 475.

Morisato, D., and K. V. Anderson. 1995. Signaling pathways that establish the dorsal–ventral pattern of the *Drosophila* embryo. *Annual Review of Genetics* 29: 371.

Nüsslein-Volhard, C. 1996. Gradients that organize embryo development. *Scientific American,* August.

Riverapomar, R., and H. Jäckle. 1996. From gradients to stripes in *Drosophila* embryogenesis: Filling in the gaps. *Trends in Genetics* 12: 478.

Sternberg, D. W. 1990. Genetic control of cell type and pattern formation in *Caenorhabditis elegans. Advances in Genetics* 27: 63.

Weigel, D. 1995. The genetics of flower development: From floral induction to ovule morphogenesis. *Annual Review of Genetics* 29: 19.

Wieschaus, E. 1996. Embryonic transcription and the control of developmental pathways. *Genetics* 142: 5.

Wolpert, L. 1996. One hundred years of positional information. *Trends in Genetics* 12: 359.

Wood, W. B., ed. 1988. *The Nematode* Caenorhabditis elegans. Cold Spring Harbor, NY: Cold Spring Harbor Laboratory.

This proud peacock lacks pigment in its feathers, but its cells are nevertheless capable of producing pigment as can be seen by the pigment in the eyes, which otherwise would be pink. Such animals are often called albinos, even though the genetic mutation is not associated with lack of pigmentation, but rather with failure to deposit pigment in the skin, hair, or in this case feathers. [© Photos.com]

key concepts

- Substitution of one base for another is an important mechanism of spontaneous mutation. A single-base substitution in a coding region may result in an amino acid replacement; a single-base deletion or insertion results in a shifted reading frame.

- In the human genome, some inherited diseases are associated with a sudden, dramatic increase in the number of copies of a trinucleotide repeat.

- Transposable element insertion can be an important mechanism of spontaneous mutation.

- Mutations can be induced by various agents, including some classes of chemicals and various types of radiation.

- Cells contain enzymatic pathways for the repair of different types of damage to DNA. Among the most important repair systems is mismatch repair of duplex DNA, in which a nucleotide containing a mismatched base is excised and replaced with the correct nucleotide.

- Most agents that cause mutation also cause cancer.

12

Molecular Mechanisms of Mutation and DNA Repair

chapter organization

12.1 Mutations are classified in a variety of ways.

12.2 Mutations result from changes in DNA sequence.

12.3 Transposable elements are agents of mutation.

12.4 Mutations are statistically random events.

12.5 Spontaneous and induced mutations have similar chemistries.

12.6 Many types of DNA damage can be repaired.

12.7 Genetic tests are useful for detecting agents that cause mutations and cancer.

the human connection
Damage Beyond Repair

Chapter Summary
Issues & Ideas
Key Terms & Concepts
Solutions: Step by Step
Concepts in Action: Problems for Solution
Further Readings

geNETics on the web

In the preceding chapters, numerous examples were presented in which the information contained in the genetic material had been altered by mutation. A **mutation** is any heritable change in the genetic material. In this chapter, we examine the nature of mutations at the molecular level. You will learn how mutations are created, how they are detected phenotypically, and the means by which many mutations are corrected by special DNA repair enzymes almost immediately after they occur. You will see that mutations can be induced by radiation and a variety of chemical agents that produce strand breakage and other types of damage to DNA.

12.1

Mutations are classified in a variety of ways.

The principal ways in which mutations are classified are listed in **Table 12.1**. The first five categories pertain to any type of mutation, whereas the last pertains only to mutations in regions of DNA that code for proteins.

■ Mutagens increase the chance that a gene undergoes mutation.

Most mutations are **spontaneous,** which means that they are statistically random, unpredictable events. Nevertheless, each gene has a characteristic

rate of mutation, measured as the probability of undergoing a change in DNA sequence in the time span of a single generation. Rates of mutation can be increased by treatment with a chemical **mutagen** or radiation, in which case the mutations are said to be **induced.** However, some of the mutations that take place in the presence of a mutagen would have taken place anyway, so it is usually impossible to state positively whether a particular mutation was or was not induced by a mutagen. For example, if treatment with a mutagen increases the spontaneous mutation rate by a factor of 10, then for every mutation that would have occurred anyway, there will now be 10. This means that 1 out of 10, or 10 percent, of all the mutations that take place in the presence of the mutagen would have occurred even in its absence.

■ Germ-line mutations are inherited; somatic mutations are not.

In multicellular organisms, one important distinction is based on the type of cell in which a mutation first occurs. Mutations that arise in cells that ultimately form gametes are **germ-line mutations;** all others are **somatic mutations.** A somatic mutation yields an organism that is genotypically a mixture (*mosaic*) of normal and mutant tissue. We shall see in Chapter 13 that most common cancers result from somatic-cell mutations. In animals, a somatic mutation cannot be transmitted to the progeny. In higher plants, somatic mutations can

Table 12.1

Major types of mutations and their distinguishing features

Basis of classification	Major types of mutations	Major features
Origin	Spontaneous	Occurs in absence of known mutagen
	Induced	Occurs in presence of known mutagen
Cell type	Somatic	Occurs in nonreproductive cells
	Germ-line	Occurs in reproductive cells
Expression	Conditional	Expressed only under restrictive conditions (such as high temperature)
	Unconditional	Expressed under permissive conditions as well as restrictive conditions
Effect on function	Loss-of-function (knockout, null)	Eliminates normal function
	Hypomorphic (leaky)	Reduces normal function
	Hypermorphic	Increases normal function
	Gain-of-function (ectopic expression)	Expressed at incorrect time or in inappropriate cell types
Molecular change	Base substitution	One base pair in duplex DNA replaced with a different base pair
	Transition	Pyrimidine (T or C) to pyrimidine, or purine (A or G) to purine
	Transversion	Pyrimidine (T or C) to purine, or purine (A or G) to pyrimidine
	Insertion	One or more extra nucleotides present
	Deletion	One or more missing nucleotides
Effect on translation	Synonymous (silent)	No change in amino acid encoded
	Missense (nonsynonymous)	Change in amino acid encoded
	Nonsense (termination)	Creates translational termination codon (UAA, UAG, or UGA)
	Frameshift	Shifts triplet reading of codons out of correct phase

often be propagated by vegetative means without going through seed production, such as by grafting or the rooting of stem cuttings. Vegetative propagation is typical of many commercially important fruits, such as the "Delicious" apple and the "Florida" navel orange.

■ **Conditional mutations are expressed only under certain conditions.**

Among the mutations that are most useful for genetic analysis are those whose effects can be turned on or off by the experimenter. These are called **conditional mutations** because they produce changes in phenotype in one set of environmental conditions (called the **restrictive conditions**) but not in another (called the **permissive conditions**). For example, a **temperature-sensitive mutation** is a conditional mutation whose expression depends on temperature. Usually, the restrictive temperature is high (in *Drosophila*, 29° C), and the organism exhibits a mutant phenotype above this critical temperature; the permissive temperature is lower (in *Drosophila*, 18° C), and under permissive conditions the phenotype is wildtype or nearly wildtype. Proteins containing amino acid replacements are often temperature-sensitive: The protein folds properly and functions nearly normally under permissive conditions, but it is unstable and denatures under restrictive conditions. Temperature-sensitive amino acid replacements are frequently used to block particular biochemical pathways under restrictive conditions, in order to test the importance of the pathways in various cellular processes, such as DNA replication.

An example of temperature sensitivity is found in the Siamese cat, with its black-tipped paws, ears, and tail (Figure 12.1). In this breed, an enzyme in the pathway for deposition of the black pigment

Figure 12.1 A Siamese cat showing the characteristic pattern of pigment deposition. [Courtesy of Jen Vertullo.]

melanin is temperature-sensitive. The pathway is blocked at normal body temperature, and pigment is not deposited over most of the body. Pigment is deposited in the tips of the legs, ears, snout, and tail because these extremities are cooler than the rest of the body.

■ **Mutations can affect the amount or activity of the gene product, or the time or tissue specificity of expression.**

Mutations can also be classified according to their effects on gene function. The major categories are described in Table 12.1.

• A mutation that results in complete gene inactivation or in a completely nonfunctional gene product is a **loss-of-function mutation,** also called a *knockout* or *null* mutation. Examples include a deletion of all or part of a gene, and an amino acid replacement that inactivates the protein.

• A mutation that reduces, but does not eliminate, the level of expression of a gene or the activity of the gene product is called a **hypomorphic mutation.** Typically resulting from a nucleotide substitution that reduces the level of transcription, or from an amino acid replacement that impairs protein function, this type of mutation is sometimes referred to as *leaky*. The basis of the term is that because the level of expression or activity differs from individual to individual by chance, a few individuals have enough enzyme activity to "leak through" to produce a quasi-normal phenotype.

• The opposite of a hypomorphic mutation is a **hypermorphic mutation.** As the prefix *hyper* implies, a hypermorphic mutant produces a greater-than-normal level of gene expression, typically because the mutation changes the regulation of the gene so that the gene product is overproduced.

• A **gain-of-function mutation** is one that qualitatively alters the action of a gene. For example, a gain-of-function mutation may cause a gene to become active in a type of cell or tissue in which the gene is not normally active. Or it may result in the expression of a gene in development at a time during which the wildtype gene is not normally expressed. Whereas most loss-of-function and hypomorphic mutations are recessive, many gain-of-function mutations are dominant. Expression of a wildtype gene in an abnormal location is also called *ectopic expression*. For example, expression of the wildtype gene product of the *Drosophila* gene *eyeless* in tissues that do not normally form eyes results in the development of parts of com-

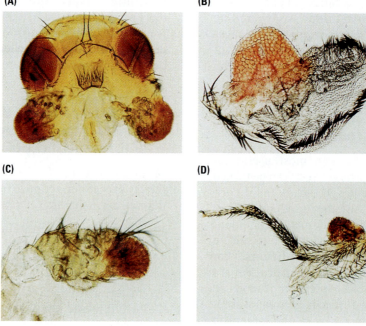

(A) (B) (C) (D)

Figure 12.2 Ecoptic expression of the wildtype allele of the *eyeless* gene in *Drosophila* results in misplaced eye tissue. (A) An adult head in which both antennae form eye structures. (B) A wing with eye tissue growing out from it. (C) A single antenna in which most of the third segment consists of eye tissue. (D) Middle leg with an eye outgrowth at the base of the tibia. [Photos courtesy of G. Halder and W. J. Gehring. Reprinted with permission from G. Halder, P. Callaerts, and W. J. Gehring. 1995. *Science* 267: 1788. © AAAS.]

pound eyes, complete with eye pigments, in abnormal locations (Figure 12.2). The locations can be in any tissues with ectopic expression, including on the legs or mouthparts, in the abdomen, or on the wings.

12.2

Mutations result from changes in DNA sequence.

All mutations result from changes in the nucleotide sequence of DNA or from deletions, insertions, or rearrangement of DNA sequences in the genome. The major types of chromosomal rearrangements have been discussed in Chapter 5. Here we examine mutations at the molecular level.

■ A base substitution replaces one nucleotide pair with another.

The simplest type of mutation is a **base substitution** (Table 12.1), in which a nucleotide pair in a DNA duplex is replaced with a different nucleotide pair. For example, in an $A \rightarrow G$ substitution, an A is replaced with a G in one of the DNA strands. This substitution temporarily creates a mis-

matched $G-T$ base pair, but at the very next replication the mismatch is resolved as a proper $G-C$ base pair in one daughter molecule and as a proper $A-T$ base pair in the other daughter molecule. In this case, the $G-C$ base pair is mutant and the $A-T$ base pair is nonmutant. Similarly, in an $A \rightarrow T$ substitution, an A is replaced with a T in one strand, creating a temporary $T-T$ mismatch, which is also resolved by replication as $T-A$ in one daughter molecule and $A-T$ in the other. In this example, the $T-A$ base pair is mutant and the $A-T$ base pair is nonmutant. The $T-A$ and the $A-T$ are not equivalent, as may be seen by considering the polarity. If the original unmutated DNA strand has the sequence 5'-GAC-3', for example, then the mutant strand has the sequence 5'-GTC-3' (which we have written as $T-A$), and the nonmutant strand has the sequence 5'-GAC-3' (which we have written as $A-T$).

Some base substitutions replace one pyrimidine base with the other or one purine base with the other. These are called **transition mutations.** The four possible transition mutations are

$$T \rightarrow C \quad \text{or} \quad C \rightarrow T$$
$$(\text{pyrimidine} \rightarrow \text{pyrimidine})$$

$$A \rightarrow G \quad \text{or} \quad G \rightarrow A$$
$$(\text{purine} \rightarrow \text{purine})$$

Other base substitutions replace a pyrimidine with a purine or the other way around. These are called **transversion mutations.** The eight possible transversion mutations are

$$T \rightarrow A \quad T \rightarrow G \quad C \rightarrow A \quad \text{or} \quad C \rightarrow G$$
$$(\text{pyrimidine} \rightarrow \text{purine})$$

$$A \rightarrow T \quad A \rightarrow C \quad G \rightarrow T \quad \text{or} \quad G \rightarrow C$$
$$(\text{purine} \rightarrow \text{pyrimidine})$$

Because there are four possible transitions and eight possible transversions, if base substitutions were strictly random, one would expect a 1 : 2 ratio of transitions to transversions. However,

> **key concept**
>
> Spontaneous base substitutions are biased in favor of transitions. Among spontaneous base substitutions, the ratio of transitions to transversions is approximately 2 : 1.

■ Mutations in protein-coding regions can change an amino acid, truncate the protein, or shift the reading frame.

Most base substitutions in coding regions result in one amino acid being replaced with another; these are called **missense mutations** or *nonsynonymous mutations* (Table 12.1). A single amino acid replacement in a protein may alter the biological properties of the protein. An example is the R408W amino acid replacement in phenylalanine hydroxylase, which results from the base-pair substitution C—G to T—A at the first position in codon 408 of the gene; codon 408 in the mRNA is thereby changed from CGG (R, arginine) to UGG (W, tryptophan). This change inactivates the enzyme and results in phenylketonuria.

Examination of the genetic code shows that not all base substitutions cause amino acid replacements, particularly if they occur in the third codon position. In all codons with a pyrimidine in the third position, the particular pyrimidine present does not matter; likewise, in most codons ending in a purine, either purine will do. This means that most transition mutations in the third codon position do not change the amino acid that is encoded. Such mutations change the nucleotide sequence without changing the amino acid sequence; they are called **synonymous substitutions** or **silent substitutions** (Table 12.1) because they are not detectable by changes in phenotype.

Occasionally a base substitution creates a new stop codon UAA, UAG, or UGA. For example, a G → A change at the third position of the normal tryptophan codon UGG converts the codon into

UGA. The result is that translation is terminated at the position of the mutant codon, and the polypeptide is truncated. A base substitution that creates a new stop codon is called a **nonsense mutation.** Nonsense mutations almost always result in loss of gene function.

Small insertions or deletions, when they take place in coding regions, can add or delete amino acids, provided that the number of nucleotides added or deleted is an exact multiple of three (the length of a codon). Otherwise, the insertion or deletion shifts the phase in which the ribosome reads the triplet codons and, consequently, alters all of the amino acids downstream from the site of the mutation. Mutations that shift the reading frame of the codons in the mRNA are called **frameshift mutations.** A common type of frameshift mutation is a single-base addition or deletion. The consequences of a frameshift can be illustrated by the insertion of an adenine at the position of the arrow in the following mRNA sequence:

Because of the frameshift, all of the amino acids downstream from the insertion are different from the original. Any addition or deletion that is not a multiple of three nucleotides will produce a frameshift. Unless it is very near the carboxyl terminus of a protein, a frameshift mutation usually results in the synthesis of a nonfunctional protein.

Courtesy of Colin Meiklejohn

This cat is an eye-color mosaic. Such mosaicism usually results from a somatic mutation in a gene affecting eye color that occurs in a cell lineage giving rise to the pigmented iris.

■ Sickle-cell anemia results from a missense mutation that confers resistance to malaria.

A classic example of the sometimes profound phenotypic effects of a single amino acid replacement is the mutation responsible for the human hereditary disease **sickle-cell anemia.** The molecular basis of sickle-cell anemia is a mutant gene for β-globin, one component of the hemoglobin present in red blood cells (Figure 12.3). The sickle-cell mutation changes the sixth codon in the coding sequence from the normal GAG, which codes for glutamic acid, into the codon GUG, which codes for valine. In the DNA, the mutant has an A−T base pair (transcribed as the middle A in the codon) replaced with a T−A base pair (transcribed as the middle U in the mutant codon). One consequence of the seemingly simple Glu → Val replacement is that hemoglobin containing the defective β polypeptide chain has a tendency to form long, needle-like crystals. Red blood cells in which crystallization happens become deformed into crescent, sickle-like

shapes. Some of the deformed red blood cells are destroyed immediately (reducing the oxygen-carrying capacity of the blood and causing the anemia), whereas others may clump together and clog the blood circulation in the capillaries. The impaired circulation affects the heart, lungs, brain, spleen, kidneys, bone marrow, muscles, and joints. Patients suffer bouts of severe pain. The anemia causes impaired growth, weakness, jaundice, and other symptoms. Affected people are so generally weakened that they are susceptible to bacterial infections, and infections are the most common cause of death in children with the disease.

Sickle-cell anemia is a severe genetic disease that often results in premature death. Yet it is a relatively common disease in areas of Africa and the Middle East in which malaria, caused by the protozoan parasite *Plasmodium falciparum,* is widespread. The association between sickle-cell anemia and malaria is not coincidental: It results from the ability of the mutant hemoglobin to afford some protection against malarial infection. In the life cycle of the parasite, it passes from a mosquito to a human being through the mos-

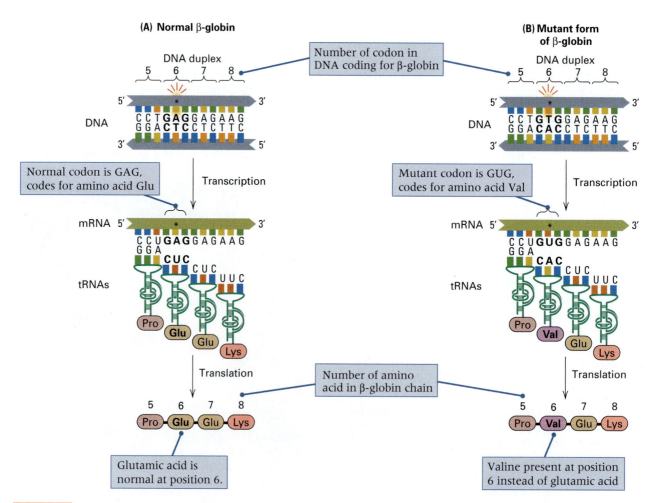

Figure 12.3 Molecular basis of sickle-cell anemia. (A) Part of the DNA in the normal β-globin gene. (B) Mutation of the normal A−T base pair to a T−A base pair results in the codon GUG (valine) instead of GAG (glutamic acid).

quito's bite. The initial stages of infection take place in cells in the liver where specialized forms of the parasite are produced that are able to infect and multiply in red blood cells. Widespread infection of red blood cells impairs the ability of the blood to carry oxygen, causing the weakness, anemia, and jaundice characteristic of malaria. In people with the mutant hemoglobin, however, infection with malaria is less likely and also less severe.

There is consequently a genetic balancing act between the prevalence of the genetic disease sickle-cell anemia and the parasitic disease malaria. If the mutant hemoglobin becomes too frequent, more lives are lost from sickle-cell anemia than are saved by the protection it affords against malaria; on the other hand, if the mutant hemoglobin becomes too rare, fewer lives are lost from sickle-cell anemia but the gain is offset by more deaths from malaria. The end result of this kind of genetic balancing act is considered in quantitative terms in Chapter 14.

In contrast to the situation with sickle-cell anemia, an amino acid replacement does not always create a mutant phenotype. For instance, replacement of one amino acid by another with the same charge (say, lysine for arginine) may in some cases have no effect on either protein structure or phenotype. Whether the substitution of a similar amino acid for another produces an effect depends on the precise role of that particular amino acid in the structure and function of the protein. Any change in the active site of an enzyme usually decreases enzymatic activity.

■ In the human genome, some trinucleotide repeats have high rates of mutation.

Genetic studies of an X-linked form of mental retardation revealed an unexpected class of mutations called **dynamic mutations** because of the extraordinary genetic instability of the region of DNA involved. The X-linked condition, one of at least

12 genetic disorders associated with dynamic mutation, is associated with a class of X chromosomes that tends to fracture in cultured cells that are starved for DNA precursors. The position of the fracture is in region Xq27.3, near the end of the long arm. The X chromosomes containing this site are called *fragile-X* chromosomes, and the associated form of mental retardation is the *fragile-X syndrome*. The fragile-X syndrome affects about 1 in 2500 children. It accounts for about half of all cases of X-linked mental retardation and is second only to Down syndrome as a cause of inherited mental impairment.

The fragile-X syndrome is highly variable in severity. Males are usually more severely affected than females. Developmental delays in speech and communication skills are common, as well as delays in gross motor skills such as sitting up and walking. Physical symptoms may include a long face with protruding ears, weakness in connective tissues resulting in poor muscle tone and extremely flexible joints, and enlarged testicles in males past puberty. Mental retardation is usually moderate in males and mild in females. Behavioral effects may include anxiety, poor concentration, trouble coping with sensory stimuli, avoidance of eye contact, and tantrums or emotional outbursts. These symptoms are nonspecific and overlap with such conditions as autism and attention deficit-hyperactivity disorder.

A hint of something unusual about the fragile-X syndrome was the paradoxical pattern of its inheritance, key features of which are illustrated in Figure 12.4 . Approximately 1 in 5 males who carry the fragile-X chromosome are themselves phenotypically normal and also have phenotypically normal children. The oddity is that the heterozygous daughters of such a "transmitting male" often have affected children of both sexes. In Figure 12.4, the transmitting male denoted I-2 is not affected, but the X chromosome that he transmits to his daugh-

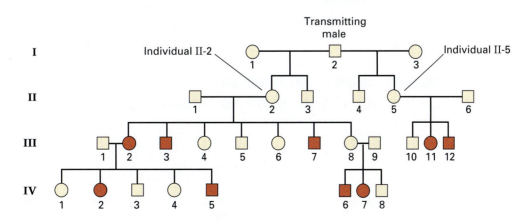

Figure 12.4 Pedigree showing transmission of the fragile-X syndrome. Male I-2 is not affected, but his daughters (II-2 and II-5) have affected children and grandchildren. [After C. D. Laird. 1987. *Genetics* 117: 587.]

ters (II-2 and II-5) somehow becomes altered in the females in such a way that sons and daughters in the next generation (III) are affected. Both affected and normal granddaughters of the transmitting male may have affected progeny (generation IV).

The molecular basis of the fragile-X chromosome has been traced to a **trinucleotide repeat** of the form CGG (or, equivalently, CCG on the other strand) present in the DNA at the site where the breakage takes place (Figure 12.5). Normal X chromosomes have 6 to 54 tandem copies of the repeating unit, with an average of about 30, whereas affected persons have 230 to 2300 or more copies of the repeat. The trinucleotide repeat in the X chromosome in transmitting males is called the *premutation* and has an intermediate number of copies, ranging from 52 to 230. Approximately 1 in 250 females and 1 in 800 males carries an X chromosome with the premutation. The unprecedented feature of the trinucleotide premutation is that when transmitted by females (and only by females), it often increases in copy number (called *trinucleotide expansion*) to a level of 230 copies or more, at which stage the chromosome causes the

fragile-X syndrome. The amplification does not take place in transmission through a male. The functional basis of the disorder is that an excessive number of copies of the CGG repeat cause loss of function of a gene designated *FMR1* (fragile-site mental retardation-1) in which the CGG repeat is present. Most fragile-X patients exhibit no *FMR1* messenger RNA, whereas normal persons and carriers do show expression. The *FMR1* gene is expressed primarily in brain and testes, which explains the strange association between mental and testes abnormalities in affected males.

There is about an 80 percent chance that a premutation transmitted by a female will undergo amplification. Surprisingly, the amplification does not take place in the mother's germ line, but in somatic cells of the early embryo. Amplification occurs to a different extent in different somatic cells, and so individuals with the fragile-X syndrome are somatic mosaics for cells with different numbers of copies of the CGG repeat in the X chromosome. This accounts for the great variation in severity of the fragile-X syndrome from one affected individual to the next.

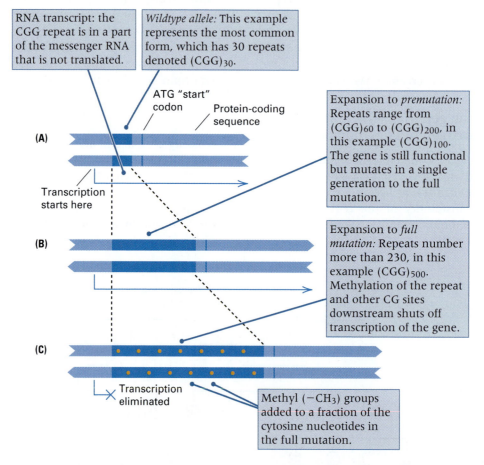

Figure 12.5 Dynamic mutation in the CGG repeat present in the *FMR1* gene implicated in the fragile-X syndrome. (A) The wildtype allele typically has 30 copies of the repeat. (B) The premutation has 60 to 200 copies, which predisposes to further amplification when transmitted through a female. (C) The full mutation, containing > 230 copies. In the full mutation, absence of transcription of the gene is associated with methylation of certain CG dinucleotides in the region.

Other genetic diseases associated with dynamic mutation include the neurological disorders myotonic dystrophy (with an unstable repeat of CTG), Kennedy disease (AGC), Friedreich ataxia (AAG), spinocerebellar ataxia type 1 (AGC), and Huntington disease (AGC). For the fragile-X syndrome and myotonic dystrophy, the trinucleotide expansions occur primarily or exclusively when transmitted by females, but for spinocerebellar ataxia type I and Huntington disease, they occur primarily or exclusively when transmitted by males. Some trinucleotide repeats can undergo amplification when transmitted by either sex.

The molecular mechanism of trinucleotide expansion is illustrated in Figure 12.6. The process is called **replication slippage** (also called *slipped-strand mispairing*). As replication is proceeding along a template strand containing the repeats (part A), the replication complex momentarily dissociates from the template strand. In reassociating with the template, the 3' end of the new strand backtracks along the template and pairs with an upstream set of repeats (part B). Replication continues normally from this point (part C), but some of the repeats will be replicated twice (expanded), the level of expansion depending on how far the replication complex backtracked in reassociating. The template

and the daughter strand cannot pair properly because they have a different number of repeats, but this situation is corrected by the mismatch repair system, and one outcome of repair is that the expanded region is introduced into the template strand (part D). Although the mechanism of dynamic mutation is known, it is not known why some trinucleotide repeats in the genome are genetically unstable whereas others are stable, or why the trinucleotide premutation state is uniquely prone to expansion whereas chromosomes that may have only somewhat fewer copies are genetically stable.

The molecular mechanism of *FMR1* inactivation is associated with the enzymatic addition of a methyl (–CH3) group to each of certain of the cytosine nucleotides in the 5' region of the *FMR1* gene (Figure 12.5). As we have seen in Chapter 9, cytosine methylation occurs at a fraction of the cytosine nucleotides in many higher eukaryotes. In mammals it occurs preferentially at CG dinucleotides, and each CGG repeat in the amplified region of *FMR1* includes a potential methylation site. A high density of methylated CG dinucleotides is usually associated with repression of transcription of the affected gene. In the case of *FMR1*, the lack of transcription of the gene in affected individuals is asso-

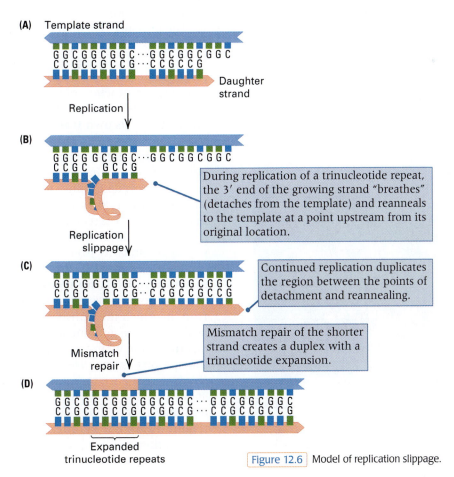

Figure 12.6 Model of replication slippage.

ciated with the methylation of the expanded CGG repeat as well as increased methylation of other CG dinucleotides nearby.

What does the *FMR1* protein do? The protein, called FMRP (for fragile-X mental retardation protein), is an RNA-binding protein that binds with the 5' end of certain messenger RNAs and regulates either their translation into protein, their localization in the cytoplasm, or both. FMRP does not bind all mRNA molecules, but only a specific subset that encode proteins that function in the development of the facial bones and the nervous system or that function in learning and memory. Many of them are proteins that function in the communication between neurons.

12.3

Transposable elements are agents of mutation.

In a 1940s study of the genetics of kernel mottling in maize (Figure 12.7), Barbara McClintock discovered a genetic element that not only regulated the mottling, but also caused chromosome breakage. She called this element *Dissociation (Ds)*. Genetic mapping showed that the chromosome breakage always occurs at or very near the location of *Ds*. McClintock's critical observation was that *Ds* does not have a constant location but occasionally moves to a new position (**transposition**), causing chromosome breakage at the new site. Furthermore, *Ds* moves only if a second element, called *Activator (Ac)*, is also present in the same genome. In addition, *Ac* itself moves within the genome and can also cause modification in the expression of genes at or near its insertion site. Since McClintock's original discovery,

Figure 12.7 Sectors of purple and yellow tissue in the endosperm of maize kernels resulting from the presence of the transposable elements *Ds* and *Ac*. The different level of sectoring in some ears results from dosage effects of *Ac*. [Courtesy of Jerry L. Kermicle.]

many other **transposable elements** have been discovered. They can be grouped into "families" based on similarity in DNA sequence. The genomes of most organisms contain multiple copies of each of several distinct families of transposable elements. Once situated in the genome, transposable elements can persist for long periods and undergo multiple mutational changes. Approximately 50 percent of the human genome consists of transposable elements; as we shall see later, most of these are evolutionary remnants no longer able to transpose.

■ Some transposable elements transpose via a DNA intermediate, others via an RNA intermediate.

The molecular mechanism of *Ds* transposition is illustrated in Figure 12.8. In this example the insertion goes into the wildtype *shrunken* gene in maize chromosome 9, causing a knockout mutation. To initiate the process, the target site for insertion is cleaved with a staggered cut, leaving a 3' overhang of 8 nucleotides on each strand (part A). The overhanging 3' ends are ligated with the 5' ends of the *Ds* element to be inserted, leaving an 8-nucleotide gap in each strand (part B). When the gap is filled by repair enzymes, the result is a new insertion of *Ds* flanked by an 8-base-pair duplication of the target sequence (part C). The *Ds* element can insert in either orientation.

The presence of a target-site duplication is characteristic of most transposable element insertions, and it results from asymmetrical cleavage of the target sequence. For elements like *Ds*, target-site cleavage is a function of a transposase protein that catalyzes transposition. Each family of transposable elements has its own **transposase** that determines the distance between the cuts made in the target DNA strands. Depending on the particular transposable element, the distance may be 1 to 12 base pairs, and this determines the length of the target-site duplication. Most transposable elements have many potential target sites scattered throughout the genome, and they usually show little or no sequence similarity from one site to the next.

The *Ds* element is one of a large class of elements called **DNA transposons** that transpose via a mechanism known as **cut-and-paste transposition,** in which the transposon is cleaved from one position in the genome and the same molecule is inserted somewhere else. Characteristic of DNA transposons is the presence of *terminal inverted repeats,* a sequence repeated in inverted orientation at each end of the element. In the case of *Ds*, the terminal inverted repeats are 11 bp in length (Figure 12.8), but in other families of DNA transposons, they can be up to a few hundred base pairs long. The terminal repeats are usually essential for efficient transposition, because they contain binding sites for the transposase that allow the element to be recognized and ligated into the cleaved target site. Many transposable elements encode their own transposase in sequences located in the central region between the terminal inverted repeats, so these elements are able to promote their own transposition. Elements in which the transposase gene has been deleted or inactivated by mutation are transposable only if another member of the family,

encoding a functional transposase, is present in the genome to provide this activity. The inability of the maize Ds element to transpose without Ac results from the absence of a functional transposase gene in Ds. The presence of an Ac element provides *transactivation* that enables a Ds element to transpose.

Another large class of transposable elements possess terminal direct repeats, typically 200 to 500 bp in length, called *long terminal repeats,* or LTRs. As the name implies, terminal direct repeats are present in the same orientation at both ends of the

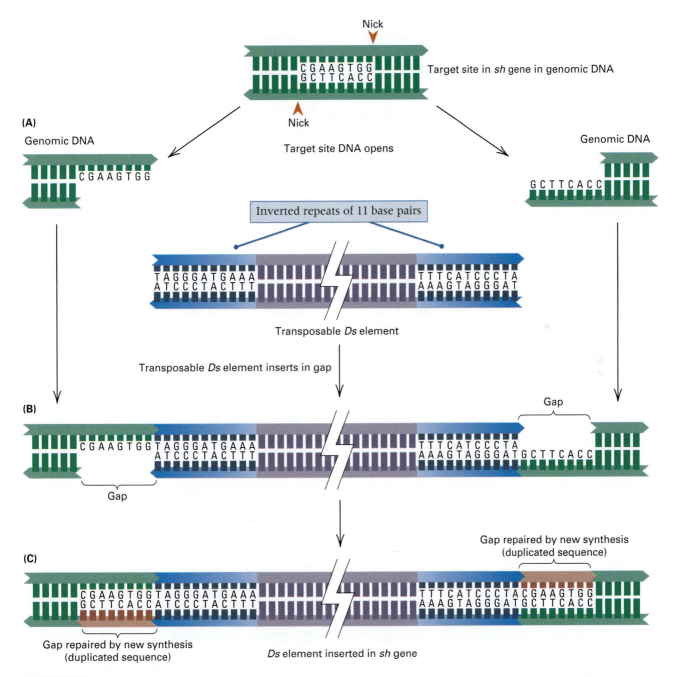

Figure 12.8 The sequence arrangement of a cut-and-paste transposable element (in this case the Ds of maize) and the changes that take place when it inserts into the genome. Ds is inserted into the maize sh gene at the position indicated. In the insertion process, a sequence of eight base pairs next to the site of insertion is duplicated and flanks the Ds element.

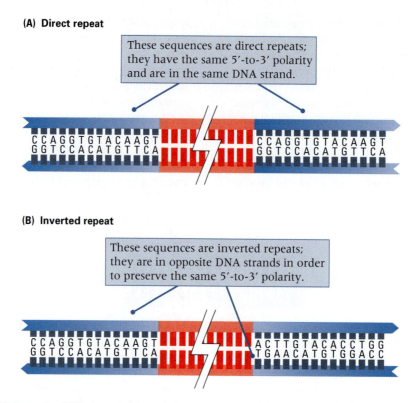

(A) Direct repeat

These sequences are direct repeats; they have the same 5'-to-3' polarity and are in the same DNA strand.

```
CCAGGTGTACAAGT          CCAGGTGTACAAGT
GGTCCACATGTTCA          GGTCCACATGTTCA
```

(B) Inverted repeat

These sequences are inverted repeats; they are in opposite DNA strands in order to preserve the same 5'-to-3' polarity.

```
CCAGGTGTACAAGT          ACTTGTACACCTGG
GGTCCACATGTTCA          TGAACATGTGGACC
```

Figure 12.9 (A) In a direct repeat, a DNA sequence is repeated in the same left-to-right orientation. (B) In an inverted repeat, the sequence is repeated in the reverse left-to-right orientation *in the opposite strand.* The opposite strand is necessary in order to maintain the correct 5'-to-3' polarity.

element (Figure 12.9, part A), whereas terminal inverted repeats are present in reverse orientation (Figure 12.9, part B). Transposable elements with long terminal repeats are called **LTR retrotransposons** because they transpose using an RNA transcript as an intermediate. A typical LTR retrotransposon is the *copia* element from *Drosophila* illustrated in Figure 12.10; in this case each LTR is itself flanked by short terminal inverted repeats. Transposition of a retrotransposon begins with transcription of the element into an RNA copy. Among the encoded proteins is an enzyme known as *reverse transcriptase*, which can "reverse-transcribe," using the RNA transcript as a template for making a complementary DNA daughter strand. A primer is needed for reverse transcrip-

tion. For retrotransposons the primer is usually a cellular transfer RNA molecule whose 3' end is complementary to part of the LTR. The reverse transcriptase adds successive deoxyribonucleotides to the 3' end of the tRNA, using the original RNA transcript as a template. Single-stranded cleavage of the RNA template by an element-encoded RNase provides a primer for second-strand DNA synthesis using the first DNA strand as a template. In this way a double-stranded DNA copy is made of the RNA transcript, and this is inserted into the target site.

Some retrotransposable elements have no terminal repeats and are called **non-LTR retrotransposons.** This class includes elements denoted **LINE elements** (long interspersed elements) and

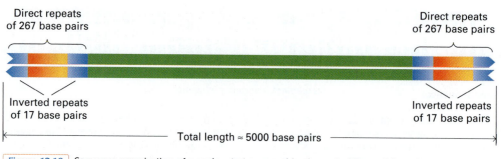

Direct repeats of 267 base pairs

Direct repeats of 267 base pairs

Inverted repeats of 17 base pairs

Inverted repeats of 17 base pairs

Total length ≈ 5000 base pairs

Figure 12.10 Sequence organization of a copia retrotransposable element of *Drosophila melanogaster*.

SINE elements (short interspersed elements). LINE and SINE elements are the most abundant types of transposable elements in mammalian genomes, although DNA and LTR retrotransposons are also found. An example of a SINE in the human genome is a set of related sequences called the AluI family because its members contain a characteristic restriction site for the restriction endonuclease *AluI*. The AluI sequences are about 300 base pairs in length and are present in approximately one million copies in the human genome. The AluI family alone accounts for about 11 percent of human DNA. In many organisms, transposable elements of various families constitute a significant part of the total genome size.

■ Transposable elements can cause mutations by insertion or by recombination.

Transposable elements can cause mutations. For example, in some genes in *Drosophila*, approximately half of all spontaneous mutations that have visible phenotypic effects result from insertions of transposable elements. We have already seen (in Figure 12.8) an example of mutation associated with the insertion of *Ds* into the *shrunken* gene in maize. The wrinkled-seed mutation in Mendel's peas, discussed in Chapter 2, is another good example. In this case, the transposable element is a DNA element related to the maize *Ac* element that also

produces an 8-bp target-site duplication. The insertion site is in the gene for starch-branching enzyme I (SBEI), and the insertion creates a loss-of-function allele. Most transposable elements are present in nonessential regions of the genome and usually cause no detectable phenotypic change. But when an element transposes, it can insert into an essential region and cause a mutant phenotype. If transposition inserts an element into a coding region of DNA, then the inserted element interrupts the coding region. Because most transposable elements contain coding regions of their own, either transcription of the transposable element interferes with transcription of the gene into which it is inserted, or transcription of the gene terminates within the transposable element. The insertion therefore causes a knockout mutation. Even if transcription proceeds through the element, the phenotype will be mutant because the coding region then contains incorrect sequences.

Genetic aberrations can also be caused by recombination between different (nonallelic) copies of a transposable element. Figure 12.11 illustrates two possible outcomes of recombination between copies present in the same DNA molecule. In part A the copies are present in direct orientation. In this case, pairing between the copies forms a loop. Recombination between the copies results in the formation of a free circle of DNA that contains the region between the elements, whereas the remaining part of the molecule has a deletion of

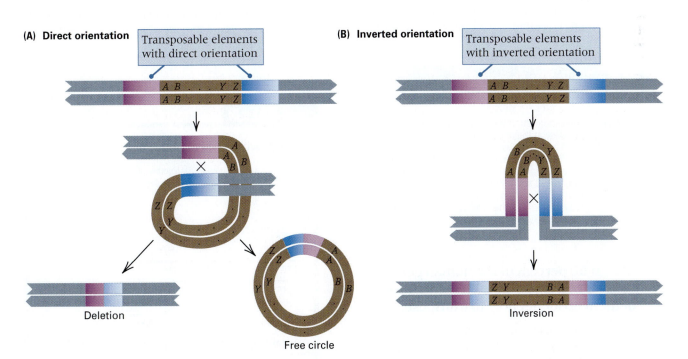

Figure 12.11 Recombination between transposable elements (or other repeated sequences) in the same chromosome. (A) If the repeats are in direct orientation, then recombination results in a deleted duplex and a circular molecule containing the deleted region. (B) If the repeats are in inverted orientation, then recombination results in an inversion of the region between them.

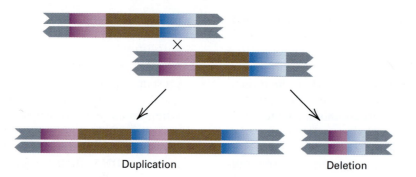

Duplication   Deletion

Figure 12.12 Unequal crossing-over between homologous transposable elements present in the same orientation in different chromatids results in products that contain either a duplication or a deletion of the DNA sequences between the elements. Crossing-over takes place at the four-strand stage of meiosis, but only the two strands participating in the crossover are shown.

this region. In part B the copies are present in inverted orientation. In this case, pairing between the copies creates a *foldback* or *hairpin* structure instead of a loop, and recombination between the copies results in an *inversion* in which the order of the genes between the elements is reversed.

Recombination between copies of a transposable element in different DNA molecules is shown in Figure 12.12 . In this case, the DNA molecules are assumed to be those in homologous chromosomes. Recombination between nonallelic copies produces one product with a duplication of the region between the copies and a reciprocal product with a deletion of the same region. A similar picture could be drawn for ectopic recombination between copies in nonhomologous chromosomes. In this case, the result would be an interchange of terminal segments between nonhomologous chromosomes, which is called a *reciprocal translocation* (Chapter 5).

dances are shown in **Table 12.2**. The general categories are shown in black, and specific examples in red. Altogether, the human genome consists of almost 50 percent transposable elements. The largest single category consists of SINE elements, of which the AluI family are the most abundant. Although transposable elements were long regarded as "selfish DNA"—a sort of genomic parasite—evidence is beginning to suggest that at least the AluI family may benefit the human genome. First, AluI elements are disproportionately represented in gene-rich regions of the genome that are high in G + C content, which suggests that they play some functional role. Second, in human beings, as in some other organisms, SINE elements are transcribed when the organism is under stress. The resulting transcripts can bind to a particular protein kinase that normally blocks translation under stress. In this way SINE elements may be able to promote translation under organismic stress.

Q A Moment to Think

Problem: A metacentric chromosome (one with its centromere approximately in the middle) has a copy of a transposable element in each arm located very close to the telomere. The two copies are present in the same left-to-right orientation. Draw a diagram of a single chromatid in which the two copies are paired and undergo a crossover within the transposable element. What kind of chromosomes result from the reciprocal products of the crossover? (The answer can be found on page 436.)

■ **Almost 50 percent of the human genome consists of transposable elements, most of them no longer able to transpose.**

Part of the reason why the human genome is so relatively large yet contains only about 32,000 genes is that it includes a high proportion of transposable elements. The principal categories and their abun-

Table 12.2

Transposable elements in the human genome

Type	Number of copies	Percentage of total genome
SINEs	1,558,000	13.1
AluI	1,090,000	10.6
LINEs	868,000	20.4
LINE1	516,000	16.9
LTR elements	443,000	8.3
DNA elements	294,000	2.8
mariner	14,000	0.1
Unclassified	3,000	0.1
Total of all types		**44.7**

Source: Data from E. S. Lander et al. 2001. *Nature* 409: 860.

The second major class of human transposable elements consists of LINE elements, of which LINE1 is the most abundant. Third on the list are the LTR retrotransposable elements, followed a distant fourth by DNA elements, of which the *mariner* transposon is an example. The *mariner* transposon is of some interest because it is widespread among eukaryotic genomes. About 14 percent of all insect species carry *mariner*, for example. One reason for its wide distribution is that *mariner* is relatively efficient in being transferred from one species to another, even unrelated, species, but the mechanisms by which this *horizontal transmission* takes place are largely matters of speculation.

The human genomic DNA sequence implies that most transposons in the genome are no longer capable of transposition. One type of evidence derives from comparing sequences of different copies of the same element throughout the genome. Because a transpositionally active element will give rise to new copies that are identical or nearly identical in sequence from one to the next, close sequence similarity among copies suggests active transposition. On the other hand, copies of transposons that can no longer move are free to change in sequence as successive mutations take place and are incorporated into the population, and so large sequence differences among copies suggest a low rate of transposition. Because the average rate of nucleotide substitution per base pair in the human genome is roughly constant through time, the amount of sequence divergence between copies can be used to estimate the time since transposition.

The analysis of sequence differences among human transposable elements suggests that the overall activity of transposable elements in the human genome has decreased substantially, and quite steadily, over the past 35 to 50 million years. The ancient times mean that the decrease in transposition was taking place in the hominid lineage long before human beings existed as a species. Other mammals that have been studied show greater and more typical rates of transposition. In the mouse, for example, the rate of transposition of SINE and LINE elements, relative to that in the human genome, has increased from 1.7-fold higher in the past 100 million years to 2.6-fold higher in the past 25 million years. This comparison is consistent with the finding that about 1 in 10 new mutations in the mouse is due to transposition, whereas only about 1 in 600 new mutations in the human genome is due to transposition. LTR retrotransposons exhibit no convincing evidence of ongoing transposition in the human genome, and DNA transposons seem to have lost their ability to transpose about 50 million years ago. Hence, human

beings stand in contrast to many other organisms, including other mammals, in which transposition is a major source of mutation as well as evolutionary innovation.

12.4

Mutations are statistically random events.

There is no way of predicting when, or in which cell, a mutation will take place, but because every gene mutates spontaneously at a characteristic rate, it is possible to assign probabilities to particular mutational events. In other words, there is a definite probability that a specified gene will mutate in a particular cell, and likewise there is a definite probability that a mutant allele of a specified gene will appear in a population of a designated size. The various kinds of mutational alterations in DNA differ substantially in complexity, so their probabilities of occurrence are quite different. A fundamental principle concerning mutation is that:

> **key concept**
>
> The mutational process is also random in the sense that whether a particular mutation happens is unrelated to any adaptive advantage it may confer on the organism in its environment. A potentially favorable mutation does not arise *because* the organism has a need for it.

The experimental basis for this conclusion is presented in the next section.

■ Mutations arise without reference to the adaptive needs of the organism.

The concept that mutations are spontaneous, statistically random events unrelated to adaptation was not widely accepted until the late 1940s. Before that time, it was believed that mutations occurred in bacterial populations *in response to* particular selective conditions. The basis for this belief was the observation that when antibiotic-sensitive bacteria are spread on a solid growth medium containing the antibiotic, some colonies form that consist of cells having an inherited resistance to the drug. The initial interpretation of this observation (and similar ones) was that these adaptive variations were *induced* by the selective agent itself.

Several types of experiments showed that adaptive mutations take place spontaneously and hence were present at low frequency in the bacterial population even *before* it was exposed to the antibiotic. One experiment utilized a technique developed by Joshua and Esther Lederberg called **replica plating**

A Moment to Think

Answer to Problem: The accompanying diagram shows the pairing configuration and the position of the crossover. Tracing along the chromatid, you can see that one product is a ring chromosome, which bears the centromere; the other product is an acentric fragment bearing a telomere at each end.

(Figure 12.13). In this procedure, a suspension of bacterial cells is spread on a solid medium. After colonies have formed, a piece of sterile velvet mounted on a solid support is pressed onto the surface of the plate. Some bacteria from each colony stick to the fibers, as shown in part A of Figure 12.13. Then the velvet is pressed onto the surface of fresh medium, transferring some of the cells from each colony, which give rise to new colonies that have positions identical to those on the first plate. Part B of Figure 12.13 shows how this method was used to demonstrate the spontaneous origin of phage T1-r mutants. A master plate

containing about 10^7 cells growing on nonselective medium (lacking phage) was replica-plated onto a series of plates that had been spread with about 10^9 T1 phages. After incubation for a time sufficient for colony formation, a few colonies of phage-resistant bacteria appeared in the same positions on each of the selective replica plates. This meant that the T1-r cells that formed the colonies must have been transferred from corresponding positions on the master plate. Because the colonies on the master plate had never been exposed to the phage, the mutations to resistance must have been present, by chance, in a few original cells not exposed to the phage.

The replica-plating experiment illustrates the following principle:

key concept

Selective techniques merely select mutants that preexist in a population.

This principle is the basis for understanding how natural populations of rodents, insects, and disease-causing bacteria become resistant to the chemical substances used to control them. A familiar example is the high level of resistance to insecticides, such as DDT, that now exists in many insect populations, the result of selection for spontaneous mutations affecting behavioral, anatomical, and enzymatic traits that enable the insect to avoid or resist the chemical. Similar problems are encountered in controlling

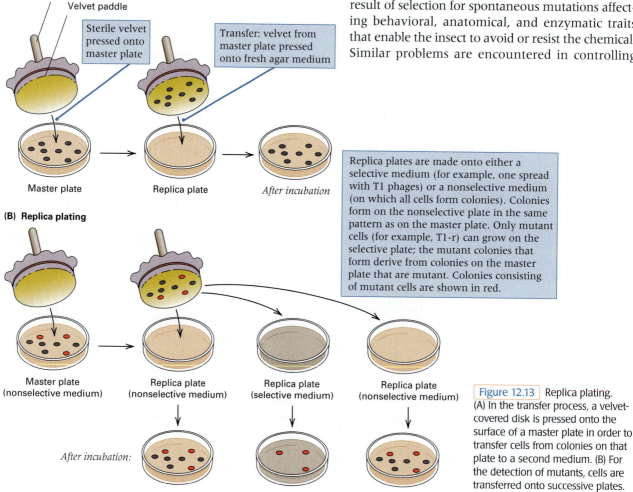

Replica plates are made onto either a selective medium (for example, one spread with T1 phages) or a nonselective medium (on which all cells form colonies). Colonies form on the nonselective plate in the same pattern as on the master plate. Only mutant cells (for example, T1-r) can grow on the selective plate; the mutant colonies that form derive from colonies on the master plate that are mutant. Colonies consisting of mutant cells are shown in red.

Figure 12.13 Replica plating. (A) In the transfer process, a velvet-covered disk is pressed onto the surface of a master plate in order to transfer cells from colonies on that plate to a second medium. (B) For the detection of mutants, cells are transferred onto successive plates.

436 **CHAPTER 12** Molecular Mechanisms of Mutation and DNA Repair

plant pathogens. For example, the introduction of a new variety of a crop plant resistant to a particular strain of disease-causing fungus results in only temporary protection against the disease. The resistance inevitably breaks down because of the occurrence of spontaneous mutations in the fungus that enable it to attack the new plant genotype. Such mutations confer a clear selective advantage, and the mutant alleles rapidly become widespread in the fungal population.

■ The discovery of mutagens made use of special strains of *Drosophila*.

The first evidence that external agents could increase the mutation rate was presented in 1927 by Hermann Muller, who showed that x rays are mutagenic in *Drosophila*. (He later was awarded the Nobel Prize for this work.) One of the techniques he used for measuring the mutation rate was the ***ClB* method.** *ClB* is a special X chromosome of *Drosophila melanogaster;* it has a large inversion (*C*) that prevents the recovery of crossover chromosomes in the progeny from a female heterozygous for the chromosome, a recessive lethal (*l*), and the dominant marker *Bar* (*B*), which reduces the normal round eye to a bar shape. The presence of a recessive lethal in the X chromosome means that males with that chromosome and females homozygous for it cannot survive. The technique is designed to detect mutations that arise in a normal X chromosome.

In the *ClB* procedure, females heterozygous for the *ClB* chromosome are mated with males carrying a normal X chromosome (Figure 12.14). From the F_1 progeny produced, females with the Bar phenotype are selected and then individually mated with normal males. (The presence of the Bar phenotype indicates that the females are heterozygous for the *ClB* chromosome and the normal

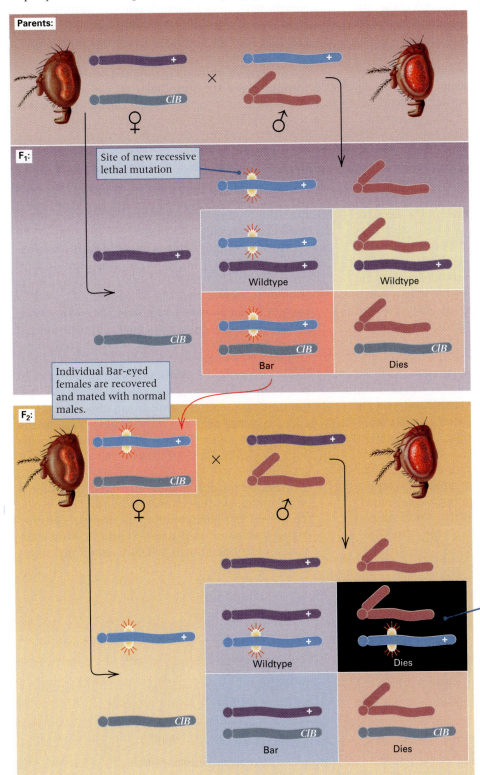

Figure 12.14 The *ClB* method for estimating the rate at which spontaneous recessive lethal mutations arise on the *Drosophila* X chromosome.

X chromosome from the male parent.) The critical observation in determining the mutation rate is the fraction of males produced in the F_2 generation. Because the *ClB* males die, all of the males in this generation must contain an X chromosome derived from the X chromosome of the initial normal male (top row of illustration). Furthermore, it must be a nonrecombinant X chromosome because of the inversion (*C*) in the *ClB* chromosome. If the wildtype X chromosome present in the original male sustained a mutation that created a new recessive lethal, then all of the males in the F_2 generation die, as shown in the lower part of the illustration. On the other hand, if the wildtype X chromosome did not undergo mutation to a new recessive lethal, a ratio of two females to one male is expected among the F_2 progeny. Hence the proportion of matings in which no male progeny are detected in the F_2 generation is a measure of the frequency with which the X chromosome present in the original sperm underwent a mutation somewhere along its length to yield a new recessive lethal. This method provides a quantitative estimate of the rate at which mutation to a recessive lethal allele occurs *at any of the large number of genes* in the X chromosome. About 0.15 percent of the X chromosomes acquire new recessive lethals in spermatogenesis. Alternatively, we could say that the mutation rate is 1.5×10^{-3} recessive lethals per X chromosome per generation. Note that the *ClB* method tells us nothing about the mutation rate for a particular gene, because the method does not reveal how many genes on the X chromosome would cause lethality if they were mutant.

Since the time that the *ClB* method was devised, a variety of other methods have been developed for determining mutation rates in *Drosophila* and other organisms. Of significance is the fact that mutation rates vary widely from one gene to another. For example, the yellow-body mutation in *Drosophila* occurs at a frequency of 10^{-4} per gamete per generation, whereas mutations to streptomycin resistance in *E. coli* occur at a frequency of 10^{-9} per cell per generation. Furthermore, the frequency can differ enormously within a single organism, ranging in *E. coli* from 10^{-5} for some genes to 10^{-9} for others.

■ Mutations are nonrandom with respect to position in a gene or genome.

Certain DNA sequences are called mutational **hotspots** because they are more likely to undergo mutation than others. Mutational hotspots include unstable trinucleotide repeats that can expand by replication slippage (Figure 12.6). Hotspots are found at many sites throughout the genome and within genes. For genetic studies of mutation, the existence of hotspots means that a relatively small number of sites account for a disproportionately large fraction of all mutations.

Sites of cytosine methylation are usually highly mutable, and the mutations are usually $G-C \rightarrow A-T$ transitions. In many organisms, including bacteria, maize, and mammals (but not *Drosophila*), a few percent of the cytosine bases are methylated at the carbon-5 position, yielding 5-methylcytosine instead of ordinary cytosine (Figure 12.15). A special enzyme adds the methyl group to the cytosine base in certain target sequences of DNA. In DNA replication, the 5-methylcytosine pairs with guanine and replicates normally.

Cytosine methylation is an important contributor to mutational hotspots, as illustrated in Figure 12.15. Both 5-methylcytosine and cytosine are subject to occasional loss of an amino group, a process called *deamination*. When 5-methylcytosine is deaminated, it becomes converted into normal thymine (part A). In duplex DNA this creates a temporary $G-T$ mismatch, which has a chance of being repaired by the mismatch repair system discussed in Chapter 4. If it is repaired to $A-T$, then the duplex has undergone a transition mutation; and if it is not repaired immediately, then in the next generation the T-bearing strand pairs normally with A, yielding a mutant $A-T$ base pair in this generation.

Whereas the deamination of 5-methylcytosine is often mutagenic, that of normal cytosine is not. The reason is that deamination of normal cytosine changes cytosine into uracil (part B). Most cells have an enzyme known as **DNA uracil glycosylase,** which recognizes the incorrect $G-U$ base pair

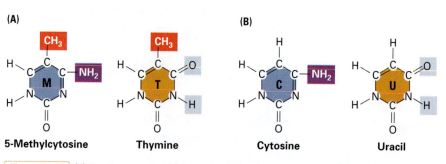

Figure 12.15 (A) Spontaneous loss of the amino group from 5-methylcytosine yields thymine. (B) Loss of the amino group from normal cytosine yields uracil.

and cleaves the offending uracil from the deoxyribose sugar to which it is attached. This enzyme works by scanning along duplex DNA until a uracil nucleotide is encountered, at which point a specific arginine residue in the enzyme intrudes into the DNA through the minor groove. The intrusion compresses the DNA backbone flanking the uracil and results in the flipping out of the uracil base so that it sticks out of the helix. The flipping out makes it accessible to cleavage by the uracil glycosylase. The cleavage leaves behind a deoxyribose sugar that lacks a base, but we shall see later in this chapter that there is a special system to repair such a defect.

12.5

Spontaneous and induced mutations have similar chemistries.

Almost any kind of mutation that can be induced by a mutagen can also occur spontaneously, but mutagens bias the types of mutations that occur according to the type of damage to the DNA that they produce. For the geneticist, the use of mutagens is a means of greatly increasing the number of mutants that can be isolated in an experiment. But mutagens are also of great importance in public health because many environmental contaminants are mutagenic, as are numerous chemicals found in tobacco products.

■ Purine bases are susceptible to spontaneous loss.

Some of the principal agents that damage DNA are listed in Table 12.3, along with the major types of damage they produce. At the head of the list is

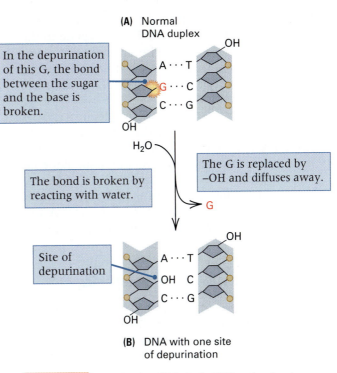

(A) Normal DNA duplex

In the depuration of this G, the bond between the sugar and the base is broken.

The bond is broken by reacting with water.

The G is replaced by —OH and diffuses away.

Site of depurination

(B) DNA with one site of depurination

Figure 12.16 Depurination. (A) Part of a DNA molecule prior to depurination. The bond between the labeled G and the deoxyribose to which it is attached is about to be hydrolyzed. (B) Hydrolysis of the bond releases the G purine, which diffuses away from the molecule and leaves a hydroxyl (−OH) in its place in the depurinated DNA.

water. In purine nucleotides, the sugar–purine bonds are relatively labile and subject to hydrolysis. The loss of the purine base, called **depurination** is illustrated in Figure 12.16 . Depurination is not always mutagenic, because the site lacking the base can be corrected by the same system that repairs sites from which uracil has been removed. If, however, the replication fork reaches the apurinic site

Table 12.3

Major agents of mutation and their mechanisms of action

Agent of mutation	Examples	Principal mechanism of mutagenesis
Water	Hydrolysis	Depurination (A or G detached from its deoxyribose sugar)
Oxidizing agent	Nitrous acid	Deamination ($-NH_2 \rightarrow =O$): C → U, 5-MeC → T, A→ Hypoxanthine
Base analog	5-Bromodeoxyuridine	Increased rate of base mispairing
Alkylating agent	Ethylmethane sulfonate Nitrogen mustard	Bulky attachments made to side groups on bases
Intercalating agent	Proflavin	Causes topoisomerase II to leave a nick in DNA strand; misrepair results in the insertion or deletion of one or a few nucleotides
Ultraviolet light	Natural sunlight UV lamps	Forms pyrimidine dimers (covalent bonds between adjacent pyrimidines, primarily T) present in the same DNA strand
Ionizing radiation	x rays Radon gas Radioactive materials	Single- and double-stranded breaks in DNA; damage to nucleotides

before repair has taken place, then replication almost always inserts an adenine nucleotide in the daughter strand opposite the apurinic site. After another round of replication, what was originally a G–C pair becomes a T–A pair, which is an example of a transversion mutation.

In air, the rate of spontaneous depurination is approximately 3×10^{-9} depurinations per purine nucleotide per minute. This rate is at least tenfold greater than any other single source of spontaneous DNA degradation. At this rate, the half-life of a purine nucleotide exposed to air is about 300 years. This sets a practical limit to how long DNA can persist in the environment before losing its biological activity.

■ Some weak acids are mutagenic.

Many mutagens are chemicals that react with DNA and change the hydrogen-bonding properties of the bases. An example is **nitrous acid**, which acts as a mutagen by deamination of the bases adenine, cytosine, and guanine.

Deamination alters the hydrogen-bonding specificity of each base. As we have seen in Figure 12.15, deamination of 5-methylcytosine results in thymine, and deamination of cytosine results in uracil. The result of deamination of adenine is illustrated in Figure 12.17 . The product is a base called *hypoxanthine*, which pairs with cytosine rather than thymine, so the result of deamination of A is an A–T → G–C transition.

■ A base analog masquerades as the real thing.

A **base analog** is a molecule sufficiently similar to one of the four DNA bases that it can be incorporated into a DNA duplex in the course of normal replication. Such a substance must be able to pair with a base in the template strand. Some base

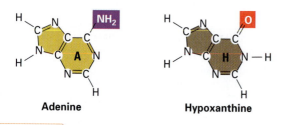

Adenine   **Hypoxanthine**

Figure 12.17   Deamination of adenine results in hypoxanthine.

analogs are mutagenic because they are more prone to mispairing than are the normal nucleotides. The molecular basis of the mutagenesis can be illustrated with 5-bromouracil (Bu), a commonly used base analog that is efficiently incorporated into the DNA of bacteria and viruses.

The base 5-bromouracil is an analog of thymine, and the bromine atom is about the same size as the methyl group of thymine (Figure 12.18 , part A). Normally, 5-bromouracil is in the *keto form*, in which it pairs with adenine (part B), but it occasionally shifts its configuration to the *enol form*, in which it pairs with guanine (part C). The shift is influenced by the bromine atom and takes place in 5-bromouracil more frequently than in thymine.

There are two pathways by which 5-bromouracil can be mutagenic. These are illustrated in Figure 12.19 .

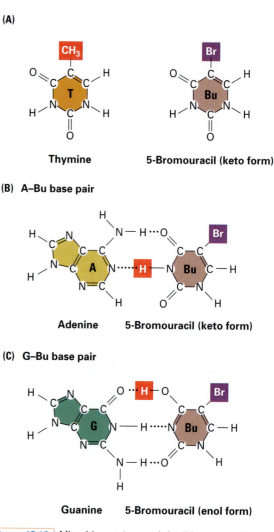

Figure 12.18   Mispairing mutagenesis by 5-bromouracil. (A) Structures of thymine and 5-bromouracil. (B) A base pair between adenine and the keto form of 5-bromouracil. (C) A base pair between guanine and the rare enol form of 5-bromouracil. One of the hydrogen atoms (shown in red) changes position when the molecule is in the keto form.

440   **CHAPTER 12** Molecular Mechanisms of Mutation and DNA Repair

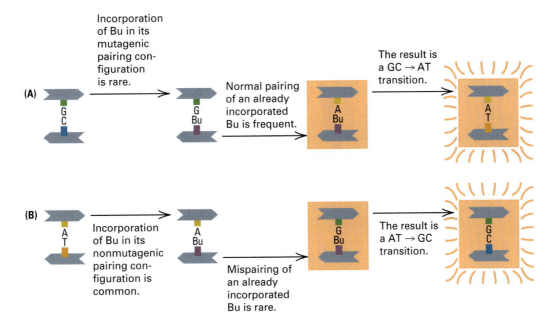

(A) Incorporation of Bu in its mutagenic pairing configuration is rare. → Normal pairing of an already incorporated Bu is frequent. → The result is a GC → AT transition.

(B) Incorporation of Bu in its nonmutagenic pairing configuration is common. → Mispairing of an already incorporated Bu is rare. → The result is a AT → GC transition.

Figure 12.19 Shown are two pathways for mutagenesis by 5-bromouracil (Bu). The position of the arrow shows which strand of each DNA duplex is being followed through the next round of replication. (A) Incorporation paired with G is rare. In this case the mutagenic base pair is formed in the first round of replication. (B) Incorporation paired with A is frequent. In this case the mutagenic base pair is formed after the first round of replication.

In pathway A, the 5-bromouracil is incorporated in its enol form, paired with G. This mode of incorporation is rare, but the mutagenic base pair is created in the first round of replication. In the next round of replication, the Bu will usually pair with A, which leads to a $G-C \rightarrow A-T$ transition. In pathway B, the 5-bromouracil is incorporated in its keto form, paired with A. This is by far the more frequent mode of incorporation, but the mutagenic base pair is not formed until a later round of replication when Bu pairs with G. In this case the result is an $A-T \rightarrow G-C$ transition.

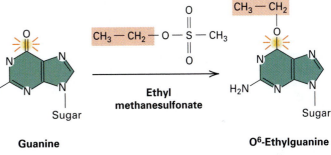

Guanine   Ethyl methanesulfonate O^6-Ethylguanine

Figure 12.21 Mutagenesis of guanine by ethyl methanesulfonate (EMS).

■ Highly reactive chemicals damage DNA.

Some mutagens react with DNA in a variety of different ways and produce a broad spectrum of effects. Among these are the **alkylating agents,** which are highly reactive chemicals that act as potent mutagens in both prokaryotes and eukaryotes. Examples of alkylating agents are ethyl

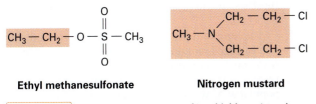

Ethyl methanesulfonate   **Nitrogen mustard**

Figure 12.20 The chemical structures of two highly mutagenic alkylating agents; the alkyl groups are in the pink rectangles.

methanesulfonate (EMS) and nitrogen mustard, the structures of which are shown in Figure 12.20. Nitrogen mustard is a gas causing extreme pain and extensive lung damage when inhaled, and was used for chemical warfare in Europe in the First World War (1914–1918). Ethyl methane sulfonate is a soluble solid and has been used widely to induce mutations for genetic research. The alkylating agents add bulky side groups to the DNA bases that either alter their base-pairing properties or cause structural distortion of the DNA molecule. For example, the reaction of EMS with guanine results in O^6-ethylguanine (Figure 12.21). Alkylation of either guanine or thymine causes mispairing, leading to the transitions $A-T$ to $G-C$ or $G-C$ to $A-T$. EMS reacts less readily with adenine and cytosine than with thymine and guanine.

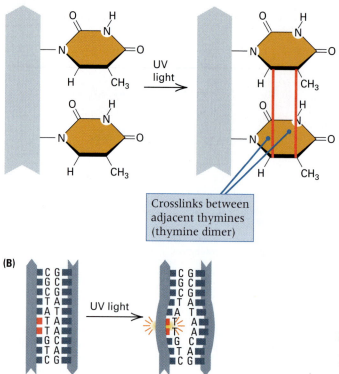

Crosslinks between adjacent thymines (thymine dimer)

Figure 12.22 (A) Structural view of the formation of a thymine dimer. Adjacent thymines in a DNA strand that have been subjected to ultraviolet (UV) irradiation are joined by formation of the bonds shown in red. Other types of bonds between the thymine rings also are possible. Although they are not drawn to scale, these bonds are considerably shorter than the spacing between the planes of adjacent thymines, so that the double-stranded structure becomes distorted. The shape of each thymine ring also changes. (B) The distortion of the DNA helix caused by two thymines moving closer together when joined in a dimer.

■ Some agents cause base-pair additions or deletions.

The **acridine** molecules are planar three-ringed molecules whose dimensions are roughly the same as those of a purine–pyrimidine pair. Acridine orange is an example whose structure is shown below.

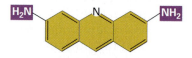

Once thought to insert between the base pairs in DNA, the acridines actually do their damage by interfering with topoisomerase II, which relieves torsional stress in DNA by making a double-stranded break, rotating the free ends, and then sealing the break. In the presence of acridine, the enzyme leaves the DNA nicked. Failure of prompt repair results in the addition or deletion of one or a few base pairs at the site. The result of a single-base addition or deletion in a coding region is a frameshift mutation.

■ Ultraviolet radiation absorbed by DNA is mutagenic.

Ultraviolet (UV) light is mutagenic in all viruses and cells. The effects are caused by chemical changes in the bases resulting from absorption of the energy of the light. The major products formed in DNA after UV irradiation are covalently joined pyrimidines (**pyrimidine dimers**), primarily thymine (Figure 12.22, part A), that are adjacent in the same polynucleotide strand. This chemical linkage brings the bases closer together, causing a distortion of the helix (part B), which blocks transcription and transiently blocks DNA replication. Pyrimidine dimers can be repaired in ways discussed later in this chapter. Nevertheless, excessive exposure of the skin to the UV rays in sunlight increases the risk of skin cancer.

■ Ionizing radiation is a potent mutagen.

Ionizing radiation includes x rays and the particles and radiation released by radioactive elements (α and β particles and γ rays). When x rays were first discovered late in the nineteenth century, their power to pass through solid materials was regarded as a harmless entertainment and source of great amusement. Witness this account from one history of the period:

By 1898, personal x rays had become a popular status symbol in New York. The *New York Times* reported that "there is quite as much difference in the appearance of the hand of a washerwoman and the hand of a fine lady in an x-ray picture as in reality." The hit of the exhibition season was Dr. W. J. Morton's full-length portrait of *"the x-ray lady,"* a "fashionable woman who had evidently a scientific desire to see her bones." The portrait was said to be a "fascinating and coquettish" picture, the lady having agreed to be photographed without her stays and corset, the better to satisfy the "longing to have a portrait of well-developed ribs." Dr. Morton said women were not afraid of x rays: "After being assured that there is *no danger* they take the rays without fear."

The titillating possibility of using x rays to see through clothing or to invade the privacy of locked rooms was a familiar theme in popular discussions of x rays and in cartoons and jokes. Newspapers carried advertisements for *"x ray proof underclothing"* for those seeking to protect themselves from x ray inspection.

The luminous properties of radium soon produced a full-fledged radium craze. A famous woman dancer performed *radium dances* using veils dipped in fluorescent salts containing radium. *Radium roulette* was popular at New York casinos, featuring a "roulette wheel washed with a radium solution, such that it glowed brightly in the darkness; an unseen hand cast the ball on the turning wheel and sparks marked its course as it bounded from pocket to glimmery pocket." A patent was issued for a process for making women's gowns luminous with radium, and Broadway producer Florenz Ziegfeld snapped up the rights for his stage extravaganzas.

Even while the unrestrained use of x rays and radium was growing, evidence was accumulating that the new forces might not be so benign after all. Hailed as tools for fighting cancer, they could also cause cancer. Doctors using x rays were the first to learn this bitter lesson. [Quoted from S. Hilgartner, R. C. Bell, and R. O'Connor. 1982. *Nukespeak*. Sierra Club Books, San Francisco, 1982.]

Doctors were indeed the first to learn the lesson. Many suffered severe x-ray burns or required amputation of overexposed hands or arms. Many others died from radiation poisoning or from radiation-induced cancer. By the mid-1930s, the number of x-ray deaths had grown so large that a monument to the "x-ray martyrs" was erected in a hospital courtyard in Germany. Yet the full hazards of x-ray exposure were not widely appreciated until the 1960s.

When ionizing radiation interacts with water or with living tissue, highly reactive ions called *free radicals* are formed. The free radicals react with other molecules, including DNA, which results in the carcinogenic and mutagenic effects. The intensity of a beam of ionizing radiation can be described quantitatively in several ways. There are, in fact, a bewildering variety of units in common use (**Table 12.4**).

Some of the units (becquerel, curie) deal with the number of disintegrations emanating from a material, others (roentgen) with the number of ionizations the radiation produces in air, still others (gray, rad) with the amount of energy imparted to material exposed to the radiation, and some (rem, sievert) with the effects of radiation on living tissue. The types of units have proliferated through the years in attempts to encompass different types of radiation, including nonionizing radiation, in a common frame of reference. The units in Table 12.4 are presented only as an aid in interpreting the multitude of units found in the literature on the health effects of radiation.

Genetic studies of ionizing radiation support the following general principle:

key concept

Over a wide range of x-ray doses, the frequency of mutations induced by x rays is proportional to the radiation dose.

One type of evidence supporting this principle is the frequency with which X-chromosome recessive lethals are induced in *Drosophila* (Figure 12.23). The mutation rate increases linearly with increasing x-ray dose. For example, an exposure of 10 sieverts increases the frequency from the spontaneous value of 0.15 percent to about 3 percent. The mutagenic and lethal effects of ionizing radiation at low to moderate doses result primarily from damage to DNA. Three types of damage in DNA are produced by ionizing radiation: single-strand

Table 12.4

Units of radiation

Unit (abbreviation)	Magnitude
Becquerel (Bq)*	1 disintegration/second = 2.7×10^{-11} Ci
Curie (Ci)	3.7×10^{10} disintegrations/second = 3.7×10^{10} Bq
Gray (Gy)*	1 joule/kilogram = 100 rad
Rad (rad)	100 ergs/gram = 0.01 Gy
Rem (rem)	Damage to living tissue done by 1 rad = 0.01 Sv
Roentgen (R)	Produces 1 electrostatic unit of charge per cubic centimeter of dry air under normal conditions of pressure and temperature. (By definition, 1 electrostatic unit repels with a force of 1 dyne at a distance of 1 centimeter.)
Sievert (Sv)	100 rem

*Units officially recognized by the International System of Units as defined by the General Conference on Weights and Measures.

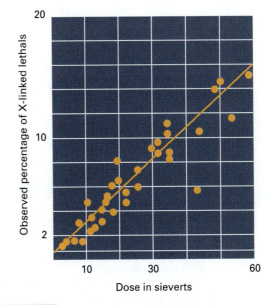

Figure 12.23 The relationship between the percentage of X-linked recessive lethals in *D. melanogaster* and x-ray dose. The frequency of spontaneous X-linked lethal mutations is 0.15 percent per X chromosome per generation.

breakage (in the sugar–phosphate backbone), double-strand breakage, and alterations in nucleotide bases. The single-strand breaks are usually efficiently repaired, but the other damage is responsible for mutation. In eukaryotes, ionizing radiation also results in chromosome breaks. Although systems exist for repairing the breaks, the repair often leads to translocations, inversions, duplications, and deletions. In human cells in culture, a dose of 0.2 sievert results in an average of one visible chromosome break per cell.

Ionizing radiation is widely used in tumor therapy. The basis for the treatment is the increased frequency of chromosomal breakage (and the consequent lethality) in cells undergoing division. Tumors usually contain many more mitotic cells than do most normal tissues, so more tumor cells than normal cells are destroyed. Because all tumor cells are not in mitosis at the same time, irradiation is carried out at intervals of several days to allow interphase tumor cells to enter mitosis. Over a period of time, most tumor cells are destroyed.

Figure 12.24 gives representative values of doses of ionizing radiation received by human beings in the United States in the course of a year. The unit of measure is the millisievert, which equals 0.1 rem. The exposures in Figure 12.24 are on a yearly basis, so over the course of a generation, the total exposure is approximately 100 millisieverts. Note

that, with the exception of diagnostic x rays, which yield important compensating benefits, most of the total radiation exposure comes from natural sources, particularly radon gas. Less than 20 percent of the average radiation exposure comes from artificial sources. Nevertheless, there are dangers inherent in any exposure to ionizing radiation, particularly an increased risk of leukemia and certain other cancers in the exposed persons. In regard to increased genetic diseases in future generations resulting from the mutagenic effects of radiation, the risk of a small amount of additional radiation is low enough that most geneticists are currently more concerned about the effects of the many mutagenic (as well as carcinogenic) chemicals that are introduced into the environment from a variety of sources.

The National Academy of Sciences of the United States regularly updates the estimated risks of radiation exposure. The latest estimates are summarized in Table 12.5. The message is that an additional 10 millisieverts of radiation per generation (about a 10 percent increase in the annual exposure) is expected to cause a relatively modest increase in diseases that are wholly or partly due to genetic factors. The most common conditions in the table are heart disease and cancer. No estimate for the radiation-induced increase is given for either of these traits because the genetic contribution to the total is still uncertain.

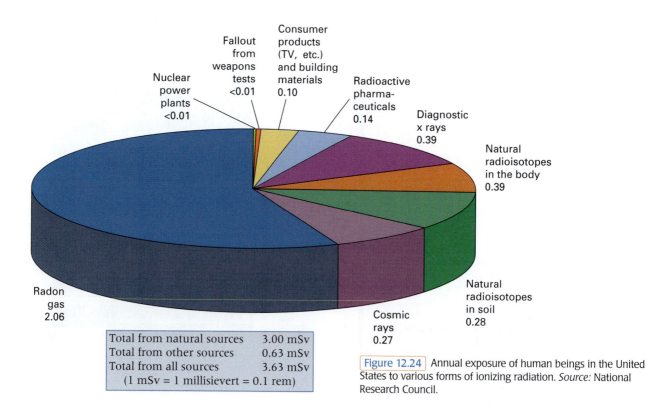

Total from natural sources 3.00 mSv
Total from other sources 0.63 mSv
Total from all sources 3.63 mSv
(1 mSv = 1 millisievert = 0.1 rem)

Figure 12.24 Annual exposure of human beings in the United States to various forms of ionizing radiation. *Source:* National Research Council.

Table 12.5

Estimated genetic effects of an additional 10 millisieverts per generation

Type of disorder	Current incidence per million liveborn	Additional cases per million liveborn per 10 mSv per generation	
		First generation	At equilibrium
Autosomal dominant			
Clinically severe	2500	5–20	25
Clinically mild	7500	1–15	75
X-linked	400	<1	<5
Autosomal recessive	2500	<1	Very slow increase
Chromosomal			
Unbalanced translocations	600	<5	Very little increase
Trisomy	3800	<1	<1
Congenital abnormalities			
(multifactorial)	20,000–30,000	10	10–100
Other multifactorial disorders			
Heart disease	600,000	Unknown	Unknown
Cancer	300,000	Unknown	Unknown
Others	300,000	Unknown	Unknown

Source: Health Effects of Exposure to Low Levels of Ionizing Radiation (BEIR V), National Research Council, Washington, D.C.

■ The Chernobyl nuclear accident had unexpectedly large genetic effects.

The city of Chernobyl in the Ukraine is a symbol for nuclear disaster. On April 26, 1986, a nuclear power plant near the city exploded, heavily contaminating the immediate area and sending clouds of radioactive debris over long distances. It was the largest publicly acknowledged nuclear accident in history. The meltdown is estimated to have released between 50 and 200 million curies of radiation, ten times greater than that released by the atomic bomb over Hiroshima in 1945. Iodine-131 and cesium-137 were the principal radioactive contaminants. More than 200,000 people living in the area were evacuated almost immediately, but many were heavily exposed to radiation. Some children absorbed an amount of radiation equal to that of a thousand chest x rays. Within a short time there was a notable increase in the frequency of thyroid cancer in children, and other health effects of radiation exposure were also detected.

At first, little attention was given to possible genetic effects of the Chernobyl disaster because relatively few people were acutely exposed and because the radiation dose to people outside the immediate area was considered too small to worry about. In the district of Belarus, some 200 kilometers north of Chernobyl, the average exposure to iodine-131 was estimated at approximately 0.185 sievert per person. This is a fairly high dose, but the exposure is brief because the half-life of iodine-131 is only 8 days. Radiation from the much longer-lived cesium-137, with a half-life of 30 years, was estimated at less than 5 millisieverts per year. On the basis of data from laboratory animals and studies of the survivors of the Hiroshima and Nagasaki atomic bombs, little detectable genetic damage was expected from these exposures.

Nevertheless, 10 years after the meltdown, studies of people living in Belarus did indicate a remarkable increase in the mutation rate. The observations focused on short tandem repeats because of their intrinsically high mutation rate due to replication slippage and other factors. Each locus was examined in parents and their offspring to detect any DNA fragments that increased or decreased in size, which would indicate an increase or decrease in the number of repeating units con-

the human connection

Damage Beyond Repair

Frederick S. Leach and 34 other investigators 1993
Johns Hopkins University, Baltimore, MD, and ten other research institutions
Mutations of a mutS Homolog in Hereditary Nonpolyposis Colorectal Cancer

Hereditary nonpolyposis colorectal cancer (HNPCC) is one of the most common conditions predisposing to colon cancer; it affects as many as 1 in 200 individuals in the Western world. There are several forms of the disease, but one form had been traced to a mutant gene in chromosome 2. Certain short, repeating nucleotide sequences (dinucleotide repeats) were known to be genetically unstable during DNA replication in cells with this form of familial colorectal cancer. The accurate replication of such sequences involves the mismatch-repair system, which had been studied extensively in bacteria. Mutants defective in the process were identified as having high rates of spontaneous mutation. Two gene products, *mutL* and *mutS*, recognize and bind to a mismatched base pair. The binding triggers the excision of a tract of nucleotides from the newly synthesized strand. Eukaryotes have similar mismatch-repair systems, and yeast enzymes involved in mismatch repair have amino acid sequences similar to those of bacterial enzymes. Yeast researchers had shown that di-

nucleotide repeats in DNA are a thousandfold less stable in yeast mutants defective in mismatch repair. Recognizing the relevance of their work to the genetic instability in HNPCC, they suggested that high-risk families might be segregating for an allele causing a mismatch-repair deficiency. This paper provides the proof.

••••••••••••••••••••••••••••••••

Thirteen new polymorphic markers were identified in the 25-cM interval [in which the gene for HNPCC was thought to lie].... They were then used to analyze six large HNPCC kindreds previously linked to chromosome 2p [the short arm].... Combined with the data from the patients with single recombinations, the double recombinants suggested that the HNPCC gene most likely resided between [genetic markers defining a 0.8-Mb region].... On the basis of the mapping results, ... we could determine whether a given gene was a candidate for HNPCC by determining its position.... A human homolog of the yeast *mutL*-related gene does not appear to reside on chromosome 2p.... To identify homologs of *mutS*, we used PCR to amplify [and clone] cDNA from cancer cell lines.... A subset of the clones contained sequences similar to that of the yeast *MSH2* gene.... The results [of FISH-fluorescence-labeled *in situ* hybridization] demonstrated that [the]

human *MSH2* gene (hMSH2) lies within the HNPCC locus defined by genetic linkage analysis.... The physical mapping of *hMSH2* to the HNPCC locus is intriguing but could not prove that this gene was responsible for the disease. To obtain more compelling evidence, we determined whether germline mutations of *hMSH2* were present in the two HNPCC kindreds that originally established linkage to chromosome 2.... The DNA from an individual afflicted with colon cancer was found to contain one

> **All eleven affected individuals contained one allele with the C to T transition, while all ten unaffected members contained two normal alleles.**

allele with a C to T transition, resulting in a substitution of leucine for proline.... Twenty-one members of the kindred were then analyzed.... All eleven affected individuals contained one allele with the C to T transition, while all ten unaffected members contained two normal alleles, thus documenting perfect segregation with the disease. Importantly, this proline was at a highly conserved position, the identical residue being found in all known *mutS*-related genes from prokaryotes and eukaryotes. These results strongly suggest that mutations of *hMSH2* are responsible for HNPCC.

Source: Cell 75: 1215–1225.

tained in the DNA fragment. Five loci were studied, with the results summarized in Figure 12.25. Two of the loci (blue dots) showed no evidence of an increase in the mutation rate. This is the expected result. The unexpected finding was that three of the loci (red dots) did show a significant increase, and the level of increase was consistent with an approx-

imate doubling of the mutation rate. At the present time, it is still not known whether the increase was detected because replication slippage is more sensitive to radiation than other types of mutations or because the effective radiation dose of the Belarus population was much higher than originally thought.

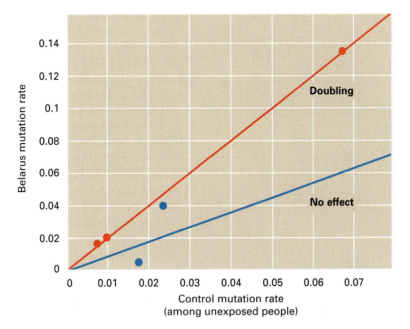

Belarus mutation rate (y-axis): 0, 0.02, 0.04, 0.06, 0.08, 0.1, 0.12, 0.14

Control mutation rate (among unexposed people) (x-axis): 0, 0.01, 0.02, 0.03, 0.04, 0.05, 0.06, 0.07

Doubling

No effect

Figure 12.25 Mutation rates of five tandem repeats among people of Belarus who were exposed to radiation from Chernobyl and among unexposed British people. Three of the loci (red dots) show evidence of an approximately twofold increase in the mutation rate. [Data from Y. E. Dubrova, V. N. Nesterov, N. G. Krouchinsky, V. A. Ostapenko, R. Neumann, D. L. Neil, and A. J. Jeffreys. 1996. *Nature* 380: 183.]

12.6

Many types of DNA damage can be repaired.

Spontaneous damage to DNA in human cells takes place at a rate of approximately 1 event per billion nucleotide pairs per minute (or, expressed per nucleotide pair, at a rate of 1×10^{-9} per nucleotide pair per minute). This may seem quite a small rate, but it implies that every 24 hours, in every human cell, the DNA is damaged at approximately 10,000 different sites. Fortunately for us, and for all living organisms, much of the damage done to DNA by spontaneous chemical reactions in the nucleus, by chemical mutagens, and by radiation can be repaired. **Table 12.6** summarizes some of the most important mechanisms of DNA repair. The first on the list is straightforward. A nick in a DNA strand is a site at which one of the phosphodiester bonds along the backbone is broken. Nicks are repaired by the enzyme DNA ligase, which restores the covalent bond. The second mechanism in the list is mediated by DNA uracil glycosylase. As we have already seen in Section 12.4, this enzyme acts to remove uracil nucleotides that get into DNA, such as by the deamination of cytosine. In the following sections, we examine other key molecular mechanisms for the repair of aberrant or damaged DNA.

■ Mismatch repair fixes incorrectly matched base pairs.

We have already encountered the mismatch repair system in the context of gene conversion and genetic recombination (Chapter 4). However, the

Table 12.6

Types of DNA damage and mechanism of repair

Type of damage	Major mechanism of repair
Nicks in DNA strand	Repaired by DNA ligase
Uracil present in DNA	Uracil removed by DNA uracil glycosylase
Mismatched bases	Corrected by mismatch repair (excision and resynthesis)
Apurinic or apyrimidinic site	Fixed by AP endonuclease repair system
Pyrimidine dimers (from UV light)	Enzymatically reversed
Damaged region of DNA	Excision repair (excision and resynthesis across partner strand)
Damaged region of DNA	Postreplication repair (sequence in damaged region recovered via recombination)

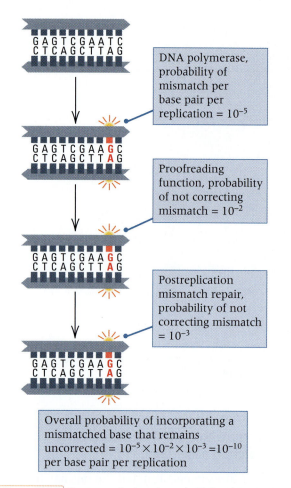

DNA polymerase, probability of mismatch per base pair per replication = 10^{-5}

Proofreading function, probability of not correcting mismatch = 10^{-2}

Postreplication mismatch repair, probability of not correcting mismatch = 10^{-3}

Overall probability of incorporating a mismatched base that remains uncorrected = $10^{-5} \times 10^{-2} \times 10^{-3} = 10^{-10}$ per base pair per replication

Figure 12.26 Summary of rates of error in DNA polymerization, proofreading, and postreplication mismatch repair. The overall rate of misincorporated nucleotides that are not repaired is 10^{-10} per base pair per replication.

most important role of mismatch repair is as a "last chance" error-correcting mechanism in replication. During DNA replication, mismatched nucleotides are incorporated at the rate of about 10^{-5} per template nucleotide per round of replication. As shown in Figure 12.26, approximately 99 percent of these are immediately corrected by the proofreading function (3′-to-5′ exonuclease activity) of the major replication polymerase. This leaves a mismatch rate of 10^{-7} per template nucleotide per round of replication; 99.9 percent of the remaining mismatches are corrected by the mismatch repair system, yielding an overall mismatch rate of

$$10^{-5} \times 10^{-2} \times 10^{-3} = 10^{-10}$$

The operation of **mismatch repair** is illustrated in Figure 12.27. When a mismatched base is detected, one of the strands is cut in two places and a region around the mismatch is removed. The excised region is variable in size. In *E. coli,* an enzyme first cleaves the DNA at the nearest GATC sequence on the unmethylated (newly synthesized) strand, provided that the site is within about 1 kb of the mismatch.

Then an exonuclease degrades the cleaved DNA strand to a position on the other side of the mismatch (the excision step), creating a single-stranded gap. After excision, DNA polymerase fills in the gap by using the remaining strand as a template, thereby eliminating the mismatch. The mismatch-repair system also corrects most small insertions or deletions.

If the DNA strand that is removed in mismatch repair is chosen at random, then the repair process sometimes creates a mutant molecule by cutting the strand that contains the correct base and using the mutant strand as a template. However, this is prevented from happening in newly synthesized DNA because the daughter strand is less methylated than the parental strand. The mismatch-repair system recognizes the degree of methylation of a strand and *preferentially excises nucleotides from the undermethylated strand*. This helps ensure that incorrect nucleotides incorporated into the daughter strand in replication will be removed and repaired. The daughter strand is always the undermethylated strand because its methylation lags somewhat behind the moving replication fork, whereas the parental strand was fully methylated in the preceding round of replication.

The mechanism of mismatch repair has been studied extensively in bacteria. Mutants defective in the process were identified as having high rates of spontaneous mutation. The products of two genes, *mutL* and *mutS*, recognize and bind to a mismatched base pair. This triggers the excision of a tract of nucleotides from the newly synthesized strand. Experiments carried out in yeast revealed that tandem repeats of short nucleotide sequences are a thousandfold less stable in mutants deficient in mismatch repair than in wildtype yeast. By that time it was already known that some forms of human hereditary colorectal cancer result in decreased stability of simple repeats, and the yeast researchers suggested that these high-risk families might be segregating for an allele causing a mismatch-repair deficiency. Within less than two years, scientists in several laboratories had identified four human genes homologous to *mutL* or *mutS*, any one of which, when mutated, results in hereditary non-polyposis colorectal cancer (HNPCC). Most cases of this type of cancer may be caused by mutations in one of these four mismatch-repair genes.

■ The AP endonuclease system repairs nucleotide sites at which the base has been lost.

We have already seen how deamination of cytosine creates uracil (Figure 12.15), which is removed by DNA uracil glycosylase from deoxyribose sugar to which it is attached. The result is a site in the DNA that lacks a pyrimidine base (an *apyrimidinic* site). We have also seen that purines in DNA are some-

Figure 12.27 Mismatch repair consists of the excision of a segment of a DNA strand that contains a base mismatch, followed by repair synthesis. In *E. coli*, cleavage takes place at the nearest methylated GATC sequence in the unmethylated strand. An exonuclease removes successive nucleotides until just past the mismatch, and the resulting gap is repaired. Either strand can be excised and corrected, but in newly synthesized DNA, methylated bases in the template strand often direct the excision mechanism to the newly synthesized strand that contains the incorrect nucleotide.

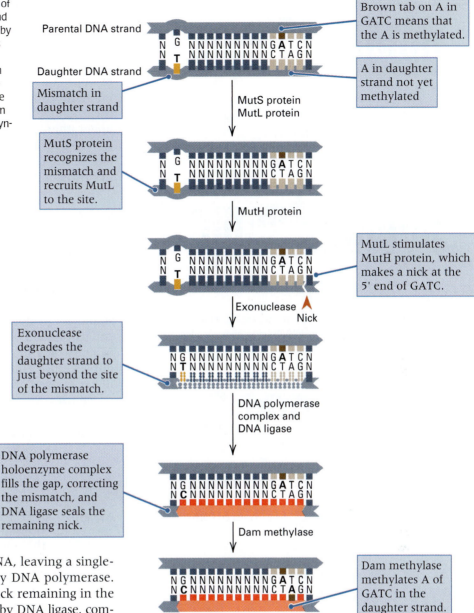

Parental DNA strand

Daughter DNA strand

Mismatch in daughter strand

Brown tab on A in GATC means that the A is methylated.

A in daughter strand not yet methylated

MutS protein
MutL protein

MutS protein recognizes the mismatch and recruits MutL to the site.

MutH protein

MutL stimulates MutH protein, which makes a nick at the 5' end of GATC.

Exonuclease
Nick

Exonuclease degrades the daughter strand to just beyond the site of the mismatch.

DNA polymerase complex and DNA ligase

DNA polymerase holoenzyme complex fills the gap, correcting the mismatch, and DNA ligase seals the remaining nick.

Dam methylase

Dam methylase methylates A of GATC in the daughter strand.

what prone to hydrolysis (Figure 12.16), which leave a site that is lacking a purine base (an *apurinic* site).

Both apyrimidinic and apurinic sites are repaired by a system that depends on an enzyme called **AP endonuclease.** The mechanism is illustrated in Figure 12.28 . The AP endonuclease cleaves the base-less sugar from the DNA, leaving a single-stranded gap that is repaired by DNA polymerase. The gap is filled, leaving one nick remaining in the repaired strand, which is closed by DNA ligase, completing the repair.

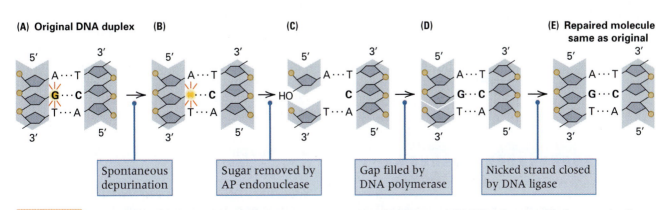

(A) Original DNA duplex **(B)** **(C)** **(D)** **(E) Repaired molecule same as original**

Spontaneous depurination

Sugar removed by AP endonuclease

Gap filled by DNA polymerase

Nicked strand closed by DNA ligase

Figure 12.28 Action of AP endonuclease. (A) Original DNA duplex. (B) Spontaneous hydrolysis of guanine results in loss of the base. (C) AP endonuclease excises the empty deoxyribose from the DNA strand. (D) DNA polymerase fills the gap using the continuous strand as a template. (E) The remaining nick is closed by DNA ligase, restoring the original sequence.

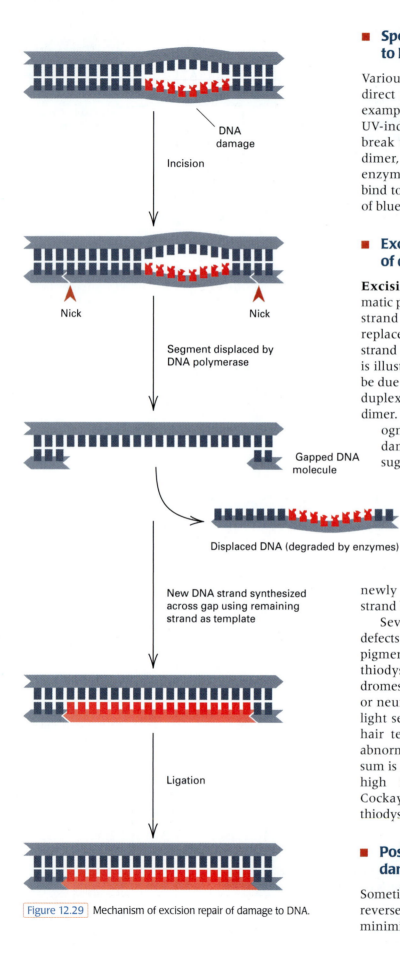

DNA damage

Incision

Nick | Nick

Segment displaced by DNA polymerase

Gapped DNA molecule

Displaced DNA (degraded by enzymes)

New DNA strand synthesized across gap using remaining strand as template

Ligation

Figure 12.29 Mechanism of excision repair of damage to DNA.

■ Special enzymes repair damage caused to DNA by ultraviolet light.

Various enzymes can recognize and catalyze the direct reversal of specific DNA damage. A classic example found in some organisms is the reversal of UV-induced pyrimidine dimers by enzymes that break the bonds that join the pyrimidines in the dimer, thereby restoring the original bases. Some enzymes of this type require light to work. They bind to the dimers in the dark but need the energy of blue light to cleave the bonds.

■ Excision repair works on a wide variety of damaged DNA.

Excision repair is a ubiquitous, multistep enzymatic process by which a stretch of a damaged DNA strand is removed from a duplex molecule and replaced by resynthesis using the undamaged strand as a template. The process of excision repair is illustrated in Figure 12.29. The DNA damage can be due to anything that produces a distortion in the duplex molecule—for example, a pyrimidine dimer. In excision repair, a repair endonuclease recognizes the distortion produced by the DNA damage and makes one or two cuts in the sugar–phosphate backbone, several nucleotides away from the damage on either side. A 3'-OH group is produced at the 5' cut, which DNA polymerase uses as a primer to synthesize a new strand while displacing the DNA segment that contains the damage. The final step of the repair process is the joining of the newly synthesized segment to the contiguous strand by DNA ligase.

Several disease syndromes are associated with defects in excision repair. These include xeroderma pigmentosum, Cockayne syndrome, and trichothiodystrophy. Although patients with these syndromes share such symptoms as skin abnormalities or neurological defects, they differ dramatically in light sensitivity, predisposition to cancer, stature, hair texture, and presence or absence of facial abnormalities. In particular, xeroderma pigmentosum is associated with severe light sensitivity and a high incidence of early-onset skin cancer, Cockayne syndrome with dwarfism, and trichothiodystrophy with sulfur-deficient, brittle hair.

■ Postreplication repair skips over damaged bases.

Sometimes DNA damage persists rather than being reversed or removed, but its harmful effects may be minimized. This often requires replication across damaged areas, so the process is called **post-replication repair.** For example, when DNA polymerase reaches a damaged site (such as a pyrimidine dimer), synthesis of the new strand stalls (Figure 12.30, part A). However, after a brief time, synthesis is reinitiated beyond the damage, and chain growth continues, producing a gapped strand with the damaged spot in the gap (part B). The gap can be filled by strand exchange with the parental strand having the same polarity (part C), and then the secondary gap produced in the undamaged strand is filled by repair synthesis (part D). The products of this exchange and resynthesis are two intact single strands, each of which can then serve in the next round of replication as a template for the synthesis of an undamaged DNA molecule.

12.7

Genetic tests are useful for detecting agents that cause mutations and cancer.

A genetic test for mutations in bacteria is widely used for the detection of chemical mutagens. In view of the increased number of chemicals used and present as environmental contaminants, tests for the mutagenicity of these substances has become important. Furthermore, most agents that cause cancer (*carcinogens*) are also mutagens, and so mutagenicity provides an initial screening for potential hazardous agents.

In the **Ames test** for mutation, histidine-requiring (His$^-$) mutants of the bacterium *Salmonella typhimurium,* containing either a base substitution or a frameshift mutation, are tested for backmutation **reversion** to His$^+$. In addition, the bacterial strains have been made more sensitive to mutagenesis by the incorporation of several mutant alleles that inactivate the excision-repair system and that make the cells more permeable to foreign molecules. Because some mutagens act only on replicating DNA, the solid medium used contains enough histidine to support a few rounds of replication but not enough to permit the formation of a visible colony. The medium also contains the potential mutagen to be tested and an extract of rat liver. The role of the liver extract is to permit identification of substances that are not directly mutagenic (or carcinogenic) but are converted into mutagens by enzymatic reactions that take place in the livers of animals. The normal function of these enzymes is to protect the organism from various naturally occurring harmful substances by converting them into soluble nontoxic substances that can be disposed of in the urine. However, when the

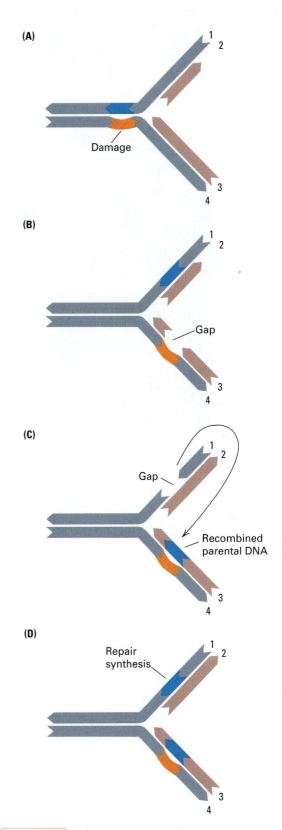

Figure 12.30 Postreplication repair. (A) A molecule with DNA damage in strand 4 is being replicated. (B) By reinitiating synthesis beyond the damage, a gap is formed in strand 3. (C) A segment of parental strand 1 is excised and inserted in strand 3. (D) The gap in strand 1 is next filled in by repair synthesis.

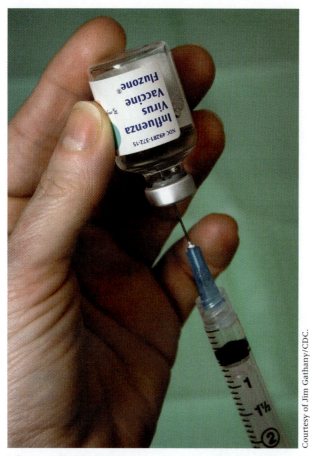

Influenza outbreaks recur because of mutations that change the amino acid sequence of proteins on the viral surface and allow the virus to escape detection by the immune system of people exposed to nonmutant forms of the virus that circulated previously. Early detection of these mutant forms enables flu vaccines to be prepared that can prevent outbreaks from turning into major epidemics.

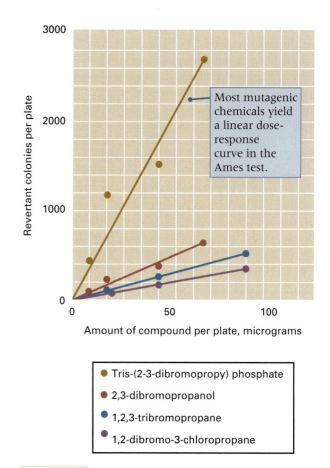

Most mutagenic chemicals yield a linear dose-response curve in the Ames test.

- Tris-(2-3-dibromopropy) phosphate
- 2,3-dibromopropanol
- 1,2,3-tribromopropane
- 1,2-dibromo-3-chloropropane

Figure 12.31 Linear dose–response relationships obtained with various chemical mutagens in the Ames test. [Data from B. N. Ames. 1974. *Science* 204: 587–593.]

enzymes encounter certain artificial and natural compounds, they convert these substances, which may not be harmful in themselves, into mutagens or carcinogens.

In the Ames test, if the test substance is a mutagen or is converted into a mutagen, some colonies are formed. A quantitative analysis of reversion frequency can also be carried out by incorporating various amounts of the potential mutagen in the medium. The reversion frequency generally depends on the concentration of the substance being tested and, for a known carcinogen or mutagen, correlates roughly with its carcinogenic potency in animals. The Ames test is simple, rapid, inexpensive, and exquisitely sensitive. Some chemicals can be detected to be mutagenic in amounts as small as 10^{-9} g, and a condensate of as little as 1/100 of a cigarette can be shown to be mutagenic. The test is also highly quantitative (Figure 12.31). Chemicals need not be classified simplistically as "mutagenic" or "nonmutagenic." They can be classified according to their potency as mutagens, because more than a millionfold range in potency can be detected in the *Salmonella* test.

An example of the importance of adequate testing involves a chemical known as tris-BP. This substance was used as a flame retardant in children's polyester pajamas from 1972 to 1977. Then it was discovered that tris-BP is a potent mutagen in *Salmonella* and also in *Drosophila*. Further studies showed that Tris-BP interacts with human DNA and damages mammalian chromosomes. It was also found to be a carcinogen in experimental tests in rats and mice, and it was found to be capable of causing sterility in laboratory animals. Moreover, the substance was shown to be absorbed through the skin, and its breakdown products could be detected in the urine of children who were wearing the treated sleepwear. Before this information became known and use of the substance in clothing was discontinued, more than 50 million children were exposed to the chemical through contact with their nightclothes.

The Ames test has been used with thousands of substances and mixtures (such as industrial chemicals, food additives, pesticides, hair dyes, and cosmetics), and numerous unsuspected substances have been found to stimulate reversion in this test. A high frequency of reversion does not necessarily indicate that the substance is definitely a carcinogen but only that it has a high probability of being so. As a result of these tests, many industries have reformulated their products; for example, the cosmetics industry has changed the formulation of many hair dyes and cosmetics to render them nonmutagenic.

Ultimate proof of carcinogenicity is determined by testing for tumor formation in laboratory animals. However, only a few percent of the substances known from animal experiments to be carcinogens failed to increase the reversion frequency in the Ames test.

chapter summary

12.1 Mutations are classified in a variety of ways.

- Mutagens increase the chance that a gene undergoes mutation.
- Germ-line mutations are inherited; somatic mutations are not.
- Conditional mutations are expressed only under certain conditions.
- Mutations can affect the amount or activity of the gene product, or the time or tissue specificity of expression.

12.2 Mutations result from changes in DNA sequence.

- A base substitution replaces one base with another.
- Mutations in protein-coding regions can change an amino acid, truncate the protein, or shift the reading frame.
- Sickle-cell anemia results from a missense mutation that confers resistance to malaria.
- In the human genome, some trinucleotide repeats have high rates of mutation.

Mutations can be classified in a variety of ways such as in terms of (1) how they come about, (2) the cell type they occur in, (3) how they affect gene function, (4) the nature of the molecular change, and (5) how they affect translation. Conditional mutations cause a change in phenotype under restrictive but not permissive conditions. For example, temperature-sensitive mutations are expressed only above a particular temperature. Spontaneous mutations are of unknown origin, whereas induced mutations result from exposure to chemical reagents or radiation. All mutations ultimately result from changes in the sequence of nucleotides in DNA, including base substitutions, insertions, deletions, and rearrangements. Base-substitution mutations are single-base changes that may be silent (for example, in a noncoding region or a synonymous codon position), may change an amino acid in a polypeptide chain (a missense mutation), or may cause chain termination by producing a stop codon (a nonsense mutation). A transition is a base-substitution mutation in which a pyrimidine is substituted for another pyrimidine or a purine for another purine. In a transversion, a pyrimidine is substituted for a purine, or the other way around. A mutation may consist of an insertion or deletion of one or more bases; in a coding region, if the number of bases is not a multiple of 3, the mutation is a frameshift. In the human genome, some trinucleotide repeats are prone to expansion (increase in copy number), and the probability of expansion of the premutation state can approach 100 percent.

12.3 Transposable elements are agents of mutation.

- Some transposable elements transpose via a DNA intermediate, others via an RNA intermediate,
- Transposable elements can cause mutations by insertion or by recombination.
- Almost 50 percent of the human genome consists of transposable elements, most of them no longer able to transpose.

Transposable elements are capable of moving from one location in the genome to another. The main classes of transposable elements are DNA transposons, which typically have inverted repeats and transpose via a transposase using a cut-and-paste mechanism; LTR retrotransposable elements, which have long direct or inverted repeats and transpose via a reverse transcriptase; and non-LTR retrotransposable elements, which also transpose via a reverse transcriptase but have no direct or inverted repeats. Insertions of transposable elements are the cause of many visible mutations.

12.4 Mutations are statistically random events.

- Mutations arise without reference to the adaptive needs of the organism.
- The discovery of mutagens made use of special strains of *Drosophila*.
- Mutations are nonrandom with respect to position in a gene or genome.

12.5 Spontaneous and induced mutations have similar chemistries.

- Purine bases are susceptible to spontaneous loss.
- Some weak acids are mutagenic.
- A base analog masquerades as the real thing.
- Highly reactive chemicals damage DNA.
- Some agents cause base-pair additions or deletions.
- Ultraviolet radiation absorbed by DNA is mutagenic.
- Ionizing radiation is a potent mutagen.
- The Chernobyl nuclear accident had unexpectedly large genetic effects.

Mutations can be induced chemically by direct alteration of DNA—for example, by nitrous acid, which deaminates bases. Base analogs, such as 5-bromouracil, are incorporated into DNA in replication. They undergo mispairing more often than the normal bases, giving rise to transition mutations. Alkylating agents react with bases and cause mispairing. Acridine molecules interfere with the activity of

chapter summary *continued*

topoisomerase II and give rise to frameshift mutations, usually of one or two bases. Ionizing radiation results in oxidative free radicals that cause a variety of alterations in DNA, including single-stranded and double-stranded breaks. Although the amount of genetic damage from normal background and other sources of radiation is believed to be low, studies of persons exposed to fallout from the Chernobyl nuclear meltdown indicate an increased mutation rate, at least for simple tandem repeats.

12.6 Many types of DNA damage can be repaired.

- Mismatch repair fixes incorrectly matched base pairs.
- The AP endonuclease system repairs nucleotide sites at which the base has been lost.
- Special enzymes repair damage caused to DNA by ultraviolet light.
- Excision repair works on a wide variety of damaged DNA.
- Postreplication repair skips over damaged bases.

A variety of systems exist for repairing damage to DNA. The mismatch-repair system corrects mispaired nucleotides missed by the proofreading function of DNA polymerase. Methylation of parental DNA strands and delayed methylation of daughter strands allow the mismatch-repair system to operate selectively on the daughter DNA strand. The AP repair system excises nucleotides that have lost a base (creating an apurinic or apyrimidinic site) as a result of spontaneous hydrolysis or chemical mutagenesis. In photoreactivation, pyrimidine dimers produced by ultraviolet radiation are restored by cleavage of the connecting bond or by excision repair. In excision repair, a stretch of a distorted DNA strand is excised and replaced with a newly synthesized copy of the undamaged strand. Postreplication repair is an exchange process in which gaps in one daughter strand produced by aberrant replication across damaged sites are filled by nondefective segments from the parental strand of the other branch of the newly replicated DNA. Thus a new template is produced by this system.

12.7 Genetic tests are useful for detecting agents that cause mutations and cancer.

The Ames test measures reversion as an indicator of mutagens and carcinogens; it uses an extract of rat liver, which in mammals occasionally converts intrinsically harmless molecules into mutagens and carcinogens.

issues & ideas

- If a mutation is a conditional mutation, what determines whether the mutant phenotype will be expressed?
- How can an organism with a temperature-sensitive, recessive-lethal mutation survive as a homozygous genotype?
- What does it mean to say that a particular allele has a mutation rate of 10^{-6} per gene per generation?
- How does replica plating demonstrate that mutations to antibiotic resistance can arise even in cells that have never been exposed to the antibiotic?

- Mutations in genes whose products are involved in DNA repair are often associated with an increased risk of cancer. What does this observation imply about the role of spontaneous mutation in the development of cancer?
- What is "mismatched" in the process of mismatch repair?
- Why were the genetic effects of the Chernobyl nuclear meltdown initially expected to be undetectable?

key terms & concepts

acridine
alkylating agent
Ames test
AP endonuclease
base analog
base substitution
ClB method
conditional mutation
cut-and-paste transposition
depurination
DNA transposon
DNA uracil glycosylase
dynamic mutation
excision repair

forward mutation
frameshift mutation
gain-of-function mutation
germ-line mutation
hotspot
hypermorphic mutation
hypomorphic mutation
induced mutation
ionizing radiation
LINE elements
loss-of-function mutation
LTR retrotransposon
mismatch repair
missense mutation

mutagen
mutation
nitrous acid
non-LTR retrotransposon
nonsense mutation
permissive condition
postreplication repair
pyrimidine dimer
rate of mutation
replica plating
replication slippage
restrictive condition
reversion
sickle-cell anemia

silent substitution
SINE elements
somatic mutation
spontaneous mutation
synonymous substitution
temperature-sensitive
  mutation
transposable element
transposase
transposition
transition mutation
transversion mutation
trinucleotide repeat

1. _____ Any agent that causes mutation.

2. _____ Refers to the probability of a new mutation occurring per unit time or per generation.

3. _____ Change from the wildtype form of a gene into a form that yields a mutant phenotype.

4. _____ Substitution of a purine nucleotide with a pyrimidine nucleotide or the other way around.

5. _____ Environment in which the mutant phenotype of a conditional mutation is expressed.

6. _____ A mutation that cannot be passed on to future generations.

7. _____ Dynamic mutation of this type of sequence is associated with the fragile-X syndrome and Huntington disease.

8. _____ A class of RNA-based transposable elements with long terminal repeats.

9. _____ A nucleotide site that is particularly prone to undergo mutation; in mammalian DNA, such sites are often the positions of 5-methyl cytosine.

10. _____ Procedure that reproduces the geometrical arrangement of bacterial colonies on a series of agar plates.

11. _____ Experiment using a rearranged *Drosophila* X chromosome bearing a recessive lethal mutation as well as a dominant *Bar* eye mutation, similar to that originally used to show that x-rays cause mutations.

12. _____ Key enzyme in the repair of apurinic or apyrimidinic sites in double-stranded DNA.

solutions: step by step

Problem 1

The molecule 2-aminopurine (Ap) is an analog of adenine that pairs with thymine. It also occasionally pairs with cytosine. What pathways of mutation are possible, and what types of mutations are formed?

■ **Solution** In problems of this sort, first note the base pair that is usually formed (the nonmutagenic base pair) and then the base pair that is rarely formed (the mutagenic base pair). In this case, the nonmutagenic base pair is Ap−T and the mutagenic base pair is Ap−C. There are two pathways of mutation, analogous to those for 5-bromouracil illustrated in Figure 12.19. If the Ap is incorporated in its mutagenic mode, it is incorporated opposite C, creating an Ap−C base pair. In the subsequent round of replication, the Ap will usually pair with T, forming an Ap−T base pair. The end result is a G−C → A−T transition. On the other hand, if the Ap is incorporated in its nonmutagenic mode, it is incorporated opposite T, creating an Ap−T base pair. In a subsequent round of replication, the Ap may pair with C, forming an Ap−C base pair. The end result is an A−T → G−C transition.

Problem 2

Among genomes as diverse as the bacteriophages λ and T4, the bacteria *Escherichia coli*, and the fungi *Neurospora crassa* and *Saccharomyces cerevisiae*, the spontaneous mutation rate is approximately 0.003 mutation per genome per DNA replication. The genome sizes of λ, *E. coli*, and *S. cerevisiae* are approximately 50 thousand, 4.6 million, and 13.5 million base pairs, respectively.

(a) What proportion of genomes escape spontaneous mutation in each replication?

(b) What is the mutation rate per base pair per DNA replication in each of these genomes?

(c) What might the differences in mutation rate per base pair imply about the evolution of mutation rates?

■ **Solution** **(a)** If the mutation rate is 0.003 mutation per genome per DNA replication, then $1 − 0.003 = 99.7$ percent of the genomes contain no new mutations after one round of replication. **(b)** The rate of mutation per base pair per DNA replication for λ is $0.003/50,000 = 6 \times 10^{-8}$. For the other genomes, the values are calculated similarly and equal 6.5×10^{-10} for *E. coli* and 2.2×10^{-10} for *S. cerevisiae*. **(c)** The great variation in spontaneous mutation rate per base pair among these genomes suggests that the larger genomes have evolved lower mutation rates per nucleotide pair through either greater fidelity of DNA replication or greater efficiency of repair mechanisms.

concepts in action: problems for solution

12.1 Occasionally, a person is found who has one blue eye and one brown eye or who has a sector of one eye a different color from the rest. Can these phenotypes be explained by new mutations? If so, in what types of cells must the mutations occur?

12.2 Occasionally a mutation is isolated that cannot be induced to reverse-mutate to wildtype. What types of molecular changes might be responsible?

12.3 Does the nucleotide sequence of a mutation tell you anything about its dominance or recessiveness?

12.4 For mutations that result from spontaneous missense substitutions, the rate of forward mutation (from wildtype allele to mutant allele) is always much greater than the rate of reverse mutation (from mutant allele back to wildtype). Suggest a hypothesis that explains why this should be expected.

12.5 If spontaneous depurination of DNA occurs at the rate of approximately 3×10^{-9} depurinations per purine nucleotide per minute, then considering that a diploid human cell has a genome size of 6×10^9 base pairs, approximately how many spontaneous depurinations must be repaired in each cell per day?

12.6 How many different codons can result from a single-base substitution in DNA coding for the cysteine codon UGC? Classify each as synonymous (silent), nonsynonymous (missense), or nonsense (chain termination).

12.7 Weedy plants that are resistant to the herbicide atrazine have a single amino acid substitution in the gene *psbA* that results in the replacement of a serine with an alanine in the polypeptide. Is the base change in the *psbA* gene that results in this amino acid replacement a transition or a transversion?

12.8 What is the minimum number of single-nucleotide substitutions that would be necessary for each of the following amino acid replacements?
(a) Trp → Lys   (b) Tyr → Gly
(c) Met → His   (d) Ala → Asp

12.9 How many amino acids can replace tyrosine by a single-base substitution in the DNA? (Do not assume that you know which tyrosine codon is being used.)

12.10 A *Drosophila* male carries an X-linked temperature-sensitive recessive allele that is viable at 18°C but lethal at 29°C. What sex ratio would be expected among the progeny if the progeny were reared at 29°C and the male were mated to:
(a) A normal XX female?
(b) An attached-X female?

12.11 Mutations caused by the insertion of a DNA transposon are often genetically unstable, reverting to wildtype (or a phenotype resembling wildtype) at a relatively high rate. Suggest a reason why this might be expected.

12.12 A population of 1×10^6 bacterial cells undergoes one round of DNA replication and cell division. The forward mutation rate of a gene is 1×10^{-6} per replication.
(a) What is the expected number of mutant cells after cell division?
(b) What is the probability that the population contains no mutant cells?

12.13 In the mouse, a dose of approximately 1 sievert (Sv) of x rays produces a rate of induced mutation equal to the rate of spontaneous mutation. Expressed as a multiple of the spontaneous mutation rate, what is the total mutation rate at 1 Sv? Assuming that the total mutation rate is proportional to the x-ray dose, what dose of x rays will increase the mutation rate by 50 percent? What dose will increase the mutation rate by 10 percent?

12.14 Among several hundred missense mutations in the gene for the A protein of tryptophan synthase in *E. coli,* fewer than 30 of the 268 amino acid positions are affected by one or more mutations. Suggest a hypothesis to explain why the number of positions affected by amino acid replacements is so low.

12.15 If every human gamete contains approximately 30,000 genes, and if the forward mutation rate is between 1×10^{-5} and 1×10^{-6} new mutations per gene per generation, what is the average number of new mutations per gamete per generation?

12.16 Human hemoglobin C is a variant in which a lysine in the beta-hemoglobin chain is substituted for a particular glutamic acid. What single-base substitution can account for the hemoglobin-C mutation?

12.17 Gene conversion results from mismatch repair in a heteroduplex DNA molecule that gives rise to fungal asci

with ratios of alleles such as 3 *A* : 1 *a* and 1 *A* : 3 *a* , instead of 2 *A* : 2 *a.*
(a) Why is a ratio of 2 *A* : 2 *a* expected?
(b) Why, among a large number of gametes chosen at random, is the Mendelian segregation ratio of 1 *A* : 1 *a* still observed in spite of gene conversion?

12.18 The accompanying diagrams show two nonhomologous chromosomes, each containing a copy of a transposable element (shaded), that can be oriented either (a) in the same direction or (b) in opposite directions with respect to the centromere. For clarity, the length of the transposable element is greatly exaggerated relative to the length of the chromosome. (In reality, the average transposable element in *Drosophila* is about 0.01 percent of the length of a chromosome.) Draw diagrams illustrating the consequences of ectopic recombination between the transposable elements.

12.19 This problem illustrates how conditional mutations can be used to determine the order of genetically controlled steps in a developmental pathway. A certain organ undergoes development in the sequence of stages A → B → C, and both gene *X* and gene *Y* are necessary for the sequence to proceed. A conditional mutation X' is sensitive to heat (the gene product is inactivated at high temperatures), and a conditional mutation Y' is sensitive to cold (the gene product is inactivated at cold temperatures). The double mutant X'/X' Y'/Y' is created and reared at either high or low temperatures. To what stage would development proceed in each of the following cases at the high temperature and at the low temperature?
(a) Both *X* and *Y* are necessary for the A → B step.
(b) Both *X* and *Y* are necessary for the B → C step.
(c) *X* is necessary for the A → B step, and *Y* is necessary for the B → C step.
(d) *Y* is necessary for the A → B step, and *X* is necessary for the B → C step.

12.20 You carry out a large-scale cross of genotypes $A\ m\ B \times a + b$ of two strains of *Neurospora crassa* and observe a number of aberrant asci, some of which are shown here.

A m B	A m B	A m B	A m B	A m B
A m B	A m B	A m B	A m B	A m B
A m b	A m b	A m b	A m b	A + b
A + b	A m b	A + b	A m b	A + b
a m B	a m B	a + B	a m B	a + B
a + B	a + B	a + B	a m B	a + B
a + b	a + b	a + b	a + b	a + b
a + b	a + b	a + b	a + b	a + b

Your colleague who provided the strains insists that they are both deficient in the same gene in the DNA mismatch-repair pathway. The results of your cross seem to contradict this assertion.
(a) Which of the asci depicted here offer evidence that your colleague is incorrect?
(b) Which ascus (or asci) would you exhibit as definitive evidence that DNA mismatch repair in these strains is not 100 percent efficient?

geNETics on the web

GeNETics on the Web will introduce you to some of the most important sites for finding genetic information on the Internet. To explore these sites, visit the Jones and Bartlett home page at

http://www.jbpub.com/genetics

For the book *Essential Genetics: A Genomics Perspective,* choose the link that says **Enter GeNETics on the Web**. You will be presented with a chapter-by-chapter list of highlighted keywords. Select any highlighted keyword and you will be linked to a Web site containing genetic information related to the keyword.

- The **Chernobyl** nuclear accident was by far the most devastating in the history of nuclear power. It released into Earth's atmosphere about 400 times more radioactive material than the atomic bomb dropped on Hiroshima, but less than 1 percent of the amount from atmospheric weapons tests conducted in the 1950s and 1960s. This keyword site summarizes what is known about the nuclear accident and its aftermath. This site contains information about what happened, how it happened, how many people were significantly exposed to radiation, the immediate health effects, and the anticipated long-term health and environmental effects of this environmental disaster.

- Among 400 people, about 200 will develop at least one tumor of the colon (large bowel), 20 of these cases will progress to malignancy, and 1 will be associated with an inherited genetic abnormality. Among the inherited forms of **colon cancer**, two important types are familial adenomatous polyposis (FAP) and hereditary nonpolyposis colorectal cancer (HNPCC). They demonstrate the difference between tumor initiation and tumor progression. Patients with FAP develop thousands of colon polyps, only a few of which progress to cancer, whereas patients with HNPCC have only a few polyps, but each has a high probability of progressing to cancer. Consult this keyword site to learn more about these and other types of colon cancer.

further readings

Ames, B. W. 1979. Identifying environmental chemicals causing mutations and cancer. *Science* 204: 587.

Berg, D. E., and M. M. Howe. 1989. *Mobile DNA.* American Association for Microbiology. Washington, DC.

Bhatia, P. K., Z. G. Wang, and E. C. Friedberg. 1996. DNA repair and transcription. *Current Opinion in Genetics & Development* 6: 146.

Buermeyer, A. B., S. M. Deschenes, S. M. Baker, and R. M. Liskay. 1999. Mammalian DNA mismatch repair. *Annual Review of Genetics* 33: 533–564.

Chu, G., and L. Mayne. 1996. Xeroderma pigmentosum, Cockayne syndrome and trichothiodystrophy: Do the genes explain the diseases? *Trends in Genetics* 12: 187.

Cox, E. C. 1997. *mutS,* proofreading, and cancer. *Genetics* 146: 443.

Crow, J. F., and C. Denniston. 1985. Mutation in human populations. *Advances in Human Genetics* 14: 59.

Drake, J. W. 1991. Spontaneous mutation. *Annual Review of Genetics* 25: 125.

Hickson, I. D. 1997. *Base Excision Repair of DNA Damage.* Austin, TX: Landes Bioscience.

Hoeijmakers, J. H. J. 1993. Nucleotide excision repair I: From *E. coli* to yeast. *Trends in Genetics* 9: 173.

Jackson, S. P. 1996. The recognition of DNA damage. *Current Opinion in Genetics & Development* 6: 19.

Nickoloff, J. A., and M. F. Hoekstra, eds. 1998. *DNA Damage and Repair.* Totowa, NJ: Humana Press.

Shcherbak, Y. M. 1996. Confronting the nuclear legacy. I. Ten years of the Chernobyl era. *Scientific American,* April.

Singer, B., and J. T. Kusmierek. 1982. Chemical mutagenesis. *Annual Review of Biochemistry* 51: 655.

Smith, P. J., and C. J. Smith, eds. *DNA Recombination and Repair.* 1999. New York: Oxford University Press.

Cancer is a genetic disease resulting from mutations in somatic cells. Usually a cell lineage requires multiple mutations to become cancerous. Some of these mutations disable the molecular mechanisms that regulate cell division, whereas others inactivate the process of programmed cell death that normally eliminates genetically stressed or damaged cells. [Courtesy of National Cancer Institute.]

key concepts

- Progression from one stage of the cell cycle to the next is controlled by protein complexes called cyclin-dependent kinase (CDK) complexes. These are made up of a cyclin component and a cyclin-dependent protein kinase. Protein degradation is also important, especially in later stages of the cell cycle.

- Checkpoints monitor a dividing cell for DNA damage, cellular defects, other abnormalities in the cell cycle, and cell size. Detection of abnormalities elicits a response that arrests the cell cycle, allowing time for repair of defects (or cell death—apoptosis) and ensuring that the phases remain in the correct order.

- Cancer cells show uncontrolled growth and proliferation and loss of contact inhibition.

- Progression from a normal to a cancerous state requires several genetic changes. Most cancers are sporadic (not inherited).

- A small proportion of cancers are associated with mutations transmitted through the germ line, which predispose the somatic cells of people carrying them to undergo cancer progression.

- The genetic changes that take place in cancer progression often involve defects or overexpression of genes that function in cell-cycle regulation or checkpoint control.

13

Molecular Genetics of the Cell Cycle and Cancer

chapter organization

13.1 The cell cycle is under genetic control.

13.2 Checkpoints in the cell cycle allow damaged cells to repair themselves or to self-destruct.

13.3 Cancer cells have a small number of mutations that prevent normal checkpoint function.

13.4 Mutations that predispose to cancer can be inherited through the germ line.

13.5 Acute leukemias are proliferative diseases of white blood cells and their precursors.

the human connection
Two Hits, Two Errors

Chapter Summary
Issues & Ideas
Key Terms & Concepts
Solutions: Step by Step
Concepts in Action: Problems for Solution
Further Readings

geNETics **on** the web

ancer is a disease characterized by the uncontrolled proliferation of cells. The normal mechanisms that regulate cellular growth and division break down. Cancer is a genetic disease. It results from mutations that overcome the normal limits to the number of cell divisions that can take place before a cell dies. These mutations usually occur in somatic cells, and full-blown malignant cancer usually requires a number of sequential mutations to get started. But occasionally a mutation affecting cell-cycle regulation is inherited through the germ line, and persons who inherit the mutation have a greatly increased risk (sometimes approaching 100 percent) of developing malignancies due to additional somatic mutations. To understand cancer, therefore, one must understand the normal mechanisms that control the cell cycle and that prevent the proliferation of genetically damaged cells. This is where we begin.

13.1

The cell cycle is under genetic control.

As we saw in Chapter 3, the cell cycle is divided into a three-part *interphase* composed of G_1 *(gap 1)*, *S (DNA synthesis)*, and G_2 *(gap 2)*, which occur in that order, followed by *M (mitosis)* proper, in which the sister chromatids are physically separated into the two daughter nuclei. The essential functions of the mitotic cell cycle are:

1. To ensure that each chromosomal DNA molecule is replicated once and only once per cycle.

2. To ensure that the identical replicas of each chromosome (the sister chromatids) are distributed equally to the two daughter cells.

Some of the key events that must take place for the proper duplication and distribution of chromosomes are highlighted in Figure 13.1 . The spindle pole is organized around a small region of clear cytoplasm near the interphase nucleus called the **centrosome;** in many organisms this role is played by a pair of **centrioles,** which are more particulate in appearance. Both are microtubule organizing centers that must be duplicated and positioned. In most cells, centrosome duplication begins late in G_1 and is completed during S phase. The duplicated poles then slowly begin to migrate to positions on opposite sides of the nucleus. Meanwhile, within the nucleus during the S phase, DNA replication takes place. Completion of DNA replication marks the beginning of G_2. Soon after the M phase commences, the centrosomes reach their final destinations and the chromosomes begin to condense. Each centrosome organizes one pole of the spindle by nucleating the formation of spindle and astral microtubules, and the condensed chromosomes

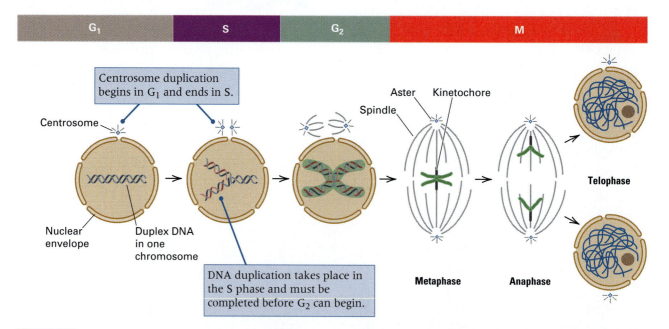

Figure 13.1 Major events in the cell cycle. In yeast, the spindle pole body serves the same function as the centrosome in many other organisms: Both are microtubule organizing centers from which the spindle emerges. In most other organisms, nuclear division takes place and the pinching off of cells (cytokinesis) follows. In yeast, the "shell" of the daughter cell forms and enlarges before nuclear division takes place. The nucleus (its membrane never breaks down) moves into the bridge between the mother and daughter cell, and nuclear division occurs there. After the daughter nuclei move into the two cell bodies, a septum is laid down between the two cells. [Adapted from L. H. Hartwell and M. Kastan. 1994. *Science* 266: 1821.]

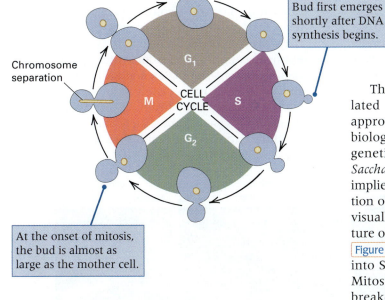

After separation, the bud must grow to reach a minimal size before a new cycle can begin.

Bud first emerges shortly after DNA synthesis begins.

Chromosome separation

At the onset of mitosis, the bud is almost as large as the mother cell.

CELL CYCLE

G₁ S G₂ M

Figure 13.2 | The cell cycle of budding yeast, *Saccharomyces cerevisiae*.

The mechanisms by which cell growth is regulated and the cell cycle controlled have been approached via the methods of biochemistry, cell biology, and genetics. Some of the most extensive genetic studies have focused on budding yeast *Saccharomyces cerevisiae*. As its common name implies, *S. cerevisiae* multiplies by budding. The position of a cell within the cell cycle can be monitored visually by the size of the bud. This convenient feature of the cell cycle is depicted diagrammatically in Figure 13.2. The bud first emerges shortly after entry into S phase and grows throughout the cell cycle. Mitosis takes place *within* the nucleus without breakdown of the nuclear envelope. The spindle poles are actually embedded within the nuclear envelope, and one pole moves to a position exactly opposite the second pole across the way. Mitosis occurs when the bud has grown to a size nearly as large as the mother cell. Shortly after chromosome separation, a barrier is laid down between the mother and daughter cells. At cell separation, the daughter cell is only slightly smaller than the mother cell, but it typically must grow this extra bit to achieve the optimal size before it can start its own cell cycle. The various stages of bud growth from beginning to end can be identified in the electron micrograph in Figure 13.3.

become attached to spindle microtubules on both sides of the kinetochore (centromere) as the nuclear envelope breaks down. Each chromosome is thus physically attached to both spindle poles and is maneuvered to a position approximately halfway between them. As anaphase begins, the spindle elongates, the centromeres separate, and the sister chromatids migrate toward opposite poles. Once the daughter chromosomes reach the poles, the spindle disassembles, the chromosomes decondense, and the nuclear membrane is formed again. These events return the cell to the G₁ phase.

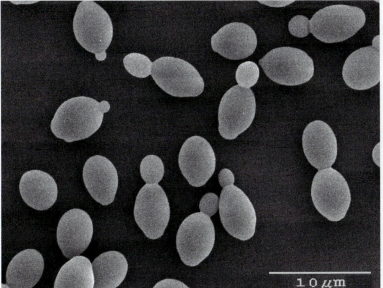

Courtesy of Elizabeth W. Jones

10 μm

Figure 13.3 | Scanning electron micrograph of cells of diploid budding yeast in various stages of the cell cycle. Bud size is correlated with the position of the cell within the cell cycle. Here the buds range in size from quite small to nearly as large as the mother cell.

13.1 The cell cycle is under genetic control   461

◼ Many genes are transcribed during the cell cycle just before their product is needed.

Because the DNA sequence of the entire genome is known for budding yeast, it is possible to analyze the transcription patterns of all of the 5538 genes in the cell in a single experiment using the type of high-density gene microarrays discussed in Chapter 10. Such experiments have shown that the transcript levels of about 800 genes vary in a periodic or cyclic pattern through the cell cycle. This result is dramatically illustrated in Figure 13.4 . Each horizontal stripe represents the expression pattern of a single gene through two or more cell cycles in synchronized cells. (In a culture of *synchronized cells*, all cells are at the same stage in the cell cycle.) The colored bands across the top indicate the stage of the cell cycle (using the same colors for the stages as labeled on the right), and the designations Alpha, *cdc15*, and *cdc28* refer to different ways in which the cells were synchronized. For each gene, red indicates overexpression relative to the level of expression of the same gene in a nondividing cell, and green indicates underexpression. The genes have been grouped according to the similarity in their pattern of transcription through the cell cycle. Transcription of each of

these 800 genes is initiated once per cell cycle, an event indicated by the red portion of the stripe as the cell cycle progresses. Some genes are transcribed in G_1, some in G_2, a few in S, and some in M. The genes are arranged in the order in which they are transcribed, and their principal stage of expression is indicated at the far right. Typically, genes encoding proteins that are needed in one part of the cycle are transcribed in the immediately preceding period. For example, enzymes needed for synthesis of the trinucleotide precursors of DNA and for DNA replication are made in G_1 immediately prior to their use in S phase. Similarly, the histone proteins are synthesized during S phase immediately prior to their incorporation into chromatin and their use in chromosome condensation.

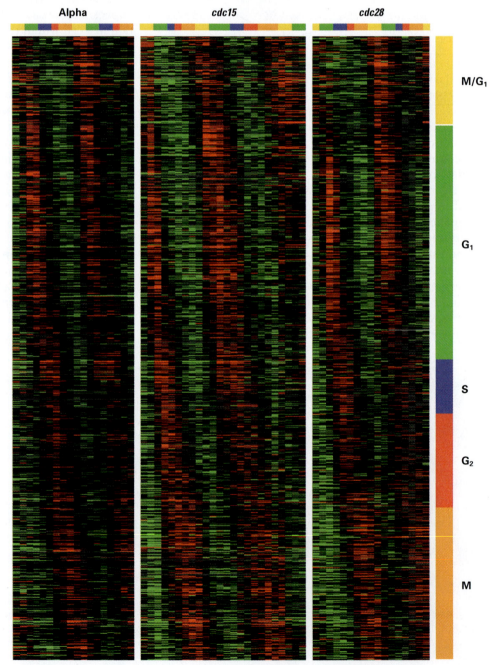

Figure 13.4 Expression patterns of about 800 yeast genes whose level of transcription varies systematically through the cell cycle. Each horizontal stripe indicates the transcription pattern of a different gene through the cell cycle. Green indicates underexpression relative to a nondividing cell, red overexpression. The color bars on the right indicate the phase in which each gene is maximally expressed: M/G_1 transition, G_1, S, G_2, or M. These same colors indicate the cell-cycle phase along the top. Alpha, *cdc15*, and *cdc28* indicate different ways in which the cell cycle has been synchronized. [Courtesy of Patrick O. Brown, Reprinted from *Molecular Biology of the Cell (Mol. Biol. Cell* 1998 9: 3273-3297) with permission of The American Society for Cell Biology.]

■ Mutations affecting the cell cycle have helped to identify the key regulatory pathways.

The ready assignment of a yeast cell to a position in the cell cycle on the basis of the relative sizes of the mother cell and its bud has made possible extensive genetic analyses of the cell cycle through the isolation and study of temperature-sensitive mutants. Temperature-sensitive **cell division cycle (*cdc*) mutants** are typically wildtype at 23°C (the *permissive* temperature) but unable to complete the cell cycle at 36°C (the *restrictive* temperature). At the higher temperature, mutant cells accumulate at a characteristic stage in the cell cycle. This is the stage at which their progression in the cell cycle is halted. The stage-specific stop is exceptional among mutations affecting cellular functions. For example, temperature-sensitive mutants with defects in protein synthesis do not cease growing abruptly when the temperature is raised. Each cell continues along until it runs out of functional protein, and new protein needs to be synthesized for continued progression in the cell cycle. The stopping point differs from cell to cell, so the mutant cells stop growing at different stages in the cell cycle. In the microscope one sees cells with no bud, cells with small buds, and cells with large buds—essentially the same distribution one sees in an asynchronous cell population. But for each *cdc* mutant the stop is at a specific stage, which differs from one mutant to the next and is related to the function of the mutant gene product. Once it has been established where in the cell cycle a *cdc* mutant is blocked, further analysis can be used to determine whether specific processes can occur in the mutants, such as DNA synthesis or spindle formation.

One mutant designated *cdc13* that causes a block, or **arrest,** at the G_2/M boundary is shown in the micrograph in Figure 13.5 . Part A shows mutant cells grown in an unsynchronized culture at the permissive temperature of 23°C. The morphology and variation in size are normal; some unbudded cells are present, as well as dividing cells with buds ranging in size from very small to nearly as large as the mother cell. Part B shows the result when this culture was placed at the restrictive temperature of 36°C for 6 hours. Two types of configurations are seen. One consists of a pair of large cells. These are cells that were not dividing or had small buds at 23°C; after the temperature shift they continued in the cell cycle until they arrested at the large-bud stage. The other configuration consists of four large cells (*quartets*). These derive from cells that, prior to the temperature shift, were nearing the end of division and had large buds; after the temperature shift

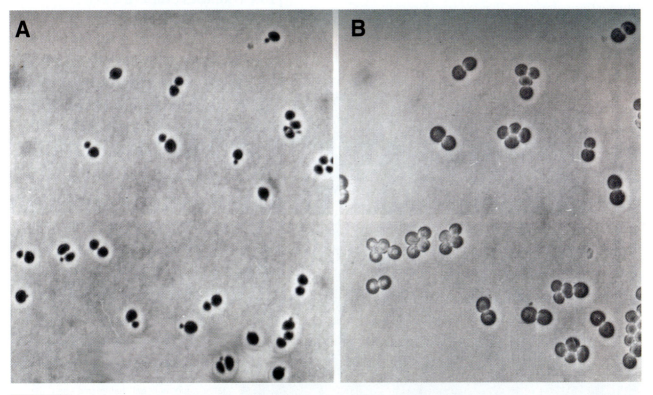

Figure 13.5 Cell-cycle arrest in temperature-sensitive *cdc13* mutants. (A) An asynchronous culture of a *cdc13* mutant grown at 23°C (the permissive temperature). The cells either are unbudded or have buds ranging in size from very small to nearly the size of the mother cell. (B) The same cells after incubation for 6 hours at 36°C (the restrictive temperature). Some cells are arrested with very large buds. Others have completed their current division and started another. These are the cells arrested in quartets. [From J. Culotti and L. H. Hartwell. 1971. *Experimental Cell Research* 67: 389.]

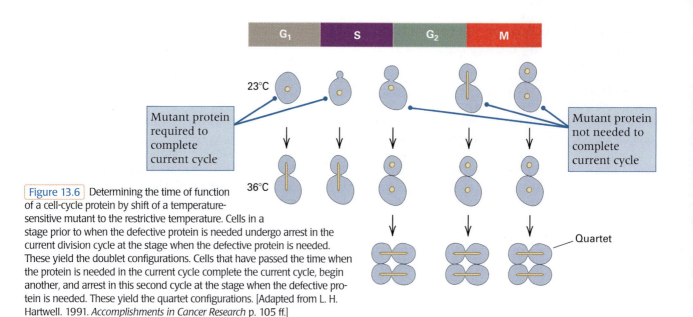

23°C

Mutant protein required to complete current cycle

Mutant protein not needed to complete current cycle

36°C

Quartet

Figure 13.6 Determining the time of function of a cell-cycle protein by shift of a temperature-sensitive mutant to the restrictive temperature. Cells in a stage prior to when the defective protein is needed undergo arrest in the current division cycle at the stage when the defective protein is needed. These yield the doublet configurations. Cells that have passed the time when the protein is needed in the current cycle complete the current cycle, begin another, and arrest in this second cycle at the stage when the defective protein is needed. These yield the quartet configurations. [Adapted from L. H. Hartwell. 1991. *Accomplishments in Cancer Research* p. 105 ff.]

they completed their current cycle, and both daughter cells, still side by side, initiated one more cell cycle and arrested at the large-bud stage, yielding the quartet.

Figure 13.6 illustrates how these results can be interpreted. Single mutant cells that are transferred to 36°C prior to the time that the affected protein is needed arrest as large budded cells, because in the absence of functional protein they cannot proceed beyond this point. Cells that are transferred to 36°C when the protein is no longer needed in the current cycle complete the current cycle, and initiate a new cycle that is terminated when both daughter cells arrest with large buds because the affected

protein is now unavailable. The *cdc13* mutant showing the pattern in Figure 13.5 has a defect in telomere processing, which occurs late in the S phase. When *cdc13* cells are grown at the restrictive temperature, abnormally extended single-stranded DNA is found at the telomeres.

■ Cyclins and cyclin-dependent protein kinases propel the cell through the cell cycle.

In the early stages of the cell cycle, progression from one phase to the next is controlled by characteristic protein complexes that are called **cyclin–CDK complexes** because they are composed of **cyclin** subunits combined with **cyclin-dependent protein kinase (CDK)** subunits (**Figure 13.7**). All eukaryotes regulate progression through the cell cycle by means of cyclin–CDK complexes, although the details of their structure and their mechanisms of action may differ slightly from one organism to another.

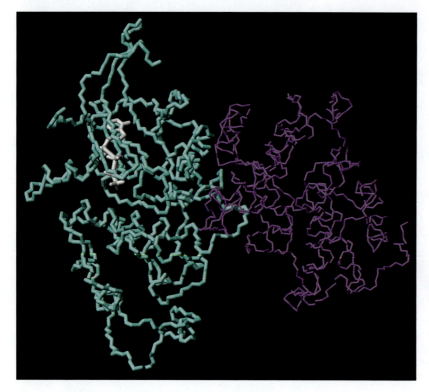

Figure 13.7 Structure of cyclin A (right) complexed with Cdk2 (left). Cyclin A binds to one side of the catalytic cleft in Cdk2, inducing a large conformational change that opens the active site and activates the kinase domain. The ATP (white) bound to Cdk2 is the phosphate donor in the kinase reaction. [Courtesy of Carlos Bustamante, from coordinate data of P. D. Jeffrey et al. 1995. *Nature* 376: 294.]

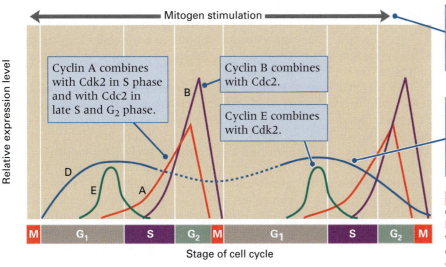

Mitogen stimulation

Relative expression level

Cyclin A combines with Cdk2 in S phase and with Cdc2 in late S and G_2 phase.

Cyclin B combines with Cdc2.

Cyclin E combines with Cdk2.

A mitogen is a chemical or treatment that stimulates the mitotic cell cycle.

D-type cyclins combine with Cdk4 and Cdk6. The mitogen keeps the level of D cyclins elevated (dotted portion of line) from one cycle to the next.

Stage of cell cycle

M | G_1 | S | G_2 | M | G_1 | S | G_2 | M

Figure 13.8 Fluctuations of cyclin levels during the cell cycle. Expression of cyclins E, A, and B (mitotic cyclins) are periodic, whereas cyclin D is expressed throughout the cell cycle in response to mitosis-stimulating drugs (*mitogens*). [Adapted from C. Sherr. 1996. *Science* 274: 1672.]

The term *cyclin* is apt because the abundance of these proteins changes cyclically in the cell cycle, and some are present only at specific times. Most cyclins appear abruptly and disappear a short time later (Figure 13.8). More than one cyclin may be present in cells at the same time in the cell cycle, and some CDKs are able to form complexes with different cyclins. Each cyclin appears at its characteristic time in the cell cycle because its transcription is linked to the cell cycle via a previously expressed cyclin. Typically, the presence of an active cyclin–CDK complex results in the activation of a transcription factor that leads to transcription of the next cyclin needed in the cell cycle. The cyclin disappears when its gene is no longer transcribed and the previously made mRNA and protein are degraded. Active cyclin–CDK complexes also cause the transcriptional activation of genes other than cyclins.

The cyclin–CDK complexes control progression through the cell cycle through their activity as pro-

tein kinases. They attach phosphate groups to the hydroxyl groups present in the amino acids serine, threonine, and tyrosine found in certain proteins. The cyclin component of the complex binds to the protein substrate and tethers it, allowing the CDK component to phosphorylate the tethered substrate. Once the targeted protein is phosphorylated, it dissociates from the cyclin–CDK complex. The activities of the phosphorylated form of a protein often differ dramatically from those of the unphosphorylated form. Phosphorylation may activate one enzyme but inactivate another. Even in a single enzyme molecule, phosphorylation at different sites may have opposite effects.

Another route of cell-cycle regulation is mediated through *phosphatase* enzymes that dephosphorylate proteins that the cyclin-dependent kinases have phosphorylated. Reversing the effects of CDKs, the phosphatases activate enzymes that are inactive or deactivate ones that are active. In mammalian cells, specific cyclin–CDK complexes can be detected at different times in the cell cycle, as illustrated in Figure 13.9. The cyclin D–Cdk4 and cyclin D–Cdk6 complexes appear in the early or middle part of G_1, whereas cyclin E–Cdk2 and cyclin A–Cdk2 appear later in G_1. The cyclin A–Cdk2 complex is present throughout the S phase and into M (mitosis), and the cyclin B–Cdc2 complex carries the cell through the G_2/M transition.

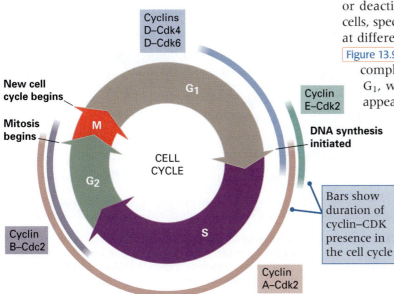

Cyclins D–Cdk4 D–Cdk6

New cell cycle begins

Mitosis begins

G_1

Cyclin E–Cdk2

DNA synthesis initiated

M

CELL CYCLE

G_2

S

Cyclin B–Cdc2

Cyclin A–Cdk2

Bars show duration of cyclin–CDK presence in the cell cycle

Figure 13.9 The temporal expression pattern of activities of the cyclin–CDKs in mammalian cells.

13.1 The cell cycle is under genetic control   465

■ The retinoblastoma protein controls the initiation of DNA synthesis.

Our first connection between cell-cycle control and cancer comes from a protein known to be involved in tumors of the retina. The type of cancer is called *retinoblastoma*, and we will examine the condition in some detail in Section 13.4. The normal role of the **retinoblastoma (RB) protein** is to maintain cycling cells at a point in G_1 called the **G_1 restriction point** or *start*, until the cell has attained proper size. The RB protein acts by binding to the transcription factor E2F, which is needed for further progression in the cell cycle (Figure 13.10, part A). If the cycling cell is growing properly, cyclin D is produced in the middle of G_1. The RB protein then begins to be phosphorylated by both the cyclin D–Cdk4 kinase and the cyclin D–Cdk6 kinase. Late in G_1, the RB phosphorylation is completed by the cyclin E–Cdk2 kinase as cells approach the **G_1/S transition,** when they become committed to DNA sysnthesis. Phosphorylation of RB eliminates its ability to bind the E2F transcription factor (part A). Release of the E2F results in transcription of the genes and translation of the enzymes responsible

for DNA synthesis, including DNA polymerase. E2F also activates transcription of the gene for E2F itself (an example of positive autoregulation), as well as those for the cyclins E and A and the Cdk2 kinase subunit.

Once cells have entered S phase, the cyclin A–Cdk2 phosphorylates E2F and inhibits its binding to DNA, thus inactivating its function as a transcription factor. The cyclin A–Cdk2 activity is required throughout S phase, apparently to keep the RB protein heavily phosphorylated. The progression from G_2 to M (the **G_2/M transition**) is controlled by a cyclin B–Cdc2 complex also known as *maturation-promoting factor*. After synthesis and assembly in the cytoplasm, the cyclin B–Cdc2 complex remains in the cytoplasm in an inactive form, because its rate of import into the nucleus is smaller than its rate of export from the nucleus. The balance is tipped in favor of import when cyclin B is phosphorylated just prior to the G_2/M transition, which masks the nuclear export signal. Reducing nuclear export ensures that cyclin B–Cdc2 accumulates in the nucleus. The cyclin B–Cdc2 complex is activated by dephosphorylation at different sites, and it then carries out phosphorylation of its sub-

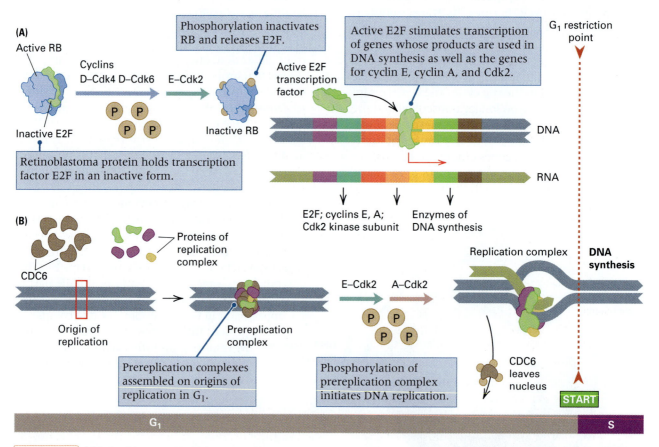

Figure 13.10 (A) Role of the retinoblastoma protein RB in controlling the transition from G_1 phase to S phase. The cyclin D-dependent kinases Cdk4 and Cdk6 initiate phosphorylation of RB in mid-G_1 phase, a process completed by cyclin E–Cdk2; this frees the transcription factor E2F, which activates transcription of enzymes for DNA synthesis. (B) The free E2F also activates transcription of the genes for cyclin A, cyclin E, and Cdc2, which help convert prereplication complexes to replication complexes for transition to S phase.

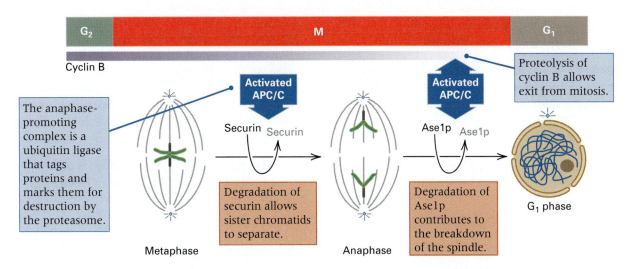

The anaphase-promoting complex is a ubiquitin ligase that tags proteins and marks them for destruction by the proteasome.

Activated APC/C

Securin → Securin

Degradation of securin allows sister chromatids to separate.

Activated APC/C

Ase1p → Ase1p

Degradation of Ase1p contributes to the breakdown of the spindle.

Proteolysis of cyclin B allows exit from mitosis.

G_2 | M | G_1

Cyclin B

Metaphase — Anaphase — G_1 phase

Figure 13.11 Role of activated anaphase-promoting complex (APC/C) in controlling the proteolysis necessary for the transition to anaphase and the exit from mitosis. APC/C is a ubiquitin ligase that marks proteins for proteasome destruction. Its substrates include securin, which unless destroyed inhibits breakdown of the protein "glue" (Scc1p) that holds the sister chromatids together, and Ase1p, a microtubule-associated protein that binds antiparallel microtubules from the spindle pole bodies in the spindle midzone. Destruction of cyclin B somewhat later allows the exit from mitosis.

strates to bring about events that complete the G_2/M transition. At this time the chromosomes condense and assemble onto the spindle, and the chromosome segregation machinery becomes active.

■ Protein degradation also helps regulate the cell cycle.

A fundamental feature of the cell cycle is that it is a true cycle: It is not reversible. The cycle is propelled forward by a process of protein degradation that complements the periodic activation of cyclin–CDK complexes. Protein degradation (*proteolysis*) eliminates proteins that were used in the preceding phase as well as proteins that would inhibit progression into the next phase. In the early stages of the cell cycle, progression requires the sequential activation of cyclin–CDKs. Entry into each new phase also requires destruction of the cyclins used in the preceding phase. In the later stages of the cell cycle, progression is propelled by proteolysis alone. This process is best understood in yeast, and we describe it here using the yeast terminology. In the completion of mitosis and the return to G_1 phase, two key regulatory events must occur:

1. The sister chromatids must separate (marking the onset of anaphase).

2. The cells must exit from mitosis, which entails chromosome decondensation, spindle disassembly, inactivation of the chromosome segregation machinery, and cytokinesis.

Both of these key events are triggered by protein degradation, as indicated in Figure 13.11 for yeast.

Exit from mitosis requires the destruction of cyclin B. Cyclin B is most abundant when cells enter mitosis, but it disappears after chromosome disjunction has occurred at the transition from metaphase to anaphase. Cyclin B is marked for destruction by the **anaphase-promoting complex (APC/C),** which is a ubiquitin–protein ligase responsible for adding the 76-amino-acid protein *ubiquitin* to its target proteins and marking them for destruction in the *proteasome*—a large, multifunctional, multi-subunit complex responsible for most of the cytoplasmic proteolysis in the cell beyond that which takes place in lysosomes. Other substrates for the APC/C are the protein securin, which inhibits a protease called separase, and Ase1p, a microtubule-associated protein that binds antiparallel microtubules from the spindle pole bodies in the midzone of the spindle. As the securin is degraded, the separase protein becomes free to degrade Scc1p, a component of the cohesin complex that condenses chromosomes and also holds sister chromatids together. Cleavage of Scc1p results in its dissociation from the chromatin, which allows the sister chromatids to come apart and be pulled toward opposite poles of the spindle.

13.2

Checkpoints in the cell cycle allow damaged cells to repair themselves or to self-destruct.

Cells monitor their external environment as well as their internal physiological state and functions. In the absence of needed nutrients or growth factors, animal cells may exit from the cell cycle and become quiescent. Upon growth stimulation, they reenter the cell cycle in a process that requires cyclin D–CDK activity. Cells also have mechanisms that respond to symptoms of stress, including DNA damage, oxygen depletion, inadequate pools of nucleoside triphosphates, and (in the case of animal cells) loss of intercellular adhesion. Inside the cell, several key events in the cell cycle are monitored. When defects are identified, progression through the cell cycle is halted at a **checkpoint.**

> **key concept**
>
> Checkpoints in the cell cycle serve to maintain the correct order of steps with respect to each other as the cycle progresses; they do this by causing the cell cycle to pause while defects are corrected or repaired.

Three principal checkpoints that function to maintain the genetic integrity of cells are

- A DNA damage checkpoint
- A centrosome duplication checkpoint
- A spindle checkpoint

These checkpoints are summarized in Figure 13.12 . All three types of checkpoints are important in maintaining the stability of the chromosome complement. When there is an error in any of the three processes monitored by the checkpoints, failure to stop at the checkpoint may lead to aneuploidy (extra or missing chromosomes), polyploidy, or an increased number of mutations. In the following sections, we shall examine how some of these checkpoints work.

■ The p53 transcription factor is a key player in the DNA damage checkpoint.

A **DNA damage checkpoint** arrests the cell cycle when DNA is damaged or replication is not completed. DNA damage includes either modification of nitrogenous bases or breakage of the phosphodiester backbone. When modified nucleotides are repaired by an excision repair pathway (Chapter 12), the repair entails removal of the affected nucleotides, followed by resynthesis with a repair polymerase and finally ligation. DNA molecules broken in the backbone are repaired by homology-based recombination, nonhomologous end joining, or addition of new telomeres. In animal cells, a DNA damage checkpoint acts at three stages in the cell cycle: at the G_1/S transition, in the S period (DNA synthesis), and at the G_2/M boundary. The S-period checkpoint continuously monitors the pro-

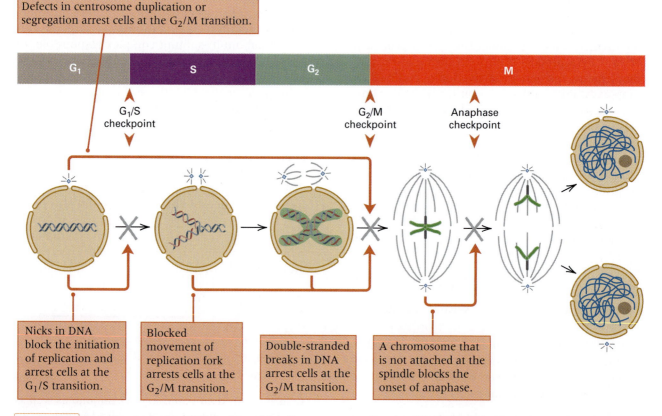

Figure 13.12 Key cell-cycle checkpoints that act to maintain the genetic stability of cells. The text in the boxes explains the events monitored and the steps affected. [Excerpted with permission from L. H. Hartwell and M. Kastan. 1994. *Science* 266: 1821. © AAAS.]

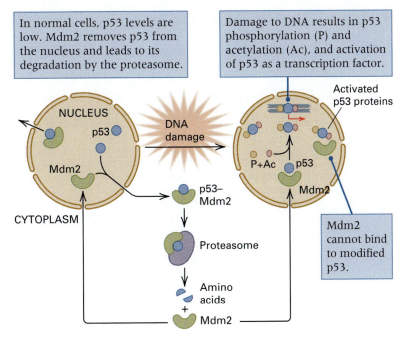

In normal cells, p53 levels are low. Mdm2 removes p53 from the nucleus and leads to its degradation by the proteasome.

Damage to DNA results in p53 phosphorylation (P) and acetylation (Ac), and activation of p53 as a transcription factor.

Activated p53 proteins

NUCLEUS

p53

DNA damage

Mdm2

CYTOPLASM

p53–Mdm2

Proteasome

Amino acids + Mdm2

P+Ac   p53

Mdm2

Mdm2 cannot bind to modified p53.

Figure 13.13 The effect of DNA damage on transcription factor p53. In normal cells the level of p53 is low, in part because Mdm2 shuttles it to the cytoplasm where it is destroyed by the proteasome. DNA damage results in phosphorylation and acetylation of p53, rendering it unable to bind with Mdm2. Hence p53 levels in the nucleus increase.

gression of DNA synthesis as it takes place. Although there are three DNA damage checkpoints, each monitors DNA damage or incomplete replication. If either type of problem is detected, the DNA checkpoint acts to block the cell cycle at multiple points.

A key protein in the mammalian cell's response to stress in general, and to DNA damage in particular, is a protein called the **p53 transcription factor.** In normal cells, the level of activated p53 protein is very low, even though substantial amounts of p53 mRNA and protein are present. The activity of p53 is kept low by another protein called Mdm2. To function as a transcription factor, p53 protein must be activated first by phosphorylation and then by acetylation. Mdm2 binds to p53 and prevents phosphorylation and subsequent steps in the activation of p53 as a transcription factor. In addition, Mdm2 continuously shuttles between the nucleus and the cytoplasm, and in this process it continuously exports p53 from the nucleus for degradation in the cytoplasm. The p53–Mdm2 export cycle is illustrated in Figure 13.13.

When cells are treated with agents that damage DNA, some cells arrest in G_1 and others arrest in G_2. The DNA damage signal is sensed and transmitted, and this causes p53 protein to become activated by protein kinases and acetylases, overriding the inhibitory effect of Mdm2. Activation of p53 causes its release from Mdm2, which stabilizes the active p53 transcription factor and results in increased levels of the p53 protein. The activated p53 then triggers the transcription of a number of genes and the repression of others. Some of the key proteins whose genes are transcriptionally activated by p53 are listed in **Table 13.1**, and a flow chart of how these

Table 13.1

Products of genes transcriptionally activated by p53

Gene product	Function
p21	Inhibits several cyclin-dependent kinases; arrests cells at G_1/S boundary.
14-3-3σ	Predicted to bind to and sequester phosphorylated Cdc25C phosphatase in the cytoplasm, which prevents Cdc25C from activating the cyclin B–Cdc2 kinase; arrests cells at the G_2/M boundary.
GADD45	Binds to proliferating cell nuclear antigen (PCNA), blocking its role as a processivity factor for DNA polymerase and hence blocking DNA replication; functions directly in DNA repair.
Bax	Acts as a positive regulator of apoptosis (programmed cell death).
Apaf1	Scaffold protein that, when activated by cytochrome c, oligomerizes caspases into the complex that promotes apoptosis.
Maspin	Acts as an inhibitor of serine proteases, and is an inhibitor of angiogenesis (formation of blood vessels) and metastasis.

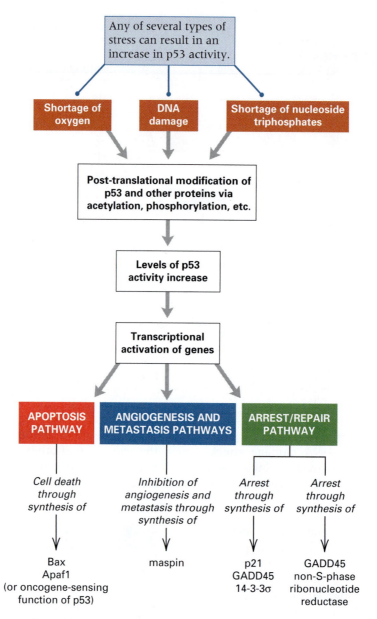

Figure 13.14 Downstream events triggered by activation of p53 protein include transcriptional activation of the genes for p21, GADD45, 14-3-3σ, Bax, maspin, and Apaf1. Activation of the apoptosis pathway by p53 follows transcriptional activation of Bax and, possibly, direct sensing of activated oncogenes (cancer-causing genes) of either cellular or viral origin.

proteins affect processes in the cell cycle and other cellular properties is given in Figure 13.14.

Increased transcription of the genes for p21, GADD45, and 14-3-3σ, and decreased transcription of the gene for cyclin B, all serve to block the cell cycle at particular points, as illustrated in Figure 13.15. The G_1/S transition checkpoint is mediated by the increased level of activated p21, which results in inhibition of the G_1 cyclin–CDKs and in this way blocks the G_1/S transition. Hence if DNA damage is detected in G_1, the cell is blocked in the G_1/S transition. The S-period response to DNA damage is mediated by the p21 protein and GADD45; these form a complex with another protein, which results in a reduction in the processivity of DNA polymerase. The **processivity** of a DNA polymerase is the number of consecutive nucleotides in the template strand that are replicated before the polymerase detaches from the template. Decreasing the processivity of DNA polymerase therefore slows down DNA synthesis, in effect buying time for the cell to repair DNA damage. The G_2/M checkpoint is mediated by the 14-3-3σ protein, which hinders activation of cyclin B–Cdc2, thus blocking the G_2/M transition. At the same time, the decrease in the level of cyclin B reduces the level of the active cyclin B–Cdc2 complex, which also ensures that the cell remains in G_2. Hence if DNA damage is detected in the S period or in G_2, the cell cycle arrests at the G_2/M checkpoint.

DNA damage also triggers activation of another pathway, a pathway for **apoptosis,** or pro-

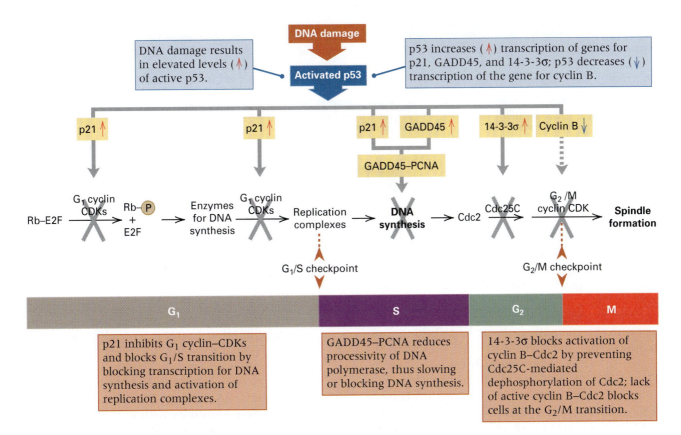

DNA damage

DNA damage results in elevated levels ($\uparrow$) of active p53.

Activated p53

p53 increases ($\uparrow$) transcription of genes for p21, GADD45, and 14-3-3σ; p53 decreases ($\downarrow$) transcription of the gene for cyclin B.

p21 $\uparrow$ p21 $\uparrow$ p21 $\uparrow$ GADD45 $\uparrow$ 14-3-3σ $\uparrow$ Cyclin B $\downarrow$

GADD45–PCNA

Rb–E2F → G_1 cyclin CDKs ✗ → Rb–P + E2F → Enzymes for DNA synthesis → G_1 cyclin CDKs ✗ → Replication complexes → DNA synthesis ✗ → Cdc2 → Cdc25C ✗ → G_2/M cyclin CDK ✗ → Spindle formation

G_1/S checkpoint G_2/M checkpoint

G_1	S	G_2	M

p21 inhibits G_1 cyclin–CDKs and blocks G_1/S transition by blocking transcription for DNA synthesis and activation of replication complexes.

GADD45–PCNA reduces processivity of DNA polymerase, thus slowing or blocking DNA synthesis.

14-3-3σ blocks activation of cyclin B–Cdc2 by preventing Cdc25C-mediated dephosphorylation of Cdc2; lack of active cyclin B–Cdc2 blocks cells at the G_2/M transition.

Figure 13.15 Role of activated p53 protein in the DNA damage checkpoint. DNA damage results in an increase in the level of activated p53, which increases transcription of the genes for p21, GADD45, and 14-3-3σ and decreases transcription of the gene for cyclin B.

grammed cell death. When the apoptotic pathway is activated, a cascade of proteolysis is initiated that culminates in cell suicide. The proteases involved are called *caspases*. Their activation ultimately results in destruction of the cellular DNA, internal organelles, and the actin cytoskeleton (Figure 13.16), and it is accompanied by nuclear condensation and usually followed by engulfment of the cellular remnants by phagocytes. The p53 transcription factor also

Early detection saves lives! Studies in various countries demonstrate that women over the age of 40 who undergo routine mammograms have an increased chance of surviving breast cancer. The American Cancer Society recommends an annual mammogram for women over age 40.

© Photodisc.

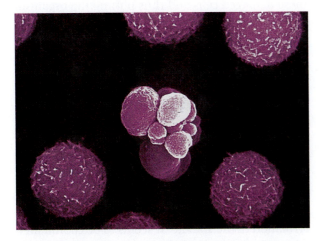

Figure 13.16 The formation of large blebs ("blisters") from the cell membrane is a characteristic of cells undergoing apoptosis. [© Dr. Gopal Murti/Visuals Unlimited.]

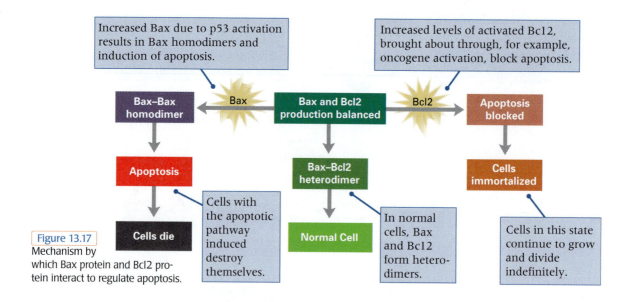

| Increased Bax due to p53 activation results in Bax homodimers and induction of apoptosis. | | Increased levels of activated Bcl2, brought about through, for example, oncogene activation, block apoptosis. |

Figure 13.17 Mechanism by which Bax protein and Bcl2 protein interact to regulate apoptosis.

Bax–Bax homodimer → Apoptosis → Cells die — Cells with the apoptotic pathway induced destroy themselves.

Bax ← Bax and Bcl2 production balanced → Bcl2

Bax and Bcl2 production balanced → Bax–Bcl2 heterodimer → Normal Cell — In normal cells, Bax and Bcl2 form heterodimers.

Apoptosis blocked → Cells immortalized — Cells in this state continue to grow and divide indefinitely.

activates the apoptotic pathway by activating transcription of *Bax* and *Apaf1*. As illustrated in Figure 13.17, Bax protein normally exists in a heterodimer with an inhibitor of apoptosis called Bcl2. When p53 activates transcription of *Bax,* the balance is tilted in favor of Bax homodimers and against the Bcl2 heterodimer, which promotes apoptosis and self-destruction of the cell. On the other hand, activation of **oncogenes,** which are genes associated with cancers, can increase the level of activated (phosphorylated) Bcl2, which prevents apoptosis and allows the affected cells to grow and divide indefinitely. Cellular immortality does not always follow oncogene activation, because in some cases it activates the apoptotic pathway, possibly through an oncogene-sensing function of p53 illustrated in Figure 13.13. In all of its functions, p53 acts to protect the integrity of the genome with respect to nucleotide sequence and strand integrity and with respect to euploidy. Instability in the genome, unbalanced genomes, and damaged DNA pose a hazard to the organism. Apoptosis is a mechanism for killing such damaged—and thus dangerous—cells.

To gain an appreciation of the importance of the DNA checkpoint, look again at Figure 13.14 and consider what would happen if part or all of this failsafe mechanism should malfunction. Suppose, for example, that cells lacked p21 protein but retained the rest of the mechanism. If DNA damage were properly sensed, and p53 functioned as usual to increase levels of 14-3-3σ and GADD45, cells would accumulate in G_2 and be unable to undergo mitosis. However, DNA would continue to be synthesized. In addition, lacking p21 protein, the cells would have a defective G_1/S checkpoint and so could embark on additional rounds of DNA synthesis. The expected result of p21 malfunction would therefore be polyploid cells. This, in fact, is the phenotype of cells that are mutant for *p21*.

To take another example, consider the consequence of loss of p53 function. Even if DNA damage were detected, the cell would be unable to mount a response—it would be unable to buy time to repair the DNA damage. The cells could initiate new rounds of synthesis with damaged chromosomes. Such synthesis would not only result in an increased frequency of mutation but would also permit gene amplification. Amplification of genes encoding cyclin D or Cdk4 would allow cells to escape the normal controls on DNA synthesis and proliferation. Cells already in S or G_2 would enter mitosis with damaged chromosomes, because not enough time would have elapsed for repair of lesions. In addition, the organism would have lost its ultimate protection against such damaged cells, apoptosis. The absence of p53 means that transcription of *Bax* and *Apaf1* would not be increased and that the apoptotic pathway would not be turned on; the normal balance between Bax and Bcl2 would be maintained, ensuring the survival of these damaged cells and thus putting the organism at risk. If the description of these runaway cells reminds you of the unchecked proliferation of cancer cells, this is because cancer cells *become* cancer cells by subverting the checkpoint mechanisms.

■ **The centrosome duplication checkpoint and the spindle checkpoint function to maintain the normal complement of chromosomes.**

Recall that the *centrosome* is the cellular organelle around which the bipolar spindle is organized. A **centrosome duplication checkpoint** monitors formation of the spindle. It seems to be coordinated with entry into mitosis, because activation of cyclin B–Cdc2 kinase is correlated with centrosome duplication and formation of the spindle. In some

Figure 13.18 The spindle checkpoint. All chromosomes must form stable, bipolar attachments to spindle microtubules to activate the anaphase-promoting complex, which promotes the onset of anaphase and separation of the sister chromatids.

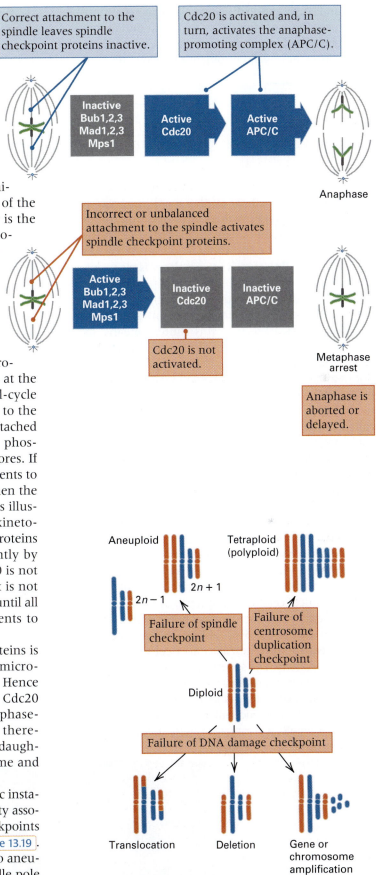

organisms, the centrosome duplication checkpoint may also be coordinated with the spindle checkpoint and the exit from mitosis.

The **spindle assembly checkpoint** monitors assembly of the spindle and attachment of the kinetochores to the spindle. (The *kinetochore* is the spindle-fiber attachment site on the chromosome.) Improper or incomplete spindle assembly triggers a block in the separation of the sister chromatids by preventing activation of the anaphase-promoting complex needed for entry into anaphase.

Cells can detect a single unattached or misattached chromosome and delay anaphase. Studies on insect and mammalian chromosomes suggest that the absence of tension at the kinetochore is the initiating signal for cell-cycle arrest. Tension on the kinetochore is related to the level of phosphorylation at kinetochores; unattached kinetochores have relatively more of a phosphorylated entity than do attached kinetochores. If all chromosomes form stable, bipolar attachments to spindle microtubules at their kinetochores, then the anaphase-promoting complex is activated. As illustrated in Figure 13.18, when an unattached kinetochore is detected, the Bub, Mad, and Mps1 proteins act to block the onset of anaphase, apparently by inhibiting the protein Cdc20. Because Cdc20 is not activated, the anaphase-promoting complex is not activated, and the cells remain at metaphase until all chromosomes form stable, bipolar attachments to spindle microtubules.

If any one of the Mad, Bub, or Mps1 proteins is defective, Cdc20 is still activated even if a microtubule has failed to attach to a kinetochore. Hence the spindle checkpoint is not activated. The Cdc20 is activated normally, and, in turn, the anaphase-promoting complex is activated. Anaphase therefore takes place, resulting in two aneuploid daughter cells, one lacking a particular chromosome and the other endowed with an extra copy.

Failure of any checkpoint results in genetic instability. The particular kinds of genomic instability associated with defects in the three checkpoints we have discussed are summarized in Figure 13.19. Malfunctioning of the spindle itself can lead to aneuploidy, whereas a failure to duplicate a spindle pole can lead to polyploidy (Chapter 9). Defects in the DNA damage checkpoints can result in chromosomal aberrations of various kinds, including translocations, dele-

Figure 13.19 Contributions of checkpoint failures to genetic instability. [Excerpted with permission from L. H. Hartwell and M. Kastan. 1994. *Science* 266: 1821. © AAAS.]

13.2 Checkpoints in the cell cycle allow damaged cells to repair themselves or to self-destruct **473**

tions, and amplification of genes or subregions of chromosomes. The amplified genes may be found as tandem repeats within a chromosome or as extrachromosomal circles lacking a centromere and telomeres.

13.3

Cancer cells have a small number of mutations that prevent normal checkpoint function.

Cancer is not one disease but rather many diseases that share similar cellular attributes. All cancer cells show uncontrolled growth as a result of mutations that affect a relatively small number of genes. It is a disease of somatic cells. About 1 percent of cancer cases are **familial,** which means there is clear evidence for segregation of a gene in the pedigree that predisposes cells in affected individuals to progress to the cancerous state. The other 99 percent of cases are called **sporadic,** which in this context means *not familial,* and are the result of genetic changes in somatic cells. Cancer cells share several properties not found in normal cells. One of these is the effect of cell-to-cell contact. In normal cells, cell-to-cell contact inhibits further growth and division, a process called **contact inhibition.** Cancer cells have lost contact inhibition: They continue to grow and divide, and they even pile on top of one another. This phenomenon is shown in Figure 13.20. The normal cells in panel A are flat and form a sheet; they have ceased to grow where they touch. In contrast, the tumor cells in panel B fail to show contact inhibition. The cells look very different from the normal cells, particularly in the protruding bubbles that indicate membrane activity. When badly damaged or starved for nutrients, normal cells undergo apoptosis and self-destruct; cancer cells do not undergo apoptosis. Even in the absence of damage, normal cells cease to divide in culture after about 50 doublings (a process called **cell senescence**), whereas cancer cells in culture divide indefinitely. The senescent behavior of normal cells is associated with a loss of *telomerase* activity: The telomeres are no longer elongated, which contributes to the onset of senescence and cell death. Cancer cells have high levels of telomerase, which help to protect them from senescence, making them immortal.

Within an organism, tumor cells are *clonal,* which means that they are descendants from a single ancestral cell that became cancerous. This conclusion is based in part on the observation that tumor cells in a female express the genes in only one of the X chromosomes, as a result of the normal inactivation of the other X chromosome (Chapter 5). If the tumor cells were not clonal, gene expression from both X chromosomes would be expected, reflecting the random inactivation of one X chromosome in each cell lineage. The finding of expression from a single X chromosome must therefore reflect clonality.

Luckily for most people, the conversion of normal somatic cells into cancer cells is a process that requires multiple mutational steps. An important contributor to cancer conversion is genetic instability in the cell population that serves as the precursor to cancer cells. This genetic instability may occur at the level of nucleotide sequences in the DNA or at the level of chromosome structure or number. The genetic instability is manifested as an increased number of mutations, gene amplification, chromosomal rearrangements, or aneuploidy. The mutations result in a cell population that is genetically heterogeneous. In such a mixed population, any cell that has a proliferative advantage will contribute a greater fraction of descendants to the future cell population than will its neighbors, and because of this advantage, its clone expands at the expense of others. Subsequent mutations in the descendants of this cell, and further clonal expansions of the derivative cells, can give rise to a clone of cells with the proliferative capacity typical of cancer cells.

Many cancers are the result of alterations in cell-cycle control, particularly in control of the G_1-to-S transition, and in the G_1/S checkpoint associated with this transition. These alterations also affect apoptosis through their interactions with p53. Figure 13.21 summarizes the key elements of the regulatory circuitry that governs the G_1/S transition, including the proteins that promote cell-cycle progression, that activate the checkpoint, and that govern the apoptotic pathway. Tumor cells commonly show altered expression or inactivation of function of one or more of these genes. Some genes for which alterations have been detected in cancer cells are listed in **Table 13.2.** The major mutational targets

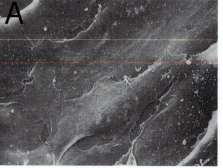

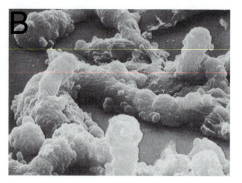

Figure 13.20 Scanning electron micrographs of (A) normal and (B) adenovirus-transformed (tumorigenic) hamster cells. [From R. Goldman et al. 1974. *Cold Spring Harbor Symposium on Quantitative Biology* 39: 601.]

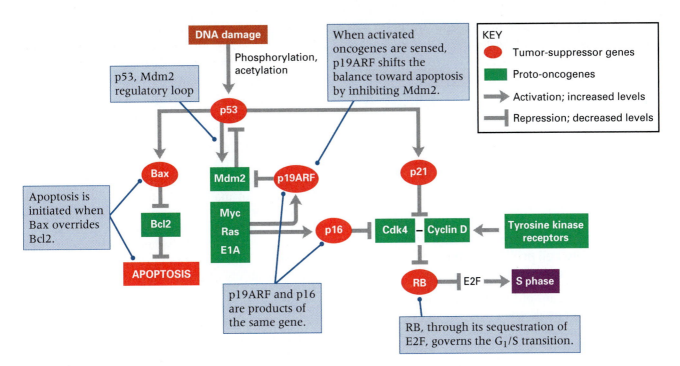

KEY
- 🔴 Tumor-suppressor genes
- 🟩 Proto-oncogenes
- → Activation; increased levels
- ⊣ Repression; decreased levels

DNA damage

Phosphorylation, acetylation

p53, Mdm2 regulatory loop

When activated oncogenes are sensed, p19ARF shifts the balance toward apoptosis by inhibiting Mdm2.

p53

Bax

Apoptosis is initiated when Bax overrides Bcl2.

Bcl2

APOPTOSIS

Mdm2   p19ARF

Myc Ras E1A

p16

p19ARF and p16 are products of the same gene.

p21

Cdk4 — Cyclin D ← Tyrosine kinase receptors

RB ⊣ E2F → S phase

RB, through its sequestration of E2F, governs the G_1/S transition.

Figure 13.21 Interactions between the p53 pathway and the RB (retinoblastoma protein) pathway in controlling apoptosis or DNA synthesis.

Table 13.2

Cell cycle regulatory genes affected in tumors

	Alteration	Consequence
Proto-oncogenes		
Cyclin D	amplification or overexpression	promotes entry into S phase
Cdk4	amplification	promotes entry into S phase
Cdk4	mutation	resistant to inhibition by p16; promotes entry into S phase
EGFR (epidermal growth-factor receptor, a tyrosine kinase receptor)	amplification	promotes proliferation by constitutive activation of pathway from growth-factor receptor
FGFR (fibroblast growth-factor receptor)	amplification	promotes proliferation by constitutive activation of pathway from growth-factor receptor
Ras	amplification	promotes proliferation by constitutively transmitting growth signal
Ras	mutation	inactivates GTPase activity; constitutive activation of pathway from growth-factor receptors
Bcl2	overexpression by translocation next to strong enhancer	blocks apoptosis
Mdm2	amplification	mimics loss of p53 with loss of G_1/S, S, and G_2/M checkpoint functions
Tumor-suppressor genes		
p53	mutation	loss of G_1/S, S, and G_2/M checkpoint functions
p21	mutation	loss of G_1/S and S checkpoint functions
RB	mutation	promotes proliferation; E2F uninhibited
Bax	mutation	failure to promote apoptosis of damaged cells
Bub1p	mutation	loss of spindle assembly checkpoint function

for the multistep cancer progression are of two types: *proto-oncogenes* and *tumor-suppressor genes*.

> **key concept**
>
> The normal function of proto-oncogenes is to promote cell division or to prevent apoptosis; the normal function of tumor-suppressor genes is to prevent cell division or to promote apoptosis.

These types of genes are discussed in the following sections.

■ Proto-oncogenes normally function to promote cell proliferation or to prevent apoptosis.

Oncogenes are gain-of-function mutations associated with cancer progression. They are derived from normal cellular genes called **proto-oncogenes.** Oncogenes are gain-of-function mutations because they improperly enhance the expression of genes that promote cell proliferation or inhibit apoptosis. In Figure 13.21, the proteins enclosed in rectangles are the products of proto-oncogenes, which when mutated can give rise to oncogenes. In this section, some of the oncogenes identified in Figure 13.21 and Table 13.2 are discussed in more detail.

Cyclin D and Cdk4 Overexpression of cyclin D promotes unscheduled entry into S phase. Overexpression is often the result of gene amplification. Amplification of the cyclin D gene is found in 35 percent of esophageal carcinomas, 15 percent of bladder cancers, and 15 percent of breast cancers. Other mechanisms of overexpression must also occur, because although only 15 percent of breast cancers show amplification of the cyclin D gene, more than 50 percent of breast cancers show overexpression of cyclin D. Paralleling the increase in cyclin D expression, the gene for the CDK partner of cyclin

D(Cdk4) is amplified in some tumors, including 12 percent of gliomas (brain tumors) and 11 percent of sarcomas (muscle or connective-tissue cancer).

Growth-factor receptors and Ras Cellular growth factors stimulate growth by binding to a growth-factor receptor at the membrane. The binding activates a signal transduction pathway that acts through Ras (discussed below), cyclin D, and its partner CDKs. The gene that encodes the receptor for epidermal growth factor (EGFR) is amplified in 45 percent of malignant astrocytomas (a kind of brain tumor), 35 to 50 percent of glioblastomas (another kind of brain tumor), 20 percent of breast cancers, 15 to 30 percent of ovarian cancers, and 10 percent of head and neck cancers and melanomas. The genes for fibroblast growth-factor receptor (FGFR) are amplified in about 20 percent of breast cancers.

Activated tyrosine kinase receptors such as EGFR and FGFR activate a signal transduction pathway that activates a small signaling protein called a **G protein,** as illustrated in Figure 13.22 for the G protein Ras. Ras is usually bound to GDP, and in this state it is inactive. Receptor activation results in downstream activation of a protein that stimulates the exchange of GTP to replace GDP in the Ras protein, and the Ras protein is active when it is bound to GTP. The Ras–GTP, in turn, activates another downstream protein, propagating the growth signal. The activation of Ras as Ras–GTP is transient. Yet another protein (GAP) activates the intrinsic GTPase activity of Ras, so Ras hydrolyzes the GTP and returns itself to the inactive state of Ras–GDP. Mutations in the *RAS* gene occur that inactivate its GTPase function; in these mutants, the GTPase activity cannot be activated by GAP, and RAS remains in its active form of Ras–GTP, whether or not the cell is receiving signals from growth-factor receptors. The signal for cellular growth is transmitted constitutively, and therefore unrestrained growth and division take place.

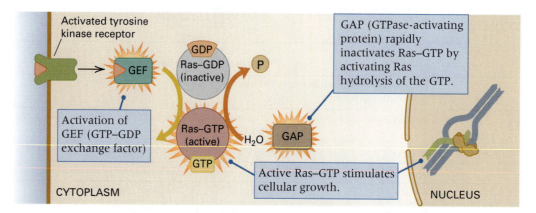

Figure 13.22 | Function of the Ras protein, which acts as a switch in stimulating cellular growth in the presence of growth factors. Ras is the product of a proto-oncogene. Certain mutant Ras proteins, such as a G19V (valine replacement for glycine at position 19), lack GTPase activity and remain in the form of Ras–GTP; hence a growth-promoting signal is present even in the absence of growth factors.

EGFR and FGFR Amplification or overexpression of EGFR or FGFR results in self-activation of the receptor and transmission of a constitutive growth signal acting through the Ras pathway. Overexpression of Ras also leads to enhanced signal transduction and renewed cycles of proliferation.

Mdm2 Amplification of the gene that encodes Mdm2 acts to promote cell proliferation, but it does so indirectly. Overexpression of Mdm2 effectively overwhelms any increase in production of p53, because the Mdm2 shuttles the extra p53 from the nucleus to the cytoplasm, where it is destroyed by the proteasome (Figure 13.13). This process effectively prevents functioning of the G_1/S checkpoint. Hence, amplification of the *Mdm2* gene and the resultant overexpression of Mdm2 protein are equivalent in their effects to inactivation of the gene for p53. Amplification of the *Mdm2* gene has been found in 19 of 28 tumor types examined, most commonly in tumors of adipose tissue (42 percent), soft-tissue sarcomas (20 percent), osteosarcoma (16 percent), and esophageal carcinoma (13 percent).

■ **Tumor-suppressor genes normally act to inhibit cell proliferation or to promote apoptosis.**

Tumor-suppressor genes are genes that normally negatively control cell proliferation or that activate the apoptotic pathway. Loss-of-function mutations in tumor-suppressor genes contribute to cancer progression. Some of the key functions of tumor-suppressor genes are illustrated in Figure 13.21 and listed in Table 13.2. They are examined in more detail in this section.

p53 Loss of function of p53 eliminates the DNA checkpoint that monitors DNA damage in G_1 and S, as we saw in Figures 13.13 and 13.14. In the absence of functional p53, the proteins responsible for arresting cells in G_1 or G_2 (p21, 14-3-3σ, and GADD45) are not synthesized in response to DNA damage. There is consequently no block to a cell's proceeding into S phase or into M phase even if it has damaged chromosomes, altered chromosome number, or amplified genes. In addition, loss of p53 function costs the organism its ultimate defense against aberrant cells, because DNA damage is no longer able to trigger enhanced expression of Bax and Apaf1 and hence the self-destruction of aberrant cells. The damaged cells survive and proliferate. Furthermore, their genetic instability increases the probability of additional genetic changes and thus progression toward the cancerous state. Given this scenario, and the key role of p53 in protecting the cell against the consequences of DNA damage, it is not surprising that p53 proves to be nonfunctional in more than half of all cancers. In addition, as we

have noted, Mdm2 overexpression—equivalent to loss of p53 function—occurs in other tumors. Because loss of p53 and overexpression of Mdm2 rarely occur in the same tumor, the actual loss of p53 function by one mechanism or another is probably substantially greater than 50 percent. Mutant p53 proteins are found frequently in melanomas, several kinds of lung cancers, colorectal tumors, bladder and prostate cancers, and astrocytomas.

One exception to the widespread involvement of p53 in cancers is testicular embryonal carcinoma. This is one of the most curable of cancers, with a cure rate of 90 to 95 percent. A striking observation is that these carcinomas never include mutations in the p53 gene. The curability and the lack of p53 involvement may be causally related. The p53 gene is not expressed in embryonal testicular cells, possibly because the process of recombination with breakage and rejoining of DNA strands would be sensed as DNA damage, and imposition of the checkpoints would block meiosis. Because the p53 gene is not expressed in embryonal testicular cells, no selective advantage can be conferred on cells by mutations in the p53 gene. The current interpretation of the high success rate with treatment using certain drugs (for example, *cisplatin*) is that the drug induces the expression of p53, which activates the apoptotic pathway and leads to suicide of the cancer cells.

p21 Loss of p21 function results in renewed rounds of DNA synthesis without accompanying mitosis, and the level of ploidy of the cell increases. Mutations in the gene that encodes p21 occur in some prostate cancers.

p16/p19ARF The p16 and p19ARF proteins are products of the same gene; they derive from two distinct transcripts from different promoters, each with a different 5′ exon. The p16 product can inhibit the cyclin D–Cdk4 complex and help control entry into S phase. The exons encoding p16 are a very frequent target for inactivation during tumor progression. The gene is deleted in 55 percent of gliomas (a form of brain tumor), 55 percent of mesotheliomas, more than 50 percent of melanomas, 40 percent of nasopharyngeal carcinomas, 50 percent of biliary-tract carcinomas, and 30 percent of esophageal carcinomas. Figure 13.21 shows the functional interactions of p19ARF as a tumor suppressor. If cellular or viral oncogenes are being overexpressed, the synthesis of p19ARF is elevated, and at this higher concentration p19ARF binds to Mdm2 and prevents it from binding with p53 and shuttling p53 from the nucleus. Given this mode of action, one might expect that some tumors might lack p19ARF. In fact, no mutation has yet been reported that inactivates p19ARF without also inactivating p16, although the reverse kind of mutation has been observed.

RB The retinoblastoma protein controls the transition from G_1 to S phase by controlling the activity of the transcription factor E2F, as illustrated in Figure 13.10 on page 466. Loss of RB function frees E2F to initiate transcription of the enzymes for DNA synthesis at all times in the cell cycle; hence excessive rounds of DNA synthesis are continuously being initiated. Loss of RB function is found in melanomas, in small-cell lung carcinoma, in osteosarcoma, and in liposarcomas. Just as loss of p53 and overexpression of Mdm2 appear to be equivalent in their effects, so do loss of either of the tumor-suppressor proteins RB or p16 and overexpression of either cyclin D or Cdk4. One basis for this conclusion is the observation that tumor cells are nearly always altered for RB, p16, cyclin D, or Cdk4, but rarely for more than one of them. This finding suggests that loss of function of either p16 or RB is equivalent to overexpression of either cyclin D or Cdk4. Any one of these defects leads to the same result: The cells lose control of the G_1/S transition and embark upon unscheduled rounds of DNA synthesis.

Bax The Bax tumor-suppressor protein promotes apoptosis (Figure 13.17). Loss of Bax function is found particularly in gastric adenocarcinomas and in colorectal carcinomas, particularly those associated with microsatellite instability because of defective mismatch repair. Cells that are defective in mismatch repair are especially prone to undergo replication slippage leading to deletions or additions of nucleotides in runs of short tandem repeats (discussed in Chapter 12). One of the sequences prone to replication slippage is a run of eight consecutive G nucleotides within the *Bax* gene, which makes this gene especially susceptible to frameshift mutation by the addition or deletion of a base pair in this region.

Bub1 Bub1 is a protein that is primarily involved in the spindle checkpoint (Figure 13.18). A subset of colon cancers show chromosomal instability: They do not maintain a constant karyotype but are aneuploid. Some of these unstable lines are defective in Bub1.

13.4

Mutations that predispose to cancer can be inherited through the germ line.

As we have noted, approximately 99 percent of cancers are sporadic: All of the genetic changes resulting in the cancer take place in somatic cells. Furthermore, the change from a normal cell into a tumor cell is progressive. It goes step-by-step as each new somatic mutation along the way compromises a different mechanism of cell-cycle control.

■ **Cancer initiation and progression occurs through mutations that allow affected cells to evade normal cell-cycle checkpoints.**

A small minority of cancers—approximately 1 percent—are familial. In these cases, one of the mutations associated with cancer progression is inherited through the germ line. The presence of this mutation predisposes the individual to cancer, because it reduces the number of additional somatic mutations necessary for a precancerous cell to progress to malignancy. Figure 13.23 illustrates some of the genetic mutations and changes in cell morphology that take place in the progression of a type of familial colon cancer called *adenomatous polyposis*. In this case, the inherited

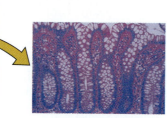

Normal colon tissue

Loss of *APC* gene

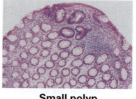

Small polyp (benign)

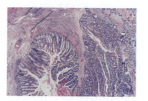

Invasive colon cancer (malignant)

Mutation or loss of p53 function

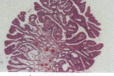

Large polyp (benign)

Oncogenic mutation in *Ras* gene

Figure 13.23 Somatic mutations and cell morphology in the progression of invasive colon cancer. [Photographs courtesy of Kathleen R. Cho.]

Table 13.3

Inherited cancer syndromes

Syndrome	Primary tumors	Associated tumors	Chromosome	Gene	Proposed function
Li–Fraumeni syndrome	sarcomas, breast cancer	brain tumors, leukemia	$17p13$	$p53$	transcription factor
familial retinoblastoma	retinoblastoma	osteosarcoma	$13q14$	$RB1$	cell-cycle regulator
familial melanoma	melanoma	pancreatic cancer	$9p21$	$p16$	inhibitor of Cdk4 and Cdk6
hereditary nonpolyposis colorectal cancer (HNPCC)	colorectal cancer	ovarian cancer, glioblastoma	$2p22$ $3p21$ $2q32$ $7p22$	$MSH2$ $MLH1$ $PMS1$ $PMS2$	DNA-mismatch repair
familial breast cancer	breast cancer	ovarian cancer	$17q21$	$BRCA1$	repair of DNA double-strand breaks
familial adenomatous polyposis of the colon	colorectal cancer	other gastrointestinal tumors	$5q21$	APC	regulation of β-catenin
xeroderma pigmentosum	skin cancer		Several complementation groups	XPB XPD XPA	DNA-repair helicases, nucleotide excision repair

mutation is in the gene APC in chromosome 5, which is a tumor-suppressor gene whose normal function is to transduce the signal of contact inhibition into the cell to inhibit further growth and division. Subsequent mutations in the progression to malignancy include an oncogenic mutation in a Ras gene and mutation or loss of the gene encoding p53. This scenario is only one possible route of progression. Different tumors may progress by different pathways, depending on what mutations occur and in what order.

Familial adenomatous polyposis of the colon and some other important familial cancer syndromes are identified in **Table 13.3**. The Li–Fraumeni syndrome, familial retinoblastoma, and familial melanoma are all associated with germ-line mutations in genes that are also found to be mutant in some sporadic tumors. The genes affected in these syndromes are the $p53$ gene, the $RB1$ gene, and the $p16$ gene, respectively. A pedigree of a cancer-prone family segregating for a mutation in the $p53$ gene (Li–Fraumeni syndrome) is given in Figure 13.24.

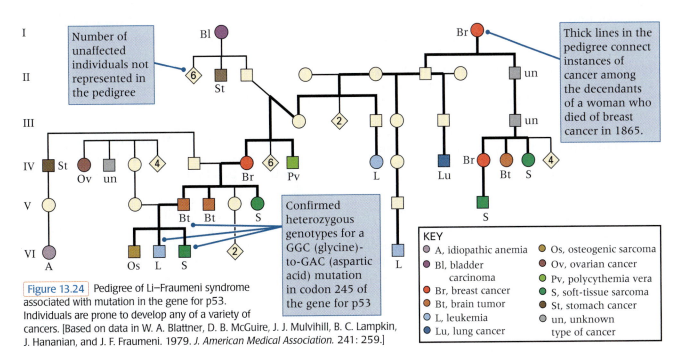

Figure 13.24 Pedigree of Li–Fraumeni syndrome associated with mutation in the gene for p53. Individuals are prone to develop any of a variety of cancers. [Based on data in W. A. Blattner, D. B. McGuire, J. J. Mulvihill, B. C. Lampkin, J. Hananian, and J. F. Fraumeni. 1979. *J. American Medical Association*. 241: 259.]

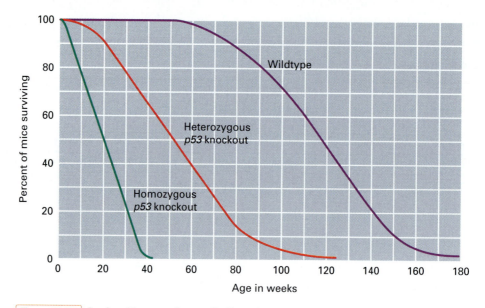

Figure 13.25 Survivorship curves for genetically engineered mice with either a heterozygous or a homozygous knockout (complete loss of function) mutation in the *p53* gene.

This syndrome shows clear autosomal dominant inheritance. However, the affected individuals show a range of different tumors and often have more than one, including osteosarcoma, leukemia, breast cancer, lung cancer, soft-tissue sarcoma, and brain tumors. A large fraction of Li–Fraumeni families show segregation for a mutation in the *p53* gene. For this family, affected members are heterozygous for a single nucleotide pair in the *p53* gene, which changes codon 245 from GGC (glycine) to GAC (aspartic acid). The observation that the affected individuals are heterozygous is consistent with the dominant autosomal inheritance.

A situation analogous to the human Li–Fraumeni syndrome has been created in mice by experimental knockout (loss of function) of the *p53* gene via the germ-line transformation methods discussed in Chapter 11. Animals heterozygous or homozygous for the *p53* knockouts were compared to wildtype mice (Figure 13.25). Normal mice do not develop tumors; 50 percent of them live to the age of 120 weeks, and about 5 percent live for at least 160 weeks. In contrast, half of the heterozygous *p53* knockout mice are dead by 55 weeks, and almost none of them live as long as 120 weeks. Furthermore, 50 percent of these mice develop tumors (mostly osteosarcomas) by 18 months of age. The homozygous knockout mice fare even worse: 75 percent develop tumors by 6 months, 50 percent are dead by 20 weeks, and all are dead by 40 weeks. In the homozygous genotype, the tumors are lymphomas rather than osteosarcomas, possibly because normal development of the immune system requires massive apoptosis within the thymus gland, and in the *p53* homozygous knockouts, apoptosis is abol-

ished because *Bax* is not up-regulated as in normal mice and *p53* knockout heterozygotes.

The fact that animals heterozygous for the *p53* knockout are severely affected for both longevity and incidence of tumors does not necessarily imply that the *p53* mutation is dominant at the level of the individual cell. In fact, in the case of *p53*, the effects of the mutation are manifested only in somatic cells that have become homozygous for the knockout mutation or in those that have lost (or have undergone somatic mutation at) the wildtype allele in the homologous chromosome. In a heterozygous animal, only one copy of the wildtype *p53* gene is present to protect the cell; inactivation of the lone wildtype allele disables the checkpoints that depend on the p53 protein. Because there are so many somatic cells in which such a rare aberration can occur, it is nearly certain that inactivation or loss of *p53* will take place somewhere in the heterozygous organism, initiating the sequence of mutations that results in cancer.

■ Retinoblastoma is an inherited cancer syndrome associated with loss of heterozygosity in the tumor cells.

In the cell cycle, the cells monitor their internal and external conditions. Until the conditions are suitable to initiate a division cycle, the cells accumulate at a point in G_1 called the G_1 restriction point, or *start*. In animal cells, a protein called the retinoblastoma (RB) protein holds cells at the restriction point by binding to and sequestering the transcription factor E2F, which is needed for further progression. The RB protein was first identified in

Figure 13.26 A retinoblastoma tumor first appears as a white mass in the retina but may grow to a size that it becomes visible through the pupil, as in this case. [© Chris Barry/Phototake, Inc./Alamy Images.]

human pedigrees because mutant forms of the protein are associated with the formation of malignant tumors in the retina, which can require surgical removal of the eyes (Figure 13.26).

Cellular growth is triggered by a growth factor combining with its receptor. This event activates a signal transduction pathway that culminates in the production of cyclin D. In the middle of G_1 phase, the RB protein begins to be phosphorylated by both the cyclin D–Cdk4 kinase and the cyclin D–Cdk6 kinase. Late in G_1, the RB phosphorylation is completed by the cyclin E–Cdk2 kinase as cells approach the G_1/S transition when they become committed to DNA replication. Phosphorylation of RB inactivates the protein and frees the bound E2F transcription factor (see Figure 13.10 on page 466). Release of the E2F results in transcription of the genes and translation of the enzymes responsible for DNA replication, including DNA polymerase. The cells therefore begin to accumulate the precursors and enzymes required for DNA synthesis. E2F also activates transcription of the gene for E2F itself (an example of positive autoregulation), as well as those for the cyclins E and A and the Cdk2 kinase subunit.

The idea that loss of the wildtype allele of a tumor-suppressor gene might be the triggering event at the cellular level for tumors in heterozygous genotypes was first suggested from studies on retinoblastoma by Alfred Knudson in 1971, long before the function of the *RB* gene was identified. As the Human Connection on page 484 indicates, Knudson noted that sporadic cases of retinoblastoma usually had a single tumor, whereas familial retinoblastoma cases usually had bilateral tumors and more than one tumor. On the basis of a statistical analysis of the time of tumor diagnosis, he suggested that genesis of a tumor in familial cases required a "single hit" in a somatic cell, whereas genesis of a tumor in sporadic cases required "two hits," which would happen only rarely. The two-hit model would also explain why sporadic cases rarely have a second tumor: Each individual "hit" is a rare event.

Retinoblastoma, like the *p53* deficiency, is inherited in pedigrees as a simple Mendelian dominant. But Knudson's hypothesis implied that even in familial cases, there must be another mutational event that triggers tumor development. Once the gene itself, called *RB1*, was located, analysis of genetic markers around the gene in tumor cells revealed that the triggering event is the loss of the wildtype *RB1* allele. Any of several mechanisms can uncover the mutant allele, including chromosome loss, mitotic recombination, deletion, and inactivating nucleotide substitutions. These can be distinguished from one another on the basis of the genetic markers that are lost or retained in the tumor cells; several of the mechanisms known to occur are diagrammed in Figure 13.27. At the organismic level, expression of the mutant gene is dominant. However, at the cellular level, expression of the mutant gene is recessive. Uncovering of the recessive allele by various mechanisms is called **loss of heterozygosity.**

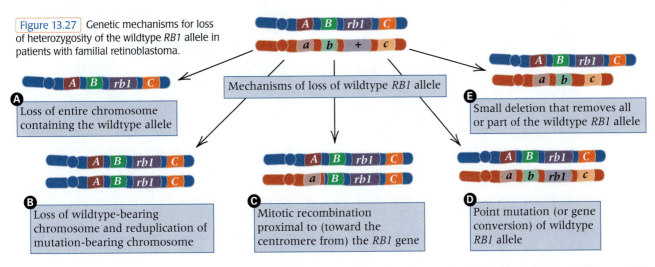

Figure 13.27 Genetic mechanisms for loss of heterozygosity of the wildtype *RB1* allele in patients with familial retinoblastoma.

Mechanisms of loss of wildtype *RB1* allele

A Loss of entire chromosome containing the wildtype allele

B Loss of wildtype-bearing chromosome and reduplication of mutation-bearing chromosome

C Mitotic recombination proximal to (toward the centromere from) the *RB1* gene

D Point mutation (or gene conversion) of wildtype *RB1* allele

E Small deletion that removes all or part of the wildtype *RB1* allele

13.4 Mutations that predispose to cancer can be inherited through the germ line   481

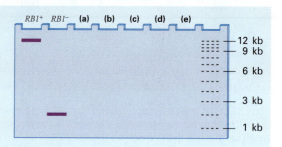

Q A Moment to Think

Problem: This problem demonstrates how DNA electrophoresis can be used to sort out different causes of loss of heterozygosity in retinoblastoma and other inherited cancer syndromes. The accompanying gel diagram shows the pattern of bands observed in a Southern blot for a 12-kb *Eco*RI restriction fragment from a nonmutant $RB1^+$ allele and a mutant allele $RB1^-$ in which there is an internal 10-kb deletion in the *Eco*RI fragment. A scale showing the electrophoretic mobility of DNA fragments of various sizes appears at the right. Show what pattern of bands would be expected in each of the following:
(a) Normal retinal cells from a heterozygous $RB1^+/RB1^-$ individual.
(b) Cells of retinoblastoma tumors caused by loss of the wildtype $RB1^+$-bearing chromosome.
(c) Cells of retinoblastoma tumors caused by mitotic recombination.
(d) Cells of retinoblastoma tumors caused by a new missense substitution in $RB1^+$.
(e) Cells of retinoblastoma tumors caused by a 6-kb deletion in the *Eco*RI fragment (not including the region that hybridizes with the probe). (The answer can be found on page 485.)

The inherited cancer syndromes listed in Table 13.3 result from mutations in tumor-suppressor genes. All show autosomal dominant inheritance except for xeroderma pigmentosum, which is inherited as a recessive. In all cancer syndromes that show dominant inheritance, loss of heterozygosity is required for manifestation of the tumor phenotype. Expression of the mutant allele *predisposes* the cell to become cancerous but is not in itself sufficient for generation of a cancer cell. Tumor progression still occurs only when additional somatic mutations and clonal expansions take place. The germ-line mutations do not themselves cause cancer; they merely make it much more likely that the progression will occur, because the tumor, in effect, has been given a "head start."

■ Some inherited cancer syndromes result from defects in processes of DNA repair.

Genetic instability clearly contributes to the origin of tumor cells. We know from studies on yeast and bacteria that cells have extensive mechanisms for repairing DNA lesions (see Chapter 12). Defects in these processes result in greatly elevated mutation rates and genetic instability. It was therefore not surprising when hereditary cancer syndromes that result from inherited defects in DNA repair were discovered.

Several of the inherited cancer syndromes listed in Table 13.3 are the consequence of defects in DNA repair. Defects in any of four genes that encode proteins involved in DNA mismatch repair cause hereditary nonpolyposis colorectal cancer, which shows autosomal dominant inheritance. Mutant cells have higher mutation rates, which promote progression toward the cancerous state. Inherited breast cancer syndromes prove to be associated with mutations in either of two genes, *BRCA1* or *BRCA2*. Less is known about the function of these genes. Current evidence suggests that *BRCA1* is

somehow involved in repair of double-strand breaks. Inherited skin cancer syndromes are called xeroderma pigmentosum. Xeroderma pigmentosum cells are defective in nucleotide excision repair; they are unable to repair defects such as thymine dimers that are induced by ultraviolet light. Individuals with this syndrome are very sensitive to the ultraviolet light that is present in sunlight and emitted by fluorescent lights.

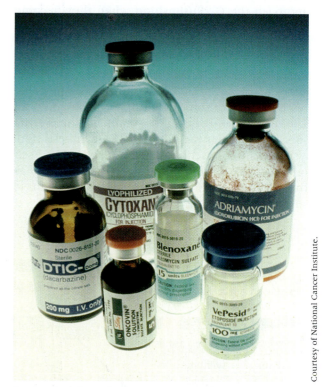

An arsenal of chemotherapy drugs is available for use singly or in combination to kill cancer cells. Among the drugs are alkylating agents that damage DNA, nitrosoureas that interfere with DNA repair, anthracyclines or topoisomerase II inhibitors that disrupt DNA replication, mitotic inhibitors that block cell division, and antimetabolites that affect DNA or RNA polymerase. While the drugs kill cancer cells, they also affect normal cells that divide rapidly such as cells in the bone marrow, hair follicles, digestive tract, and reproductive organs.

13.5

Acute leukemias are proliferative diseases of white blood cells and their precursors.

Acute *leukemia* is a malignant disease of the bone marrow, spleen, and lymph nodes associated with uncontrolled proliferation of white blood cells (**leukocytes**) and their precursors in the bone marrow. The initial genetic events that result in the acute leukemias are quite different from those indicated in the cancers examined previously. They do not arise as a consequence of alterations in cell cycle regulation or checkpoints, nor are they familial.

■ **Some acute leukemias result from a chromosomal translocation that fuses a transcription factor with a leukocyte regulatory sequence.**

Up to 65 percent of the acute leukemias arise as a consequence of chromosomal translocations involving genes that encode transcription factors that play a role in blood-cell development (hematopoiesis). The translocations are of the two types illustrated in Figure 13.28 . In a **promoter fusion** (part A), the coding region for a gene that encodes a

transcription factor is translocated near an enhancer for an immunoglobulin heavy-chain gene or a T-cell receptor gene. The result is overexpression of the transcription factor, derangement of normal hematopoiesis, and overproduction of lymphocytes. One important aspect of normal hematopoiesis is the apoptotic destruction of progenitor cells that fail to rearrange their antigen-receptor genes productively (Chapter 9). Estimates are that 75 percent of B-cell and 95 percent of T-cell precursors self-destruct during normal development. *Bcl2* (*Bcl* stands for B-cell lymphoma) is an example of an oncogene in Table 13.3 that was originally discovered in a promoter fusion that placed *Bcl2* next to an immunoglobulin heavy-chain gene enhancer. Overexpression of Bcl2 blocks apoptosis by preventing formation of Bax homodimers (Figure 13.21). These "undeservedly alive" lymphocytes proliferate and come to dominate the population of white blood cells, but they are useless in fighting infection because they have not rearranged their immunoglobulin genes properly. In such cases the bone marrow becomes almost totally occupied by the cancerous leukocytes, leading to severe anemia and bleeding. Chemotherapy can sometimes be quite effective in treating acute leukemias in young children by activating the *p53* gene, provided that the aberrant leukocytes still carry the wildtype alleles encoding p53.

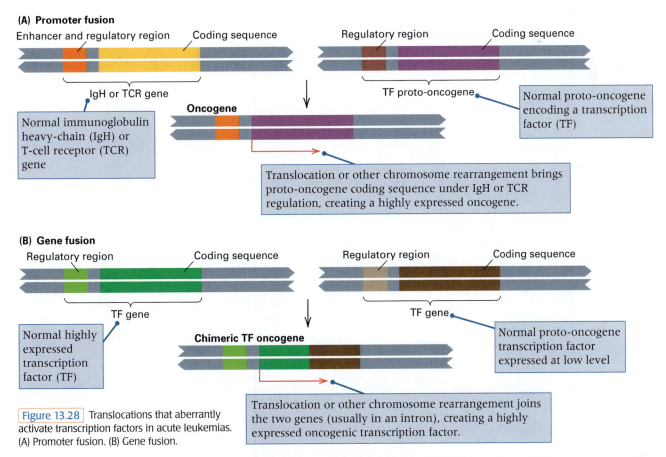

Figure 13.28 Translocations that aberrantly activate transcription factors in acute leukemias. (A) Promoter fusion. (B) Gene fusion.

Two Hits, Two Errors

Alfred G. Knudson 1971
M. D. Anderson Hospital
The University of Texas
Houston, Texas
Mutation and Cancer: Statistical Study of Retinoblastoma

This landmark paper was a turning point in thinking about genetics and cancer. Pedigree studies had already shown that there were hereditary predispositions to certain cancers, such as retinoblastoma. But "sporadic" cases also occurred, in which only one member of a kindred was affected. What is the relationship between these two forms? By an ingenious statistical analysis, Knudson showed that the sporadic cases exhibit "two-hit" kinetics as a function of age, indicating that two independent mutations in the same retinal cell are involved, whereas familial cases exhibit "one-hit" kinetics. The simplest interpretation is that in sporadic cases, the first mutation knocks out a key gene and the second knocks out its allele, whereas in familial cases, the initial mutant allele is inherited, and consequently only one mutation (in the remaining functional allele) is needed to cause the disease. At the level of the individual cell, therefore, a mutation in the retinoblastoma gene (*RB1*) is recessive, whereas at the familial level, the disease is inherited as a dominant, because almost every person who inherits one mutant allele will undergo the second mutation in at least one retinal cell.

. . . .

The hypothesis is developed that retinoblastoma is a cancer caused by two mutational events.

In the dominantly inherited form, one mutation is inherited via the germinal cells and the second occurs in somatic cells. In the nonhereditary form, both mutations occur in somatic cells. . . . Several authors have concluded that retinoblastoma may be caused by either a germinal or a somatic mutation. . . . All bilateral cases [25 to 30 percent] should be counted as hereditary because the proportion of affected offspring closely approximates the 50 percent expected with dominant inheritance. . . . If a second, single event is involved [in the hereditary cases], the distribution of bilateral cases with time should be an exponential function, i.e., the fraction of the total cases that develop should be constant, as expressed in the relationship $dS/dt = -kS$, and $\ln(S) = -kt$, where S is the fraction of survivors not yet diagnosed at time t, and dS is the change in this fraction in the interval dt.

As shown in the figure, this is indeed the case. By contrast, the fractional decrease in unilateral cases per unit time does not show this relationship. . . . The exponential decline in new hereditary cases with time reflects the occurrence of a second event at a

The two-mutation hypothesis is consistent with current thought that the common cancers are produced by about 3 to 7 mutations.

constant rate . . . of the order of 2×10^{-7} per year. . . . The two-mutation hypothesis is consistent with current thought . . . that the common cancers are produced by about 3 to 7 mutations. Interestingly, one of the lowest estimates [is] for brain tumors, which are, like retinoblastoma, derived from neural elements.

Source: Proceedings of the National Academy of Sciences USA. 68: 820–823.

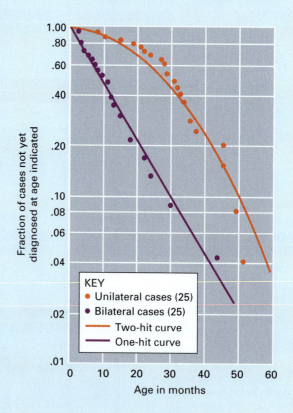

A Moment to Think

Answer to Problem: To deduce the expected pattern of bands in each case, first think through the consequences of each type of event with respect to the $RB1$ alleles present in the cells with loss of heterozygosity.
(a) Normal heterozygous cells will contain both DNA fragments.
(b) Loss of the wildtype chromosome eliminates the $RB1^+$ allele.
(c) Mitotic recombination makes the $RB1^-$ allele homozygous in the tumor cells.
(d) A missense mutation in the $RB1^+$ allele does not change the size of the EcoRI fragment (except in the extremely unlikely case that the mutation hits one of the EcoRI sites).
(e) In the new $RB1$ deletion mutation, the size of the EcoRI fragment will be 6 kb instead of 12 kb. Therefore, the expected gel patterns are as shown in the illustration.

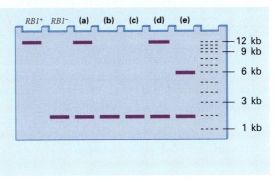

■ Other acute leukemias result from a chromosomal translocation that fuses two genes to create a novel chimeric gene.

The second type of translocation associated with acute leukemia is a **gene fusion** (part B of Figure 13.28). This type of rearrangement is found more frequently in acute leukemia than a promoter fusion. Most commonly, the translocation breakpoints occur in introns of genes that encode transcription factors in two different chromosomes. The result is a fusion gene called a **chimeric gene** composed of parts of the original genes. The chimeric gene produces a chimeric protein. In the case of acute leukemia, the chimeric protein is a transcription factor with an altered function that interferes with normal hematopoiesis. The uniqueness of these chimeric proteins, and the fact that they are present only in cancer cells and not in normal cells, make them inviting targets for drugs or chemotherapy. If one could successfully attack cells expressing the chimeric protein, one could selectively kill the cancer cells.

Chronic myeloid leukemia (CML) accounts for 15 to 20 percent of all cases of leukemia. The hallmark of CML is the presence of the *Philadelphia chromosome*, which results from a reciprocal translocation between chromosomes 9 and 22 in hematopoietic stem cells of the bone marrow. These are the cells that differentiate into various specialized types of cells that become part of the blood and immune sys-

tem. The molecular result of the *t(9, 22)* translocation is replacement of the first exon of *c-abl* with sequences from the *bcr* gene, resulting in a Bcr-Abl fusion protein. Because the N-terminal region of Abl normally inhibits function of the catalytic domain, the fusion protein, having lost the inhibitory domain, is a constitutively active protein kinase.

The tyrosine kinase inhibitor STI571 (imatinib mesylate) is a therapeutic agent specifically directed toward the Bcr-Abl tyrosine kinase. By inhibiting the Bcr-Abl kinase, the drug not only inhibits the kinase activity but also inhibits proliferation of cells expressing the kinase in vivo and in vitro. Continuous exposure to the drug eradicates most cancer cells. It is thought to be very effective because Bcr-Abl may be the sole molecular abnormality early in the course of chronic myeloid leukemia. STI571 also inhibits the Kit tyrosine kinase and has proven effective against gastrointestinal stromal tumors, in which mutations that constitutively activate the *c-kit* gene encoding Kit are involved.

In one study of colorectal cancers, 182 cancers were tested for mutations in 138 genes encoding protein kinases, out of a total of 518 known protein-kinase genes in the human genome. About 30 percent of the colorectal cancers proved to have a mutation in one of the genes encoding tyrosine kinases. These results, along with the positive outcomes of treatment with drugs like STI571, have led to an increased focus on protein kinases as therapeutic possibilities.

chapter summary

13.1 The cell cycle is under genetic control.

- Many genes are transcribed during the cell cycle just before their product is needed.
- Mutations affecting the cell cycle have helped to identify the key regulatory pathways.
- Cyclins and cyclin-dependent protein kinases propel the cell through the cell cycle.

- The retinoblastoma protein controls the initiation of DNA synthesis.
- Protein degradation also helps regulate the cell cycle.

The essential function of the cell cycle is to ensure that each chromosomal DNA molecule is replicated once in the cell cycle and that the genetically identical daughter chromosomes are distributed equally to daughter cells. The cell cycle comprises four phases: G_1 (gap/growth 1), S (DNA synthesis), G_2 (gap/growth 2), and M (mitosis). Key events in the cycle that ensure proper replication and segregation of chromosomes include centrosome duplication and positioning, replication of the DNA, formation of the bipolar spindle, attachment of each centromere to spindle fibers from each pole, proper functioning of the spindle, and disjunction of the chromosomes.

The cell cycle has been studied extensively in budding yeast. Conditional mutants defective in the cell division cycle (*cdc* mutants) arrest at particular places in the cell cycle under restrictive conditions. Arrest of the cell cycle in *cdc* mutants takes place at a limited number of points, which correspond to the position of checkpoints within the cell cycle.

Progression from one stage of the cell cycle to the next is controlled by protein complexes called cyclin-dependent kinase (CDK) complexes. These are made up of a cyclin component, which is stage-specific, and a cyclin-dependent protein kinase. The CDK complexes phosphorylate stage-specific protein substrates to change their properties and thus make possible particular reaction sequences. Cyclins are synthesized immediately prior to their use and degraded immediately thereafter.

Prior to the G_1/S transition in animal cells, the cyclin–CDKs phosphorylate the retinoblastoma protein, RB, which changes its conformation and thereby frees the E2F transcription factor, which until this time is rendered inactive by binding with RB. The free E2F activates transcription of enzymes needed for synthesis of DNA. The event responsible for the onset of S phase is phosphorylation of one or more proteins in the protein complexes assembled at the origins of replication, and DNA replication ensues.

The G_2/M transition involves phosphorylation of cyclin B, resulting in its retention in the nucleus. Cyclin B–CDK phosphorylates proteins required for assembly of the spindle and dissolution of the nuclear membrane. The onset of anaphase and the exit from mitosis are controlled by proteolysis via activity of the anaphase-promoting complex (APC/C). Proteolysis ensures that sister chromatids separate and that cyclin B is degraded, propelling the cell into the next G_1 phase.

13.2 **Checkpoints in the cell cycle allow damaged cells to repair themselves or to self-destruct.**

- The p53 transcription factor is a key player in the DNA damage checkpoint.
- The centrosome duplication checkpoint and the spindle checkpoint function to maintain the normal complement of chromosomes.

Cells monitor external and internal conditions. If defects are detected, checkpoint mechanisms halt cell-cycle progression, allowing time for repair or correction. Major checkpoints monitor for DNA damage in G_1 and S, duplication of the centrosome, and attachment of chromosomes to the spindle.

In animal cells, DNA damage activates a signal transduction pathway that results in phosphorylation and acetylation of p53 protein, preventing its being bound by Mdm2 and shuttled from the nucleus and destroyed by the proteasome. This results in both stabilization of p53 and an increase in the level of p53.

The p53 protein activates transcription of several genes whose products block progression of the cell cycle at the G_1/S and G_2/M boundaries, in effect through inhibition of the cyclin–CDKs at these steps. The p53 protein also activates transcription of the gene for Bax, which propels the cell toward programmed cell death (apoptosis). Detection of a defect in attachment of a chromosome to the spindle blocks activation of the anaphase-promoting complex and thereby prevents sister chromatid separation and exit from mitosis.

13.3 **Cancer cells have a small number of mutations that prevent normal checkpoint function.**

- Proto-oncogenes normally function to promote cell proliferation or to prevent apoptosis.
- Tumor-suppressor genes normally act to inhibit cell proliferation or to promote apoptosis.

Cancer cells exhibit uncontrolled proliferation. Most cancers are sporadic; they involve a sequence of genetic changes in somatic cells. Tumors are clonal, deriving from a single cell. The genetic changes in cancer cells include the dominant oncogenes, which derive from normal cellular genes called proto-oncogenes by mutation or overexpression. Other genetic changes include the recessive tumor-suppressor genes. Many of the genetic changes involved in progression from the normal to the cancerous state are changes in cell-cycle regulatory genes or checkpoint genes. These changes often result in genetic instability, which enhances the probability of further changes.

13.4 **Mutations that predispose to cancer can be inherited through the germ line.**

- Cancer initiation and progression occur through mutations that allow affected cells to evade normal cell-cycle checkpoints.
- Retinoblastoma is an inherited cancer syndrome associated with loss of heterozygosity in the tumor cells.
- Some inherited cancer syndromes result from defects in processes of DNA repair.

About 1 percent of cancer syndromes are heritable. These often can be traced to germ-line mutations in the same cell-cycle regulatory or checkpoint genes that are mutated in sporadic cancers. Such germ-line changes often show a dominant inheritance pattern at the organismic or pedigree level. At the cellular level, expression of the mutant phenotype is recessive. For expression, the cell must undergo a loss of heterozygosity through any of several mechanisms. Other inherited cancer syndromes can be traced to defects in DNA-repair genes. Defects in mismatch repair show dominant inheritance, whereas defects in nucleotide excision repair show recessive inheritance.

13.5 Acute leukemias are proliferative diseases of white blood cells and their precursors.

- Some acute leukemias result from a chromosomal translocation that fuses a transcription factor with a leukocyte regulatory sequence.
- Other acute leukemias result from a chromosomal translocation that fuses two genes to create a novel chimeric gene.

A large fraction of the acute leukemias involve chromosomal translocations of either of two types. In the promoter fusion type of translocation, a proto-oncogene is translocated near the strong enhancer of an immunoglobulin heavy-chain gene (IgH) or a T-cell receptor gene, resulting in high-level expression of the oncogene. In the gene fusion type of translocation, genes for two different transcription factors become fused, giving rise to a chimeric transcription factor.

issues & ideas

- What role does the centriole (in some organisms, the centrosome) play in cell division?
- What molecular process that takes place in the nucleus defines the S period?
- In yeast, what phenotype defines a *cdc* mutant? For a temperature-sensitive *cdc* mutant grown at the restrictive temperature, why do cells arrest at a particular stage in the cell cycle?
- What is a cell-cycle checkpoint? Which checkpoints are emphasized in this chapter, and what does each "check" for?
- What is apoptosis and what role does it play in preserving the integrity of the genome of a multicellular organism?

- Mutations in genes whose products are involved in DNA repair are often associated with an increased risk of cancer. What does this observation imply about the role of spontaneous mutations in the development of cancer?
- How can one reconcile the following statements: "Cancer is a genetic disease" and "Most cancers are sporadic (not inherited)."
- Distinguish between proto-oncogenes and tumor-suppressor genes, and give one example of each. In which class of genes does a loss-of-function mutation predispose to cancer? A gain-of-function mutation? Explain your answer.
- What is "loss of heterozygosity," and how is this phenomenon related to the progression of some types of cancer?

key terms & concepts

anaphase-promoting complex (APC/C)
apoptosis
arrest
cancer
cell cycle arrest
cell division cycle (*cdc*) mutant
cell senescence
centriole
centrosome

centrosome duplication checkpoint
checkpoint
chimeric gene
contact inhibition
cyclin
cyclin–CDK complex
cyclin-dependent protein kinase (CDK)
DNA damage checkpoint

familial
gene fusion
G protein
G_1 restriction point (start)
G_1/S transition
G_2/M transition
leukocyte
loss of heterozygosity
oncogene
p53 transcription factor

processivity
programmed cell death
promoter fusion
proto-oncogene
retinoblastoma (RB) protein
spindle assembly checkpoint
sporadic
tumor-suppressor gene

1. _____ Opposite of familial.

2. _____ Component of the cyclin-CDK complex that determines which proteins will be phosphorylated.

3. _____ The normal function of this protein is to bind the transcription factor E2F to prevent premature initiation of DNA synthesis.

4. _____ Key protein in the DNA damage checkpoint.

5. _____ Term for the average number of template nucleotides processed by a polymerase before spontaneous disassociation from the template.

6. _____ Activation of this protein-tagging complex allows entry into anaphase.

7. _____ Technical term for programmed cell death.

8. _____ Any of a large family of GTP-associated proteins involved in signal transduction.

9. _____ The process in which a new gene is formed from parts of two or more other genes.

10. _____ Any process by which a heterozygous somatic cell becomes homozygous for the recessive allele.

11. _____ A gene whose normal function is to restrict cell proliferation or to promote apoptosis.

12. _____ A gene whose normal function is to promote cell proliferation or to prevent apoptosis.

Problem 1

Indicate which of the checkpoints in the cell cycle discussed in this chapter would be expected to be the "first line of defense" in preventing changes in the genome due to:
(a) The effects of x rays and other forms of ionizing radiation.
(b) Endoreduplication as a cause of polyploidy.
(c) Nondisjunction as a cause of polysomy.

■ Solution (a) Because x rays and other forms of ionizing radiation produce single-stranded and double-stranded breaks in DNA molecules, the DNA damage checkpoint is primarily responsible for preventing (not always successfully) mutations due to these agents. (b) Endoreduplication occurs when a cell undergoes chromosome replication without chromosome separation, which represents a failure of the spindle to form properly. In this case the centrosome duplication checkpoint is the first line of defense. (c) Polysomy, in contrast to polyploidy, results from the misbehavior of a single chromosome at anaphase, rather than the whole set of chromosomes. Therefore, the spindle assembly checkpoint is expected to be the first line of defense against nondisjunction.

Problem 2

Distinguish between an oncogene and a tumor-suppressor gene. Why are oncogene mutations associated with cancer gain-of-function mutations, whereas tumor-suppressor genes are associated with cancer loss-of-function mutations?

■ Solution Oncogenes are mutant genes whose normal counterparts, called proto-oncogenes, encode products that promote cell proliferation or inhibit apoptosis. The products of tumor-suppressor genes inhibit cell proliferation or activate the apoptotic pathway. Oncogenes are gain-of-function mutations because their enhanced expression makes uncontrolled cell division possible or prevents the apoptosis pathway. Tumor-suppressor mutations associated with cancer are loss-of-function mutations because absence of a functional gene product fails in inhibiting cell division or in activating the apoptosis pathway.

Problem 3

The accompanying pedigree includes individuals affected with retinoblastoma, and the diagram of the gel shows a restriction fragment from a number of alleles of the *RB1* gene, mutant forms of which are associated with this cancer. Three sizes of restriction fragments (a, b, and c) are observed. Individuals in generations I and II are old enough to have developed the cancer if they carry a mutant *RB1* allele, but the individuals in generation III are all too young to have developed the disease. Identify the high-risk individuals in generation III and those who are not at risk.

■ Solution First compare the pedigree with the gel patterns to identify which restriction fragment is associated with the mutant *RB1* allele in this pedigree. Because the father in generation I (individual 10) is affected, the mutant allele must be associated with either band a or band b. All of the affected individuals in generation II have band b, so this band must mark the mutant *RB1* allele. Therefore, the individuals at high risk in generation III are those with band b, whereas the others are not at risk. The high-risk individuals are numbers 1, 2, 8, 11, 13, and 18; those not at risk are individuals 3, 6, 7, 12, 16, and 17.

Problem 3

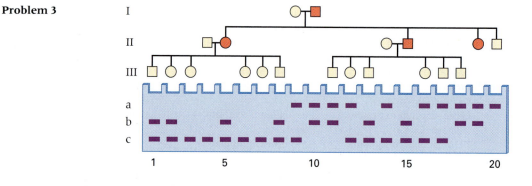

13.1 If cancer is a "genetic disease," how can it be true that most cases are sporadic (that is, not familial)?

13.2 A human cell in culture is homozygous for a temperature-sensitive mutation in a gene necessary to repair double-stranded breaks in DNA. If cells were irradiated at the restrictive temperature, at what stage of the cell division cycle would you expect the mutant cells to accumulate?

13.3 The p53 protein is defective in more than half of all cancers. Why might this be expected?

13.4 What role does Ras–GTP play in intracellular signaling that makes *Ras* a proto-oncogene?

13.5 What does it mean to say that mutations in the retinoblastoma gene *RB1* are "dominant at the organismic level but recessive at the cellular level"?

13.6 Draw a diagram showing how recombination between homologous chromosomes during mitosis can result in a cell lineage with loss of heterozygosity for a *p53* mutation. What other genes also lose heterozygosity in this process?

13.7 How does the normal retinoblastoma protein function to hold mammalian cells at the G_1 restriction point ("start")?

13.8 What are the roles of Bax and Bcl2 proteins in programmed cell death? How is the balance in the amounts of these proteins affected by activated p53? By certain cellular oncogenes?

13.9 What protein is the major player in activating a DNA damage checkpoint? How is this protein normally kept from triggering the checkpoint? What happens to the protein when there is DNA damage?

13.10 Many types of cancer cells have defects in the G_1/S checkpoint. These also tend to have abnormalities in chromosome number or structure. Why would chromosome abnormalities be expected in such cases?

13.11 A woman has a mammogram (breast x ray) that reveals a suspicious lump of tissue. Cytological analysis of the lump reveals cells with a highly variable chromosome number and many chromosome rearrangements. What does this finding suggest about the malignant or nonmalignant nature of the suspicious growth? Explain your answer.

13.12 In familial retinoblastoma, there is an average of 3 retinal tumors per heterozygous carrier of the mutation. Assuming that the number of retinal cells at risk is 2×10^6 in each eye and that each tumor results from an independent loss of heterozygosity, what is the estimated rate of loss of heterozygosity per cell? Should this be regarded as a "mutation rate"? Why or why not?

13.13 The protein GADD45, for which p53 is a transcription factor, reduces the processivity of the DNA polymerase. What does *processivity* mean? How does a reduction in processivity help in the repair of damaged DNA?

13.14 DNA from cells of a patient with retinoblastoma was analyzed using a Southern blot with a probe for a particular restriction fragment in the *RB1* gene. The result from nontumor cells and the results from cells taken from a tumor in each eye are shown in the accompanying diagram.
(a) Which band should be associated with the mutant allele and which with the nonmutant allele?
(b) How is it possible for the bands from the tumor in the left eye to be different from those from the tumor in the right eye?
(c) Explain how cells in the tumor in the right eye can have a "loss of heterozygosity" even though the bands are indistinguishable from those observed from nontumor cells.

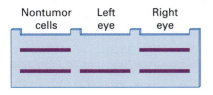

13.15 In an inherited cancer syndrome for pancreatic cancer there is an average of 2 independent tumors resulting from loss of heterozygosity, and 8×10^6 cells are at risk of causing this type of cancer. What is the penetrance of the familial form of the disease?

13.16 The accompanying gel diagram shows the pattern of bands observed in a Southern blot for three linked RFLPs in and flanking the *p53* gene in a family in which Li–Fraumeni syndrome is found. Shown are an affected mother, an unaffected father, and one affected son. The bands AL and AS are for one RFLP (with alleles *AL* and *AS*), and the bands BL and BS are for another RFLP (with alleles *BL* and *BS*). The $p53^+$ and $p53^-$ bands are for the RFLP for *p53*, where the wildtype $p53^+$ allele yields the band $p53^+$, and a mutant p53 allele, which has a 1-kb insertion in the restriction fragment, yields the band $p53^-$. (The blot was probed with a fragment from each of the three RFLPs, so all three can be visualized in the same blot.) The order of the genes along the chromosome arm is *centromere–A–p53–B–telomere*. Considering the mother and the father, determine which RFLP alleles are linked with $p53^+$ in the son. If DNA from tumor cells in the son were assayed for these RFLPs, what pattern of bands would be expected from each of the following cells?
(a) Cells that had lost the $p53^+$-bearing chromosome and had a reduplication of the $p53^-$-bearing homolog.
(b) Cells that had undergone mitotic recombination in the *centromere–A* interval.
(c) Cells that had undergone mitotic recombination in the *A–p53* interval.
(d) Cells that had a new nonsense mutation in the $p53^+$ allele.
(e) Cells that had undergone gene conversion of $p53^+$ to $p53^-$.

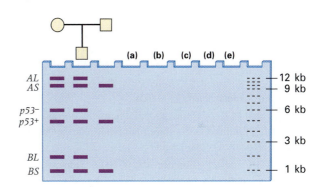

13.17 In patients with bilateral retinoblastoma, would the mechanism of loss of heterozygosity in tumors in different eyes be expected to be the same or different? Explain your answer.

13.18 Mutagenesis of a *RAS* gene of budding yeast yields a temperature-sensitive conditional mutation.
(a) Would you expect a cell that is carrying a mutation that prevents Ras from exchanging GTP for GDP at the restrictive temperature to continue to divide at this temperature?
(b) Would the mutation be dominant or recessive?
(c) Would you expect a cell that is carrying a mutation that inactivates the GTPase activity of Ras at 36°C to continue to divide at the restrictive temperature?
(d) Would this mutation be dominant or recessive?

13.19 Human papilloma virus (HPV) is present in greater than 90 percent of cervical cancers. HPV encodes two proteins, E6 and E7, that are potent contributors to its tumorigenicity. E7 is known to disable RB; E6 binds to p53 and targets it for degradation. Discuss how these activities might contribute to the development of the cancerous state in infected cells.

13.20 The accompanying pedigree includes individuals affected with adenomatous polyposis, and the diagram of the gel shows a restriction fragment from a number of alleles of *APC,* mutant forms of which are associated with this cancer. Four sizes of restriction fragments (a through d) are observed. Individuals in generations I and II are old enough to have developed the cancer if they carry a mutant *APC* allele, but the individuals in generation III are all too young to have developed the disease. Identify the high-risk individuals in generation III and those who are not at risk. (Note that a mutant allele and a nonmutant allele can yield the same size restriction fragment.)

Problem 13.20

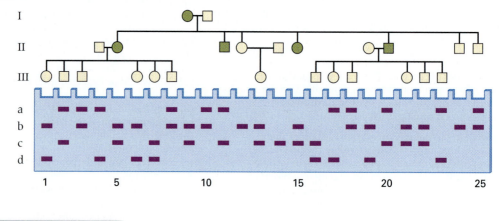

further readings

Cahill, D. P., K. W. Kinzler, B. Vogelstein, and C. Lengauer. 1999. Genetic instability and Darwinian selection in tumours. *Trends in Biochemical Sciences* 24: M57.

Cavenee, W. K., and R. L. White. 1995. The genetic basis of cancer. *Scientific American,* March.

Chin, L., and R. A. DePinho. 2000. Flipping the oncogene switch: Illumination of tumor maintenance and regression. *Trends in Genetics* 16: 147.

Clarke, D. J., and J. F. Gimenez-Abian. 2000. Checkpoints controlling mitosis. *Bioessays* 22: 351.

Fearon, E. R. 1997. Human cancer syndromes: Clues to the origin and nature of cancer. *Science* 278: 1043.

Fearon, E. R., and C. V. Dang. 1999. Cancer genetics: Tumor suppressor meets oncogene. *Current Biology* 9: R62.

Golub, T. R., D. K. Slonim, P. Tamayo, et al. 1999. Molecular classification of cancer: Class discovery and class prediction by gene expression monitoring. *Science* 286: 531.

Hartwell, L. H., and T. A. Weinert. 1989. Checkpoints: Controls that ensure the order of cell cycle events. *Science* 246: 629.

Jacks, T. 1996. Tumor suppressor gene mutations in mice. *Annual Review of Genetics* 30: 603.

Kaelin, W. G. 1999. Functions of the retinoblastoma protein. *Bioessays* 21: 950.

Levine, A. J. 1997. p53, the cellular gatekeeper for growth and division. *Cell* 88: 323.

Lew, D. J., T. A. Weinert, and J. R. Pringle. 1997. Cell cycle control in *Saccharomyces cerevisiae*. In *The Molecular and Cellular Biology of the Yeast Saccharomyces*. Vol. 3: *Cell Cycle and Cell Biology,* ed. J. R. Pringle, J. R. Broach, and E. W. Jones. Cold Spring Harbor, NY: Cold Spring Harbor Laboratory Press, p. 67.

Look, A. T. 1997. Oncogenic transcription factors in the human acute leukemias. *Science* 278: 1059.

MacNeill, S. A., and P. Nurse. 1997. Cell cycle control in fission yeast. In *The Molecular and Cellular Biology of the Yeast Saccharomyces*. Vol. 3: *Cell Cycle and Cell Biology*, ed. J. R. Pringle, J. R. Broach, and E. W. Jones. Cold Spring Harbor, NY: Cold Spring Harbor Laboratory Press, p. 697.

McCormick, F. 1999. Signalling networks that cause cancer. *Trends in Biochemical Sciences* 24: M53.

Murray, A. W., and T. Hunt. 1993. *The Cell Cycle: An Introduction*. New York: Oxford University Press.

Rahman, N., and M. R. Stratton. 1998. The genetics of breast cancer susceptibility. *Annual Review of Genetics* 32: 95.

Song, Z. W., and H. Steller. 1999. Death by design: Mechanism and control of apoptosis. *Trends in Biochemical Sciences* 24: M49.

Vogelstein, B., and K. W. Kinzler., eds. 1998. *The Genetic Basis of Human Cancer*. New York: McGraw-Hill.

Weinberg, R. A. 1996. How cancer arises. *Scientific American,* September.

Welcsh, P. L., K. N. Owens, and M. C. King. 2000. Insights into the functions of *BRCA1* and *BRCA2*. *Trends in Genetics* 16: 69.

geNETics on the web

GeNETics on the Web will introduce you to some of the most important sites for finding genetic information on the Internet. To explore these sites, visit the Jones and Bartlett home page at

http://www.jbpub.com/genetics

For the book *Essential Genetics: A Genomics Perspective,* choose the link that says **Enter GeNETics on the Web**. You will be presented with a chapter-by-chapter list of highlighted keywords. Select any highlighted keyword and you will be linked to a Web site containing genetic information related to the keyword.

- Among 400 people, about 200 will develop at least one tumor of the colon (large bowel), 20 of these cases will progress to malignancy, and 1 will be associated with an inherited genetic abnormality. Among the inherited forms of **colon cancer**, two important types are familial adenomatous polyposis (FAP) and hereditary nonpolyposis colorectal cancer (HNPCC). They demonstrate the difference between tumor initiation and tumor progression. Patients with FAP develop thousands of colon polyps, only a few of which progress to cancer, whereas patients with HNPCC have only a few polyps, but each has a high probability of progressing to cancer. Consult this keyword site to learn more about these and other types of colon cancer.

- In the United States, the frequency of **retinoblastoma** is about 1 in 23,000 live births. It is a disease of young children, the great majority of cases being diagnosed prior to the age of 5. In many cases there are treatment options other than surgical removal of the eye. These include cryotherapy to destroy the cancer cells with extreme cold, laser light to target the tumors and the surrounding blood vessels that feed them, and radiation and chemotherapy. This keyword site aims to provide information and links for anyone interested in any aspect of the disease.

Called *tahtanka* by the Lakota Sioux, the American bison (*Bison bison*) was a source of meat, hide, hair, horn, and fuel for the Native Americans of the Great Plains. Although the size of the bison population was once about 60 million, the coming of the railroads and grooved rifle barrels to the American West resulted in their unprecedented slaughter until, in 1890, only about 1,000 animals remained in all of North America. From this population *bottleneck*, as a drastic reduction in population size is known, the herd has since recovered to about 350,000. [Courtesy of S. Schmidt/Yellowstone National Park/NPS]

key concepts

- Because DNA and protein sequences change through time, sequence comparisons reveal how genes and proteins evolve, and make it possible to infer the evolutionary relationships among different species.

- Many genes in natural populations are polymorphic; they have two or more common alleles.

- With random mating, the alleles in gametes are combined at random to form the zygotes of the next generation.

- Genetic polymorphisms can be used as genetic markers in pedigree studies and for individual identification (DNA typing).

- Relative to the frequencies of genotypes expected with random mating, inbreeding results in an excess of homozygous genotypes.

- Mutation and migration introduce new alleles into populations.

- Natural selection and random genetic drift are the usual causes of change in allele frequency; selection changes allele frequency in a systematic direction, whereas random genetic drift changes allele frequency in an unpredictable direction.

- Analysis of human mitochondrial DNA sequences is consistent with a scenario in which people migrating out of subsaharan Africa approximately 100,000 years ago became the progenitors of all present-day populations existing outside of Africa.

14

Molecular Evolution and Population Genetics

chapter organization

14.1 DNA and protein sequences contain information about the evolutionary relationships among species.

14.2 Genotypes may differ in frequency from one population to another.

14.3 Random mating means that mates pair without regard to genotype.

14.4 Highly polymorphic sequences are used in DNA typing.

14.5 Inbreeding means mating between relatives.

14.6 Evolution is accompanied by genetic changes in species.

14.7 Mutation and migration bring new alleles into populations.

14.8 Natural selection favors genotypes that are better able to survive and reproduce.

14.9 Some changes in allele frequency are random.

14.10 Mitochondrial DNA is maternally inherited.

the human connection
Resistance in the Blood

Chapter Summary
Issues & Ideas
Key Terms & Concepts
Solutions: Step by Step
Concepts in Action: Problems for Solution
Further Readings

 geNETics on the web

Molecular genetics has brought about major changes in the way biologists study genetic differences within and among natural populations of organisms. Traditionally, the study of genetic differences among organisms required controlled matings and analysis of the progeny. This requirement made it almost impossible to study genetic differences between species, because all but the most closely related species either do not mate or they yield progeny that are inviable or sterile. But the discovery that DNA is the genetic material made it possible to compare corresponding genes even in distantly related species. As we shall see in this chapter, studies of DNA sequences from different species can reveal important information not only about how genes evolve, but also about the evolutionary relationships among the species. Comparative study of macromolecules within and among species constitutes the field of *molecular evolution.*

Molecular methods have also transformed the study of genetic variation within species. Traditionally, genetic differences among individuals of a species were undetectable unless they caused a difference in phenotype. The primacy of phenotypic differences severely limited the types of population studies that could be carried out, because most genetic differences among individuals in a population cause no detectable difference in any aspect of the phenotype. But the study of molecular markers revealed that natural populations contain abundant genetic variation at the molecular level. This type of genetic variation can be used to investigate population history, subdivision, and the genealogical relationships among individuals, as well as to identify the chromosomal locations of genetic risk factors for inherited diseases. The application of genetic principles to entire populations of organisms constitutes the subject of *population genetics.*

14.1

DNA and protein sequences contain information about the evolutionary relationships among species.

Macromolecules such as DNA, RNA, and protein are linear polymers of subunits. The specific sequence of subunits along each molecule determines its information content or function. With the vast outpouring of data from large-scale genomic sequencing, there is great interest in comparing the sequences of related molecules among species, motivated in part by the hope of correlating differences in sequence with differences in function, especially in proteins.

Although the sequences of macromolecules contain information about function, they also contain information about evolutionary history. Sequences change through time even among macromolecules whose function remains identical. In fact, it is often difficult to distinguish which differences in sequence between species are important to the function of a molecule, and which differences have such small effects that they simply reflect changes that take place by chance over evolutionary time.

The study of how (and why) the sequences of macromolecules change through time constitutes **molecular evolution.** In this section we consider several aspects of molecular evolution, beginning with reconstructing the evolutionary history of a set of sequences and ending with the origin of new genes.

■ A gene tree is a diagram of the inferred ancestral history of a group of sequences.

Because sequences change through time, it follows that sequence differences accumulate through time. The accumulation of differences is the basis of **molecular systematics,** which is the analysis of molecular sequences in order to infer their evolutionary relationships.

The principles of reconstructing evolutionary history from molecular sequences can best be illustrated by example. We will use the data in Figure 14.1 , part A, which depicts the sequence of the first 50 amino acids of the beta globin chain of adult hemoglobin from each of seven organisms. The technical term used for the source of each sequence is **taxon** (plural **taxa**), and in this case the taxa are species of vertebrates. In evolutionary studies, the sequences to be compared are normally much longer, and there can be many more species, but this small data set will serve to illustrate the methods. Each raised dot means that the amino acid at the site is identical to that in the molecule listed along the top (in this case human). The cow and sheep sequences can be aligned with the others only if gaps are introduced at positions 2 and 3. These are indicated by dashes and correspond to deletions of 6 bp in the cow and sheep genes relative to the others. (Alternatively, there could be a 6-bp insertion in each of the other sequences, but this interpretation is less likely.)

Because the sequences of biological macromolecules change through time, the amount of difference between any two sequences can be taken as a measure of how long they have been evolving along separate evolutionary paths. There are two problems in assuming such an equivalence. The first problem is that changes in sequence are a matter of

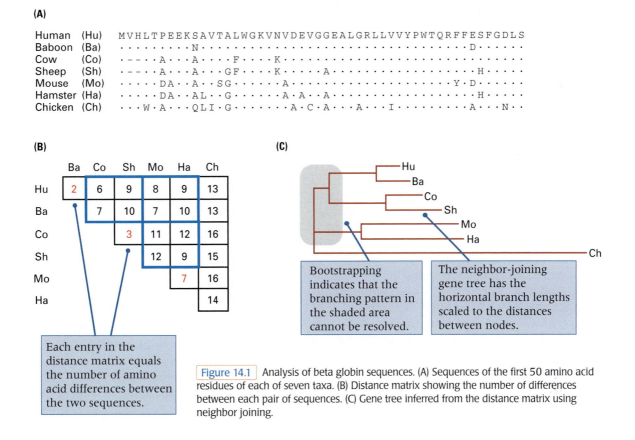

(A)

Human	(Hu)	MVHLTPEEKSAVTALWGKVNVDEVGGEALGRLLVVYPWTQRFFESFGDLS
Baboon	(Ba)	· · · · · · · · · · N · D · · · · · ·
Cow	(Co)	· — · · · A · · A · · · · F · · · K ·
Sheep	(Sh)	· — · · · A · · A · · · GF · · · K · · · · A · · · · · · · · · · · · · · · · · H · · · · ·
Mouse	(Mo)	· · · · · DA · · A · · SG · · · · · A · · · · · · · · · · · · · · · Y · D · · · · ·
Hamster	(Ha)	· · · · · DA · · AL · · G · · · · · · A · A · · A · · · · · · · · · · · H · · · · ·
Chicken	(Ch)	· · · W · A · · · QLI · G · · · · · · · A · C · A · · · A · · · I · · · · · · · A · · · N · ·

(B)

	Ba	Co	Sh	Mo	Ha	Ch
Hu	2	6	9	8	9	13
Ba		7	10	7	10	13
Co			3	11	12	16
Sh				12	9	15
Mo					7	16
Ha						14

Each entry in the distance matrix equals the number of amino acid differences between the two sequences.

(C)

Bootstrapping indicates that the branching pattern in the shaded area cannot be resolved.

The neighbor-joining gene tree has the horizontal branch lengths scaled to the distances between nodes.

Figure 14.1 Analysis of beta globin sequences. (A) Sequences of the first 50 amino acid residues of each of seven taxa. (B) Distance matrix showing the number of differences between each pair of sequences. (C) Gene tree inferred from the distance matrix using neighbor joining.

chance depending on which mutations take place and their likelihood of being fixed in the population. (A mutant allele is said to be **fixed** if it replaces all other alleles in the population.) Because new mutations occur at random times, any two molecules separated by the same length of time may differ at more or fewer sites purely by chance. In practice, this problem can be minimized by avoiding sequences that are so closely related that their expected number of differences is of the same magnitude as the random variation. The second problem is that some mutations may increase the ability of the organism to survive and reproduce, and these mutations have a much better chance of being fixed in the population. This problem can be minimized by studying regions of DNA, RNA, or protein that are less likely to include favorable mutation, but it is often difficult to know where these regions are.

One frequent goal of molecular evolution is to estimate the pattern of evolutionary relationships among a set of sequences, which is called a *gene tree*. It is a gene tree because it is based on a single gene (or, in this example, part of a gene). As will be explained below, the pattern of evolutionary relationships among the sequences in a gene tree is not necessarily the same as the pattern of evolutionary relationships among the species. A number of methods can be used to estimate a gene tree, and standard software packages such as PAUP and

PHYLIP are available for doing the calculations. For example, one method assumes that the best gene tree is the tree that minimizes the number of mutations. Among the simplest methods are trees based on the distances between the sequences taken in pairs. The sequences are converted into a **distance matrix,** which provides the data for the analysis. A number of distance measures are in use, but for the distance matrix shown in part B the distance measure is straightforward: for each pair of sequences, the distance is the number of amino acid sites that differ between the sequences. If the proportion of sites that differ is large, then it is necessary to make a correction for the possibility that an amino acid site may have changed over the course of time, then changed back again to what it was originally. In this example the correction is unnecessary.

One way to estimate a gene tree from a distance matrix is known as **neighbor joining.** First the pair of taxa with the smallest distance is combined into a single group, and then a new distance matrix is calculated in order to identify the new pair of nearest neighbors that should be joined. Neighbor joining is conceptually simple and often yields a gene tree that is identical or very similar to the best trees produced by more complex methods. For the hemoglobin data, the gene tree based on neighbor joining is shown in part C, where the length of each horizontal line is scaled to distance.

■ Bootstrapping is a method of assigning a level of confidence to each node in a gene tree.

A gene tree is an estimate of the true pattern of evolutionary relationships among a set of sequences. It is only an estimate because there is random variation in the number of substitutions, and the true gene tree is unknown. As might be expected, shorter branches in a gene tree are less reliable than longer branches. But what criterion can be used to assess how reliable a particular branching order is? For example, in part C, do the data really imply that the lineage leading to mouse and hamster lineage split off from the common ancestor prior to the lineage leading to humans, baboons, cows, and sheep?

A common technique for assessing the reliability of a node in a gene tree is called **bootstrapping.** In this procedure, 1000 or more different data sets are constructed from the actual data by choosing sites at random. Bootstrap sampling is carried out "with replacement," which means that the same site can, by chance, be chosen two or more times. A bootstrap sample from the sequence in part A would therefore consist of a sample of 50 sites chosen at random with replacement. In a particular bootstrap sample of 50, 18 sites are expected to be present once, 9 twice, and 5 three or more times—and 18 sites not at all. Hence, if a branching in the gene tree is supported by many of the sites in the sequences, then gene trees from most bootstrap samples will include the same branching, but if a branching is supported by relatively few sites, the gene trees from many bootstrap samples will not include it. In the gene-tree in part C, fewer than 50 percent of 1000 bootstrap samples support the branching order included in the shaded area. In practical terms, what this means is that, insofar as this small region of the protein is concerned, the lineages became separated so closely in time that it cannot be resolved which taxon split off first.

■ A gene tree does not necessarily coincide with a species tree.

It seems reasonable to suppose that the evolutionary relationships among a set of genes from different species (the **gene tree**) must be the same as the evolutionary relationships among the species themselves (the **species tree**), because the genes are present in the genomes of the species. But this is not the case. The gene tree is not necessarily the same as the species tree.

One way in which the gene tree can differ from the species tree is shown in Figure 14.2. The ancestral population is originally fixed for the A_1 allele (top), but then an A_2 mutation occurs and the population becomes polymorphic. The polymorphism is maintained even as species 1 splits off and as species 2 and 3 become separated. However, loss of one allele (and fixation of the other) eventually takes place, and in this example species 1 and 2 become fixed for allele A_2 whereas species 3 becomes fixed for allele A_1. This means that the gene tree would group the alleles from species 1 and 2 as being the most closely related, whereas the species tree shows that species 2 and 3 are actually the most closely related.

The situation becomes even more complicated because of recombination. When genes can undergo recombination, it implies that different genes can have different evolutionary histories. Because recombination within a gene can occur, it is even possible for different parts of the same gene to have different evolutionary histories. This is one reason why the nonrecombining part of the Y chromosome and the entirety of mitochondrial DNA are

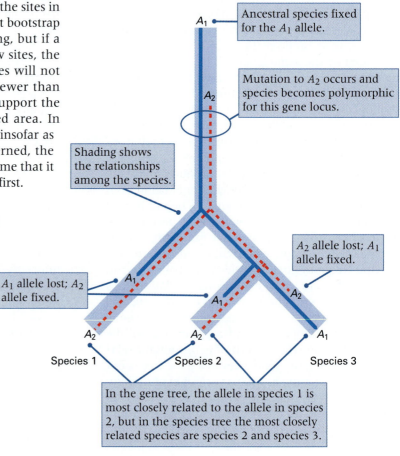

Figure 14.2 A gene tree may not coincide with a species tree because of the sorting of polymorphic alleles in the different lineages.

of such utility in tracing the recent history of human populations. However, these studies yield estimates of either the Y chromosome tree or the mitochondrial DNA tree, and it is important to bear in mind that the gene trees for nuclear genes will not necessarily be the same, and in fact may differ from one nuclear gene to the next.

On the other hand, for genes with polymorphisms that persist for relatively short times or for species that are sufficiently old, gene trees often do coincide with species trees. Discordance between gene trees and species trees is a potential problem primarily for genes that can maintain polymorphisms for long periods of time or for species that are closely related. The ancient polymorphism problem is exemplified by many genes that function in the immune system that are highly polymorphic and have been polymorphic for periods that are as long or longer than the time required for the formation of new species. In such cases, the allele sorting process in Figure 14.2 can result in discrepancies between the gene tree and the species tree. The young taxa problem is exemplified by the branching order between human, chimpanzee, and gorilla. The species are still so relatively young that polymorphisms in the ancestral populations have been sorted so that some genes support one branching order and other genes support another. (As data have accumulated, however, the majority of gene trees have supported a branching order in which the gorilla splits off first.)

■ Rates of evolution can differ dramatically from one protein to another.

By the *rate* of sequence evolution of a molecule, we mean the fraction of sites that undergo a change in some designated interval of time. For example, in the entire beta globin molecule between mouse and human, the rate of sequence evolution is 1.23 amino acid replacements per amino acid site per billion years, or 1.23×10^{-9} amino acid replacements per amino acid site per billion years. The mouse and human lineages have been separated for an estimated 80 million years, which means that they last shared a common ancestor about 80 million years ago. The total time separating the mouse and human molecules is therefore 160 million years (80 million years in the mouse lineage and 80 million years in the human lineage), and the entire beta hemoblogin chain is 147 amino acids. Therefore, a rate of 1.23×10^{-9} amino acid replacements per amino acid site per billion years implies that the expected number of differences between the molecules is $(1.23 \times 10^{-9}) \times (160 \times 10^6) \times 147 = 29$. If these differences were uniformly distributed along the molecule, the expected number of differences in the first 50 amino acids would be 10, with which the observed number 8 (Figure 14.1, part A) agrees quite well.

Different proteins (and sometimes different parts of proteins) evolve at very different rates. Molecules such as the histones H3 and H4 evolve very slowly. For example, among the 103 amino acids in histone H4, the molecules in rice and humans differ at only two sites. Other molecules evolve relatively rapidly. For example, between mouse and humans the antiviral protein gamma interferon has evolved at a rate of about 5×10^{-9} amino acid replacements per amino acid site per billion years, a rate approximately 4 times faster than beta globin and very much faster than histone H4 (there are no differences in histone H4 between mouse and human).

Although protein sequences evolve at very different rates, some proteins in some taxa show a rough constancy in their rate of amino acid replacement over long periods of evolutionary time. The apparently constant rate of sequence change has been called a **molecular clock,** and affords a basis for attaching a time scale to a gene tree and therefore a time scale for the branching of species independent of the fossil record. There is an elegant theoretical argument that explains why a constant rate of sequence evolution might be expected, as least in the simplest cases. Consider a gene in a population of N diploid individuals, so that in the entire population there are $2N$ copies of the gene. Suppose that some new mutations are **selectively neutral,** which means that they have no effects on the ability of the organisms to survive and reproduce, and that the rate of mutation to selectively neutral alleles is μ per gene per generation. Then in any one generation, the expected number of new selectively neutral alleles is $2N\mu$. As time goes on, because the population is finite in size, some of the lineages of these genes will become extinct by chance and they will be replaced with other gene lineages. Eventually a time will come when all the gene lineages will have become extinct except one. The probability that any particular gene lineage replace all others is $1/(2N)$, and because there are $2N\mu$ new mutations, the expected number of new mutations (sequence changes) that become fixed in each generation is:

$$\text{Rate of neutral evolution} = \frac{2N\mu}{2N} = \mu$$

The expected rate of neutral evolution is therefore equal to the rate of neutral mutation, which constitutes a sort of molecular clock whose ticks are mutations that become fixed.

In applying this model there are some important caveats. One warning is that the neutral mutation rate can differ from one gene to the next according to what fraction of new mutations is neutral or nearly neutral. For genes that evolve extremely slowly, like histone H4, most amino acid replacements are probably very deleterious, and so the neutral mutation rate will be very low. On the other hand, for genes that

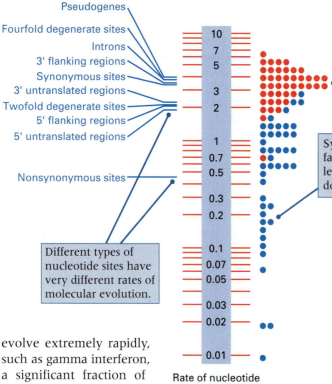

Pseudogenes
Fourfold degenerate sites
Introns
3' flanking regions
Synonymous sites
3' untranslated regions
Twofold degenerate sites
5' flanking regions
5' untranslated regions

Nonsynonymous sites

10
7
5
3
2
1
0.7
0.5
0.3
0.2
0.1
0.07
0.05
0.03
0.02
0.01

Rate of nucleotide
substitution per site
per billion years

● Synonymous sites
● Nonsynonymous sites

Figure 14.3 Rates of nucleotide substitution in mammalian genes. The left side shows the average rates for different classes of sequence; the right side shows the rates for synonymous and nonsynonymous sites in a sample of 43 genes.

Synonymous sites evolve much faster on the average, and with less variation among genes, than do nonsynonymous sites.

Different types of nucleotide sites have very different rates of molecular evolution.

evolve extremely rapidly, such as gamma interferon, a significant fraction of the amino acid replacements that become fixed may be favorable mutations, which violate the assumption of neutrality. A second caveat is that the molecular clock is not like a timepiece that ticks at reproducible intervals. It is a random or stochastic clock, in which only the average interval between ticks is predictable. An analogy may be made with radioactive decay, which is a random but clocklike process, but even this analogy has the shortcoming that the random variation in a molecular clock is much larger, relative to the mean, than the random variation in an atomic clock. A third warning is that in many cases molecular clocklike behavior is not observed, because there are different rates of sequence evolution along different branches of the gene tree. For example, in the lineages of humans and mice, the rate of sequence evolution along the branch leading to mice is about two times faster than that along the branch leading to human, which is thought to be due to the shorter generation time of organisms in the mouse lineage, resulting in more generations along that lineage. The best methods for reconstructing genes trees take this source of variation into account.

■ **Rates of evolution of nucleotide sites differ according to their function.**

The general principles of the molecular evolution of protein sequences also apply to DNA sequences, but there are some important differences. Proteins consist of 20 amino acids, but DNA consists of only 4 nucleotides. The smaller number of subunits means

that independent but identical changes at nucleotide sites are much more likely than at amino acid sites, and also that a DNA site that undergoes two sequential mutations is more likely to return to the original state than is an amino acid site. In analyzing DNA sequences, these kinds of events often require that corrections be made to the distance matrix.

There are also different kinds of nucleotide sites depending on their position and function in the genome, and these evolve at different rates (Figure 14.3). Note that the vertical scale in Figure 14.3 is logarithmic and covers three orders of magnitude, so the differences are very large. On the right are shown the rates of synonymous substitution (red) and nonsynonymous substitution (blue) in each of 43 genes. A *synonymous substitution* in a coding sequence does not result in an amino acid replacement. Synonymous sites are sites at which synonymous substitutions can occur, primarily at the third codon position. A *nonsynonymous substitution* does result in an amino acid replacement. Nonsynonymous nucleotide sites occur primarily at first and second codon positions. Reflecting the great variation in rate of amino acid replacement among different proteins, the rates of nonsynonymous substitution in Figure 14.3 are highly variable among genes. The rates of synonymous substitution in the same genes are much less variable and also much faster.

Plotted on the left in Figure 14.3 are the average rates of nucleotide substitution for different classes of DNA sequence. The fastest evolving DNA sequences are those of **pseudogenes,** which are duplicate genes that have lost their function because of mutation. Introns and fourfold degenerate sites also evolve very rapidly. (A *fourfold degenerate site* is a synonymous nucleotide site at which the same amino acid is specified whatever the identity of the nucleotide; a *twofold degenerate site* is one at which the encoded amino acid depends only on whether the nucleotide at the site is a pyrimidine or a purine.) The differences in the average rate of nucleotide substitution among the different classes of DNA sequence are thought to reflect differing tolerance for nucleotide substitutions. DNA sequences in which most nucleotide substitutions are deleterious are relatively intolerant to nucleotide substitutions, and the rate of nucleotide substitution is relatively low. Conversely, in sequences in which many nucleotide substitutions are equivalent or nearly equivalent in their effects on survival and reproduction, the rate of nucleotide substitution is relatively high. The high rate of nucleotide substitution in pseudogenes is understandable from this point of view, as are the high rates in fourfold degenerate sites and introns.

■ New genes usually evolve through duplication and divergence.

In the course of evolution, new genes usually come from preexisting genes, and new gene functions evolve from previous gene functions. The raw material for new genes comes from duplications of regions of the genome, which may include one or more genes. Duplications take place relatively frequently. Analysis of the genomic sequences of a wide variety of eukaryotes suggests that a eukaryotic genome containing 30,000 genes may be expected to undergo roughly 60 to 600 duplications per million years.

From an evolutionary standpoint, two types of duplications need to be distinguished. The first is typified by beta globin in the gene tree in Figure 14.1. Each time a speciation event took place, which is represented by a branching of the tree, the beta globin gene became duplicated in the sense that each derived species has a copy of the beta globin gene that existed in the parental species. Genes that are duplicated as an accompaniment to speciation and that retain the same function are known as **orthologous genes.**

New gene functions can arise from duplications that take place in the genome of a single species. Duplications within a genome result in **paralogous genes.** An example of paralogous genes are the beta globin gene and the alpha globin gene. Although the genes are distinct now, they are sufficiently similar in sequence to show without ambiguity that they originated from a single gene in a remote common ancestor. Hence, the beta globin and alpha globin genes are paralogs. (To take this discussion one step further, the beta genes in any two mammalian species are orthologs, as are the alpha genes. However, the beta gene in one species and the alpha gene in another are paralogs.)

When a gene duplication has taken place, the paralogs are redundant and one is free to evolve along any path. Probably the most common event is that one of the paralogs undergoes a mutation that destroys its function, or a deletion that eliminates it. But occasionally mutations take place that cause the functions of the paralogs to diverge. They may

evolve different pH optima, for example, so that one gene product performs optimally in compartments of the cell that are relatively basic and the other in compartments that are relatively acidic. Or genetic rearrangements can fuse two unrelated genes and yield a new activity. We have already seen examples in Chapter 13 of how translocations that fuse a transcription factor with an oncogene can lead to acute leukemia. These rearrangements are extremely deleterious, but the creation of chimeric genes affords an example of how new functions can be acquired.

Gene duplications also allow the paralogous copies to evolve more specialized functions. An example is shown in Figure 14.4. At the top is a gene in a multicellular eukaryote that has enhancers for

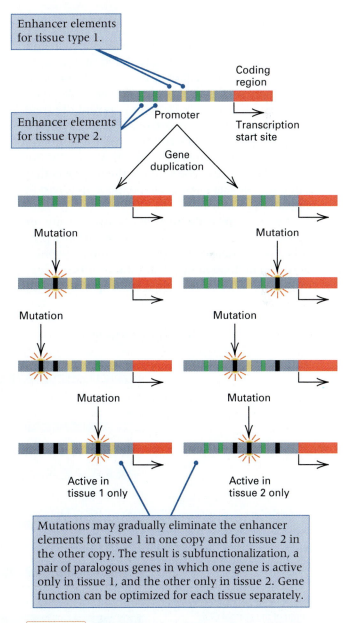

Mutations may gradually eliminate the enhancer elements for tissue 1 in one copy and for tissue 2 in the other copy. The result is subfunctionalization, a pair of paralogous genes in which one gene is active only in tissue 1, and the other only in tissue 2. Gene function can be optimized for each tissue separately.

Figure 14.4 Specialization of paralogous genes by subfunctionalization. In this example, the paralogs become specialized for tissue-specific expression.

expression in tissue types 1 (orange) and 2 (green). The other panels show a gene duplication followed by sequential mutations that knock out either the type-2 enhancer elements (left) or the type-1 enhancer elements (right). The result is that the paralog on the left is expressed only in tissue type 1 and that on the right only in tissue type 2. Specialization of paralogs accompanying loss of functional capabilities is known as **subfunctionalization.** It may often be advantageous because each of the specialized genes is free to evolve toward optimal function in its own domain of expression.

14.2

Genotypes may differ in frequency from one population to another.

Our discussion of molecular evolution focused on new mutations that become fixed in lineages because the rate of sequence evolution depends on how rapidly such fixations can take place. Between the time that a new mutation occurs and the time that it ultimately becomes fixed or lost are important processes that determine what the ultimate fate of the mutation will be and how rapidly the fate will be realized. The study of the processes that determine the fate of alleles in populations is part of population genetics.

The term **population** refers to a group of organisms of the same species living within a prescribed geographical area. Most widespread populations are subdivided by geographical or other features into smaller units called **subpopulations** or *local populations*. In the human population, for example, we may distinguish the subpopulation of people who live in the United States. Although matings occur primarily within each subpopulation, because of occasional migration between subpopulations they all share a common pool of genetic information called the **gene pool.** This section begins with an analysis of local populations with respect to a phenotype determined by two alleles. The phenotype is of considerable interest because it is associated with genetic resistance to AIDS (acquired immune deficiency syndrome).

■ Allele frequencies are estimated from genotype frequencies.

The genetic composition of a population can often be described in terms of the frequencies, or relative abundances, in which alternative alleles are found. The concepts can be illustrated with respect to an AIDS-resistance phenotype determined by a relatively common mutant allele. The gene in question is a chemokine receptor gene, *CCR5*, found in the human population. (*Chemokines* are molecules that white blood cells of the immune system use to attract one another.) The *CCR5* receptor enables the HIV virus to combine with the plasma membrane and infect the CD4(+) class of T cells of the immune system, which is necessary for an HIV infection to progress to full-blown AIDS. Most human subpopulations contain a *CCR5* allele known as $\Delta 32$ because it has a 32-bp deletion within the coding region. The molecular consequences of the deletion are shown in Figure 14.5 . The relevant part of the normal *CCR5* DNA

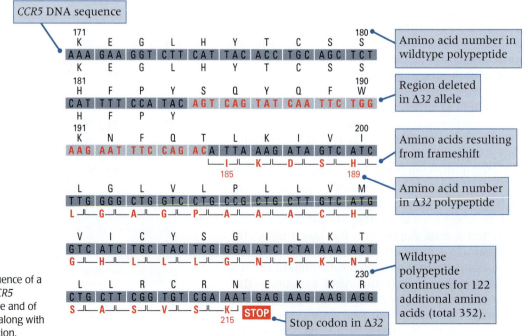

Figure 14.5 DNA sequence of a portion of the normal *CCR5* chemokine receptor gene and of the $\Delta 32$ deletion allele, along with their products of translation.

sequence is shown, grouped into codons, along with the amino acid sequence of the polypeptide, given in the single-letter abbreviations. The nucleotides missing in the $\Delta 32$ deletion are highlighted in red. Above the DNA sequence is the normal CCR5 polypeptide; below is the $\Delta 32$ mutant polypeptide. The deletion creates a frameshift in translation following codon 184, which results in the insertion of 31 incorrect amino acids until a termination codon is encountered after amino acid 215. The truncated protein is nonfunctional and does not support HIV entry into CD4(+) cells. The $\Delta 32$ mutation was originally discovered among persons infected with HIV-1 who had remained free of AIDS for at least 10 years. The usual frequency of heterozygous genotypes, as we shall see, is about 20 percent, but among AIDS nonprogressors the frequency is approximately 40 percent. The homozygous $\Delta 32$ genotypes, which are much less frequent, seem to have even greater protection.

Why is the $\Delta 32$ homozygous genotype much less frequent than the $\Delta 32$ heterozygous genotype? And what is the relationship between the homozygous and heterozygous frequencies? To begin to answer these questions, we should look at some data. For convenience, we will represent the normal CCR5 allele as A and the $\Delta 32$ deletion allele as a. In one study of 1000 French people whose DNA was genotyped for CCR5, the numbers of homozygous normal (AA), heterozygous $\Delta 32$ (Aa), and homozygous $\Delta 32$ (aa) individuals were as follows:

$$795\ AA \quad 190\ Aa \quad 15\ aa$$

These numbers contain a great deal of information about the population, such as whether it is a single homogeneous population or a mixture of genetically somewhat different subpopulations. To interpret this information, first note that the sample contains two types of data: (1) the number of each of the three genotypes, and (2) the number of individual CCR5 wildtype (A) and $\Delta 32$ (a) alleles.

Furthermore, the 1000 genotypes represent 2000 alleles of the CCR5 gene, because each human genome is diploid. These alleles break down as shown in Figure 14.6. Each homozygous AA genotype represents two A alleles, each homozygous aa genotype two a alleles, and each heterozygous Aa genotype one allele of each type. By the kind of allele counting shown in Figure 14.6, the sample of 1000 people therefore represents 1780 CCR5 normal (A) alleles and 220 $\Delta 32$ (a) alleles.

Usually, it is more convenient to analyze the data in terms of the relative frequencies of genotypes and alleles than in terms of the observed numbers. For genotypes, the **genotype frequency** in a population is the proportion of organisms that have the particular genotype. For any specified allele, the **allele frequency** is the proportion of all alleles that are of the specified type. In the CCR5 example, the genotype frequencies are obtained by dividing the observed numbers by the total sample size—in this case, 1000. Therefore, the genotype frequencies are

$$0.795\ AA \quad 0.190\ Aa \quad 0.015\ aa$$

Similarly, the allele frequencies are obtained by dividing the observed number of each allele by the total number of alleles (in this case, 2000):

$$\text{Allele frequency of } A = \frac{1780}{2000} = 0.89$$

$$\text{Allele frequency of } a = \frac{220}{2000} = 0.11$$

Note that the genotype frequencies add up to 1.0, as do the allele frequencies. This is a consequence of their definition in terms of proportions: They must add up to 1.0 when all of the possibilities are taken into account. An allele with a frequency of 1.0 is *fixed*, and an allele whose frequency has reached 0 is *lost*.

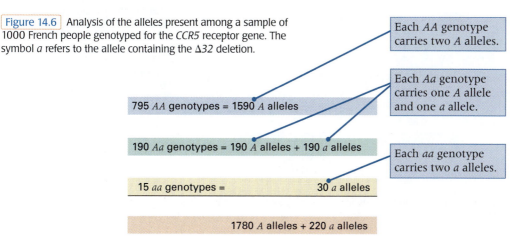

Figure 14.6 Analysis of the alleles present among a sample of 1000 French people genotyped for the CCR5 receptor gene. The symbol a refers to the allele containing the $\Delta 32$ deletion.

Each AA genotype carries two A alleles.

Each Aa genotype carries one A allele and one a allele.

Each aa genotype carries two a alleles.

795 AA genotypes = 1590 A alleles

190 Aa genotypes = 190 A alleles + 190 a alleles

15 aa genotypes = 30 a alleles

1780 A alleles + 220 a alleles

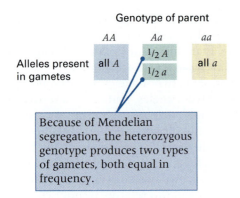

Genotype of parent

	AA	Aa	aa
Alleles present in gametes	all A	$1/2\ A$ $1/2\ a$	all a

Because of Mendelian segregation, the heterozygous genotype produces two types of gametes, both equal in frequency.

Figure 14.7 Mendelian considerations in population genetics. Each homozygous genotype produces gametes containing a single allele, but a heterozygous genotype, because of Mendelian segregation, produces two types of gametes in equal frequency.

■ The allele frequencies among gametes equal those among reproducing adults.

One question we can ask at this stage is whether the genotypes in the sample have the frequencies we might expect if the alleles were joined in pairs at random. Two A alleles (AA) would then be paired with a frequency of $0.89 \times 0.89 = 0.7921$, an A allele would be paired with an a allele (Aa) with a frequency of $2 \times 0.89 \times 0.11 = 0.1958$, and two a alleles (aa) would be paired with a frequency of $0.11 \times 0.11 = 0.0121$. These values are, in fact, quite close to the observed genotype frequencies. It is also noteworthy that the allele frequencies among adults, as we have calculated them in Figure 14.8, are the same as those among the gametes produced by the adults. This is true because Mendelian segregation ensures that each heterozygous genotype will produce equal numbers of each type of gamete (Figure 14.7). Thus when we calculate the gametic frequencies, the heterozygous genotypes contribute equally to both classes of gametes, whereas the

homozygous AA and homozygous aa genotypes contribute only A-bearing or only a-bearing gametes, respectively. Consequently, when the genotype frequencies among the parents are taken into account, as well as Mendelian segregation in the heterozygous Aa genotypes, the gametes produced in the population have the composition deduced in Figure 14.8. The equality between the allele frequencies among the adults and those among the gametes produced by the adults—namely 0.89 A and 0.11 a—must be true whenever each adult in the population produces the same number of functional gametes. The apparent random pairing of alleles in the adult genotypes is also important, because it is a clue as to how the gametes are united in fertilization. This issue is examined in the next section.

14.3

Random mating means that mates pair without regard to genotype.

When a local population undergoes **random mating,** it means that organisms in the local population form mating pairs independently of genotype. Each type of mating pair is formed as often as would be expected by chance encounters. Random mating is by far the most prevalent mating system for most species of animals and plants, except for plants that regularly reproduce through self-fertilization. Self-fertilization is an extreme example of another important type of mating system, which is called *inbreeding,* or mating between relatives. Figure 14.9 represents an example of inbreeding. In this case, the female I is the offspring of a first-cousin mating (G with H). The closed loop in the pedigree is diagnostic of inbreeding, and the individuals designated A and B are called *common ancestors* of I, because they are ancestors of both of the parents of I.

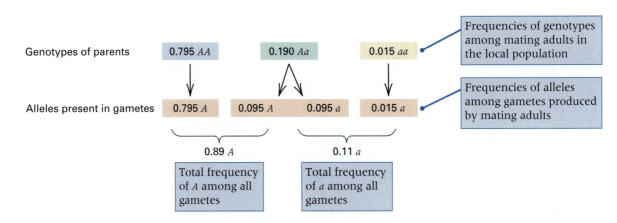

Figure 14.8 Calculation of the allele frequencies in gametes. In any population, the frequency of gametes containing any particular allele equals the frequency of the genotypes that are homozygous for the allele, plus one-half the frequency of all genotypes that are heterozygous for the allele.

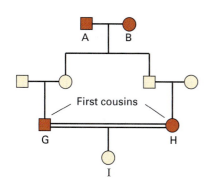

First cousins

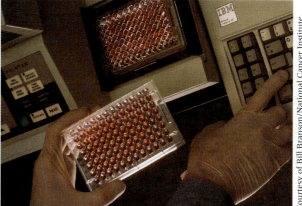

Figure 14.9 An inbreeding pedigree in which individual I is the result of a mating between first cousins.

Genetic variation in cellular responses to drugs can be tested by means of automated plate readers such as this one. In this particular test, each well contains growth medium plus a different drug, and a change in color is used to estimate the proportion of cancer cells that are killed.

Because A and B are common ancestors, a particular allele present in A (or in B) could, by chance, be transmitted in inheritance down both sides of the pedigree, to meet again in the formation of I. This possibility is the most important and characteristic consequence of inbreeding, and it will be discussed later in this chapter.

Random mating implies a pleasingly simple relationship between allele frequency and genotype frequency. This is summarized in the following principle:

> **key concept**
>
> Random mating of individuals is equivalent to the random union of gametes.

On the basis of this principle, we may imagine random mating to be equivalent to drawing female and male gametes at random from a large container. Each zygote genotype is formed from the pairing of one female and one male gamete. To be specific, consider two alleles A and a with allele fre-

quencies p and q, respectively, where $p + q = 1$. In the sample of $CCR5$ receptor genotypes considered earlier in this chapter, we calculated

$$p = 0.89 \quad \text{for the } A \text{ allele}$$
$$q = 0.11 \quad \text{for the } a \text{ allele}$$

The genotype frequencies expected with random mating can be deduced from the tree diagram in Figure 14.10. The gametes at the left represent the eggs and those in the middle the sperm. The genotypes that can be formed with two alleles are shown at the right, and with random mating, the frequency of each genotype is calculated by multiplying the allele frequencies of the corresponding gametes. However, the genotype Aa can be formed in two ways: The A allele could have come from the mother (top part of diagram) or from the father

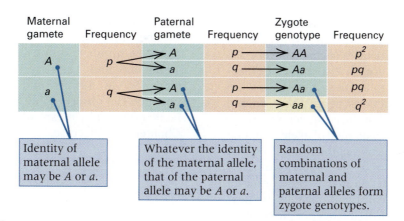

Figure 14.10 When gametes containing either of two alleles unite at random to form the next generation, the genotype fre- quencies among the zygotes are given by the ratio $p^2 : 2pq : q^2$. This constitutes the Hardy–Weinberg principle.

14.3 Random mating means that mates pair without regard to genotype   **503**

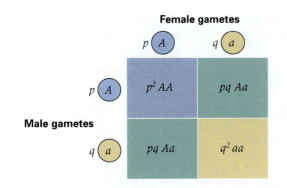

Female gametes

	p A	q a
p A	p^2 AA	pq Aa
q a	pq Aa	q^2 aa

Male gametes

Figure 14.11 A Punnett square showing, in a cross-multiplication format, the ratio $p^2 : 2pq : q^2$ that is characteristic of random mating.

(bottom part of diagram). In each case, the frequency of the Aa genotype is pq; considering both possibilities, we find that the frequency of Aa is $pq + pq = 2pq$. Consequently, the overall genotype frequencies expected with random mating are

$$AA: p^2 \quad Aa: 2pq \quad aa: q^2 \quad (14.1)$$

The frequencies p^2, $2pq$, and q^2 result from random mating for a gene with two alleles; they constitute the **Hardy–Weinberg principle,** named after Godfrey Hardy and Wilhelm Weinberg, who derived it independently of each other in 1908. Sometimes the Hardy–Weinberg principle is demonstrated by the type of Punnett square illustrated in Figure 14.11 . Such a square is completely equivalent to the tree diagram used in Figure 14.10. Although the Hardy–Weinberg principle is exceedingly simple, it has a number of important implications that are not obvious. These are described in the following sections.

■ The Hardy–Weinberg principle has important implications for population genetics.

One important implication of the Hardy–Weinberg principle is that *the allele frequencies remain constant from generation to generation.* To understand why, consider again a gene with two alleles, A and a, having frequencies p and q, respectively ($p + q = 1$). With random mating, the frequencies of genotypes AA, Aa, and aa among zygotes are p^2, $2pq$, and q^2, respectively. Assuming equal ability to survive among the genotypes, these frequencies equal those among adults. If, in addition, all of the adult genotypes are equally fertile, then the frequency of allele A among gametes that form the zygotes of the next generation can be calculated in terms of the frequency of the A allele in the previous generation. If we use a prime (') to denote the frequency of the A allele in the next generation, then the allele frequency is p'. In terms of the alleles present in the previous generation, p' includes all of the A alleles in homozygous AA geno-

types (frequency p^2) plus half of the alleles in heterozygous Aa genotypes (frequency $2pq$). The Aa heterozygotes are multiplied by $1/2$ because of Mendelian segregation; only half of the gametes from Aa genotypes carry A. Putting all this together, the frequency p' of the A allele in the next generation is

$$p' = p^2 + 2pq/2 = p(p + q) = p$$

The final equality follows because $p + q = 1$. We have therefore shown that the frequency of allele A remains constant at the value of p through the passage of one or more complete generations. This principle depends on certain assumptions, of which the most important are the following:

- Mating is random; there are no subpopulations that differ in allele frequency.
- Allele frequencies are the same in males and females.
- All the genotypes are equal in survival and fertility (*selection* does not operate).
- Mutation does not occur.
- Migration into the population is absent.
- The population is sufficiently large that the frequencies of alleles do not change from generation to generation because of chance.

As a practical application of the Hardy–Weinberg principle, consider again the *CCR5* receptor discussed in Section 14.2. The frequencies of the *CCR5* normal allele (A) and the $\Delta32$ allele (a) among 1000 adults were 0.89 and 0.11, respectively. Assuming random mating, the expected genotype frequencies can be calculated from Equation (14.1) as

$$AA: \quad (0.89)^2 = 0.7921$$
$$Aa: \quad 2(0.89)(0.11) = 0.1958$$
$$aa: \quad (0.11)^2 = 0.0121$$

Note that these are the same frequencies calculated earlier on the basis of the supposition that the alleles were joined in pairs at random. Now we know this means that the genotype frequencies are given by the Hardy–Weinberg principle. Because the total number of people in the sample is 1000, the expected numbers of the genotypes are 792.1 AA, 195.8 Aa, and 12.1 aa. The *observed* numbers were 795 AA, 190 Aa, and 15 aa. Goodness of fit between the observed and expected numbers can be determined by means of the χ^2 test described in Chapter 3. In this case,

$$\chi^2 = \frac{(795 - 792.1)^2}{792.1}$$
$$+ \frac{(190 - 195.8)^2}{195.8} + \frac{(15 - 12.1)^2}{12.1}$$
$$= 0.877$$

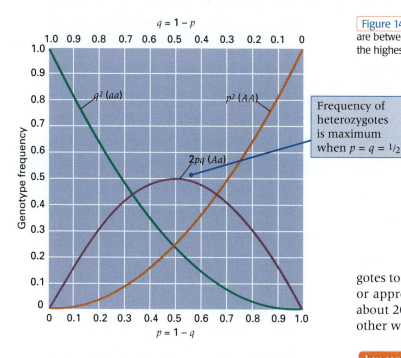

Figure 14.12 Graphs of p^2, $2pq$, and q^2. If the allele frequencies are between 1/3 and 2/3, the heterozygote is the genotype with the highest frequency.

The fit is obviously satisfactory, but to evaluate it quantitatively we need to use the chart in Figure 3.34 on page 109. There is also an adjustment needed in the degrees of freedom. Normally, with 3 classes of data, we would have 2 degrees of freedom. But when any quantity, such as an allele frequency, is estimated from the data, 1 degree of freedom must be deducted for each quantity estimated. In this case we estimated the allele frequency p ($q = 1 - p$ follows automatically), so we lose 1 of the 2 degrees of freedom we would otherwise have, and thus the appropriate number of degrees of freedom for the test is 1. For this we obtain a P value of 0.35, which means that the hypothesis of random mating can account for the data. On the other hand, the χ^2 test detects only deviations that are rather large, so a good fit to Hardy–Weinberg frequencies should not be over interpreted, because

key concept

It is entirely possible for one or more assumptions of the Hardy–Weinberg principle to be violated, including the assumption of random mating, and still not produce deviations from the expected genotype frequencies that are large enough to be detected by the χ^2 test.

■ **If an allele is rare, it is found mostly in heterozygous genotypes.**

Another important implication of the Hardy–Weinberg principle is that *for a rare allele, the frequency of heterozygotes far exceeds the frequency of the rare homozygote.* For example, when the frequency of the rarer allele is $q = 0.1$, the ratio of heterozygotes to homozygotes equals $2pq/q^2 = 2(0.9)/(0.1)$, or approximately 20; when $q = 0.01$, this ratio is about 200; and when $q = 0.001$, it is about 2000. In other words,

key concept

When an allele is rare, there are many more heterozygotes than there are homozygotes for the rare allele.

The reason for this perhaps unexpected relationship is shown in Figure 14.12, which plots the frequencies of homozygous and heterozygous genotypes with random mating. Note that at allele frequencies near 0 or 1, the frequency of the heterozygous genotype goes to 0 much more slowly than does the frequency of the rarer homozygous genotype.

One practical implication of this principle is seen in the example of cystic fibrosis, an inherited secretory disorder of the pancreas and lungs. Cystic fibrosis is one of the most common recessively inherited severe disorders among Caucasians; it affects about 1 in 1700 newborns. In this case, the heterozygotes cannot readily be identified by phenotype, so a method of calculating allele frequencies that is different from the gene-counting method used earlier must be applied. This method is straightforward because with random mating, the frequency of recessive homozygotes must correspond to q^2. In the case of cystic fibrosis,

$$q^2 = 1/1700 = 0.00059$$
$$\text{or}$$
$$q = (0.00059)^{1/2} = 0.024$$

and consequently,

$$p = 1 - q = 1 - 0.024 = 0.976$$

The frequency of heterozygous genotypes that carry a mutant allele for cystic fibrosis is calculated as

$$2pq = 2(0.976)(0.024) = 0.047 = 1/21$$

This calculation implies that for cystic fibrosis, although only 1 person in 1700 is affected with the disease (homozygous), about 1 person in 21 is a carrier (heterozygous). The calculation should be regarded as approximate because it is based on the assumption of Hardy–Weinberg genotype frequencies. Nevertheless, considerations like these are important in predicting the outcome of population screening for the detection of carriers of harmful recessive alleles, which is essential in evaluating the potential benefits of such programs.

■ Hardy–Weinberg frequencies can be extended to multiple alleles.

Extension of the Hardy–Weinberg principle to multiple alleles of a single autosomal gene can be illustrated by a three-allele case. Figure 14.13 shows the results of random mating in which three alleles are considered. The alleles are designated A_1, A_2, and A_3, where the uppercase letter represents the gene and the subscript designates the particular allele. The allele frequencies are p_1, p_2, and p_3, respectively. With three alleles (as with any number of alleles), the allele frequencies of all alleles must sum to 1; in this case, $p_1 + p_2 + p_3 = 1.0$. As in Figure 14.11, the entry in each square is obtained by multiplying the frequencies of the alleles at the corresponding margins; any homozygous genotype (such as A_1A_1) has a random-mating frequency equal to the square of the corresponding allele frequency (in this case, p_1^2). Any heterozygous genotype (such as A_1A_2) has a random-mating frequency equal to twice the product of the corresponding allele frequencies (in this case, $2p_1p_2$). The extension to any number of alleles is straightforward:

Frequency of any homozygous genotype
= square of allele frequency

Frequency of any heterozygous genotype
= 2 × product of allele frequencies   (14.2)

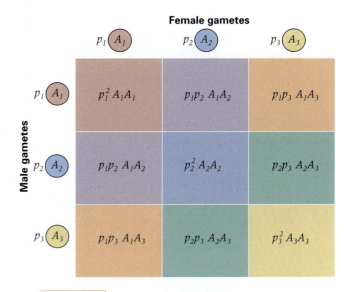

Multiple alleles determine the human ABO blood groups (Chapter 2). The gene has three principal alleles, designated I^A, I^B, and I^O. In one study of 3977 Swiss people, the allele frequencies were found to be 0.27 I^A, 0.06 I^B, and 0.67 I^O. Applying the rules for multiple alleles, we can expect the genotype frequencies that result from random mating to be as shown in Figure 14.14. Because both I^A and I^B are dominant to I^O, the expected frequency of blood-group *phenotypes* is that shown at the right in the illustration. Note that the majority of A and B phenotypes are actually heterozygous for the I^O allele; this is because the I^O allele has such a high frequency in the population.

■ X-linked genes are a special case because males have only one X-chromosome.

The implications of random mating for two X-linked alleles (H and h) are illustrated in Figure 14.15. The principles are the same as those

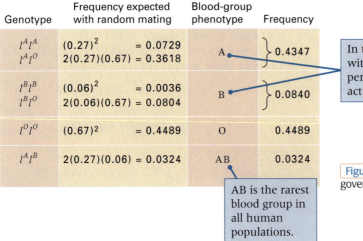

In this population, 83% of persons with blood group A and 96% of persons with blood group B are actually heterozygous for I^O.

AB is the rarest blood group in all human populations.

Figure 14.14 Random-mating frequencies for the three alleles governing the ABO blood groups.

(A)

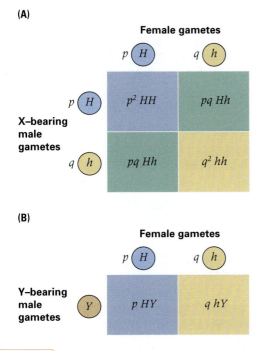

Female gametes

Figure layout: X-bearing male gametes ($p\,H$, $q\,h$) crossed with Female gametes ($p\,H$, $q\,h$):
- $p^2\,HH$, $pq\,Hh$
- $pq\,Hh$, $q^2\,hh$

(B)

Female gametes

Y-bearing male gametes (Y) crossed with Female gametes ($p\,H$, $q\,h$):
- $p\,HY$, $q\,hY$

Figure 14.15 The results of random mating for an X-linked gene. (A) Genotype frequencies in females. (B) Genotype frequencies in males.

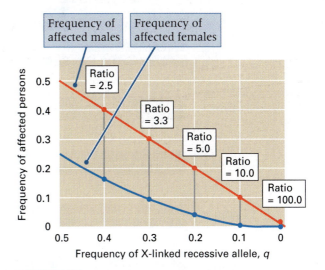

Figure 14.16 For a recessive X-linked allele, the upper curve gives the frequency of affected males (q), and the lower curve the frequency of affected females (q^2), for values of $q < 0.5$. Although both frequencies decrease as the recessive allele becomes rare, the frequency of affected females decreases more rapidly. The result is that the ratio of affected males to affected females increases as the allele frequency decreases.

considered earlier, but male gametes carrying the X chromosome (part A) must be distinguished from those carrying the Y chromosome (part B). When the male gamete carries an X chromosome, the Punnett square is exactly the same as that for the two-allele autosomal gene in Figure 14.11. However, because the male gamete carries an X chromosome, all the offspring in question are female. Consequently, among females, the genotype frequencies are

Frequency of HH females $= p^2$
Frequency of Hh females $= 2pq$
Frequency of hh females $= q^2$

When the male gamete carries a Y chromosome, the outcome is quite different (Figure 14.15, part B). All the offspring are male, and each has only one X chromosome, which is inherited from the mother. Therefore, each male receives only one copy of each X-linked gene, and the genotype frequencies among males are the same as the allele frequencies:

Frequency of H males $= p$
Frequency of h males $= q$

An important implication of Figure 14.15 is that if h is a rare recessive allele, then there will be many more males exhibiting the trait than females because the frequency of affected females (q^2) will be much smaller than the frequency of affected males (q). This principle is illustrated in Figure 14.16.

As the allele frequency of the recessive decreases toward 0, the frequencies of affected males and females both decrease, but the frequency of affected females decreases faster. The result is that the ratio of affected males to affected females increases. At an allele frequency of $q = 0.3$, for example, the ratio of affected males to affected females is 3.33; but for an allele frequency of $q = 0.1$, the ratio of affected males to affected females is 10.0. In general, the ratio of affected males to affected females is q/q^2, or $1/q$.

For an X-linked recessive trait, the frequency of affected males provides an estimate of the frequency of the recessive allele. A specific example is found in the common form of X-linked color blindness in human beings. This trait affects about 1 in 20 males among Caucasians, so $q = 1/20 = 0.05$. The expected frequency of color-blind females is therefore estimated as $q^2 = (0.05)^2 = 0.0025$, or about 1 in 400.

14.4

Highly polymorphic sequences are used in DNA typing.

Many genes in human populations are **polymorphic**, which means that they have two or more alleles that are common in the population. (Quantitatively speaking, a gene is generally regarded as polymorphic if the frequency of heterozygous genotypes is 10 percent or greater.) Averaged across the human genome, about 1 per thousand nucleotide sites differs from one individual to the next at a frequency high enough to be

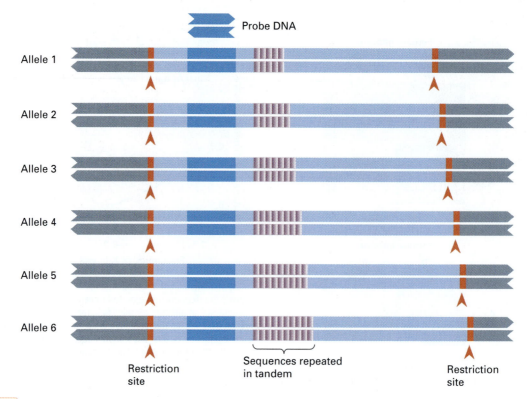

Probe DNA

Allele 1

Allele 2

Allele 3

Allele 4

Allele 5

Allele 6

Restriction
site

Sequences repeated
in tandem

Restriction
site

Figure 14.17 Allelic variation resulting from a variable number of units repeated in tandem in a nonessential region of a gene. The probe DNA detects a restriction fragment for each allele. The length of the fragment depends on the number of repeating units present.

regarded as polymorphic. Owing to the high level of genetic variation, it is virtually impossible for two human beings to be genetically identical. The only exceptions are identical twins, identical triplets, and so forth. Each human genotype is unique. The theoretical principle of genetic uniqueness has become a practical reality through the study of DNA polymorphisms. Small samples of human material from an unknown person (for example, material left at the scene of a crime) often contain enough DNA that the genotype can be determined for a number of polymorphisms and matched against those present among a group of suspects. Typical examples of crime-scene evidence include blood, semen, hair roots, and skin cells. Even a small number of cells is sufficient, because predetermined regions of DNA can be amplified by the polymerase chain reaction (Chapter 6).

If a suspect's DNA contains sequences that are clearly not present in the crime-scene sample, then the sample must have originated from a different person. On the other hand, if a suspect's DNA *does* match that of the crime-scene sample, then the suspect could be the source. The strength of the DNA evidence depends on the number of polymorphisms that are examined and the number of alleles present in the population. The greater the number of polymorphisms that match, especially if they are highly polymorphic, the stronger the evidence linking the suspect to the sample taken from the scene of the crime. The use of polymorphisms in DNA to link suspects with samples of human material is called **DNA typing.** This method of identifying individuals is generally regarded as the most important innovation in criminal investigation since the development of fingerprinting more than a century ago.

Figure 14.17 illustrates one type of polymorphism used in DNA typing. The restriction fragments corresponding to each allele differ in length because they contain different numbers of units repeated in tandem. A polymorphism of this type is called an **SSR,** which stands for **simple sequence repeat.** SSRs are abundant in the human genome (Table 14.1). They account for about 3 percent of the total DNA and have a density of about one SSR per 2 kb. Genetic markers based on SSRs have been the workhorse of mapping human disease genes, particularly those based on repeats of the dinucleotide 5'-CA-3'.

The SSRs used in DNA typing usually have longer repeat units (14 to 500 bases) than those used in genetic mapping (2 to 11 bases). SSRs are of value in DNA typing because many alleles are possible, owing to the variable number of repeats present at the site from one chromosome to the next. Although many alleles may be present in the population as a whole, each person can have no more than two alleles for each SSR polymorphism. An example of an SSR used in DNA typing is shown in Figure 14.18. The lanes in the gel labeled M contain

Table 14.1

Simple sequence repeats (SSRs) in the human genome

Length of repeat unit	Average number of occurrences per Mb	Average number of repeats per occurrence
1	36.7	45
2	43.1	59
3	11.8	29
4	32.5	26
5	17.6	31
6	15.2	15
7	8.4	15
8	11.1	13
9	8.6	12
10	8.6	18
11	8.7	8

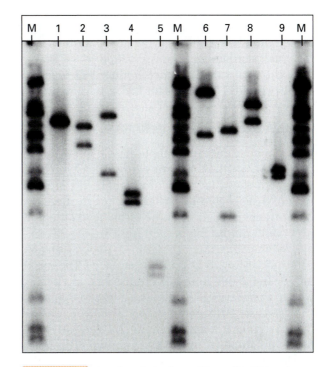

Figure 14.18 Genetic variation in an SSR used in DNA typing. Each numbered lane contains DNA from a single person; the DNA has been cleaved with a restriction enzyme, separated by electrophoresis, and hybridized with radioactive probe DNA. The lanes labeled M contain molecular-weight markers. [Courtesy of R. W. Allen.]

multiple DNA fragments of known size to serve as molecular-weight markers. Each numbered lane 1 through 9 contains DNA from a single person. Two typical features of SSRs are to be noted:

1. Most people are heterozygous for SSR alleles that produce restriction fragments of different sizes. Heterozygosity is indicated by the presence of two distinct bands. In Figure 14.18, only the person numbered 1 appears to be homozygous for a particular allele.

2. The restriction fragments from different people cover a wide range of sizes. The variability in

size indicates that the population as a whole contains many SSR alleles.

The reason why SSRs are useful in DNA typing is also evident in Figure 14.18: Each of the nine people tested has a different pattern of bands and thus could be identified uniquely by means of this SSR. On the other hand, the uniqueness of each

 A Moment to Think

Problem: In the early-morning hours of July 17, 1918, on orders of Bolshevik leader Vladmir Lenin, the last Russian Tsar, Nicholas II, was machine gunned to death in the basement of a home in Ekaterinburg, Russia. Executed with him were his wife, Tsarina Alexandra, their four daughters Olga, Tatiana, Marie, and Anastasia, and their hemophiliac son the Tsarevitch Alexis. Also executed were the family's personal physician, a male, and three female servants. In 1979 the remains of nine human beings were found in a shallow grave in the area where the bodies were rumored to have been disposed of, and in 1991 they became available for DNA testing. The gel pattern obtained for one SSR probe is shown in the accompanying diagram. The bands are numbered in order of increasing electrophoretic mobility. Degree of bone development and presence of Y-chromosome-specific DNA indicated that the remains included those of four adult females (A–D), two adult males (E–F), and three juvenile females (G–I).
(a) Are the DNA typing results consistent with the presence of six unrelated adult persons?
(b) Are they consistent with including a mother, father, and three of their daughters?
(c) Identify the lanes containing DNA from Tsar Nicholas II and Tsarina Alexandra.
(The answers can be found on page 516.)

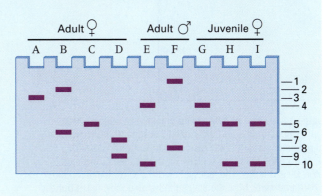

14.4 Highly polymorphic sequences are used in DNA typing   509

person in Figure 14.18 is due in part to the high degree of polymorphism of the SSR and in part to the small sample size. If more people were examined, then pairs that matched in their SSR types by chance would certainly be found.

■ DNA exclusions are definitive.

DNA typing cannot only implicate the guilty, it can also exonerate the innocent whose DNA type does *not* match the evidence. For example, if the DNA type of a suspected rapist fails to match the DNA type of semen taken from the victim, then the suspect could not be the perpetrator. Another example, which illustrates the use of DNA typing in paternity testing, is given in Figure 14.19 . The numbers 1 and 2 designate different cases, in each of which a man was alleged to have fathered a child. In each case, DNA was obtained from the mother (M), the child (C), and the man (A). The pattern of bands obtained for one SSR is shown. The lanes labeled A + C contain a mixture of DNA from the

man and the child. In case 1, the lower band in the child was inherited from the mother and the upper band from the father; because the upper band is the same size as one of those in the alleged father, he could have contributed this allele to the child. This finding does not prove that this individual is the father; it says only that he cannot be excluded on the basis of this particular SSR. (However, if enough SSRs are studied, and the man cannot be excluded by any of them, it does make it more likely that he really is the father.) Case 2 in Figure 14.19 is an exclusion. The small band lowest in the gel is the band inherited by the child from the mother. The other band in the child does not match either of the bands in the accused father, so the accused man could not be the biological father. (In theory, mutation could be invoked to explain the result, but this is extremely unlikely.)

14.5

Inbreeding means mating between relatives.

Inbreeding means mating between relatives, such as first cousins. The principal consequence of inbreeding is that the frequency of heterozygous offspring is smaller than it is with random mating. This effect is seen most dramatically in repeated self-fertilization, which occurs naturally in certain plants. To understand why the frequency of heterozygous genotypes decreases with self-fertilization, consider a hypothetical initial population consisting exclusively of *Aa* heterozygotes (Figure 14.20). With self-fertilization, each plant would produce offspring in the proportions 1/4 *AA*, 1/2 *Aa*, and 1/4 *aa*. Thus one generation of self-fertilization reduces the proportion of heterozygotes from 1 to 1/2. In the second generation, only the heterozygous plants can again produce hetero-

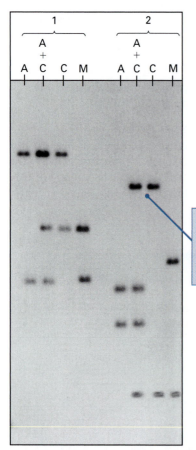

Figure 14.19 Use of DNA typing in paternity testing. The sets of lanes numbered 1 and 2 contain DNA samples from two different paternity cases. In each case, the lanes contain DNA fragments from the following sources: M, the mother; C, the child; A, the alleged father. The lanes labeled A + C contain a mixture of DNA fragments from the alleged father and the child. [Courtesy of R. W. Allen.]

AA	Aa	aa	Genotype
0	1	0	Initial frequency
$\frac{1}{4}$	$\frac{1}{2}$	$\frac{1}{4}$	After one generation of self-fertilization
$\frac{3}{8}$	$\frac{1}{4}$	$\frac{3}{8}$	After two generations of self-fertilization
$\frac{7}{16}$	$\frac{1}{8}$	$\frac{7}{16}$	After three generations of self-fertilization
$\frac{15}{32}$	$\frac{1}{16}$	$\frac{15}{32}$	After four generations of self-fertilization

With repeated self-fertilization, the frequency of heterozygous genotypes is reduced by half in each successive generation.

Figure 14.20 Effects of repeated self-fertilization on the genotype frequencies. In each generation, the proportion of heterozygous genotypes decreases to half of its value in the previous generation.

zygous offspring, and only half of their offspring will again be heterozygous. Heterozygosity is therefore reduced to 1/4 of what it was originally. Three generations of self-fertilization reduce the heterozygosity to $1/4 \times 1/2 = 1/8$, and so forth. The remainder of this section demonstrates how the reduction in heterozygosity because of inbreeding can be expressed quantitatively.

Repeated self-fertilization is a particularly intense form of inbreeding, but weaker forms of inbreeding are qualitatively similar in that they also lead to a reduction in heterozygosity. A convenient measure of the effect of inbreeding is based on the reduction in heterozygosity. Suppose that H_I is the frequency of heterozygous genotypes in a population of inbred organisms. The most widely used measure of inbreeding is called the **inbreeding coefficient,** symbolized F, which is defined as the proportionate reduction in H_I compared with the value of $2pq$ that would be expected with random mating:

$$F = (2pq - H_I)/2pq$$

This equation can be rearranged as

$$H_I = 2pq(1 - F)$$

which says that in a population of organisms having an inbreeding coefficient of F, the proportion of heterozygous genotypes is reduced by the fraction F relative to what it would be with random mating. As the proportion of heterozygous genotypes de-

creases in frequency, the proportion of homozygous genotypes increases correspondingly. The overall genotype frequencies in an inbred population are given by

Frequency of AA genotype $= p^2(1 - F) + pF$

Frequency of Aa genotype $= 2pq(1 - F)$ (14.3)

Frequency of aa genotype $= q^2(1 - F) + qF$

These frequencies are a modification of the Hardy–Weinberg principle that takes inbreeding into account. When $F = 0$ (no inbreeding), the genotype frequencies are the same as those given in the Hardy–Weinberg principle in Equation (14.1), namely p^2, $2pq$, and q^2. At the other extreme, when $F = 1$ (complete inbreeding), the inbred population consists entirely of AA and aa genotypes in the frequencies p and q, respectively. Whatever the value of F, however, the allele frequencies remain at the values of p and q because

$$p^2(1 - F) + pF + (1/2)[2pq(1 - F)]$$
$$= p(p + q)(1 - F) + pF = p$$

A graphical representation of the genotype frequencies in Equation (14.3) is shown in Figure 14.21 . For each organism, the population is divided conceptually into two groups. In one group (amounting to a proportion F of the population), the gene in question has been affected by the inbreeding, which means that the two alleles present in the organism are identical by descent, owing

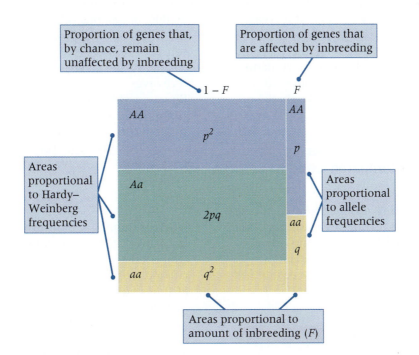

Figure 14.21 **Figure 14.21** Effect of inbreeding on genotype frequencies. The large rectangles on the left pertain to alleles whose ancestries, by chance, are not affected by inbreeding and for which the genotype frequencies remain in Hardy–Weinberg proportions. The narrow rectangles on the right pertain to alleles whose ancestries are affected by the inbreeding, and in this case the genotype frequencies of AA and aa are related as $p : q$. (Note that there are no heterozygous genotypes in the latter case.)

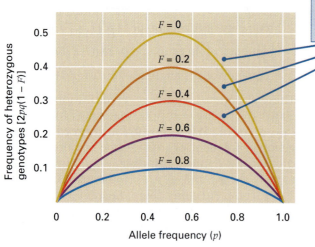

Inbreeding reduces the frequency of heterozygous genotypes, and increases the frequency of homozygous genotypes, relative to random mating.

$F = 0$
$F = 0.2$
$F = 0.4$
$F = 0.6$
$F = 0.8$

Frequency of heterozygous genotypes [$2pq(1 - F)$]

Allele frequency (p)

Figure 14.22 Frequency of heterozygous genotypes in an inbred population (y-axis) against allele frequency (x-axis). As the inbreeding becomes more intense (greater inbreeding coefficient F), the proportion of heterozygous genotypes decreases.

to DNA replication in a common ancestor of the inbred organism. In the other group (amounting to a proportion $1 - F$ of the population), the gene in question has, by chance, escaped the effects of inbreeding, which means that the genotype frequencies are those expected with random mating. Taking both groups into account results in the genotype frequencies in Equation (14.3).

The reduction in heterozygosity due to inbreeding is shown in Figure 14.22 . Each curve is given by an equation of the form $H_I = 2pq(1 - F)$. The case $F = 0$ corresponds to random mating. The curves show that as the inbreeding coefficient increases, the frequency of heterozygous genotypes decreases proportionately until, when $F = 1$, there are no heterozygotes remaining in the inbred population. In other words, in a highly inbred population with $F = 1$, the genotypes consist of AA and aa in the relative proportions p and q, respectively.

■ Inbreeding results in an excess of homozygotes compared with random mating.

The effects of inbreeding differ according to the normal mating system of an organism. At one extreme, in regularly self-fertilizing plants, inbreeding is already so intense and the organisms are so highly homozygous that additional inbreeding has virtually no effect. However,

key concept

In most species, inbreeding is harmful, and much of the effect is due to rare recessive alleles that would not otherwise become homozygous.

Among human beings, close inbreeding is usually uncommon because of social conventions. The closest type of inbreeding usually found is mating between first cousins. In small, isolated populations (such as aboriginal groups and religious communities), some inbreeding is inevitable as a result of matings between remote relatives. The effect of inbreeding is always an increase in the frequency of genotypes that are homozygous for rare, usually harmful recessives. For example, among American whites, the frequency of albinism among offspring of matings between nonrelatives is approximately 1 in 20,000, but among offspring of first-cousin matings, the frequency is approximately 1 in 2000. The reason for the increased risk may be understood by comparing the genotype frequencies of homozygous recessives in Equation (14.3) (for inbreeding) and Equation (14.1) (for random mating). In the most common form of inbreeding among human beings (mating between first cousins), $F = 0.062$ ($= 1/16$) among the offspring. Therefore, the frequency of homozygous recessives produced by first-cousin mating is

$$q^2(1 - 0.062) + q(0.062)$$

whereas, among the offspring of nonrelatives, the frequency of homozygous recessives is q^2. For albinism, $q = 0.007$ (approximately), and the calculated frequencies are 5×10^{-4} for the offspring of first cousins and 5×10^{-5} for the offspring of nonrelatives.

The effect of first-cousin mating in causing an increased frequency of offspring homozygous for a rare recessive allele is shown in Figure 14.23 . The increase occurs for any allele frequency, but the relative increase is greater for alleles that are more rare. Considering the curves in Figure 14.23, with

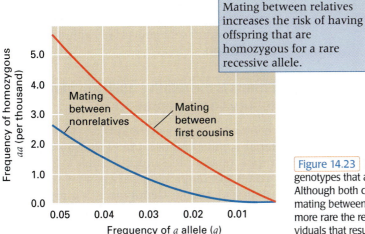

Mating between relatives increases the risk of having offspring that are homozygous for a rare recessive allele.

Mating between nonrelatives

Mating between first cousins

Frequency of homozygous aa (per thousand)

Frequency of a allele (q)

Figure 14.23 Effect of first-cousin mating on the frequency of offspring genotypes that are homozygous for a rare autosomal recessive allele. Although both curves decrease as the allele becomes rare, the curve for mating between nonrelatives decreases more rapidly. As a consequence, the more rare the recessive allele, the greater the proportion of all affected individuals that result from first-cousin matings.

an allele frequency $q = 0.05$, for example, the relative risk of producing a homozygous offspring is $0.0335938/0.0025 = 13.4$; whereas with an allele frequency $q = 0.01$, the relative risk of a homozygous offspring is $0.00634375/0.0001 = 63.4$.

14.6

Evolution is accompanied by genetic changes in species.

In its broadest sense, the term **evolution** means any change in the gene pool of a population or in the allele frequencies present in a population. Evolution is possible because genetic variation exists in populations. Four processes account for most of the evolutionary changes. They form the basis of cumulative change in the genetic characteristics of populations, leading to the descent with modification that characterizes the process of evolution. Although the point has yet to be proved, most evolutionary biologists believe that these same processes, when carried out continuously over geological time, also account for the formation of new species and higher taxonomic categories. These processes are

1. **Mutation,** the origin of new genetic capabilities in populations by means of spontaneous heritable changes in genes.

2. **Migration,** the movement of organisms among subpopulations within a larger population.

3. **Natural selection,** resulting from the different abilities of organisms to survive and reproduce in their environment. Natural selection is the primary process by which populations of organisms become progressively better adapted to their environments.

4. **Random genetic drift,** the random, undirected changes in allele frequency that happen

by chance in all populations, but particularly in small ones.

Evolution is a population phenomenon, so it is conveniently discussed in terms of allele frequencies. In the following sections, we consider some of the population genetic implications of the major evolutionary processes.

14.7

Mutation and migration bring new alleles into populations.

Mutation is the ultimate source of genetic variation. It is an essential process in evolution, but it is a relatively weak force for changing allele frequencies, primarily because typical mutation rates are so low. Moreover, most newly arising mutations with phenotypic effects are harmful to the organism.

Migration is similar to mutation in that new alleles can be introduced into a local population, although the alleles derive from another subpopulation rather than from new mutations. In the absence of migration, the allele frequencies in each local population can change independently, so local populations may undergo considerable genetic differentiation. Genetic differentiation among subpopulations means that there are differing frequencies of common alleles among the local populations or that some local populations possess certain rare alleles not found in others. The accumulation of genetic differences among subpopulations can be minimized if some migration of organisms among the subpopulations is possible. In fact, only a relatively small amount of migration among subpopulations (on the order of just a few migrant organisms in each local population in each generation) is usually sufficient to prevent high levels of genetic differentiation. On the other hand, some genetic differentiation can accumulate in spite of migration if other evolutionary forces, such as natural selection for adaptation to the local environments, are sufficiently strong.

14.8

Natural selection favors genotypes that are better able to survive and reproduce.

The driving force of adaptive evolution is natural selection, which is a consequence of hereditary differences among organisms in their ability to survive and reproduce in the prevailing environment. Since it was first proposed by Charles Darwin in 1859, the concept of natural selection has been revised and extended, most notably by the incorporation of genetic concepts. In its modern formulation, the concept of natural selection rests on three premises:

- In all organisms, more offspring are produced than survive and reproduce.
- Organisms differ in their ability to survive and reproduce, and some of these differences are due to genotype.
- In every generation, genotypes that promote survival in the prevailing environment (favored genotypes) are present in excess at the reproductive age, and hence they contribute disproportionately to the offspring of the next generation. In this way, the alleles that enhance survival and reproduction increase in frequency from generation to generation, and the population becomes progressively better able to sur-

vive and reproduce in its environment. This progressive genetic improvement in populations constitutes the process of evolutionary **adaptation.**

■ Fitness is the relative ability of genotypes to survive and reproduce.

Selection over many generations can be studied in bacterial populations because of the short generation time (about 30 minutes). Figure 14.24 shows the result of competition between two bacterial genotypes, A and B. Genotype A is the superior competitor under the particular conditions. In the experiment, the competition was allowed to continue for 120 generations, during which time the proportion of A genotypes (p) increased from 0.60 to 0.9995 and that of B genotypes decreased from 0.40 to 0.0005. The data points give a satisfactory fit to an equation of the form

$$\frac{p_n}{q_n} = \left(\frac{p_0}{q_0}\right)\left(\frac{1}{w}\right)^n \qquad (14.4)$$

in which p_0 and q_0 are the initial frequencies of A and B (in this case 0.6 and 0.4, respectively), p_n and q_n are the frequencies after n generations of competition, n is the number of generations of competition, and w is a measure of the competitive ability of B when competing against A under the conditions of the experiment. The theoretical derivation of Equation (14.4) is based on the definition of w as the rate of survival and/or reproduction of genotype B, relative to genotype A, under the conditions of the experiment, and assuming that w is constant through all generations. What this means is that if

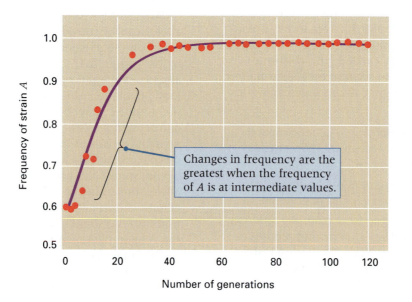

Changes in frequency are the greatest when the frequency of A is at intermediate values.

Figure 14.24 | Increase in frequency of a favored strain of *E. coli* resulting from selection in a continuously growing population. The y value for each point is the number of A cells at any time divided by the total number of cells ($A + B$). Note that the changes in frequency are greatest when the frequency of the favored strain, A, is at intermediate values.

the relative frequencies of $A : B$ in generation 0 are $p_0 : q_0$, then in the next generation they will be in the relative frequencies $p_0 : q_0 w$; which is to say that $p_1/q_1 = (p_0/q_0)(1/w)$. Likewise, in the next generation, the relative frequencies will be given by $p_2/q_2 = (p_1/q_1)(1/w)$, and by substituting for p_1/q_1, we obtain $p_2/q_2 = (p_0/q_0)(1/w)^2$. Continuing in this manner, generation after generation, we obtain Equation (14.4). Because $q = 1 - p$ in every generation, Equation (14.4) implies that

$$p_n = \frac{p_0}{p_0 + q_0 w^n} \qquad (14.5)$$

This is the smooth curve plotted in Figure 14.24 for the values $p_0 = 0.6$, $q_0 = 0.4$, and $w = 0.96$. The dots show the good fit with the experimental data.

The value of $w = 0.96$ is called the **relative fitness** of the B genotype relative to the A genotype under these particular conditions. As we have seen, the relative fitness measures the comparative contribution of each parental genotype to the pool of offspring genotypes produced in each generation. A value $w = 0.96$ means that for each offspring cell produced by an A genotype, a B genotype produces, on the average, 0.96 offspring cell.

In population genetics, relative fitnesses are usually calculated with the superior genotype (A in this case) taken as the standard with a fitness of 1.0. However, the selective disadvantage of a genotype is often of greater interest than its relative fitness. The selective disadvantage of a disfavored genotype is called the **selection coefficient** associated with the genotype, and it is calculated as the difference between the fitness of the standard (taken as 1.0) and the relative fitness of the genotype in question.

In the present example, the selection coefficient against B, denoted s, is

$$s = 1.000 - 0.96 = 0.04 \qquad (14.6)$$

The specific meaning of s in this example is that the selective disadvantage of strain B is 4 percent per generation. When the selection coefficient is known, Equation (14.5) also makes it possible to predict the allele frequencies in any future generation, given the original frequencies. Alternatively, it can be used to calculate the number of generations required for selection to change the allele frequencies from any specified initial values to any later ones. For example, from the relative fitnesses of A and B just estimated, what is the number of generations necessary to change the frequency of A from 0.1 to 0.8? In this particular problem, $p_0/q_0 = 0.1/0.9$, $p_n/q_n = 0.8/0.2$, and $w = 0.96$. A little manipulation of Equation (14.4) gives

$$n = [\log(0.1/0.9) - \log(0.8/0.2)]/\log(0.96)$$
$$= 87.8 \text{ generations}$$

■ Allele frequencies change slowly when alleles are either very rare or very common.

Selection in diploids is analogous to that in haploids, but dominance and recessiveness create additional complications. Figure 14.25 shows the change in allele frequencies for both a favored dominant and a favored recessive. The striking feature of the figure is that the frequency of the favored dominant allele changes very slowly when the allele is common, and the frequency of the favored reces-

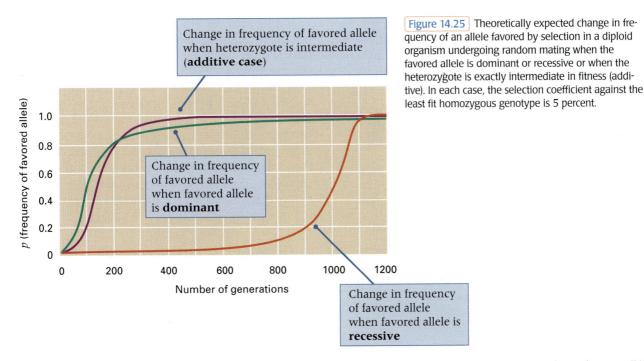

Change in frequency of favored allele when heterozygote is intermediate (**additive case**)

Change in frequency of favored allele when favored allele is **dominant**

Change in frequency of favored allele when favored allele is **recessive**

p (frequency of favored allele)

Number of generations

Figure 14.25 Theoretically expected change in frequency of an allele favored by selection in a diploid organism undergoing random mating when the favored allele is dominant or recessive or when the heterozygote is exactly intermediate in fitness (additive). In each case, the selection coefficient against the least fit homozygous genotype is 5 percent.

sive allele changes very slowly when the allele is rare. The reason is that rare alleles are found much more frequently in heterozygotes than in homozygotes. With a favored dominant at high frequency, most of the recessive alleles are present in heterozygotes, and the heterozygotes are not exposed to selection and hence do not contribute to change in allele frequency. Conversely, with a favored recessive at low frequency, most of the favored alleles are in heterozygotes, and again the heterozygotes are not exposed to selection and do not contribute to change in allele frequency. The principle is quite general:

key concept

Selection for or against recessive alleles is very inefficient when the recessive allele is rare.

One simple example of this principle is selection against a recessive lethal. In this case, it can be shown that the number of generations required to reduce the frequency of the recessive allele from q to $q/2$ equals $1/q$ generations. For example, if $q = 0.01$, then successive halvings of the frequency of the recessive allele require 100, 200, 400, 800, 1600, . . . generations, so a recessive lethal allele is eliminated very slowly.

The inefficiency of selection against rare recessive alleles has an important practical implication. There is a widely held belief that medical treatment to save the lives of persons with rare recessive disorders will cause a deterioration of the human gene pool because of the reproduction of persons who carry the harmful genes. This belief is unfounded. With rare alleles, the proportion of homozygotes is so small that reproduction of the homozygotes contributes a negligible amount to change in allele frequency. Considering their low frequency, it matters little whether homozygotes reproduce or not. Similar reasoning applies to eugenic proposals to

Answer to Problem: (a) The DNA samples from the six adults have no bands in common, so the data are consistent with the presence of six unrelated adults. **(b)** Comparing the bands from the DNA from the juveniles with those from the adults indicates that only male E could have contributed DNA for fragment size 4 in individual G and for fragment size 10 in individuals H and I. Similarly, only female C could have contributed DNA for fragment size 5 in all three juveniles. **(c)** Consequently, the remains C are consistent with being those of Tsarina Alexandra, who was evidently homozygous for this SSR, and the remains E are consistent with being those of Tsar Nicholas II. The remains of the Tsarevitch Alexis and one of the daughters (thought to be Anastasia) have not been recovered.

"cleanse" the human gene pool of harmful recessives by preventing the reproduction of affected persons. People with severe genetic disorders rarely reproduce anyway and, even when they do, they have essentially no effect on allele frequency. In other words,

key concept

The largest reservoir of harmful recessive alleles is in the genomes of heterozygous carriers, who are phenotypically normal.

■ Selection can be balanced by new mutations.

It is apparent from Figure 14.25 that selection tends to eliminate harmful alleles from a population. However, harmful alleles can never be eliminated totally because recurrent mutation of the normal allele continually creates new harmful alleles. These new mutations tend to replenish the harmful alleles eliminated by selection. Eventually the population will attain a state of *equilibrium* in which the new mutations exactly balance the selective eliminations. Two important cases, pertaining to a complete recessive and to partial dominance, must be considered. In both cases, equilibrium results from the balance of new mutations against those alleles eliminated by selection.

The allele frequency of a harmful allele maintained at equilibrium depends strongly on whether the allele is completely recessive. Because selection against a complete recessive is so inefficient when the allele is rare, even a small amount of selection against heterozygous carriers results in a dramatic reduction in the equilibrium allele frequency. For example, consider a homozygous-lethal allele that has an equilibrium frequency of 0.01 when completely recessive; if the heterozygous carriers had a relative fitness of 0.99, instead of 1.0, the equilibrium frequency would decrease to 0.001. In other words, a mere 1 percent decrease in the fitness of heterozygous carriers results in a tenfold decrease in the equilibrium frequency. The decrease is so dramatic because there are many more heterozygotes than homozygotes for the rare allele; therefore, a small amount of selection against heterozygotes affects such a relatively large number of organisms that the result is very great.

■ Occasionally the heterozygote is the superior genotype.

So far, we have considered only cases in which the heterozygote is intermediate in fitness between the homozygotes (or possibly equal in fitness to one homozygote). In these cases, the allele associated

the human connection

Resistance in the Blood

Anthony C. Allison 1954
Radcliffe Infirmary, Oxford, England
Protection Afforded by Sickle-Cell Trait Against Subtertian Malarial Infection

Malaria is the most prevalent infectious disease in tropical and subtropical regions of the world, infecting up to 300 million people each year and causing as many as 1 million deaths. The disease is characterized by recurrent episodes of fever with alternating shivering and sweating. Patients suffer anemia because of the destruction of red blood cells, as well as enlargement of the spleen, inflammation of the digestive system, bronchitis, and many other severe complications. The type of malaria caused by the protozoan parasite *Plasmodium falciparum* is called "subtertian" malaria because there is less than a 3-day interval between bouts of fever. This parasite is spread through bites by the mosquito *Anopheles gambiae*. Upon transmission, the parasites multiply in the liver for about a week and then begin to infect and multiply in red blood cells (parasitemia), which are destroyed after a few days. In parts of Africa, the Middle East, the Mediterranean region, and India where *falciparum* malaria is endemic, there is also a relatively high frequency of the sickle-cell mutation affecting the amino acid sequence of the beta chain of hemoglobin. Heterozygous carriers have no severe clinical symptoms, but they have the so-called "sickle-cell trait," in which the red blood cells collapse into half-moon, or sickle, shapes after 1 to 3 days when sealed under a cover slip on a microscope slide. Homozygous affected persons have "sickle-cell anemia," in which many red blood cells sickle spontaneously while still in the bloodstream, causing severe complica-

tions and often death. Why would a genetic disease that is effectively lethal when homozygous be maintained at a frequency of 10 percent or more in a population? Allison noted the correlation between the sickle-cell mutation and malaria and speculated that the sickle-cell trait gives heterozygous carriers some protection against malaria. Key evidence supporting this hypothesis is presented here. Later work showed that the heterozygous carriers have an approximately 15 percent selective advantage as a result of their malaria resistance.

....

During the course of field work in Africa in 1949 I was led to question the view that the sickle-cell trait is neutral from the point of view of natural selection and to reconsider the possibility that it is associated with a selective advantage. I noted that the incidence of the sickle-cell trait was higher in regions where malaria was prevalent than elsewhere. . . . It became imperative, then, to ascertain whether sickle cells can afford some degree of protection against malarial infection. . . . Children were chosen rather than adults as subjects for the observations so as to minimize the effect of acquired immunity to malaria. The recorded incidence of parasitemia in a group of 290 Ganda children [living near Kampala, Uganda] is presented in the accompanying table (below). . . .

It is apparent that the incidence of parasitemia is lower in the sickle-

cell group than in the group without sickle cells. The difference is statistically significant ($\chi^2 = 4.8$ for 1 degree of freedom). . . . The parasite density in the two groups also differed: of 12 sicklers with malaria, 66.7 percent had only slight parasitemia while 33.3 percent had a moderate parasitemia. Of the 113 non-sicklers with malaria, 34 percent had slight parasitemia, the parasite density in the remainder being moderate or severe. . . . [Among a group of adult males who volunteered to be bitten by heavily infected *Anopheles gambiae* mosquitoes], an infection with *Plasmodium falciparum* was established in 14 out of 15 men without the sickle-cell trait, whereas in a comparable group of 15 men with the trait only 2 developed parasites.

> **It became imperative . . . to ascertain whether sickle cells can afford some degree of protection against malarial infection.**

It is concluded that the abnormal erythrocytes of individuals with the sickle-cell trait are less easily parasitized by *P. falciparum* than are normal erythrocytes. Hence those who are heterozygous for the sickle-cell gene will have a selective advantage in regions where malaria is hyperendemic. This fact may explain why the sickle-cell gene remains common in these areas in spite of the elimination of genes in patients dying of sickle-cell anemia.

Source: British Medical Journal 1: 290–294.

	With parasitemia	Without parasitemia	Total
Sicklers	12 (27.9%)	31 (72.1%)	43
Non-sicklers	113 (45.7%)	134 (53.3%)	247

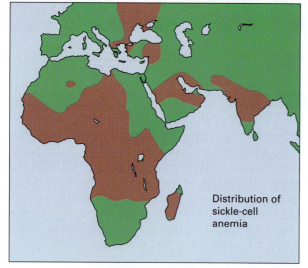

(A)

Distribution of
sickle-cell
anemia

(B)

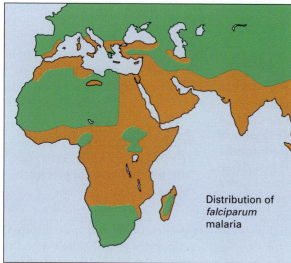

Distribution of
falciparum
malaria

Figure 14.26 Geographic distribution of sickle-cell anemia (A) and *falciparum* malaria (B) in the 1920s, before extensive malaria-control programs were launched.

with the superior homozygote eventually becomes fixed, unless the selection is opposed by mutation. In this section, we consider the possibility of **heterozygote superiority,** in which the fitness of the heterozygote is greater than that of both homozygotes.

When there is heterozygote superiority, neither allele can be eliminated by selection. In each generation, the heterozygotes produce more offspring than the homozygotes, and the selection for heterozygotes keeps both alleles in the population. Selection eventually produces an equilibrium in which the allele frequencies no longer change. If the relative fitnesses of the genotypes AA, Aa, and aa are $1 - s : 1 : 1 - t$, then it can be shown that at equilibrium, the values of p and q are given by the ratio $p/q = t/s$. This formula makes good intuitive sense, because the greater the selection coefficient

against aa (t), the larger the equilibrium ratio of p/q, and vice versa.

Heterozygote superiority does not appear to be a particularly common form of selection in natural populations. However, there are several well-established cases, the best known of which is the sickle-cell hemoglobin mutation (Hb^S) and its relationship to a type of malaria caused by the parasitic protozoan *Plasmodium falciparum* (Figure 14.26). The mutation in sickle-cell anemia is in the gene for β-globin, and it changes the sixth codon in the coding sequence from the normal 5'-GAG-3', which codes for glutamic acid, into the codon 5'-GUG-3', which codes for valine. In the absence of effective medical care, the Hb^S allele is virtually lethal when homozygous, yet in certain parts of Africa and the Middle East, the allele frequency reaches 10 percent or even higher. The Hb^S allele is maintained because heterozygous persons are less susceptible to malaria, and when they are infected have milder infections, than homozygous-normal persons; the heterozygous genotypes therefore have the highest fitness.

14.9

Some changes in allele frequency are random.

Random genetic drift comes about because populations are not infinitely large, as we have been assuming all along, but finite (limited in size). The breeding organisms in any one generation produce a potentially infinite pool of gametes. Barring differences in fertility, the allele frequencies among gametes would equal the allele frequencies among adults. However, because of the finite size of the population, only relatively few of the gametes participate in fertilization to form the zygotes of the next generation. In other words, a process of *sampling* takes place from one generation to the next; because there is chance variation among samples, the allele frequencies among gametes and zygotes may differ.

The concrete example in Figure 14.27 illustrates the essential features of random genetic drift. The graphs in part A show the number of A alleles in each of 12 hypothetical subpopulations, each consisting of 8 diploid individuals and all initially containing equal numbers of A and a alleles. Within each subpopulation, mating is random. For each subpopulation in each generation, a computer program was used to choose 16 gametes from among the pool of gametes produced by the subpopulation in the previous generation. The dispersion of allele frequencies resulting from random genetic drift is apparent. In larger populations than the very small populations illustrated here, the changes in allele frequency would be less pronounced and would

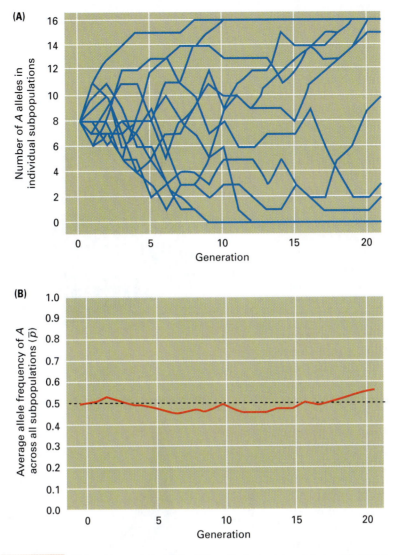

Figure 14.27 (A) Random genetic drift in 12 hypothetical subpopulations of 8 diploid organisms. (B) Average allele frequency among the subpopulations in part A.

require more generations, but the overall effect would be the same. The extent of the dispersion of allele frequency resulting from random genetic drift depends on population size; the smaller the population, the greater the dispersion and the more rapidly it takes place.

In Figure 14.27, part A, the principal effect of random genetic drift is evident even in the first few generations: The allele frequencies begin to spread out over a wider range. By the seventh generation, the spreading is extreme, and the number of A alleles ranges from 1 to 15. This spreading out means that the allele frequencies among the subpopulations become progressively more different. In general,

key concept

Random genetic drift causes differences in allele frequency among subpopulations and therefore causes genetic differentiation among subpopulations.

Although allele frequencies among subpopulations spread out over a wide range because of random genetic drift, the *average* allele frequency among subpopulations remains approximately constant. This point is illustrated in Figure 14.27, part B. The average allele frequency stays close to 0.5, its initial value. If an infinite number of subpopulations were being considered instead of only 12 subpopulatons, then the average allele frequency would be exactly 0.5 in every generation. This principle implies that in spite of the random drift of allele frequency in individual subpopulations, the average allele frequency among a large number of subpopulations remains constant and equal to the average allele frequency among the original subpopulations.

After a sufficient number of generations of random genetic drift, some of the subpopulations become fixed for A and others for a. Because we are excluding the occurrence of mutation, a popu-

14.9 Some changes in allele frequency are random 519

lation that becomes fixed for an allele remains fixed thereafter. After 21 generations in Figure 14.27, only four of the populations are still segregating; eventually, these too will become fixed. Because the average allele frequency of A remains constant, it follows that a fraction p_0 of the populations will ultimately become fixed for A and a fraction $1 - p_0$ will become fixed for a. (The symbol p_0 represents the allele frequency of A in the initial generation.) Therefore,

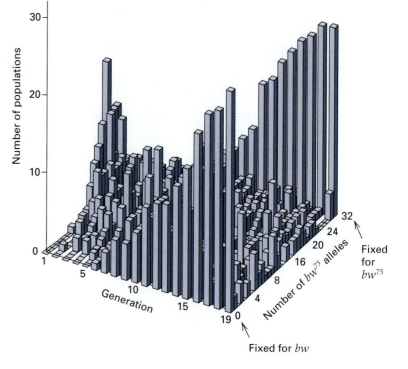

(A) **Actual result of random genetic drift**

key concept

With random genetic drift, the probability of ultimate fixation of a particular allele is equal to the frequency of the allele in the original population.

In Figure 14.27, five of the fixed populations become fixed for A and three for a, which is not very different from the equal numbers expected theoretically with an infinite number of subpopulations.

A real example of random genetic drift in small experimental populations of *Drosophila* that exhibits the characteristics pointed out in connection with Figure 14.27 is illustrated in part A of Figure 14.28. The figure is based on 107 subpopulations, each initiated with eight bw^{75} / bw females (bw = brown eyes) and eight bw^{75} / bw males and maintained at a constant size of 16 by randomly choosing 8 males and 8 females from among the progeny of each generation. Note how the allele frequencies among subpopulations spread out because of random genetic drift and how subpopulations soon begin to be fixed for either bw^{75} or bw. Although the data are somewhat rough because there are only 107 subpopulations, the overall pattern of genetic differentiation has a reasonable resemblance to that expected from the theory based on the binomial distribution (Figure 14.28, part B).

If random genetic drift were the only force at work, all alleles would become either fixed or lost and there would be no polymorphism. On the other hand, many factors can act to retard or prevent the effects of random genetic drift. The most important of these factors are (1) large population size; (2) mutation and migration, which impede fixa-

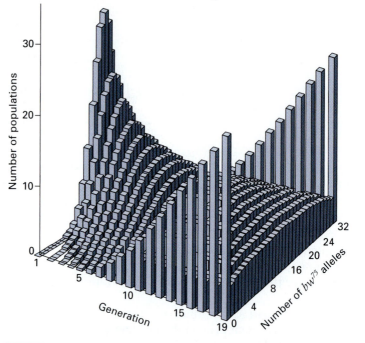

(B) **Predicted result of random genetic drift**

Figure 14.28 (A) Random genetic drift in 107 experimental populations of *Drosophila melanogaster,* each consisting of 8 females and 8 males. (B) Theoretical expectation of the same situation, calculated from the binomial distribution. [Data in part A from P. Buri. 1956. *Evolution* 10: 367. Graphs from D. L. Hartl and A. G. Clark. 1989. *Principles of Population Genetics.* Sunderland, MA: Sinauer Associates.]

tion because alleles lost by random genetic drift can be reintroduced by either process; and (3) natural selection, particularly those modes of selection that tend to maintain genetic diversity, such as heterozygote superiority.

14.10

Mitochondrial DNA is maternally inherited.

Mitochondrial DNA (**mtDNA**) has a number of features that make it useful for studying the genetic relationships among organisms. In higher animals, mitochondria usually show **maternal inheritance,** which means genetic transmission only through the female. The mitochondria are typically maternally inherited because the egg is the major contributor of cytoplasm to the zygote. A typical pattern of maternal inheritance of mitochondria is shown in the human pedigree in Figure 14.29. When human mitochondrial DNA is cleaved with the restriction enzyme *Hae*II, the cleavage products include either one fragment of 8.6 kb or two fragments of 4.5 kb and 4.1 kb (Figure 14.29, part A). The pattern with two smaller fragments is typical, and it results from the presence of a *Hae*II cleavage site within the 8.6-kb fragment. Maternal inheritance of the 8.6-kb mitochondrial DNA fragment is indicated in part B, because males (I-2, II-7, and II-10) do not transmit the pattern to their progeny, whereas females (I-3, I-5, and II-8) transmit the pattern to all of their progeny. Although the mutation in the *Hae*II site yielding the 8.6-kb fragment is not associated with any disease, a number of other mutations in mitochondrial DNA do cause diseases and have similar patterns of mitochondrial inheritance. Most of these conditions decrease the ATP-generating capacity of the mitochondria and affect the function of muscle and nerve cells, particularly in the central nervous system, leading to blindness, deafness, or stroke. Many of the conditions are lethal in the absence of some normal mitochondria, and there is variable expressivity because of differences in the proportions of normal and mutant mitochondria among affected persons. The condition in which two or more genetically different types of mitochondria are present in the same cell is unusual among animals. For example, a typical human cell contains from 1000 to 10,000 mitochondria, all of them genetically identical.

■ Human mtDNA evolves changes in sequence at an approximately constant rate.

Another important feature of mitochondrial DNA is that it does not undergo genetic recombination. Hence the DNA molecule in any mitochondrion derives from a single ancestral molecule. Mitochondrial DNA is also a good genetic marker for tracing human ancestry, because it evolves considerably more rapidly than that of nuclear genes. Differences in mitochondrial DNA sequences accumulate among human lineages at a rate of approximately 1 change per mitochondrial lineage every 3800 years. For example, the people of Papua New Guinea have been relatively isolated genetically from the aboriginal people of Australia ever since these areas were colonized approximately 40,000 years ago and 30,000 years ago, respectively. The total time for evolution between the populations is therefore

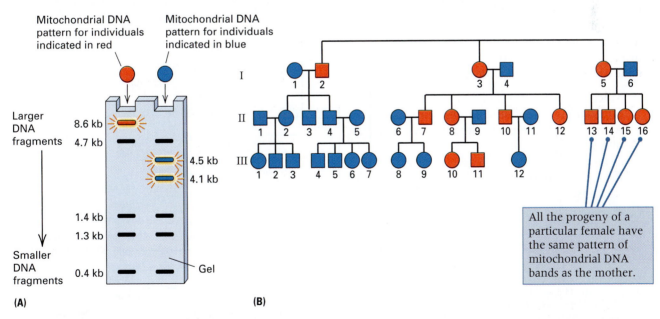

Figure 14.29 Maternal inheritance of human mitochondrial DNA. (A) Pattern of DNA fragments obtained when mitochondrial DNA is digested with the restriction enzyme *Hae*II. The DNA type at the left includes a fragment of 8.6 kb (red). The DNA type at the right contains a cleavage site for *Hae*II within the 8.6-kb fragment, which results in smaller fragments of 4.5 kb and 4.1 kb (blue). (B) Pedigree showing maternal inheritance of the DNA pattern with the 8.6-kb fragment (red symbols). The mitochondrial DNA type is transmitted only through the mother. [After D. C. Wallace. 1989. *Trends in Genetics* 5: 9.]

70,000 years (40,000 years in Papua New Guinea and 30,000 years in Australia). If one nucleotide change accumulates every 3800 years, the average number of differences in the mitochondrial DNA of modern-day Papua New Guineans and Australian aborigines is expected to be 18.4 nucleotides (calculated as 70,000/3800), with of course some statistical variation from one pair of persons to the next. This example shows how the rate of mitochondrial DNA evolution can be used to predict the number of differences between populations. In practice, the calculation is usually done the other way around, and the observed number of differences between pairs of populations is used to estimate the number of years since the populations have been geographically separated.

■ Modern human populations originated in subsaharan Africa approximately 100,000 years ago.

Nucleotide differences in mitochondrial DNA have been used to reconstruct the probable historical relationships among human populations. Figure 14.30 shows the gene tree of mtDNA based on the complete mtDNA sequences from 53 persons representing human populations from throughout the world. Because mtDNA undergoes no recombination, the tree is the genetic history of only a single locus; and because mtDNA is maternally inherited, it is a genetic history of females. Nevertheless, the phylogenetic tree shows several remarkable features:

• Much of the mtDNA diversity in African populations is not found among non-Africans; on the average, the mtDNA of Africans shows about twice as much genetic variation as the mtDNA among non-Africans.

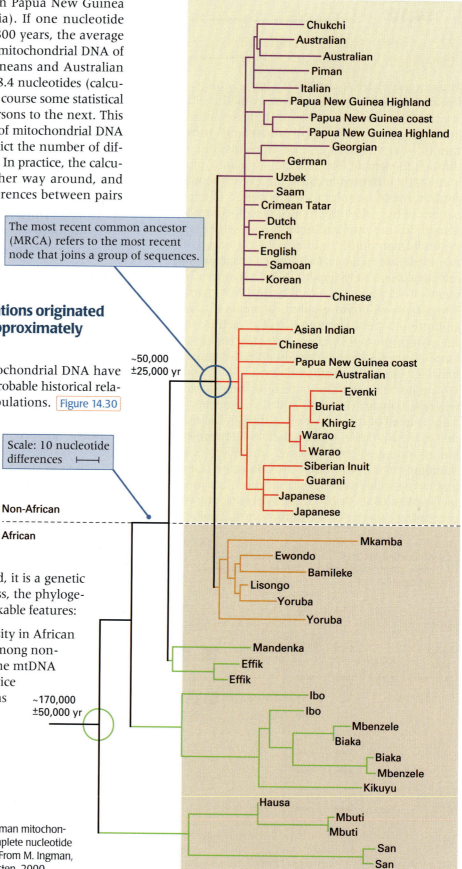

Figure 14.30 Phylogenetic tree of human mitochondrial DNA based on analysis of the complete nucleotide sequence of mtDNA from 53 persons. [From M. Ingman, H. Kaessmann, S. Pääbo, and U. Gyllensten. 2000. *Nature* 408: 708.]

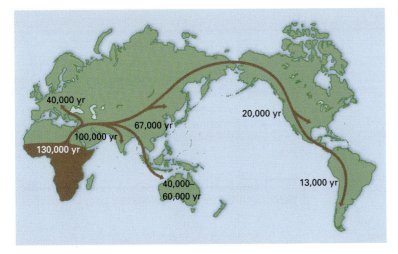

Figure 14.31 The dispersal of modern human populations from subsaharan Africa began approximately 100,000 years ago. The dates on the map are based on the earliest fossil and archaeological evidence from each continent. [From S. B. Hedges. 2000. *Nature* 408: 652.]

est fossils of the modern subspecies of human beings (*Homo sapiens sapiens*) found in Africa is 130,000 years. This age is consistent with that of the MRCA of all of the mtDNA lineages in Figure 14.30 (170,000 ± 50,000 years). The age of the Great Migration out of Africa—100,000 years—is based on DNA sequences of nuclear genes and archaeological evidence. At first sight, this date is inconsistent with the age of the MRCA of the non-African mtDNA lineages in Figure 14.30. However, the genetic history of mtDNA is the genetic history of females, and it is reasonable to suppose that some of the mtDNA diversity among the original migrants was lost in the first few tens of thousands of years in the subsequent expansion. It is not known whether the original migrants came from a restricted geographical locality, but it is worth noting that the contemporary mtDNA lineages in orange in Figure 14.30 are present in geographically dispersed individuals.

The Great Migration 100,000 years ago went initially to the Middle East and Northern Europe (first fossil and archaeological evidence ~40,000 years), then east and south to Asia (~67,000 years) and Australasia (40,000–60,000 years), and finally from East Asia across the Bering Strait to North America (~20,000 years) and South America (~13,000 years). At a rate of sequence evolution averaging 1 nucleotide substitution every 3800 years, the average number of differences between any two mtDNA sequences descended from the Great Migration is about 50 in a molecule of total length 16.5 kb. In other words, our mtDNA molecules are 99.7 percent identical.

- Three of the four major lineages of mtDNA are found only in subsaharan Africans (green); the age of the **most recent common ancestor (MRCA)** of these sequences (green circle) is approximately 170,000 ± 50,000 years.

- A restricted subset of mtDNA lineages is found among non-Africans (purple and red); these sequences share a more recent MRCA with five African mtDNAs (orange) than these African mtDNAs share with other Africans.

- The age of the MRCA of the mtDNA lineage joining African and non-African populations (blue circle) is approximately 50,000 ± 25,000 years.

These features of the mtDNA tree are consistent with a widely accepted scenario in which all modern human populations are derived from a migration out of Africa that took place approximately 100,000 years ago (Figure 14.31). There were earlier migrations out of Africa as well, but the descendants of these earlier migrants were apparently displaced by those who came later. In Figure 14.31, the dates are those of the earliest fossil and archaeological finds on each continent. The age of the old-

chapter summary

14.1 DNA and protein sequences contain information about the evolutionary relationships among species.

- A gene tree is a diagram of the inferred ancestral history of a group of sequences.
- Bootstrapping is a method of assigning a level of confidence to each node in a gene tree.
- A gene tree does not necessarily coincide with a species tree.
- Rates of evolution can differ dramatically from one protein to another.
- Rates of evolution of nucleotide sites differ according to their function.

- New genes usually evolve through duplication and divergence.

The sequences of protein and nucleic acid molecules change through evolutionary time as new mutations occur and some eventually become fixed. The accumulation of sequence differences is the basis for estimating a gene tree depicting the ancestral history of a group of molecules. Proteins and different classes of DNA sequences evolve at very different rates. Among the fastest evolving DNA sequences are those of pseudogenes, introns, and fourfold degenerate sites. New genes usually evolve from existing genes through the acquisition of novel or specialized functions among duplicate copies (paralogs).

14.2 Genotypes may differ in frequency from one population to another.

- Allele frequencies are estimated from genotype frequencies.
- The allele frequencies among gametes equal those among reproducing adults.

Population genetics is the application of Mendel's laws and other principles of genetics to populations of organisms. A subpopulation, or local population, is a group of organisms of the same species living within a geographical region of such size that most matings are between members of the group. In most natural populations, many genes have two or more common alleles. One of the goals of population genetics is to determine the nature and phenotypic consequences of genetic variation in natural populations.

14.3 Random mating means that mates pair without regard to genotype.

- The Hardy–Weinberg principle has important implications for population genetics.
- If an allele is rare, it is found mostly in heterozygous genotypes.
- Hardy–Weinberg frequencies can be extended to multiple alleles.
- X-linked genes are a special case because males have only one X chromosome.

The relationship between the relative proportions of particular alleles (allele frequencies) and genotypes (genotype frequencies) is determined in part by the frequencies with which particular genotypes form mating pairs. In random mating, mating pairs are independent of genotype. When a population undergoes random mating for an autosomal gene with two alleles, the frequencies of the genotypes are given by the Hardy–Weinberg principle. If the alleles of the gene are A and a, and their allele frequencies are p and q, respectively, then the Hardy–Weinberg principle states that the genotype frequencies with random mating are AA with frequency p^2, Aa with frequency $2pq$, and aa with frequency q^2. These are often good approximations for genotype frequencies within subpopulations. An important implication of the Hardy–Weinberg principle is that rare alleles are found much more frequently in heterozygotes than in homozygotes ($2pq$ versus q^2).

14.4 Highly polymorphic sequences are used in DNA typing.

- DNA exclusions are definitive.

14.5 Inbreeding means mating between relatives.

- Inbreeding results in an excess of homozygotes compared with random mating.

Inbreeding means mating between relatives, and the extent of inbreeding is measured by the inbreeding coefficient, F. The main consequence of inbreeding is that a rare, harmful allele present in a common ancestor may be transmitted to both parents of an inbred organism in a later generation and become homozygous in the inbred offspring. Among the progeny produced by inbreeding, the frequency of heterozygous genotypes is smaller, and that of homozygous genotypes greater, than it would be with random mating.

14.6 Evolution is accompanied by genetic changes in species.

14.7 Mutation and migration bring new alleles into populations.

Evolution is a progressive change in the gene pool of a population or in the allele frequencies present in a population. One principal mechanism of evolution is natural selection, in which genotypes superior in survival or reproductive ability in the prevailing environment contribute a disproportionate share of genes to future generations, thereby gradually increasing the frequency of the favorable alleles in the whole population. By this process, a species becomes genetically better adapted to its environment. At least three other processes also can change allele frequency: mutation (heritable change in a gene), migration (movement of organisms among subpopulations), and random genetic drift (resulting from restricted population size). Spontaneous mutation rates are generally so low that the effect of mutation on changing allele frequency is minor, except for rare alleles. Migration can have significant effects on allele frequency because migration rates may be very large. The main effect of migration is the tendency to equalize allele frequencies among the local populations that exchange migrants. Selection operates through differences in viability (the probability of survival of a genotype) and in fertility (the probability of successful reproduction).

14.8 Natural selection favors genotypes that are better able to survive and reproduce.

- Fitness is the relative ability of genotypes to survive and reproduce.
- Allele frequencies change slowly when alleles are either very rare or very common.
- Selection can be balanced by new mutations.
- Occasionally the heterozygote is the superior genotype.

Populations maintain harmful alleles at low frequencies as a result of a balance between selection, which tends to eliminate the alleles, and mutation, which tends to increase their frequencies. When there is a balance between selection and mutation, the allele frequency at equilibrium is usually greater if the allele is completely recessive than if it is partially dominant. This difference arises because selection is quite ineffective in affecting the frequency of a completely recessive allele when the allele is rare, because the allele appears almost exclusively in heterozygotes.

A few examples are known in which the heterozygous genotype has a greater fitness than either of the homozygous genotypes (heterozygote superiority). Heterozygote superiority results in an equilibrium in which both alleles are maintained in the population. An example is sickle-cell anemia in regions of the world where *falciparum* malaria is endemic. Heterozygous persons have an increased resistance to malaria and only a mild anemia, which results in fitness greater than that of either homozygote.

14.9 Some changes in allele frequency are random.

Random genetic drift is a statistical process of change in allele frequency in small populations, resulting from the inability of every organism to contribute equally to the offspring of successive generations. In a subdivided population, random genetic drift results in differences in allele frequency among the subpopulations. In an isolated population, barring mutation, an allele will ultimately become fixed or lost as a result of random genetic drift.

14.10 Mitochondrial DNA is maternally inherited.

- Human mtDNA evolves changes in sequence at an approximately constant rate.
- Modern human populations originated in subsaharan Africa approximately 100,000 years ago.

Human mitochondrial DNA sequences have been organized into a phylogenetic tree depicting their ancestral relationships. The most recent common ancestor of human mtDNA has an estimated age of 170,000 years. Much of the mtDNA diversity is found in populations in subsaharan Africa. A subset of African mtDNA lineages appears to have given rise to all mtDNA lineages present in populations outside Africa, as well as to some African lineages. These mtDNA molecules are thought to be descendants of a migration of people out of subsaharan Africa that took place about 100,000 years ago.

issues & ideas

- What is a gene tree? A species tree? Must they always agree? Explain why or why not.
- Distinguish between orthologous genes and paralogous genes. Which type of duplication provides the raw material for the evolution of new gene functions?
- What does the Hardy–Weinberg principle imply about the relative frequencies of heterozygous carriers and homozygous affected organisms for a rare, harmful recessive allele?
- Traits due to recessive alleles in the X chromosome are usually much more prevalent in males than in females. Explain why this discrepancy is expected with random mating.
- Why are the effects of inbreeding more easily observed with a rare recessive allele than with a rare dominant allele?

- What is the fitness of an organism that dies before the age of reproduction? What is the fitness of an organism that is sterile?
- Many recessive alleles are extremely harmful when homozygous, so there is selection in every generation that tends to reduce the allele frequency. Yet harmful recessive alleles are maintained at a low frequency for almost every gene. What process prevents harmful recessive alleles from being completely eliminated?
- Heterozygote superiority of the type observed with sickle-cell anemia is sometimes called *balancing selection*. Do you think this is an appropriate term? Explain why or why not.
- What is random genetic drift and why does it occur? Explain why this process implies that in the absence of other forces, the ancestry of all alleles present at a locus in a population can eventually be traced back to a single allele present in some ancestral population.

key terms & concepts

adaptation
allele frequency
bootstrapping
distance matrix
DNA typing
evolution
fixed allele
gene pool
gene tree
genotype frequency
Hardy–Weinberg principle

heterozygote superiority
inbreeding
inbreeding coefficient (F)
lost allele
maternal inheritance
migration
molecular clock
molecular evolution
molecular systematics
most recent common
 ancestor (MRCA)

mtDNA
mutation
natural selection
neighbor joining
orthologous genes
paralogous genes
polymorphic gene
population
pseudogene
random genetic drift

random mating
relative fitness
selection coefficient
selectively neutral mutation
simple sequence repeat
 (SSR)
species tree
subfunctionalization
subpopulation
taxon

1. _____ A method for assigning confidence to the nodes in a gene tree.

2. _____ Genes that are duplicates in the sense that they derive from a single gene in the most recent common ancestor of a pair of species.

3. _____ Refers to the observation that some DNA or protein sequences change at an approximately constant rate through evolutionary time.

4. _____ Genetic information shared among all members of a species.

5. _____ A gene that has two or more relatively common alleles in a population.

6. _____ When pairs of genotypes undergo mating in proportion to their relative frequencies in the population.

7. _____ Relationship among the fitnesses of beta hemoglobin genotypes that accounts for the polymorphism of sickle-cell anemia in Africa.

8. _____ Darwin's proposed mechanism for the evolution of adaptations.

9. _____ Measure of the difference in relative fitness between two genotypes.

10. _____ Process that can result in the fixation of a selectively neutral mutation.

11. _____ Measures the reduction in the frequency of heterozygous genotypes brought about by mating between relatives.

12. _____ Maternally inherited DNA molecule.

solutions: step by step

Problem 1

The accompanying illustration shows a human pedigree, along with the pattern of restriction fragments observed in a DNA sample from each of the individuals when hybridized with a labeled probe. Examine the pedigree and the band patterns, and suggest a mode of inheritance.

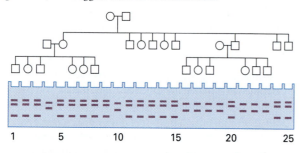

■ **Solution** The first step is to examine the pedigree carefully, looking for signs of Mendelian segregation to indicate whether the DNA fragment is inherited on an autosome, on the X chromosome, or on the Y chromosome. In this case the pattern of inheritance fits none of these possibilities. In each sibship, the phenotypes (band patterns) of all the progeny are identical to each other. Furthermore, when the maternal and paternal phenotypes differ, the progeny phenotype is exactly like that of the mother. Therefore, the pedigree implies that the DNA fragment detected by the probe shows maternal inheritance, and the most likely explanation is that the probe is a fragment of mitochondrial DNA.

Problem 2

In human populations, a locus called *secretor* determines whether the carbohydrate A and B antigens of the ABO blood groups are secreted into the saliva and other body fluids. The genotypes *Se Se* and *Se se* are secretors, and the genotype *se se* is not. Among Caucasians known to have blood group A, B, or AB because of the presence of these antigens on the red blood cells, the frequency of the nonsecretor phenotype is about 33 percent.
(a) Assuming random mating, what is the allele frequency of the *se* allele?
(b) Among Caucasians with blood group A, B, or AB, what are the expected proportions of homozygous secretors and heterozygous secretors?

■ **Solution** This is a typical problem that makes use of the Hardy–Weinberg principle for random mating. **(a)** With Hardy–Weinberg proportions, the frequency of homozygous recessive genotypes equals q^2, which in this case is given as 0.33. Hence $q = \sqrt{0.33} = 0.574$, so $p = 1 - q = 0.426$.
(b) The expected proportions of *Se Se* and *Se se* genotypes are $p^2 = (0.426)^2 = 0.181$ and $2pq = 2(0.574)(0.426) = 0.489$, respectively. Note that approximately half of the genotypes in the population are heterozygous for the *secretor* locus.

Problem 3

Xeroderma pigmentosum (XP) is an often fatal skin cancer resulting from a recessive mutant allele that affects DNA excision repair. In the United States, the frequency of homozygous-recessive affected people is approximately 1 in 250,000. (The mutant allele can be in any one of about eight different genes, but for the purposes of this problem it makes no difference.)
(a) What is the expected frequency of XP among the offspring of first-cousin matings?
(b) What is the ratio of XP among the offspring of first-cousin matings to that among the offspring of nonrelatives?

■ **Solution** First we must calculate the frequency of the recessive allele, q, using the information that the frequency of homozygous recessives is 1 in 250,000. Assuming random mating frequencies, $q = \sqrt{(1/250,000)} = 0.002$. **(a)** Among the offspring of first-cousin matings, the expected frequency of XP equals $q^2(1 - F) + qF$, where $F = 1/16$ for the offspring of first cousins. In this case the formula yields

$(0.002)^2(15/16) + (0.002)(1/16) = 0.00012875$, or about 1 in 7767. **(b)** The ratio is given in the following formula, where $q = 0.002$ and $F = 1/16$.

$$\text{Ratio} = \frac{q^2(1 - F) + qF}{q^2} = \frac{(1/7,767)}{(1/250,000)} = 32.2$$

In other words, the offspring of first cousins have more than a 32-fold greater risk of being homozygous for a recessive mutant allele causing XP.

concepts in action: problems for solution

14.1 If the genotype AA is an embryonic lethal and the genotype aa is fully viable but sterile, what genotype frequencies would be found in adults in an equilibrium population containing the A and a alleles? Is it necessary to assume random mating?

14.2 The accompanying illustration shows the gel patterns observed with a probe for an SSR (simple sequence repeat) locus among four pairs of parents (1–8) and four children (A–D). One child comes from each pair of parents.
(a) Why does each person have two bands?
(b) How would it be possible for a person to have only one band?
(c) What type of dominance is illustrated by the SSR alleles?
(d) Which pairs of people who have the DNA in lanes 1–8 are the parents of each child?

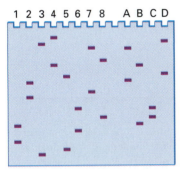

14.3 A trait due to a harmful recessive X-linked allele in a large, randomly mating population affects 1 in 50 males. What is the frequency of carrier females? What is the expected frequency of affected females?

14.4 How many A and a alleles are present in a sample of organisms consisting of 10 AA, 15 Aa, and 4 aa genotypes? What are the allele frequencies in this sample?

14.5 DNA from 100 unrelated people was digested with the restriction enzyme HindIII, and the resulting fragments were separated and probed with a sequence for a particular gene. Four fragment lengths that hybridized with the probe were observed—namely 5.7, 6.0, 6.2, and 6.5 kb—where each fragment defines a different restriction-fragment allele. The accompanying illustration shows the gel patterns observed; the number of individuals with each gel pattern is shown across the top. Estimate the allele frequencies of the four restriction-fragment alleles.

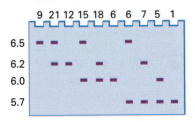

14.6 A randomly mating population of dairy cattle contains an autosomal recessive allele causing dwarfism. If the frequency of dwarf calves is 10 percent, what is the frequency of heterozygous carriers of the allele in the entire herd? What is the frequency of heterozygotes among nondwarf cattle?

14.7 In a Pygmy group in Central Africa, the frequencies of alleles determining the ABO blood groups were estimated as 0.74 for I^O, 0.16 for I^A, and 0.10 for I^B. Assuming random mating, what are the expected frequencies of ABO genotypes and phenotypes?

14.8 Suppose a randomly mating diploid population has n equally frequent alleles of an autosomal locus. What is the expected frequency of:
(a) Any specified homozygous genotype?
(b) Any specified heterozygous genotype?
(c) All homozygous genotypes together?
(d) All heterozygous genotypes together?

14.9 Which of the following genotype frequencies of AA, Aa, and aa, respectively, satisfy the Hardy–Weinberg principle?
(a) 0.25, 0.50, 0.25   **(d)** 0.64, 0.27, 0.09
(b) 0.36, 0.55, 0.09   **(e)** 0.29, 0.42, 0.29
(c) 0.49, 0.42, 0.09

14.10 A man is known to be a carrier of the cystic fibrosis allele. He marries a phenotypically normal woman. In the general population, the incidence of cystic fibrosis at birth is approximately 1 in 1700. Assume Hardy–Weinberg proportions.
(a) What is the probability that the wife is also a carrier?
(b) What is the probability that their first child will be affected?

14.11 How does the frequency of heterozygotes in an inbred population compare with that in a randomly mating population with the same allele frequencies?

14.12 Galactosemia is an autosomal recessive condition associated with liver enlargement, cataracts, and mental retardation. Among the offspring of unrelated individuals, the frequency of galactosemia is 8.5×10^{-6}. What is the expected frequency among the offspring of first cousins ($F = 1/16$) and among the offspring of second cousins ($F = 1/64$)?

14.13 A man with normal parents whose brother has phenylketonuria marries a phenotypically normal woman. In the general population, the incidence of phenylketonuria at birth is approximately 1 in 10,000. Assume Hardy–Weinberg proportions.
(a) What is the probability that the man is a carrier?
(b) What is the probability that the wife is also a carrier?
(c) What is the probability that their first child will be affected?

concepts in action: problems for solution *continued*

14.14 For an X-linked gene with two alleles in a large, randomly mating population, the frequency of carrier females equals one-half of the frequency of the males carrying the recessive allele. What are the allele frequencies?

14.15 Electrophoretic differences in alcohol dehydrogenase in the flowering plant *Phlox drummondii* are determined by codominant alleles of a single gene. In one sample of 35 plants, the following data were obtained:

Genotype	AA	AB	BB	BC	CC	AC
Number	2	5	12	10	5	1

What are the allele frequencies in this sample? With random mating, what are the expected numbers of each of the genotypes?

14.16 A population of maple trees contains a lethal allele that allows homozygous recessive seeds to germinate, but the plants produce no chlorophyll and die shortly after germination. The allele frequency of the recessive allele for this condition in the population is 0.01.
(a) If a randomly mating population produces one million seeds, how many plants will lack chlorophyll?
(b) Will natural selection eliminate the allele after one generation? Why or why not?

14.17 Hartnup disease is an autosomal recessive disorder of intestinal and renal transport of amino acids. The frequency of affected newborn infants is about 1 in 14,000. Assuming random mating, what is the frequency of heterozygotes?

14.18 Self-fertilization in the annual plant *Phlox cuspidata* results in an average inbreeding coefficient of $F = 0.66$.
(a) What frequencies of the genotypes for the enzyme phosphoglucose isomerase would be expected in a population with alleles A and B at respective frequencies 0.43 and 0.57?
(b) What frequencies of the genotypes would be expected with random mating?

14.19 Two strains of bacteria, A and B, are placed into direct competition in a chemostat. A is favored over B. If the selection coefficient per generation is constant, what is its value if, in an interval of 100 generations:
(a) The ratio of A cells to B cells increases by 10 percent?
(b) The ratio of A cells to B cells increases by 90 percent?
(c) The ratio of A cells to B cells increases by a factor of 2?

14.20 An allele A undergoes mutation to the allele a at the rate of 10^{-5} per generation. If a very large population is fixed for A (generation 0), what is the expected frequency of A in the following generation (generation 1)? What is the expected frequency of A in generation 2? Deduce the rule for the frequency of A in generation n.

further readings

Allison, A. C. 1956. Sickle cells and evolution. *Scientific American,* August.

Ayala, F. 1978. The mechanisms of evolution. *Scientific American,* September.

Bittles, A. H., W. M. Mason, J. Greene, and N. A. Rao. 1991. Reproductive behavior and health in consanguineous marriages. *Science* 252: 789.

Bodmer, W. F., and L. L. Cavalli-Sforza. 1976. *Genetics, Evolution, and Man.* New York: Freeman.

Cavalli-Sforza, L. L. 1974. The genetics of human populations. *Scientific American,* September.

Cavalli-Sforza, L. L. 1998. The DNA revolution in population genetics. *Trends in Genetics* 14: 60.

Crow, J. F., and M. Kimura. 1970. *An Introduction to Population Genetics Theory.* New York: Harper & Row.

Dobzhansky, T. 1941. *Genetics and the Origin of the Species.* New York: Columbia University Press.

Falconer, D. S. and T. F. C. Mackay. 1996. *Introduction to Quantitative Genetics.* 4th ed. Essex, England: Longman.

Graur, D. and W.-H. Li. 2000. *Fundamentals of Molecular Evolution.* 2d ed. Sunderland, MA: Sinauer.

Grivell, L. A. 1983. Mitochondrial DNA. *Scientific American,* March.

Hartl, D. L. 2000. *A Primer of Population Genetics.* 3d ed. Sunderland, MA: Sinauer.

Hartl, D. L., and A. G. Clark. 1997. *Principles of Population Genetics.* 3d ed. Sunderland, MA: Sinauer.

Hoffmann, A. A., C. M. Sgro, and S. H. Lawler. 1995. Ecological population genetics: The interface between genes and the environment. *Annual Review of Genetics* 29: 349.

Ingman, M., H. Kaessmann, S. Pääbo, and U. Gyllensten. 2001. Mitochondrial genome variation and the origin of modern humans. *Nature* 408: 708.

Kimura, M. 1983. *The Neutral Theory of Molecular Evolution.* Cambridge, England: Cambridge University Press.

Kline, J., N. Takahata, and F. J. Ayala. 1993. MHC polymorphism and human origins. *Scientific American,* November.

Lewontin, R. C. 1974. *Genetic Basis of Evolutionary Change.* New York: Columbia University Press.

Li, W.-H. 1997. *Molecular Evolution.* Sunderland, MA: Sinauer.

Mitton, J. F. 1997. *Selection in Natural Populations.* New York: Oxford University Press.

Nei, M. 1987. *Molecular Evolutionary Genetics.* New York: Columbia University Press.

Nei, M. and S. Kumar. 2000. *Molecular Evolution and Phylogenetics.* New York: Oxford University Press.

Wright, S. 1978. *Evolution and the Genetics of Populations* (4 volumes). Chicago: University of Chicago Press.

geNETics on the web

GeNETics on the Web will introduce you to some of the most important sites for finding genetic information on the Internet. To explore these sites, visit the Jones and Bartlett home page at

http://www.jbpub.com/genetics

For the book *Essential Genetics: A Genomics Perspective,* choose the link that says **Enter GeNETics on the Web**. You will be presented with a chapter-by-chapter list of highlighted keywords. Select any highlighted keyword and you will be linked to a Web site containing genetic information related to the keyword.

- At the keyword site **CFTR** you can learn about geographical variation in common cystic fibrosis mutations in the human population. The high frequency of the disease has been difficult to explain in view of the seriousness of the condition in affected individuals. Supposing that mutant alleles are completely recessive, and that the frequency of homozygous recessives (1/1700) represents an equilibrium between mutation and selection, then the rate of mutation to new alleles would have to be 6×10^{-4} per generation, which is much higher than can easily be explained. Another alternative is that heterozygotes have a selective advantage of some kind, so the mechanism of selection is actually overdominance. Recent data support the hypothesis that heterozygotes are more resistant to typhoid fever. The bacterium *Salmonella typhi,* the pathogen for typhoid fever, invades the gastrointestinal system through the CFTR membrane channel encoded in the CFTR gene. Cells expressing wildtype CFTR internalize more *S. typhi* than do mutant cells, and transgenic mice that are heterozygous for CFTR mutations internalize 86 percent fewer bacteria than do wildtype homozygotes. The resistance hypothesis is also consistent with the high frequency of particular CFTR mutations that are found in patients.

- The occasionally extreme effects of random genetic drift are dramatically illustrated by **Pingelap disease**, a form of total color blindness combined with extreme nearsightedness and cataract, which affects 4 to 10 percent of the Pingelapese people of the eastern Caroline Islands in the Pacific, who are blind from infancy. Consult this keyword site to learn the molecular and cellular basis of this condition and why its frequency is so high in this particular human population.

The Shetland Sheepdog is a herding breed that originated in the Shetland Islands near Scotland. Friendly, intelligent, inquisitive, and agile, the "Sheltie" is one of about 150 breeds of dog recognized by the American Kennel Club. Studies of mitochondrial DNA sequences from 67 dog breeds and 162 wild wolves indicate a very close genetic relationship. The results suggest that domestication began at least 100,000 years ago, that there were at least two separate domestication events, and that during domestication there were repeated episodes of genetic admixture between dogs and wolves. [Courtesy Anne Spencer]

key concepts

- Multifactorial traits are determined by multiple genetic and environmental factors acting together.

- The relative contributions of genotype and environment to a trait are measured by the variance due to genotype (genotypic variance) and the variance due to environment (environmental variance).

- Correlations between relatives are used to estimate various components of variation, such as genotypic variance, additive variance, and dominance variance.

- Additive variance accounts for the parent–offspring correlation; dominance variance accounts for the sib–sib correlation over and above that expected from the additive variance.

- Narrow-sense heritability is the ratio of additive (transmissible) variance to the total phenotypic variance; it is widely used in plant and animal breeding.

- Genes that affect quantitative traits (QTLs) can be identified and genetically mapped using various kinds of genetic polymorphisms.

- Many complex human behaviors are affected by multiple genetic and environmental factors and the interactions among them.

15

The Genetic Basis of Complex Inheritance

chapter organization

15.1 Multifactorial traits are determined by multiple genes and the environment.

15.2 Variation in a trait can be separated into genetic and environmental components.

15.3 Artificial selection is a form of "managed evolution."

15.4 Genetic variation is revealed by correlations between relatives.

15.5 Pedigree studies of genetic polymorphisms are used to map loci for quantitative traits.

the human connection
Win, Place, or Show?

Chapter Summary
Issues & Ideas
Key Terms & Concepts
Solutions: Step by Step
Concepts in Action: Problems for Solution
Further Readings

geNETics **on** the web

Earlier chapters have emphasized traits in which differences in phenotype result from alternative genotypes of a single gene. Examples include green versus yellow peas, red eyes versus white eyes in *Drosophila,* normal versus sickle-cell hemoglobin, and the ABO blood groups. These traits are particularly well suited for genetic analysis through the study of pedigrees, because there is a small number of genotypes and phenotypes and because there is a relatively simple correspondence between genotype and phenotype. However, many traits that are important in medical genetics, animal breeding, and plant breeding are influenced by multiple genes as well as by the effects of environment. These are known as **multifactorial traits** because of the multiple genetic and environmental factors implicated in their causation. With a multifactorial trait, a single genotype can have any one of many possible phenotypes (depending on the environment), and similar phenotypes can result from many different genotypes.

Multifactorial traits are often called **complex traits** because each factor that affects the trait contributes, at most, a modest amount to the total variation in the trait observed in the entire population. Most traits that vary in populations of humans and other organisms, including common human diseases that have a genetic component, are complex traits. For a complex trait, the **genetic architecture** consists of a description of all of the genetic and environmental factors that affect the trait, along with the magnitudes of their individual effects and the magnitudes of interactions among the factors. It is, in principle, possible to define the genetic components in terms of Mendelian segregation and locations along a genetic map. Environmental factors are much less easily partitioned into separate factors whose individual effects and interactions can be sorted out. The genetic analysis of complex traits requires special concepts and methods, which are introduced in this chapter.

15.1

Multifactorial traits are determined by multiple genes and the environment.

Multifactorial traits are often called **quantitative traits** because the phenotypes in a population differ in quantity rather than in type. A trait such as height is a quantitative trait. Heights are not found in discrete categories but differ merely in quantity from one person to the next. The opposite of a quantitative trait is a "discrete trait," in which the phenotypes differ in kind—for example, brown eyes versus blue eyes.

Quantitative traits are typically influenced not only by the alleles of two or more genes but also by

the effects of environment. Therefore, with a quantitative trait, the phenotype of an organism is potentially influenced by

- **Genetic factors** in the form of alternative genotypes of one or more genes.

- **Environmental factors** in the form of conditions that are favorable or unfavorable for the development of the trait. Examples include the effect of smoking on the development of lung cancer in human beings, that of nutrition on the growth rate of animals, and that of fertilizer, rainfall, and planting density on the yield of crop plants.

With some quantitative traits, differences in phenotype result largely from differences in genotype, and the environment plays a minor role. With other quantitative traits, differences in phenotype result largely from the effects of environment, and genetic factors play a minor role. Most quantitative traits fall between these extremes, so both genotype and environment must be taken into account in their analysis.

In a genetically heterogeneous population, many genotypes are formed by the processes of segregation and recombination. Variation in genotype can be eliminated by studying *inbred lines*, which are homozygous for most genes, or the F_1 progeny from a cross of inbred lines, which are uniformly heterozygous for all genes in which the parental inbreds differ (Figure 15.1). In contrast, complete elimination of environmental variation is impossible, no matter how hard the experimenter may try to render the environment identical for all members of a population. With plants, for example, small variations in soil quality or exposure to the sun will produce slightly different environments, sometimes even for adjacent plants. Similarly, highly inbred *Drosophila* still show variation in phenotype (for example, in body size) brought about by environmental differences among animals within the same culture bottle. Therefore, traits that are susceptible to small environmental effects will never be uniform, even in inbred lines.

Important quantitative traits in human genetics include infant growth rate, adult weight, blood pressure, serum cholesterol, and length of life. In plant and animal breeding, traits of key economic importance are often quantitative traits. One economically important quantitative trait in crop plants is yield per unit area—whether it be the yield of corn, tomatoes, soybeans, or grapes. In domestic animals, important quantitative traits include meat quality, milk production per cow, egg laying per hen, fleece weight per sheep, and litter size per sow. In evolutionary studies, fitness is the preeminent quantitative trait. Most quantitative traits cannot be studied by means of the usual pedigree methods,

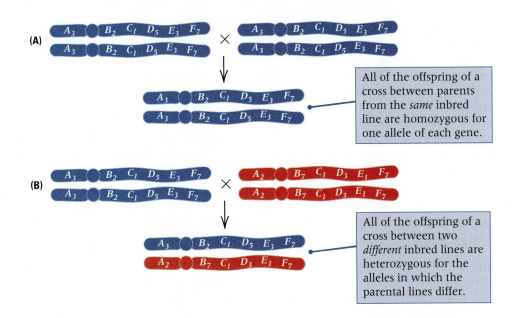

All of the offspring of a cross between parents from the *same* inbred line are homozygous for one allele of each gene.

All of the offspring of a cross between two *different* inbred lines are heterozygous for the alleles in which the parental lines differ.

Figure 15.1 By definition, a completely inbred line is homozygous for every gene. A population of organisms, all identical in genotype, can be created by crossing inbred parents. (A) If the parents are from the same inbred line, the progeny are all genetically identical and homozygous. (B) If the parents are from different inbred lines, the progeny are all genetically identical but heterozygous for alleles at which the parental inbred lines differ.

because the effects of segregation of alleles of one gene may be concealed by effects of other genes, and because environmental effects may cause identical genotypes to have different phenotypes. Therefore, individual pedigrees of quantitative traits do not fit any simple pattern of dominance, recessiveness, or X linkage. Nevertheless, genetic effects on quantitative traits can be assessed by comparing the phenotypes of relatives who, because of their familial relationship, must have a certain proportion of their genes in common. Such studies utilize many of the concepts of population genetics discussed in Chapter 14.

Three categories of traits are frequently found to have quantitative inheritance. They are described in the following section.

■ Continuous, categorical, and threshold traits are usually multifactorial.

Most phenotypic variation in populations is not manifested in a few easily distinguished categories. Instead, the traits vary continuously from one phenotypic extreme to the other, with no clear-cut breaks in between. Human height is a prime example of such a trait. Other examples include milk production in cattle, growth rate in poultry, yield in corn, and blood pressure in human beings. Such traits are called **continuous traits** because there is a continuous gradation from one phenotype to the next. The range of phenotypes is continuous, from minimum to maximum, with no clear categories. Weight is an example of a continuous trait because the weight of an organism can fall anywhere along a continuous scale of weights, so the number of possible phenotypes is virtually unlimited.

Two other types of quantitative traits are not continuous:

Categorical traits are traits in which the phenotype corresponds to any one of a number of discrete categories. Typically the phenotype corresponds to a count of, for example, the number of skin ridges forming the fingerprints, number of kernels on an ear of corn, number of eggs laid by a hen, number of bristles on the abdomen of a fly, and number of puppies in a litter. An example of a categorical trait is the number of ears on a stalk of corn, which typically has the value 1, 2, 3, or 4 ears on a given stalk.

Threshold traits are traits that have only two, or a few, phenotypic classes, but their inheritance is determined by the effects of multiple genes together with the environment. Examples of threshold traits include twinning in cattle and parthenogenesis (development of unfertilized eggs) in turkeys. In a threshold trait, each organism has an underlying and not directly observable predisposition to express the trait, such as a predisposition for a cow to give birth to twins. If the underlying predisposition is high enough (above a "threshold"), the cow will actually give birth to twins; otherwise, she will give birth to a single calf. Many human diseases with a genetic component are threshold-trait disorders, and the phenotypic classes are "affected" versus "not affected." Examples of such disorders include adult-onset diabetes, schizophrenia, and many congenital abnormalities, such as spina bifida. Threshold traits can be interpreted as continuous traits by imagining that each individual has an underlying risk or *liability* toward manifestation of the condition. A liability above a certain cutoff, or *threshold*, results in expression of the condition; a liability below the threshold results in normality. The liability of an individual to a threshold trait cannot be observed directly, but inferences about liability can be drawn from the incidence of the condition among individuals and their relatives. The manner in which this is done is discussed in Section 15.4.

■ The distribution of a trait in a population implies nothing about its inheritance.

The **distribution** of a trait in a population is a description of the population in terms of the proportion of individuals that have each of the possible phenotypes. Characterizing the distribution of some traits is straightforward because the number of phenotypic classes is small. For example, the distribution of progeny in a certain pea cross may consist of 3/4 green seeds and 1/4 yellow seeds, and the distribution of ABO blood groups among one sample of Greeks may consist of 42 percent O, 39 percent A, 14 percent B, and 5 percent AB. However, with continuous traits, the large number of possible phenotypes makes such summaries impractical. Often, it is convenient to reduce the number of phenotypic classes by grouping similar phenotypes together. Data for an example pertaining to the distribution of height among 4995 British women are given in **Table 15.1** and in Figure 15.2 . You can imagine each bar in the graph in Figure 15.2

being built step-by-step, as each of the women is measured, by placing a small square along the x-axis at the location corresponding to the height of each woman. As sampling proceeds, the squares begin to pile up in certain places, leading ultimately to the bar graph shown.

It is worth taking a moment to consider a more vivid example of what a histogram like that in Figure 15.2 really represents. The "real thing" is shown in Figure 15.3 , which is a picture of 162 scholars in genetics at the University of Connecticut, Storrs, who have arranged themselves in order of their height. The women are in white, the men in blue. This is a "living histogram," in which each building block not merely represents a person but actually *is* a person.

Displaying a distribution completely, either in tabular form, as in Table 15.1, or in graphical form, as in Figure 15.2 (or even Figure 15.3), is always helpful but often unnecessary. In many cases a description of the distribution in terms of two major features is sufficient. These features are the *mean* and the *variance*. To discuss the mean and the variance in quantitative terms, we shall use the data in Table 15.1. The height intervals are numbered from 1 (53 to 55 inches) to 11 (73 to 75 inches). The symbol x_i designates the midpoint of the height interval numbered i; for example, $x_1 = 54$ inches, $x_2 = 56$ inches, and so on. The number of women in height interval i is designated f_i; for example, $f_1 = 5$ women, $f_2 = 33$ women, and so forth. The total size of the sample—in this case 4995—is denoted N. The mean and variance serve to characterize the distribution of height among

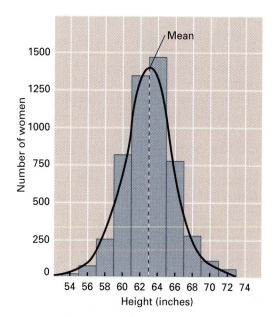

Figure 15.2 Distribution of height among 4995 British women and the smooth normal distribution that approximates the data.

Table 15.1

Distribution of height among British women

Interval number, i	Height interval (inches)	Midpoint, x_i	Number of women, f_i
1	53–55	54	5
2	55–57	56	33
3	57–59	58	254
4	59–61	60	813
5	61–63	62	1340
6	63–65	64	1454
7	65–67	66	750
8	67–69	68	275
9	69–71	70	56
10	71–73	72	11
11	73–75	74	4
		Total $N =$	4995

Figure 15.3 Genetics scholars at the University of Connecticut in Storrs, who have helpfully arranged themselves by height to form a "living histogram." [Photo by Peter Morenus/UCONN, courtesy of Linda Strausbaugh.]

these women as well as the distribution of many other quantitative traits.

- The **mean,** or average, is the peak of the distribution. The mean of a population is estimated from a sample of individuals from the population as follows:

$$\bar{x} = \frac{\sum f_i x_i}{N} \qquad (15.1)$$

where $\bar{x}$ is the estimate of the mean and Σ symbolizes summation over all classes of data (in this example, summation over all 11 height intervals). In Table 15.1, the mean height in the sample of women is 63.1 inches.

- The **variance** is a measure of the spread of the distribution and is estimated in terms of the squared *deviation* (difference) of each observation from the mean. The variance is estimated from a sample of individuals as follows:

$$s^2 = \frac{\sum f_i (x_i - \bar{x})^2}{N - 1} \qquad (15.2)$$

where s^2 is the estimated variance and x_i, f_i, and N are as in Table 15.1. Note that $(x_i - \bar{x})$ is the

difference from the mean of each height category, and that the denominator is the total number of individuals minus 1. The variance describes the extent to which the phenotypes

© Philip Regan.

Lambs of the Shropshire breed, named after its origin in the county of Shropshire in west-central England where Charles Darwin was born. Once very popular in the United States, the breed was said to have "wool from the tip of the nose to the tip of the toes." By the 1950s the trait had become so extreme that the animals required frequent hair trimming around the eyes to prevent "wool blindness," and they began to lose popularity.

15.1 Multifactorial traits are determined by multiple genes and the environment   535

Figure 15.4 Graphs showing that the variance of a distribution measures the spread of the distribution around the mean. The area under each curve covering any range of phenotypes equals the proportion of individuals having phenotypes within the range.

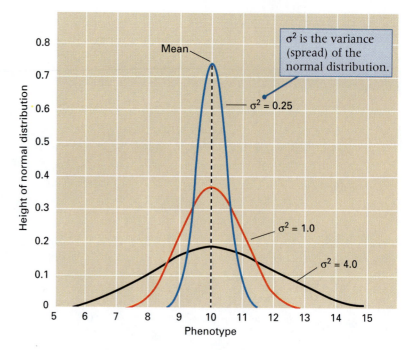

are clustered around the mean, as shown in Figure 15.4. A large value implies that the distribution is spread out, and a small value implies that it is clustered near the mean. From the data in Table 15.1, the variance of the population of British women is estimated as $s^2 = 7.24 \text{ in}^2$.

A quantity closely related to the variance—the **standard deviation** of the distribution—is defined as the square root of the variance. For the data in Table 15.1, the estimated standard deviation s is obtained from Equation 15.2 as $s = (s^2)^{1/2} = (7.24 \text{ in}^2)^{1/2} = 2.69$ inches. The standard deviation has the useful feature of having the same units of dimension as the mean—in this example, inches.

When the data are symmetrical, or approximately symmetrical, the distribution of a trait can often be approximated by a smooth, bell-shaped curve of the type shown in Figure 15.4. The bell curve is called the **normal distribution.** Because the normal curve is symmetrical, half of its area is determined by points with values greater than the mean and half by points with values less than the mean, and thus the proportion of phenotypes that exceed the mean is 1/2. One important characteristic of the normal distribution is that the entire distribution is completely determined by the value of the mean and the variance.

The mean and standard deviation (square root of the variance) of a normal distribution provide a great deal of information about the distribution of phenotypes in a population, as is illustrated in Figure 15.5. Specifically, for a normal distribution,

1. Approximately 68 percent of the population have a phenotype within *one* standard deviation of the mean (in the symbols of Figure 15.5, between $\mu - \sigma$ and $\mu + \sigma$).

2. Approximately 95 percent lie within *two* standard deviations of the mean (between $\mu - 2\sigma$ and $\mu + 2\sigma$).

3. Approximately 99.7 percent lie within *three* standard deviations of the mean (between $\mu - 3\sigma$ and $\mu + 3\sigma$).

Applying these rules to the data in Figure 15.2, in which the mean and standard deviation are 63.1 and 2.69 inches, reveals that approximately 68 percent of the women are expected to have heights in the range from $63.1 - 2.69$ inches to $63.1 + 2.69$ inches (that is, 60.4 to 65.8), and approximately 95 percent are expected to have heights in the range from $63.1 - 2 \times 2.69$ inches to $63.1 + 2 \times 2.69$ inches (that is, 57.7 to 68.5).

Real data frequently conform to the normal distribution. Normal distributions are usually the rule when the phenotype is determined by the cumulative effect of many individually small independent factors. This is the case for many multifactorial traits.

15.2

Variation in a trait can be separated into genetic and environmental components.

In considering the genetics of multifactorial traits, an important objective is to assess the relative importance of genotype versus environment. In some cases in experimental organisms, it is possible to separate genotype and environment with respect to their effects on the mean. For example, a plant breeder may study the yield of a series of inbred lines grown in a group of environments that differ

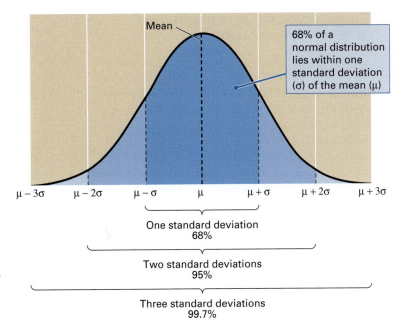

Mean

68% of a normal distribution lies within one standard deviation (σ) of the mean (μ)

$\mu - 3\sigma$ $\mu - 2\sigma$ $\mu - \sigma$ μ $\mu + \sigma$ $\mu + 2\sigma$ $\mu + 3\sigma$

One standard deviation
68%

Two standard deviations
95%

Three standard deviations
99.7%

Figure 15.5 Features of a normal distribution. The proportions of individuals lying within one, two, and three standard deviations from the mean are approximately 68 percent, 95 percent, and 99.7 percent, respectively. In this normal distribution, the mean is symbolized μ and the standard deviation σ.

in planting density or amount of fertilizer. It would then be possible

1. to compare yields of the same genotype grown in different environments and thereby rank the *environments* relative to their effects on yield, or

2. to compare yields of different genotypes grown in the same environment and thereby rank the *genotypes* relative to their effects on yield.

Such a fine discrimination between genetic and environmental effects is not usually possible, particularly in human quantitative genetics. For example, with regard to the height of the women in Figure 15.2, environment could be considered favorable or unfavorable for tall stature only in comparison with the mean height of a genetically identical population reared in a different environment. This reference population does not exist. Likewise, the genetic composition of the population could be judged as favorable or unfavorable for tall stature only in comparison with the mean of a genetically different population reared in an identical environment. This reference population does not exist, either.

Without such standards of comparison, it is impossible to determine the genetic versus environmental effects on the mean. However, it is still possible to assess genetic versus environmental contributions to the *variance*, because instead of comparing the means of two or more populations, we can compare the phenotypes of individuals within the *same* population. Some of the differences in phenotype result from differences in genotype and others from differences in environment, and it is often possible to separate these effects.

In any distribution of phenotypes, such as the one in Figure 15.2, four sources contribute to phenotypic variation:

1. Genotypic variation

2. Environmental variation

3. Variation due to genotype-by-environment interaction

4. Variation due to genotype-by-environment association

Each of these sources of variation is discussed in the following sections.

© Courtesy of Mary Hollinger, NODC biologist, NOAA

The modern world presents many environmental hazards to which organisms have not been adapted by natural selection. In people, for example, a high salt, high fat, high calorie fast-food diet contributes to hypertension and obesity. As another example, a modern water hazard—an oil slick—has soiled this unfortunate swan.

The genotypic variance results from differences in genotype.

The variation in phenotype caused by differences in genotype among individuals is termed **genotypic variance.** Figure 15.6 illustrates the genotypic variation expected among the F_2 generation from a cross of two inbred lines differing in genotype for three unlinked genes. The alleles of the three genes are represented as A / a, B / b, and C / c, and the genetic variation in the F_2 generation caused by segregation and recombination is evident in the differences in color. Relative to a categorical trait (one whose phenotype is determined by counting, such as ears per stalk in corn), if we assume that each uppercase allele is favorable for the expression of

the trait and adds one unit to the phenotype, whereas each lowercase allele is without effect, then the *aa bb cc* genotype has a phenotype of 0 and the *AA BB CC* genotype has a phenotype of 6. There are seven possible phenotypes (0 through 6) in the F_2 generation. The distribution of phenotypes in the F_2 generation is shown in the colored bar graph in Figure 15.7 . The normal distribution approximating the data has a mean of 3 and a variance of 1.5. In this case, we are assuming that *all* of the variation in phenotype in the population results from differences in genotype among the individuals.

Figure 15.7 also includes a bar graph with diagonal lines representing the theoretical distribution when the trait is determined by 30 unlinked genes segregating in a randomly mating population,

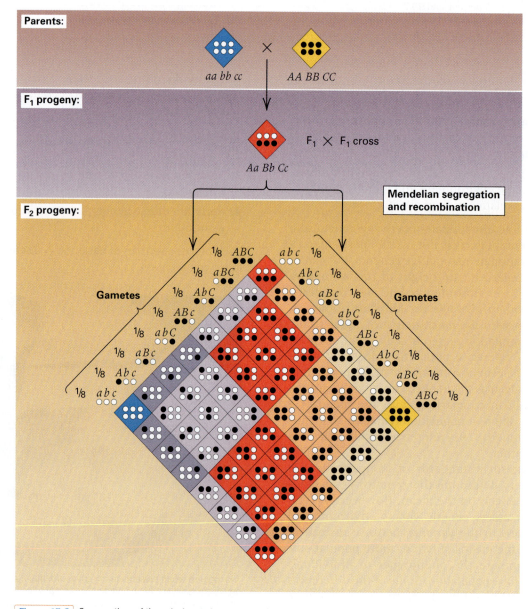

Figure 15.6 Segregation of three independent genes affecting a quantitative trait. Each uppercase allele in a genotype contributes one unit to the phenotype.

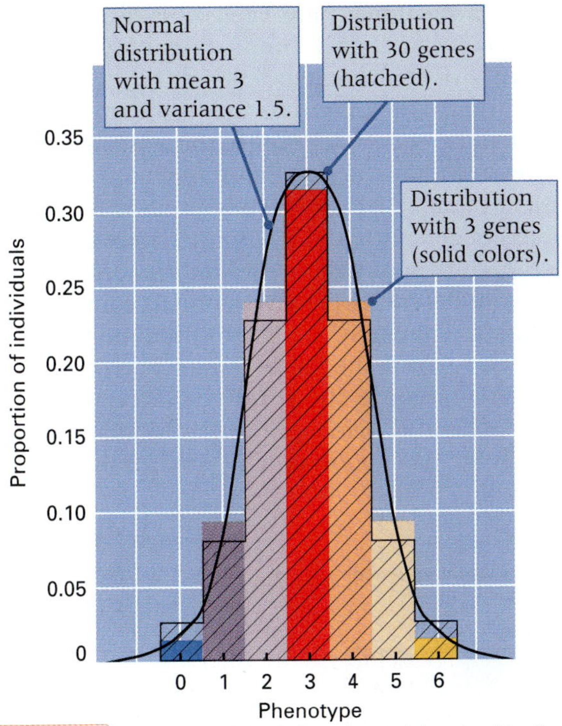

Normal distribution with mean 3 and variance 1.5.
Distribution with 30 genes (hatched).
Distribution with 3 genes (solid colors).
0.35
0.30
0.25
0.20
0.15
0.10
0.05
0
Proportion of individuals
0 1 2 3 4 5 6
Phenotype

Figure 15.7 The distribution of phenotypes determined by the segregation of 3 and 30 independent genes. Both distributions are approximated by the same normal distribution (black curve).

grouped into the same number of phenotypic classes as the 3-gene case. We assume that 15 of the genes are nearly fixed for the favorable allele and that 15 are nearly fixed for the unfavorable allele. The contribution of each favorable allele to the phenotype has been chosen to make the mean of the distribution equal to 3 and the variance equal to 1.5. Note that the distribution with 30 genes is virtually identical to that with three genes and that both are approximated by the same normal curve. If such distributions were encountered in actual research, the experimenter would not be able to distinguish between them. The key point is that

key concept

Even in the absence of environmental variation, the distribution of phenotypes, by itself, provides no information about the number of genes influencing a trait and no information about the dominance relations of the alleles.

However, the number of genes influencing a quantitative trait is important in determining the potential for long-term genetic improvement of a population by means of artificial selection. For example, in the 3-gene case in Figure 15.7, the best possible genotype would have a phenotype of 6, but in the 30-gene case, the best possible genotype (homozygous for the favorable allele of all 30 genes) would have a phenotype of 60.

Later in this chapter, some methods for estimating the number of genes affecting a quantitative trait will be presented. All the methods depend on comparing the phenotypic distributions in the F_1 and F_2 generations of crosses between nearly or completely homozygous lines.

The environmental variance results from differences in environment.

The variation in phenotype caused by differences in environment among individuals is termed **environmental variance.** Figure 15.8 is an example showing the distribution of seed weight in edible beans. The mean of the distribution is 500 mg and the standard deviation is 95 mg. All of the beans in this population are genetically identical and homozygous because they are highly inbred. Therefore, in this population, *all* of the phenotypic variation in seed weight results from environmental variance. A comparison of Figures 15.7 and 15.8 demonstrates the following principle:

key concept

The distribution of a trait in a population provides no information about the relative importance of genotype and environment. Variation in the trait can be entirely genetic, entirely environmental, or a combination of both influences.

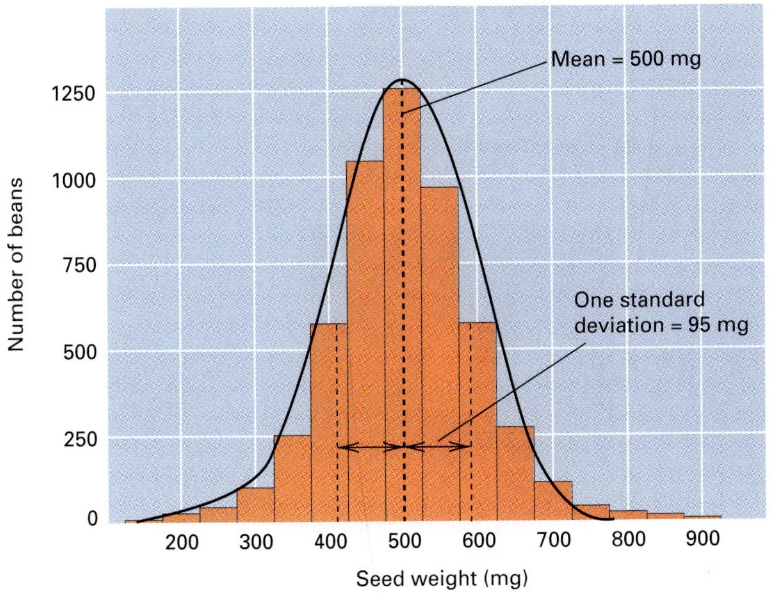

Mean = 500 mg
One standard deviation = 95 mg
1250
1000
750
500
250
0
Number of beans
200 300 400 500 600 700 800 900
Seed weight (mg)

Figure 15.8 Distribution of seed weight in a homozygous line of edible beans. All variation in phenotype among individuals results from environmental differences.

Genotypic and environmental variance are seldom separated as clearly as in Figures 15.7 and 15.8, because usually they work together. Their combined effects are illustrated for a simple hypothetical case in Figure 15.9. At the upper left is the distribution of phenotypes for three genotypes assumed not to be influenced by environment. As depicted, the trait can have one of three distinct and nonoverlapping phenotypes determined by the effects of two additive alleles. The genotypes are in random-mating proportions for an allele frequency of 1/2, and the distribution of phenotypes has mean 5 and variance 2. Because it results solely from differences in genotype, this variance is *genotypic variance*, which is symbolized σ_g^2. The three panels at the upper right depict the distribution of phenotypes of each of the three genotypes in the presence of environmental variation. In each case the variance in phenotype that is due to environment alone equals 1. Because this variance results solely from differences in environment, it is *environmental variance*, which is symbolized σ_e^2. When the effects of genotype and environment are combined in the same population, both differences in genotype and differences in environment contribute to variation in the trait, and the distribution shown in the lower part of the figure results. The variance of this distribution is the **total variance** in phenotype, which is symbolized σ_p^2. Because we are assuming that genotype and environment have

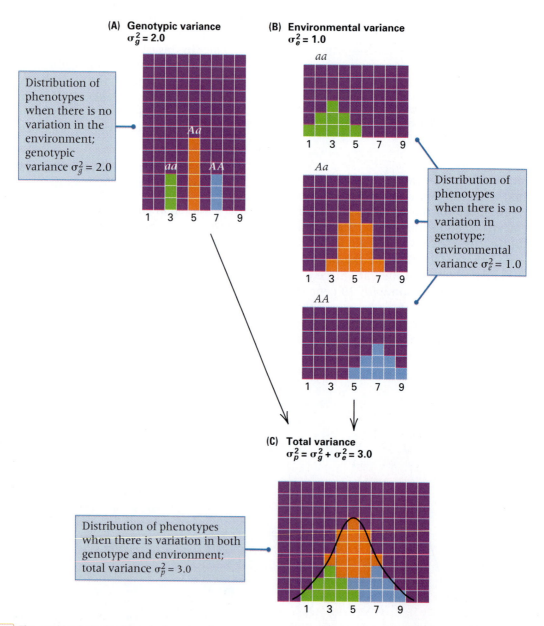

(A) **Genotypic variance** $\sigma_g^2 = 2.0$

Distribution of phenotypes when there is no variation in the environment; genotypic variance $\sigma_g^2 = 2.0$

(B) **Environmental variance** $\sigma_e^2 = 1.0$

Distribution of phenotypes when there is no variation in genotype; environmental variance $\sigma_e^2 = 1.0$

(C) **Total variance** $\sigma_p^2 = \sigma_g^2 + \sigma_e^2 = 3.0$

Distribution of phenotypes when there is variation in both genotype and environment; total variance $\sigma_p^2 = 3.0$

Figure 15.9 The combined effects of genotypic and environmental variance. (A) Population affected only by genotypic variance σ_g^2. (B) Populations of each genotype separately showing the effects of environmental variance σ_e^2. (C) Population affected by both genotypic and environmental variance; the total phenotypic variance σ_p^2 equals the sum of σ_g^2 and σ_e^2.

separate and independent effects on phenotype, we expect σ_p^2 to be greater than either σ_g^2 or σ_e^2 alone. In fact,

$$\sigma_p^2 = \sigma_g^2 + \sigma_e^2 \qquad (15.3)$$

In words, Equation (15.3) states that

> **key concept**
>
> When genetic and environmental effects contribute independently to phenotype, the total variance equals the sum of the genotypic and environmental variance.

Equation (15.3) is one of the most important relations in quantitative genetics. How it can be used to analyze data will be explained shortly. Although the equation serves as an excellent approximation in very many cases, it is valid in an exact sense only when genotype and environment are independent in their effects on phenotype. The two most important departures from independence are discussed in the next section.

■ Genotype and environment can interact, or they can be associated.

In the simplest cases, environmental effects on phenotype are additive, and each environment adds (or detracts) the same amount to (or from) the phenotype, independent of the genotype. When this is not true, the environmental effects on phenotype differ according to genotype, and a **genotype-by-environment interaction (G-E interaction)** is said to be present. In some cases, G-E interaction can even change the relative ranking of the genotypes, so a genotype that is superior in one environment may become inferior in another.

An example of genotype-by-environment interaction in maize is illustrated in Figure 15.10 . The two strains of corn are hybrids formed by crossing different pairs of inbred lines, and their overall means, averaged across all of the environments, are approximately the same. However, the strain designated A clearly outperforms strain B in the negative, stressful environments, whereas the performance is reversed when the environment is of high quality. (Environmental quality is judged on the basis of soil fertility, moisture, and other factors.) In some organisms, particularly plants, experiments like those illustrated in Figure 15.10 can be carried out to determine the contribution of G-E interaction to the total observed variation in phenotype. In other organisms, particularly human beings, the effect cannot be evaluated separately.

Interaction of genotype and environment is common and is very important in both plants and animals. Because of this interaction, no one plant variety outperforms all others in all types of soil and climate, and therefore plant breeders must develop special varieties that are suited to each growing area.

Another important type of interaction is **genotype-by-sex interaction,** in which the same genotype results in a different phenotype according to the sex of the organism. Genotype-by-sex interactions are very common in quantitative genetics. One example is seen in the living histogram in Figure 15.3. The distribution of height among the women (dressed in white) is clearly shifted to the left relative to the distribution of height among men (dressed in blue). The averages differ by more than 5 inches (64.8 inches for the women, 70.1 inches for the men). Yet there is no reason to think that the genes that affect height are distributed differently in women and men. The genes simply have somewhat different effects depending on sex—an example of genotype-by-sex interaction.

When different genotypes in a population are not distributed at random among all the possible environments, there is **genotype-by-environment association (G-E association).** In these circumstances, certain genotypes are preferentially associated with certain environments, which may either increase or decrease the phenotype of these genotypes compared with what would result in the absence of G-E association. An example of deliberate genotype-by-environment association can be found in dairy husbandry, in which some farmers feed each of their cows in proportion to its level of

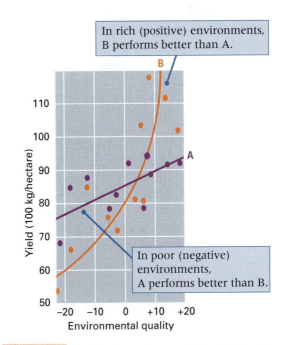

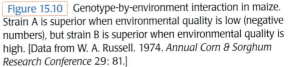

Figure 15.10 Genotype-by-environment interaction in maize. Strain A is superior when environmental quality is low (negative numbers), but strain B is superior when environmental quality is high. [Data from W. A. Russell. 1974. *Annual Corn & Sorghum Research Conference* 29: 81.]

milk production. Because of this practice, cows with superior genotypes with respect to milk production also have a superior environment in that they receive more feed. In plant breeding, genotype-by-environment association can often be eliminated or minimized by appropriate randomization of genotypes within the experimental plots. In other cases, human genetics again being a prime example, the possibility of G-E association cannot usually be controlled.

■ There is no genotypic variance in a genetically homogeneous population.

Equation (15.3) can be used to separate the effects of genotype and environment on the total phenotypic variance. Two types of data are required:

1. The phenotypic variance of a genetically uniform population, which provides an estimate of σ_e^2 because a genetically uniform population has a value of $\sigma_g^2 = 0$

2. The phenotypic variance of a genetically heterogeneous population, which provides an estimate of $\sigma_g^2 + \sigma_e^2$

An example of a genetically uniform population is the F_1 generation from a cross between two highly homozygous strains, such as inbred lines (Figure 15.1). An example of a genetically heterogeneous population is the F_2 generation from the same cross, as indicated in Figure 15.6. If the environments of both populations are the same, and if there is no G-E interaction, then the estimates may be combined to deduce the value of σ_g^2.

To take a specific numerical illustration, consider variation in the size of the eyes in the cave-dwelling fish *Astyanax*, reared in the same environments (Figure 15.11). The variances in eye diameter in the F_1 and F_2 generations from a cross of two highly homozygous strains were estimated as 0.057 and 0.563, respectively. Written in terms of the components of variance, these are

$$F_2:\ \sigma_p^2 = \sigma_g^2 + \sigma_e^2 = 0.563$$
$$F_1:\ \sigma_e^2 = 0.057$$

The estimate of genotypic variance σ_g^2 is obtained by subtracting the second equation from the first; that is,

$$\sigma_g^2 = 0.563 - 0.057 = 0.506$$

because

$$(\sigma_g^2 + \sigma_e^2) - \sigma_e^2 = \sigma_g^2$$

Hence the estimate of σ_g^2 is 0.506, whereas that of σ_e^2 is 0.057. In this example, the genotypic variance is much greater than the environmental variance, but this is not always the case.

Figure 15.11 Reduced eye size and pigmentation in a cave-dwelling *Astyanax* (above), compared with a surface-dwelling relative (below). [Courtesy of Richard Borowsky.]

The next section shows what information can be obtained from an estimate of the genotypic variance.

■ The number of genes affecting a quantitative trait need not be large.

When the number of genes influencing a quantitative trait is not too large, knowledge of the genotypic variance can be used to estimate the number of genes. All that is needed is the means and variances of two phenotypically divergent strains and their F_1, F_2, and backcrosses. In ideal cases, the data appear as in Figure 15.12 , in which P_1 and P_2 represent the divergent parental strains (for example, inbred lines). The points lie on a triangle, with increasing variance according to the increasing genetic heterogeneity (genotypic variance) of the populations. If the F_1 and backcross means lie exactly between their parental means, then these means will lie at the midpoints along the sides of the triangle, as shown in Figure 15.12. This finding implies that the alleles affecting the trait are *additive*; that is, for each gene, the phenotype of the heterozygote is the average of the phenotypes of the corresponding homozygotes. In such a simple situation, it may be shown that the number, n, of genes contributing to the trait is

$$n = \frac{D^2}{8\sigma_g^2} \qquad (15.4)$$

in which D represents the difference between the means of the original parental strains, P_1 and P_2. This equation can be verified by applying it to the ideal case in Figure 15.9. In this case, the parental strains would be the homozygous genotypes AA, with a mean phenotype of 7, and aa, with a mean phenotype of 3. Consequently, $D = 7 - 3 = 4$. The geno-

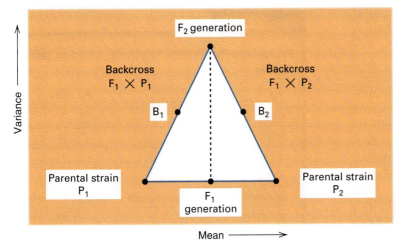

Figure 15.12 Means and variances of parents (P) and of backcross (B) and hybrid (F) progeny of inbred lines for an ideal quantitative trait affected by unlinked and completely additive genes. [After R. Lande. 1981. *Genetics* 99: 541.]

example, after a population of the flour beetle *Tribolium* was bred for increased pupa weight, the mean value for pupa weight was found to be 17 standard deviations above the mean of the original population. Determination of traits by a large number of genes implies that

typic variance is given in Figure 15.9 as $\sigma_g^2 = 2$. Substituting D and σ_g^2 into Equation (15.4), we obtain $n = 16/(8 \times 2) = 1$, which is correct because there is only one gene, with alleles A and a, affecting the trait.

Applied to actual data, Equation (15.4) requires several assumptions that are not necessarily valid. In addition to the assumption that all generations are reared in the same environment, the theory also makes the genetic assumptions that (1) the alleles of each gene are additive, (2) the genes contribute equally to the trait, (3) the genes are unlinked, and (4) the original parental strains are homozygous for alternative alleles of each gene. However, when the assumptions are invalid, the outcome is that the calculated n is smaller than the actual number of genes affecting the trait. The estimated number is a minimum, because almost any departure from the genetic assumptions leads to a smaller genotypic variance in the F_2 generation and so, for the same value of D, would yield a larger value of n in Equation (15.4). This is why the estimated n is the *minimum* number of genes that can account for the data. For the cave-dwelling *Astyanax* fish discussed in the preceding section, the parental strains had average phenotypes of 7.05 and 2.10, giving $D = 4.95$. The estimated value of $\sigma_g^2 = 0.506$, so the minimum number of genes affecting eye diameter is $n = (4.95)^2/(8 \times 0.506) = 6.0$. Therefore, at least six different genes affect the diameter of the eye of the fish.

The number of genes that affect a quantitative trait is important because it influences the amount by which a population can be genetically improved by selective breeding. With traits determined by a small number of genes, the potential for change in a trait is relatively small, and a population consisting of the best possible genotypes may have a mean value that is only two or three standard deviations above the mean of the original population. However, traits determined by a large number of genes have a large potential for improvement. For

key concept

Selective breeding can create an improved population in which the value of *every* individual greatly exceeds that of the *best* individuals that existed in the original population.

This principle at first seems paradoxical, because in a large enough population, every possible genotype should be created at some low frequency. The explanation of the paradox is that real populations subjected to selective breeding typically consist of a few hundred organisms (at most), and thus many of the theoretically possible genotypes are never formed because their frequencies are much too rare. As selection takes place and the allele frequencies change, these genotypes become more common and allow the selection of superior organisms in future generations.

■ The broad-sense heritability includes all genetic effects combined.

Estimates of the numbers of genes that determine quantitative traits are frequently unavailable because the necessary experiments are impractical or have not been carried out. Another attribute of quantitative traits, which requires less data to evaluate, makes use of the ratio of the genotypic variance to the total phenotypic variance. This ratio of σ_g^2 to σ_p^2 is called the **broad-sense heritability,** symbolized H^2, and it measures the importance of genetic variation, relative to environmental variation, in causing variation in the phenotype of a trait of interest. Broad-sense heritability is a ratio of variances, specifically

$$H^2 = \frac{\sigma_g^2}{\sigma_p^2} = \frac{\sigma_g^2}{\sigma_g^2 + \sigma_e^2} \qquad (15.5)$$

Substitution of the data for eye diameter in *Astyanax*, in which $\sigma_g^2 = 0.506$ and $\sigma_g^2 + \sigma_e^2 = 0.563$, into Equation (15.5) yields $H^2 = 0.506/0.563 = 0.90$ for the estimate of broad-sense heritability. This value implies that 90 percent of the variation in eye diameter in this population results from differences in genotype among the fish.

Knowledge of heritability is useful in the context of plant and animal breeding because heritability can be used to predict the magnitude and speed of population improvement. The broad-sense heritability defined in Equation (15.5) is used in predicting the outcome of selection practiced among clones, inbred lines, or varieties. Analogous predictions for randomly bred populations utilize another type of heritability (the narrow-sense heritability), which will be discussed in the next section. Broad-sense heritability measures how much of the total variance in phenotype results from differences in genotype. For this reason, H^2 is often of interest in regard to human quantitative traits.

■ Twin studies are often used to assess genetic effects on variation in a trait.

In human beings, twins would seem to be ideal subjects for separating genotypic and environmental variance because **identical twins,** who arise from the splitting of a single fertilized egg, are genetically identical and are often strikingly similar in such traits as facial features and body build. **Fraternal twins,** who arise from two fertilized eggs, have the same genetic relationship as ordinary siblings, so only half of the genes in either twin are identical with those in the other. Theoretically, the variance between members of an identical-twin pair would be equivalent to σ_e^2, because the twins are genetically identical. The variance between members of a fraternal-twin pair would include not only σ_e^2 but also part of the genotypic variance (approximately $\sigma_g^2/2$ because of the identity of half of the genes in fraternal twins). Consequently, both σ_g^2 and σ_e^2 could be estimated from twin data and combined as in Equation (15.5) to estimate H^2. Table 15.2 summarizes estimates of H^2 based on twin studies of several traits.

Unfortunately, twin studies are subject to several important sources of error, most of which increase the similarity of identical twins, so the numbers in **Table 15.2** should be considered very approximate and probably too high. Here are four of the potential sources of error:

1. Genotype-by-environment interaction, which increases the variance in fraternal twins but not in identical twins

2. Frequent sharing of embryonic membranes between identical twins, resulting in a more similar intrauterine environment

3. Greater similarity in the treatment of identical twins by parents, teachers, and peers, resulting

Table 15.2

Broad-sense heritability, in percent, based on twin studies

Trait	Heritability, H^2
Longevity	29
Height	85
Weight	63
Amino acid excretion	72
Serum lipid levels	44
Maximum blood lactate	34
Maximum heart rate	84
Verbal ability	63
Numerical ability	76
Memory	47
Sociability index	66
Masculinity index	12
Temperament index	58

in a decreased environmental variance in identical twins

4. Different sexes in half of the pairs of fraternal twins, in contrast with the same sex of identical twins

These pitfalls and others imply that data from human twin studies should be interpreted with caution and reservation.

15.3

Artificial selection is a form of "managed evolution."

The practice of breeders to choose a select group of organisms from a population to become the parents of the next generation is termed **artificial selection.** When artificial selection is carried out either by choosing the best organisms in a species that reproduces asexually or by choosing the best among subpopulations propagated by close inbreeding (such as self-fertilization), broad-sense heritability permits an assessment of how rapidly progress can be achieved. Broad-sense heritability is important in this context, because with clones, inbred lines, or varieties, superior genotypes can be perpetuated without disruption of favorable gene combinations by Mendelian segregation. An example is the selection of superior varieties of plants that are propagated asexually by means of cuttings or grafts, or of animals that are reproduced by cloning. Because there is no sexual reproduction, each offspring has exactly the same genotype as its parent.

In sexually reproducing populations that are genetically heterogeneous, broad-sense heritability is not relevant in predicting progress resulting from artificial selection, because superior genotypes must necessarily be broken up by the processes of segregation and recombination. For example, if the best genotype is heterozygous for each of two unlinked loci, $A/a; B/b$, then because of segregation and independent assortment, among the progeny of a cross between parents with the best genotypes—$A/a; B/b \times A/a; B/b$—only 1/4 will have the same favorable $A/a; B/b$ genotype as the parents. The rest of the progeny will be genetically inferior to the parents. For this reason, to the extent that high genetic merit may depend on particular combinations of alleles, each generation of artificial selection results

in a slight setback in that the offspring of superior parents are generally not quite so good as the parents themselves. Progress under selection can still be predicted, but the prediction must make use of another type of heritability, the narrow-sense heritability, which is discussed in the next section.

■ The narrow-sense heritability is usually the most important in artificial selection.

Figure 15.13 illustrates a typical form of artificial selection and its result. The organism is *Nicotiana longiflora* (tobacco), and the trait is the length of the corolla tube (*corolla* is a collective term for all the petals of a flower). Part A shows the distribution of phenotypes in the parental generation, and part B shows the distribution of phenotypes in the offspring generation. The parental generation is the

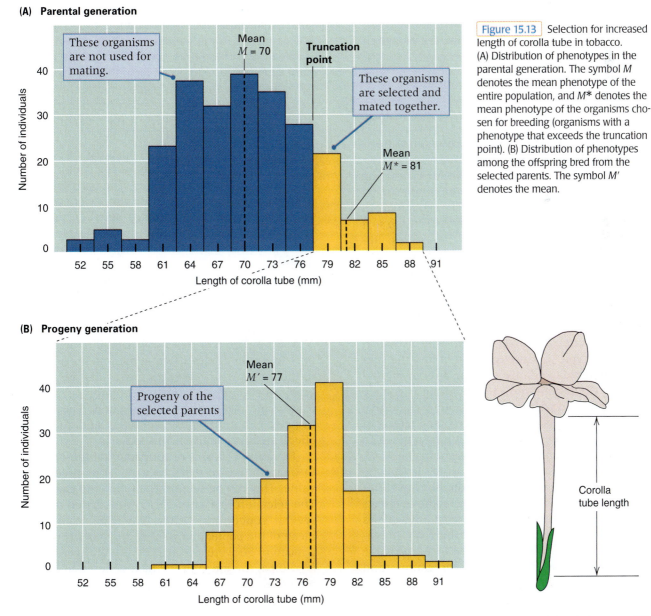

Figure 15.13 Selection for increased length of corolla tube in tobacco. (A) Distribution of phenotypes in the parental generation. The symbol M denotes the mean phenotype of the entire population, and M^* denotes the mean phenotype of the organisms chosen for breeding (organisms with a phenotype that exceeds the truncation point). (B) Distribution of phenotypes among the offspring bred from the selected parents. The symbol M' denotes the mean.

(A) Parental generation

These organisms are not used for mating.

Mean $M = 70$

Truncation point

These organisms are selected and mated together.

Mean $M^* = 81$

Length of corolla tube (mm)

(B) Progeny generation

Mean $M' = 77$

Progeny of the selected parents

Length of corolla tube (mm)

Corolla tube length

population from which the parents were chosen for breeding. The type of selection is called **individual selection,** because each member of the population to be selected is evaluated according to its own individual phenotype. The selection is practiced by choosing some arbitrary level of phenotype—called the **truncation point**—that determines which individuals will be saved for breeding purposes. All individuals with a phenotype above the threshold are randomly mated among themselves to produce the next generation.

In evaluating progress through individual selection, three distinct phenotypic means are important. In Figure 15.13, these means are symbolized as M, M^*, and M', and they are defined as follows:

1. M is the mean phenotype of the entire population in the parental generation, including both the selected and the nonselected individuals.

2. M^* is the mean phenotype among those individuals selected as parents (those with a phenotype above the truncation point).

3. M' is the mean phenotype among the progeny of selected parents.

The relationship among these three means is given by

$$M' = M + h^2(M^* - M) \qquad (15.6)$$

in which the symbol h^2 is the **narrow-sense heritability** of the trait in question.

Later in this chapter, a method for estimating narrow-sense heritability from the similarity in phenotype among relatives will be explained. In Figure 15.13, h^2 is the only unknown quantity, so it can be estimated from the data themselves. Rearranging Equation (15.6) and substituting the values for the means from Figure 15.13, we get

$$h^2 = \frac{M' - M}{M^* - M} = \frac{77 - 70}{81 - 70} = 0.64$$

In a manner analogous to the way in which total phenotypic variance can be split into the sum of the genotypic variance and the environmental variance (Equation 15.3), the genotypic variance can be split into parts resulting from the additive effects of alleles, dominance effects, and effects of interaction between alleles of different genes. The difference between the broad-sense heritability, H^2, and the narrow-sense heritability, h^2, is that H^2 includes all of these genetic contributions to variation, whereas h^2 includes only the additive effects of alleles. In ordinary language, the additive effects of alleles are the genetic effects that are transmissible from parent to offspring; dominance effects are not transmissible because of segregation, and epistatic (interaction) effects are not transmissible

because of independent assortment and recombination. It follows that from the standpoint of animal or plant improvement, h^2 is the heritability of interest because

<table>
<tr><td style="background:#e8631a;color:white">**key concept**</td></tr>
<tr><td>The narrow-sense heritability, h^2, is the proportion of the variance in phenotype that is transmissible from parents to offspring and that can be used to predict changes in the population mean with individual selection, according to Equation (15.6).</td></tr>
</table>

The distinction between the broad-sense heritability and the narrow-sense heritability can be appreciated intuitively by considering a population in which there is a rare recessive gene. In such a case, most homozygous recessive genotypes come from matings between heterozygous carriers. Some such kindreds have more than one affected offspring. Hence affected siblings can resemble each other more than they resemble their parents. For example, if a is a recessive allele, the mating $Aa \times Aa$, in which both parents show the dominant phenotype, may yield two offspring that are aa, which both show the recessive phenotype; in this case, the offspring are more similar to each other than they are to either of their parents. It is the dominance of the wildtype allele that causes this paradox; dominance and epistasis contribute to the broad-sense heritability of the trait but not to the narrow-sense heritability. The narrow-sense heritability includes only those genetic effects that contribute to the resemblance between parents and their offspring, because narrow-sense heritability measures how similar offspring are to their parents.

In general, the narrow-sense heritability of a trait is smaller than the broad-sense heritability. For example, in the parental generation in Figure 15.13, the broad-sense heritability of corolla tube length is $H^2 = 0.82$. The two types of heritability are equal only when the alleles affecting the trait are additive in their effects; *additive effects* means that each heterozygous genotype shows a phenotype that is exactly intermediate between the phenotypes of the respective homozygous genotypes and that the effects are also additive across loci.

Equation (15.6) is of fundamental importance in quantitative genetics because of its predictive value. This can be seen in the following example. The selection in Figure 15.13 was carried out for several generations. After two generations, the mean of the population was 83, and parents having a mean of 90 were selected. By use of the estimate $h^2 = 0.64$, the mean in the next generation can be predicted. The information provided is that $M = 83$

and $M^* = 90$. Therefore, Equation (15.6) implies that the predicted mean is

$$M' = 83 + (0.64)(90 - 83) = 87.5$$

This value is in good agreement with the observed value of 87.9.

A Moment to Think

Problem: The raised skin ridges on the fingers of each person form a fingerprint pattern that is unique. Each finger can have a ridge pattern in the shape of an arch, a loop, or a whorl, and in fingerprint identification certain conventions are used to define the center and edge of the pattern. The number of ridges between the center and the edge constitutes the fingerprint ridge count for any finger, and the sum of the ridge counts for all 10 fingers constitutes the total fingerprint ridge count. (Arches have no well-defined edge, so fingers with arches are ignored in calculating the total ridge count.) Remarkably, whereas the fingerprint pattern differs even among identical twins, the total ridge count is a highly heritable multifactorial trait with a narrow-sense heritability (h^2) of 86 percent. Suppose a man whose fingerprint ridge count is 1.5 standard deviations above the mean marries a woman whose fingerprint ridge count is 0.5 standard deviation below the mean. What is the expected average fingerprint ridge count among their offspring? (Total ridge count is normally distributed with a mean of 125 and a standard deviation of 50.) (The answer can be found on page 550.)

■ There are limits to the improvement that can be achieved by artificial selection.

Artificial selection is analogous to natural selection in that both types of selection cause an increase in the frequency of alleles that improve the selected trait (or traits). The principles of natural selection discussed in Chapter 14 also apply to artificial selection. For example, artificial selection is most effective in changing the frequency of alleles that are in an intermediate range of frequency ($0.2 < p < 0.8$). Selection is less effective for alleles with frequencies outside this range, and it is least effective for rare recessive alleles. For quantitative traits, including fitness, the total selection is shared among all the genes that influence the trait, and the selection coefficient affecting each allele is determined by (1) the magnitude of the effect of the allele, (2) the frequency of the allele, (3) the total number of genes affecting the trait, (4) the narrow-sense heritability of the trait, and (5) the proportion of the population that is selected for breeding.

The value of heritability is determined both by the magnitude of effects and by the frequency of alleles. If all favorable alleles were fixed ($p = 1$) or lost ($p = 0$), the heritability of the trait would be 0. Therefore, the heritability of a quantitative trait is expected to decrease over many generations of artificial selection as a result of favorable alleles becoming nearly fixed. For example, ten generations of selection for less fat in a population of Duroc pigs decreased the heritability of fatness from 73 to 30 percent because of changes in allele frequency resulting from the selection.

Population improvement by means of artificial selection cannot continue indefinitely. A population may respond to selection until its mean is many standard deviations different from the mean of the original population, but eventually the population reaches a **selection limit** at which successive generations show no further improvement. (An exception to this generalization is found in traits that are affected by a very large number of genes and in which selection is carried out in a very large population. In such a case, selective advance can be continued indefinitely because new mutations continue to add genetic variation.) In a few cases, progress under selection ceases because all alleles that affect the trait are either fixed or lost, and so the narrow-sense heritability of the trait becomes 0. On the other hand, fixation or loss of all relevant alleles is rare. The usual reason for a selection limit is that natural selection counteracts artificial selection. Many genes that respond to artificial selection as a result of their favorable effect on a selected trait also have indirect harmful effects on fitness. For example, selection for increased size of eggs in poultry results in a decrease in the number of eggs, and selection for extreme body size (large or small) in most animals results in a decrease in fertility. When one trait (for example, number of eggs) changes in the course of selection for a different trait (for example, size of eggs), the unselected trait is said to have undergone a **correlated response** to selection. Correlated response of fitness is typical in long-term artificial selection. Each increment of progress in the selected trait is partially offset by a decrease in fitness because of correlated response. Eventually, artificial selection for the trait of interest is exactly balanced by natural selection against the trait, so a selection limit is reached and no further progress is possible without changing the strategy of selection.

■ Inbreeding is generally harmful, and hybrids may be the best.

Inbreeding can have harmful effects on economically important traits such as yield of grain and egg production. This decline in performance is called **inbreeding depression,** and it results principally from rare harmful recessive alleles becoming homozygous because of inbreeding (Chapter 14). The degree of inbreeding is measured by the inbreeding coefficient F discussed in Chapter 14.

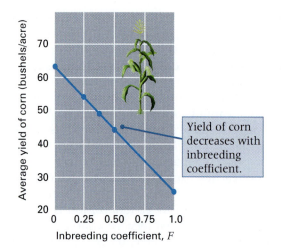

Yield of corn decreases with inbreeding coefficient.

Figure 15.14 Inbreeding depression for yield in corn. [Data from M. Neal. 1935. *J. Amer. Soc. Agronomy* 27: 666.]

Figure 15.14 is an example of inbreeding depression in yield of corn, in which the yield decreases linearly as the inbreeding coefficient increases.

Most highly inbred strains suffer from many genetic defects, as might be expected from deleterious recessive alleles becoming homozygous. One would also expect that if two different inbred strains were crossed, the F_1 would show improved features, because a harmful recessive allele inherited from one parent would be likely to be complemented by a normal dominant allele from the other parent. In many organisms, the F_1 generation of a cross between inbred lines is superior to the parental lines, and the phenomenon is called **heterosis,** or **hybrid vigor.** This phenomenon, which is widely used in the production of corn and other agricultural products, yields genetically identical hybrid plants with traits that are sometimes more favorable than those of the ancestral plants from which the inbreds were derived. The features that most commonly distinguish hybrid plants from their inbred parents are their rapid growth, larger size, and greater yield. Furthermore, the F_1 plants have a fairly uniform phenotype; because the progeny of inbred parents all have the same genotype (Figure 15.1), the genetic variance $\sigma_g^2 = 0$, and so all of the variance in phenotype is due to variation in the environment. Genetically heterogeneous crops with high yields or certain other desirable features can also be produced by traditional plant-breeding programs, but growers often prefer hybrid plants because of this relative uniformity in phenotype. For example, uniform height and time of maturity facilitate machine harvesting, and plants that all bear fruit at the same time accommodate picking and shipping schedules.

Hybrid varieties of corn are used almost exclusively in the United States for commercial crops. A farmer cannot plant the seeds from his crop because the F_2 generation consists of a variety of genotypes, most of which do not show hybrid vigor. The production of hybrid seeds is a major industry in corn-growing sections of the United States. Since the 1930s, when hybrid corn was first introduced into the United States, average yield has increased about fivefold. About 70 percent of this increase is due to the greater productivity of hybrids, the rest to agricultural practices such as the use of fertilizer and irrigation.

15.4

Genetic variation is revealed by correlations between relatives.

Quantitative genetics relies extensively on similarity among relatives to assess the importance of genetic factors. Particularly in the study of complex behavioral traits in human beings, genetic interpretation of familial resemblance is not always straightforward because of the possibility of nongenetic, but nevertheless familial, sources of resemblance. In plant and animal breeding, the situation is usually less complex because genotypes and environments are under experimental control.

■ Covariance is the tendency for traits to vary together.

Genetic data about families are frequently pairs of numbers: pairs of parents, pairs of twins, or pairs consisting of a single parent and a single offspring. An important issue in quantitative genetics is the degree to which the numbers in each pair are associated. The usual way to measure the association is to calculate a statistical quantity called the *correlation coefficient* between the members of each pair.

The correlation coefficient among relatives is based on the covariance in phenotype among them. Much as the variance describes the tendency of a set of measurements to vary [Equation (15.2)], the covariance describes the tendency of pairs of numbers to vary together (co-vary). Calculation of the covariance is similar to that of the variance in Equation (15.2) except that the squared deviation term $(x_i - \bar{x})^2$ is replaced with the product of the deviations of the pairs of measurements from their respective means—that is, $(x_i - \bar{x})(y_i - \bar{y})$.

For example, $(x_i - \bar{x})$ could be the deviation of a particular father's height from the overall father mean, and $(y_i - \bar{y})$ could be the deviation of his son's height from the overall son mean. In symbols, let f_i be the number of pairs of relatives with phenotypic measurements x_i and y_i. Then the estimated **covariance (*Cov*)** of the trait among the relatives is

$$Cov(x,y) = \frac{\sum f_i (x_i - \bar{x})(y_i - \bar{y})}{N - 1} \quad (15.7)$$

where N is the total number of pairs of relatives studied.

The **correlation coefficient (r)** of the trait between the relatives is calculated from the covariance as follows:

$$r = \frac{Cov(x,y)}{s_x s_y} \qquad (15.8)$$

where s_x and s_y are the standard deviations of the measurements, estimated from Equation (15.2). The correlation coefficient can range from -1.0 to $+1.0$. A value of $+1.0$ means perfect association. When $r = 0$, x and y are not associated.

■ The additive genetic variance is transmissible; the dominance variance is not.

Covariance and correlation are important in quantitative genetics because the correlation coefficient of a trait between individuals with various degrees of genetic relationship is related fairly simply to the narrow-sense or broad-sense heritability, as shown in Table 15.3. The table gives the theoretical values of the correlation coefficient for various pairs of relatives; h^2 represents the narrow-sense heritability and H^2 the broad-sense heritability. Considering parent–offspring, half-sibling, or first-cousin pairs, narrow-sense heritability can be estimated directly by multiplication. Specifically, h^2 can be estimated as twice the parent–offspring correlation, four times the half-sibling correlation, or eight times the first-cousin correlation.

With full siblings, identical twins, and double first cousins, the correlation coefficient is related to the broad-sense heritability, H^2, because phenotypic resemblance depends not only on additive effects but also on dominance. In these relatives,

dominance contributes to resemblance because the relatives can share *both* of their alleles as a result of their common ancestry, whereas parents and offspring, half siblings, and first cousins can share at most a single allele of any gene because of common ancestry. Therefore, to the extent that phenotype depends on dominance effects, full siblings can resemble each other more than they resemble their parents.

■ The most common disorders in human families are multifactorial.

The most common human diseases are caused by genetic and environmental factors acting together. Each of the factors adds to the risk or **liability** that a person has for manifesting the trait. For example, genetic risk factors for heart disease are revealed in the family history of the disease, and environmental risk factors include being overweight and such behaviors as cigarette smoking. A person's overall risk for manifesting a disease is a multifactorial trait determined by numerous genetic and environmental factors and the interactions among these factors. Such a trait is a *threshold trait;* although the underlying risk itself is not directly observable, the trait will be either present or absent according to whether the risk is above or below a critical (threshold) value.

As with other multifactorial traits, the risk of manifesting a threshold trait has a broad-sense and a narrow-sense heritability that may differ among populations according to the allele frequencies and the distribution of environmental risk factors. The heritabilities cannot be estimated directly, because the risk is not observable directly, but the heritabilities can be inferred from the incidences of the trait

Table 15.3

Theoretical correlation coefficient in phenotype between relatives

Degree of relationship	Correlation coefficient*
Offspring and one parent	$h^2/2$
Offspring and average of parents	$h^2/2$
Half siblings	$h^2/4$
First cousins	$h^2/8$
Monozygotic twins	H^2
Full siblings	$\sim H^2/2$

*Contributions from interactions among alleles of different genes have been ignored. For this and other reasons, H^2 correlations are approximate.

© Ablestock

Identical twins share the same genotype, and their degree of physical and behavioral resemblance is often impressive. But they can differ genetically owing to such phenomena as the randomness of X-chromosome inactivation, genetic imprinting, and somatic mutation. They also differ in their environmental influences. The result is that identical twins often differ in phenotype, especially for complex traits.

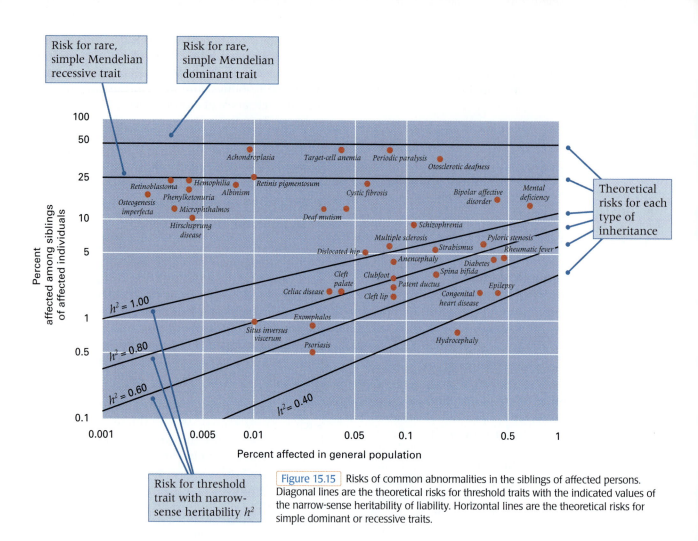

Risk for rare, simple Mendelian recessive trait

Risk for rare, simple Mendelian dominant trait

Theoretical risks for each type of inheritance

Risk for threshold trait with narrow-sense heritability h^2

Figure 15.15 Risks of common abnormalities in the siblings of affected persons. Diagonal lines are the theoretical risks for threshold traits with the indicated values of the narrow-sense heritability of liability. Horizontal lines are the theoretical risks for simple dominant or recessive traits.

among individuals and their relatives. The statistical techniques for doing this are quite specialized, but some of the theoretical results are shown in Figure 15.15, along with observed data for the most common congenital abnormalities in Caucasians. The x-axis gives the incidence of the trait in the general population, and the y-axis gives the risk of the trait in brothers or sisters of affected persons. The two horizontal lines near the top yield the proportions for simple Mendelian dominance and simple Mendelian recessive inheritance, which are 50 percent (dominance) and 25 percent (recessiveness), respectively. Note that these proportions do not depend on the incidence of the trait in the population. The other curves pertain to threshold traits with the narrow-sense heritabilities of liability as noted. Now the proportion of affected siblings *does* depend on the incidence of the trait in the general population, as well as on the heritability of liability. Consider a trait with a population incidence of 0.05 percent (1 in 2000). If the heritability of liability is 0.40, the proportion of affected siblings among affected persons is a little less than 0.5 percent (1 in 200), or about a tenfold increase over that of the general population. If the heritability is 1.00, the proportion of affected siblings

among affected persons is a little less than 5 percent (1 in 20), or about a hundredfold increase over that of the general population. Note in Figure 15.15 that the most common traits tend to be threshold traits; with a few exceptions, the proportion of affected siblings tends to be moderate or low, corresponding to the heritabilities indicated.

A Moment to Think

Answer to Problem: Equation (15.6) [$M' = M + h^2(M^* - M)$] applies to this situation, but the values given are deviations from the mean rather than ridge counts. No matter: The equation still applies using the deviations. In this case, $M = 0$ (that is, the mean of the population deviates by 0 standard deviations from its own mean), $M^* =$ the average of the parents (or in this case the average deviation), which is $[1.5 + (-0.5)]/2 = 0.5$. Now we apply Equation 15.6:

$$M' = M + h^2(M^* - M)$$
$$= 0 + (0.86)(0.5 - 0)$$
$$= 0.43$$

This means that the expected progeny average is 0.43 standard deviation above the mean of the population. Because the mean of the population is 125 and the standard deviation is 50, the expected progeny mean for total fingerprint ridge count is $125 + 0.43 \times 50 = 125 + 21.5 = 146.5$.

15.5

Pedigree studies of genetic polymorphisms are used to map loci for quantitative traits.

Genes affecting quantitative traits cannot usually be identified in pedigrees because their individual effects are obscured by the segregation of other genes and by environmental variation. Even so, genes affecting quantitative traits can be localized if they are genetically linked with genetic markers, such as simple sequence repeat polymorphisms (SSRs), discussed in Chapter 4, because the effects of the genotype affecting the quantitative trait are correlated with the genotype of the linked SSR. A gene that affects a quantitative trait is a **quantitative-trait locus (QTL).** Locating QTLs in the genome is important to the manipulation of genes in breeding programs and to the cloning and study of genes in order to identify their functions.

SSRs result from variation in number, n, of repeating units of a simple-sequence repeat, such as $(TG)_n$. The number of repeats can be determined by PCR amplification of the SSR using oligonucleotide primers to unique-sequence DNA flanking the repeats. The size of the amplified DNA fragment is a function of the number of repeats in the SSR. Such SSR markers are abundant, are distributed throughout the genome, and often have multiple codominant alleles—qualities that make them ideally suited for linkage studies of quantitative traits. In SSR studies, as many widely scattered SSRs as possible are monitored, along with the quantitative trait, in successive generations of a

genetically heterogeneous population. Statistical studies are then carried out to identify which SSR alleles are the best predictors of the phenotype of the quantitative trait because their presence in a genotype is consistently accompanied by superior performance with respect to the quantitative trait. These SSRs identify regions of the genome that contain one or more genes with important effects on the quantitative trait, and the SSRs can be used to trace the segregation of the important regions in breeding programs and even as starting points for cloning genes with particularly large effects.

An example of genetic mapping of quantitative trait loci for several quantitative traits in tomatoes is illustrated in Figure 15.16 . More than 300 highly polymorphic genetic markers have been mapped in the tomato genome, with an average spacing between markers of 5 map units. The chromosome maps in Figure 15.16 show a subset of 67 markers

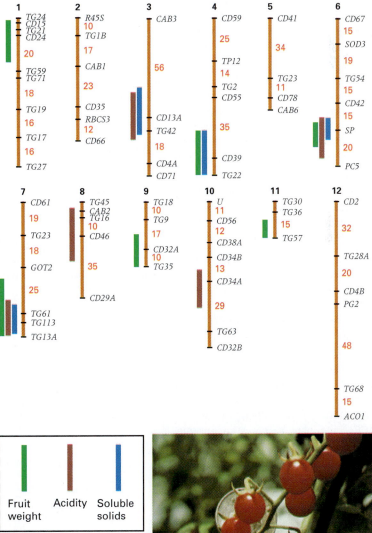

Figure 15.16 Location of QTLs for several quantitative traits in the tomato genome. The genetic markers are shown for each of the 12 chromosomes. The numbers in red are distances in map units between adjacent markers, but only map distances of 10 or greater are indicated. The regions in which the QTLs are located are indicated by the bars: green bars, QTLs for fruit weight; blue bars, QTLs for content of soluble solids; and dark red bars, QTLs for acidity (pH). The data are from crosses between the domestic tomato (*Lycopersicon esculentum*) and a wild South American relative with small, green fruit (*Lycopersicon chmielewskii*). The photograph is of fruits of wild tomato. Note the small size in comparison with the coin (a U.S. quarter). The F_1 generation was backcrossed with the domestic tomato, and fruits from the progeny were assayed for the genetic markers and each of the quantitative traits. [Data from A. H. Paterson et al. 1988. *Nature* 335: 721. Photograph courtesy of Steven D. Tanksley.]

Win, Place, or Show?

Hugh E. Montgomery and 18 other investigators 1998
University College, London, and 6 other research institutions
Human Gene for Physical Performance

A professional basketball executive once said of legendary superstar Michael Jordan, "Gene for gene, DNA for DNA, nobody can replace him." The implication that Jordan's phenomenal success was not due exclusively to his work ethic, practice, commitment, consistency, and character is interesting. But nobody doubts that Jordan's success in basketball was associated in part with his being tall. And height is a multifactorial trait partly under genetic control. About 50 percent of the variation in height among people is due to genetic differences. But might invisible, metabolic differences also influence physical performance? Some people find this possibility unsettling, because it potentially raises all sorts of issues about the value of effort and practice, the fairness of competition, genetic testing for entry into elite physical training programs, and so forth. The extent to which genotype influences physical performance is not known, but some genetic differences may be important. In this paper, a polymorphism in the angiotensin-converting enzyme (ACE) is reported to have a major effect on physical endurance. It is not difficult to see how ACE could play a role. The enzyme degrades a class of vasodilator proteins and also converts angiotensin I into a vasoconstrictor. Note, however, that the authors are cautious in their conclusions and are careful not to speculate that the gene for ACE is a QTL (quantitative trait locus) affecting physical performance. It is still too early to reach firm conclusions.

A specific genetic factor that strongly influences human physical performance has not so far been reported, but here we show that a polymorphism in the gene encoding angiotensin-converting enzyme does just that. An "insertion" allele of the gene is associated with elite endurance performance among high-altitude mountaineers. Also, after physical training, repetitive weight-lifting is improved elevenfold in individuals homozygous for the "insertion" allele compared with those homozygous for the "deletion" allele. The endocrine renin–angiotensin system is important in controlling the circulation. . . . A polymorphism in the human *ACE* gene for angiotensin-converting enzyme has been described in which the deletion (*D*) rather than insertion (*I*) allele is associated with higher activity by tissue ACE. . . . Our initial studies suggested that the *I* allele was associated with improved endurance performance. . . . High-altitude moutaineers perform extreme-endurance exercise. [The *ACE* genotypes of 25] elite unrelated male British mountaineers . . . were compared with those of 1906 British males. . . . Both the genotype distribution and allele frequency differed significantly between climbers and controls (Figure A).

> We show that a polymorphism in the gene encoding angiotensin-converting enzyme [influences human physical performance].

In a second study, *ACE* genotype was determined in 123 Caucasian males recruited into the United Kingdom army consecutively. . . . The maximum duration (in seconds) for which they could perform repetitive elbow flexion while holding a 15-kg barbell was assessed both before and after a 10-week physical training period. . . . Duration of exercise improved significantly for those of *II* and *ID* genotypes, but not for those of *DD* genotype (Figure B). . . . Increased performance is likely to be due to an improvement in the endurance characteristics of the tested muscles. . . . Further work will be needed to determine whether this correlation holds beyond the limited group studied here.

Source: Nature 393: 221.

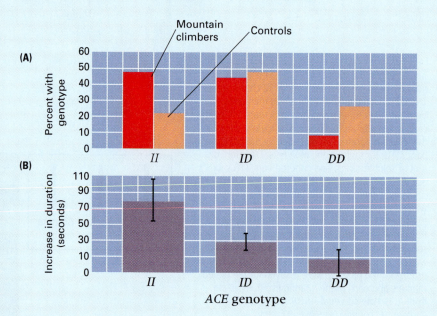

that were segregating in crosses between the domestic tomato and a wild South American relative. The average spacing between the markers is 20 map units. Backcross progeny of the cross $F_1 \times$ domestic tomato were tested for the genetic markers, and the fruits of the backcross progeny were assayed for three quantitative traits—fruit weight, content of soluble solids, and acidity. Statistical analysis of the data was carried out in order to detect marker alleles that were associated with phenotypic differences in any of the traits; a significant association indicates genetic linkage between the marker gene and one or more QTLs affecting the trait. A total of six QTLs affecting fruit weight were detected (green bars), as well as four QTLs affecting soluble solids (blue bars) and five QTLs affecting acidity (dark red bars). Although additional QTLs of smaller effects undoubtedly remained undetected in these types of experiments, the effects of the mapped QTLs are substantial: The mapped QTLs account for 58 percent of the total phenotypic variance in fruit weight, 44 percent of the phenotypic variance in soluble solids, and 48 percent of the phenotypic variance in acidity. The genetic markers linked to the QTLs with substantial effects make it possible to trace the transmission of the QTLs in pedigrees and manipulate them in breeding programs by following the transmission of the linked marker genes. Figure 15.16 also indicates a number of chromosomal regions containing QTLs for two or more of the traits—for example, the QTL regions on chromosomes 6 and 7, which affect all three traits. Because the locations of the QTLs can be specified only to within 20 to 30 map units, it is unclear whether the coincidences result from pleiotropy, in which a single gene affects several traits simultaneously, or from the independent effects of multiple, tightly linked genes. However, the locations of QTLs for different traits coincide frequently enough that in most cases, pleiotropy is likely to be the explanation.

■ Some QTLs have been identified by examining candidate genes.

Besides genetic mapping, another approach to the identification of QTLs is through the use of candidate genes. A **candidate gene** for a trait is a gene for which there is some *a priori* basis for suspecting that the gene affects the trait. In human behavior genetics, for example, if a pharmacological agent is known to affect a personality trait, and the molecular target of the drug is known, then the gene that codes for the target molecule and any gene whose product interacts with the drug or with the target molecule are candidate genes for affecting the personality trait.

One example of the use of candidate genes in the study of human behavior genetics is in the identification of a naturally occurring genetic polymorphism associated with serious depression in response to stressful life experiences. The neurotransmitter substance serotonin (5-hydroxytryptamine) is known to influence a variety of psychiatric conditions, such as anxiety and depression. Among the important components in serotonin action is the serotonin transporter protein. Neurons that release serotonin to stimulate other neurons also take it up again through the serotonin transporter. This uptake terminates the stimulation and also recycles the molecule for later use.

The serotonin transporter became a strong candidate gene for depression when it was discovered that the transporter is the target of a class of antidepressants known as selective serotonin reuptake inhibitors. The widely prescribed antidepressant Prozac is an example of such a drug.

Motivated by the strong suggestion that the serotonin transporter might be involved in depression, researchers looked for evidence of genetic polymorphisms affecting the transporter gene in human populations. Such a polymorphism was found in the promoter region of the transporter. About 1 kb upstream from the transcription start site is a series of 16 tandem repeats of a nearly identical DNA sequence of about 15 base pairs. This is the L (*long*) allele, which has an allele frequency of 57 percent among Caucasians. There is also an S (*short*) allele, in which three of the repeated sequences are not present. The S allele has an allele frequency of 43 percent. The genotypes L/L, L/S, and S/S are found in the Hardy–Weinberg proportions of 32 percent, 49 percent, and 19 percent, respectively (Chapter 14). Further studies revealed that the polymorphism does have a physiological effect. In cells grown in culture, L/L cells had approximately 50 percent more mRNA for the transporter than L/S or S/S cells, and L/L cells had approximately 35 percent more membrane-bound transporter protein than L/S or S/S cells.

On the basis of these results, the researchers predicted that the serotonin transporter polymorphism would be found to be associated with depression. Such an association was found in a study of 845 people who were genotyped for the transporter polymorphism, but the association occurs only as a result of genotype-by-environment interaction in response to multiple high-stress life events. In this study, each individual's life history from age 21 to 26 years was assessed in terms of employment, finance, housing, health, and personal relationships to identify stressful events involving threat, defeat, humiliation, or death of a loved one. Each individual was examined at age 26 for depression during the previous year

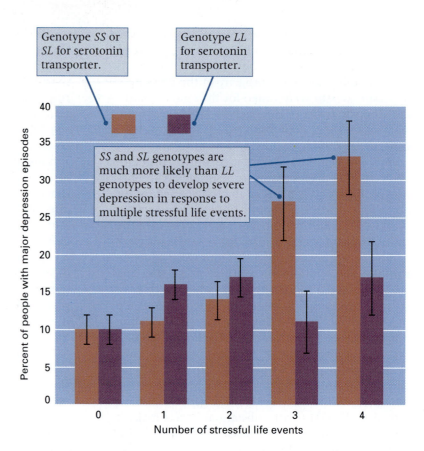

Figure 15.17 Risk of severe depression in response to stressful life events in relation to the serotonin reuptake transporter polymorphism. The strong genotype-by-environment interaction is clear in comparing relative risk of depression with 0 to 2 stressful events as against 3 to 4 stressful events. [Data from A. Caspi, et al. 2003. *Science* 274: 1527.]

Genotype *SS* or *SL* for serotonin transporter.

Genotype *LL* for serotonin transporter.

SS and *SL* genotypes are much more likely than *LL* genotypes to develop severe depression in response to multiple stressful life events.

Percent of people with major depression episodes

Number of stressful life events

with the use of diagnostic interviews. The results comparing the *S*– and *LL* transporter genotypes are shown in Figure 15.17 . There is clearly a much higher risk of severe depression in the *SS* and *SL* genotypes, but only in response to 3 or 4 stressful life events occurring in the previous 5 years. (About 25 percent of the individuals experienced such high and repeated levels of stress.) Among the individuals

with 2 or fewer stressful events in the previous 5 years, there was no difference between the serotonin transporter genotypes. The high level of genotype-by-environment interaction is emphasized by noting that, among individuals with 3 or more stressful life events, the risk of severe depression is approximately twice as high among *SS* and *SL* genotypes and among *LL* genotypes.

chapter summary

15.1 Multifactorial traits are determined by multiple genes and the environment.

- Continuous, categorical, and threshold traits are usually multifactorial.
- The distribution of a trait in a population implies nothing about its inheritance.

Many traits that are important in human genetics and in agriculture are determined by the effects of multiple genes and by the environment. Such traits are multifactorial traits or quantitative traits, and their analysis is known as quantitative genetics. There are three types of multifactorial traits: continuous, categorical, and threshold. Continuous traits are expressed according to a continuous scale of measurement, such as height. Categorical traits are traits that are expressed in whole numbers, such as the number of grains on an ear of corn. Threshold traits have an underlying risk and are either expressed or not expressed in each individual; an example is diabetes. The genes affecting quantitative traits are no different from those affecting simple Mendelian traits, and the

genes can have multiple alleles, partial dominance, and so forth. When several genes affect a trait, the pattern of genetic transmission need not fit a simple Mendelian pattern, because the effects of one gene can be obscured by other genes or the environment. However, the number of genes can be estimated, and many of them can be mapped using linkage to molecular polymorphisms.

15.2 Variation in a trait can be separated into genetic and environmental components.

- The genotypic variance results from differences in genotype.
- The environmental variance results from differences in environment.
- Genotype and environment can interact, or they can be associated.
- There is no genotypic variance in a genetically homogeneous population.

- The number of genes affecting a quantitative trait need not be large.
- The broad-sense heritability includes all genetic effects combined.
- Twin studies are often used to assess genetic effects on variation in a trait.

Many quantitative and categorical traits have a distribution that approximates the bell-shaped curve of a normal distribution. A normal distribution can be completely described by two quantities: the mean and the variance. The standard deviation of a distribution is the square root of the variance. In a normal distribution, approximately 68 percent of the individuals have a phenotype within one standard deviation from the mean, and approximately 95 percent of the individuals have a phenotype within two standard deviations from the mean.

15.3 Artificial selection is a form of "managed evolution."

- The narrow-sense heritability is usually the most important in artificial selection.
- There are limits to the improvement that can be achieved by artificial selection.
- Inbreeding is generally harmful, and hybrids may be the best.

Variation in phenotype of multifactorial traits among individuals in a population derives from four principal sources: (1) variation in genotype, which is measured by the genotypic variance; (2) variation in environment, which is measured by the environmental variance; (3) variation resulting from the interaction between genotype and environment (G-E interaction); and (4) variation resulting from nonrandom association of genotypes and environments (G-E association). The ratio of genotypic variance to the total phenotypic variance of a trait is called the broad-sense heritability; this quantity is useful in predicting the outcome of artificial selection among clones, inbred lines, or varieties. When artificial selection is carried out in a randomly mating population, the narrow-sense heritability is used for prediction. The value of the narrow-sense heritability can be determined from the correlation in phenotype among groups of relatives.

One common type of artificial selection is called truncation selection, in which only those individuals that have a phenotype above a certain value (the truncation point) are saved for breeding the next generation. Artificial selection usually results in improvement of the selected population. The general principle is that the deviation of the progeny mean from the mean of the previous generation equals the narrow-sense heritability times the deviation of the parental mean from the mean of their generation: $M' = M + h^2(M^* - M)$. When selection is carried out for many generations, progress often slows or ceases because (1) some of the favorable genes become nearly fixed in the population, which decreases the narrow-sense heritability, and (2) natural selection may counteract the artificial selection.

15.4 Genetic variation is revealed by correlations between relatives.

- Covariance is the tendency for traits to vary together.
- The additive genetic variance is transmissible; the dominance variance is not.
- The most common disorders in human families are multifactorial.

15.5 Pedigree studies of genetic polymorphisms are used to map loci for quantitative traits.

- Some QTLs have been identified by examining candidate genes.

A gene affecting a quantitative trait is a quantitative-trait locus, or QTL. The locations of QTLs in the genome can be determined by genetic mapping with respect to simple sequence repeat polymorphisms or other types of genetic markers. The mapping of QTLs aids in the manipulation of genes in breeding programs and in the identification of the genes. An alternative approach to identifying QTLs is the study of candidate genes known to function in the biochemical pathway that affects the trait. The Human Genome Project has generated a dense map of genetic markers, both simple sequence repeats (SSRs) and single nucleotide polymorphisms (SNPs), for QTL mapping, and the human genome sequence affords a large number of potential candidate genes for a variety of inherited traits.

issues & ideas

- Give an example of a trait in human beings that is affected both by environmental factors and by genetic factors. Specify some of the environmental factors that affect the trait.

- Does the distribution of phenotypes of a trait in a population tell one anything about the relative importance of genes and environment in causing differences in phenotype among individuals? Does it tell one anything about the number of genes that may affect the trait? Explain.

- In regard to the genotypic variance, what is special about an inbred line or about the F_1 progeny of a cross between two inbred lines?

- The distribution of bristle number on one of the abdominal segments in a population of *Drosophila* ranges from 12 to 26. The narrow-sense heritability of the trait is approximately 50 percent. Do you think that it would be possible, by practicing artificial selection over a number of generations, to obtain a population in which the *mean* bristle number was greater than 26? A population in which the *minimum* bristle number was greater than 26? Explain why or why not.

- Why are correlations between relatives of interest to the quantitative geneticist?

- In the context of multifactorial inheritance, what is a QTL? Does a QTL differ from any other kind of gene? How can QTLs be detected?

- In the context of multifactorial inheritance, what is a candidate gene? Would you regard the human *PAH* gene for phenylalanine hydroxylase as a natural candidate gene for mental retardation? Why or why not?

key terms & concepts

artificial selection
broad-sense heritability
candidate gene
categorical trait
complex trait
continuous trait
correlated response
correlation coefficient (*r*)
covariance (*Cov*)
distribution

environmental variance
fraternal twins
genetic architecture
genotype-by-sex interaction
genotype-by-environment
  association (G-E)
genotype-by-environment
  interaction (G-E)
genotypic variance

heterosis
hybrid vigor
identical twins
inbreeding depression
individual selection
liability
mean
multifactorial trait
narrow-sense heritability

normal distribution
quantitative trait
quantitative-trait locus (QTL)
selection limit
standard deviation
threshold trait
total variance
truncation point
variance

1. _____ A trait for which the observed phenotypes can be assigned to any one of a relatively small number of discrete categories.

2. _____ Synonym for average.

3. _____ For a normal distribution of phenotypes, this measure of dispersion defines a region around the mean that includes approximately 68 percent of the population.

4. _____ Measure of dispersion defined as the average of the squared deviations of the individual values from the overall mean in the population.

5. _____ Term for a situation in which combinations of genotypes and environments do not occur in the frequencies expected at random.

6. _____ Measure of heritability used to predict the change in mean phenotype expected with truncation selection.

7. _____ The mean phenotype observed when a population no longer responds to artificial selection.

8. _____ The component of the total phenotypic variance that is zero in a genetically uniform population.

9. _____ When artificial selection for one trait also changes a different trait.

10. _____ Interchangeable terms referring to the superiority that hybrids typically show relative to their inbred parents.

11. _____ An informed guess as to the identity of a gene that might influence a quantitative trait.

12. _____ In theory, these individuals have a genotypic variance equal to zero.

solutions: step by step

Problem 1

The weight in pounds at the time of weaning in a representative sample of lambs taken from a large flock is shown in the accompanying table. Estimate the mean, the variance, and the standard deviation of weaning weight in this population.

37	37	38	46	39
30	31	35	30	42
43	39	48	27	41
43	41	37	29	26

■ **Solution** The mean and variance of weaning weight are estimated from the sum of the individual values and the sum of the deviations from the mean, as expressed in the formulas. Note that the variance has units of pounds squared. The original units of pounds are restored by taking the square root, which is the standard deviation $s = \sqrt{40.16 \text{ lb}^2} = 6.24 \text{ lb}$. The calculations are shown below.

$$\bar{x} = \frac{\Sigma x_i}{N}$$

$$= \frac{739}{20} = 36.95 \text{ pounds}$$

$$s^2 = \frac{\Sigma(x_i - \bar{x})^2}{N - 1}$$

$$= \frac{762.95}{19} = 40.16 \text{ pounds}^2$$

An equivalent formula for the variance, which is much easier to use in calculations, is shown below. The symbol Σx_i^2 means the sum of the squares of the individual values, and $(\Sigma x_i)^2$ means the square of the sum of the individual values.

$$s^2 = \frac{\Sigma x_i^2}{N - 1} - \frac{(\Sigma x_i)^2}{N(N - 1)}$$

$$= 1477.32 - 1437.16$$

$$= 40.16 \text{ pounds}^2$$

Problem 2

Artificial selection for fruit weight is carried out in a tomato population. From each plant, all the fruits are weighed, and the average weight per tomato is regarded as the phenotype of the plant. In the population as a whole, the average fruit weight per plant is 75 grams. Plants whose fruit weight averages 100 grams per tomato are selected and mated at random to produce the next generation. The narrow-sense heritability of average fruit weight in this population is estimated to be 20 percent.

(a) What is the expected average fruit weight among plants after one generation of selection?

(b) What is the expected average fruit weight among plants after five consecutive generations of selection when, in each generation, the chosen parents have an average fruit weight 25 grams above the population average in that generation? Assume that the narrow-sense heritability remains constant during this time.

■ Solution In the first generation of selection, the mean of the population is $M = 75$ grams, that of the selected parents is $M^* = 100$ grams, and the narrow-sense heritability h^2 is given as 0.20. **(a)** The formula below gives the prediction

equation for this case, and the predicted average fruit weight is 80 grams per tomato. **(b)** It is clear from the formula that if $M^* - M = 25$ in each generation, and h^2 remains at 0.20, then the expected gain per generation is 5 grams. After five generations the expected gain is 25 grams, yielding an expected average fruit weight after five generations of 100 grams per tomato. In the supermarket, this represents a decrease from about 6 tomatoes per pound to about 4 to 5 tomatoes per pound.

$$M' = M + h^2(M^* - M)$$
$$= 75 + (0.20)(100 - 75) = 75 + 5$$
$$= 80 \text{ grams}$$

concepts in action: problems for solution

15.1 Two varieties of corn, A and B, are field-tested in Indiana and North Carolina. Strain A is more productive in Indiana, but strain B is more productive in North Carolina. What phenomenon in quantitative genetics does this example illustrate?

15.2 Distinguish between the broad-sense heritability of a quantitative trait and the narrow-sense heritability. If a population is fixed for all genes that affect a particular quantitative trait, what are the values of the narrow-sense and broad-sense heritabilities?

15.3 The following questions pertain to a normal distribution.

(a) What term applies to the value along the x-axis that corresponds to the peak of the distribution?

(b) If two normal distributions have the same mean but different variances, which is the broader?

(c) What proportion of the population is expected to lie within one standard deviation of the mean?

(d) What proportion of the population is expected to lie within two standard deviations of the mean?

15.4 When we compare a quantitative trait in the F_1 and F_2 generations obtained by crossing two highly inbred strains, which set of progeny provides an estimate of the environmental variance? What determines the variance of the other set of progeny?

15.5 Some estimates of broad-sense heritabilities of human traits are 0.85 for stature, 0.62 for body weight, 0.57 for systolic blood pressure, 0.44 for diastolic blood pressure, 0.50 for twinning, and 0.1 to 0.2 for overall fertility. Which of these characteristics is most likely to "run in families"? If one of your parents and one of your grandparents has high blood pressure, should you be concerned about the likelihood of your having the same problem?

15.6 Tabulated below are the numbers of eggs laid by 50 hens over a 2-month period. The hens were selected at random from a much larger population. Estimate the mean, variance, and standard deviation of the distribution of egg

numbers in the entire population from which the sample was drawn.

48	50	51	47	54	45	50	38	40	52
58	47	55	53	54	41	59	48	53	49
51	37	31	47	55	46	49	48	43	68
59	51	52	66	54	37	46	55	59	45
44	44	57	51	50	57	50	40	63	33

15.7 Two highly inbred strains of mice are crossed, and the F_1 generation has a mean tail length of 5 cm and a standard deviation of 1.5 cm. The F_2 generation has a mean tail length of 5 cm and a standard deviation of 4 cm. What are the environmental variance, the genetic variance, and the broad-sense heritability of tail length in this population?

15.8 For the difference between the domestic tomato, *Lycopersicon esculentum*, and its wild South American relative, *Lycopersicon chmielewskii*, the environmental variance (σ_e^2) accounts for 13 percent of the total phenotypic variance (σ_p^2) of fruit weight, for 9 percent of σ_p^2 of soluble-solid content, and for 11 percent of σ_p^2 of acidity. What are the broad-sense heritabilities of these traits?

15.9 Two homozygous genotypes of *Drosophila* differ in the number of abdominal bristles. In genotype *AA*, the mean bristle number is 20 with a standard deviation of 2. In genotype *aa*, the mean bristle number is 23 with a standard deviation of 3. Both distributions conform to the normal curve, in which the proportions of the population with a phenotype within an interval defined by the mean ± 1, ± 1.5, ± 2, and ± 3 standard deviations are 68, 87, 95, and 99.7 percent, respectively.

(a) In genotype *AA*, what is the proportion of flies with a bristle number between 20 and 23?

(b) In genotype *aa*, what is the proportion of flies with a bristle number between 20 and 23?

(c) What proportion of *AA* flies have a bristle number greater than the mean of *aa* flies?

(d) What proportion of *aa* flies have a bristle number greater than the mean of *AA* flies?

concepts in action: problems for solution *continued*

15.10 The narrow-sense heritability of withers height in a population of quarterhorses is 30 percent. (Withers height is the height at the highest point of the back, between the shoulder blades.) The average withers height in the population is 17 hands. (A "hand" is a traditional measure equal to the breadth of the human hand, now taken to equal 4 inches.) From this population, studs and mares with an average withers height of 16 hands are selected and mated at random. What is the expected withers height of the progeny? How does the value of the narrow-sense heritability change if withers height is measured in meters rather than hands?

15.11 Consider a complex trait in which the phenotypic values in a large population are distributed approximately according to a normal distribution with mean 100 and standard deviation 15. What proportion of the population has a phenotypic value above 130? Below 85? Above 85?

15.12 In an experimental population of the flour beetle *Tribolium castaneum*, the pupal weight is distributed normally with a mean of 2.0 mg and a standard deviation of 0.2 mg. What proportion of the population is expected to have a pupal weight between 1.8 and 2.2 mg? Between 1.6 and 2.4 mg? Would you expect to find an occasional pupa weighing 3.0 mg or more? Explain your answer.

15.13 Estimate the minimum number of genes affecting fruit weight in a population of the domestic tomato produced by crossing two inbred strains. Measured as the logarithm of fruit weight in grams, the inbred lines have average fruit weights of -0.137 and 1.689. The F_1 generation has a variance of 0.0144, and the F_2 generation has a variance of 0.0570.

15.14 In human beings, the narrow-sense heritability of the total fingerprint ridge count is 86 percent. On the basis of this value, what is the estimated correlation coefficient between first cousins in the total fingerprint ridge count?

15.15 Maternal effects are nongenetic influences on offspring phenotype that derive from the phenotype of the mother. For example, in many mammals, larger mothers have larger offspring, in part because of a maternal effect on birth weight. What result would a maternal effect have on the correlation in birth weight between mothers and their offspring compared with that between fathers and their offspring? How would such a maternal effect influence the estimate of narrow-sense heritability?

15.16 In terms of the narrow-sense heritability, what is the theoretical correlation coefficient in phenotype between first cousins who are the offspring of monozygotic twins?

15.17 If the correlation coefficient of a trait between first cousins is 0.09, what is the estimated narrow-sense heritability of the trait?

15.18 A mouse population has an average weight gain between ages 3 and 6 weeks of 12 g, and the narrow-sense heritability of the weight gain between 3 and 6 weeks is 20 percent.
(a) What average weight gain would be expected among the offspring of parents whose average weight gain was 16 g?
(b) What average weight gain would be expected among the offspring of parents whose average weight gain was 8 g?

15.19 To estimate the heritability of maze-learning ability in rats, a selection experiment was carried out. From a population in which the average number of trials necessary to learn the maze was 10.8, with a variance of 4.0, animals were selected that managed to learn the maze in an average of 5.8 trials. Their offspring required an average of 8.8 trials to learn the maze. What is the estimated narrow-sense heritability of maze-learning ability in this population?

15.20 For a phenotype determined by a single, completely penetrant recessive allele at frequency q in a random-mating population, it can be shown that the narrow-sense heritability is $2q/(1 + q)$.
(a) Calculate the narrow-sense heritability for $q = 1.0$, 0.5, 0.1, 0.05, 0.01, 0.005, and 0.001.
(b) Note that the narrow-sense heritability goes to 0 as q goes to 0, yet the phenotype is completely determined by heredity. How can this be explained?

geNETics on the web

GeNETics on the Web will introduce you to some of the most important sites for finding genetic information on the Internet. To explore these sites, visit the Jones and Bartlett home page at

http://www.jbpub.com/genetics

For the book *Essential Genetics: A Genomics Perspective,* choose the link that says **Enter GeNETics on the Web**. You will be presented with a chapter-by-chapter list of highlighted keywords. Select any highlighted keyword and you will be linked to a Web site containing genetic information related to the keyword.

- This keyword will open a table of frequencies of the most common types of **birth defects** in the U.S. population. The overall frequency of these conditions is about 1 in 25 births. Most of the conditions that have genetic risk factors are complex traits affected by multiple genes; these include disorders of the heart and circulation, muscles and skeleton, genital and urinary tract, and nervous system and eye, which together have an incidence of 1 in 35.

- The debate over the growing and marketing of **genetically modified organisms** (GMO) has been fierce, especially in Europe. Opponents call them Frankenstein's plants.

They argue that it is foolhardy to plant millions of acres of novel crop plants and let their products enter the food supply without evaluating the possible long-term effects on human health and the environment. Proponents argue that the maintenance of an affordable food supply depends on the increased productivity expected from genetically modified plants. This keyword site provides a balanced discussion of the issues and points out that although the debate continues, genetically modified crops already constitute a large portion of the food supply in the United States.

further readings

Black, D. M., and E. Solomon. 1993. The search for the familial breast/ovarian cancer gene. *Trends in Genetics* 9: 22.

Bouchard, T. J., Jr., D. T. Lykken, M. McGue, N. L. Segal, and A. Tellegen. 1990. Sources of human psychological differences: The Minnesota study of twins reared apart. *Science* 250: 223.

Comprehensive Human Genetic Linkage Map Center. 1994. A comprehensive human linkage map with centimorgan density. *Science* 265: 2049.

Crow, J. F. 1993. Francis Galton: Count and measure, measure and count. *Genetics* 135: 1.

Devor, E. J., and C. R. Cloninger. 1989. The genetics of alcoholism. *Annual Review of Genetics* 23: 19.

Falconer, D. S., and T. F. C. Mackay. 1996. *Introduction to Quantitative Genetics,* 4th ed. Essex, England: Longman.

Feldman, M. W., and R. C. Lewontin. 1975. The heritability hang-up. *Science* 190: 1163.

Gottesman, I. I., 1997. Twins: En route to QTLs for cognition. *Science* 276: 1522.

Gottesman, I. I., and J. Shields. 1982. *Schizophrenia: The Epigenetic Puzzle.* Cambridge, England: Cambridge University Press.

Greenspan, R. J. 1995. Understanding the genetic construction of behavior. *Scientific American,* April.

Haley, C. S. 1996. Livestock QTLs: Bringing home the bacon. *Trends in Genetics* 11: 488.

Hartl, D. L. 2000. *A Primer of Population Genetics.* 3d ed. Sunderland, MA: Sinauer.

Lander, E. S., and N. J. Schork. 1994. Genetic dissection of complex traits. *Science* 265: 2037.

Lesch, K.-P., D. Bengel, A. Heils, S. Z. Sabol, B. D. Greenberg, S. Petri, J. Benjamin, C. R. Müller, D. H. Hamer, and D. L. Murphy. 1996. Association of anxiety-related traits with a polymorphism in the serotonin transporter gene regulatory region. *Science* 274: 1527.

Paterson, A. H., E. S. Lander, J. D. Hewitt, S. Peterson, S. E. Lincoln, and S. D. Tanksley. 1988. Resolution of quantitative traits into Mendelian factors by using a complete linkage map of restriction fragment length polymorphisms. *Nature* 335: 721.

Pericakvance, M. A., and J. L. Haines. 1995. Genetic susceptibility to Alzheimer disease. *Trends in Genetics* 11: 504.

Pirchner, F. 1983. *Population Genetics in Animal Breeding.* 2d ed. New York: Plenum.

Plomin, R., and J. C. Defries. 1998. The genetics of cognitive abilities and disabilities. *Scientific American,* May.

Risch, N. 2000. Searching for genetic determinants in the new millenium. *Nature* 405: 847.

Risch, N., and H. Zhang. 1995. Extreme discordant sib pairs for mapping quantitative trait loci in humans. *Science* 268: 1584.

Stigler, S. M. 1995. Galton and identification by fingerprints. *Genetics* 140: 857.

Stuber, C. W. 1996. Mapping and manipulating quantitative traits in maize. *Trends in Genetics* 11: 477.

Tanksley, S. D. and S. R. McCouch. 1997. Seed banks and molecular maps: Unlocking genetic potential from the wild. *Science* 277: 1063.

answers
to even-numbered problems

Chapter 1

1.2 The importance of the nucleus in inheritance was implied by its prominence in fertilization. The discovery of chromosomes inside the nucleus, their behavior during cell division, and the observation that each species has a characteristic chromosome number made it likely that chromosomes were the carriers of the genes. Microscopic studies showed that DNA and proteins are both present in chromosomes, but whereas nearly all cells of a given species contain a constant amount of DNA, the amount and kinds of proteins differ greatly in different cell types.

1.4 Because the mature T2 phage contains only DNA and protein, the labeled RNA was left behind in material released by the burst cells.

1.6 RNA differs from DNA in that the sugar–phosphate backbone contains ribose rather than deoxyribose, RNA contains the base uracil (U) instead of thymine (T), and RNA usually exists as a single strand (although any particular molecule of RNA may contain short regions of complementary base pairs that can come together to form duplexes).

1.8 Because A $\neq$ T and G $\neq$ C in this DNA, it seems likely that the DNA molecule present in this particular virus is single-stranded.

1.10 The repeating Asn results from translation in the reading frame

$$5'\text{-AAUAAUAAUAAU}\cdots\text{-}3'$$

and the repeating Ile results from translation in the reading frame

$$5'\text{-AUAAUAAUAAUA}\cdots\text{-}3'$$

There is no product corresponding to the third reading frame ($5'$-UAAUAAUAAUAA $\cdots$ -$3'$) because $5'$-UAA-$3'$ is a stop codon.

1.12 It is the case because each codon is exactly three nucleotides in length. In a protein-coding region, an insertion or deletion of anything other than an exact multiple of three nucleotides would shift the reading frame (phase) in which the mRNA is translated. All amino acids downstream of the site of the mutation would be translated incorrectly.

1.14 Transcription takes place from left to right, and the mRNA sequence is $5'$-UGUCGUAUUUGCAAG-$3'$.

1.16 Cys−His−Ile−Cys−Lys or, using the single-letter abbreviations, CHICK.

1.18 (a) Met−Ser−Thr−Ala−Val−Leu−Glu−Asn−Pro−Gly. (b) The mutation changes the initiation codon into a noninitiation codon, so translation will not start with the first AUG; translation will start either with the next AUG farther along the mRNA or, if this is too distant, not at all. (c) Met−Ser−Thr−Ala−Val−Leu−Glu−Asn−Pro−Gly; there is no change, because both $5'$-UCC-$3'$ and $5'$-UCG-$3'$ code for serine. (d) Met−Ser−Thr−Ala−Ala−Leu−Glu−Asn−Pro−Gly; there is a Val $\rightarrow$ Ala amino acid replacement because $5'$-GUC-$3'$ codes for Val, whereas $5'$-GCC-$3'$ codes for Ala.

(e) Met−Ser−Thr−Ala−Val−Leu; translation is terminated at the UAA because $5'$-UAA-$3'$ is a "stop" (termination) codon.

1.20 The finding that the cells can grow in the presence of Y implies that step C is functional. The finding that the cells cannot grow in the presence of X implies that step B is blocked. The results with W imply that a downstream step is blocked, but do not reveal which one.

Chapter 2

2.2

	Progeny		
Parents	**AA**	**Aa**	**aa**
$AA \times AA$	1	0	0
$AA \times Aa$	1/2	1/2	0
$AA \times aa$	0	1	0
$Aa \times Aa$	1/4	1/2	1/4
$Aa \times aa$	0	1/2	1/2
$aa \times aa$	0	0	1

2.4 The birth of an affected offspring implies that parents are at risk of having another affected offspring. Once the genotypes of the parents have been deduced from their own phenotypes and the fact that they have an affected offspring, the recurrence risk is calculated as the probability of an affected offspring in the next birth. (a) The mating must be $Aa \times aa$, and so the risk of an Aa offspring in any birth (and therefore the recurrence risk) is 1/2. (b) The mating must be $Aa \times Aa$, and so the risk of an aa offspring in any birth is 1/4. (c) The mating must be $Aa \times aa$, and so the risk of an aa offspring in any birth is 1/2.

2.6 (a) II-2 is the affected person, and so the genotype must be aa. (b) I-1 and I-2 are the parents of the affected person, and because neither is affected, their genotypes must both be Aa. (c) II-1 and II-3 are siblings of the affected person. Because they are not affected, and because they result from the mating $Aa \times Aa$, their possible genotypes are AA and Aa. (d) From the mating $Aa \times Aa$, the ratio of $AA : Aa$ is 1 : 2; hence the probability that II-3 is a carrier (genotype Aa) is 2/3.

2.8 The DNA fragment from a_1 with the 2-kb insertion results in a band at 3 kb + 2 kb = 5 kb, and the DNA fragment from a_2 with the 1-kb deletion results in a band at 3 kb − 1 kb = 2 kb. DNA from each homozygous genotype produces only one band, whereas that from each heterozygous genotype produces two bands.

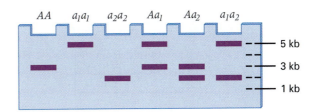

561

2.10 The existing data enable us to group the mutants into three complementation groups as follows: {a, c, d}, {b, f}, and {e}. The missing entries are shown in the accompanying table.

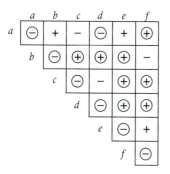

2.12 We expect a 3 : 1 ratio of the dominant to the recessive phenotype when two heterozygous gentypes are crossed, but in this case we observe 2 : 1. A cross of two $Cy/+$ heterozygotes is expected to yield a progeny genotypic ratio of 1 Cy/Cy : 2 $Cy/+$: 1 $+/+$. Because we see only two curly-winged flies for every wildtype fly, one possible explanation is that the Cy/Cy homozygotes die. In other words, Cy is dominant with respect to wing phenotype but recessive with respect to lethality.

2.14 Let AABB represent the wildtype genotype with a phenotype of disc. We are told that genotypes $aaBB$ and $AAbb$ both have spherical fruit and that $aabb$ has elongate fruit. The F2 progeny are expected in the ratio $9\,A{-}B{-} : 3\,A{-}bb : 3\,aaB{-} : 1\,aabb$, which implies that the expected ratio of fruit-shaped phenotypes is 9 disc : 6 sphere : 1 elongate.

2.16 (a) Because the trait is rare, it is reasonable to assume that the affected father is heterozygous $HD\ hd$, where hd represents the normal allele. Half of the father's gametes contain the mutant HD allele, so the probability is $1/2$ that the son received the allele and will later develop the disorder. **(b)** We do not know whether the son is heterozygous $HD\ hd$, but the probability is $1/2$ that he is; and if he is heterozygous, half of his gametes will contain the HD allele. Therefore, the overall probability that the grandchild carries the HD allele is $(1/2) \times (1/2) = 1/4$.

2.18 In the functional female gametes, the ratio of $A : a$ is $1/2 : 1/2$ because of Mendelian segregation. In males, the products of meiosis in an Aa individual also consist of $A + A + a + a$, but as stated in the problem, half of the A-bearing products are nonfunctional. Hence each male meiosis produces, on the average, three functional products, namely $A + a + a$. The ratio of $A : a$ among functional male gametes is therefore 1 : 2 or, converting to proportions, $1/3\ A : 2/3\ a$. The Punnett square shown here indicates that the F$_2$ ratio of $AA : Aa : aa$ is $1/6 : 3/6 : 2/6$ (or, reducing the fractions, $1/6 : 1/2 : 1/3$).

		Eggs	
		$1/2\ A$	$1/2\ a$
Pollen	$1/3\ A$	$1/6\ AA$	$1/6\ Aa$
	$2/3\ a$	$2/6\ Aa$	$2/6\ aa$

2.20 The accompanying Punnett squares indicate that the expected ratios of genotypes in **(a)** are $9/16\ DD : 6/16\ Dd : 1/16\ dd$, and in **(b)** are $3/8\ DD : 4/8\ Dd : 1/8\ dd$.

(a)

		Eggs	
		$3/4\ D$	$1/4\ d$
Sperm	$3/4\ D$	$9/16\ DD$	$3/16\ Dd$
	$1/4\ d$	$3/16\ Dd$	$1/16\ dd$

(b)

		Eggs	
		$3/4\ D$	$1/4\ d$
Sperm	$1/2\ D$	$3/8\ DD$	$1/8\ Dd$
	$1/2\ d$	$3/8\ Dd$	$1/8\ dd$

Chapter 3

3.2 After one cell cycle carried out in the presence of colchicine, a human cell would be expected to have $46 \times 2 = 92$ chromosomes.

3.4 After the centromeres have split, each former sister chromatid is counted as a chromosome in its own right, so at anaphase there are 48 chromosomes.

3.6 The chromosome tips (telomeres) would be expected to become shorter in each cell cycle. The reason is that DNA replication cannot begin at the extreme 3' end of a template strand. The function of telomerase is to restore the unreplicated region by elongating the 3' end after each round of DNA replication.

3.8 Panel C is anaphase of mitosis, because the homologous chromosomes are not paired. Panel A is anaphase I of meiosis, because the homologous chromosomes are paired. Panel B is anaphase II of meiosis, because the chromosome number has been reduced by half.

3.10 (a) Ratio X/A = 0.5, phenotype male. **(b)** Ratio X/A = 1.0, phenotype female. **(c)** Ratio X/A > 1.0, phenotype female. **(d)** Ratio X/A = 0.66, phenotype intersex.

3.12 (a) $1/2 \times 1/2 = 1/4$. **(b)** $1/2 \times 1/2 = 1/4$.

3.14 The chance that the second gene is on a different chromosome than the first is 6/7; the chance that the third gene is on a different chromosome than either of the first two is 5/7. Continuing in this manner, we find that the overall chance is $6/7 \times 5/7 \times 4/7 \times 3/7 \times 2/7 \times 1/7 = 0.00612$, or less than 1 percent. The likelihood of having every gene on a different chromosome is quite small, and for this reason Mendel has been accused of deliberately choosing unlinked genes. In fact, it is now known that not all of Mendel's genes are on different chromosomes. Three of the genes (*fa*, determining axial versus terminal flowers; *le*, determining tall versus short plants; and *v*, determining smooth versus constricted pods) are all located on chromosome 4. The *le* and *v* genes undergo recombination at the rate of about 12 percent, but Mendel apparently did not study this particular combination for independent assortment. The *fa* gene is very distant from the other two and shows independent assortment with them.

3.16 The chi-square value equals

$$(188 - 186)^2/186 + (203 - 186)^2/186 + (175 - 186)^2/186 + (178 - 186)^2/186 = 2.57$$

It has 3 degrees of freedom (four classes of data), and the P value is about 0.5. We therefore conclude that there is no reason, on the basis of these data, to reject the hypothesis of 1 : 1 : 1 : 1 segregation.

3.18 The chi-square value equals

$$(462 - 450)^2/450 + (167 - 150)^2/150 + (127 - 150)^2/150 + (44 - 50)^2/50 = 6.49$$

It has 3 degrees of freedom (four classes of data), and the P value is about 0.08. Although this is close to the 5% level of significance, it is not sufficient reason to reject the hypothesis of 9 : 3 : 3 : 1 segregation.

3.20 DNA from males exhibits one band and that from females two bands, which suggests X-linked inheritance. This can be verified further by examining the pattern of inheritance. Sons exhibit a band present in the mother, and daughters exhibit one band present in the mother and one in the father (which may be identical if the daughter is homozygous). The genotypes are: I-1, A_2A_2; I-2, A_1Y; II-1, A_2Y; II-2, A_1A_2; II-3, A_2Y; II-4, A_2Y; II-5, A_1A_2; II-6, A_1Y; III-1, A_1Y; III-2, A_1A_2; III-3, A_1A_2; III-4, A_2Y.

Chapter 4

4.2 Females produce gametes of types A B, a b, A b, and a B in the proportions $(1 - 0.08)/2 = 0.46$, $(1 - 0.08)/2 = 0.46$, $0.08/2 = 0.04$, and $0.08/2 = 0.04$, respectively. Males produce gametes of types A B and a b only (owing to the absence of crossing-over), in the proportions 0.50 and 0.50.

4.4 Only **(b)** is true; all the others are false.

4.6 This is a testcross, so the phenotypes of the progeny reveal the ratio of meiotic products from the doubly heterozygous parent. The expected progeny are 86 percent parental (43 percent wildtype and 43 percent scarlet, spineless) and 14 percent recombinant (7 percent scarlet and 7 percent spineless).

4.8 The data include $59 + 52 = 111$ progeny with either both mutations or neither, and $46 + 43 = 89$ with one mutation or the other. One of these groups consists of parental chromosomes and the other of recombinant chromosomes, and so the appropriate chi-square test compares the ratio 111 : 89 against an expected 100 : 100 with no linkage. The chi-square value equals 2.42 and there is 1 degree of freedom, from which P equals approximately 0.12. There is no evidence of linkage, even though both genes are in the X chromosome.

4.10 (a) They are not alleles because they fail to segregate; if they were alleles, all the progeny would be resistant to one insecticide or the other. **(b)** The alleles are not linked. The chi-square of 177 parental : 205 recombinant types against 191 : 191 equals 2.05 with 1 degree of freedom, which is not significant. **(c)** The two genes must be far apart on the chromosome.

4.12 (a) The parental types are evidently v^+ pr bm and v pr^+ bm^+ and the double-recombinant types v^+ pr^+ bm^+ and v pr bm. This puts v in the middle. The $pr-v$ recombination frequency is $(69 + 76 + 36 + 41)/1000 = 22.2$ percent, and the $v-bm$ recombination frequency is $(175 + 181 + 36 + 41)/1000 = 43.3$ percent. The expected number of double crossovers equals $0.222 \times 0.433 \times 1000 = 96.13$, so the coincidence is $(36 + 41)/96.13 = 0.80$. The interference is therefore $1 - 0.80 = 0.20$. **(b)** The true map distances are certainly larger than 22.2 and 43.3 centimorgans. The frequencies of recombination between these genes are so large that there are undoubtedly many undetected double crossovers in each region.

4.14 (a) The three-allele hypothesis predicts that the matings will be $FS \times FS$ and should yield $FF : FS : SS$ offspring in a ratio of 1 : 2 : 1. **(b)** These ratios are not observed, and furthermore, some of the progeny show the "null" pattern. **(c)** One possibility is that the F and S bands are from unlinked loci and that there is a "null" allele of each—say, f and s. Because f and s are common and F and S are rare, most persons with two bands would have the genotype Ff Ss. The cross should therefore yield $9/16$ $F-$ $S-$ (two bands), $3/16$ $F-$ ss (fast band only), $3/16$ ff $S-$ (slow band only), and $1/16$ ff ss (no bands). **(d)** The data are consistent with this hypothesis (chi-square = 2.67 with 3 degrees of freedom, P value approximately 0.50).

4.16 In this case one must start by calculating the number of double-crossover gametes that would be observed, given the interference. The number of observed doubles equals the number of expected doubles times the coincidence, or $0.15 \times 0.20 \times 0.20 = 0.006$, or among 1000 gametes, 3 each of $o + +$ and $+ ci$ p. The single recombinants in the $o-ci$ interval would therefore number $0.15 \times 1000 - 6 = 144$, or 72 each of $o + p$ and $+ ci +$. The single recombinants in the $ci-p$ interval would number $0.20 \times 1000 - 6 = 194$, or 97 each of o ci p and $+ + +$. The remaining 656 gametes are nonrecombinant, 328 each of o ci $+$ and $+ + p$.

4.18 The gene-centromere map distance equals $1/2$ the frequency of second-division segregation, which also equals the frequency of crossing-over in the region. In this problem it is easiest to answer the questions by taking the cases out of order, considering the second-division segregations at the beginning. **(c)** The frequency of second-division segregation of $cys-1$ must be 14 percent, because the map distance is 7 cM. Because of the complete interference, a crossover on one side of the centromere precludes a crossover on the other side, so these asci must have first-division segregation for $pan-2$. **(b)** Similarly, the frequency of second-division segregation of $pan-2$ must be 6 percent, because the map distance is 3 cM; these asci must have first-division segregation for $cys-1$. **(d)** Because of the complete interference, second-division segregation of both markers is not possible. **(a)** The only remaining possibility is first-division segregation of both markers, which must have a frequency of $1 - 0.14 - 0.06 = 80$ percent. **(e)** First-division segregation of both markers yields a PD tetrad, and second-division segregation for one of the markers yields a TT tetrad. Because there are no double crossovers, there can be no NPD tetrads. Hence the frequencies are PD = 80 percent and TT = 20 percent.

4.20 (a) This is like any other 3-point cross problem once the genotypes of the progeny have been inferred from the gel patterns. One of the parents is triply heterozygous, and one is homozygous a b c. The most frequent classes of gametes from the triply heterozygous parent are A B C and a b c, and the least frequent are a B c and a B C. Hence, gene C is in the middle. **(b)** A C B/a c b. **(c)** $(54 + 45 + 5 + 3)/1000 = 0.107$, or 10.7 map units. **(d)** $(80 + 59 + 5 + 3)/1000 = 0.147$, or 14.7 map units. **(e)** The coincidence = $8/(0.107 \times 0.147 \times 1000) = 0.509$. **(f)** The interference = $1 -$ coincidence = $1 - 0.509 = 0.491$.

Chapter 5

5.2 For 47 chromosomes: 47, +21 (trisomy 21 or Down syndrome), 47,XXX, 47,XXY (Klinefelter), or 47,XYY; for 45 chromosomes: 45, X.

5.4 It is true but unexpected, because 47,XXX would be expected to produce many XX eggs and 47,XXY many XY or XX sperm. Apparently the extra X chromosome is eliminated from the nucleus prior to meiosis.

5.6 The chromosomes underwent endoreduplication, resulting in an autotetraploid.

5.8 The inversion has the sequence A B E D C F G, the deletion A B F G. The possible translocated chromosomes are (1) A B C D E T U V and M N O P Q R S F G or (2) A B C D E S R Q P O N M and V U T F G. One of these possibilities includes two monocentric chromosomes, and the other includes a dicentric and an acentric. Only the translocation with two monocentrics is genetically stable.

5.10 The gene order is $(a$ $e)$ d f c b or the reverse, where the parentheses mean that the gene order cannot be determined from these data.

5.12 Gene a is present in band 2, b in band 1, c in band 3, d in band 5, e in band 4, and f in band 6.

5.14 The Cy gene is now linked with the Y chromosome, the most likely reason being because of a reciprocal translocation between chromosome 2 and the Y chromosome.

5.16 (a) In homozygotes, there is no impediment to crossing-over. Hence, for a map distance of 12 map units, one should expect to observe 12 percent recombination, because over this length of genetic interval, multiple crossing-over can be neglected. **(b)** In heterozygotes, the products of recombination would not be recovered, and if the whole region were involved in the inversion, the frequency of recombination would be 0. Because only $1/3$ of the interval is inverted in this case, the recombination frequency is expected to be $(2/3) \times 12 = 8$ percent.

5.18 Species A ($n = 6$) hybridizes with species B ($n = 6$). The F_1 progeny will have 12 chromosomes and will be sterile. The sterility can be overcome by endoreduplication in an F_1 organism, which creates a fertile new species with a chromosome number of 24. This new species ($n = 12$) hybridizes with a third species, C ($n = 6$), yielding other sterile F_1 progeny with 18 chromosomes. Endoreduplication in one of these sterile F_1 organisms gives rise to a fully fertile new species with 36 chromosomes. Note that this scenario is very similar to that which produced hexaploid wheat.

5.20 The original strain carried an inversion with a breakpoint between A and the centromere. Single crossovers within the inversion loop created chromosomally abnormal ascospores, which were the ascospores that failed to survive. The ascospore that was recovered from second-division segregation carried the inversion as well as the a allele. When crossed with the original wildtype strain, the inversion became homozygous, restoring the normal 20 percent second-division segregation, which corresponds to 10 map units from the centromere.

Chapter 6

6.2 Because of semiconservative replication, equal amounts of $^{14}N^{14}N$ and $^{14}N^{15}N$ would be expected.

6.4 Ten percent of the haploid genome is 10,000 kb, or 10^7 base pairs. Each base pair in the double helix extends 3.4 angstrom units, so 10^7 base pairs equals 3.4×10^7 angstrom units. Because there are 10^{-4} micrometers per angstrom unit, the required length is 3.4×10^3 micrometers or, dropping the scientific notation, 3400 micrometers. This length is 3.4 mm, which, if the molecule were not so thin, would be long enough to see with the naked eye.

6.6 (a) The distance between nucleotide pairs is 3.4 Å. The length of the molecule is 68×10^4 Å. Therefore, the number of nucleotide pairs equals $(68 \times 10^4)/3.4 = 2 \times 10^5$. **(b)** There are 10 nucleotide pairs per turn of the helix, so the total number of turns is $2 \times 10^5/10 = 2 \times 10^4$.

6.8 Every BamHI site is also a Sau3A site, but not every Sau3A site is a BamHI site. The reason is that a Sau3A site is included within any BamHI site but not the other way around.

6.10 Replication is bidirectional from a single origin of replication.

6.12 In a double-stranded DNA molecule, A = T and G = C because of complementary base pairing. Therefore, if A/C = 1/3, then C = G = 3 × A. Because A + T + G + C = 1, everything can be put in terms of A as follows: A + A + 3A + 3A = 1. Hence A = T = 0.125. This makes C = G = (1 − 0.25)/2 = 0.375. In other words, the DNA is 12.5 percent A, 12.5 percent T, 37.5 percent G, and 37.5 percent C.

6.14 The difference in size of the restriction fragment could be due either to the insertion of DNA between the two restriction sites or to the elimination of one of the two flanking XhoI restriction sites. The pattern of hybridization of the 10-kb fragment to multiple locations along the polytene chromosomes, which differ between strains, suggests not only that the mutation is caused by insertion of DNA but also that the inserted DNA is a transposable element. Transposable elements are often present in multiple copies in a genome, and because of the lack of specificity for insertion sites that many transposable elements display, one would not expect to find elements inserted at the same genomic locations in two independent wildtype isolates.

6.16 The leading strands are c d and e f, the lagging strands are a b and g h. The 3′ ends are c, b, e, and h; the 5′ ends are a, d, g, and f.

6.18 The sequence, read from left to right, is

3′-CTTGACCGATACACAACTGAGACT-5′

6.20

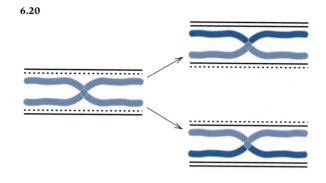

Chapter 7

7.2 For a mean of 1 phage per cell, the expected proportion of uninfected cells is given by the Poisson distribution as $e^{-1} = 0.368$. If you do not know the Poisson distribution, you can work out the answer as follows. The probability that a particular bacterium is infected by a particular phage is 1×10^{-6}, hence the probability of a particular bacterium escaping a particular phage is 0.999999. The probability of a particular bacterium escaping all 1,000,000 phage is therefore $(0.999999)^{1000000} = 0.368$, which equals the expected proportion of uninfected cells.

7.4 Yes, as long as it has a replication origin.

7.6 One wants 50 cells in 0.1 ml, so the concentration in the final dilution should be 5×10^2 viable cells per ml. Dilution of the original 5×10^7 to a final 5×10^2 would require two dilutions of 100-fold each and one dilution of 10-fold.

7.8 *E. coli* has a genome size of 4.6×10^6 bp and there are 100 minutes in the map; hence the length of 1 minute in the map is 4.6×10^4 bp = 46 kb. The genetic length of lambda prophage is therefore approximately 1 minute.

7.10 Apparently h and *tet* are closely linked, so recombinants that contain the h^+ allele of the Hfr tend also to contain the *tet-s* allele of the Hfr, and these recombinants are eliminated by the counterselection for *tet-r*.

7.12 The attachment site, *att*, must be between genes *f* and *g*.

7.14 The three possible orders are (1) $pur-pro-his$, (2) $pu-his-pro$, and (3) $pro-pur-his$. The predictions of the three orders are as follows: (1) Virtually all $pur^+ his^-$ transductants should be pro^-, but this is not true. (2) Virtually all $pur^+ pro^-$ transductants should be his^-, but this is not true. (3) Some $pur^+ pro^-$ transductants will be his^-, and some $pur^+ his^-$ transductants will be pro^- (depending on the locations of the exchanges). Therefore, order (3) is the only one that is not contradicted by the data.

7.16 Any phage transduces one small, contiguous piece of DNA. Cotransduction therefore indicates very close linkage. Hence G and H are close, and G and I are close, but H and I are not close (because they do not cotransduce), so the order of these three genes must be $I G H$ (or the reverse). The locations of the other three genes can be deduced similarly: A is close to H, B is close to I, and T must be close to A but not close to H. Hence the gene order is $B I G H A T$ or, equivalently, $T A H G I B$.

7.18 The map implied by the time-of-entry data is shown here. The entries that correspond to the question marks should be as follows: for *his* in Hfr1, 87 min; for *phe* in Hfr2, 4 min; at 35 min in Hfr3, *trp*; and for *bio* in Hfr3, 66 min.

7.20 The times of entry correspond to the *x*-intercepts in the accompanying graph. For the genes *a, b, c,* and *d* the times of entry are 10, 15, 20, and 30 minutes, respectively.

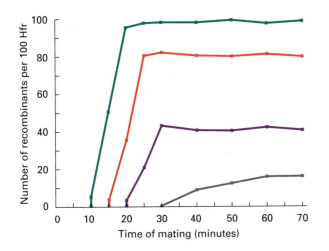

Chapter 8

8.2 The 5' end of the RNA transcript is transcribed first, so **(a)** through **(c)** are all wrong. The 5' end of the RNA transcript is transcribed from the 3' end of the DNA template, so **(e)** is wrong. The additional statement in **(e)** that the amino end of the polypeptide is synthesized first serves as confirmation of the correctness of statement **(d)**.

8.4 The mRNA sequence is

5'-AAAAUGCCCUUAAUCUCAGCGUCCUAC-3'

and the first AUG is the initiation codon, so the resulting amino acid sequence is Met−Pro−Leu−Ile−Ser−Ala−Ser−Tyr, or, in the single-letter abbreviations, M−P−L−I−S−A−S−Y.

8.6 (a) Methionine has 1 codon, histidine 2, and threonine 4, so the total number is $1 \times 2 \times 4 = 8$. The sequence AUGCAYACN encompasses them all. **(b)** Because arginine has six codons, the possible number of sequences coding for Met−Arg−Thr is $1 \times 6 \times 4 = 24$. The sequences are AUGCGNACN (16 possibilities) and AUGAGRACN (8 possibilities).

8.8 With a G at the 5' end, the first codon is GUU, which codes for valine; and with a G at the 3' end, the last codon is UUG, which codes for leucine.

8.10 27/64 Lys (AAA), 9/64 Asn (AAC), 12/64 Thr (9/64 ACA + 3/64 ACC), 9/64 Gln (CAA), 3/64 His (CAC), 4/64 Pro (3/64 CCA + 1/64 CCC).

8.12 (a) With four kinds of nucleotides and only five kinds of amino acids, the smallest codon size is 2, which gives 16 possible different combinations of four bases taken two at a time ($4^2 = 16$). **(b)** If the codon size is 2, it would take $2 \times 10 = 20$ bases to code for one protein, or $20 \times 100 = 2000$ to code for all the proteins the organism makes. Therefore, the minimum size of the genome of the organism is 2000 base pairs, assuming that the genes are nonoverlapping.

8.14 (a) The deletion must have fused the amino-coding terminus of the *B* gene with the carboxyl-coding terminus of the

A gene. The nontranscribed strand must therefore be oriented 5'-*B*−*A*-3'. **(b)** The number of bases deleted must be a multiple of 3; otherwise, the carboxyl terminus would not have the correct reading frame.

8.16 The wildtype and double-mutant mRNA codons are

Wildtype:

5'-AAR AAR UAY CAY CAR UGG ACN UGY AAY-3'

Double-mutant:

5'-AAR CAR AUH CCN CCN GUN GAY AUG AAY-3'

Comparing these sequences, it can be deduced that the first frameshift addition is the C at position 4 in the double-mutant gene. Realigning the sequences to take this into account yields

Wildtype:

5'-AAR AAR UAY CAY CAR UGG ACN UGY AAY-3'

Double-mutant:

5'-AARC ARA UHC CNC CNG UNG AYA UGA AY-3'

Now it is also clear that the single-nucleotide deletion in the wildtype gene must be the Y in the fourth nucleotide position from the 3' end. Realigning again to take this into account, we have

Wildtype:

5'-AAR AAR UAY CAY CAR UGG ACN UGY AAY-3'

Double-mutant:

5'-AARC ARA UHC CNC CNG UNG AYA UG AAY-3'

Hence the complete sequences, as thoroughly as they can be resolved from the data, are

Wildtype:

5'-AAR AAA UAC CAC CAG UGG ACA UGY AAY-3'

Double-mutant:

5'-AAR CAA AUA CCA CCA GUG GAC AUG AAY-3'

8.18 (a) Lys has the codon formula AAR, hence the next amino acid must have the formula ARN; this includes Asn (AAY), Lys (AAR), Ser (AGY), and Arg (AGR). **(b)** Met has one codon, AUG, and could be followed by Cys (UGY) or Trp (UGG). **(c)** Tyr has the formula UAY and could be followed by Ile (AUU, AUC, AUA), Met (AUG), or Thr (ACN). **(d)** Tyr has one codon, UGG, and could be followed only by Gly (GGN).

8.20 The 25 ribonucleotides encode a polypeptide of 22 amino acids because a shift of reading frame occurs each time the circle is traversed until a stop codon is finally encountered. The complete amino acid sequence of the product is given below.

Problem 8.20—Complete Amino Acid Sequence

Met−Ala−Asp−Ser−Ala−Asp−Leu−Ala−Asp−Gly−Arg−Phe−Gly−Arg−Phe−Ser−Arg−Trp−Gln−Ilu−Arg−Glu−Ile,

or, in the single-letter abbreviations,

M−A−D−S−A−D−L−A−D−G−R−F−G−R−F−S−R−W−Q−I−R−E−I.

Chapter 9

9.2 The repressor gene need not be near because the repressor is diffusible. The *trp* repressor is one example in which the repressor gene is located quite far from the structural genes.

9.4 In the presence of glucose, bacterial cells have very low levels of cAMP. Without cAMP, the binding of the CRP protein cannot take place, transcription cannot be activated, and no *lac* proteins are made. Lack of induction is true even if lactose is present. The system ensures that primary carbon sources are exhausted before the cell expends energy for the synthesis of new enzymes to utilize secondary carbon sources.

9.6 In the absence of lactose or glucose, two proteins are bound: the *lac* repressor and CRP–cAMP. In the presence of glucose, only the repressor is bound.

9.8 The mutant gene should bind more of the activator protein or bind it more efficiently. Therefore, the mutant gene should be induced with lower levels of the activator protein or expressed at higher levels than the wildtype gene, or both.

9.10 In the first cross, active repressor is already present in the recipient cells, so no activation of *lac* genes is possible. In the second cross, when the *lac* genes enter the recipient, no repressor is present, so *lac* mRNA is produced and β-galactosidase is synthesized. The effect is transient, because after a short period of time, repressor is also made using the donor *lacI*$^+$ gene, and transcription of *lac* genes is repressed.

9.12 An inversion not only inverts the sequence but, because the DNA strands in a duplex are antiparallel, exchanges the strands. The inversion of a promoter sequence would have a profound effect because the mutant promoter is in the wrong strand pointing in the wrong direction. One would expect little or no effect of the inversion of an enhancer element, because enhancers are targets for DNA binding proteins irrespective of orientation in the DNA.

9.14 (a) Constitutive. Deletion of A causes a frameshift that switches the reading frame from AUG AUG, and so forth, to UGA UGA, and so forth. UGA is a stop codon; therefore, the attenuator will not be translated and transcription will continue unstopped. **(b)** Met$^-$. Deletion of the two A's changes the reading frame from AUG AUG, and so forth, to GAU GAU, and so forth. GAU is an Asp codon. Therefore, translation will proceed through even when the concentration of methionine is very low, and transcription will terminate prematurely. **(c)** Wildtype. A deletion of all three A's does not change the reading frame.

9.16 The predicted levels are shown in the table.

Genotype	Uninduced level		Induced level	
	Z	**Y**	**Z**	**Y**
(a)	100	100	100	100
(b)	0.1	0.1	100	100
(c)	25	25	100	100
(d)	25	25	100	100

9.18 (a) Yes; the repressor is functional, and the presence of lactose activates transcription of the *lac* genes. **(b)** and **(d)** Yes; at 42°C, the repressor cannot bind the operator, which means that the *lac* operon is transcribed whether or not the inducer is present. **(c)** No; at this temperature, the repressor functions normally, and because lactose is absent, the *lac* operon is repressed.

9.20 Recessive mutations will fail to complement if they are in the same exon, or if one mutation is in an exon that is shared among all of the alternatively spliced transcripts. However, recessive mutations will complement if they are in different exons and these are not shared among all of the alternatively spliced transcripts. **(a)** The complementation matrix is as shown in the accompanying diagram. **(b)** This matrix is unusual in that some mutations are shared between complementation groups. In particular, mutations *1, 2, 3*, and *4* fail to complement in all combinations, and mutations *3, 4, 5*, and *6* fail to complement in all combinations. One would therefore expect mutations *1* and *2* to fail to complement mutations *5* and *6*, but in fact these mutations do complement. In other words, mutations *3* and *4* appear to belong to two different complementation groups.

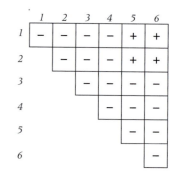

Chapter 10

10.2 It occurs 23 times there also, because the sequence is a "palindrome."

10.4 One must use cDNA whose transcription is driven by a suitable promoter that functions in a prokaryotic system. Eukaryotic genomic DNA includes regulatory elements that will not work in bacteria and introns that cannot be spliced out of RNA in prokaryotic cells.

10.6 The primer sequences are included in the PCR product at both ends of the gene fragment. The total length of the amplified piece of DNA is therefore 19 + 303 + 19 = 341 base pairs.

10.8 The 3.1-kb fragment has inverted repeats at the ends; it is probably a DNA transposable element.

10.10 The $E-B$ fragment contains a repressor-binding site and the $H-S$ fragment contains a repressor-binding site; these act cooperatively to produce full repression.

10.12 Six-base cutters, such as *Bam*HI, find their recognition sequences every 4096 (4^6) bases, on the average. In a genome of 4,600,000 bases, about 1123 fragments are expected, assuming that A, G, C, and T are in equal proportions in the genome and are randomly distributed.

10.14 The hint says that if the genome were represented x times in the library, the probability that a particular sequence would be missing is e^{-x}, which we want to equal 0.01. Hence $x = 4.6$. Because one haploid representation of the genome equals $(6 \times 10^9)/2 = 3 \times 10^9$ base pairs, and the average insert size is 2×10^4 base pairs, one requires $[(3 \times 10^9)/(2 \times 10^4)] \times 4.6 = 6.9 \times 10^5$ clones.

10.16 The only restriction map consistent with the data is a circular one:

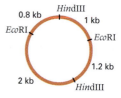

10.18 The resulting electrophoresis gel would have bands in the positions indicated in the accompanying diagram.

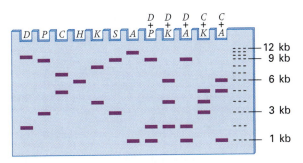

10.20 The tryptophan operon is expressed equally under all growth conditions. The *lacI* and *crp* genes are expressed constitutively. Although the cyclic AMP receptor protein co-regulates the *lac* operon, it does so through variation in the levels of cAMP under various growth conditions rather than through variation in the expression of *crp*. Transcription of *lacZ* and *lacY* into a polycistronic mRNA is induced by growth in lactose medium.

Experimental minimal medium	Control minimal medium	Transcript				
		trpE	*lacI*	*lacZ*	*lacY*	*crp*
Glucose	Glucose	○	○	○	○	○
Glucose	Glycerol	○	○	○	○	○
Glycerol	Glucose	○	○	○	○	○
Lactose	Glucose	○	○	●	●	○
Glucose	Lactose	○	○	●	●	○
Lactose	Glycerol	○	○	●	●	○
Glycerol	Lactose	○	○	●	●	○

Chapter 11

11.2 The first experiment suggests that the genetic program of cell death is already activated and cannot be turned off. The second experiment shows that intercellular signaling must be responsible for inducing the cell-death fate, because when uninduced cells are transplanted to a new site, they do not die.

11.4 Perform reciprocal crosses. A maternal-effect mutation results in phenotypically mutant progeny when the mother is homozygous.

11.6 The proteins required for cleavage and blastula formation are translated from mRNAs present in the mature oocyte, but transcription of zygotic genes is required for gastrulation.

11.8 Distinct sets of genes are activated in the different target cells.

11.10 A loss-of-function mutation in a gene that is necessary for a developmental fate to be expressed prevents the fate from being realized; such a mutation is expected to be recessive.

11.12 Petal–petal–stamen–carpel.

11.14 An *X Y* double mutant has only *Z* function, and gene product Z expands into all four whorls. The mutant flower would therefore have carpels induced in all four whorls.

11.16 The 20 mutations need not be mutations of 20 different genes; indeed, many of them proved to be alleles of one or another of a small set of genes.

11.18 **(a)** *agamous* mutants have the phenotype sepal–petal–petal–sepal; unless *ap1* expression expands into whorls 3 and 4, the phenotype would be sepal–petal–leaf–leaf; **(b)** *apetala-1* mutants have a phenotype carpel–stamen–stamen–carpel; unless *ag* expression expanded into whorls 1 and 2, the phenotype would be leaf–leaf–stamen–carpel.

11.20 a^+ is necessary for development of the medial and distal segments, b^+ is necessary for the development of the proximal segment, and c^+ is necessary for the development of the proximal and medial segments. When c^+ is missing from the medial segment, but a^+ is present in this segment, the segment is transformed to proximal, which implies that it is the interaction between a^+ and c^+ that yields the medial segment.

Chapter 12

12.2 A deletion, or possibly such a complex insertion or rearrangement that the original nucleotide sequence cannot be restored.

12.4 For forward missense mutations, there are a large number of amino acid replacements that can reduce or eliminate protein function, but once a particular amino acid replacement has been made, there are only a small number of ways in which it can be reversed.

12.6 There are nine different mutant codons: (1) UGC → CGC (Arg), (2) UGC → AGC (Ser), (3) GGC → GGC (Gly), (4) UGC → UUC (Phe), (5) UGC → UCC (Ser), (6) UGC → UAC (Tyr), (7) UGC → UGU (Cys), (8) UGC → UGA (Stop), and (9) UGC → UGG (Trp). Number 7 is synonymous, number 8 is nonsense (termination codon), and the rest are nonsynonymous (missense).

12.8 **(a)** 2; **(b)** 2; **(c)** 3; **(d)** 1.

12.10 **(a)** 1/2 females and 1/2 males (the females are heterozygous); **(b)** all female (the males die).

12.12 **(a)** There are 1×10^6 replications and a rate of mutation of 1×10^{-6} new mutants per replication, so the expected number of new mutants is $(1 \times 10^6) \times (1 \times 10^{-6}) = 1$. **(b)** The total number of cells after division is 2×10^6. The probability of any one cell being a mutant is therefore $1/(2 \times 10^6) = 5 \times 10^{-7}$. The probability that any cell is a nonmutant is thus $1 - (5 \times 10^{-7})$, and the probability that each of the cells in the population is nonmutant is this quantity raised to the power of 2×10^6. The answer is $P = 0.368$. [The log $P = (2 \times 10^6) \times \log[1 - (5 \times 10^{-7})] = (2 \times 10^6) \times (-2.17 \times 10^{-7}) = -0.43429$, so $P = 0.368$. Students familiar with the Poisson distribution may recognize this as $e^{-1} = 0.368$, the "Poisson zero term."

12.14 Many of the amino acid replacements impair protein function to such a small extent that the mutants still have enough tryptophan synthase activity to be able to grow nearly normally under the conditions of the experiment.

12.16 The lysine codons are AAA and AAG and the glutamic codons are GAG and GAA. Hence an A → G transition in the first codon position will change a codon for lysine into one for glutamic acid.

12.18 Trace the consequences of a crossover within the transposable element. The answers are shown in the diagrams.

(a)

(b)

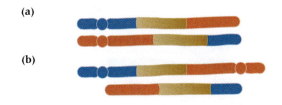

12.20 **(a)** Acsi of types (4) and (5) show typical 3 : 1 patterns of segregation expected with gene conversion, which strongly suggests that mismatch repair does take place. The ratios of $m : +$ in these asci are 6 : 2 and 2 : 6, respectively. Asci (2) and (3) also show evidence of mismatch repair because the segregation ratios are, respectively, 5 : 3 and 3 : 5. **(b)** In asci of types (1), (2), and (3), adjacent pairs of ascospores, which result from postmeiotic mitosis, do not have the same genotype; this result implies that the immediate product of meiosis was a heteroduplex, resulting in genetically different ascospores after the mitotic division. The persistence of the hetero-duplexes implies that mismatch repair is not 100 percent efficient.

Chapter 13

13.2 At the G_1/S checkpoint, because of the unrepaired double-stranded DNA breaks caused by the x rays.

13.4 Ras–GDP is activated by cellular growth factors to produce Ras–GTP, which stimulates cellular growth. A gain-of-function mutation such as G19V promotes constitutive growth; hence Ras is a proto-oncogene.

13.6 The diagram at the bottom of this page shows how mitotic recombination can cause loss of heterozygosity. The diagram also shows that all genetic markers distal to (toward the telo-mere side of) the site of recombination also lose heterozygosity.

13.8 Bax and Bcl2 form Bax–Bcl2 heterodimers. Over-production of Bax due to p53 activation promotes apoptosis. Overproduction of Bcl2 caused by certain oncogenes blocks apoptosis and the cells can progress to cancer.

13.10 A defect in the G_1/S checkpoint will allow cells with damaged chromosomes to enter DNA synthesis, when replica-tion and repair can cause chromosome rearrangements or repli-cation errors leading to aneuploidy.

13.12 3 tumors/(4×10^6 cells) = 0.75×10^{-6} = 7.5×10^{-7} per cell. This is not exactly a "mutation" rate, because it includes such events as chromosome loss, mitotic recombina-tion, and so forth, which are not, strictly speaking, mutations.

13.14 **(a)** The smaller band should be associated with the mutant RB1 allele, the larger with the nonmutant allele. **(b)** The loss of heterozygosity is independent in different tumors; hence the mechanism need not be the same in both cases. **(c)** In this case, the wildtype allele has undergone a new mutation (or perhaps a small deletion or insertion), which inactivates the gene but does not detectably change the size of the band.

13.16 The son's genotype is $AL\ p53^-\ BL/AS\ p53^+\ BS$. **(a)** The bands $AS\ p53^+$ and BS are lost. **(b)** $AL\ p53^-$ and BL all become homozygous. **(c)** $p53^-$ and BL, but AL/AS remains heterozy-gous. **(d)** No apparent change in banding pattern from non-cancerous cells. **(e)** $p53^-$ has replaced $p53^+$. The expected band patterns are shown in the illustration.

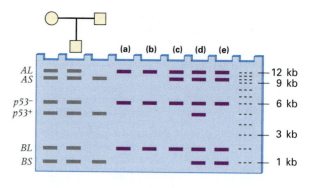

13.18 **(a)** The cells should stop dividing at the restrictive tem-perature because they would no longer sense and propagate a growth-stimulatory signal (probably the presence of nutrients rather than growth factors in this example). **(b)** Recessive. The Ras produced from the wildtype gene in the same cell would still propagate the signal. **(c)** Such a mutant cell will continue to divide even when nutrients are exhausted, but such a mutant will die when starved rather than cease growing. **(d)** The mutation would be dominant. The mutant Ras will remain in the GTP-bound form regardless of what the wildtype Ras is doing and will ceaselessly propagate a signal to grow.

13.20 For generations I and II, the band at position c is associ-ated with the disease. In generation III, individuals 2 and 7 are at risk owing to the parent 5, and individuals 21 and 22 are at risk owing to the parent 20. However, individual 13 is not at risk because the parent 14 with a band at position c is not affected; band c in this individual evidently orginates from a nonmutant allele.

Chapter 14

14.2 **(a)** Each genomic DNA sample shows two bands because the person is heterozyygous. **(b)** It would be possible if a person were homozygous. **(c)** Codominance. **(d)** 2 and 7 are the par-ents of A, 1 and 4 the parents of B, 6 and 8 the parents of C, and 3 and 5 the parents of D.

14.4 The 10 AA genotypes contribute 20 A alleles to the sample, the 15 Aa genotypes contribute 15 A and 15 a alleles to the sample, and the 4 aa genotypes contribute 8 a alleles to the sample. Altogether, there are 35 A and 23 a alleles in the sam-ple, for a total of 58. In this sample the allele frequency of A is $35/58 = 0.603$ and that of a is $23/58 = 0.397$.

14.6 Because the frequency of homozygous recessives is 0.10, the frequency of the recessive allele q is $(0.10)^{1/2} = 0.316$. This means $p = (1 - 0.316) = 0.684$, so the frequency of heterozygotes is $2 \times 0.316 \times 0.684 = 0.43$. Because 0.90 of the individuals are nondwarfs, the frequency of hetero-zygotes among nondwarfs is $0.43/0.90 = 0.48$.

14.8 **(a)** $(1/n)^2 = 1/n^2$; **(b)** $2(1/n)(1/n) = 2/n^2$; **(c)** $n(1/n^2) = 1/n$; **(d)** $[n(n-1)/2] \times (2/n^2) = 1 - (1/n)$.

Problem 13.6

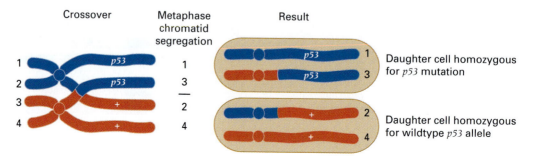

14.10 (a) The probability that the wife is a carrier is the probability that a phenotypically normal person is a carrier. This probability is $2pq/(p^2 + 2pq)$, where $q = \sqrt{(1/1700)} = 0.024$ and $p = 0.976$. The numerator $2pq = 0.047$ and the denominator $p^2 + 2pq = 0.999$ (close enough to 1 that it can be ignored). Hence the probability that the wife is a carrier equals 0.047. **(b)** If she is a carrier, the probability that the first child will be affected is 1/4, so the overall probability that the first child will be affected equals $0.047 \times 1/4 = 0.012$, or about 1 in 84.

14.12 The allele frequency is the square root of the frequency of affected offspring among unrelated individuals, or $(8.5 \times 10^{-6})^{1/2} = 2.9 \times 10^{-3}$. With inbreeding, the expected frequency of homozygous recessives is $q^2(1 - F) + qF$, in which F is the inbreeding coefficient. When $F = 1/16$, the expected frequency of homozygous recessives is $(2.9 \times 10^{-3})^2(1 - 1/16) + (2.9 \times 10^{-3})(1/16) = 1.9 \times 10^{-4}$; when $F = 1/64$, the value is 5.3×10^{-5}.

14.14 Set $q = 2 \times 2pq$; hence $p = 1/4$ and $q = 3/4$.

14.16 (a) $(0.01)^2 \times 1,000,000 = 100$ seedlings. **(b)** The lethal recessive allele will persist for many generations at a low frequency, because it is present primarily in heterozygotes and is shielded from selection by the dominant allele.

14.18 (a) Here $F = 0.66$, $p = 0.43$, and $q = 0.57$. The expected genotype frequencies are as follows: AA, $(0.43)^2(1 - 0.66) + (0.43)(0.66) = 0.347$; AB, $2(0.43)(0.57)(1 - 0.66) = 0.167$; and BB, $(0.57)^2(1 - 0.66) + (0.57)(0.66) = 0.487$. **(b)** With random mating, the expected genotype frequencies are as follows: AA, $(0.43)^2 = 0.185$; AB, $2(0.43)(0.57) = 0.490$; and BB, $(0.57)^2 = 0.325$.

14.20 Let p_n be the allele frequency of A in generation n, and for convenience, set $\mu = 10^{-5}$. The probability of a particular A allele not mutating in any one generation is $1 - \mu$; hence $p_1 = (1 - \mu)p_0$, but because $p_0 = 1$, $p_1 = 1 - \mu = 0.99999$. The relationship between p_2 and p_1 is the same as that between p_1 and p_0, so $p_2 = (1 - \mu)p_1 = (1 - \mu)(1 - \mu)p_0 = (1 - \mu)^2 p_0 = (1 - \mu)^2 = 0.99998$. In general, the rule is $p_n = (1 - \mu)^n p_0$, which, in this example, is $(0.99999)^n$.

Chapter 15

15.2 Broad-sense heritability is the proportion of the phenotypic variance attributable to all differences in genotype, which includes dominance effects and interactions between alleles. Narrow-sense heritability is the proportion of the phenotypic variance due only to additive ("transmissible") gene effects, which is used to predict the resemblance between parents and offspring in artificial selection. If all allele frequencies equal 1 or 0, both heritabilities equal 0.

15.4 The variance of the F_1 progeny is an estimate of the environmental variance. The variance of the F_2 progeny is the sum of the genotypic and environmental variance.

15.6 $\Sigma x_i = 2480$, so the estimated mean equals $2480/50 = 49.60$ eggs. To estimate the variance, subtract the mean from each value, square, sum these together, and divide by 49; in symbols, $s^2 = \Sigma(x_i - 49.60)^2/49 = 60.3265$. The standard deviation is the square root of the variance, hence $s = \sqrt{(60.3265)} = 7.76702$. (*Note:* An alternative method of calculating the estimate of the variance is to use the sum of the values and the sum of the squares as follows: $s^2 = N/(N - 1) \times [\Sigma x^2/N - (\Sigma x/N)^2]$. In the present example, this expression evaluates to $(49/50) \times [125,964/50 - (49.60)^2] = 60.3265$, the same as calculated earlier.)

15.8 For fruit weight, $\sigma_e^2/\sigma_p^2 = 0.13$ is given. Because $\sigma_p^2 = \sigma_g^2 + \sigma_e^2$, we know that $H^2 = \sigma_g^2/\sigma_p^2 = 1 - \sigma_e^2/\sigma_p^2 = 1 - 0.13 = 0.87$. Therefore, 87 percent is the broad-sense heritability of fruit weight. The values of H^2 for soluble-solid content and acidity are 91 percent and 89 percent, respectively.

15.10 The expected average height of the progeny equals $17 + 0.30 \times (16 - 17) = 16.7$ hands. The heritability is a ratio of variances and is independent of the unit of measurement. Thus, whether the withers height is measured in meters or in hands makes no difference in the heritability.

15.12 The range 1.8 to 2.2 mg is the mean $\pm$ 1 standard deviation, and 68 percent of the population is expected to fall within this range. Similarly, the range 1.6 to 2.4 mg is the mean $\pm$ 2 standard deviations, and 95 percent of the population is expected to fall within this range. A beetle with a pupal weight of 3.0 mg or more deviates by more than $+5$ standard deviations from the mean. Because only 0.15 percent of the population deviates by more than $+3$ standard deviations, animals with pupae weighing more than 3.0 mg are expected to be exceedingly rare and, in practice, are not found. (The expected proportion weighing ≥ 3.0 mg actually equals 3×10^{-7}, or roughly 1 animal in 3 million.)

15.14 The narrow-sense heritability equals 8 times the first-cousin correlation coefficient, so the first-cousin correlation coefficient for total fingerprint ridge count is $0.86/8 = 0.108$.

15.16 The offspring are genetically related as half siblings, and hence the theoretical correlation coefficient is $h^2/4$.

15.18 (a) $12 + 0.20 \times (16 - 12) = 12.8$ g. **(b)** $12 + 0.20 \times (8 - 12) = 11.2$ g.

15.20 (a) 1.0, 0.67, 0.18, 0.10, 0.02, 0.01, and 0.002, respectively. **(b)** With rare recessive alleles, most homozygous offspring come from matings between heterozygous genotypes, so there is essentially no parent–offspring correlation in phenotype.

word roots
prefixes, suffixes, and combining forms

Roots and Prefixes	Meaning	Example
a-, an-	*absence or lack*	acentric, lack of a centromere
ab-	*departing from, away from*	abnormal, departing from normal
ac-, acro-	*extreme or extremity, peak*	acrocentric, centromere near the end of a chromosome
allel-	*of one another*	alleles, alternative forms of a gene
amphi-	*on both sides, of both kinds*	amphidiploid, an organism containing diploid genomes from two different organisms
ana-	*apart, up, again*	anaphase of mitosis, when the chromosomes separate (move apart)
ant-, anti-	*opposed to, preventing or inhibiting*	antibiotic, preventing or inhibiting life
ante-	*preceding, before*	antedate, precedes a date
apo-	*former, from*	aporepressor, precursor to repressor
aut-, auto-	*self*	autogenous, self-generated
bi-	*two*	bidirectional, going in two directions
bio-	*life*	biology, the study of life and living organisms
blast-	*bud or germ*	blastoderm, structure formed in early development
carcin-	*cancer*	carcinogen, a cancer-causing agent
cata-	*down*	catabolism, chemical breakdown
caud-	*tail*	caudal (directional term)
chiasm-	*crossing*	chiasma, the cross-shaped figure that occurs at the site of crossing-over between homologous chromosomes
chrom-	*colored*	chromosome (staining body), so named because they stain darkly
circum-	*around*	circumnuclear, surrounding the nucleus
co-, con-, com-	*together*	codominant, expression of both alleles in a heterozygote
cyt-	*cell*	cytology, the study of cells
de-	*undoing, reversal, loss, removal*	deoxy, lacking an oxygen atom
di-	*twice, double*	dicentric, having two centromeres
dys-	*difficult, faulty, painful*	dysfunctional, disturbed function
ec-, ex-, ecto-	*out, outside, away from*	excise, to cut away
ectop-	*displaced*	ectopic, expression that occurs in the wrong tissue or cell type
endo-	*within, inner*	endonuclease, cleaving a nucleic acid in the interior, not the end, of a molecule
epi-	*over, above*	episome, a genetic element over (beyond) the core genome
eu-	*good, well*	eukaryote, a cell with a good or true nucleus
exo-	*outside, outer layer*	exonuclease, an enzyme that digests nucleic acids beginning at the ends of the molecule
extra-	*outside, beyond*	extracellular, outside the body cells of an organism
flagell-	*whip*	flagellum, the tail of a sperm cell
gam-, gamet-	*married, spouse*	gametes, the sex cells
gene	*beginning, origin*	genetics
gon-, gono-	*seed, offspring*	gonads, the sex organs
haplo-	*single*	haploid, gametic chromosome number
hema-, hemato-, hemo-	*blood*	hemoglobin, blood protein
hemi-	*half*	hemimethylated, methylated on one DNA strand only
hetero-	*different or other*	heterosexuality, sexual desire for a person of the opposite sex
holo-	*whole*	holoenzyme, form of an enzyme containing all subunits
hom-, homo-	*same*	homozygous, having the same allele of a gene in homologous chromosomes
homeo-	*similar*	homeotic, related structurally
hyper-	*excess*	hypermorphic, state in which a gene is expressed at levels greater than normal
hypo-	*below, deficient*	hypomorphic, state in which a gene is expressed at levels less than normal
in-	*in, into*	induce, to lead into (a new state)
inter-	*between*	intercellular, between the cells

Roots and Prefixes	Meaning	Example
intercal-	insert	intercalated dyes, dyes that insert between adjacent base pairs in DNA
intra-	within, inside	intracellular, inside the cell
iso-	equal, same	isothermal, equal, or same, temperature
juxta-	near, close to	juxtapose, place near or next to
karyo-	kernel, nucleus	karyotype, the set of the nuclear chromosomes
kin-, kines-	move	kinetic energy, the energy of motion
lact-	milk	lactose, milk sugar
lumen	light	lumen, center of a hollow structure
lys-	dissolution or loosening	lysis, disruption by dissolution
macro-	large	macromolecule, large molecule
mal-	bad, abnormal	malfunction, abnormal functioning of an organ
mater-	mother	maternal, pertaining to the mother
mega-	large	megabase, a million base pairs
meio-	less	meiosis, nuclear division that halves the chromosome number
mero-	partial	merodiploid, partial diploid (in bacteria)
meta-	beyond, between, transition	metaphase, the stage of mitosis or meiosis in which chromosomes are located between the poles of the spindle
micro-	small	microscope, an instrument used to make small objects appear larger
mito-	thread, filament	mitochondria, small, filament like structures located in cells
mono-	single	monhybrid, heterozygous for one allelic pair
morpho-	form	morphology, the study of form and structure of organisms
multi-	many	multinuclear, having several nuclei
muta-	change	mutation, change in the base sequence of DNA
nano-	dwarf	nanometer, one billionth of a meter
nucle-	pit, kernel, little nut	nucleus
nulli-, nullo-	none	nullisomic, having no copies of a particular chromosome
oligo-	few	oligonucleotide, a nucleic acid molecule containing a few nucleotides
onco-	a mass	oncology, study of cancer
oo-	egg	oocyte, precursor of female gamete
org-	living	organism
ov-, ovi-	egg	ovum, oviduct
oxy-	oxygen	oxygenation, the saturation of a substance with oxygen
para-	near, beside	paracentric, near the centromere
pater-	father	paternal, pertaining to the father
patho-	pathogen	disease-causing
pent-	five	pentose, a 5-carbon sugar
per-	through	permease, a protein that carries a small molecule through the cell membrane
peri-	around	pericentric, around the centromere
phago-	eat	bacteriophage, a virus that infects bacteria
pheno-	show, appear	phenotype, the physical appearance of an individual
pili	hair	arrector pili muscles of the skin, which make the hairs stand erect
poly-	multiple	polymorphism, multiple forms
post-	after, behind	posterior, places behind (a specific) part
pre-, pro-	before, ahead of	prenatal, before birth
proto-	first or original	prototroph, having the nutritional requirements of the wild type
pseudo-	false	pseudodominant, appears but isn't dominant
re-	back, again	reinfect
retro-	backward, behind	retrovirus, to move "backward" from RNA into DNA
sanguin-	blood	consanguineous, indicative of a genetic relationship between individuals
se-	apart	segregate, to set apart
semi-	half	semicircular, having the form of half a circle
septum	fence	septum, membrane or wall between cells
soma-	body	somatic cell
sub-	under	subunit
super-	above, over	supercoil
telo-	end	telomere, the end of a chromosome arm
tetra-	four	tetraploid, having four sets of chromosomes
thermo-	heat	thermophile, heat-loving, able to grow at high temperatures
topo-	locale, local	topoisomerase, altering the local state

Roots and Prefixes	Meaning	Example
toti-	*wholly*	totipotent, having the ability to generate all cell types
tra-, trans-	*across, through*	transgenic, placement of novel DNA into an organism
tri-	*three*	triploid, three complete sets of chromosomes in a cell
ultra-	*beyond*	ultraviolet radiation, beyond the band of visible light
vita-	*life*	vital, alive
vitre-	*glass*	in vitro, in "glass"—the test tube
viv-	*live*	in vivo
zyg-	*a yoke, twin*	zygote

Suffixes	Meaning	Example
-able	*able to, capable of*	viable, ability to live or exist
-ac	*referring to*	cardiac, referring to the heart
-age	*action, process*	cleavage, process of cleaving
-al	*relating to, pertaining to*	chromosomal
-ary	*associated with, relating to*	coronary, associated with the heart
-bryo	*swollen*	embryo
-cide	*destroy or kill*	germicide, an agent that kills germs
-ell, -elle	*small*	organelle
-emia	*condition of the blood*	anemia, deficiency of red blood cells
-gen	*an agent that initiates*	pathogen, any agent that produces disease
-gram	*data that are systematically recorded, a record*	autoradiogram, a record of where radioactive atoms decayed
-ic	*having the character of*	acidic
-logy	*the study of*	pathology, the study of changes in structure and function brought on by disease
-lysis	*loosening or breaking down*	hydrolysis, chemical decomposition of a compound into other compounds as a result of taking up water
-oid	*like, resembling*	cuboid, shaped as a cube
-oma	*tumor*	lymphoma, a tumor of the lymphatic tissues
-ory	*referring to, of*	auditory, referring to hearing
-pathy	*disease*	osteopathy, any disease of the bone
-phil, -philo	*like, love*	hydrophilic, water-attracting (e.g., molecules)
-phobia	*fear*	acrophobia, fear of heights
-phragm	*partition*	diaphragm, which separates the thoracic and abdominal cavities
-plasm	*form, shape*	cytoplasm
-scope	*instrument used for examination*	microscope, instrument used to examine small things
-some	*body*	chromosome
-stasis	*arrest, fixation, stand*	epistasis, "stand upon", the genotype at one locus affects the phenotypic expression of the genotype at a second locus
-troph	*nutrition*	prototroph, nutritional requirements of wild-type organism
-ula, -ule	*diminutive*	blastula, little "bud", sphere of cells in early development
-zyme	*ferment*	enzyme

glossary

acentric chromosome A chromosome with no centromere.

acridine A chemical mutagen that intercalates between the bases of a DNA molecule, causing single-base insertions or deletions.

acrocentric chromosome A chromosome with the centromere near one end.

adaptation Any characteristic of an organism that improves its chance of survival and reproduction in its environment; the evolutionary process by which a species undergoes progressive modification favoring its survival and reproduction in a given environment.

addition rule The principle that the probability that any one of a set of mutually exclusive events is realized equals the sum of the probabilities of the separate events.

adenine (A) A nitrogenous purine base found in DNA and RNA.

adjacent segregation Type of segregation from a heterozygous reciprocal translocation in which a structurally normal chromosome segregates with a translocated chromosome. In adjacent-1 segregation, homologous centromeres go to opposite poles of the first-division spindle; in adjacent-2 segregation, homologous centromeres go to the same pole of the first-division spindle.

albinism Absence of melanin pigment in the iris, skin, and hair of an animal; absence of chlorophyll in plants.

alkaptonuria A recessively inherited metabolic disorder in which a defect in the breakdown of tyrosine leads to excretion of homogentisic acid (alkapton) in the urine.

alkylating agent An organic compound capable of transferring an alkyl group to other molecules.

allele Any of the alternative forms of a given gene.

allele frequency The relative proportion of all alleles of a gene that are of a designated type.

allopolyploid A polyploid formed by hybridization between two different species.

alpha satellite Highly repetitive DNA sequences associated with mammalian centromeres.

alternate segregation Segregation from a heterozygous reciprocal translocation in which both parts of the reciprocal translocation separate from both nontranslocated chromosomes in the first meiotic division.

Ames test A bacterial test for mutagenicity; also used to screen for potential carcinogens.

amino acid Any one of a class of organic molecules that have an amino group and a carboxyl group; 20 different amino acids are the usual components of proteins.

aminoacylated tRNA A tRNA covalently attached to its amino acid; charged tRNA.

A (aminoacyl) site The tRNA-binding site on the ribosome to which each incoming charged tRNA is initially bound.

aminoacyl-tRNA synthetase The enzyme that attaches the correct amino acid to a tRNA molecule.

amino terminus The end of a polypeptide chain at which the amino acid bears a free amino group ($-NH_2$).

amniocentesis A procedure for obtaining fetal cells from the amniotic fluid for the diagnosis of genetic abnormalities.

anaphase The stage of mitosis or meiosis in which chromosomes move to opposite ends of the spindle. In anaphase I of meiosis, homologous centromeres separate; in anaphase II, sister centromeres separate.

anaphase-promoting complex (APC/C) A ubiquitin-protein ligase that targets proteins whose destruction enables a cell to transition from metaphase into anaphase.

aneuploid A cell or organism in which the chromosome number is not an exact multiple of the haploid number; more generally, aneuploidy is a condition in which particular genes or chromosomal regions are present in extra or fewer copies compared with wildtype.

antibiotic-resistant mutant A cell or organism that carries a mutation conferring resistance to an antibiotic.

antibody A blood protein produced in response to a specific antigen and capable of binding with the antigen.

anticodon The three bases in a tRNA molecule that are complementary to the three-base codon in mRNA.

antigen A substance able to stimulate the production of antibodies.

antiparallel The chemical orientation of the two strands of a double-stranded nucleic acid molecule; the 5′-to-3′ orientations of the two strands are opposite one another.

antisense RNA An RNA molecule complementary in nucleotide sequence to all or part of a messenger RNA.

antiterminator A sequence in RNA that allows transcription to continue through the gene.

AP endonuclease An endonuclease that cleaves a DNA strand at any site at which the deoxyribose lacks a base.

apoptosis Genetically programmed cell death.

aporepressor A protein converted into a repressor by binding with a particular molecule.

Archaea One of the three major classes of organisms; also called archaebacteria, they are unicellular microorganisms, usually found in extreme environments, that differ as much from bacteria as either group differs from eukaryotes. *See also* **Bacteria.**

artificial selection Selection, imposed by a breeder, in which organisms of only certain phenotypes are allowed to breed.

ascospore *See* **ascus.**

ascus A sac containing the spores (ascospores) produced by meiosis in certain groups of fungi, including *Neurospora* and yeast.

asexual polyploidization Formation of a polyploid through the fusion of normal gametes followed by endoreduplication of the chromosome sets in the hybrid.

attached-X chromosome A chromosome in which two X chromosomes are joined to a common centromere; also called a compound-X chromosome.

attenuation *See* **attenuator.**

attenuator A regulatory base sequence near the beginning of an mRNA molecule at which transcription can be terminated; when an attenuator is present, it precedes the coding sequences.

autonomous determination Cellular differentiation determined intrinsically and not dependent on external signals or interactions with other cells.

autopolyploid An organism whose cells contain more than two sets of homologous chromosomes.

autoregulation Regulation of gene expression by the product of the gene itself.

autosomes All chromosomes other than the sex chromosomes.

auxotroph A mutant microorganism that is unable to synthesize a compound required for its growth but is able to grow if the compound is provided.

BAC *See* **bacterial artificial chromosome.**

backcross The cross of an F_1 heterozygote with a partner that has the same genotype as one of its parents.

Bacteria One of the major kingdoms of living things; includes most bacteria. *See also* **Archaea.**

bacterial artificial chromosome (BAC) A plasmid vector with regions derived from the F plasmid that contains a large fragment of cloned DNA.

bacteriophage A virus that infects bacterial cells; commonly called a phage.

575

Barr body   A darkly staining body found in the interphase nucleus of certain cells of female mammals; consists of the condensed, inactivated X chromosome.

basal transcription factors   Transcription factors that are associated with transcription of a wide variety of genes.

base   Single-ring (pyrimidine) or double-ring (purine) component of a nucleic acid.

base analog   A chemical so similar to one of the normal bases that it can be incorporated into DNA.

base pair   A pair of nitrogenous bases, most commonly one purine and one pyrimidine, held together by hydrogen bonds in a double-stranded region of a nucleic acid molecule; commonly abbreviated bp, the term is often used interchangeably with the term *nucleotide pair*. The normal base pairs in DNA are A−T and G−C.

base-substitution   Incorporation of an incorrect base into a DNA duplex.

β-galactosidase   An enzyme that cleaves lactose into its glucose and galactose constituents; produced by a gene in the *lac* operon.

biochemical pathway   A diagram showing the order in which intermediate molecules are produced in the synthesis or degradation of a metabolite in a cell.

bioinformatics   The use of computers in the interpretation and management of biological data.

bivalent   A pair of homologous chromosomes, each consisting of two chromatids, associated in meiosis I.

blastoderm   Structure formed in the early development of an insect larva; the syncytial blastoderm is formed from repeated cleavage of the zygote nucleus without cytoplasmic division; the cellular blastoderm is formed by migration of the nuclei to the surface and their inclusion in separate cell membranes.

block in a biochemical pathway   Stoppage in a reaction sequence due to a defective or missing enzyme.

blunt ends   Ends of a DNA molecule in which all terminal bases are paired; the term usually refers to termini formed by a restriction enzyme that does not produce single-stranded ends.

bootstrapping   Analysis of multiple data sets formed by random sampling with replacement from an actual data set in order to estimate a degree of confidence in a particular branch or branching pattern in a gene tree.

broad-sense heritability   The ratio of genotypic variance to total phenotypic variance.

cAMP–CRP complex   A regulatory complex consisting of cyclic AMP (cAMP) and the CAP protein; the complex is needed for transcription of certain operons.

cancer   Any of a large number of diseases characterized by the uncontrolled proliferation of cells.

candidate gene   A gene proposed to be involved in the genetic determination of a trait because of the role of the gene product in the cell or organism.

cap   A complex structure at the 5′ termini of most eukaryotic mRNA molecules, having a 5′-5′ linkage instead of the usual 3′-5′ linkage.

carbon-source mutant   A cell or organism that carries a mutation preventing the use of a particular molecule or class of molecules as a source of carbon.

carboxyl terminus   The end of a polypeptide chain at which the amino acid has a free carboxyl group (−COOH).

carrier   A heterozygote for a recessive allele.

cassette   In bacterial genetics, a circular molecule of duplex DNA containing a target sequence for a site-specific recombinase (integrase) and usually one or more protein-coding sequences. In yeast genetics, either of two sets of inactive mating-type genes that can become active by relocating to the *MAT* locus.

categorical trait   A complex trait in which each possible phenotype can be classified into one of a number of discrete categories. Also called a meristic trait.

cDNA   *See* **complementary DNA.**

cell cycle arrest   A blockage in the progression of the cell cycle.

cell division cycle (cdc) mutant   A mutant whose phenotype is to arrest the cell cycle at a specific and reproducible point.

cell senescence   A normal process in which mammalian cells in culture cease dividing after about 50 doublings.

cell cycle   The growth cycle of a cell; in eukaryotes, it is subdivided into G_1 (gap 1), S (DNA synthesis), G_2 (gap 2), and M (mitosis).

cell fate   The pathway of differentiation that a cell normally undergoes.

cell lineage   The ancestor–descendant relationships of a group of cells in development.

cellular oncogene   A gene coding for a cellular growth factor whose abnormal expression predisposes to malignancy. *See also* **oncogene.**

centimorgan   A unit of distance in the genetic map equal to 1 percent recombination; also called a map unit.

central dogma   The concept that genetic information is transferred from the nucleotide sequence in DNA to the nucleotide sequence in an RNA transcript to the amino acid sequence of a polypeptide chain.

centriole   In animal cells, one of a pair of particulate structures composed of an array of microtubules around which the spindle is organized.

centromere   The region of the chromosome that is associated with spindle fibers and that participates in normal chromosome movement during mitosis and meiosis.

centrosome   A localized region of clear cytoplasm found near the nucleus of nondividing cells, which in dividing cells duplicates to form the centers around which the spindle is organized. *See also* **centriole.**

centrosome duplication checkpoint   A mechanism that arrests the cell cycle while the centrosome (the spindle-organizing center) remains undivided.

chain elongation   The process of addition of successive amino acids to the growing end of a polypeptide chain.

chain initiation   The process by which polypeptide synthesis is begun.

chain termination   The process of ending polypeptide synthesis and releasing the polypeptide from the ribosome; a chain-termination mutation creates a new stop codon, resulting in premature termination of synthesis of the polypeptide chain.

chaperone   A protein that assists in the three-dimensional folding of another protein.

charged tRNA   A tRNA molecule to which an amino acid is linked; acylated tRNA.

checkpoint   Any mechanism that arrests the cell cycle until one or more essential processes are completed.

chiasma   The cytological manifestation of crossing-over; the cross-shaped exchange configuration between nonsister chromatids of homologous chromosomes that is visible in prophase I of meiosis. The plural can be either *chiasmata* or *chiasmas*.

chimeric gene   A gene produced by recombination, chromosome rearrangement, or genetic engineering that is a mosaic of DNA sequences from two or more different genes.

chi-square (χ^2)   A statistical quantity calculated to assess the goodness of fit between a set of observed numbers and the theoretically expected numbers.

chromatid   Either of the longitudinal subunits produced by chromosome replication.

chromatid interference   In meiosis, the effect that crossing-over between one pair of nonsister chromatids may have on the probability that a second crossing-over in the same chromosome will involve the same or different chromatids; chromatid interference does not generally occur.

chromatin   The aggregate of DNA and histone proteins that makes up a eukaryotic chromosome.

chromatin-remodeling complex   Any of a number of complex protein aggregates that reorganizes the nucleosomes of chromatin in preparation for transcription.

chromomere   A tightly coiled, bead-like region of a chromosome most readily seen during the pachytene substage of meiosis; the beads are in register in a polytene chromosome, resulting in the banded appearance of the chromosome.

chromosome   In eukaryotes, a DNA molecule that contains genes in linear order to which numerous proteins are bound and that has a telomere at each end and a centromere; in prokaryotes, the DNA is associated with fewer proteins, lacks telomeres and a centromere, and is often circular; in viruses, the chromosome is DNA or RNA, single-stranded or double-stranded, linear or circular, and often free of bound proteins.

chromosome complement The set of chromosomes in a cell or organism.

chromosome map A diagram showing the locations and relative spacing of genes along a chromosome.

chromosome painting Use of differentially labeled, chromosome-specific DNA strands for hybridization with chromosomes to label each chromosome with a different color.

chromosome territory The three-dimensional region occupied by a particular chromosome in the nucleus of an interphase or noncycling cell.

chromosome theory of heredity The theory that chromosomes are the cellular objects that contain the genes.

cis **configuration** The arrangement of linked genes in a double heterozygote in which both mutations are present in the same chromosome—for example, $a_1 a_2 / + +$; also called coupling.

cis-**dominant** Of or pertaining to a mutation that affects the expression of only those genes on the same DNA molecule.

cis **heterozygote** *See cis* configuration.

cistron A DNA sequence specifying a single genetic function as defined by a complementation test; a nucleotide sequence coding for a single polypeptide.

ClB **method** A genetic procedure used to detect X-linked recessive lethal mutations in *Drosophila melanogaster;* so named because one X chromosome in the female parent is marked with an inversion (*C*), a recessive lethal allele (*l*), and the dominant allele for Bar eyes (*B*).

clone A collection of organisms derived from a single parent and, except for new mutations, genetically identical to that parent; in genetic engineering, the linking of a specific gene or DNA fragment to a replicable DNA molecule, such as a plasmid or phage DNA.

cloned DNA sequence A DNA fragment inserted into a vector and transformed into a host organism.

cloned gene A DNA sequence incorporated into a vector molecule capable of replication in the same or a different organism.

cloning The process of producing cloned genes.

coding region The part of a DNA sequence that codes for the amino acids in a protein.

coding sequence A region of a DNA strand with the same sequence as is found in the coding region of a messenger RNA, except that T is present in DNA instead of U.

codominant Refers to phenotypes in which the presence of both alleles in heterozygous genotypes can be detected.

codon A sequence of three adjacent nucleotides in an mRNA molecule, specifying either an amino acid or a stop signal in protein synthesis.

coefficient of coincidence An experimental value obtained by dividing the observed number of double recombinants by the expected number calculated under the assumption that the two events take place independently.

cohesive ends Single-stranded regions at the ends of otherwise double-stranded DNA molecules that are complementary in base sequence.

cointegrate A DNA molecule, usually circular and formed by recombination, that joins two replicons.

colchicine A chemical that prevents formation of the spindle in nuclear division.

colinearity The linear correspondence between the order of amino acids in a polypeptide chain and the corresponding sequence of nucleotides in the DNA molecule.

colony A visible cluster of cells formed on a solid growth medium by repeated division of a single parental cell and its daughter cells.

colony hybridization assay A technique for identifying colonies that contain a particular cloned gene; many colonies are transferred to a filter, lysed, and exposed to radioactive DNA or RNA complementary to the DNA sequence of interest, after which colonies that contain a sequence complementary to the probe are located by autoradiography.

color blindness In human beings, the usual form of color blindness is X-linked red–green color blindness. Unequal crossing-over between the adjacent red and green opsin pigment genes results in chimeric opsin genes that cause mild or severe green-vision defects (deuteranomaly or deuteranopia, respectively) and mild or severe red-vision defects (protanomaly or protanopia, respectively).

combinatorial control Strategy of gene regulation in which a relatively small number of time- and tissue-specific positive and negative regulatory elements are used in various combinations to control the expression of a much larger number of genes.

complementary base pairing Regions of nucleic acid molecules whose nucleotides can undergo Watson-Crick base pairing.

complementary DNA (cDNA) A DNA molecule made by copying RNA with reverse transcriptase.

complementation The phenomenon in which two recessive mutations with similar phenotypes result in a wildtype phenotype when both are heterozygous in the same genotype; complementation means that the mutations are in different genes.

complementation group A group of mutations that fail to complement one another.

complementation test A genetic test to determine whether two mutations are alleles (are present in the same functional gene).

complex trait A multifactorial trait influenced by multiple genetic and environmental factors, each of relatively small effect, and their interactions.

conditional mutation A mutation that results in a mutant phenotype under certain (restrictive) environmental conditions but results in a wildtype phenotype under other (permissive) conditions.

conjugation A process of DNA transfer in sexual reproduction in certain bacteria; in *E. coli,* the transfer is unidirectional, from donor cell to recipient cell. Also, a mating between cells of *Paramecium.*

conjugative plasmid A plasmid encoding proteins and other factors that make possible its transmission between cells.

consensus sequence A generalized base sequence derived from closely related sequences found in many locations in a genome or in many organisms; each position in the consensus sequence consists of the base found in the majority of sequences at that position.

conserved sequence A base or amino acid sequence that changes very slowly in the course of evolution.

constant antibody region The part of the heavy and light chains of an antibody molecule that has the same amino acid sequence among all antibodies derived from the same heavy-chain and light-chain genes.

constitutive mutant A mutant in which synthesis of a particular mRNA molecule (and the protein that it encodes) takes place at a constant rate independent of the presence or absence of any inducer or repressor molecule.

contact inhibition A phenomenon in normal mammalian cells in culture whereby cells cease to grow and divide when they are in close physical proximity.

contig A set of cloned DNA fragments overlapping in such a way as to provide unbroken coverage of a contiguous region of the genome; a contig contains no gaps.

continuous trait A trait in which the possible phenotypes have a continuous range from one extreme to the other rather than falling into discrete classes.

coordinate gene Any of a group of genes that establish the basic anterior–posterior and dorsal–ventral axes of the early embryo.

coordinate regulation Control of synthesis of several proteins by a single regulatory element; in prokaryotes, the proteins are usually translated from a single mRNA molecule.

core particle The aggregate of histones and DNA in a nucleosome, without the linking DNA.

co-repressor A small molecule that binds with an aporepressor to create a functional repressor molecule.

correlated response Change of the mean in one trait in a population accompanying selection for another trait.

correlation coefficient (r) A measure of association between pairs of numbers, equaling the covariance divided by the product of the standard deviations.

cotransduction Transduction of two or more linked genetic markers by one transducing particle.

cotransformation Transformation in bacteria of two genetic markers carried on a single DNA fragment.

counterselected marker A mutation used to prevent growth of a donor cell in an Hfr $\times$ F$^-$ bacterial mating.

coupled transcription–translation In prokaryotes, the translation of an mRNA molecule before its transcription is completed.

coupling configuration *See cis* configuration.

covariance (*Cov*) A measure of association between pairs of numbers that is defined as the average product of the deviations from the respective means.

crossing-over A process of exchange between nonsister chromatids of a pair of homologous chromosomes that results in the recombination of linked genes.

cut-and-paste element A transposable element that transposes through cleavage from double-stranded DNA and insertion elsewhere in the genome.

cut-and-paste transposition A mechanism of transposition in which the transposable element is not replicated in the process of transposition, but is cleaved ("cut") from an existing site in the genome and inserted ("pasted") at a new target site.

C-value paradox The observation that among eukaryotes, the DNA content of the haploid genome (C-value) bears no consistent relationship to the metabolic, developmental, or behavioral complexity of the organism.

cryptic splice site A potential splice site not normally used in RNA processing unless a normal splice site is blocked or mutated.

cyclin One of a group of proteins that participates in controlling the cell cycle. Different types of cyclins interact with the p34 kinase subunit and regulate the G_1/S and G_2/M transitions. The proteins are called cyclins because their abundance rises and falls rhythmically in the cell cycle.

cyclin–CDK complex Protein complex formed by the interaction between a cyclin and a cyclin-dependent protein kinase (CDK).

cyclin-dependent protein kinase (CDK) Any of a number of proteins that are activated by combining with a cyclin and that regulate the cell cycle by phosphorylation of other proteins.

cytokinesis Division of the cytoplasm.

cytological map Diagrammatic representation of a chromosome.

cytosine (C) A nitrogenous pyrimidine base found in DNA and RNA.

daughter strand A newly synthesized DNA or chromosome strand.

deficiency *See* deletion.

degeneracy *See* redundancy.

degrees of freedom An integer that determines the significance level of a particular statistical test. In the goodness-of-fit type of chi-square test in which the expected numbers are not based on any quantities estimated from the data themselves, the number of degrees of freedom is one less than the number of classes of data.

deletion Loss of a segment of the genetic material from a chromosome; also called deficiency.

deoxyribonucleic acid *See* DNA.

deoxyribose The five-carbon sugar present in DNA.

depurination Removal of purine bases from DNA.

diakinesis The substage of meiotic prophase I, immediately preceding metaphase I, in which the bivalents attain maximum shortening and condensation.

dicentric chromosome A chromosome with two centromeres.

dideoxyribose A deoxyribose sugar that lacks the 3' hydroxyl group; when incorporated into a polynucleotide chain, it blocks further chain elongation.

dideoxy sequencing method Procedure for DNA sequencing in which a template strand is replicated from a particular primer sequence and terminated by the incorporation of a nucleotide that contains dideoxyribose instead of deoxyribose; the resulting fragments are separated by size via electrophoresis.

dihybrid Heterozygous at each of two loci; progeny of a cross between true-breeding, or homozygous, strains that differ genetically at two loci.

diploid A cell or organism with two complete sets of homologous chromosomes.

diplotene The substage of meiotic prophase I, immediately following pachytene and preceding diakinesis, in which pairs of sister chromatids that make up a bivalent (tetrad) begin to separate from each other and chiasmata become visible.

direct repeat Copies of an identical or very similar DNA or RNA base sequence in the same molecule and in the same orientation.

distance matrix A matrix showing the amount of sequence divergence between all possible pairs of a set of protein or nucleic acid sequences.

distribution In quantitative genetics, the mathematical relation that gives the proportion of members in a population that have each possible phenotype.

DNA Deoxyribonucleic acid, the macromolecule, usually composed of two polynucleotide chains in a double helix, that is the carrier of the genetic information in all cells and many viruses.

DNA chip An array of tiny dots ("microarray") of DNA molecules immobilized on glass or on another solid support used for hybridization with a probe of fluorescently labeled nucleic acid.

DNA cloning *See* cloned gene.

DNA damage checkpoint A mechanism that arrests the cell cycle while damaged DNA remains unrepaired.

DNA ligase An enzyme that catalyzes formation of a covalent bond between adjacent 5'-P and 3'-OH termini in a broken polynucleotide strand of double-stranded DNA.

DNA looping A mechanism by which enhancers that are distant from the immediate proximity of a promoter can still regulate transcription; the enhancer and promoter, both bound with suitable protein factors, come into indirect physical contact by means of the looping out of the DNA between them. The physical interaction stimulates transcription.

DNA methylase An enzyme that adds methyl groups ($-CH_3$) to certain bases, particularly cytosine.

DNA microarray An array of tiny dots of DNA molecules immobilized on glass or on another solid support used for hybridization with a probe of fluorescently labeled nucleic acid.

DNA polymerase Any enzyme that catalyzes the synthesis of DNA from deoxynucleoside 5'-triphosphates, using a template strand.

DNA repair Any of several different processes for restoration of the correct base sequence of a DNA molecule into which incorrect bases have been incorporated or whose bases have been chemically modified.

DNA replication The semiconservative copying of a DNA molecule.

DNA transposon A type of transposable element that undergoes transposition without its genetic information passing through an RNA intermediate.

DNA typing Electrophoretic identification of individual persons by the use of DNA probes for highly polymorphic regions of the genome, such that the genome of virtually every person exhibits a unique pattern of bands; sometimes called DNA fingerprinting.

DNA uracil glycosylase An enzyme that removes uracil bases when they occur in double-stranded DNA.

domain Contiguous region of a polypeptide chain or protein molecule that folds into a characteristic structure relatively independently of other such regions, often contributing unique structural characteristics or binding properties to the molecule as a whole.

dominant trait Refers to an allele whose presence in a heterozygous genotype results in a phenotype characteristic of the allele.

dosage compensation A mechanism regulating X-linked genes such that their activities are equal in males and females; in mammals, random inactivation of one X chromosome in females results in equal amounts of the products of X-linked genes in males and females.

double-stranded DNA A DNA molecule consisting of two antiparallel strands that are complementary in nucleotide sequence.

double-Y syndrome The clinical features of the karyotype 47,XYY.

Down syndrome The clinical features of the karyotype 47,+21 (trisomy 21).

duplex DNA A double-stranded DNA molecule.

duplication A chromosome aberration in which a chromosome segment is present more than once in the haploid genome; if the two segments are adjacent, the duplication is a tandem duplication.

dynamic mutation Mutations in certain genetically unstable tandem repeats that increase or decrease in repeat number at a relatively high rate.

editing function The activity of DNA polymerases that removes incorrectly incorporated nucleotides; also called the proofreading function.

electrophoresis A technique used to separate molecules on the basis of their different rates of movement in response to an applied electric field, typically through a gel.

elongation (protein synthesis) Addition of amino acids to a growing polypeptide chain.

embryoid A small mass of dividing cells formed from haploid cells in anthers that can give rise to a mature haploid plant.

embryonic stem cells Cells in the blastocyst that give rise to the body of the embryo.

endonuclease An enzyme that breaks internal phosphodiester bonds in a single- or double-stranded nucleic acid molecule; usually specific for either DNA or RNA.

endoreduplication Doubling of the chromosome complement because of chromosome replication and centromere division without nuclear or cytoplasmic division.

enhancer A base sequence in eukaryotes and eukaryotic viruses that increases the rate of transcription of nearby genes; the defining characteristics are that it need not be adjacent to the transcribed gene and that the enhancing activity is independent of orientation with respect to the gene.

environmental variance The part of the phenotypic variance that is attributable to differences in environment.

enzyme A protein that catalyzes a specific biochemical reaction and is not itself altered in the process.

epigenetic Inherited changes in gene expression resulting from altered chromatin structure or DNA modification (usually methylation) rather than from changes in DNA sequence.

episome A DNA element that can persist in the cell by undergoing autonomous replication or by becoming incorporated into the genome.

epistasis A term referring to an interaction between nonallelic genes in their effects on a trait. Generally, *epistasis* means any type of interaction in which the genotype at one locus affects the phenotypic expression of the genotype at another locus. In a more restricted sense, it refers to a situation in which the genotype at one locus determines the phenotype in such a way as to mask the genotype present at a second locus.

equational division Term applied to the second meiotic division because the haploid chromosome complement is retained throughout.

EST *See* **expressed sequence tag.**

euchromatin A region of a chromosome that has normal staining properties and undergoes the normal cycle of condensation; relatively uncoiled in the interphase nucleus (compared with condensed chromosomes), it apparently contains most of the genes.

Eukarya One of the major kingdoms of living organisms, in which the cells have a true nucleus and divide by mitosis or meiosis.

eukaryote A cell with a true nucleus (DNA enclosed in a membranous envelope) in which cell division takes place by mitosis or meiosis; an organism composed of eukaryotic cells.

euploid A cell or an organism having a chromosome number that is an exact multiple of the haploid number.

evolution Cumulative change in the genetic characteristics of a species through time.

excisionase An enzyme that is needed for prophage excision; works together with an integrase.

excision repair Type of DNA repair in which segments of a DNA strand that are chemically damaged are removed enzymatically and then resynthesized, using the other strand as a template.

E (exit) site The tRNA-binding site on the ribosome that binds each uncharged tRNA just prior to its release.

exon The sequences in a gene that are retained in the messenger RNA after the introns are removed from the primary transcript.

exon shuffle The theory that new genes can evolve by the assembly of separate exons from preexisting genes, each coding for a discrete functional domain in the new protein.

exonuclease An enzyme that removes a terminal nucleotide in a polynucleotide chain by cleavage of the terminal phosphodiester bond; nucleotides are removed successively, one by one; usually specific for either DNA or RNA and for either single-stranded or double-stranded nucleic acids. A 5'-to-3' exonuclease cleaves successive nucleotides from the 5' end of the molecule; a 3'-to-5' exonuclease cleaves successive nucleotides from the 3' end.

expressed sequence tag A partial or complete cDNA sequence.

factorial Of a number, the product of all integers from 1 through the number itself; 0! is defined to equal 1.

familial Tending to be present in multiple generations of a pedigree.

F_1 **generation** The first generation of descent from a given mating.

F_2 **generation** The second generation of descent from a given mating, produced by intercrossing or self-fertilizing F_1 organisms.

first-division segregation Separation of a pair of alleles into different nuclei in the first meiotic division; happens when there is no crossing-over between the gene and the centromere in a particular cell.

first meiotic division The meiotic division that reduces the chromosome number; sometimes called the reduction division.

fitness A measure of the average ability of organisms with a given genotype to survive and reproduce.

5' end The end of a DNA or RNA strand that terminates in a free phosphate group not connected to a sugar farther along.

5' untranslated region The initial part of a messenger RNA, which does not code for protein.

fixed allele An allele whose allele frequency equals 1.0.

F^- **cell** A cell, typically of *Escherichia coli*, that lacks the F plasmid.

flower ABC model A model of floral determination in which a unique combination of gene activities present in each whorl of the floral meristem results in the differentiation of a distinct organ in the mature flower.

forward mutation A change from a wildtype allele to a mutant allele.

F factor A bacterial plasmid—often called the fertility factor or sex plasmid—that is capable of transferring itself from a host (F^+) cell to a cell not carrying an F factor (F^- cell); when an F factor is integrated into the bacterial chromosome (in an Hfr cell), the chromosome becomes transferrable to an F^- cell during conjugation.

F^+ **cell** A cell, typically of *Escherichia coli*, that contains the F plasmid.

F' plasmid An F plasmid that contains genes obtained from the bacterial chromosome in addition to plasmid genes; formed by aberrant excision of an integrated F factor, taking along adjacent bacterial DNA.

fragile-X chromosome A type of X chromosome containing a site toward the end of the long arm that tends to break in cultured cells that are starved for DNA precursors; causes fragile-X syndrome.

frameshift mutation A mutational event caused by the insertion or deletion of one or more nucleotide pairs in a gene, resulting in a shift in the reading frame of all codons following the mutational site.

fraternal twins Twins that result from the fertilization of separate ova and are genetically related as siblings; also called dizygotic twins.

frequency of cotransduction The proportion of transductants carrying a selected genetic marker that also carry a nonselected genetic marker.

frequency of recombination The proportion of gametes carrying combinations of alleles that are not present in either parental chromosome.

functional genomics The use of DNA microarrays and other methods to study the coordinated expression of many genes simultaneously.

gain-of-function mutation Mutation in which a gene is overexpressed or inappropriately expressed.

gamete A mature reproductive cell, such as a sperm or egg in animals.

gametophyte In plants, the haploid part of the life cycle that produces the gametes by mitosis.

gap gene Any of a group of genes that control the development of contiguous segments or parasegments in *Drosophila* such that mutations result in gaps in the pattern of segmentation.

gel electrophoresis *See* **electrophoresis.**

gene The hereditary unit defined experimentally by the complementation test. At the molecular level, a region of DNA containing genetic information, usually transcribed into an RNA molecule that is processed and either functions directly or is translated into a polypeptide chain; a gene can mutate to various forms called alleles.

gene amplification A process in which certain genes undergo differential replication either within the chromosome or extrachromosomally, increasing the number of copies of the gene.

gene cloning *See* **cloned gene.**

gene conversion The phenomenon in which the products of a meiotic division in an *Aa* heterozygous genotype are in some ratio other than the expected $1A : 1a$—for example, $3A : 1a$, $1A : 3a$, $5A : 3a$, or $3A : 5a$.

gene dosage Number of gene copies.

gene expression The multistep process by which a gene is regulated and its product synthesized.

gene fusion A new gene created by the joining of DNA from two preexisting genes. *See also* **chimeric gene.**

gene library A large collection of cloning vectors containing a complete (or nearly complete) set of fragments of the genome of an organism.

gene pool The totality of genetic information in a population of organisms.

gene product A term used for the polypeptide chain translated from an mRNA molecule transcribed from a gene; if the RNA is not translated (for example, ribosomal RNA), the RNA molecule is the gene product.

generalized transducing phage *See* **transducing phage.**

general transcription factor A protein molecule needed to bind with a promoter before transcription can proceed; transcription factors are necessary, but not sufficient, for transcription, and they are shared among many different promoters.

gene regulation Processes by which gene expression is controlled in response to external or internal signals.

gene targeting Disruption or mutation of a designated gene by homologous recombination.

gene therapy Deliberate alteration of the human genome for alleviation of disease.

gene tree A diagram showing the real or estimated ancestral relationships among a set of protein or nucleic acid sequences.

genetic architecture Of a complex trait, specification of the genetic and environmental factors that contribute to the trait, and their interactions.

genetic code The set of 64 triplets of bases (codons) that correspond to the 20 amino acids in proteins and the signals for initiation and termination of polypeptide synthesis.

genetic engineering The linking of two DNA molecules by *in vitro* manipulations for the purpose of generating a novel organism with desired characteristics.

genetic map *See* **linkage map.**

genetic marker Any pair of alleles whose inheritance can be traced through a mating or through a pedigree.

genetics The study of biological heredity.

genome The total complement of genes contained in a cell or virus; commonly used to refer to all genes present in one complete haploid set of chromosomes in eukaryotes.

genomic imprinting A process of DNA modification in gametogenesis that affects gene expression in the zygote; one probable mechanism is the methylation of certain bases in the DNA.

genomics The systematic study of the genome using large-scale DNA sequencing, gene-expression analysis, or computational methods.

genotype The genetic constitution of an organism or virus, typically with respect to one or a few genes of interest, as distinguished from its appearance, or phenotype.

genotype-by-sex interaction Genetic determination that differs between the sexes to result in different phenotypes for the same genotype, depending on the sex of the individual.

genotype–environment (G-E) association The condition in which genotypes and environments are not in random combinations.

genotype–environment (G-E) interaction The condition in which genetic and environmental effects on a trait are not additive.

genotype frequency The proportion of members of a population that are of a prescribed genotype.

genotypic variance The part of the phenotypic variance that is attributable to differences in genotype.

germ cell A cell that gives rise to reproductive cells.

germ line Cell lineage consisting of germ cells.

germ-line mutation A mutation that takes place in a reproductive cell.

goodness of fit The extent to which observed numbers agree with the numbers expected on the basis of some specified genetic hypothesis.

G protein One of a family of signaling proteins that is activated by binding to a molecule of guanosine triphosphate (GTP).

G_1 period *See* **cell cycle.**

G_1 restriction point The "start" point at which cells are arrested in the G_1 phase until they become committed to division.

G_1/S transition Transition from the first "growth" phase of the cell cycle to DNA synthesis.

G_2/M transition Transition between the second "growth" phase of the cell cycle to mitosis.

G_2 period *See* **cell cycle.**

guanine (G) A nitrogenous purine base found in DNA and RNA.

guide RNA The RNA template present in telomerase.

gyrase A type of topoisomerase II that cleaves and rejoins both strands of a DNA duplex to relieve torsional stress.

haploid A cell or organism of a species containing the set of chromosomes normally found in gametes.

haplotype The allelic form of each of a set of linked genes present in a single chromosome.

Hardy–Weinberg principle The genotype frequencies expected with random mating.

helicase A protein that separates the strands of double-stranded DNA.

hemophilia A One of two X-linked forms of hemophilia; patients are deficient in blood-clotting factor VIII.

heritability A measure of the degree to which a phenotypic trait can be modified by selection. *See also* **broad-sense heritability** and **narrow-sense heritability.**

heterochromatin Chromatin that remains condensed and heavily stained during interphase; commonly present adjacent to the centromere and in the telomeres of chromosomes. Some chromosomes are composed primarily of heterochromatin.

heteroduplex All or part of a double-stranded nucleic acid molecule in which the two strands have different hereditary origins; produced either as an intermediate in recombination or by the *in vitro* annealing of single-stranded complementary molecules.

heterosis The superiority of hybrids over either inbred parent with respect to one or more traits; also called hybrid vigor.

heterozygote superiority The condition in which a heterozygous genotype has greater fitness than either of the homozygotes.

580 Glossary

heterozygous Carrying dissimilar alleles of one or more genes; not homozygous.

hexaploid A cell or organism with six complete sets of chromosomes.

H4 histone *See* **histone.**

Hfr cell An *E. coli* cell in which an F plasmid is integrated into the chromosome, making possible the transfer of part or all of the chromosome to an F^- cell.

histone Any of the small basic proteins bound to DNA in chromatin; the five major histones are designated H1, H2A, H2B, H3, and H4. Each nucleosome core particle contains two molecules each of H2A, H2B, H3, and H4. The H1 histone forms connecting links between nucleosome core particles.

Holliday junction A cross-shaped configuration of two DNA duplexes formed as an intermediate in recombination.

Holliday junction-resolving enzyme An enzyme that catalyzes the breakage and rejoining of two DNA strands in a Holliday junction to generate two independent duplex molecules.

homeobox A DNA sequence motif found in the coding region of many regulatory genes; the amino acid sequence corresponding to the homeobox has a helix-loop-helix structure.

homeotic (HOX) gene Any of a group of genes in which a mutation results in the replacement of one body structure by another body structure.

homogentisic acid Substance excreted in the urine of alkaptonurics that turns black upon oxidation.

homologous Genes derived from a common ancestral gene, or proteins whose genes derive from a common ancestral gene. *See also* **orthologous** and **paralogous.**

homologous chromosomes Chromosomes that pair in meiosis and have the same genetic loci and structure; also called homologs.

homothallism The capacity of cells in certain fungi to undergo a conversion in mating type to make possible mating between cells produced by the same parental organism.

homozygous Having the same allele of a gene in homologous chromosomes.

hotspot A site in a DNA molecule at which the mutation rate is much higher than the rate for most other sites.

housekeeping gene A gene that is expressed at the same level in virtually all cells and whose product participates in basic metabolic processes.

Human Genome Project A worldwide project to map genetically and sequence the human genome.

Huntington disease Dominantly inherited degeneration of the neuromuscular system, with onset in middle age.

hybrid An organism produced by the mating of genetically unlike parents; also, a duplex nucleic acid molecule produced of strands derived from different sources.

hybrid vigor *See* **heterosis.**

hydrogen bond A weak, noncovalent linkage between two negatively charged atoms in which a hydrogen atom is shared.

hypermorphic mutation A mutation in which the wildtype gene function is overexpressed or overactive.

hypomorphic mutation A mutation in which the wildtype gene function is underexpressed or only partially active.

identical twins Twins developed from a single fertilized egg that splits into two embryos at an early division; also called monozygotic twins.

imaginal disk Structures present in the body of insect larvae from which the adult structures develop during pupation.

immunity A general term for resistance of an organism to specific substances, particularly agents of disease.

imprinting A process of DNA modification in gametogenesis that affects gene expression in the zygote; a probable mechanism is the methylation of certain bases in the DNA.

inborn error of metabolism A genetically determined biochemical disorder, usually in the form of an enzyme defect that produces a metabolic block.

inbreeding Mating between relatives.

inbreeding coefficient (F) A measure of the genetic effects of inbreeding in terms of the proportionate reduction in heterozygosity in an inbred organism compared with the heterozygosity expected with random mating.

inbreeding depression A phenomenon in which the average value of a quantitative trait in a population undergoes progressive deterioration as the level of inbreeding increases.

incomplete dominance Condition in which the phenotype of a heterozygous genotype is intermediate between the phenotypes of the homozygous genotypes.

independent assortment Random distribution of unlinked genes into gametes, as with genes in different (nonhomologous) chromosomes or genes that are so far apart on a single chromosome that the recombination frequency between them is 1/2.

individual selection Selection based on each organism's own phenotype.

induced mutation A mutation formed under the influence of a chemical mutagen or radiation.

inducer A small molecule that inactivates a repressor, usually by binding to it and thereby altering the ability of the repressor to bind to an operator.

inducible transcription Transcription of a gene, or a group of genes, only in the presence of an inducer molecule.

induction Activation of an inducible gene; prophage induction is the derepression of a prophage that initiates a lytic cycle of phage development.

initiation (protein synthesis) The process by which mRNA binds with ribosomes and other factors and protein synthesis begins.

inosine (I) One of a number of unusual bases found in transfer RNA.

insertion sequence A DNA sequence capable of transposition in a prokaryotic genome; such sequences usually code for their own transposase.

integrase An enzyme that catalyzes a site-specific exchange between two DNA sequences.

integrated phage A state in which the phage DNA molecule is inserted intact into the bacterial chromosome; the integrated phage is called a prophage.

integron In bacteria, a DNA structure consisting of a promoter, a flanking target site for a site-specific recombinase (integrase), a coding sequence for the integrase, and usually one or more protein-coding cassettes that have been "captured" by site-specific recombination.

interference The tendency for crossing-over to inhibit the formation of another crossover nearby.

interphase The interval between nuclear divisions in the cell cycle, extending from the end of telophase of one division to the beginning of prophase of the next division.

interrupted-mating technique In an Hfr × F^- cross, a technique by which donor and recipient cells are broken apart at specific times, allowing only a particular length of DNA to be transferred.

intervening sequence *See* **intron.**

intron A noncoding DNA sequence in a gene that is transcribed but is then excised from the primary transcript in forming a mature mRNA molecule; found primarily in eukaryotic cells. *See also* **exon.**

inversion A structural aberration in a chromosome in which the order of several genes is reversed from the normal order. A pericentric inversion includes the centromere within the inverted region, and a paracentric inversion does not include the centromere.

inversion loop Loop structure formed by synapsis of homologous genes in a pair of chromosomes, one of which contains an inversion.

inverted repeat Either of a pair of base sequences present in the same molecule that are identical or nearly identical but are oriented in opposite directions; often found at the ends of transposable elements.

ionizing radiation Electromagnetic or particulate radiation that produces ion pairs when dissipating its energy in matter.

IS element *See* **insertion sequence.**

J (joining) region Any of multiple DNA sequences that code for alternative amino acid sequences of part of the variable region of an antibody molecule; the J regions of heavy and light chains are different.

karyotype The chromosome complement of a cell or organism; often represented by an arrangement of metaphase chromosomes according to their lengths and the positions of their centromeres.

kilobase (kb) Unit of length of a duplex DNA molecule; equal to 1000 base pairs.

kinetochore The cellular structure, formed in association with the centromere, to which the spindle fibers become attached in cell division.

kissing complex The bipartite structure formed by base pairing between a small regulatory RNA and a messenger RNA.

Klinefelter syndrome The clinical features of human males with the karyotype 47,XXY.

lactose permease An enzyme responsible for transport of lactose from the environment into bacteria.

lagging strand The DNA strand whose complement is synthesized in short fragments that are ultimately joined together.

lariat structure Structure of an intron immediately after excision in which the 5' end loops back and forms a 5'-2' linkage with another nucleotide.

leader polypeptide A short polypeptide encoded in the leader sequence of some operons coding for enzymes in amino acid biosynthesis; translation of the leader polypeptide participates in regulation of the operon through attenuation.

leading strand The DNA strand whose complement is synthesized as a continuous unit.

leptotene The initial substage of meiotic prophase I during which the chromosomes become visible in the light microscope as unpaired, thread-like structures.

leukocyte Any of several classes of mature white blood cells.

liability Risk, particularly toward a threshold type of quantitative trait.

library *See* **gene library.**

ligand The molecule that binds to a specific receptor.

lineage diagram A diagram of cell lineages and their developmental fates.

LINE element A type of transposable element lacking long terminal repeats that undergoes transposition via an RNA intermediate; the acronym LINE stands for long interspersed element.

linkage The tendency of genes located in the same chromosome to be associated in inheritance more frequently than expected from their independent assortment in meiosis.

linkage group The set of genes present together in a chromosome.

linkage map A diagram of the order of genes in a chromosome in which the distance between adjacent genes is proportional to the rate of recombination between them; also called a genetic map.

linker DNA In genetic engineering, synthetic DNA fragments that contain restriction enzyme cleavage sites used to join two DNA molecules. *See also* **nucleosome.**

local population A group of organisms of the same species occupying an area within which most individual members find their mates; synonymous terms are *deme* and *Mendelian population*.

locus The site or position of a particular gene on a chromosome.

loss-of-function mutation A mutation that eliminates gene function; also called a null mutation.

loss of heterozygosity Loss of the presence of the wildtype allele, or loss of its function, in a heterozygous cell, enabling the phenotype of a recessive mutant allele to be expressed; mechanisms for loss of heterozygosity include chromosome loss, gene conversion, and mutation.

lost allele An allele no longer present in a population; its frequency is 0.

LTR retrotransposon A type of transposable element that transposes via an RNA intermediate and that has long terminal repeats (LTRs) in direct orientation at its ends.

lysis Breakage of a cell caused by rupture of its cell membrane and cell wall.

lysogen Clone of bacterial cells that have acquired a prophage.

lysogenic cycle In temperate bacteriophage, the phenomenon in which the DNA of an infecting phage becomes part of the genetic material of the cell.

lytic cycle The life cycle of a phage, in which progeny phage are produced and the host bacterial cell is lysed.

major groove In B-form DNA, the larger of two continuous indentations running along the outside of the double helix.

map distance The genetic distance between two marker genes expressed as the sum of the length in map units across of a set of small, nonoverlapping intervals between the marker genes; corresponds to one-half of the average number of chiasmata between the genes multiplied by 100.

mapping function The mathematical relation between the genetic map distance across an interval and the observed percentage of recombination in the interval.

map unit A unit of distance in a linkage map that corresponds to a recombination frequency of 1 percent. Technically, the map distance across an interval in map units equals one-half the average number of crossovers in the interval, expressed as a percentage. Map units are sometimes called centimorgans (cM).

maternal-effect gene A gene that influences early development through its expression in the mother and the presence of the gene product in the oocyte.

maternal inheritance Extranuclear inheritance of a trait through cytoplasmic factors or organelles contributed by the female gamete.

maternal PKU A condition that resembles phenylketonuria and results from embryonic development in the uterus of a woman deficient in phenylalanine hydroxylase.

mating-type interconversion Phenomenon in homothallic yeast in which cells switch mating type as a result of the transposition of genetic information from an unexpressed cassette into the active mating-type locus.

MCS Multiple cloning site. *See* **polylinker.**

mean The arithmetic average.

megabase (Mb) Unit of length of a duplex nucleic acid molecule; equal to 1 million base pairs.

meiocyte A germ cell that undergoes meiosis to yield gametes in animals or spores in plants.

meiosis The process of nuclear division in gametogenesis or sporogenesis in which one replication of the chromosomes is followed by two successive divisions of the nucleus to produce four haploid nuclei.

Mendelian genetics The mechanism of inheritance in which the statistical relations between the distribution of traits in successive generations result from (1) particulate hereditary determinants (genes), (2) random union of gametes, and (3) segregation of unchanged hereditary determinants in the reproductive cells.

meristem The mitotically active growing point of plant tissue.

meristic trait A trait in which the phenotype is determined by counting, such as number of ears on a stalk of corn and number of eggs laid by a hen.

messenger RNA (mRNA) An RNA molecule transcribed from a DNA sequence and translated into the amino acid sequence of a polypeptide. In eukaryotes, the primary transcript undergoes elaborate processing to become the mRNA.

metabolic pathway A set of chemical reactions that take place in a definite order to convert a particular starting molecule into one or more specific products.

metabolism The totality of chemical and physical processes that take place in cells and organisms.

metabolite Any small molecule that serves as a substrate, an intermediate, or a product of a metabolic pathway.

metacentric chromosome A chromosome with its centromere about in the middle so that the arms are equal or almost equal in length.

metaphase In mitosis, meiosis I, or meiosis II, the stage of nuclear division in which the centromeres of the condensed chromosomes are arranged in a plane between the two poles of the spindle.

metaphase plate Imaginary plane, equidistant from the spindle poles in a metaphase cell, on which the centromeres of the chromosomes are aligned by the spindle fibers.

migration Movement of organisms among subpopulations; also, the movement of molecules in electrophoresis.

minimal medium A growth medium consisting of simple inorganic salts, a carbohydrate, vitamins, organic bases, essential amino acids, and other essential compounds; its composition is precisely known. Minimal medium contrasts with complex medium or broth, which is an extract of biological material (vegetables, milk, meat) that contains a large number of compounds and the precise composition of which is unknown.

minor groove In B-form DNA, the smaller of two continuous indentations running along the outside of the double helix.

mismatch repair Removal, from duplex DNA, of a single-stranded region in which a nucleotide pair does not form proper hydrogen bonds, followed by replacement with a region of newly synthesized DNA using the intact strand as a template.

missense mutation An alteration in a coding sequence of DNA that results in an amino acid replacement in the polypeptide.

mitosis The process of nuclear division in which the replicated chromosomes divide and the daughter nuclei have the same chromosome number and genetic composition as the parent nucleus.

mobile DNA Alternative term for transposable elements.

molecular clock A condition in which a protein or nucleic acid molecule has the same probability of change per unit time in every branch of a gene tree.

molecular genetics The branch of genetics concerned with the chemistry of DNA and the molecules that participate in its replication, function, mutation, and repair.

molecular systematics A group of statistical methods for estimating gene trees and often, by inference, the evolutionary relationships among the taxa of which the genes are representative.

monohybrid A genotype that is heterozygous for one pair of alleles; the offspring of a cross between genotypes that are homozygous for different alleles of a gene.

monoploid The basic chromosome set that is reduplicated to form the genomes of the species in a polyploid series; the smallest haploid chromosome number in a polyploid series.

monosomy Condition of an otherwise diploid organism in which one member of a pair of chromosomes is missing.

mosaic An organism composed of two or more genetically different types of cells.

most recent common ancestor (MRCA) In a phylogenetic tree, the most recent node that unites a particular subset of sequences, characters, or species.

M period *See* **cell cycle.**

mRNA *See* **messenger RNA.**

mtDNA Mitochondrial DNA.

multifactorial trait A trait determined by the combined action of many factors, typically some genetic and some environmental.

multiple alleles The presence, in a population, of more than two alleles of a gene.

multiple cloning site *See* **polylinker.**

multiplication rule The principle that the probability that all of a set of independent events are realized simultaneously equals the product of the probabilities of the separate events.

mutagen An agent that is capable of increasing the rate of mutation.

mutant Any heritable biological entity that differs from wildtype, such as a mutant DNA molecule, mutant allele, mutant gene, mutant chromosome, mutant cell, mutant organism, or mutant heritable phenotype; also, a cell or organism in which a mutant allele is expressed.

mutant screen A type of genetic experiment in which the geneticist seeks to isolate multiple new mutations that affect a particular trait.

mutation A heritable alteration in a gene or chromosome; also, the process by which such an alteration happens. Used incorrectly, but with increasing frequency, as a synonym for *mutant*, even in some excellent textbooks.

mutation rate The probability of a new mutation in a particular gene, either per gamete or per generation.

narrow-sense heritability The fraction of the phenotypic variance revealed as resemblance between parents and offspring; technically, the ratio of the additive genetic variance to the total phenotypic variance.

natural selection The process of evolutionary adaptation in which the genotypes genetically best suited to survive and reproduce in a particular environment give rise to a disproportionate share of the offspring and so gradually increase the overall ability of the population to survive and reproduce in that environment.

negative regulation Regulation of gene expression in which mRNA is not synthesized until a repressor is removed from the DNA of the gene.

neighbor joining A method for estimating a gene tree in which pairs of taxa are joined sequentially according to which pair are separated by the shortest distance.

nick A single-strand break in a DNA molecule.

nitrous acid HNO_2, a chemical mutagen.

nondisjunction Failure of chromosomes to separate (disjoin) and move to opposite poles of the division spindle; the result is loss or gain of a chromosome.

non-LTR retrotransposon A type of transposable element that transposes via an RNA intermediate and that lacks terminal repeats at its ends.

nonparental ditype (NPD) An ascus containing two pairs of recombinant spores.

nonselective medium A growth medium that allows growth of wildtype and of one or more mutant genotypes.

nonsense-mediated decay In eukaryotes, a process in which a messenger RNA containing a premature chain-terminating codon is destroyed in the first round of translation owing to the presence of proteins that are bound to the exon-exon junctions.

nonsense mutation A mutation that changes a codon specifying an amino acid into a stop codon, resulting in premature polypeptide chain termination; also called a chain termination mutation.

normal distribution A symmetrical bell-shaped distribution characterized by the mean and the variance; in a normal distribution, approximately 68 percent of the observations are within 1 standard deviation of the mean, and approximately 95 percent are within 2 standard deviations.

nuclease An enzyme that breaks phosphodiester bonds in nucleic acid molecules.

nucleic acid A polymer composed of repeating units of phosphate-linked five-carbon sugars to which nitrogenous bases are attached. *See also* **DNA** and **RNA.**

nucleic acid hybridization The formation of duplex nucleic acid from complementary single strands.

nucleolus (*pl.* **nucleoli)** Nuclear organelle in which ribosomal RNA is made and ribosomes are partially synthesized; usually associated with the nucleolar organizer region. A nucleus may contain several nucleoli.

nucleoside A purine or pyrimidine base covalently linked to a sugar.

nucleosome The basic repeating subunit of chromatin, consisting of a core particle composed of two molecules each of four different histones around which a length of DNA containing about 145 nucleotide pairs is wound, joined to an adjacent core particle by about 55 nucleotide pairs of linker DNA associated with a fifth type of histone.

nucleotide A nucleoside phosphate.

nucleotide analog A molecule that is structurally similar to a normal nucleotide and that is incorporated into DNA.

nutritional mutation A mutation in a metabolic pathway that creates a need for a substance to be present in the growth medium or that eliminates the ability to utilize a substance present in the growth medium.

Okazaki fragment Any of the short strands of DNA produced during discontinuous replication of the lagging strand; also called a precursor fragment.

oligonucleotide primer A short, single-stranded nucleic acid synthesized for use in DNA sequencing or as a primer in the polymerase chain reaction.

oncogene A gain-of-function mutation in a cellular gene, called a proto-oncogene, whose normal function is to promote cellular proliferation or inhibit apoptosis; oncogenes are often associated with tumor progression.

open reading frame (ORF) In the coding strand of DNA or in mRNA, a region containing a series of codons uninterrupted by stop codons and therefore capable of coding for a polypeptide chain.

operator A regulatory region in DNA that interacts with a specific repressor protein in controlling the transcription of adjacent structural genes.

operon A collection of adjacent structural genes regulated by an operator and a repressor.

ORF *See* **open reading frame.**

orthologous Genes that share a common ancestral gene through the process of speciation. *See also* **paralogous.**

PAC *See* **P1 artificial chromosome.**

pachytene The middle substage of meiotic prophase I, in which the homologous chromosomes are closely synapsed.

pair-rule gene Any of a group of genes active early in *Drosophila* development that specifies the fates of alternating segments or parasegments. Mutations in pair-rule genes result in loss of even-numbered or odd-numbered segments or parasegments.

paracentric inversion An inversion that does not include the centromere.

paralogous Genes that share a common ancestral gene through the process of gene duplication within a species. *See also* **orthologous.**

parasegment Developmental unit in *Drosophila* consisting of the posterior part of one segment and the anterior part of the next segment in line.

parental combination Alleles present in an offspring chromosome in the same combination as that found in one of the parental chromosomes.

parental ditype (PD) An ascus containing two pairs of non-recombinant spores.

parental strand In DNA replication, the strand that served as the template in a newly formed duplex.

partial diploid A cell in which a segment of the genome is duplicated, usually in a plasmid.

Pascal's triangle Triangular configuration of integers in which the nth row gives the binomial coefficients in the expansion of $(x + y)^{n-1}$. The first and last numbers in each row are 1, and the others equal the sum of the adjacent numbers in the row immediately above.

pattern formation The creation of a spatially ordered and differentiated embryo from a seemingly homogeneous egg cell.

PCR *See* **polymerase chain reaction.**

pedigree A diagram representing the familial relationships among relatives.

penetrance The proportion of organisms having a particular genotype that actually express the corresponding phenotype. If the phenotype is always expressed, penetrance is complete; otherwise, it is incomplete.

peptide bond A covalent bond between the amino group ($-NH_2$) of one amino acid and the carboxyl group ($-COOH$) of another.

pericentric inversion An inversion that includes the centromere.

permissive condition An environmental condition in which the phenotype of a conditional mutation is not expressed; contrasts with the nonpermissive or restrictive condition.

p53 transcription factor A key protein that helps regulate a mammalian cell's response to stress, especially to DNA damage.

P_1 generation The parents used in a cross, or the original parents in a series of generations; also called the P generation if there is no chance of confusion with the grandparents or more remote ancestors.

phage *See* **bacteriophage.**

phage repressor Regulatory protein that prevents transcription of genes in a prophage.

phenotype The observable properties of a cell or an organism, which result from the interaction of the genotype and the environment.

phenylalanine hydroxylase (PAH) The enzyme, deficient in phenylketonuria, that converts phenylalanine into tyrosine.

phenylketonuria (PKU) A hereditary human condition resulting from the inability to convert phenylalanine into tyrosine; causes severe mental retardation unless treated in infancy and childhood via a low-phenylalanine diet.

phosphodiester bond In nucleic acids, the covalent bond formed between the 5'-phosphate group (5'-P) of one nucleotide and the 3'-hydroxyl group (3'-OH) of the next nucleotide in line; these bonds form the backbone of a nucleic acid molecule.

phylogenetic tree A diagram showing the genealogical relationships among a set of genes or species.

plaque A clear area in an otherwise turbid layer of bacteria growing on a solid medium, caused by the infection and killing of the cells by a phage; because each plaque results from the growth of one phage, plaque counting is a way of counting viable phage particles. The term is also used for animal viruses that cause clear areas in layers of animal cells grown in culture.

plasmid An extrachromosomal genetic element that replicates independently of the host chromosome; it may exist in one or many copies per cell and may segregate in cell division to daughter cells in either a controlled or a random fashion. Some plasmids, such as the F factor, may become integrated into the host chromosome.

pleiotropic effect Any phenotypic effect that is a secondary manifestation of a mutant gene.

pleiotropy The condition in which a single mutant gene affects two or more distinct and seemingly unrelated traits.

polarity The 5'-to-3' orientation of a strand of nucleic acid.

pole cell Any of a group of cells, set off at the posterior end of the *Drosophila* embryo, from which the germ cells are derived.

Pol II holoenzyme A large protein complex containing the type of RNA polymerase used in transcribing most protein-coding genes.

poly-A tail The sequence of adenines added to the 3' end of many eukaryotic mRNA molecules in processing.

polycistronic mRNA An mRNA molecule from which two or more polypeptides are translated; found primarily in prokaryotes.

polylinker A short DNA sequence that is present in a vector and that contains a number of unique restriction sites suitable for gene cloning.

polymerase chain reaction (PCR) Repeated cycles of DNA denaturation, renaturation with primer oligonucleotide sequences, and replication, resulting in exponential growth in the number of copies of the DNA sequence located between the primers.

polymorphic gene A gene for which there is more than one relatively common allele in a population.

polymorphism The presence, in a population, of two or more relatively common forms of a gene, chromosome, or genetically determined trait.

polynucleotide chain A polymer of covalently linked nucleotides.

polypeptide *See* **polypeptide chain.**

polypeptide chain A polymer of amino acids linked together by peptide bonds.

polyploidy The condition of a cell or organism with more than two complete sets of chromosomes.

polysome A complex of two or more ribosomes associated with an mRNA molecule and actively engaged in polypeptide synthesis; a polyribosome.

polysomy The condition of a diploid cell or organism that has three or more copies of a particular chromosome.

polytene chromosome A giant chromosome consisting of many identical strands laterally apposed and in register, exhibiting a characteristic pattern of transverse banding.

P1 artificial chromosome A plasmid vector containing regions of the bacteriophage P1 and a large inserted DNA fragment.

population A group of organisms of the same species.

population genetics Application of Mendel's laws and other principles of genetics to entire populations of organisms.

population substructure Organization of a population into smaller breeding groups between which migration is restricted. Also called population subdivision.

positional information Developmental signals transmitted to a cell by virtue of its position in the embryo.

positive regulation Mechanism of gene regulation in which an element must be bound to DNA in an active form to allow transcription. Positive regulation contrasts with negative regulation, in which a regulatory element must be removed from DNA.

postreplication repair DNA repair that takes place via recombination in nonreplicating DNA or after the replication fork is some distance beyond a damaged region.

precursor fragment *See* **Okazaki fragment.**

P (peptidyl site) The tRNA-binding site on the ribosome to which the tRNA bearing the nascent polypeptide becomes bound immediately after formation of the peptide bond.

primary transcript An RNA copy of a gene; in eukaryotes, the transcript must be processed to form a translatable mRNA molecule.

primer In nucleic acids, a short RNA or single-stranded DNA segment that functions as a growing point in polymerization.

primosome The enzyme complex that forms the RNA primer for DNA replication in eukaryotic cells.

probe A radioactive DNA and RNA molecule used in DNA–RNA and DNA–DNA hybridization assays.

processivity Refers to the number of consecutive nucleotides in a template strand of nucleic acid that are traversed before a DNA polymerase or an RNA polymerase detaches from the template.

product molecule The end result of a biochemical reaction or a metabolic pathway.

programmed cell death Cell death that happens as part of the normal cellular response to damage or as part of the normal developmental process. *See also* **apoptosis.**

prokaryote An organism that lacks a nucleus; prokaryotic cells divide by fission.

promoter A DNA sequence at which RNA polymerase binds and initiates transcription.

promoter fusion Joining of the promoter region of one gene with the protein-coding region of another.

proofreading function *See* **editing function.**

prophage The form of phage DNA in a lysogenic bacterium; the phage DNA is repressed and is usually integrated into the bacterial chromosome, but some prophages are in plasmid form.

prophage induction Activation of a prophage to undergo the lytic cycle.

prophase The initial stage of mitosis or meiosis, beginning after interphase and terminating with the alignment of the chromosomes at metaphase; often absent or abbreviated between meiosis I and meiosis II.

proteome The set of all proteins encoded in a genome.

proteomics Study of the complement of proteins present in a cell or organism in order to identify their localization, functions, and interactions.

proto-oncogene A eukaryotic gene that functions to promote cellular proliferation or inhibit apoptosis, in which gain-of-function mutations (oncogenes) are associated with cancer progression.

prototroph Microbial strain capable of growth in a defined minimal medium that ideally contains only a carbon source and inorganic compounds. The wildtype genotype is usually regarded as a prototroph.

pseudoautosomal region In mammals, a small region of the X and Y chromosome containing homologous genes.

pseudogene A DNA sequence that is not functional because of one or more mutations but that has a functional counterpart in the same organism; pseudogenes are regarded as mutated forms of ancient gene duplications.

P transposable element A *Drosophila* transposable element used for the induction of mutations, germ-line transformation, and other types of genetic engineering.

Punnett square A cross-multiplication square used for determining the expected genetic outcome of a mating.

purine An organic base found in nucleic acids; the predominant purines are adenine and guanine.

pyrimidine An organic base found in nucleic acids; the predominant pyrimidines are cytosine, uracil (in RNA only), and thymine (in DNA only).

pyrimidine dimer Two adjacent pyrimidine bases, typically a pair of thymines, in the same polynucleotide strand, between which chemical bonds have formed; the most common lesion formed in DNA by exposure to ultraviolet light.

quantitative trait A trait—typically measured on a continuous scale, such as height or weight—that results from the combined action of several or many genes in conjunction with environmental factors.

quantitative trait locus (QTL) A locus segregating for alleles that have different, measurable effects on the expression of a quantitative trait.

random genetic drift Fluctuation in allele frequency from generation to generation resulting from restricted population size.

random mating System of mating in which mating pairs are formed independently of genotype and phenotype.

random spore analysis In fungi, the genetic analysis of spores collected at random rather than from individual tetrads.

reading frame The phase in which successive triplets of nucleotides in mRNA form codons; depending on the reading frame, a particular nucleotide in an mRNA could be in the first, second, or third position of a codon. The reading frame actually used is defined by the AUG codon that is selected for chain initiation.

recessive trait Refers to an allele, or the corresponding phenotypic trait, expressed only in homozygotes.

reciprocal cross A cross in which the sexes of the parents are the reverse of those in another cross.

reciprocal translocation Interchange of parts between nonhomologous chromosomes.

recombinant A chromosome that results from crossing-over and that carries a combination of alleles differing from that of either chromosome participating in the crossover; the cell or organism that contains a recombinant chromosome.

recombinant DNA A DNA molecule composed of one or more segments from other DNA molecules.

recombination Exchange of parts between DNA molecules or chromosomes; recombination in eukaryotes usually entails a reciprocal exchange of parts, but in prokaryotes it is often nonreciprocal.

recruitment The process in which a transcriptional activator protein interacts with one or more components of the transcription complex and attracts it to the promoter.

red–green color blindness *See* **color blindness.**

reductional division Term applied to the first meiotic division because the chromosome number (counted as the number of centromeres) is reduced from diploid to haploid.

redundancy The feature of the genetic code in which an amino acid corresponds to more than one codon; also called degeneracy.

release (protein synthesis) The process by which the completed polypeptide chain is freed from the ribosome.

relative fitness The fitness of a genotype expressed as a proportion of the fitness of another genotype.

replica plating Procedure in which a particular spatial pattern of colonies on an agar surface is reproduced on a series of agar surfaces by stamping them with a template that contains an image of the pattern; the template is often produced by pressing a piece of sterile velvet upon the original surface, which transfers cells from each colony to the cloth.

replication *See* **DNA replication; θ replication.**

replication fork In a replicating DNA molecule, the region in which nucleotides are added to growing strands.

replication origin The base sequence at which DNA synthesis begins.

replication slippage The process in which the number of copies of a small tandem repeat can increase or decrease during replication.

replicon A DNA molecule that has a replication origin.

reporter gene A gene whose expression can readily be monitored.

repressible transcription A regulatory process in which a gene is temporarily rendered unable to be transcribed.

repressor A protein that binds specifically to a regulatory sequence adjacent to a gene and blocks transcription of the gene.

repulsion configuration   *See trans* configuration.

restriction endonuclease   A nuclease that recognizes a short nucleotide sequence (restriction site) in a DNA molecule and cleaves the molecule at that site; also called a restriction enzyme.

restriction enzyme   *See* **restriction endonuclease.**

restriction fragment   A segment of duplex DNA produced by cleavage of a larger molecule by a restriction enzyme.

restriction fragment length polymorphism (RFLP) Genetic variation in a population associated with the size of restriction fragments that contain sequences homologous to a particular probe DNA; the polymorphism results from the positions of restriction sites flanking the probe, and each variant is essentially a different allele.

restriction map   A diagram of a DNA molecule showing the positions of cleavage by one or more restriction endonucleases.

restriction site   The base sequence at which a particular restriction endonuclease makes a cut.

restrictive condition   A growth condition in which the phenotype of a conditional mutation is expressed.

retinoblastoma   An inherited cancer caused by a mutation in the tumor-suppressor gene located in chromosome band *13q14.* Inheritance of one copy of the mutation results in multiple malignancies in retinal cells of the eyes in which the mutation becomes homozygous—for example, through a new mutation or mitotic recombination.

retinoblastoma protein   Any of a family of proteins found in animal cells that functions to hold cells at the G_1/S restriction point ("start") by binding to and sequestering a transcription factor that initiates the cell cycle.

retrovirus   One of a class of RNA animal viruses that cause the synthesis of DNA complementary to their RNA genomes on infection.

reverse genetics   Procedure in which mutations are deliberately produced in cloned genes and introduced back into cells or the germ line of an organism.

reverse mutation   A mutation that undoes the effect of a preceding mutation.

reverse transcriptase   An enzyme that makes complementary DNA from a single-stranded RNA template.

reverse transcriptase PCR (RT-PCT)   Amplification, using an RNA template, of a duplex DNA molecule originally produced by reverse trascriptase.

reversion   Restoration of a mutant phenotype to the wildtype phenotype by a second mutation.

RFLP   *See* **restriction fragment length polymorphism.**

R group   Refers to the side chain connected to the alpha carbon that differs for each amino acid.

ribonucleic acid   *See* **RNA.**

ribose   The five-carbon sugar in RNA.

ribosomal RNA (rRNA)   RNA molecules that are components of the ribosomal subunits; in eukaryotes, there are four rRNA molecules—5S, 5.8S, 18S, and 28S; in prokaryotes, there are three—5S, 16S, and 23S.

ribosomal translocation   Movement of the ribosome along a molecule of messenger RNA in translation.

ribosome   The cellular organelle on which the codons of mRNA are translated into amino acids in protein synthesis. Ribosomes consist of two subunits, each composed of RNA and proteins. In prokaryotes, the subunits are 30S and 50S particles; in eukaryotes, they are 40S and 60S particles.

ribosome-binding site   The base sequence in a prokaryotic mRNA molecule to which a ribosome can bind to initiate protein synthesis; also called the Shine–Dalgarno sequence.

ribosome tRNA-binding sites   The tRNA-binding sites on the ribosome to which tRNA molecules are bound. The aminoacyl site receives the incoming charged tRNA, the peptidyl site holds the tRNA with the nascent polypeptide chain, and the exit site holds the outgoing uncharged tRNA.

riboswitch   A 5′ RNA leader sequence that, according to whether it is bound with a small molecule, can adopt either of two configurations, one of which permits transcription and the other of which terminates transcription.

ribozyme   An RNA molecule able to catalyze one or more biochemical reactions.

ring chromosome   A chromosome whose ends are joined; one that lacks telomeres.

RNA   Ribonucleic acid, a nucleic acid in which the sugar constituent is ribose; typically, RNA is single-stranded and contains the four bases adenine, cytosine, guanine, and uracil.

RNA interference (RNAi)   The ability of small fragments of double-stranded RNA to silence genes whose transcripts contain homologous sequences.

RNA polymerase   An enzyme that makes RNA by copying the base sequence of a DNA strand.

RNA polymerase holoenzyme   Any of several large protein complexes that includes RNA polymerase among its constituents.

RNA processing   The conversion of a primary transcript into an mRNA, rRNA, or tRNA molecule; includes splicing, cleavage, modification of termini, and (in tRNA) modification of internal bases.

RNA splicing   Excision of introns and joining of exons.

Robertsonian translocation   A chromosomal aberration in which the long arms of two acrocentric chromosomes become joined to a common centromere.

rolling-circle replication   A mode of replication in which a circular parent molecule produces a linear branch of newly formed DNA.

R plasmid   A bacterial plasmid that carries drug-resistance genes; commonly used in genetic engineering.

rRNA   *See* **ribosomal RNA.**

RT-PCR   *See* **reverse transcriptase PCR.**

satellite DNA   Eukaryotic DNA that forms a minor band at a different density from that of most of the cellular DNA in equilibrium density gradient centrifugation; consists of short sequences repeated many times in the genome (highly repetitive DNA) or of mitochondrial or chloroplast DNA.

scaffold   A protein-containing material in chromosomes, believed to be responsible in part for the compaction of chromatin.

scanning   The process in which the eukaryotic translational initiation complex moves along the mRNA until the first start codon AUG is encountered.

second-division segregation   Segregation of a pair of alleles into different nuclei in the second meiotic division, the result of crossing-over between the gene and the centromere of the pair of homologous chromosomes.

second meiotic division   The meiotic division in which the centromeres split and the chromosome number is not reduced; also called the equational division.

segment   Any of a series of repeating morphological units in a body plan.

segmentation gene   Any of a group of genes that determines the spatial pattern of segments and parasegments in *Drosophila* development.

segment-polarity gene   Any of a group of genes that determines the spatial pattern of development within the segments of *Drosophila* larvae.

segregation   Separation of the members of a pair of alleles into different gametes in meiosis.

selected marker   A genetic mutation that allows growth in selective medium.

selection   In evolution, intrinsic differences in the ability of genotypes to survive and reproduce; in plant and animal breeding, the choosing of organisms with certain phenotypes to be parents of the next generation; in mutation studies, a procedure designed in such a way that only a desired type of cell can survive, as in selection for resistance to an antibiotic.

selection coefficient   The amount by which relative fitness is reduced or increased.

selection limit   The condition in which a population no longer responds to artificial selection for a trait.

selectively neutral mutation   A mutation that has no (or negligible) effects on fitness.

selective medium   A medium that allows growth only of cells with particular genotypes.

selfish DNA   DNA sequences that do not contribute to the fitness of an organism but are maintained in the genome through their ability to replicate and transpose.

semiconservative replication   The usual mode of DNA replication, in which each strand of a duplex molecule serves as a template for the synthesis of a new complementary strand, and the daughter molecules are composed of one old (parental) strand and one newly synthesized strand.

semisterility A condition in which a significant proportion of the gametophytes produced by a plant or of the zygotes produced by an animal are inviable, as in the case of a translocation heterozygote.

sequence-tagged site (STS) A DNA sequence, present once per haploid genome, that can be amplified by the use of suitable oligonucleotide primers in the polymerase chain reaction in order to identify clones that contain the sequence.

sex chromosome A chromosome, such as the human X or Y, that plays a role in the determination of sex.

sexual polyploidization Formation of a polyploid through the fusion of unreduced gametes.

sib *See* **sibling.**

sibling A brother or sister, each having the same parents.

sibship A group of brothers and sisters.

sickle-cell anemia A severe anemia in human beings, inherited as an autosomal recessive and caused by an amino acid replacement in the β-globin chain; heterozygotes tend to be more resistant to falciparum malaria than are normal homozygotes.

significant *See* **statistically significant.**

silencer A nucleotide sequence that binds with certain proteins whose presence prevents gene expression.

silent substitution A mutation that has no phenotypic effect.

simple sequence repeat (SSR) A DNA polymorphism in a population in which the alleles differ in the number of copies of a short, tandemly repeated nucleotide sequence.

SINE element A type of transposable element lacking long terminal repeats that undergoes transposition via an RNA intermediate; the acronym SINE stands for short interspersed element.

single-active-X principle In mammals, the genetic inactivation of all X chromosomes but one in each cell lineage, except in the very early embryo.

single-nucleotide polymorphism (SNP) A site in the DNA occupied by a different nucleotide pair among a significant fraction of the individuals in a population.

single-stranded DNA A DNA molecule that consists of a single polynucleotide chain.

single-stranded DNA binding protein A protein able to bind and stabilize single-stranded DNA.

sister chromatids Chromatids produced by replication of a single chromosome.

site-specific recombinase An enzyme that catalyzes inter-molecule recombination between two duplex DNA molecules at the site of a target sequence that they have in common.

small ribonucleoprotein particles Small nuclear particles that contain short RNA molecules and several proteins. They are involved in intron excision and splicing and in other aspects of RNA processing.

snRNP Any of several classes of small ribonucleoprotein particles involved in RNA splicing.

somatic cell Any cell of a multicellular organism other than the gametes and the germ cells from which gametes develop.

somatic mutation A mutation arising in a somatic cell.

Southern blot A nucleic acid hybridization method in which, after electrophoretic separation, denatured DNA is transferred from a gel to a membrane filter and then exposed to radioactive DNA or RNA under conditions of renaturation; the radioactive regions locate the homologous DNA fragments on the filter.

specialized transducing phage *See* **transducing phage.**

species tree A real or estimated ancestral history of a group of species.

S period *See* **cell cycle.**

spindle A structure composed of fibrous proteins on which chromosomes align during metaphase and move during anaphase.

spindle assembly checkpoint A mechanism that arrests the cell division cycle until the spindle is properly deployed.

splice acceptor The 3′ end of an exon.

splice donor The 5′ end of an exon.

spliceosome An RNA-protein particle in the nucleus in which introns are removed from RNA transcripts.

spontaneous mutation A mutation that happens in the absence of any known mutagenic agent.

sporadic An instance of a disease that is solitary, lacking other affected members in the same pedigree.

spore A unicellular reproductive entity that becomes detached from the parent and can develop into a new organism upon germination; in plants, spores are the haploid products of meiosis.

sporophyte The diploid, spore-forming generation in plants, which alternates with the haploid, gamete-producing generation (the gametophyte).

standard deviation The square root of the variance.

start codon An mRNA codon, usually AUG, at which polypeptide synthesis begins.

statistically significant Said of the result of an experiment or study that has only a small probability of happening by chance on the assumption that some hypothesis is true. Conventionally, if results as bad or worse would be expected less than 5 percent of the time, the result is said to be statistically significant; if less than 1 percent of the time, the result is called statistically highly significant; both outcomes cast the hypothesis into serious doubt.

sticky ends The single-stranded ends of a DNA fragment produced by certain restriction enzymes, each capable of reannealing with a complementary sequence in another molecule.

stop codon One of three mRNA codons—UAG, UAA, and UGA—at which polypeptide synthesis stops.

STRP *See* **simple tandem repeat polymorphism.**

STS *See* **sequence-tagged site.**

subfunctionalization Evolutionary change in a gene that results in loss of one or more of its functional or regulatory motifs.

submetacentric chromosome A chromosome whose centromere divides it into arms of unequal length.

subpopulation Any of the breeding groups within a larger population between which migration is restricted.

substrate molecule A substance acted on by an enzyme.

synapsis The pairing of homologous chromosomes or chromosome regions in the zygotene substage of the first meiotic prophase.

synonymous substitution A change in a coding region that alters the nucleotide sequence of a codon without changing the amino acid that is specified.

synteny group A group of genes present in a continuous region of chromosome in two or more species.

tandem duplication A pair of identical or closely related DNA sequences that are adjacent and in the same orientation.

TATA-box–binding protein (TBP) A protein that binds to the TATA motif in the promoter region of a gene.

TATA box The base sequence 5′-TATA-3′ in the DNA of a promoter.

taxon A population, species, or other group of organisms of which a protein or nucleic acid sequence, or a set of such sequences, is regarded as representative.

TBP-associated factor (TAF) Any protein found in close association with TATA binding protein.

T DNA Transposable element found in *Agrobacterium tumefaciens*, which produces crown gall tumors in a wide variety of dicotyledonous plants.

telomerase An enzyme that adds specific nucleotides to the tips of the chromosomes to form the telomeres.

telomere The tip of a chromosome, containing a DNA sequence required for stability of the chromosome end.

telophase The final stage of mitotic or meiotic nuclear division.

temperature-sensitive mutation A conditional mutation that causes a phenotypic change only at certain temperatures.

template A strand of nucleic acid whose base sequence is copied in a polymerization reaction to produce either a complementary DNA or an RNA strand.

terminator A sequence in RNA that halts transcription.

testcross A cross between a heterozygote and a recessive homozygote, resulting in progeny in which each phenotypic class represents a different genotype.

testis-determining factor (TDF) Genetic element on the mammalian Y chromosome that determines maleness.

tetrad The four chromatids that make up a pair of homologous chromosomes in meiotic prophase I and metaphase I; also, the four haploid products of a single meiosis.

tetraploid A cell or organism with four complete sets of chromosomes; in an autotetraploid, the chromosome sets are homologous; in an allotetraploid, the chromosome sets consist of a complete diploid complement from each of two distinct ancestral species.

tetratype An ascus containing spores of four different genotypes—one each of the four genotypes possible with two alleles of each of two genes.

θ replication Bidirectional replication of a circular DNA molecule, starting from a single origin of replication.

30-nm fiber The level of compaction of eukaryotic chromatin resulting from coiling of the extended, nucleosome-bound DNA fiber.

three-point cross Cross in which three genes are segregating; used to obtain unambiguous evidence of gene order.

3' end The end of a DNA or RNA strand that terminates in a sugar and so has a free hydroxyl group on the number-3' carbon.

3' untranslated region The terminal portion of a messenger RNA, following the stop codon, which does not code for protein.

threshold trait A trait with a continuously distributed liability or risk; organisms with a liability greater than a critical value (the threshold) exhibit the phenotype of interest, such as a disorder.

thymine (T) A nitrogenous pyrimidine base found in DNA.

thymine dimer *See* **pyrimidine dimer.**

time of entry In an Hfr $\times$ F$^-$ bacterial mating, the earliest time that a particular gene in the Hfr parent is transferred to the F$^-$ recipient.

Ti plasmid A plasmid that is present in *Agrobacterium tumefaciens* and is used in genetic engineering in plants.

total variance Summation of all sources of genetic and environmental variation.

trait Any aspect of the appearance, behavior, development, biochemistry, or other feature of an organism.

trans configuration The arrangement in linked inheritance in which a genotype heterozygous for two mutant sites has received one of the mutant sites from each parent—that is, $a_1 + / + a_2$.

transcript An RNA strand that is produced from, and is complementary in base sequence to, a DNA template strand.

transcription The process by which the information contained in a template strand of DNA is copied into a single-stranded RNA molecule of complementary base sequence.

transcriptional activator protein Positive control element that stimulates transcription by binding with particular sites in DNA.

transcription complex An aggregate of RNA polymerase (consisting of its own subunits) along with other polypeptide subunits that makes transcription possible.

transducing phage A phage type capable of producing particles that contain bacterial DNA (transducing particles). A specialized transducing phage produces particles that carry only specific regions of chromosomal DNA; a generalized transducing phage produces particles that may carry any region of the genome.

transduction The carrying of genetic information from one bacterium to another by a phage.

transfer RNA (tRNA) A small RNA molecule that translates a codon into an amino acid in protein synthesis; it has a three-base sequence, called the anticodon, complementary to a specific codon in mRNA, and a site to which a specific amino acid is bound.

transformation Change in the genotype of a cell or organism resulting from exposure of the cell or organism to DNA isolated from a different genotype; also, the conversion of an animal cell, whose growth is limited in culture, into a tumor-like cell whose pattern of growth is different from that of a normal cell.

transforming rescue The ability of an introduced DNA fragment to correct a gametic defect in a mutant organism.

transgenic organism An animal or plant in which novel DNA has been incorporated into the germ line.

trans heterozygote *See* **trans configuration.**

transition mutation A mutation resulting from the substitution of one purine for another purine or that of one pyrimidine for another pyrimidine.

translation The process by which the amino acid sequence of a polypeptide is synthesized on a ribosome according to the nucleotide sequence of an mRNA molecule.

translocation Interchange of parts between nonhomologous chromosomes; also, the movement of mRNA with respect to a ribosome during protein synthesis. *See also* **reciprocal translocation.**

transmembrane receptor A receptor protein containing amino acid sequences that span the cell membrane.

transmission genetics The processes by which genes are passed from one generation to the next.

transposable element A DNA sequence capable of moving (transposing) from one location to another in a genome.

transposase Protein necessary for transposition.

transposition The movement of a transposable element.

transposon A transposable element that contains bacterial genes—for example, for antibiotic resistance; also used loosely as a synonym for *transposable element.*

transposon tagging Insertion of a transposable element that contains a genetic marker into a gene of interest.

transversion mutation A mutation resulting from the substitution of a purine for a pyrimidine or that of a pyrimidine for a purine.

trinucleotide repeat A tandemly repeated sequence of three nucleotides; genetic instability in some trinucleotide repeats is the cause of a number of human hereditary diseases.

triplet code A code in which each codon consists of three bases.

triploid A cell or organism with three complete sets of chromosomes.

trisomic A diploid organism with an extra copy of one of the chromosomes.

trisomy-X syndrome The clinical features of the karyotype 47,XXX.

trivalent Structure formed by three homologous chromosomes in meiosis I in a triploid or trisomic chromosome when each homolog is paired along part of its length with first one and then the other of the homologs.

tRNA *See* **transfer RNA.**

true-breeding Refers to a strain, breed, or variety of organism that yields progeny like itself; homozygous.

truncation point In artificial selection, the value of the phenotype that determines which organisms will be retained for breeding and which will be culled.

tumor-suppressor gene A gene that normally controls cell proliferation or that activates the apoptotic pathway, in which loss-of-function mutations are associated with cancer progression.

Turner syndrome The clinical features of human females with the karyotype 45,X.

two-hybrid analysis A method for detecting protein-protein interactions that makes use of two fused (hybrid) proteins, one including the DNA-binding domain and the other the transcriptional activation domain of a trancriptional activator protein. If the polypeptide chains attached to these components interact within the nucleus, then the interaction brings the domains together and transcription of a reporter gene takes place.

unequal crossing-over Crossing-over between nonallelic copies of duplicated or other repetitive sequences—for example, in a tandem duplication, between the upstream copy in one chromosome and the downstream copy in the homologous chromosome.

univalent Structure formed in meiosis I in a monoploid or a monosomic when a chromosome has no pairing partner.

uracil (U) A nitrogenous pyrimidine base found in RNA.

uridylate A nucleotide in RNA in which the base is uridine.

UTR The untranslated region of a messenger RNA; the 5' UTR is upstream of the protein-coding region and the 3' UTR is downstream.

variable expressivity Differences in the severity of expression of a particular genotype.

variable antibody region The portion of an immunoglobulin molecule that varies greatly in amino acid sequence among antibodies in the same subclass.

variance A measure of the spread of a statistical distribution; the mean of the squares of the deviations from the mean.

vector A DNA molecule, capable of replication, into which a gene or DNA segment is inserted by recombinant DNA techniques; a cloning vehicle.

viral oncogene A class of genes found in certain viruses that predispose to cancer. Viral oncogenes are the viral counterparts of cellular oncogenes. *See also* **cellular oncogene; oncogene.**

V–J joining DNA splicing that unites one of the V (variable) regions of an antibody light-chain gene with one of the J (joining) regions of the same gene to create a unique antibody sequence.

Watson–Crick base pairing Base pairing in DNA or RNA in which A pairs with T (or U in RNA) and G pairs with C.

wildtype The most common phenotype or genotype in a natural population; also, a phenotype or genotype arbitrarily designated as a standard for comparison.

wobble The acceptable pairing of several possible bases in an anticodon with the base present in the third position of a codon.

X chromosome A chromosome that plays a role in sex determination and that is present in two copies in the homogametic sex and in one copy in the heterogametic sex.

xeroderma pigmentosum An inherited defect in the repair of ultraviolet-light damage to DNA, associated with extreme sensitivity to sunlight and multiple skin cancers.

X inactivation In mammals, the genetic inactivation of all X chromosomes but one in each cell lineage, except in the very early embryo.

X-linked gene A gene located in the X chromosome; X-linked inheritance is usually evident from the production of nonidentical classes of progeny from reciprocal crosses.

YAC *See* **yeast artificial chromosome.**

Y chromosome The sex chromosome present only in the heterogametic sex; in mammals, the male-determining sex chromosome.

yeast artificial chromosome (YAC) In yeast, a cloning vector that can accept very large fragments of DNA; a chromosome introduced into yeast derived from such a vector and containing DNA from another organism.

zygote The product of the fusion of a female gamete and a male gamete in sexual reproduction; a fertilized egg.

zygotene The substage of meiotic prophase I in which homologous chromosomes synapse.

zygotic gene Any of a group of genes that control early development through their expression in the zygote.

Index

A

Aamodt-Leeper, Gina, 400
ABC transporters, 381
Abnormal segregation and trisomic chromosomes, 177-178
ABO blood groups, 55-58, 68, 106, 161, 166, 532, 534
ACE. *See* Angiotensin-converting enzyme
Acentric chromosomes, 167
Acquired immune deficiency syndrome (AIDS), 230, 233, 384, 500
Acridine, 442
Acrocentric chromosomes, 167
Acron, 404
Action mechanisms in mutation agents, 439
Acute leukemias, 483-485
Adaptation, evolutionary, 514
Adaptive needs of the organism, 435-437
Addition rule, 49, 51, 105-106
Additive effects, 546
Additive genetic variance, 549
Adenine (A), 7, 204, 209
Adenomatous polyposis, 478-479
Adenylyl cyclase, 324
Adjacent-1 segregation, 186
Adjacent-2 segregation, 188
Aedes aegypti (yellow-fever mosquito), 158
Agglutination, 57
Agrobacterium tumefaciens (soil bacterium), 379-380, 382
AIDS. *See* Acquired immune deficiency syndrome
Albinism, 53, 512
Albinos, 420
Alexandra (tsarina of Russia), 102, 509
Alexis (tsarevich of Russia), 102, 103, 509
Alix (tsarina of Russia), 103
Alkaptonuria, 10
Alkylating agents, 441
Alleles, 42, 74
    classes of, 63-66
    frequencies and genotype frequencies, 500-501
    frequencies in gametes, 502
    frequencies of, 501, 504, 515-516, 519
    frequency random changes, 418-420
    in heterozygous gneotypes, 505-506
    independent segregation of, 45-49
    multiple of, 62
    recessive, 516
Allelism, 62, 68
Allison, Anthony C., 517
Allopolyploids, 192
Allotetraploid, 192
Alpha satellite of DNA, 96
Alpha synchronization, 462
Alternate segregation, 188
Alternative promoters, 339-340
Alternative splicing, 26, 343
Ames test, 452
Amino acids, 10, 30, 284
    in a polypeptide chain, 282
    chemical structures of, 283
    in DNA base sequences, 284
    and mutations, 425
    triplet genetic code, 203-208
Amino terminus, 282
Aminoacyl-tRNA synthetases, 294, 306-309

Ammospermophilus harrissi (ground squirrels), 94
Amniocentesis, 177, 201
Amnion, 177
Amoebozoa, 24
Amphibian eggs, 81
Amphiuma means (salamander), 205
Amylopectin, 38
Anaphase I in first meiotic division, 86
Anaphase II in second meiotic division, 87
Anaphase in mitosis, 78
Anaphase-promoting complex (APC/C), 467
Ancestral cereal, 195
Ancestral history, 494
Anchor cells (ACs), 396
Aneuploid, 174, 175
Aneuploid zygotes, 179
Angelman syndrome, 341
Angiotensin-converting enzyme (ACE), 552
Anhidrotic ectodermal dysplasia, 199
Animals
    growth genetically engineered, 381-382
    meiosis in, 80
    viruses for gene therapy, 384
Anopheles funetus (mosquito), 294
Anopheles gambiae (mosquito), 294, 517
Answers to even-numbered problems, 561-569
Antennapedia-complex (ANT-C), 409
Anterior genes, 404
Anthrax, 250
Antibiotic-resistant bacteria, 242, 245, 250
Antibiotic-resistant cassettes, 247-250
Antibiotic-resistant mutants, 250
Antibiotics, 240, 279
Antibodies, 58, 59
Antibody classes, 346
Antibody regions, 346-348
Antidepressants, 553
Antigens, 58, 59, 68
Antiparallel, 8, 210
Antisense RNA, 344
Antiterminator conformation, 330
AP endonuclease system, 448-449
Apaf1 protein, 472
Apoptosis, 470-471, 475
    prevention of, 476-477
Aporepressors, 319
Apterous genes, 409
Apurinic sites, 449
Arabidopsis thaliana, 25, 388
    floral development, 411-414
    web site, 419
Archaea, 24
Archaea genomic sequencing, 369
Archael genomes, 204
Arrest in the cell cycle, 463
Artificial chromosomes, 362
Artificial selection as managed evolution, 544-548
Ascospore, 143-144
Ascus, 143-144
Asexual polyploidization, 191
Aspergillus ficuum, 382
Association of environment and genotype, 541-542
Astyanax, 542, 543
Attenuation, 328-330
Automated plate readers, 503

Autonomous development, 394
Autoregulation, 319
Autosomal dominant inheritance, 480
Autosomal monosomies, 197
Autosomal trisomies, 197
Auxotrophs, 251
Avery, Oswald, 4
Azacytidine, 340
AZT, 233

B

B cells, 345-348, 483
Bacillus amyloliquefaciens, 226
Bacillus subtilis, 330, 331
Backcrosses, 45
BACs. *See* Bacterial artificial chromosomes
Bacteria, 24
Bacterial and bacteriophage genetics, 241-279
    bacteria with resistance to antibiotics, 245, 247, 250
    chromosome transfers, 254-255
    cointegration of plasmids, 247
    colony formation abilities, 250-251
    E. coli genetic map unit of distance, 255-259
    episomes, 253
    F plasmids, 243-245, 253-259
    insertion sequences and transposens, 245-246
    lysogenic bacteriophages, 267-274
    mobile DNA, 242-250
    phages transferring small pieces of bacterial DNA, 250-263
    plasmids, 242-243
    recombination of bacteriophage DNA molecules, 263-267
    reproductive systems of, 242
    site-specific recombinases and integrons, 247-250
    specialized transducing phages, 273-274
    transformation from the uptake of DNA and recombination, 251-253
    transposable elements in, 245
    unidirectional DNA transfer, 253-259
Bacterial artificial chromosomes (BACs), 362
Bacterial attachment site, 271
Bacterial genomes, 205
Bacterial infections, 240
Bacterial meningitis, 25
Bacteriophage MS2, 104
Bacteriophages, 4
    see also Bacterial and bacteriophage genetics
Balanced and unbalanced translocation, 174
Balstoderm cells, 399, 401
Baltimore, David, 17
*Bam*HI enzyme, 226, 227
Barcarese-Hamilton, Monique, 400
Barr bodies, 169, 199
Basal transcription factors, 336-337
Base analogs masquerading as the real thing, 440-441
Base-pair additions or deletions, 442
Base pairing, 7
Base substitutions and mutations, 424-425
Bases, 7
Bax protein, 472, 478, 480
Bcl2 protein, 472, 483

Bcr-Abl fusion protein, 485
Beadle, George W., 261
Beijersbergen, Roderick L., 99
Benzer, Seymour, 265
Beta-scaffold factors, 355
β-galactosidase, 319, 323, 324
Beta-globin, 426
Beta-carotene rice, 382
Beta-carotene, 383
Beta-galactosidase activity, 365-367
Bicoid genes, 337, 344, 404, 405
Biconcave disks, 280
Binomial probability formula, 105-107
bio genes, 262, 263, 270
Biochemical pathway, 10-11
Bioinformatics, 370
Biological weapons, 250
Biology as commodity, 356
Birds, sex chromosomes in, 103
Birth defects, 239, 559
Bishop, Dorothy V.M., 400
Bison bison, 492
Bisphenol A, 179
Bithorax, 409
Bivalent, 85
Black Jews of South Africa, 173
Black urine disease, 10-12
Bleomycin, 245
Blocked pathways, 11
Blocks, 11
Blood-cell development (hematopoiesis), 483
Blood-group phenotypes, 506
Blunt ends, 359
Bone marrow, 280, 483
Bootstrapping, 496
Borrelia burgdorfei, 204
Bottleneck, population, 492
Boundary enhancers, 409
Brachystola magna (grasshoppers), 86
BRCA1 (inherited breast cancer in women), 192, 292, 482
BRCA2, 482
Bread mold, 12, 148-149
Breast cancer, inherited, 192, 292, 482
Breast cancer syndromes, 482
Breeding, selective, 34
Bridges, Calvin, 103-104
Broad-sense heritability, 543-544
Brooks, Mary W., 99
Bub protein, 473
Bub1 protein, 478
Budding yeast. See *Saccharomyces cerevisiae*
Butterflies, 390
    sex chromosomes in, 103

C
c-kit gene, 485
C-value paradox, 205
C1B method, 437-438
Caenorhabditis elegans (nematode worms), 25, 69
    autonomous development, 393-399
    early stages of development, 374
    reverse genetics in, 377
    RNAi in, 343
    small regulatory RNAs in, 345
Calico cat's X-chromosome inactivation, 170, 199
cAMP. See Cyclic adenosine monophosphate
cAMP-CRP, 325
Cancer
    colon, 446, 457
    defined, 458
    familial, 474, 478
    genetic tests for, 451-453
    skin, 450
    sporadic, 474
    and telomeres, 99
    and x rays, 443
Cancer and molecular genetics of the cell cycle, 459-491
    acute leukemias, 483-485
    cancer-initiation and progression, 478-480

centrosome duplication checkpoint, 472-474
    checkpoints in the cell cycle, 467-474
    cyclin-CDK complexes, 464-465
    damaged cells repair or self-destruction, 467-474
    defects in processes of DNA, 482
    DNA damage checkpoint, 468-472
    genetic control of the cell cycle, 460-467
    germ-line inheritance, 460, 478-482
    inherited cancer syndromes, 478-482
    mutations in the cell cycle and key regulatory pathways, 463-464
    mutations that prevent normal checkpoint function in cancer, 474-482
    p53 transcription factor, 468-472, 475, 477
    protein degradation, 467
    proto-oncogenes, 476-477
    retinoblastoma, 480-482, 484, 491
    retinoblastoma (RB) protein, 466-467, 475, 478
    transcription of genes in the cell cycle, 462
    tumor-suppressor genes, 477-478
Cancer syndromes, inherited, 478-482
Candidate genes, 553, 553-554
Cap, 290
Capsella bursapastoris (shepherd's purse), 70
Carbon-source mutants, 2251
Carboxyl terminus, 282
Carothers, Eleanor, 86
Carriers, 53, 70
Cassettes, 248-250
Categorical traits, 533
Caudal genes, 405
CCR5 receptors, 500-501, 504
cdc15 synchronization, 462
cdc20, 473
cdc28 synchronization, 462
Cdk4, 476
cDNA (complementary DNA), 365, 383
Cell cycle, 76, 460
    mutations in, 463-464
    see also Cancer and molecular genetics of the cell cycle
Cell cycle regulatory genes affected in tumors, 475
Cell death, programmed, 458, 470-471
Cell division, 74, 119, 394
    and cancer, 458
    and differentiation of the fertilized egg, 392
Cell division cycle (cdc) mutants, 463
Cell lineages, 394
Cell senescence, 474
Centimorgan (cM), 127
    see also Genetic mapping
Central dogma of molecular genetics, 15, 16
Centrioles, 460
Centromeres, 76, 95-96, 148, 461
    in human chromosomes, 166-168
Centrosome duplication checkpoints, 468, 472, 472-474
Centrosomes, 460
Cercozoa, 24
Cereal grasses, 195
CGG repeat, 428
Chain elongation, 287
Chain initiation, 2287
Chain termination, 287-289
Chance in genetic transmission, 105
Chance in Mendelian genetics, 49-51
Chaperones, 300
Chaperonins, 300
Charged tRNA, 294
Chase, Martha, 4-7
Checkpoint function in cancer, 474-482
Checkpoints in the cell cycle, 76, 467-474
Chemical modifications in the DNA, 340-342
Chemical mutations, 422
Chemical warfare, 441
Chemicals damaging DNA, 441

Chemicals in the environment and chromosome abnormalities, 179
Chemokines, 500
Chemotherapy, 482, 491
Chernobyl nuclear accident, 445-447, 457
Chi-square, 107-110, 117
Chiasmata, 85, 88
Chimeric genes, 183, 485
Chimpanzees, 135, 167-168, 497
Chloroplast DNAs, 369
Chocolate, 316
Chorion, 177
Chorionic villus sampling (CVS), 177, 201
Chromalveolates, 25
Chromatids, 76, 88
Chromatin, 88-92, 90
Chromatin-remodeling complexes (CRCs), 338-339
Chromomeres, 84
Chromosomal basis of heredity, 73-119
    centromeres, 95-96
    chromosome complement, 74-75
    eukaryotic chromosomes, 88-95
    genes are located in chromosomes, 99-105
    meiosis, 75, 78-88, 113, 119
    mitosis, 75-78, 113
    somatic chromosome numbers of some plant and animal species, 74
    telomeres, 96-99
    see also Chromosomes
Chromosome 21, 175, 176
Chromosome abnormalities
    plants and animals, 164
    see also Human chromosome abnormalities
Chromosome behavior
    in meiosis, 78-88
    in mitosis, 76-78
Chromosome complement, 74-75
Chromosome condensation, 93-94
Chromosome interference, 139
Chromosome maps, 126
    see also Genetic mapping
Chromosome painting, 164-167
Chromosome structure abnormalities, 179-189
    deletions of chromosomes, 180
    duplications of chromosomes, 182
    giant polytene chromosomes, 180-182
    human color-blindness mutations, 182-184
    inversions in chromosomes, 184-186
    polyploid species with multiple sets of chromosomes, 189-196
    translocations of chromosomes, 186-189
Chromosome territory, 92
Chromosome theory of heredity, 194
Chromosome transfers in bacteria, 254-255
Chromosomes, 3, 74
    calico cat's X-chromosome inactivation, 170
    and DNA, 3
    genetic effects of rearrangements in, 178-189
    homologous, 78
    pseudoautosomal regions, 170
    replication of, 87
    see also Chromosomal basis of heredity; Human chromosomes
Chrysanthemum, 189-190, 191, 192
Ciliated protozoa, 97, 98, 292
Cis configuration, 122, 124
Cis-dominant, 321
Cistron, 267
Classical genetics, 2
Clonal tumor cells, 474
Cloning, 360-361
    directional, 367
    recombinant, 368
Cloning a DNA molecule, 358-368
Co-repressors, 319
Cockayne syndrome, 450
Coding sequence of a eukaryotic gene, 364-365

Codominance in human ABO blood groups, 55-58
Codominant genes, 39
Codons, 16, 294
Codons in polypeptide synthesis, 204-205
Coefficient of coincidence, 139, 157
Cohanim legacy, 173
Cohen modal haplotypes, 173
Cohesive ends, 269
Coho salmon, 382
Cointegration of plasmids, 247
Colchicine, 116, 194
Colinearity, 284, 285
Colon cancers, 457, 478, 491
Colonies of bacteria, 250
Colony, 3
Colony formation abilities of bacteria, 250-251
Colony hybridization, 368
Color blindness, 115, 117, 166, 507
Color-blindness mutations, 182-184
Colorectal carcinomas, 478, 485
Combinatorial control of gene expression, 411-414
Common ancestors, 502-503
Complementarity and DNA replication, 104
Complementary base pairing, 7
Complementary DNA (cDNA), 365, 383
Complementation between mutations of genes, 62-66
Complementation groups, 65, 69
Complementation principle, 65
    and recombination, 155
Complementation tests, 63-66, 69, 70
Complex inheritance, 531-559
    additive genetic variance, 549
    artificial selection as managed evolution, 544-548
    broad-sense heritability, 543-544
    candidate genes, 553-554
    covariance as tendency for traits to vary together, 548-549
    distribution of a trait in a population, 534-536
    environmental variance, 539-541
    genetic and environmental components in variation of a trait, 536-544
    genetic variation revealed by correlations between relatives, 548-550
    genetically homogenous populations, 542
    genotype and environment interaction or association, 541-542
    genotypic variance, 538-539
    human disorders are multifactorial, 549-550
    inbreeding depression, 547-548
    limits to improvement in artificial selection, 547
    multifactorial traits, 532-536
    narrow-sense heritability, 545-547
    number of genes affecting a quantitative trait, 542-543
    quantitative-trait locus (QTL), 52, 551-554
    twin studies, 544
Complex traits, 532
Conditional mutations, 423
Congenital abnormalities, 533
Congenital defects, 179
Conjugation, 243, 253
Conjugative plasmids, 243
Consensus sequences, 286
Conservative replication, 212
Constant antibody regions, 346, 347
Constitutive mutants, 332
Contact inhibition, 474
Continuous traits, 533
Coordinate genes, 402, 403-404
Coordinate regulation, 318
Coppin, Brian, 400
Core particles, 91
Corn, 158, 195, 438, 541
    genetic mapping of, 130-132
    inbreeding depression, 548

life cycle, 81
    transposable elements in, 430, 431
Corolla tubes, 545
Correlated responses, 547
Correlation coefficient, 548-549
Correns, Carl, 34
Cosmetic industry, 453
Cotransduction, 263
Cotransformation, 251, 252
Cotton, 199
Counter, Christopher M., 99
Counter-selected markers, 255
Coupled processes, 290
Coupling, 122, 124, 309
Covariance, 548-549
Covariance as tendency for traits to vary together, 548-549
CRCs. See Chromatin-remodeling complexes
Cre recombinase, 248
Creation of Human Tumor Cells with Defined Genetic Elements (Hahn et al.), 99
Creswell, Catharine, 400
Crick, Francis, 7, 9, 31, 211, 302, 308
Crocus augustifolius, 193-194
Crocus flavus, 193-194
Crop plants with improved nutritional qualities, 382
Crossing-over, 85, 132-134
    multiple crossing-over and interference, 138-139
    three-point crosses, 134-139
Crown gall tumors, 379
CRP. See Cyclic AMP receptor protein
Cryotherapy, 491
Curcurbita pepo (summer squash), 70
Cut-and-paste transposition, 430
Cuytosine (C), 7
CVS. See Chorionic villus sampling
Cyclic adenosine monophosphate (cAMP), 324-325
Cyclic AMP receptor protein (CRP), 325
Cyclin-CDK complexes, 464-465, 467
Cyclin D, 476, 481
Cyclin-dependent protein kinase (CDK), 464-465
Cystic fibrosis, 239, 315, 505-506
    web site, 529
Cytokinesis, 75, 76
Cytological maps, 181
Cytosine (C), 7, 204, 209

D
D4T, 233
Dalton, Paola, 400
Damaged cells repair or self-destruction, 467-474
Darwin, Charles, 514
Daughter strands, 210, 211, 231
ddI, 233
de Vries, Hugo, 34
Deadenylation-dependent pathway, 343
Deadenylation-independent pathway, 343
Deamination, 438
Decaploid, 190
Defects in processes of DNA, 482
Deficiencies in chromosomes, 180
Degrees of freedom, 109-110, 117
Deletions of chromosomes, 180
Denaturation, 224
Deoxyribonucleic acid. See DNA
Deoxyribonucleotides, 205-207
Dependence of Cell-Free Protein Synthesis in E. coli upon Naturally Occuring or Synthetic Polyribonucleotides (Nirenberg and Matthaei), 305
Depression, 553-554
Depurination, 439-440
Deuteranomaly, 182
Deuteranopia, 182
Development. See Genetic control of development
Development alterations by mutations, 392
Deviation, 535
Diabetes, 533

Diakinesis, 84, 85, 114
Dicentric chromosomes, 167
Dideoxy sequencing method, 230-233
Dideoxynucleoside analogs for treatment of diseases, 232-233
Dideoxyribose, 230-231
Dihybrids, 36
    crosses of, 45-47
Diodon holacanthus (spiny pufferfish), 202
Diploids, 75, 120, 194, 393
Diplotene, 84, 85, 114
Directional cloning, 367
Discrete traits, 532
Disease treatments, dideoxynucleoside analogs for, 232-233
Dissociation (Ds), 430
Distal part of a chromosome, 187
Distance matrix, 495
Distribution of a trait in a population, 534-536, 539
Distribution of phenotypes, 539
Divergence in molecular evolution, 499-500
DMD. See Duchenne-type muscular dystrophy
DNA, 2
    alpha satellite, 96
    altering genetic traits with, 3-4
    and chromosome condensation, 93-94
    and chromosomes, 3
    double helix structure of, 7-8, 31
    double-stranded break in, 152-154
    double-stranded or duplex, 8
    function of telomerase in chromosomal DNA, 97
    guide, 98
    and heterochromatin, 94-95
    highly reactive chemicals damaging, 441
    and histone proteins, 89
    and meiosis, 78
    Mendel and analysis of, 38-39
    as the molecule of heredity, 3-7
    and nucleosomes, 90-91
    polymorphism in, 142
    replication in, 8-9
    satellite, 94-95
    single-stranded, 8
    transmission of, 4-7
DNA and RNA base sequences, 284-289
DNA-binding proteins, 355
DNA damage and mechanism of repair, 447
DNA damage checkpoint, 468-472
DNA exclusions, 510
DNA fragments, 363-364
DNA intermediates, 430-433
DNA ligase, 224, 363-364
DNA looping, 335-336
DNA manipulation, 224-233
    dideoxy sequencing method, 230-233
    dideoxynucleoside analogs for treatment of crosses of diseases, 232-233
    hybridization of single strands with complementary sequences, 224-225
    and knowledge of DNA structure, 204, 224
    polymerase chain reaction (PCR), 228-230
    restriction enzymes, 225-228
    Southern blot, 228
DNA methylase, 340
DNA methylation, 400
DNA microarrays (or chips), 372-375
DNA polymerase, 210, 220, 285
DNA repair, 447-451
DNA replication, 210-224
    and complementarity, 104
    eukaryotic DNA molecules contain multiple origins of replication, 216-217
    nucleotides in, 210
    polymerase with proofreading function, 220-222
    proteins participating in, 217-224
    semiconservative replication, 211-213
    short RNA primer, 218-219
    strands must be unwound, 213-216
    synthesizing in fragments, 222-224
    template for, 210

DNA sequence changes and mutations, 424-430
DNA structure, 203-210
  and DNA manipulation, 204, 224
  a double helix with hydrogen bonds, 208-210
  genome size, 204-205
  a linear polymer of four deoxyribonucleotides, 205-207
  paired, complementary, antiparallel strands, 204
DNA transposons, 430
DNA typing, 508
DNA uracil glycosylase, 438-439
Domain structure, 284
Dominance
  of a pair of alleles, 53-59
  human ABO blood groups, 55-58
  variance in, 549
Dominant traits, 36, 37, 43
dorsal gene, 404
Dosage compensation, 168-170, 323
Double crossovers and three-point crosses, 134-139
Double helix structure of DNA, 7-8
Double helix with hydrogen bonds (DNA), 208-210
Double-stranded or duplex DNA, 8
Double-Y syndrome, 179, 199
Down, John Langdon, 201
Down syndrome, 135, 199, 200
  and human chromosome abnormalities, 175-177
  and maternal age, 177
  myths and truths, 201
  and Robertsonian translocations of chromosomes, 189
Drosophila melanogaster (fruit fly), 25, 70, 96, 156, 532
  alternative promoters in, 339-340
  copia retrotransposable elements, 432
  development, 399-411
  Dscam gene in, 342-343
  ectopic expression in, 423, 424
  gene white, 380-381
  genetic mapping of, 127, 130, 132, 139, 158
  genome of, 88
  long and short wings, 117
  Morgan's study of, 100-101, 103
  mutations in, 437-438
  Notch deletions, 180
  PcG proteins in, 335
  polytene chromosomes, 181
  random genetic drift, 520
  replicating DNA of, 216, 217
  RNAi in, 343
  salivary gland nuclei, 181-182, 200
  symbolism used in genetics of, 124
  transcriptional activiation in, 337-338
  transformation in, 377-378
  web site, 419
  X-linked inhteritance in, 104, 105, 110, 116, 122, 123
  and x-ray doses, 443
  Y chromosome in, 95
Drosophila texana, 200
Drosphila virilis, 158, 200
Dscam gene, 342-343
DsrA, 345
Duchenne-type muscular dystrophy (DMD), 117, 119
Duplex DNA, 8
Duplication in molecular evolution, 499-500
Duplications of chromosomes, 182
Duroc-Jersey pigs, dyhybrid ratio in, 62
Dwarfism, 450
Dynamic mutations, 427
Dystrophin (DMD) gene, 371

E

E (exit) sites, 294
E2F transcription factor, 478

Early pachtene, 84
Ecdysone hormone, 408
EcoRI (restriction enzyme), 140-141, 358-359, 363
Ectopic expression, 423-424
Editing function, 222
EGFR. See Epidermal growth factor
Electrophoresis, 88
Electrophoresis gel, 70
Electroporation, 361
Elongation in polypeptide synthesis, 294-297
Elongation phase of polypeptide synthesis, 294-297
Embryoid forms, 194
Embryonic pattern formation, 401-403
Embryonic stem cells, 378-379
Emmer wheat, 116
EMS. See Ethyl methanesulfonate
Endoreduplication, 191
Enhancers, 335
Enol form, 440
Environment
  effect on traits, 22
  and genotype interaction or association, 541-542
Environmental components in variation of a trait, 536-544
Environmental factors in multifactorial traits, 532-533
Environmental hazards, 537
Environmental insults, 411
Environmental risk factors, 179
Environmental variance, 539-540, 539-541
Enzyme defects
  and inborn errors of metabolism, 9-13
  and mutant genes, 14
Enzymes, 9
Enzymes repairing damage, 450
Epidermal growth factor (EGFR), 397, 398, 476, 477
Epigenetic, 340
Episomes, 253
Epistasis and phenotypes, 59-62
Equational division, 87, 113
Equilibrium density-gradient centrifugation, 211
Error-correcting mechanism, 222
Erwinia uredovora, 382
Escherichia coli, 4, 5, 6
  annotated maps of, 279
  circular genetic map of, 258
  cloning in, 360
  cloning vectors in, 361, 362
  Cre recombinase in, 248
  F factor in, 245
  F pilus connecting two cells, 243
  gene regulation in, 319-328
  genetic map unit of distance, 255-259
  genome size, 204
  genome size of, 204
  initiation in, 301
  lac genes, 259
  mismatch repair in, 449
  natural isolates in, 247
  phage T4, 364
  plaques from the progeny of a mixed infection, 265
  plasmid origin of replication, 366-367
  plasmids in, 242
  promoter regions, 286, 287
  replication in, 211-216, 220
  and restriction enzyme EcoRI, 140-141, 358-359
  selection over generations, 514
  termination regions, 288
  transcription and translation, 309
  transcriptional termination in, 328-330
  tryptophan synthase gene trpA in, 284, 285
  in vitro translation experiments, 305-306
Estrogen, 179
Ethical, legal, and social implications (ELSI) in genetic discovery, 371

Ethyl methanesulfonate (EMS), 441
Eubacteria genomic sequencing, 369
Euchromatin, 94, 95
Eukarya, 24
Eukaryota genomic sequencing, 369
Eukaryotes, 24
Eukaryotic chromosomes, 88-95
Eukaryotic DNA molecules, multiple origins of replication, 216-217
Eukaryotic genomes, 205
Eukaryotic RNA polymerases, 286
Euploid, 174
European Gypsies, 173-174
Even-skipped affect, 405
Evidence from Turner's Syndrome of an Imprinted X-lined Locus Affecting Cognitive Function (Skuse et al.), 400
Evolution, 22-26
  adaptation in, 514
  in formation of new species, 513
  and genetic changes in species, 413
  managed, 544-548
  rates of, 497-498, 497-499
  relationships among species, 494-500
  relationships among the species, 494
  see also Molecular evolution
Evolutionary history, 494
Excavates, 24
Excision repair, 450
Excisionase, 272
Exon shuffle model, 293
Exons, 290, 293
Exonuclease activity, 220

F

F- cells (F minus), 243, 244
F factor (fertility), 243, 362
F' plasmid (F prime), 259
F plasmids, 243-245, 253-259
F+ cells, 243, 244
F_1 generation, 36, 40
F_2 generation
  in a dihybrid cross,43, 45-47
Factorials, 106-107
Familial adenomatous polyposis (FAP), 457, 479, 491
Familial cancer, 474, 478
Familial melanoma, 479
Familial retinoblastoma, 479
FAP. See Familial adenomatous polyposis
Fertilization, 74
FGFR. See Fibroblast growth-factor receptor
Fibroblast growth-factor receptor (FGFR), 476, 477
Fingerprint patterns, 547
First-cousin mating, 512-513
First-division segregation, 150
First meiotic division, 81, 84-87
Fisher, Ronald, 110-111
Fitness in molecular evolution, 514-515
Five' end, 206, 207
Five' untranslated region, 289
Fixed mutant alleles, 495
Flesh-eating forms of Streptococcus, 263
flhA gene, 344
Flower ABC model, 414
Flower development, 419
FMR1 gene, 428
FMR1 inactivation, 429-430
Foldback structure, 434
Folding of polypeptide chains, 299-300
Fourfold degenerate sites, 498
Foxtail millet, 195
Fragile-X syndrome, 427, 428, 429, 430
Frameshift mutations, 302, 425
"Frankenstein's plants", 559
Fraternal twins, 544
Free radicals, 443
Frequency of cotransduction, 263
Frequency of recombination, 124
Friedreich ataxia, 429
Fringe genes, 409
Fruit flies. See Drosophila melanogaster
Functional genomics, 371-376

G

G-C pair, 209
G protein, 476
G_1 restriction point, 466
G_1/S transition, 466, 470
G_2/M transition, 466, 470
Gain of function, 396
Gain-of-function mutations, 423
gal enzymes, 332-334
gal genes, 262, 263, 270, 334, 375
Galactose metabolism in yeast, 332-334
Gametes, 41
    random fertilization of, 41-42
Gametophyte organism, 80
Gamma interferon, 497
Gap genes, 402, 404
GAP protein, 476
Garden peas. *See* Peas
Garrod, Archibald, 9-12
    Inborn Errors of Metabolism, 261
Gastric adenocarcinomas, 478
Gastrointestinal stomal tumors, 485
Gastrulas, 399
Gautier, Marthe, 176
Gel electrophoresis, 38, 70
Gene Action in the X Chromosome of the Mouse (Lyon), 323
Gene cloning, 358
    see also Genetic engineering
Gene conversion, 151-152
Gene expression, steps in, 282
Gene function and mutations, 423-424
Gene fusion, 485
Gene linkage, 121-161
    in a human autosome, 161
    double crossovers in three-point crosses, 134-139
    linked alleles in meiosis, 122-126
    polymorphic DNA sequences in human genetic mapping, 140-143
    recombination and double-stranded breaks in DNA, 152-154
    recombination results and linked alleles, 126-134
    tetrads and products of meiosis, 143-152
    see also Genetic mapping
Gene Mutations in Human Hemoglobin: The Chemical difference Between Normal and Sickle-Cell Hemoglobin (Ingram), 221
Gene pools, 500
Gene products, 282
Gene regulation, 317-355
    alternative promoters, 339-340
    attenuation, 328-330
    chromatin-remodeling complexes (CRCs), 338-339
    galactose metabolism in yeast, 332-334
    heritable chemical modifications in the DNA, 340-342
    lactose metabolism, 319-328
    negative regulation, 318-319, 325
    positive regulation, 319, 324, 325
    programmed DNA rearrangements, 345-349
    RNA processing and decay, 342-344
    transcription in eukaryotes, 330-340
    transcription in prokaryotes, 318-328
    transcriptional activator proteins, 334-335
    transcriptional termination, 328-330
    translation level, 344-345
    tryptophan biosynthesis, 326-328
Gene targeting, 379
Gene therapy, 384
Gene trees, 494-495
Gene trees versus species trees, 496-497
Geneotypes in multifactorial traits, 532-533
General transcription factors, 336-337
Generalized transducing phages, 259
Genes, 2
    addition rule, 49, 51
    changing by mutation, 20-22
    in chromosomes, 99-105

codominant, 39
complementation between mutations of, 62-66
complementation principle, 65
controlling metabolic processes, 261
and evolution, 22-26
experimental definition of, 65, 69
independent assortment principle, 47-48, 49
multiplication rule, 50-51
mutations in, 20-22
pairs of, 40-41
segregation principle, 41, 49
variable gene expression, 58-59
    see also Transmission genetics
Genes code for proteins, 9-20
    enzyme defects and inborn errors of metabolism, 9-13
    enzyme defects and mutant genes, 14
    RNA molecules, 14-16
    triplet code, 18-20
Genetic aberrations, 411
Genetic architecture, 532
Genetic code table, 18-20
Genetic codes for amino acids, 302-308
Genetic components in variation of a trait, 536-544
Genetic Control of Biochemical Reactions in Neurospora (Beadle and Tatum), 261
Genetic control of development, 391-419
    Arabidopsis thaliana floral development, 411-414
    Caenorhabditis elegans autonomous development, 393-399
    Drosophila melanogaster development, 399-411
    Saccharomyces cerevisiae (yeast) mating types, 392-393
Genetic control of the cell cycle, 460-467
Genetic diseases, 429, 450
Genetic effects of rearrangements in chromosomes, 178-189
Genetic engineering, 333, 357-389
    animal growth genetically engineered, 381-382
    beta-galactosidase activity, 365-367
    cloning a DNA molecule, 358-368
    colony hybridization, 368
    crop plants with improved nutritional qualities, 382
    DNA microarrays (or chips), 372-375
    functional genomics, 371-376
    gene therapy, 384
    genomic sequencing, 368-371
    human genes, 383
    Human Genome Project, 368, 371
    in medicine, industry, agriculture, and research, 381-384
    patents issued for clinical applications, 383
    proteins produced with recombinant DNA, 382-383
    recombinant cDNA, 364-365
    recombinant DNA, 360-362, 377-379
    recombinant DNA introduced into animals, 377-379
    recombinant DNA introduced into plants, 379-380
    restriction enzymes, 358-360
    restriction fragments, 360-361
    reverse genetics, 377-381
    transformation rescue, 380-381
    vectors, 361-367
    yeast two-hybrid analysis, 375-376
    see also Genomics; Proteomics
Genetic imprinting, 400
Genetic mapping, 122, 126-127
    centimorgan, 127
    in complex inheritance, 532
    of *Drosophila melanogaster* (fruit fly), 127, 130, 132, 239
    locus of genes in, 122
    map distance, 129, 139, 155
    map unit, 127
    mapping function, 139

physical distance and map distance, 132
    of *Zea mays* (corn), 130-132
    see also Gene linkage
Genetic markers, 128-129
    in human chromosomes, 161
Genetic mosaics, 169-170
Genetic tests for mutations and cancer, 51-53
Genetic variation revealed by correlations between relatives, 548-550
Genetic variation within species, 494
Genetically homogenous populations, 542
Genetically modified organisms (GMOs), 333, 358, 559
    see also Genetic engineering
Genetics, 2
    see also Transmission genetics
Genghis Khan, 172-173
Genomes, 358
    compared with proteomes, 25-26
    from different species in polyploids, 192-194
    duplications before or after fertilization, 190-191
    sequencing projects, 389
    sizes of, 204-205
Genomic imprinting, 340-342
Genomic sequencing, 368-371, 370
Genomics, 2, 358
    see also Genetic engineering
Genotype-by-environment association (G-E association), 541-542
Genotype-by-environment interaction (G-E interaction), 541
Genotype-by-sex interaction, 541
Genotype frequencies, 501
Genotype frequency differences, 500-502
Genotypes, 42
Genotypic variance, 538-539
Germ cells, 74
Germ-line inheritance, 460, 478-482
Germ-line mutations, 422
Giant polytene chromosomes, 180-182
Giemsa, 165
Glossary, 575-589
GMOs. *See* Genetically modified organisms
Goodness of fit, 107, 109
Gorillas, 135, 168, 497
Grass family illustrates polyploidy, 195-196
Grasshoppers, 86
    X chromosome in, 99-100
Great apes, 168
Great Migration, 523
Griffith, Frederick, 3-4
Ground squirrels, 94
Growth-factor receptors, 476
Growth hormones, 382
GTP. *See* Guanosine triphosphate
GTPase activity, 476
Guanine (G), 7, 204, 209
Guanosine triphosphate (GTP), 296
Guide DNA, 98
Gyrase, 217-218

H

*Hae*II enzyme, 521
Haemantuhus, chromosome behavior in mitosis, 76, 78
Hahn, William C., 99
Hair dyes, 453
Hairpin structure, 434
Hairy genes, 405
Hamster cells, normal and adenovirus-transformed (tumorigenic), 474
Hansen-Melander, Eva, 135
Haploid, 190
Haploids, 75, 120
Haplotypes, 172-174
Hardy-Weinberg principle, 503, 504-506, 553
Heavy (H) chains, 346
Helicase protein, 217
Helix-turn-helix, 335, 355, 409
Hematopoiesis (blood-cell development), 483

Hemophilia, 102-103, 117, 509
Hemophilia A, 102
Hemophilus influenzae (bacterial meningitis), 25
Hepatitis B virus, 384
Hereditary nonpolyposis colorectal cancer (HNPCC), 446, 457, 491
Heritability
    broad-sense, 543-544
    narrow-sense, 546
Heritable chemical modifications in the DNA, 340-342
Hermaphrodites, 393
Herpes virus, 205
Hershey, Alfred, 4-7
Hershey-Chase "blender" experiment, 6
Heterochromatin, 94-95
Heteroduplex, 151
Heterosis, 548
Heterozygosity, loss of, 480-481
Heterozygote superiority, 516, 518
Heterozygous genotypes, 42, 74, 505
Heterozygous organisms, 186
Heterozygous reciprocal translocations, 187
Hexaploid, 190
Hfr cells, 253, 254, 255, 257
Hibiscus, 117
Highly reactive chemicals damaging DNA, 441
Hiroshima, 457
Histone proteins, 89
HIV. *See* Human immunodeficiency virus
HNPCC. *See* Hereditary nonpolyposis colorectal cancer
Holliday junction, 153-154
Holliday junction-resolving enzyme, 153
Holliday, Robin, 153
Homeobox, 409
Homeotic genes, 407, 409
Homo sapiens, 25
Homogentisic acid, 10
Homologous chromosomes, 78
Homothallism, 348
Homozygote excesses in inbreeding, 512-513
Homozygous genotypes, 42, 505
Homozygous reciprocal translocations, 186
Horizontal transmission, 435
Hospitals, infections in, 240
Hotspot of mutation, 267
Hotspots, 438-439
Housekeeping genes, 330
HOX genes, 409-411
Huckebein genes, 405
Human ABO blood groups, dominance and codominance, 55-58
Human beings, 497
    "cleansing" the gene pool, 516
    hybridization of human metaphase chromosomes with alpha-satellite DNA, 96
    inbreeding in, 512
    quantitative traits in, 532
    sex chromosomes, 100
    see also Human chromosomes
Human chromosome abnormalities, 135, 164
    chemicals in the environment, 179
    Down syndrome, 175-177
    extra X or Y chromosomes, 179
    in human pregnancies, 174
    and spontaneous abortions, 174-175
    trisomic chromosomes and abnormal segregation, 177-178
Human chromosomes, 74-75, 163-201
    centromeres, 166-168
    dosage compensation, 168-170
    genetic markers in, 161
    at metaphase of mitosis, 162, 165
    numbers of, 135, 164, 165
    standard human karyotype, 165-166
    x-linked genes in females, 169-170
    see also Chromosomes
Human cytogenetics, 135

Human disorders
    multifactorial, 549-550
    and threshold traits, 533-534
Human gene characteristics, 292-293
Human Gene for Physical Performance (Montgomery et al.), 552
Human genes, 383
    genetically engineered, 383
Human genetic mapping, polymorphic DNA sequences in, 140-143
Human genome
    genetic landmarks in, 166
    simple sequence repeats (SSRs), 142, 143
    transposable elements in, 434-435
    trinucleotide repeats in, 427-430
Human Genome Project, 368, 371
Human immunodeficiency virus (HIV), 230, 233, 365, 384, 500
Human karyotype, 165-166
Human pedigrees
    segregation principle in, 51-53
    segregation results in, 51-53
    simple sequence repeats (SSRs), 143
    symbols used in, 52
    X-linked inheritances in, 103
Human population origins, 522-523
Human populations and the Y chromosome, 172-174
    Cohanim legacy, 173
    European Gypsies, 173-174
    Genghis Khan legacy, 172-173
Human pregnancies, chromosome abnormalities, 175
Human proteins, 284
Hunchback genes, 337, 404, 405
Huntington disease, 52-53, 70, 71, 158, 429
Hybrid vigor, 548
Hybridization, 164
    colony, 368
    of DNA single strands with complementary sequences, 224-225
    of human metaphase chromosomes with alpha-satellite DNA, 96
Hybrids, 36, 547-548
Hydrogen bonds, 208
Hymenoptera, 120
Hypermorphic mutations, 423
Hypomorphic mutations, 423
Hypoxanthine, 440

Identical twins, 544, 549
Iionizing radiation in mutations, 442-445
Imaginal disks, 407-408
Immune system in vertebrate, 345-348
Immunoglobulin, 346
Inborn errors of metabolism, 19
Inborn Errors of Metabolism (Garrod), 12, 261
Inbred lines, 532, 533, 544
Inbreeding, 510-513
Inbreeding coefficient, 511
Inbreeding depression, 547-548
Incomplete dominance in snapdragon flower color, 54-55
Incomplete penetrance, 59
Independent assortment principle, 47-48, 59
    genes in chromosomes, 99
Independent segregation of alleles, 45-49
Individual selection, 546
Induced mutations, 422, 439-447
Inducers, 319
Inducible transcription, 319
Infertility of triploids, 190
Influenza, 452
Ingram, Vernon M., 221
Inheritance. *See* Complex inheritance
Inherited breast cancer in women, 292
Inherited cancer syndromes, 478-482
Inherited diseases, 494
Inherited skin cancer syndromes, 482
Initial Sequencing and Analysis of the Human Genome (Lander), 371
Initiation in polypeptide synthesis, 294, 295
Initiation of a new replication fork, 214

Insertion in mutations, 433-434
Insertion sequences (IS elements), 245
    and transposons, 245-246
Instars, 399
Insulin receptor gene in humans, 342
Integrase, 248
Integrons, 248-249
Interaction of environment and genotype, 541-542
Intercellular signaling, 394
Interference in chromosomes, 139
    and multiple crossing-over, 138-139
International Union of Biochemistry and Molecular Biology (IUMBMB), 313
Interphase, 76
Interrupted-mating technique, 255-256
Introns, 290, 291
Inversion loops, 184
Inversions, 434
    in chromosomes, 184-186, 199
Ionizing radiation, 442-445
    annual exposure of human beings, 444
    radiation poisoning, 442-443
    risks of exposure, 444-445
    in tumor therapy, 444
    units of radiation, 433
Iron deficiency, 382
IS elements (insertion sequences), 245
IUMBMB. *See* International Union of Biochemistry and Molecular Biology

J

Jacob, François, 322
Jacobs, Patricia A., 400
James, Rowena S., 400
Japanese pufferfish, 202, 205
Jews, 173
Joining (J) antibody regions, 347

K

Kangaroos, 169, 323
Karyotypes, 165
    standard human, 165-166
Kennedy disease, 429
Keto form, 440
Kilobase (kb), 204
Kilobase pairs (kb), 88
Kinetochore, 78, 95, 473
Kissing complex, 344
Kit tyrosine kinase, 485
Klinefelder syndrome, 179, 199
Knirps genes, 405
Knockout mutations, 423
Knudson, Alfred G., 481, 484
Koalas, 169
Krüppel gene, 404, 405

L

Labeled probes, 368
Laboratory mouse, 25
lac DNA, 325
Lac genes, 259, 288, 289
lac operon, 322-327
lac repressor, 325
*lac*A, 322, 323
*lac*I, 321, 323, 324
*lac*O, 321, 322, 323, 324
*lac*P, 321, 322, 323
Lactose intolerance, 287
Lactose metabolism, 319-328
Lactose operator, 321-322
Lactose permease, 319
Lactose promoter, 322
*lac*Y, 321, 322, 323
*lac*Z, 321, 322, 323, 366, 367
Lagging strands, 222
Lakota Sioux, 492
Lambs, 535
Lander, Eric S., 371
Landsteiner, Karl, 57
Laser light therapy, 491
Late pachtene, 84
Leach, Frederick S., 446
Leader polypeptide, 328, 330

Leading strands, 222
Leaky mutations, 423
Lederberg, Esther, 435
Lederberg, Joshua, 261, 435
Lejeune, Jerome, 176
Lemba tribe, 173
Leptotene, 84, 114
Leukemias
    acute, 483-485
    lymphoblastic, 384
Leukocytes, 483
Levan, Albert, 135
Li-Fraumeni syndrome, 479, 480
Liability, 549
Library (large set of clones), 367
Ligands, 397, 398
Light (L) chains, 346
Lilium longiflorum (lily), 84, 86
LIN-3 protein, 398
LIN-12 protein, 396-397
LINE elements, 432-433, 435
Lineage diagrams, 394
Linear polymers
    with four deoxyribonucleotides, 205-207
    in polypeptide chains, 282
Linkage, 122
    see also Gene linkage
Linkage groups, 130
Linkage maps, 126
    see also Genetic mapping
Linked alleles in meiosis, 122-126
Linked alles and recombination, 126-134
Linker DNA, 91-92
Liposarcomas, 478
LMO-2 gene, 384
Local populations, 500
Locus of genes in gene mapping, 122
Long terminal repeats (LTRs), 431-432
Loss of function, 396
Loss-of-function mutations, 423
Loss of heterozygosity, 480-481
LTR retrotransposons, 432
LTRs. *See* Long terminal repeats
Lundberg, Ante S., 99
Lycopersicon chmielewskii (wild South
    American tomato), 551
Lycopersicon esculentum (domestic tomato), 551
Lymphoblastic leukemia, 384
Lyon, Mary F., 169, 323
Lysogenic bacteriophages, 267-274
Lysogenic cycle, 267-274
Lysogens, 268
Lytic cycle, 263

M

MacLeod, Colin, 4
Mad protein, 473
MADS box factor, 355
MADS box transcription factors, 412-413,
    419
Main body axes, 403-404
Maize. *See* Corn
Major grooves, 210
Malaria, 294, 384, 427, 517
Mammograms, 471
Managed evolution, 544-548
Map distance, 129, 139, 155
    see also Genetic mapping
Map unit, 127
    see also Centimorgan
Mapping function, 139
    see also Genetic mapping
Mariner transposon, 435
Marsupials, 169, 170, 323
Maternal-age effect, 177, 179
Maternal-effect gene mutations, 401
Maternal-fetal conflict, 342
Maternal inheritance, 521
Maternal phenylketonuria (PKU), 31
Mating-type interconversion, 348-359
Matthaei, J. Heinrich, 305
Maturation-promoting factor, 466
McCarty, Maclyn, 4
McClintock, Barbara, 430

McGurk, Rhona, 400
MCSs. *See* Multiple cloning sites
Mdm2, 477
Mean, 534, 535
Megabase (Mb), 204
Megabase pairs, 88
Megaspores, 81
Meiocytes, 80
Meiosis, 75, 78-88, 113, 119
    in animals, 80
    first meiotic division, 81, 84-87
    linked alleles in, 122-126
    in plants, 80-81
    second meiotic division, 81, 87-88
Meiotic drive, 71
Melanomas, 478
Mendel, Gregor, 2, 74
    addition rule, 49, 51
    and analysis of DNA, 38-39
    background of, 42
    chance in Mendelian genetics, 49-51
    character differences in peas studied by, 37
    independent assortment principle, 47-
        48, 49
    information on the Web, 71
    multiplication rule, 50-51
    results of monohybrid experiments, 40
    results of testcross experiments, 44-45
    segregation principle, 41, 49, 86, 532, 544
    statistics of data, 110-111
    work in genetics, 34-39, 42, 43
Mendelian genetics, 34, 41
Mendel's law, 12
Mental retardation, 179
Meristems, 411
Merodiploids, 259
Meselson, Matthew, 211
Meselson-Stahl experiment, 211-213
Messenger RNA (mRNA), 15, 18, 280, 293
    synthesis, 289, 321
Metabolic pathway, 10-11
Metabolism, 9
Metabolites, 12
Metacentric chromosomes, 167
Metallothionein, 381
Metaphase chromosomes, 92-94
Metaphase I in first meiotic division, 86, 87
Metaphase II in second meiotic division, 87
Metaphase in mitosis, 76-78
Metaphase plate, 78
Methanococcus janaschi, 204
Methionine biosynthesis, 330
Methylation in the germ line, 340-342
Mice, 323
    dyhybrid ratio in color, 62
    mammary turmor virus, 335
    p53 knockouts in, 480
    transformation of the germ line using
        embryonic stem cells, 378
    vaccination of mice against malaria, 384
Microarrays (or chips), 372-375
Micrococcus luteus, 30
Migration in formation of new species, 513
Migration in molecular evolution, 513-514
Milkweed bugs, 94
Minimal mediums, 251
Minor grooves, 210
Mismatch repair, 151, 447-448, 482
Mispairing mutagenesis, 440
Missense mutations, 425, 426
Mitochondrial DNA (mtDNA) maternal
    inheritance, 521-523
Mitochrondrial DNAs, 369
Mitosis, 75-78, 113, 460, 461
Mobile DNA, 242-250
Molecular clock, 497
Molecular disease, 221
Molecular evolution, 493-529
    allele frequencies, 501, 504, 515-516, 519
    allele frequencies among gametes, 502
    allele frequencies and genotype
        frequencies, 500-501
    allele frequency random changes, 418-420
    bootstrapping, 496

DNA exclusions, 510
duplication and divergence, 499-500
evolutionary relationships among species,
    494-500
evolution and genetic changes in species,
    413
fitness, 514-515
gene trees, 494-495
genotype frequency differences, 500-502
Hardy-Weinberg principle, 503, 504-506
heterozygote as superior genotyoe, 516,
    518
homozygote excesses in inbreeding,
    512-513
inbreeding, 510-513
mitochondrial DNA (mtDNA) maternal
    inheritance, 521-523
modern human population origins,
    522-523
mutation and migration, 513-514
natural selection, 514-518
polymorphic sequences in DNA typing,
    407-410
random mating, 502-507
rare alleles in heterozygous gneotypes,
    505-506
rates of evolution, 497-498, 497-499
selection and new mutations, 516
species trees versus gene trees, 496-497
x-linked genes in males, 506-507
Molecular genetics, 2
    see also Cancer and molecular genetics of
        the cell cycle
Molecular genetics of gene expression,
    281-315
    amino acid triplet genetic code, 203-208
    amino acids, 284
    aminoacyl-tRNA synthetase, 306-309
    codons in polypeptide synthesis, 204-205
    DNA and RNA base sequences, 284-289
    elongation phase of polypeptide synthesis,
        294-297
    eukaryotic RNA polymerases, 286
    exon shuffle model, 293
    folding of polypeptide chains, 299-300
    human gene characteristics, 292-293
    human proteins, 284
    messenger RNA in the synthesis of a
        polypeptide chain, 289
    nucleotide sequences, 286-289
    polypeptide chains, 282-284
    polyuridylic acid, 305
    redundancy and near-universality in the
        genetic code, 305-306
    ribosomes moving in tandem along a
        mRNA, 308-309
    RNA chemical synthesis, 284-286
    RNA transcript and messenger RNA,
        289-293
    splicing introns from RNA transcript,
        291-292
    steps in, 282
    termination of polypeptide synthesis
        (release phase), 298-299
    translation initiation in eukaryotes, 294,
        295
    translation initiation in prokaryotes,
        301-302
    translation into a polypeptide chain,
        293-302
    wobble in codon-anticodon pairing, 308
    see also Cancer and molecular genetics of
        the cell cycle
Molecular markers, 494
Molecular mechanisms. *See* Gene regulation
Molecular systematics, 494
Molecular traits, 38-39
mom-2 gene, 394
Monod, Jacques, 322
Monohybrids, 36, 40
Monoploid plants can be cultured, 194
Monoploids, 189, 190, 194
Monosomy, 174, 175
Montgomery, Hugh E., 552

Morgan, Thomas Hunt, 100-101, 122, 127
Morphological traits, 38-39
Morton, W.J., 442
Mosaics, genetic, 169-170
Mosquitoes, 294, 517
Most recent common ancestor (MRCA), 523
Moths, sex chromosomes in, 103
Mps1 protein, 473
MRCA. See Most recent common ancestor
mRNA. See Messenger RNA
Mullis, Kary, 239
Multifactorial traits, 532-536, 549
Multiple alleles, 62
Multiple cloning sites (MCSs), 366, 367
Multiplication rule, 50-51, 105-106
Mus musculus (laboratory mouse), 25
Mutagens, 422
Mutant, 20
Mutant screen, 63-66
Mutation and Cancer: Statistical Study of
    Retinoblastoma (Knudson), 484
Mutation in formation of new species, 513
Mutation in molecular evolution, 513-514
Mutations, 421-457
    and adaptive needs of the organism,
        435-437
    agents and their mechanisms of action,
        439
    altering development, 392
    AP endonuclease system, 448-449
    base analogs masquerading as the real
        thing, 440-441
    base-pair additions or deletions, 442
    and base substitutions, 424-425
    in the cell cycle and key regulatory
        pathways, 463-464
    Chernobyl nuclear accident, 445-447, 457
    classification of, 422-424
    conditional mutations, 423
    DNA and RNA intermediates, 430-433
    DNA repair, 447-451
    and DNA sequence changes, 424-430
    in Drosophila, 437-438
    enzymes repairing damage, 450
    excision repair, 450
    and gene function, 423-424
    in genes, 20-22
    genetic tests for mutations and cancer,
        451-453
    germ-line mutations, 422
    highly reactive chemicals damaging DNA,
        441
    insertion or recombination, 433-434
    ionizing radiation, 442-445
    in maternal-effect genes, 401
    mismatch repair, 447-448
    mutagens, 422
    nonrandomness in position in a gene or
        genome, 438-439
    postreplication repair, 450-451
    preventing normal checkpoint function in
        cancer, 474-482
    in protein-coding regions, 425
    purine bases and spontanious loss,
        439-440
    sickle-cell anemia, 426-427
    similarity in spontaneous and induced
        mutations, 439-447
    somatic mutations, 422-423
    as statistically random events, 435-439
    transposable elements in, 430-435
    transposable elements in the human
        genome, 434-435
    trinucleotide repeats in the human
        genome, 427-430
    in tulip bulbs, 356
    types of, 422
    types of DNA damage and mechanism of
        repair, 447
    ultraviolet radiation, 442, 450
    weak acids, 440
Mutations of a mutS Homolog in Hereditary
    Nonpolyposis Colorectal Cancer (Leach),
    446

Mycoplasma genitalium, 370
Myotonic dystrophy, 429

N
Nägeli, Carl, 34, 75
Nail-patella syndrome (NPS), 161
Nanos, 404
Narcissus pseudonarcissus (common daffodil),
    382
Narrow-sense heritability, 546
National Academy of Sciences of the United
    States, 444
National Center for Biotechnology
    Information (NCBI), 279
Native Americans of the Great Plains,
    492
Natural populations of organisms, 494
Natural selection, 514-518
Natural selection in formation of new
    species, 513, 517
Nazi death camps, 173
NCBI. See National Center for Biotechnology
    Information
Near-universality in the genetic code,
    305-306
necdin gene, 341
Negative regulation, 318-319, 325, 393
Neighbor joining, 495
Nemotodes. See Caenorhabditis elegans
Neomycin, 245
Neurological disorders, 429
Neurospora, 144, 261
Neurospora crassa (bread mold), 12, 148-149,
    157, 159
Neurotoxins, 202
Newts, 288
Nicholas II (tsar of Russia), 102, 509
Nicotiana longiflora (tobacco), 545
Nirenberg, Marshall W., 305
Nitrogen mustard, 441
Nitrous acid, 440
Non-LTR retrotransposons, 432-433
Nondisjunction, 103-104
Nonparental ditype (NPD), 145
Nonrandomness in position in a gene or
    genome, 438-439
Nonselective mediums, 251
Nonsense mutations, 425
Nonsynonmous mutations, 425, 498
Normal distribution, 536, 537
Notch choromosomes, 180
Notch gene, 396
NPS. See Nail-patella syndrome
Nuclear division, 74, 75
Nucleic acid hybridization, 224, 2225
Nucleolus, 76
Nucleoside analogs, 233
Nucleosides, 206, 233
Nucleosome, 88-92, 90
Nucleotide sequences, 286-289
Nucleotides, 7, 30, 206
    in DNA replication, 210
Null mutations, 423
Nutritional mutants, 250-251

O
Oat seeds, genetic studies of color in, 61
Octoploid, 190, 191
Odd-skipped affect, 405
Oedipina poelzi (salamanders), 85
On Agglutination Phenomena in Normal Blood
    (Landsteiner), 57
Oncogenes, 472, 476
Oncogenesis (cancer) and telomeres, 99
Oncopeltus fasciatus (milkweed bug), 94
Oocytes, 401
Open reading frame (ORF), 289
Operators, lactose, 321-322
Operons, 323
Opisthokonts, 24
Orangutans, 168
ORF. See Open reading frame
Orthologous genes, 499
Oryza sativa (rice), 389

Osteosarcomas, 478
OxyS, 344, 345

P
P element, 377
P (peptidyl) sites, 294
P_1 generation, 36
p16 protein, 477
p19ARF protein, 477
p21 function, 477
p53 transcription factor, 468-472, 475, 477
Pachytene, 84, 85, 114
PAH protein, mutant, 20-22, 30
Pair-rule genes, 403, 404
Palindrome symmetry, 226, 358
Paracentric inversions, 185, 186
Paralogous genes, 499
Parasegments, 402, 404
Parental combinations, 122, 123
Parental ditype (PD), 144
Parental strands, 210, 211
Parental traits, 74
Partial digestion, 362
Partial diploids, 259
Pascal's triangle, 106
Patents issued for clinical applications, 383
Paternity testing, 510
Pattern formation, 392
PAUP (software), 495
PCR. See Polymerase chain reaction
Peacocks, 420
Peas, 34-39, 117, 532
    axial and terminal flowers, 68
    character differences studied, 37
    chromosomes in, 74-75
    distribution of traits in, 534
    flower color in, 60
    round and wrinkled seeds, 38, 69
    See also Mendel, Gregor
Pedigrees, 51
Penetrance, 59
Peptide bonds, 282
Peptidyl transferase, 297
Pericentric inversions, 185
Permissive conditions in mutations, 423
Permissive temperatures, 463
Petri dishes, 250
Phage repressor proteins, 273
Phage therapy, 279
Phages, 4
    attachment sites, 271
    specialized transducing, 259, 273-274
    transferring small pieces of bacterial DNA,
        250-263
Phaseolus vaugaris (French bean), 382
Phenotypes, 42
    and epistasis, 59-62
Phenotypic differences, 494
Phenotypic traits, 282
Phenotypic variation, 537
Phenyl ring, 10
Phenylalanine hydrox-13ylase (PAH), 12
Phenylketonuria (PKU), 12-13, 22, 30
    maternal, 31
Phenylthiocarbamide (PTC), 115-116
Philadelphia chromosome, 485
Phosphatase enzymes, 465
Phosphodiester bonds, 206
PHYLIP (software), 495
Phylogenetic tree of human mitochondrial
    DNA, 522
Physical performance gene, 552
Phytate, 382
Pigmentation, 420
Pingelap disease, 529
Pisum sativum (common garden pea), 34
PKU. See Phenylketonuria
Plantae, 24
Plants, meiosis in, 80-81
Plaque formation, 264
Plasmids, 242-243
    cointegration of, 247
    F plasmids, 243-245, 253-259
Plasmodium berghei, 384

Plasmodium falciparum, 294, 384, 426, 517, 518
Plastic growth patterns of higher plants, 411
Pleiotropic effects, 22
Pleiotropy, 22
Ploytene chromosomes, 180-182
Pol II holoenzyme, 337
Polar granules, 394
Polarity, 8
Pole cells, 399
Poly-A tails, 290
Polycistronic mRNA, 301, 318
Polylinkers, 366, 367
Polymerase, with proofreading function, 220-222
Polymerase chain reaction (PCR), 228-230, 239
Polymorphic DNA sequences in human genetic mapping, 140-143
Polymorphic sequences in DNA typing, 407-410
Polymorphism, 140-142
Polynucleotide chain, 206
Polypeptide chains, 14, 282, 282-284
  properties of, 283
Polyploid species with multiple sets of chromosomes, 189-196
  genome duplications before or after fertilization, 190-191
  and genomes from different species, 192-194
  the grass family illustrates polyploidy, 195-196
  monoploid plants can be cultured, 194
Polyploidy, 189-194
Polysaccharides, 3
Polysomes, 309
Polytene chromosomes, 180-182
Polyuridylic acid, 305
Population bottleneck, 492
Population genetics, 494
  see also Molecular evolution
Populations, 500
Positional information, 394
Positive regulation, 319, 324, 325, 393
Posterior genes, 404
Postreplication repair, 450-451, 451
Prader-Willi syndrome, 341
Precursor fragments, 222, 223
Precusor to cancer cells, 474
Premutation, 428
Primary transcripts, 289
Primer chains, 220
Primosome, 219
Probes, 225, 228
  labeled, 368
Processivity of DNA polymerase, 470
Product molecules, 10
Proflavin, 302
Programmed cell death and cancer, 458, 470-471
Programmed deletions in mammalian immune system, 345-348
Programmed DNA rearrangements, 345-349
Prokaryotes, 24
Promoters, 286, 289
  fusion of, 483
  lactose, 322
  mutations in, 288-289
  recognition of, 286
  sequences in, 249
Proofreading function, 222
Prophage, 268
Prophage induction, 272
Prophase I in first meiotic division, 84-85, 88
Prophase II in second meiotic division, 87
Prophase in mitosis, 76, 77
Protanopia, 182
Proteasome, 467
Protection Afforded by Sickle-Cell Trait Against Subtertian Malarial Infection (Allison), 517
Protein-coding regions and mutations, 425

Protein factors in the eukaryotic transcription complex, 335-338
Proteins
  degradation of, 467
  in DNA replication, 217-224
  of humans and invertebrates, 284
  produced with recombinant DNA, 382-383
  synthesis of, 280
Proteomes, compared with genomes, 25-26
Proteomics, 358, 371, 375-376
  see also Genetic engineering
Proto-oncogenes, 476, 476-477
Prototrophs, 251
Proximal part of a chromosome, 188
Prozac, 553
Pseudoautosomal regions, 170
  progressive shortening of the X and Y chromosomes, 170-172
Pseudogenes, 498
PTC. *See* Phenylthiocarbamide
Punnett square, 47, 70, 71, 110
Purine bases and spontanious loss, 439-440
Purines, 206, 209
Pyrimidine dimers, 442
Pyrimidines, 206, 209

Q
Quadrant enhancers, 409
Quantitative-trait locus (QTL), 52, 551-554
Quantitative traits, 532, 542-543

R
R group, 282
R plasmids, 250
Radiation therapy, 491
Radiation units, 443
Radium craze, 442-443
Radom genetic drift, 518-520
Random events, mutations as, 435-439
Random fertilization, 41
Random genetic drift in formation of new species, 513, 518-520, 529
Random mating, 502-507
Random-mating frequencies, 506
Random-spore analysis, 147
Rare alleles in heterozygous genotypes, 505-506
Ras gene, 476-477, 479
Rate of mutation, 422
Rates of evolution, 497-498, 497-499
Reading frames, 302, 425
Recessive alleles, 516
Recessive genes, 199
Recessive traits, 36, 43
Reciprocal crosses, 36-37
Reciprocal translocations in chrommomes, 186, 434
Recombinant cDNA, 364-365
Recombinant cloning, 368
Recombinant DNA, 358, 382
  see also Genetic engineering
Recombinant DNA introduced into animals, 377-379
Recombinant vectors, 365
Recombinants, 122, 123
Recombination, 122, 123
  of bacteriophage DNA molecules, 263-267
  and complementation, 155
  crossing-over between linked alleles, 126-134
  in *Drosophila* males, 125-126
  frequency of, 122-125, 129, 139, 155
  and gene conversion, 151-152
  initiated by a double-strand break in DNA, 152-154
  in mutations, 433-434
  see also Gene linkage
Recruitment, transcriptional activation by, 336
Red blood cells, 280
Red-green color blindness, 182
Reductional division, 84, 113
Redundancy, 306

Redundancy and near-universality in the genetic code, 305-306
Redundant genetic code, 306
Relative fitness, 515
Release phase, 298
Renaturation, 224
Replica plating, 435-436
Replicating circle (Greek letter theta), 214
Replication fork, 214
Replication of DNA molecules, 8-9, 460
Replication origin, 214
Replication slippage, 429
Reporter genes, 376
Repressible transcription, 319
Repressor proteins, 319
Reproductive systems of bacteria, 242
Repulsion, 122, 124
Restriction endonuclease, 140
Restriction enzymes, 225-228, 226, 358-360
Restriction fragment length polymorphisms (RFLPs), 140
Restriction fragments, 226, 360-361
Restriction maps, 226-227
Restriction site, 226
Restrictive conditions in mutations, 423
Restrictive temperatures, 463
Retinoblastoma, 480-482, 484, 491
Retinoblastoma (RB) protein, 466-467, 475, 478
Retroviruses, RNA, 378, 384
Reverse genetics, 377-381
Reverse transcriptase, 17, 364, 365, 432
Reverse transcriptase PCR (RT-PCR), 365
Reversions, 451
RFLPs. *See* Restriction fragment length polymorphisms
Rhodopsin, 182
Ribonucleic acid. *See* RNA
Ribose (sugar), 15, 218
Ribosomal RNA (rRNA), 15
Ribosome-binding site, 301
Ribosomes, 15, 293
Ribosomes moving in tandem along mRNA, 308-309
Riboswitches, 330, 331
Ribozymes, 292
Rice, 195, 382, 383, 389
rII frameshift mutations, 303, 306
rII gene, 265-267, 269
RISC (RNA-induced silencing complex), 343-344
RNA, messenger, 280
RNA and DNA base sequences, 284-289
RNA chemical synthesis, 284-286
RNA interference (RNAi), 343-344
RNA intermediates, 430-433
RNA molecules, 4, 14-16, 29, 30
  processing in, 92
RNA polymerase, 218-219, 282, 285
RNA polymerase holoenzyme, 286
RNA polymerase I, II, and III, 286
RNA primer, 219, 220
RNA processing, 282, 290
RNA processing and decay, 342-344
RNA retroviruses, 378, 384
RNA splicing, 290
RNA transcript and messenger RNA, 289-293
RNAi. *See* RNA interference
Robertsonian translocations, 188
Rolling-circle replication, 214, 215
Roma (Gypsies), 173-174
RPA (replication protein A), 223
rRNA. *See* Ribosomal RNA
RT-PCR. *See* Reverse transcriptase PCR
Rye, 116

S
S-adenosylmethionine (SAM), 330, 331
S-linked genes in males, 506-507
Saccharomyces cerevisiae (budding yeast), 25, 88, 89, 95, 159
  cell cycle of, 460-461
  genome size, 204
  life cycle of, 145

mating-type interconversion, 348-349
    mating types, 392-393
    metabolic pathway in, 332
Salamanders, 85, 205
Salmonella typhimurium, 451, 452, 529
SAM. *See* S-adenosylmethionine
Satellite DNA, 94-95
Schizophrenia, 533
Scorpions, 80
Secale cereale (rye), 116
Second-division segregation, 150
Second meiotic division, 81, 87-88
Secretors, 68
Segment identity, 407
Segment-polarity genes, 403, 4405
Segmentation genes, 402, 403
Segments, 402
Segregation principle, 41, 49, 86
    genes in chromosomes, 99
    in human pedigrees, 51-53
Selected markers, 255
Selection and new mutations in molecular
    evolution, 516
Selection coefficient, 515
Selection limits, 547
Selective breeding, 34, 333, 543
Selective mediums, 251
Selective serotonin reuptake inhibitors
    (SSRIs), 553-554
Selectively neutral mutations, 497
Self-fertilization, 502, 510, 512, 544
Self-splicing, 292
Semiconservative replication, 211-213
Semisterility, 186
Sequence logo, 315
Sequencing of bases in DNA manipulation,
    230-233
Serotinin transporter protein, 553-554
Serrate genes, 409
Sex chromatin bodies, 323
Sex chromosomes, 99-100, 164
Sexual polyploidization, 190-191
Shepherd's purse, 70
Shetland sheepdogs, 530
Shine-Dalgarno sequence, 301
Short RNA primer, 218-219
Siamese cats, 423
Siblings, 52
Sibs, 52
Sibship, 50, 69
Sickle-cell anemia, 29, 221, 239, 426-427,
    517, 518, 532
Silencers, 335
Silent substitutions, 425
Simple sequence repeats (SSRs), 142, 172,
    508-510, 551
SINE elements, 433, 434, 435
Single-active-X principle, 169
Single-nucleotide polymorphisms (SNPs),
    143
Single-stranded DNA, 8
Single-stranded DNA binding proteins
    (SSBs), 217
Sister chromatids, 78, 460
Site-specific recombinases and integrons,
    247-250
Skin cancer, 450
Skin cancer syndromes, inherited, 482
Skuse, David H., 400
Slipped-strand mispairing, 429
Small-cell lung carcinomas, 478
Small regulatory RNAs, 344-345
Snapdragon flower color, incomplete
    dominance, 54-55
SNPs. *See* Single-nucleotide polymorphisms
snRNPs, 291
SNRPN gene, 341
Social implications of advances in genomics,
    389
Sockeye salmon, 382
Somatic cell mutations and cancer, 458, 460
Somatic cells, 74
Somatic chromosome numbers of some plant
    and animal species, 74

Somatic mutations, 422-423
Sorghum, 195
Southern blot, 228
Specialized transducing phages, 259,
    273-274
Specialized vectors, 362-363
Species trees versus gene trees, 496-497
Spina bifida, 533
Spindle, 76-78, 116
Spindle assembly checkpoint, 473
Spindle checkpoint function, 472
Spindle checkpoints, 468
Spiny pufferfish, 202
Spliceosomes, 291
Splicing introns from RNA transcript,
    291-292
Sponocerebellar ataxia type 1, 429
Spontaneous abortions, and human
    chromosome abnormalities, 174-175
Spontaneous mutations, 422, 439-447
Sporadic cancer, 474
Spore killers, 157
Spores in meiosis, 80
Sporophyte plants, 81
SRY (sex-determining region), 166
SSBs. *See* Single-stranded DNA binding
    proteins
SSRIs. *See* Selective serotonin reuptake
    inhibitors
SSRs. *See* Simple sequence repeats
Stahl, Franklin, 211
Standard deviation, 536
Standard human karyotype, 165-166
Starch-branching enzyme I (SBEI), 38, 53
Start codons, 294
Statistically highly significant, 109
Statistically random events, mutations as,
    435-439
Statistically significant, 109
Statistics, 534-536
Statistics and probability in genetic data
    analysis, 105-111
    binomial probability formula, 105-107
    chi-square, 107-110, 117
    degrees of freedom, 109-110, 117
    factorials, 106-107
    goodness of fit, 107, 109
    Pascal's triangle, 106
    statistically highly significant, 109
    statistically significant, 109
STI571, 485
Sticky ends, 358, 360, 363, 364
Stop codons, 298-299
Streptococcus pneumoniae, 3-4
Streptococcus pyrogenes bacteria, 263
Streptomycin, 245
Structural genes, 322-323
Sturtevant, A.H., 139
Sublineages, 395
Submetacentric chromosomes, 167
Subpopulations, 500
Substrate molecules, 10
Sugar cane, 195
Summer squash, 70
Supernumerary telocentric chromosome,
    176
Superrepressors, 321
Surface exclusion, 245
Sutosomes, 100
SV40 virus, 104
Sweat glands, 199
Synapsis, 84
Synchronized cells, 462
Synonymous substitutions, 425, 498
Synteny group, 196
Synthesizing in fragments in DNA
    replication, 222-224

T

T cells, 345-348, 483
T DNA, 380
TAFs. *See* TBP-associated factors
Tahtanka, 492
Tailless genes, 405

Takifugu rubripes (Japanese pufferfish), 202,
    205
Tandem duplication in chromosomes, 182
TAPs. *See* Transcriptional activator proteins
Tasters, 115-116
TATA-box-binding protein (TBP), 337, 339
TATA boxes, 286, 315
Tatum, Edward L., 261
Taxa, 494, 497
Tay-Sachs disease, 239
TBP. *See* TATA-box-binding protein
TBP-associated factors (TAFs), 337
TDF. *See* Testis-determining factor
Telocentric chromosomes, 176
Telomerase, 97
Telomerase activity, 474
Telomeres, 96-99
    and cancer, 99
Telophase I in first meiotic division, 86-87
Telophase II in second meiotic division,
    87-88
Telophase in mitosis, 78
Temin, Howard, 17
Temperate phage, 268
Temperature-sensitive mutations, 423, 463
Template strands, 282
Templates, 9, 210
    for DNA replication, 210
Terminal genes, 404
Terminal inverted repeats, 430
Termination in polypeptide synthesis, 294,
    298-299
Terminator conformation, 330
Testcrosses, 44-45
    independent assortment in progeny of,
    47-48
Testis-determining factor (TDF), 171
Tetracycline-resistance gene (tet-r), 242
Tetrads, 85, 155
    analysis of, 145-147
    four products of meiosis, 143-152
    meiosis geometry in, 148-150
    tetratype, 145
    unordered, 144-145
Tetrahymena (ciliated protozoa), 97, 98, 292
Tetraploids, 174, 189
Tetratype (TT), 145
Tetrodotoxin, 202
Thermophiles, 230
Thermus aquaticus, 230, 232, 286
Thermus thermophilus, 308
Thirty-nm chromatin fibers, 91-92
Three' end, 206, 207
Three-point crosses, 134-139
    reciprocal products in, 138
Three' untranslated region, 289
Threshold traits, 533-534, 549
Thymine (T), 7, 204, 209
Ti plasmid, 379-380
Tijo, Joe Hin, 135
Time of entry, 256
Toll gene, 344
Tomato plants, 200
Topoisomerases, 218
torso gene, 344, 404
Tortoiseshell cats, 323
Total variance, 540-541
Traits
    altering with DNA, 3-4
    biologically inherited, 2
    categorical, 533
    complex, 532
    continuous, 533
    discrete, 532
    dominant, 36, 37, 43, 53-59
    environment affecting, 22
    Mendel's choice of, 35-36
    molecular, 38-39
    morphological, 38-39
    multifactorial, 532-536, 549
    parental, 74
    phenotypic, 282
    quantitative, 532, 542-543
    recessive, 36, 43

Traits, *Continued*
  threshold, 533-534, 549
  wildtype, 38
  see also Complex inheritance
Trans configuration, 122, 124
Transactivation, 431
Transcript, 16
Transcription, 15, 16, 17, 282
  in eukaryotes, 330-340
  of genes in the cell cycle, 462
  inactivation, 340
  in prokaryotes, 318-328
  termination, 328-330
Transcription factors, 171, 336-337, 355
Transcriptional activator proteins (TAPs),
    319, 334-335, 339
Transducing phages, 259, 260
Transduction, 250
Transfer RNA (tRNA), 15, 18, 293-294,
    306-308, 328, 329
Transformation, 4
  of DNA, 251
  from the uptake of DNA and
      recombination, 251-253
Transformation rescue, 380-381, 381
Transgenic organisms, 358, 378, 381
  see also Genetic engineering
Translation, 15, 16, 282
  initiation in eukaryotes, 294, 295
  initiation in prokaryotes, 301-302
  into a polypeptide chain, 293-302
  level of, 344-345
Translocations in chromosomes, 186-189
Transmembrane receptors, 295-297, 396-397
Transmission genetics, 33-71, 74
  see also Genes; Genetics
Transmission of DNA, 4-7
Transposable elements, 38, 430, 431
  in bacteria, 245
  in the human genome, 434-435
  in mutations, 430-435
Transposase protein, 245
Transposition, 38, 430
Transposons, 245
  and insertion sequences, 245-246
Transverse bands (G-bands), 165
Tribolium (flour beetle), 543
Trichothiodystrophy, 450
Trigger factors, 300
Trihybrids, 36, 70
Trinucleotide expansion, 428, 429
Trinucleotide repeats in the human genome,
    427-430
Triplet code, 18-20, 302
Triploids, 174, 190
Tris-BP, 452
Trisomic, 174, 175
  meiotic synapsis in a, 178
Trisomic chromosomes and abnormal
    segregation, 177-178
Trisomy-21 (Down syndrome), 175
Trisomy-X syndrome, 179, 199
Triticum aestivum (wheat), 195
Triticum dicoccum (Emmer wheat), 116
Triticum turgidum (wheat), 193

Triturus viridescens (newt), 288
Trivalent, 178
tRNA. *See* Transfer RNA
trp attenuator, 327, 329
trp enzymes, 328
trp leader mRNA, 329
trp mRNA, 328
trp operon, 326-328
trpR gene, 327
True-breeding, 35
Truncation points, 546
Tryptophan biosynthesis, 326-328
Tryptophan tRNA, 330
Tulip craze in Holland, 356
Tumor cells, 474-476
Tumor-suppressor genes, 476, 477-478, 479,
    482
Tumor therapy and ionizing radiation, 444
Turner syndrome, 179, 199, 400
Turpin, Raymond, 176
Twin studies, 544
Twins, identical and fraternal, 544, 549
Two-hybrid analysis, 376
Twofold degenerate sites, 498
Tyrosine kinase inhibitor STI571, 485

U

UBE3A gene, 341
Ubiquitin, 467
UBX (*Ultrabithorad*) genes, 409-411
Ultrabithorad (UBX) genes, 409-411
Ultracentrifuge, 211
Ultraviolet radiation and mutations, 442,
    450
Unequal crossing-over in chromosomes, 182
Unidirectional DNA transfer in bacteria,
    253-259
Uninducible mutants, 332
Univalent, 178
Universal donors, 58
Universal recipients, 58
Uracil (U), 15, 218

V

V-J joining, 347
Vaccines, 452
Variable antibody regions, 346
Variable expressivity of genes, 58-59
Variable (V) antibody regions, 347
Variance, 534, 535
Variance of a distribution, 536, 537
Vectors, 360-361, 361-367
  recombinant, 365
  specialized, 362-363
Vegetative propagation, 423
Ventral uterine pecursor cells (VUs), 396
Vesicular stomatitis virus, 384
Vestigial genes, 408, 409
Victoria (queen of England), 102-103
Vir genes, 380
Virulent phage, 268
Vitamin B6 (pyridoxine), 261
von Tschermak, Erich, 34
Vulval development, 396
Vulval differentiations, 397

W

Watson-Crick base pairing, 7, 8, 204, 211,
    212, 312
Watson, James, 7, 9, 31, 211
Weak acids and mutations, 440
Weaponizing biological agents, 250
Weinberg, Robert A., 99
Wexler, Nancy, 52
Wheat, 116, 193, 195
White blood cells and acute leukemias,
    483-485
White Leghorn and White Wyandotte
    chickens, epistasis in, 61-62
Wildtype alleles, 62-63
Wildtype traits, 38
Wobble in codon-anticodon pairing, 308
Wolves, 530
Wombats, 169
Word roots, prefixes, suffixes, and combining
    forms, 571-573
Wright, Sewall, 111

X

X chromosomes, 100
  extra, 179
  inactivation in the calico cat, 170
  inactivation of, 168-170
  pseudoautosomal region of, 170-172
  single-active-X principle, 169
X-inactivation center, 323
X-linked genes, 101, 506-507
  in females, 169-170
x rays, 442-443
Xeroderma pigmentosum, 450, 482
XIC (X-inactivation center), 169
Xist (X-inactivation-specific transcript), 169

Y

Y chromosomes, 100
  extra, 179
  history of human populations traced
      through, 172-174
  pseudoautosomal region of, 170-172
  SRY (sex-determining region), 166
Yeast. *See Saccharomyces cerevisiae*
Yeast cell cycle, 460-461
Yeast centromere, 95
Yeast mating-type interconversion, 348-359
Yeast phenylalanine tRNA, 307
Yeast two-hybrid analysis, 375-376
Yellow-fever mosquito, 158
Yellowstone National Park, 230, 232
yitJ gene, 330, 331

Z

Zea mays (corn)
  genetic mapping of, 130-132, 195
  life cycle, 81
Zebras, 147
Zeste (z) mutation, 182
Ziegfeld, Florenz, 443
Zinc fingers, 335, 355
Zygotene, 84, 114
Zygotes, 42
Zygotic genes, 401